Sandra Dow.

Statistical Analysis for Business: A Conceptual Approach Second Edition

Lee H. Smith
Southwest Texas State University
Donald R. Williams
North Texas State University

Wadsworth Publishing Company, Inc. Belmont, California

ISBN 0–534–00399–0
L.C. Cat. Card No. 74–16896
Printed in the United States of America

1 2 3 4 5 6 7 8 9 10—80 79 78 77 76

Preface

This new edition of *Statistical Analysis for Business: A Conceptual Approach* benefits from extensive experience, by us and others, in using the text in both undergraduate and graduate business statistics classes. All of our changes have the common purpose of helping the student master the subject more efficiently.

Changes in Pedagogy

We have:

1. simplified the numerical system for identifying such elements as sections, equations, and rules.
2. selectively moved some of the more involved mathematical proofs to chapter supplements.
3. replaced many of the examples and problems involving dice and balls in urns with problems that relate to business.
4. included problem sets *within the text,* following their sections, and added Review Problems at the ends of chapters. (Answers to selected problems are in a special section at the end of the text.)
5. extensively rewritten parts of chapters and entire chapters to add some concepts, delete others, shift the placement of some topics, and to get rid of verbiage that can slow the student down without enhancing understanding.

Changes in Content

1. Bayes' theorem and elementary Bayesian probability material have been promoted to an earlier chapter (Chapter 4 on general probability).
2. Also in Chapter 4, we have added an important section on the use of tree diagrams in solving complex probability problems and a section on the notation and concept of set theory as related to probability theory.
3. We have eliminated the treatment of the negative binomial distribution in connection with discrete probability distributions.
4. We have added a section involving the Kolmogorov-Smirnov goodness-of-fit test to our treatment of the Chi-Square test.
5. Chapter 12, on survey sampling is entirely new.
6. Finally, we have made the discussion of the abbreviated Doolittle method supplemental to the main chapter dealing with multiple regression analysis (15).

Rationale for this text

Like its predecessor, this edition is designed to be used in either a one- or two-semester course at the undergraduate level or in a fast-moving one-semester course for beginning graduate students. Anyone with a good background in college algebra will find the material digestible. Our guiding objectives have been to

1. set forth very clearly the mathematical ideas we find necessary for comprehension of statistical concepts, including the underlying definitions and assumptions.

   For example, even at the risk of being too elementary, we are careful to enlighten the student about what is meant by the entity called a statistic; and we make a clear distinction between distributions of random variables and sampling distributions.

2. augment this foundation with a thorough treatment of the intuitive aspects of statistics.

   We believe, for example, that a student cannot tie together the theoretical and intuitive aspects of statistics without a firm background in probability. Since we feel so deeply that a keen understanding of probability can simplify the study of statistics, we take every opportunity to use probability statements when dealing with random variables.

3. clarify the mechanics associated with well-known statistical procedures.

   For example, we include a detailed look at the results of the normal distribution and operations related to it. We believe that an understanding of the relation between the normal distribution and probability can carry a student a long way toward mastery of more advanced concepts. We also avoid being over-specific in treating confidence intervals and tests of hypotheses (including sampling distributions). While one certainly needs to know how to construct confidence intervals for the population mean, the student also needs the more general approach involving other parameters and statistics.

4. tie together the mathematical ideas, intuitive aspects, and mechanics with appropriate examples.

   For example, we shun the categorization of estimation as either completely intuitive or completely mathematical. Our approach is structured to tie intuition and mathematics together as equal components of estimation.

Adapting the Text to Your Course

An important aim in designing the order of presenting topics has been to provide flexibility in terms of possible course-content combinations. However, we feel strongly that, regardless of the length or level of the course, much of the material of Chapters 4 through 9 should be included. A rather fast-moving, one-semester introductory undergraduate course can be fashioned very nicely through a selective process that incorporates only the high points of these chapters with relevant material from other chapters. For example, a traditional one-semester course would include Chapters 1, 2, and 3, the high points of Chapters 4 through 9, selected topics from Chapters 11 and 14, and much of Chapters 18 and 19.

There are almost countless possibilities for the two-semester undergraduate sequence. The last four chapters have been written so that they do not need to be read sequentially; Chapters 16 and 17 can be covered immediately after Chapter 9 with no loss of continuity. With a responsive class, one can cover the entire text in two semesters. Our experience shows that Chapters 1–3, 18, and 19 can be covered quite rapidly (although Chapter 19 assumes the reader is familiar with regression analysis, described in 14). We recommend a detailed coverage of most of the topics of Chapters 4–9 to enable the student to move rapidly through Chapters 10–17.

Acknowledgements

We are indebted to the literary executor of the late Sir Ronald A. Fisher, F.R.S., to Dr. Frank Yates, F.R.S., and to Oliver & Boyd, Edinburgh, for permission to reprint Tables II and III from their book *Statistical Tables for Biological, Agricultural, and Medical Research;* and to F. J. Massey, Jr., and the *Journal of the American Statistical Association* for permission to reprint Table 11.

For their helpful suggestions, we would like to thank many users and reviewers of both editions; for their help on this second edition, we particularly thank Robert L. Andrews, Virginia Polytechnic Institute and State University; James D. Dilworth, University of Alabama; George L. Miller, Northern Illinois University; Joseph G. Van Matre, University of Alabama; and Othmar W. Winkler, Georgetown University.

We sincerely appreciate the invaluable assistance afforded by Charlotte Bowling and Marinelle Skretteberg of the University of Texas at Dallas, Dorothy McCommack, College of Business, University of Houston, and Karen Whaley of Oklahoma State University. Finally, we owe a special vote of thanks to our families, particularly our wives, for their support, patience, and tolerance.

Lee H. Smith
Donald R. Williams

Contents

One

Introduction

1.1 On Statistics and Business Statistics

The procedures of statistics can be said to parallel the general procedures of the scientific method. But misconceptions about statistics abound, largely because, until recently, a very limited presentation of the subject has been the rule in textbooks. Statistics has been defined roughly as the collection, description, summarization, and presentation of large amounts of data. However, such a definition is incomplete because these factors represent only part of the discipline. A more appropriate first definition for our purposes would be:

Statistics deals with description, analysis, and interpretation of data that are subject to errors of measurement and prediction.

By this definition, it should be obvious that statistics is concerned with practically *all* data. We shall look at a more rigorous definition in Chapter 7. Since errors of measurement and prediction create uncertainty, it follows that statistics is useful for aspects of decision making that involve uncertainty. It includes the design and analysis of experiments and surveys, as well as every intermediate phase related to these tasks—for example, the collection and reduction of data. In fact, statistics is applicable to a wide variety of problems in almost every field imaginable, including aerospace, biology, business, economics, medicine, sociology, physics, psychology, and so forth. While some statistical tools are used across the board, so to speak, others are more specialized. For example, that special part of statistics known as "time-series analysis" (Chapter 19) was developed specifically for problems arising in business (although it has since been adapted to other fields).

Since business is not yet exploiting many of the vital tools of statistical analysis, there is an implied emphasis on the word *statistics* when we refer to business statistics. Only when you have a genuine appreciation of the underlying principles can you apply statistical tools to their best advantage in business analysis.

Among the various ways of classifying statistics, one breaks the subject matter into two classes: *descriptive statistics* and *inferential statistics.* Descriptive statistics, which involves the summarization and presentation of data, represents that part which has traditionally been the concern of students of business statistics. In inferential statistics one draws inferences about an entire set of data by considering only part of the data. As such, it represents the essence of a logical approach to decision making under uncertainty. As one might suppose, the concepts and tools of inferential statistics involve more rigor than those of descriptive statistics. Mastery of it requires a keen understanding of the mathematical field of probability.

We can also classify statistics by fields of interest, dividing the discipline into *theory* and *methodology*. The theory is based almost entirely upon probability, extending it to the development of such theoretical principles as estimation (Chapter 8), testing of hypotheses (Chapter 9), and analysis of variance (Chapter 13). Statistical methodology involves applying the tools and techniques that stem from statistical theory.

This second classification, into theory and methodology, opens an even wider chasm between statistics and business statistics than the classification based on description and inference. It puts the business statistician and the mathematical statistician at opposite poles. A glance into the history of statistics probably explains why. Until recently, most statisticians could be best described as "pure statisticians." Their primary objective was to develop and extend statistical theory, rather than develop statistical tools for the research worker or decision maker, who, in our context, may be viewed as special breeds of the same thing: Both are interested in applying the methodological tools that have developed as "spin offs" from the mathematical statisticians' theoretical endeavors. As statistical theory deepened over the years, it became harder and harder for the research worker to interpret its results well enough to apply them to his own pursuits. And the kit of statistical tools available became so big it was hard for him even to keep track of the techniques.

This widening of the chasm between the pure statistician and the research worker (or decision maker) finally brought about a new breed, known as the "applied statistician," whose primary concern is to apply statistics to various areas of research. Since he is necessarily thoroughly trained in both theory and methodology of statistics, he has helped slow down the growth of the chasm. What caused the chasm to diminish has been the recognition in the field of education that aspiring research workers and decision makers must learn enough statistics, both theory and methodology, to be able to collaborate with the applied statisticians. The purpose of this text, therefore, is to help you do just that.

1.2 Why Study Statistics?

The statistics professor is often asked, "Why should I study statistics?" From the foregoing discussion you probably have an inkling why. This reasonable question deserves some specific answers.

1. Quantitative measurement and evaluation, for centuries the watchwords of scientists, have been adopted by other disciplines—largely because of the development of the computer. Thus, social scientists and humanists also are moving more and more to quantitative analysis. It follows that statistics, as a specific branch of mathematics designed primarily to handle errors of measurement and prediction, is now considered essential to the *general* preparation of the student.
2. Even though you may not use the specific methods of statistics directly, you will benefit indirectly through a knowledge of the conceptual framework and the logical approaches of statistical thinking.
3. As a future executive, you should be aware of the vital role statistics plays in all walks of business, because you will be the user of statisticians' reports. The decision makers of a company cannot get full benefit from statistical studies unless they understand and appreciate statistics.

While most people readily understand that the statistician must develop his skills to a great extent, many fail to recognize that the businessman must understand what the statistician does. So many decisions are statistically based today that the decision maker who is ill-equipped in statistics may find himself in one or more of the following predicaments: (1) making inadequate decisions "by the seat of the pants" because he cannot interpret the wealth of information offered by statistical studies; (2) following the recommendations of co-workers and subordinates because he cannot challenge the assumptions and techniques used for the statistical studies; (3) making poor decisions based on improper interpretations of statistical studies; and (4) the unkindest blow of all—after operating on the "management is an art, not a science" theory, being passed up for promotion in favor of a "new young tiger" who knows both the art of management *and* the science of statistics and quantitative analysis.

Let us take a more specific look at the significance of statistical tools in decision making. All basic functions of business lend themselves to statistical consideration. For example:

Planning involves projections; statistical concepts help the planner identify alternative possibilities and determine the likelihood of success under each possibility, given different sets of circumstances that might arise.

Marketing involves statistical evaluations ranging from market research to sales analyses; probabilistic studies help the marketing experts determine optimal distribution systems.

Finance uses extensive and sophisticated statistical techniques for capital budgeting.

Accounting uses many basic aspects of statistics in its projections of cash flow and operating statements.

Production's "on-line," "real-time" control systems are rooted in statistics.

These five examples give only an indication of the areas of business and the applications within areas; for today's businessman, a keen working knowledge of statistics is not a luxury, but a necessity.

1.3 Scope, Level, and Objectives of the Book

In this book, we have tried to bridge the chasm between statistical theory and methodology. To be sure, we cover most of the methodology that constitutes the major portion of traditional business statistics texts. In addition, we relate methods to theory whenever the relation can enhance your comprehension of a particular application. Further, a primary purpose is to continually emphasize the conditions and assumptions under which the various statistical tools and techniques may be used.

While we hope you will become proficient in certain fundamental techniques and concepts, our aim is not to turn you into a statistician. Rather, you should become familiar enough with the problems, techniques, and language of statistics to be able to recognize when a problem warrants the help of a professional statistician and to communicate .intelligently and effectively with him. Thus, the theoretical developments you encounter will depend more on a logical approach to general problem solving than on mathematical rigor.

References

1.1 American Statistical Association, *Careers in Statistics*, Washington, D.C., 1962.

1.2 Judith Tanur, *Statistics: A Guide to the Unknown*, American Statistical Association, NCTM—Joint Committee on Statistics, Holden-Day, San Francisco, 1972.

Review Problems

1. Explain the difference between *descriptive* and *inferential* statistics.
2. Give three reasons why business students should study statistics.
3. Name four situations that a businessman might experience if he is ill equipped in statistics.
4. Name at least one area in which statistics is needed for each of the following basic functions of business:

   a. Accounting b. Finance c. Marketing d. Planning e. Production
5. What purpose is this textbook designed to serve?

Two

Frequency Distributions

2.1 Introduction

Many business situations require that someone organize large masses of data and present the results in a form that is readily meaningful. An effective method for meeting such a challenge, traditionally termed *descriptive statistics*, has, until recently, been the chief consideration in courses in business statistics. However, in such courses today, statistical inference is generally afforded a much more detailed treatment than descriptive statistics. Although we feel that this shift of emphasis is justified, we cannot condone the current tendency toward the exclusion of descriptive statistics. Consequently, we devote this and the following chapter to what we consider very important elements of descriptive statistics. For additional related information, see Croxton, Cowden, and Klein (2.1) and Huff and Geis (2.2).

2.2 Constructing Frequency Distributions

When an analyst receives a set of data, it usually is accompanied by a request for his conclusions or inferences. However, the data more often than not come from a survey or experiment and are delivered to the analyst in a state of disarray. Thus, before he can make inferences, the analyst must organize and summarize the data. One means of organizing data, known as a frequency distribution, is the subject of this section.

The sets of data the analyst receives are viewed in statistics as *values of a variable.* At the moment we need not be concerned with whether or not the data represent all of the values that might exist for the variable. A set might contain only part of the possible values. For our purposes, we may consider a variable simply as something that varies. To put it another way, a variable is not constant in value—it can assume more than one value.

To explore this difference between a variable and a constant, let us consider a group of circular objects of various sizes—metal washers, for example. If we measure the objects of a set selected randomly from the group, we will obtain different values for the radius for different washers. Thus, in this example, the radius is a variable. On the other hand, if we could determine very accurately the ratio of the circumference to the diameter for each washer, we would obtain the same value for every washer measured—namely, the constant π. ($\pi = 3.14159$, accurate to five decimal places).

Obviously, it is the variable that interests statisticians. Other examples of variables are: weights of animals, volumes of containers, incomes of families, populations of cities, and gross profits of companies.

When making a general reference to a variable, such as "diameters of metal washers," we shall use a capital letter to denote it, usually X. When referring to specific observed values of that variable, such as the diameters of several individual washers, we shall use subscripted lowercase letters corresponding to the capital letter used for the general variable—for example, x_1, x_2, x_3. The subscripts serve to distinguish among the various observations.

For example, let us suppose we have observed that six families on a city block have respective yearly incomes of \$5,800, \$6,200, \$4,900, \$7,100, \$5,400, and \$5,800. We would let X represent the variable "yearly income in dollars." The observed values of X, that is, the individual incomes, would be denoted as

$$X: x_1 = 5{,}800,\ x_2 = 6{,}200,\ x_3 = 4{,}900,\ x_4 = 7{,}100,\ x_5 = 5{,}400,\ x_6 = 5{,}800.$$

Note that in the foregoing discussion we implied a distinction between the *possible* and the *observed* values of the variable X.

*The **possible** values of the variable X are all the values that X may assume.*

*The **observed** values of the variable X are those values that actually have been observed.*

Theoretically at least, a *continuous* variable can take on any real number within the entire range of possible values. A variable that can assume only specific values is said to be *discrete*. Thus, we have the following nonrigorous definitions.

*A **continuous variable** is one whose possible values can be represented by an unbroken line within the relevant range.*

*A **discrete variable** is one whose possible values skip or jump within the relevant range and thus cannot be represented by a continuous line.*

Example 1

Suppose we have the following observations on the variable "number of employees" of companies in a given area:

$$X: 38, 79, 253, 176, 98, 135. \tag{1}$$

Since it makes no sense to consider fractions of employees, the variable X may assume only positive integer values. Thus, the possible values (1, 2, 3, etc.) have breaks, and the variable is discrete. The observed values of X are those given by (1).

Example 2

Let X be the weight of watermelons in a truckload. If we assume we can measure the weights of each melon as accurately as we desire, the possible values of X theoretically include all positive real numbers, and we call X a continuous variable. In other words, there is no break in the possible values that X may assume. The observed values would be the weights of the melons actually weighed.

Let us now return to the problem of organizing and summarizing data. Suppose we wish to study the variable "number of minutes per sales call" (measured to the nearest minute) for salesmen in a given sales territory. The raw data might be tabulated in the order received and hence might initially be presented as shown by Table 2.1.

Table 2.1

Minutes per sales call

25	44	30	59	23	56	48	63	39	61
85	68	41	51	87	67	58	56	43	50
40	28	53	61	32	24	82	42	72	59
65	58	63	84	47	46	52	58	53	49
33	72	42	76	54	33	74	86	36	22
54	88	26	48	75	83	57	78	47	37
73	38	70	67	60	45	64	46	63	79
58	48	64	37	72	35	53	56	85	47

Obviously, it would be difficult to draw any meaningful conclusions from the table about such questions as

1. What is the average length of time spent in a sales call?
2. On approximately what percentage of calls should a salesman expect to spend an hour or more?
3. How do the salesmen of this territory compare with those in other territories with regard to time per sales call?

What can we do to provide easy answers to these questions and countless others that might arise in a statistical study of the data?

One method is simply to arrange the numerical values in ascending (or descending) order—that is, in an *ordered array.* However, this method is quite tedious. Worse, it adds very little to the description of the set of data.

An effective method for summarizing data involves a *frequency distribution.* To construct a frequency distribution, we divide the range of values—the interval between the smallest and largest value—of the data under consideration into a number of *classes*; then, we simply note the number of values that fall into each class. A frequency distribution provides a much clearer description of the overall pattern of the distribution of the data than do tables such as Table 2.1 or ordered arrays. However, note that information is necessarily sacrificed when the raw data are summarized in a frequency distribution; in the process, the identities of the individual values are lost.

A frequency distribution may be recorded as either a table or a graph. In tabular form, it is called a *frequency table*. To illustrate, let us use the data of Table 2.1. Our first step is to answer the following questions.

1. How many classes shall we use?
2. How shall we determine *class boundaries* (dividing lines between classes) so that no ambiguity exists about which class contains any individual value?
3. How shall we label the classes?

The answer to the first question is somewhat arbitrary. However, one of the primary objectives in constructing a frequency distribution is to provide as meaningful a picture of the data as possible; Sturges (2.3) has set forth an empirical relationship for obtaining a guideline figure for an "optimum" number of classes. He suggests that the approximate number of classes be determined by

$$k = 1 + 3.3 \log n, \tag{2}$$

where k is the number of classes, n is the number of data values under consideration, and $\log n$ is the common (base 10) logarithm of n. (Common logarithms for numbers are given in Table A.9 of the Appendix). The formula is by no means intended to substitute for judgment; when n is extremely large (or small), its use leads to groupings which are perhaps too fine (or too coarse). For the values of Table 2.1, we have

$$k = 1 + 3.3 \log 80 = 1 + 3.3(1.9031) \cong 1 + 6.28 = 7.28,$$

where the symbol $\cong$ means "approximately equal."

Thus, we should use approximately seven classes in constructing a frequency table for the data on "number of minutes per sales call." However, it is not necessary to use exactly seven classes; Equation (2) merely provides a guideline, not an inflexible rule. Finally, even though the choice of the number of classes is somewhat arbitrary, there are other guidelines to note: The number of classes should be large enough to demonstrate the major characteristics of the data. Yet, the number should not be so large that the advantages of summarization are lost.

To determine class boundaries, we first note that all classes need not be of equal size. Sometimes a better picture results from classes of unequal sizes; however, it is easier to work with classes of equal size. Approximate class sizes can be determined simply by dividing the range of values by the number of classes to be used. From Table 2.1, we see that the values of the data range from 22 to 88 minutes. Thus, the size of the combined classes must be at least 66 ($88 - 22 = 66$). We could use six classes of 11 units each, seven classes of 10 units each, or eight classes of 9 units each. Since Equation (2) indicates that seven classes will be adequate, let us select seven classes of 10 units each.

Avoiding ambiguity in assigning values to classes can cause problems if one tries to use one set of rules for continuous data and another set for discrete data. It is easy to choose unambiguous class boundaries for discrete data, since there are natural breaks in the data. However, at least from a practical standpoint, this essentially holds true for continuous data. *There is a practical limit to how precise measure-*

ments can be, even in the case of continuous data. Whether dealing with discrete or continuous data, we can avoid ambiguity by selecting class boundaries that leave no doubt into which class any given observed value will be distributed. *We can accomplish this merely by including one more decimal place in the representation of the class boundary than we do in the measurement and representation of the data under consideration.*

To clarify the point, let us refer again to Table 2.1. To select classes with boundaries 20 and 30, 30 and 40, etc., would cause doubt about where to assign an observed value of 30. However, if we select as class boundaries values which cannot possibly be observed, we eliminate the ambiguity. For the data of Table 2.1, we can avoid the difficulty by setting the boundaries at 19.5, 29.5, etc. Since the values of Table 2.1 are measured only to the nearest minute, no ambiguity will develop in subsequent classification of the data among the various classes.

The proper manner for labeling classes is somewhat flexible; if the data are obviously discrete, it is usually easier to designate the classes in terms of *class limits,* that is, the largest and smallest possible values that can be attained by any observed value in a given class. For example, for the data of Table 2.1, where we established class boundaries of 19.5, 29.5, etc., the class-limit type of designation would result in the classes 20–29, 30–39, etc. (When the classes are designated by class limits, the class boundaries are located halfway between the upper and lower limits of adjacent classes.)

The length (in units) of a class interval is often referred to as *class-interval length.* Regardless of the methods chosen for determining class boundaries and labeling classes, the *class-interval length is, by definition, the difference between the upper and lower class boundaries.* For example, referring again to Table 2.1, the class 20–29 will have the boundaries 19.5 and 29.5, and the class-interval length will be 29.5 − 19.5 = 10 minutes.

When the analyst has his answers to the three questions, he is prepared to construct a frequency table. Table 2.2 presents the raw unorganized data of Table 2.1 in the form of a frequency table. Actually, the first and third columns of Table 2.2 comprise a frequency table, with the middle column (tally) merely representing an intermediate step in its construction. Note in Column 1 that, although statisticians use the respective labels 20–29, 30–39, etc., the labels should be thought of as

$$20 \leq x \leq 29, \qquad 30 \leq x \leq 39,$$

and so forth. The frequency table is constructed merely by assigning each value of Table 2.1 to one of the classes of Table 2.2. Each such assignment is indicated by a

Table 2.2
Frequency table

Class (minutes per sales call)	*Tally*	*Class frequency (number of values per class)*
20–29	~~IIII~~ I	6
30–39	~~IIII~~ ~~IIII~~	10
40–49	~~IIII~~ ~~IIII~~ ~~IIII~~ I	16
50–59	~~IIII~~ ~~IIII~~ ~~IIII~~ III	18
60–69	~~IIII~~ ~~IIII~~ II	12
70–79	~~IIII~~ ~~IIII~~	10
80–89	~~IIII~~ III	8
		80

tally mark for the proper class. The total tally for each class represents the number of values assigned to that class. This number is called the *class frequency*.

For ease of computation, one value in each class is often selected as representative of all the values in such class. Ordinarily, the midpoint of the class, or the class mark, is selected. The *class mark* may be defined as the value which lies exactly halfway between the class boundaries (or class limits). For example, the class mark of the first class in Table 2.2 is

$$(20 + 29)/2 = 24.5$$

(using class limits), or

$$(19.5 + 29.5)/2 = 24.5$$

(using class boundaries).

Since it is not necessary for class intervals to be of equal length, we can sometimes present a better picture by collecting all extreme values into one *open class* (a class with only one boundary) and then defining our other classes to cover the range of all values except the extreme values. For example, consider a set of data wherein a few of the values are either much larger or much smaller than the others. If we construct a frequency distribution for such data in the manner discussed above (using Equation (2) as a guideline for the number of classes and constructing class intervals of equal length), it may turn out that several of the class frequencies will be very small, some possible even zero. Furthermore, classes of unequal length are permissible at other than the extreme classes.

Let us consider an example. Suppose we observe a production employee perform a given operation repeatedly (say 200 times) and measure the time required for each performance. We might note that the range of values is from 8.35 to 37.56 minutes. The guideline of Equation (2) suggests approximately eight classes ($k = 1 + 3.3 \log 200 = 7.59$) to describe such data. However, suppose the resulting frequency distribution were as given by Table 2.3, where the great majority of the observations lie in the range $8 \leq x < 24$. In this case, we have failed to present a clear picture in the frequency distribution. It appears that observations are about equally likely to occur at any value within the range $8 \leq x < 24$. However, if we collect the four extreme values into one open class and divide the range 8–24 into 8 classes of equal size, we see from Table 2.4 that we can extract much better information.

Table 2.3

Frequency distribution

Time to perform operation (minutes)	*Class frequency*
8 but less than 12	40
12 but less than 16	54
16 but less than 20	58
20 but less than 24	44
24 but less than 28	2
28 but less than 32	1
32 but less than 36	0
36 but less than 40	1
	200

Table 2.4

Frequency distribution

Time to perform operation (minutes)	*Class frequency*
8 but less than 10	17
10 but less than 12	23
12 but less than 14	26
14 but less than 16	28
16 but less than 18	32
18 but less than 20	26
20 but less than 22	25
22 but less than 24	19
24 or greater	4
	200

For example, we note that values occur much more frequently toward the center of the range than at each end, a fact that is less evident in Table 2.3. This type of presentation in Table 2.4 also has disadvantages. For instance, in collecting all extreme values into one open class, we sacrifice information regarding how much larger such values might be than the other values. We also cause computational difficulties that we shall consider in Chapter 3.

Once we know the frequency of each class and the total frequency, we can convert each class frequency to a relative frequency by dividing it by the total frequency. Table 2.5 combines this information with that of Table 2.2 to construct what is called *a relative frequency distribution.* Columns 1 and 3 comprise a *relative frequency table.*

Table 2.5

Relative frequency table

Class	*Class frequency*	*Relative frequency*
20–29	6	0.075
30–39	10	0.125
40–49	16	0.200
50–59	18	0.225
60–69	12	0.150
70–79	10	0.125
80–89	8	0.100
	80	1.000

Problems

1. Explain the difference between

   a. a *discrete variable* and a *continuous variable.*

   b. the *possible values* of a variable and the *observed values* of a variable.

2. A drug manufacturer has measured the contents of five bottles of cough syrup. The observed values of the variable "volume of contents" (call it X) in cubic centimeters are:

   X: 149.6, 150.2, 148.9, 150.3, 149.9.

   a. If X can be measured to any desired degree of accuracy, is the variable X discrete or continuous?

   b. If the bottles have a capacity of exactly 151 cubic centimeters, what are the possible values of X?

3. A hardware manufacturer has counted the number of $\frac{1}{2}$-inch machine screws contained in each of eight packages. Each package has a capacity of 15 screws. He made the following observations on the variable (Y):

   Y: 15, 15, 14, 12, 15, 14, 15, 15

   a. If the variable Y can be measured to any desired degree of accuracy, is it discrete or continuous?

   b. What are the possible values of Y?

4. Suppose we have the following observations on the variable "length of screws" (X) obtained by measuring the lengths in inches of the screws contained in a package with a maximum internal dimension of 2 inches:

$$X: 0.51, 0.50, 0.50, 0.48, 0.52, 0.50, 0.49, 0.50,$$
$$0.36, 0.50, 0.47, 0.51, 0.50, 0.50, 0.49.$$

a. If the variable X can be measured to any desired degree of accuracy, is it discrete or continuous?

b. What are the possible values of X?

5. Suppose we have the following continuous data on the "maximum torque in foot-pounds of electric motors" produced by an electrical equipment manufacturer:

3.90	4.06	4.02	4.13	3.97	4.01
4.32	3.85	4.01	4.19	4.23	4.31
3.94	4.13	4.16	3.89	4.04	4.00
4.11	4.02	4.28	4.09	4.27	4.18

a. Calculate the approximate number of classes which should be used in constructing a frequency distribution of the data by Sturges' formula (2). Round your answer to the *nearest* whole number.

b. Assuming equal class-interval lengths, calculate the approximate class sizes by dividing the range of values by the number of classes calculated in Part a. Round your answer to two decimal places.

c. Construct a frequency table. Use class intervals of equal length such that each is open to the right. Use 3.85 as the lower class boundary of the first class with the number and size of the classes determined by the results of Parts a and b. Include columns for the class and class frequency in the table.

6. Suppose we have 100 observations on the variable Y. Calculate the approximate number of classes to be used in constructing a frequency distribution of this data according to Sturges' formula. Round your answer to the *nearest* whole number.

7. Sturges' formula yields a result of 8.5933 classes for a set of observations on the variable X. Approximately how many observations on X are there? Round your answer to the *nearest* whole number.

8. Explain the difference between *class boundaries* and *class limits*.

9. Define *class interval length* and *class mark*.

10. Suppose we have the following *discrete* data for the variable X.

31	47	35	51	38	62	54
60	40	117	55	43	71	43
50	58	48	67	129	51	69

Sturges' formula yields approximately 5 classes with class-interval lengths of approximately 20 for equal-size classes.

a. From this information, construct a frequency table and label the classes 30–49, 50–69, etc.

b. Construct a frequency table with 5 classes, the last one longer than the other four. Label them 30–39, 40–49, 50–59, 60–69, and 70 or greater.

c. Compare the tables of Parts a and b and determine which offers a more meaningful or more useful presentation of the data.

11. Explain the difference between *frequency* and *relative frequency*.

12. Construct a relative frequency table from your frequency table of Problem 5. Show the relative frequencies as natural fractions. Check that the total of the relative frequencies is $24/24 = 1$. Include columns for the class and the relative frequency.

2.3 Cumulative Frequency Distributions

We often need to show the number of values in a given set that are less than, or greater than or equal to, a certain number, or perhaps the number of values between two given numbers. Such information is best presented in the form of a *cumulative frequency distribution.* We shall first construct a *cumulative frequency table* from information given in the frequency table.

Cumulative frequency distributions may be of the *less-than* type or of the *greater-than-or-equal-to* type, each providing the kind of information implied by the respective names. For practical purposes, both provide the same information. Thus, we shall generally deal with only one type, the less-than cumulative frequency distribution. To construct it from a frequency table, we merely tabulate the number of values in the entire set of data that are less than the lower class limit of each class, including the lower limit of the hypothetical class that is to the right of the rightmost class of the frequency distribution. For example, in Table 2.6, the last two columns represent a less-than cumulative frequency table constructed from the first two columns (from Table 2.2). Note that each value in column 4 (cumulative frequencies) is the sum of all the values (frequencies) of column 2 corresponding to classes (column 1) for which

Table 2.6

A "less-than" cumulative frequency table

Frequency table		*Cumulative frequency table*	
Class (minutes per sales call)	*Class frequency*	*Minutes per sales call*	*Cumulative frequency (less-than)*
		$X < 20$	0
20–29	6	$X < 30$	6
30–39	10	$X < 40$	16
40–49	16	$X < 50$	32
50–59	18	$X < 60$	50
60–69	12	$X < 70$	62
70–79	10	$X < 80$	72
80–89	8	$X < 90$	80
	80		

the upper limits are less than the relevant lower limit given in column 3. For example, 32 (the cumulative frequency corresponding to the possible value 50) is obtained by adding 6 (frequency of the 20–29 class), 10 (frequency of the 30–39 class), and 16 (frequency of the 40–49 class).

In a greater-than-or-equal-to cumulative frequency distribution (sometimes refered to as a *decumulative frequency distribution*), we tabulate the number of values in the entire set of data which are greater than or equal to the lower class limit of each class, again including the lower limit of the hypothetical class which is after the last class $80 \leq X \leq 89$. The last two columns of Table 2.7 present a greater-than-or-equal-to

Table 2.7

A "greater-than-or-equal-to" frequency table

Frequency table		*Decumulative frequency table*	
Class (minutes per sales call)	*Class frequency*	*Minutes per sales call*	*Decumulative frequency (greater-than-or-equal-to)*
		$X \geq 20$	80
20–29	6	$X \geq 30$	74
30–39	10	$X \geq 40$	64
40–49	16	$X \geq 50$	48
50–59	18	$X \geq 60$	30
60–69	12	$X \geq 70$	18
70–79	10	$X \geq 80$	8
80–89	8	$X \geq 90$	0
	80		

cumulative frequency table constructed from the frequency table of Table 2.2. Note that each value in column 4 (decumulative frequencies) is the sum of all the values (frequencies) of column 2 corresponding to classes (column 1) for which the lower limits are greater than or equal to the relevant lower limit in column 3. For example, 30 (the decumulative frequency corresponding to the possible value 60) is obtained by adding 12, 10, and 8—in other words, all the sales calls that took 60 minutes or more.

Tables 2.6 and 2.7 show that, for any possible value of the variable X, the sum of the corresponding cumulative frequency and decumulative frequency is equal to the total number of values in the entire set of data.

The method for constructing cumulative distributions for relative frequencies is completely analogous to the procedure described for absolute frequencies.

Problems

1. Suppose we have the following frequency table for a *discrete* variable:

Class (units sold/week)	*Frequency*
0–19	4
20–39	19
40–59	8
60–79	12
80–99	7
	50

Construct:

a. A table of relative frequencies accurate to two decimal places.

b. A *less-than* cumulative frequency table.

c. A *greater-than-or-equal-to* cumulative frequency table.

d. A *less-than* cumulative relative frequency table.

e. A *greater-than-or-equal-to* cumulative relative frequency table.

2. Suppose we have the following frequency table for a *continuous* variable:

Class (diameter in millimeters)	*Frequency*
86 but less than 87	7
87 but less than 88	5
88 but less than 89	10
89 but less than 90	32
90 but less than 91	29
91 but less than 92	11
92 but less than 93	6
	100

Construct:

a. A table of relative frequencies accurate to two decimal places.

b. A *less-than* cumulative frequency table.

c. A *greater-than-or-equal-to* cumulative frequency table.

d. A *less-than* cumulative relative frequency table.

e. A *greater-than-or-equal-to* cumulative relative frequency table.

3. a. Construct a *less-than-or-equal-to* cumulative frequency table from the frequency table of Problem 1.

b. Construct a *greater-than* cumulative frequency table from the frequency table of Problem 1.

c. Check that the sum of the cumulative frequencies of each corresponding class of the *less-than-or-equal-to* and *greater-than* cumulative frequency tables constructed in Parts a and b, respectively, is equal to the total number of observations. Explain why this relation must hold.

4. Assume that there are no observations exactly equal to any of the class boundaries.

a. Construct a *less-than-or-equal-to* cumulative relative frequency table from the relative frequency table of Problem 2. Use upper class boundaries as limits.

b. Construct a *greater-than* cumulative relative frequency table from the relative frequency table of Problem 2. Use upper class boundaries as limits.

c. Check that the sum of the cumulative relative frequencies of each corresponding class of the *less-than-or-equal-to* and *greater-than* cumulative frequency tables constructed in Parts a and b, respectively, is equal to unity. Explain why this relation must hold. Why must we assume that there are no observations exactly equal to any of the class boundaries in this problem?

2.4 Graphic Representation

Although frequency tables are useful for summarizing and presenting data, many people lack the time or inclination to cope with numerical data. Analysts often solve this problem by presenting a picture, or a graph, that tells the same story. Their general method, known as *graphic representation*, may be used either to supplement or replace a tabular presentation. Of the many forms for presenting data graphically, we shall discuss three: *histograms*, *frequency polygons* (or curves), and *cumulative frequency polygons* (or curves). The last is often called an *ogive* (pronounced ō-jive).

A histogram is a graphic presentation of a frequency distribution, constructed by drawing rectangles with *areas* proportional to the frequencies of the class intervals. Values of the variable under consideration are represented on a horizontal scale, while, in the case of class intervals of equal length, frequencies are represented on the vertical scale. When the class intervals are not of equal length, it is appropriate to represent *class density* on the vertical scale. (Density is defined here as the ratio of class frequency to class interval length.) To show the proper method and clarify the reason for the two kinds of vertical scales, let us construct histograms showing the data of Table 2.2 and a modified version of such data.

We obtain the histogram of Figure 2.1 from the frequencies given in Table 2.2. Note that the class intervals are separated in the graph by class boundaries (19.5, 29.5, etc.). Careful examination of Figure 2.1 reveals that the areas of the rectangles are in fact proportional to the corresponding frequencies of Table 2.2. In this particular example, the heights of the rectangles are also proportional to the corresponding frequencies, but only because we are using class intervals of equal length.

When dealing with class intervals of unequal length, any attempt to make the *heights* of the rectangles proportional to corresponding frequencies will preclude

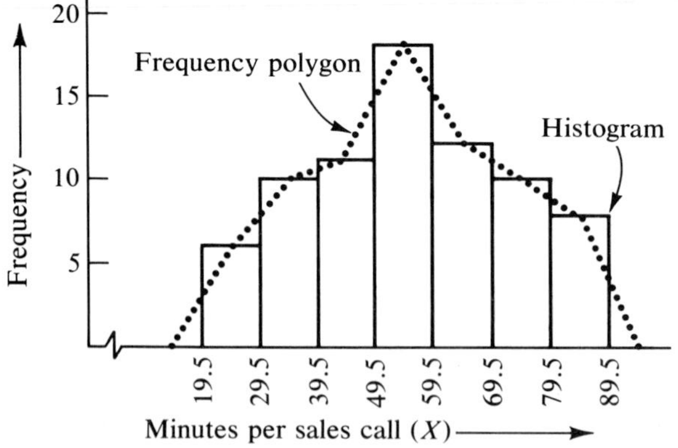

Figure 2.1
Histogram and frequency polygon of frequencies from Table 2.2

making the *areas* proportional. Thus, we will, by definition, have something other than a histogram. Worse yet, the resulting graph will distort the information.

Consider a modified version of the data of Table 2.2. Suppose we are given a frequency distribution from the data of Table 2.1, in which the observations in the last three classes (60–69, 70–79, and 80–89) have been grouped into one larger class (60–89). Table 2.8 shows the resulting frequency table.

Table 2.8
Frequency table

Class (minutes/sales call)	*Class frequency*
20–29	6
30–39	10
40–49	16
50–59	18
60–89	30

Now, let us look at the distortion that occurs if the graph is constructed so that the heights of the rectangles are proportional to the corresponding frequencies. The result is shown in Figure 2.2. Although the 30 calls represented there are only $36\frac{1}{2}\%$ of the total number of calls in the data, they are given an area more than twice the area given to all of the other 52 calls. One would infer from the relative areas that there are far more values between 59.5 and 89.5 than between 19.5 and 59.5!

To depict the information correctly, we must construct the histogram so that the area of the corresponding rectangle is in fact proportional to the number of values which lie between 59.5 and 89.5, as in Figure 2.3. Thus, the use of class density is necessary to prevent a distorted graphic presentation. To do this, we must average the total frequencies of the three combined classes over their total area.

A *frequency polygon* provides essentially the same information as a histogram. In fact, it is quite easy to construct a frequency polygon from a histogram. To do so, we merely connect by straight line segments the midpoints of the tops of the rectangles of the histogram. So that the curve will touch the horizontal axis at each end (thus forming a closed polygon), we ordinarily include a class with zero frequency at each

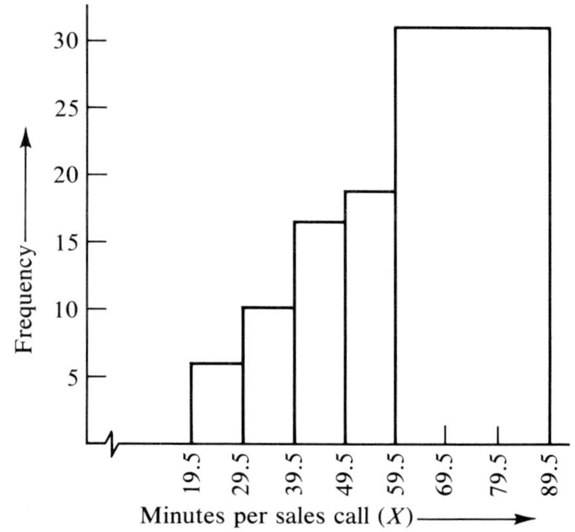

Figure 2.2
Incorrect histogram of frequencies, Table 2.8

end of the distribution. A frequency polygon for the raw data of Table 2.2 is given by the dotted lines in Figure 2.1. Note that the frequency polygon can also be constructed directly from a frequency table even when there is no reason to build a histogram as an intermediate step. In such case, we merely plot densities (or frequencies when using classes of equal length) in the vertical direction perpendicular to the midpoints of corresponding class intervals in the horizontal direction.

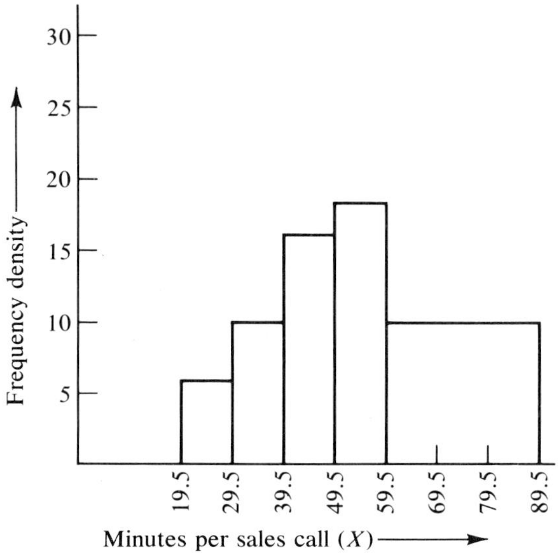

Figure 2.3
Correct histogram of frequencies, Table 2.8

Interpreting a frequency polygon requires caution. It is incorrect to interpret the height of the polygon at any given point along the horizontal axis as the frequency associated with any single value of the variable. Rather, we simply say that *the height of the polygon at each midpoint represents the density for the class interval associated with that midpoint*. In effect, the polygon is the same as the histogram in that we represent the frequency associated with a given class interval by the area under the frequency polygon between the boundary points of the interval. In other words, once again, corresponding areas are proportional to frequencies.

Why is this fact so important? Consider what happens if we associate frequencies with intermediate points along the frequency polygon of Figure 2.1. We know that the total number of observed values is 80. Yet by choosing only a few points along the horizontal axis and interpreting the corresponding heights as frequencies, we see that our total frequency would quickly exceed 80—an absurdity.

The point takes on added significance when we extend our thinking slightly. If we were dealing with a very large number of observed values and could select very narrow class intervals (so narrow that each essentially has zero length), the frequency polygon would appear as a smooth polygon, or *curve*. Such a curve is commonly labeled a *frequency curve*, a graphic representation of a theoretical frequency distribution.

Even in this instance, it is not proper to interpret the height of the curve as frequency associated with a given possible value. Rather, as with histograms and frequency polygons, we interpret "the area under the curve between any two possible values" as representing "the frequency (or relative frequency) of values which are between the two possible values." In succeeding chapters, we shall find this type of thinking highly important, even though the curves which we shall be discussing may be obtained or derived in a manner quite different from that described above.

A cumulative frequency distribution may be depicted by a *cumulative frequency polygon* (or curve). The cumulative frequency polygon is constructed by (1) plotting cumulative frequencies on the vertical axis versus appropriate class boundaries (or limits) on the horizontal axis and (2) connecting the resulting points by straight-line segments. Figure 2.4 presents both the cumulative and decumulative frequency

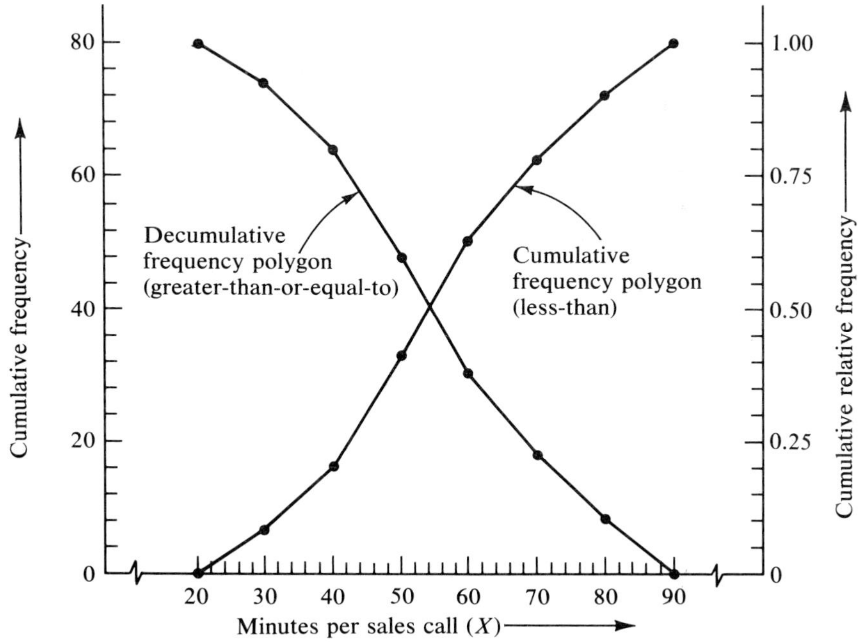

Figure 2.4
Cumulative and decumulative frequency polygons from data of Tables 2.6 and 2.7

polygons constructed from the corresponding distributions of Table 2.6 (less-than) and Table 2.7 (greater-than-or-equal-to). In Figure 2.4, the vertical axis is recorded in "cumulative frequency" on the left and "cumulative relative frequency" on the right, while the horizontal scale records possible values of the variable under consideration (minutes per sales call).

It is permissible (once one realizes results are approximate) to read information from cumulative frequency polygons from points between the points actually plotted. A cumulative frequency polygon can be used to answer many questions, as the following examples demonstrate.

Example 1

Use the cumulative frequency polygon of Figure 2.4 to find:
(a) The valuc (say x^*) below which 45 percent of the values of X lie.
(b) The value (say x^*) such that 65 of the values of X are greater than or equal to it.
(c) The number of values of X which are less than 45.
(d) The proportion of values of X which are greater than or equal to 74.

Solution

(a) Find 0.45 (45 percent) on the right vertical scale and draw a horizontal line from there to the cumulative frequency polygon. Draw a vertical line from the point the horizontal line intersects the polygon to the horizontal axis. Read x^* at the point where your vertical line intersects the horizontal axis. The result is approximately $x^* = 52$. (See Figure 2.5.)

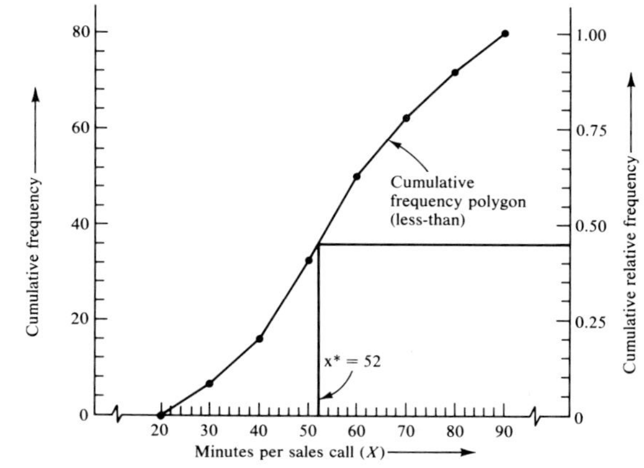

Figure 2.5

(b) If 65 of the values of X are greater than or equal to x^*, 15 must be less than x^* ($80 - 65 = 15$). Thus, we find x^* as follows. Find 15 on the left vertical scale. Draw a horizontal line from 15 to the cumulative frequency polygon. Draw a vertical line to the horizontal axis from the intersection of the horizontal line and the cumulative frequency polygon. Read x^* at the point where the vertical line intersects the horizontal axis. (See Figure 2.6.)

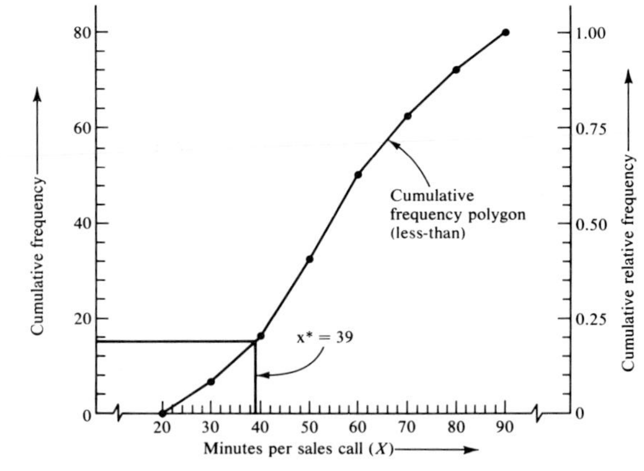

Figure 2.6

(c) Find 45 on the horizontal axis. Draw a vertical line from 45 to the cumulative frequency polygon. Draw a horizontal line from the intersection of the vertical line with the cumulative frequency polygon to the left vertical axis. Read the answer at the point where the constructed horizontal line segment intersects the left vertical axis. The result is that approximately 24 values are less than 45.

(d) To find the proportion (or percentage) of values of X greater than or equal to 74, we subtract the proportion (or percentage) less than 74 from 1 (or 100 percent). To do this, first find 74 on the horizontal axis. Draw a vertical line from 74 to the polygon and a horizontal line from there to the right vertical axis. The value at the point is 0.825, the proportion of values less than 74. Finally, the proportion of values greater than or equal to 74 becomes $1 - 0.825 = 0.175$. □

Example 2

In the cumulative frequency polygon of Figure 2.4 find the number of values of X greater than or equal to 38 but less than 79.

Solution

Read (as in Example 1) the number of values less than 79 to be 71. Read the number of values less than 38 to be 14. Subtract the number of values less than 38 (14) from the number of values less than 79 (71) to obtain the number of values greater than or equal to 38 but less than 79. We obtain $71 - 14 = 57$. □

Note that any of the information of Examples 1 and 2 could have been obtained just as easily by using the decumulative frequency polygon of Figure 2.4.

Problems

1. Define *density* and explain why it should be used for constructing histograms when dealing with intervals of unequal length.
2. Using the frequency table of Problem 5-c, page 12:
   a. Construct a histogram.
   b. Construct a frequency polygon on the same set of axes used in Part a.
3. Using the frequency and relative frequency tables of Problem 1, page 14:
   a. Construct a frequency polygon. Label the vertical axis "frequency."
   b. Construct a relative frequency polygon. Label the vertical axis "relative frequency."
   c. Group the first three classes into one class and construct a frequency polygon.
   d. Group the last two classes into one class and construct a relative frequency polygon.
4. Using the frequency and relative frequency tables of Problem 2, page 15:
   a. Construct a frequency polygon. Label the vertical axis "frequency."
   b. Construct a relative frequency histogram. Label the vertical axis "relative frequency."
   c. Group the last two classes into one class and construct a histogram.

d. Group the second and third classes into one class and construct a relative frequency polygon.

5. In a relative frequency polygon, we interpret "the area under the curve between any two possible values" as representing "the relative frequency of values which are between the two possible values." What number does "the area under the curve" between the extreme limits of a relative frequency polygon represent? Why?

6. a. Construct a *less-than* cumulative frequency polygon from the cumulative absolute and cumulative relative frequency tables of Problem 1, page 14. Label the variable of interest "X".

   b. From the same tables, construct a *greater-than-or-equal-to* cumulative frequency polygon on the same axes used for Part a.

7. Construct a *less-than* cumulative frequency polygon from the cumulative absolute and cumulative relative frequency tables of Problem 2, page 15. Label the variable of interest "X."

8. Use the cumulative frequency polygon of Problem 7 to find:

   a. The number of values such that $X < 91$; such that $X \geq 89$.

   b. The percentage of values such that $X \geq 90$; such that $X < 87$.

   c. The number of values such that $X < 89.5$.

   d. The percentage of values such that $X \geq 87.4$.

   e. The number of values such that $88 \leq X < 92$.

   f. The percentage of values such that $89.5 \leq X < 91$.

   g. The value x^* such that there are 54 values such that $X < x^*$.

   h. The value x^* such that 62% of the values of X are equal to or greater than x^*.

   i. The value x_1^* such that there are 21 values of X such that $x_1^* \leq X < 89.5$.

   j. The value x_2^* such that 77% of the values of X satisfy $88.8 \leq X < x_2^*$.

9. Use the appropriate cumulative frequency polygon of Problem 6 to find:

   a. The percentage of values such that $X \geq 40$.

   b. The number of values such that $X < 70$.

   c. The value x^* such that 62% of the values of X are less than x^*.

   d. The value x^* such that 36.5 values of X are equal to or greater than x^*.

   e. The number of values such that $57.5 \leq X < 80$.

   f. The value x_2^* such that 34% of the values of X satisfy $40 \leq X < x_2^*$.

   g. The percentage of values such that $45 \leq X < 65$.

   h. The value x_1^* such that 35 values of X satisfy $x_1^* \leq X < 70$.

   Check your answers to Parts a through h by applying linear interpolation to the appropriate cumulative frequency and cumulative relative frequency tables of Problem 1, page 14. The answers obtained by the graphical and numerical methods should be approximately equal.

10. Explain why, given a frequency distribution (frequency table, histogram, polygon; cumulative frequency table or polygon), we can ascertain (at least approximately) the number of observations *between* any two given values, but not the number of observations *equal* to a given value.

11. Suppose we have the following *less-than* cumulative frequency table for the continuous variable "man days per job":

Man days/job (X)	*Cumulative frequency* (*less-than*)
$X < 5$	0
$X < 10$	7
$X < 15$	12
$X < 20$	25
$X < 25$	34
$X < 30$	40

a. Construct a frequency table.
b. Draw a histogram with "frequency" on the vertical axis.
c. Construct a relative frequency table. Round numbers to three decimal places.
d. Draw a relative frequency polygon with "relative frequency density" on the vertical axis.
e. Calculate the relative areas of every class from the relative frequency polygon by geometric methods and check to see that they approximately equal the corresponding relative frequencies. Show your work. Round all numbers to three decimal places.

12. Suppose we have the following *greater-than-or-equal-to* cumulative relative frequency table for 25 observations on the discrete variable "cost per unit" in cents:

Cost/unit (X) (*cents*)	*Cumulative relative frequency* (*greater-than-or-equal-to*)
$X \geq 2815$	1.00
$X \geq 2830$	0.88
$X \geq 2845$	0.56
$X \geq 2860$	0.32
$X \geq 2875$	0.00

a. Construct a relative frequency table. Round all numbers to two decimal places.
b. Construct a relative frequency polygon. Label the vertical axis "relative frequency."
c. Calculate the relative areas of all classes from the relative frequency polygon by geometric methods and check to see that they are approximately equal to the corresponding relative frequencies. Show your work. Round all numbers to three decimal places.

13. a. Use information from the histogram in Figure 2.7 to construct a frequency table for the distribution. Assume that all data are continuous.

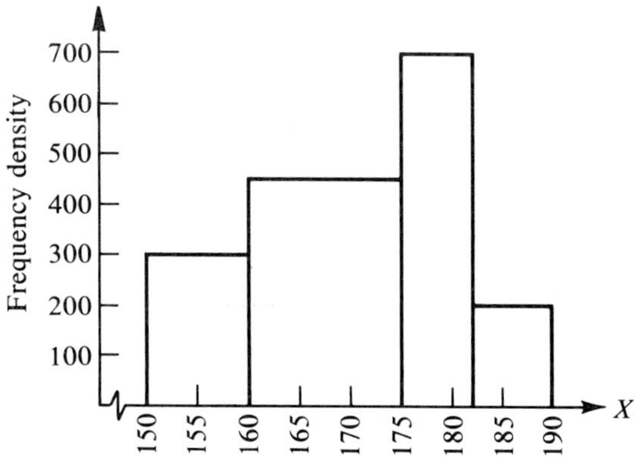

Figure 2.7

b. Use information from the frequency polygon in Figure 2.8 to construct a frequency table for the distribution. Assume that all data are continuous.

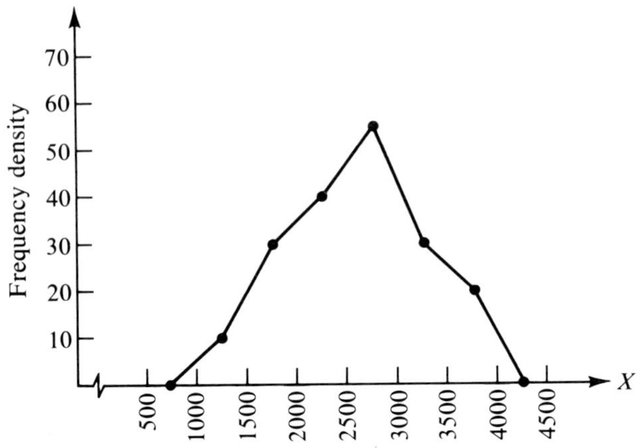

Figure 2.8

14. Read the following information from the cumulative frequency polygon in Figure 2.9.

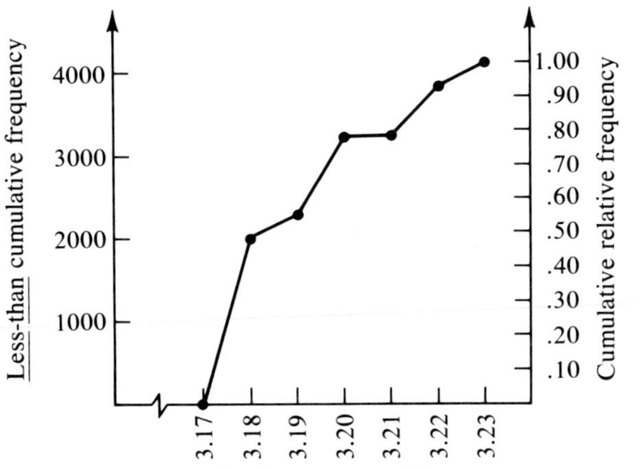

Figure 2.9

a. The number of values such that $X < 3.205$.

b. The percentage of values such that $X \geq 3.215$.

c. The value x^* such that 51% of the values of X satisfy $X < x^*$.

d. The number of values of X between 3.178 and 3.196.

e. The value x_2^* such that 37% of the values of X satisfy $3.192 \leq X < x_2^*$.

15. Construct a frequency table from the cumulative frequency polygon in Figure 2.9. Assume that all data are continuous.

2.5 *Summary Remarks*

The most basic and important point in the construction of frequency distributions is to present the data in such a way that it can be easily understood. Therefore, the foregoing rules and guidelines should always be tempered with an ample dosage of common sense and judgment. After all, the type of data description required in any given situation depends on its ultimate purpose. The analyst can make this decision better than anyone else.

Important Terms and Concepts

class
class boundary
class interval
class interval length
class limit
class mark
constant
continuous variable
cumulative frequency
cumulative frequency distribution
cumulative frequency polygon
decumulative frequency distribution
density
descriptive statistics
discrete variable
frequency
frequency curve
frequency distribution
frequency polygon
frequency table
histogram
intervals open on the right
observed value
ogive
possible value
relative frequency
relative frequency distribution
variable

References

2.1 Frederick Emery Croxton, Dudley J. Cowden & Sydney Klein, *Applied General Statistics*, 5th Ed. Prentice Hall, New Jersey, 1967.

2.2 Darrell Huff & Irving Geis, *How to Lie with Statistics*, Norton Publishing, New York, 1954.

2.3 H. A. Sturges, "The Choice of a Class Interval," *J. Am. Stat. Assoc.*, March 1926, pp. 65 66.

Review Problems

1. Suppose we have the following observations on the variable "number of customers per hour" in a small retail store:

34	19	27	44	31	29	38	17
26	30	35	21	36	24	31	27
25	41	15	26	30	28	21	32
25	39	18	22	29	32	26	24

   a. Arrange the data in an ascending ordered array.

   b. Calculate the approximate number of classes which should be used in constructing a frequency distribution of the data by Sturges' formula (2). Round the result to the *nearest* whole number.

   c. Assuming class intervals of equal length, calculate the approximate class sizes by dividing the range of values by the number of classes calculated in Part b. Round your answer to the *nearest* whole number.

   d. Construct a frequency table, using class intervals of equal length and determining the number and size of the classes from your answers to Parts b and c. Use 14.5 as the lower class boundary of the first class and label the classes by using class limits. Include columns for the class, tally, and class frequency in the table.

   e. Construct a histogram.

   f. Draw a frequency polygon on the same set of axes used for Part e.

2. Construct a relative frequency table from your frequency table of Problem 1. Include columns for the class, class frequency, and relative frequency. Compute the relative frequencies accurate to 5 decimal places. Check that the total of the relative frequencies is 1.00000.

3. A set of data consists of 500 observations on the variable X ranging in value from 618 to 707.

   a. Calculate the approximate number of classes to be used in constructing a frequency distribution of this data according to Sturges' formula. Round your answer to the *nearest* whole number.

   b. Calculate the approximate length of each class interval. Assume classes of equal size and determine the number of classes from your answer to Part a. Round your answer to the *nearest* whole number.

4. Sturges' formula yields a result of 10.9000 classes for a set of observations on the variable Y. Approximately how many observations on Y are there? Round your answer to the *nearest* whole number.

5. The Internal Revenue Service wishes to construct a frequency distribution of the variable "gross personal annual income," in dollars, for every U.S. citizen for the year 1970. The values range from 0 to 100,000,000.

   a. If there are 200,000,000 U.S. citizens, approximately how many classes should be used according to Sturges' formula? Round your answer to the *nearest* integer.

   b. If equal class interval lengths are used and the results of Part a are applied, approximately how long will the class intervals be? Round your answer to the *nearest* whole number.

   c. If the distribution is constructed according to the results of Parts a and b, how useful will it be for depicting the characteristics of the variable of interest? Explain. Would the use of appropriate unequal class intervals improve its usefulness?

6. Read the following information from the cumulative frequency polygon in Figure 2.10.

   a. The percentage of values such that $X \geq 715$.

   b. The number of values such that $X < 724$.

   c. The value x^* such that 8 values of X are equal to or greater than x^*.

   d. The percentage of values such that $725 \leq X < 739$.

   e. The value x_2^* such that 34 values of X satisfy $711 \leq X < x_2^*$.

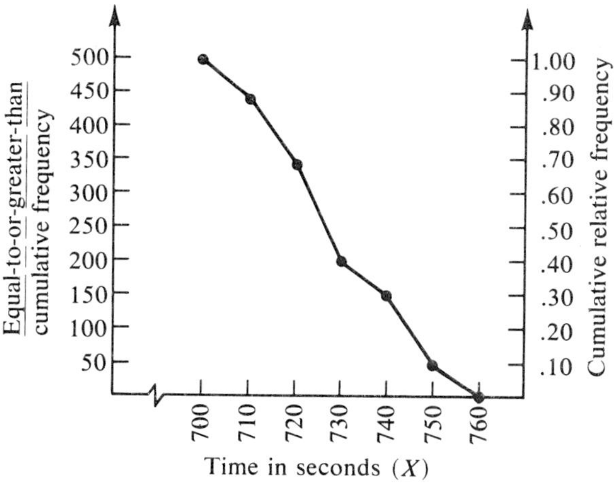

Figure 2.10

Three

Measures of Location and Dispersion

3.1 Introduction

We saw in the previous chapter that a set of raw data can be presented more meaningfully by organizing it into a frequency distribution. Sometimes we want to characterize data even more compactly, by calculating certain descriptive measures. Naturally, we cannot characterize it as completely by only two or three measures as we can by using an entire frequency distribution, particularly if we stay strictly in the domain of descriptive statistics. However, if we calculate appropriate measures, we can extract a great deal of valuable information. What measures should we calculate?

A graphical representation of a frequency distribution allows us to determine very quickly

1. where on the horizontal axis the central tendency of the distribution lies and
2. how much spread or variation exists within the data.

For example, Figure 2.1 (page 16) shows clearly that the observed values are concentrated and distributed around a central value somewhere between 39 and 69; the values cover the range 20–90; and the values are much more closely concentrated around the possible values toward the center of the distribution than around possible values near the end points.

We shall find in the present chapter that we can obtain similar information by calculating and interpreting two kinds of descriptive measures—*measures of location* and *measures of dispersion.*

These measures are important for several reasons. First, it is much easier to calculate them than to build an entire frequency distribution. Second, in some situations, particularly when the number of values is fairly small, measures of location and dispersion actually provide more useful information than does a frequency distribution.

Finally, and by far most important, such measures play a vital role in statistical inference, a topic of later chapters.

Before formally discussing measures of location and dispersion, we must deal with several concepts that are quite important in related calculations. In Chapter 2, we were not too descriptive about the nature of the data. We avoided discussing whether the data represented all of the possible values associated with the variable of interest or only a part of the possible values. It was not necessary, because all the techniques of Chapter 2 are applicable in both situations. However, we must now begin to recognize this difference and the necessary terminology to do so.

If a set of data consists of the totality of objects (values) under consideration, we shall label it a *population*; if it consists of only a part of the objects under consideration, we shall label it a *sample*. The interpretation of any given set of data as a population or a sample depends entirely on the circumstances which give rise to the data. In this respect, the phrase "totality of objects under consideration" is the key; what we intend to do with a set of data essentially determines whether the set is a population or sample. Let us clarify this idea with an example.

Suppose we have a set of data comprised of the ages of all 25 executives of one public accounting firm. With only this information, we cannot determine whether to interpret the set as a population or as a sample. It may be either. However, if we find that the set is to be used by the firm in designing a retirement plan for its executives, we know that our set of data represents the totality of objects under consideration; thus, it should be interpreted as a population. On the other hand, suppose the American Accounting Association is conducting a study from which will be drawn inferences about the distribution of ages of all public accounting firm executives in the United States. If the ages of the 25 executives in our hypothetical firm were to be incorporated into such study, our set of data would be a sample; the "totality of objects under consideration"—that is, the population—would be the ages of all public accounting firm executives in the country.

The reason for differentiating between a sample and a population will become more apparent as we move along. We introduce the matter in this chapter, because the symbolism used when working with a sample is different from that used for a population. It is customary to use letters of the Roman alphabet (a, b, c, ...) to represent sample descriptive measures. In this chapter we shall treat all data sets as if they were samples—that is, we shall use Roman letters to represent descriptive measures. In later chapters, when it becomes critical to distinguish between populations and samples, we shall use Greek letters (α, β, γ, ...) to represent population descriptive measures.

We must also distinguish between ungrouped data and grouped data, primarily because computational formulas vary according to the form of the set of data under consideration. If all individual values (observations) of the data under consideration can be identified, we describe them as *ungrouped data*. For example, the set of data of Table 2.1 represents ungrouped data, since, in this form, we are able to attach a numerical value to every observed value (25, 44, ..., 47). On the other hand, if the data are presented in the form of a frequency distribution, we describe the set as *grouped data*. That is, the various values are available only in groups. For example, if we were given the data of Table 2.2 without knowledge of the information of Table 2.1, we would have a set of grouped data. We know, for instance, that there are six observations between 20 and 29 inclusive, but we have no way of knowing the exact numerical values of these six observations.

3.2 Summation Notation

Since the steps required in calculating descriptive measures for data involve so many summations, products, summations of products, products of summations, and so forth, it is very important for you to be adept at manipulating summation notation. Recall that we have decided to represent a general variable (measurement of a certain characteristic of an object) by an uppercase letter (say X) and individual observations on such variable by subscripted lowercase letters ($x_1, x_2, \ldots$). Under this scheme of notation, we usually designate any arbitrary observed value of X by x_i, and call it the "ith observed value of X." Thus, if the subscript i takes on the value 5, for the fifth observed value of x, we specify it as x_5.

Suppose we have a set of n (where n is any positive integer) observed values on each of three variables, X, Y and Z. We would denote the variables and sets of observations by

$$X: x_1, x_2, x_3, \ldots, x_n;$$

$$Y: y_1, y_2, y_3, \ldots, y_n;$$

and

$$Z: z_1, z_2, z_3, \ldots, z_n.$$

To facilitate the discussion of sums of various combinations of observations, we introduce a summation operator, denoted by $\sum$ (the capital Greek letter *sigma*). Henceforth, $\sum$ will indicate that a summation is to be performed. For example, to denote the sum of n values of the variable X, we write

$$\sum_{i=1}^{n} x_i, \tag{1}$$

where $\sum$ is the summation operator, the subscript i is the index of summation, and the 1 and n designate the range of summation. Relation (1) is read "the summation of all the values of the variable X which have subscripts between 1 and n, inclusive." That is,

$$\sum_{i=1}^{n} x_i = x_1 + x_2 + \cdots + x_n. \tag{2}$$

In general, the value appearing below the summation sign represents the beginning value for i, and the value appearing above the summation sign represents the ending value of i. Thus, to denote a partial sum, say, the four values from 5 to 8, inclusive, we write

$$\sum_{i=5}^{8} x_i = x_5 + x_6 + x_7 + x_8.$$

If it is clear from the context that the summation is to be taken over the entire range of data, the range of summation is sometimes omitted, a practice we shall follow quite often.

If we wish to denote the sum of products of the n observations on two variables, X and Y, we write $\sum_{i=1}^{n} x_i y_i$, which is read "the summation of the products of all observations having equal subscripts between 1 and n, inclusive." That is,

$$\sum_{i=1}^{n} x_i y_i = x_1 y_1 + x_2 y_2 + \cdots + x_n y_n. \tag{3}$$

As a general rule, the summation operator followed by any algebraic expression whatsoever will mean that the expression should be computed for every value indicated for i (the range of i), and then the resulting values should be summed. For example, given the summation

$$\sum_{i=3}^{4} (x_i^2 - y_i)^3,$$

we would

1. subtract y_3 from the square of x_3 and raise the difference to the third power;
2. subtract y_4 from the square of x_4 and raise the difference to the third power;
3. add the results of 1 and 2.

The three algebraic rules that follow greatly facilitate computations with summations.

Rule 3.1

The summation of a sum (or difference) is the sum (or difference) of the individual summations; that is,

$$\sum_{i=1}^{n} (y_i - x_i + z_i) = \sum_{i=1}^{n} y_i - \sum_{i=1}^{n} x_i + \sum_{i=1}^{n} z_i.$$

This rule follows easily from the fact that

$$\begin{aligned}\sum_{i=1}^{n} (y_i - x_i + z_i) &= (y_1 - x_1 + z_1) + (y_2 - x_2 + z_2) + \cdots + (y_n - x_n + z_n)\\ &= (y_1 + y_2 + \cdots + y_n) - (x_1 + x_2 + \cdots + x_n)\\ &\quad + (z_1 + z_2 + \cdots + z_n)\\ &= \sum_{i=1}^{n} y_i - \sum_{i=1}^{n} x_i + \sum_{i=1}^{n} z_i.\end{aligned}$$

Rule 3.2

The summation of the product of a constant and a variable is equal to the product of the constant and the summation of the variable; that is,

$$\sum_{i=1}^{n} c x_i = c \sum_{i=1}^{n} x_i.$$

(Note that we indicate a *constant* by omitting a subscript. In other words, because c is a constant, it takes on the same value regardless of the value of i.) For example,

$$\sum_{i=1}^{n} 3x_i = 3x_1 + 3x_2 + \cdots + 3x_n = 3(x_1 + x_2 + \cdots + x_n) = 3\sum_{i=1}^{n} x_i.$$

Rule 3.3

The summation of a constant is the constant multiplied by the number of terms in the summation; that is,

$$\sum_{i=1}^{n} c = nc.$$

It should be noted that Rule 3.3 is a special case of a more general rule. If the summation begins with i assuming a value other than 1 (say $i = m$), the rule becomes

$$\sum_{i=m}^{n} c = (n - m + 1)c.$$

If you master the foregoing rules, and adhere to algebraic tradition, by reading summation and product indicators (parentheses, brackets, braces, etc.) from the innermost grouping to the outermost grouping, interpretation of notation involving the summation operator will be quite easy. For example, note the difference between what we mean by

$$\sum_{i=1}^{n} x_i y_i \quad \text{and} \quad \left(\sum_{i=1}^{n} x_i\right)\left(\sum_{i=1}^{n} y_i\right).$$

The first case indicates a *sum of products*, while the second indicates a *product of sums*.

Example 1

Suppose we have the following observations on two variables, X and Y.

$$X: x_1 = 50; \quad x_2 = 25; \quad x_3 = 40; \quad x_4 = 60; \quad x_5 = 65,$$

and

$$Y: y_1 = 20; \quad y_2 = 20; \quad y_3 = 40; \quad y_4 = 30; \quad y_5 = 10.$$

Find:

(a) $\displaystyle\sum_{i=1}^{5} \frac{x_i}{y_i}$, (b) $\displaystyle\frac{\sum_{i=1}^{5} x_i}{\sum_{i=1}^{5} y_i}$, (c) $\displaystyle\sum_{i=3}^{5} \left(\frac{x_i - 35}{5}\right)^2$.

Solution

(a) We calculate the sum of quotients as

$$\sum_{i=1}^{5} \frac{x_i}{y_i} = \frac{x_1}{y_1} + \frac{x_2}{y_2} + \frac{x_3}{y_3} + \frac{x_4}{y_4} + \frac{x_5}{y_5}$$

$$= \frac{50}{20} + \frac{25}{20} + \frac{40}{40} + \frac{60}{30} + \frac{65}{10} = 2.5 + 1.25 + 1.0 + 2.0 + 6.5 = 13.25.$$

(b) We calculate the quotient of sums as

$$\frac{\sum_{i=1}^{5} x_i}{\sum_{i=1}^{5} y_i} = \frac{50 + 25 + 40 + 60 + 65}{20 + 20 + 40 + 30 + 10} = \frac{240}{120} = 2.$$

(c) Here, we want the sum of the quotients squared. Working from the inside out, so to speak, we have

$$\sum_{i=3}^{5}\left(\frac{x_i - 35}{5}\right)^2 = \left(\frac{x_3 - 35}{5}\right)^2 + \left(\frac{x_4 - 35}{5}\right)^2 + \left(\frac{x_5 - 35}{5}\right)^2$$

$$= \left(\frac{40 - 35}{5}\right)^2 + \left(\frac{60 - 35}{5}\right)^2 + \left(\frac{65 - 35}{5}\right)^2$$

$$= \left(\frac{5}{5}\right)^2 + \left(\frac{25}{5}\right)^2 + \left(\frac{30}{5}\right)^2$$

$$= (1)^2 + (5)^2 + (6)^2 = 1 + 25 + 36 = 62.$$ □

Problems

In all problems for this chapter, intermediate calculations should be accurate to the number of decimal places required in the answer unless otherwise specified. If no degree of accuracy is specified, round your answer to one more decimal place than that of the data used. If the formula to be used in calculating a measure is not specified, use the formula which *defines* the measure (not the computational formula) for the type of data given.

In Problems 1 through 16, use the following data and round your answers to one decimal place:

W	X	Y	Z
$w_1 = 8$	$x_0 = 4.7$	$y_1 = 6$	$z_1 = 31$
$w_2 = 11$	$x_1 = 3.9$	$y_2 = 1$	$z_2 = 62$
$w_3 = 3$	$x_2 = 7.2$	$y_3 = 5$	$z_3 = 7$
$w_4 = 2$	$x_3 = 0.5$	$y_4 = 4$	$z_4 = 15$
$w_5 = 17$	$x_4 = 2.8$	$y_5 = 4$	$z_5 = 53$
$w_6 = 9$	$x_5 = 5.1$	$y_6 = 8$	$z_6 = 16$
$w_7 = 6$	$x_6 = 7.9$	$y_7 = 1$	$z_7 = 94$
$w_8 = 11$	$x_7 = 4.6$	$y_8 = 7$	$z_8 = 59$

1. Find $\sum_{i=2}^{6} w_i$.
2. Find $\sum_{j=3}^{8} z_j$.
3. Find $\sum_{i=1}^{4} y_{(2i)}$.
4. Find $\sum_{i=4}^{7} x_{(2i-8)}$.
5. Find $\sum_{j=3}^{7} (w_j - y_{(j-2)})$.
6. Find $\sum_{i=1}^{4} x_{(i-1)} y_i$.
7. Find $\sum_{j=2}^{5} \frac{z_j}{y_{(j+1)}}$.
8. Find $\sum_{i=1}^{3} \frac{4w_i x_{(i-1)}}{w_{(i+1)} - y_{(2i)}}$.

In Problems 9 and 10 $v_{i,j}$ is defined by $v_{i,j} = w_i y_j$. For example, $v_{2,3} = w_2 y_3 = 11 \cdot 5 = 55$.

9. Find $\sum_{i=1}^{3} \sum_{j=5}^{7} v_{i,j}$.

10. Find $\sum_{i=2}^{5} \sum_{j=3}^{4} v_{(i+3),(j-2)}$.

11. a. Show that $\sum_{i=1}^{5} 3w_i = 3 \sum_{i=1}^{5} w_i$.

b. Given $u_1, u_2, u_3, \ldots, u_n$ and any constant k, show that, in general,

$$\sum_{i=1}^{n} ku_i = k \sum_{i=1}^{n} u_i.$$

12. a. Show that $\sum_{j=1}^{4} 13 = 4 \cdot 13$.

b. Given any constant k, show that, in general, $\sum_{i=1}^{n} k = n \cdot k$.

13. a. Show that $\sum_{j=1}^{4} (3w_j - 2y_j) = 3 \sum_{j=1}^{4} w_j - 2 \sum_{j=1}^{4} y_j$.

b. Given $u_1, u_2, u_3, \ldots, u_n$, and $v_1, v_2, v_3, \ldots, v_n$, and the constants k and m, show that, in general,

$$\sum_{i=1}^{n} (ku_i - mv_i) = k \sum_{i=1}^{n} u_i - m \sum_{i=1}^{n} v_i.$$

14. a. Show that $\sum_{j=1}^{4} (z_j - 6) = \sum_{j=1}^{4} z_j - 4 \cdot 6$.

b. Given $v_1, v_2, v_3, \ldots, v_n$ and any constant k, show that, in general,

$$\sum_{i=1}^{n} (v_i + k) = \sum_{i=1}^{n} v_i + n \cdot k.$$

15. a. Show that $\sum_{j=1}^{3} (w_j - 3)^2 = \sum_{j=1}^{3} w_j^2 - 6 \sum_{j=1}^{3} w_j + 3(3)^2$.

b. Given $u_1, u_2, u_3, \ldots, u_n$ and any constant k, show that, in general,

$$\sum_{i=1}^{n} (u_i + k)^2 = \sum_{i=1}^{n} u_i^2 + 2k \sum_{i=1}^{n} u_i + n \cdot k^2.$$

16. a. Find $\left[\sum_{i=1}^{4} w_i\right] \cdot \left[\sum_{i=1}^{4} y_i\right]$.

b. Find $\sum_{i=1}^{4} w_i y_i$.

c. Given $u_1, u_2, u_3, \ldots, u_n$ and $v_1, v_2, v_3, \ldots, v_n$, prove that the relation

$$\sum_{i=1}^{n} u_i v_i = \left[\sum_{i=1}^{n} u_i\right] \cdot \left[\sum_{i=1}^{n} v_i\right]$$

is not true in general.

In Problems 17 through 20, a, b, c, and d represent *any* functions of those variables. Expand the given expressions by eliminating the sigma (summation) notation and indicating the actual summations involved.

Example: $\sum_{i=2}^{4} (b_i^2 + c_i) = (b_2^2 + c_2) + (b_3^2 + c_3) + (b_4^2 + c_4)$.

17. $\sum_{i=3}^{6} (a_i^2 - 4b_{(j-1)}) \cdot \sum_{j=2}^{4} (2d_j + 4)^2$.

18. $\sum_{i=1}^{3} \left(b_i^3 - \frac{2b_i}{3}\right) - \sum_{i=4}^{6} (c_i^2 + a_{2i})^3$.

19. $\dfrac{\sum_{j=2}^{3} \left(\frac{d_j}{3} + 2b_j^2 - a_{(j+1)}^3\right) \cdot \sum_{j=2}^{4} (3b_j c_{2j} - 1)}{\sum_{j=1}^{2} (d_j^3 - 2b_j d_j)^2}$.

20. $\sum \left(b_i^3 - \frac{2b_i}{3i}\right) - \sum (i \cdot c_i^2 + a_{2i})^3$.

3.3 *Measures of Location for Ungrouped Data*

Recall that in Chapter 2 we were concerned with how the set of data of Table 2.1 (minutes per sales call for salesmen of a given territory) might compare with a similar set of data for salesmen in other territories. Obviously, a comparison could be made by building a frequency distribution for each set of data. However, suppose we simply want to compare values of a single descriptive measure for each set of data. We can calculate a central tendency value for each set and then compare these values. (Note that such a simple comparison may not give meaningful results unless the distributions of the two sets of data are similar.)

The most frequently used measures of location are the *arithmetic mean*, the *median*, and the *mode*. Particularly in elementary statistics, the mean usually has the greatest utility. In fact, when we use either the median or mode, it sometimes is simply because they are easier to determine. However, even this advantage is less pronounced when the data are grouped or when the set consists of a relatively large number of values. In the present section we consider *ungrouped data*.

Arithmetic Mean

The *arithmetic mean* is sometimes referred to simply as the *mean*, a practice which we shall often follow. However, since there are other kinds of means, we shall always imply an arithmetic mean when we utilize the shorter term mean. Suppose X is a variable and $x_1, x_2, \ldots, x_n$ are n observed values of X. The arithmetic mean of such data is the sum of the values divided by the number of values; that is,

$$\bar{x} = \frac{1}{n} \sum_{i=1}^{n} x_i, \qquad (1)$$

where $\bar{x}$ represents the arithmetic mean of the set of data, x_i is the ith observed value of X, and n is the number of observed values of X.

Example 1

Suppose X is a variable representing the yearly salaries (in dollars) earned by the five officers of a given company and that the salaries are

$$X: 28{,}000;\ 34{,}000;\ 19{,}000;\ 26{,}000;\ 37{,}000.$$

What is the arithmetic mean of the officers' salaries?

Solution

The mean salary is

$$\bar{x} = \frac{1}{n}\sum_{i=1}^{n} x_i = \frac{1}{5}(28{,}000 + 34{,}000 + 19{,}000 + 26{,}000 + 37{,}000)$$

$$= \frac{1}{5}(144{,}000) = 28{,}800. \qquad \square$$

Example 2

Find the mean of the values of Table 2.1.

Solution

To find the mean, we simply add the 80 values of Table 2.1 and divide by 80. The solution is

$$\bar{x} = \frac{1}{n}\sum_{i=1}^{n} x_i = \frac{1}{80}(25 + 44 + 30 + \cdots + 47) = \frac{4{,}403}{80} = 55.0375. \qquad \square$$

Equation (1) provides a formula for one of the most important descriptive measures in all of statistics. Moreover, it tells us the relationship among the number of observations (n), the sum of the values ($\sum x$), and the mean of the values ($\bar{x}$). We can manipulate it just as we might any algebraic relation. Indeed, in some situations we might be given the mean of a set of data and be required to determine the corresponding sum. In such case, we simply solve Equation (1) to obtain

$$\sum x_i = n\bar{x}. \qquad (2)$$

Example 3

A company payroll consists of 10 (n_1) officers, 200 (n_2) salaried employees, and 1500 (n_3) hourly workers. The mean monthly salaries (in dollars) are (1) officers: 2,000 ($\bar{x}_1$); (2) salaried: 900 ($\bar{x}_2$); and (3) hourly: 400 ($\bar{x}_3$). What is the mean monthly salary of all employees of the company?

Solution

To calculate the overall mean by Equation (1), we need the sum of all salaries and the total number of employees. Yet, we have not been given all the individual salaries, so we cannot arrive at these values directly. We do, however, know that the sum of all salaries can be obtained by adding the sums of the salaries of the three different types of employees. Similarly, we can obtain the total number of employees. We have

$$\begin{aligned}\text{total of all salaries} &= n_1\bar{x}_1 + n_2\bar{x}_2 + n_3\bar{x}_3\\ &= 10(2{,}000) + 200(900) + 1{,}500(400)\\ &= 20{,}000 + 180{,}000 + 600{,}000\\ &= 800{,}000,\end{aligned}$$

and

$$\text{total number of employees} = n_1 + n_2 + n_3 = 10 + 200 + 1{,}500 = 1{,}710.$$

Finally,

$$\begin{aligned}\text{overall mean (average)} &= \frac{\text{total of all salaries}}{\text{total number of employees}} = \frac{800{,}000}{1{,}710}\\ &\cong 467.84.\end{aligned}$$

□

The important result is as follows. If we know the means of several sets of data (along with the number of values in each set), we can calculate the overall mean of the data without knowing the individual observations. That is, if we have k sets of data with $n_1, n_2, \ldots, n_k$ observations, respectively, and respective means of $\bar{x}_1, \bar{x}_2, \ldots,$ and $\bar{x}_k$, we obtain the overall mean by

$$\bar{x} = \frac{n_1\bar{x}_1 + n_2\bar{x}_2 + \cdots + n_k\bar{x}_k}{n_1 + n_2 + \cdots + n_k}. \tag{3}$$

Weighted Mean

Note that, when we calculate the mean by using Equation (1), we essentially attach equal importance to all observations. However, there are times when some observations are more important than others. In this situation, the mean is more valuable as a measure of central tendency if we compute it by weighting the various values according to their importance. In other words, we are to compute a *weighted mean.* For example, suppose a person invests in three stocks, A, B, and C. After the first year, we find that the values of the respective stocks have increased by 17 percent, 3 percent, and 4 percent. If we treat these increases as equally important, we conclude that the average increase is

$$\frac{0.17 + 0.03 + 0.04}{3} = 0.08 = 8 \text{ percent} \tag{4}$$

and, thus, that the person has not had a bad year with regard to the growth of his investments. However, if we know additionally that the investor has held \$1,000 worth of stock A, \$10,000 worth of stock B, and \$9,000 worth of stock C, we see that his real investment growth has been (simply dividing the total dollar increase by the total amount invested)

$$\frac{\$1{,}000(0.17) + \$10{,}000(0.03) + \$9{,}000(0.04)}{\$1{,}000 + \$10{,}000 + \$9{,}000}$$

$$= \frac{\$170 + \$300 + \$360}{\$20{,}000} = \frac{\$830}{\$20{,}000} = 0.0415 = 4.15 \text{ percent} \qquad (5)$$

Thus, we correctly conclude that his investment record is not so successful as it first appeared to be. Obviously, the first answer is wrong because we failed to weight the percent increases by their relative importances.

In general, a weighted mean is calculated by

$$\bar{x}_w = \frac{\sum_{i=1}^{k} x_i w_i}{\sum_{i=1}^{k} w_i}, \qquad (6)$$

where $\bar{x}_w$ is the weighted mean of k observations, x_i represents the ith value, and w_i is the weight attached to the ith value.

There is a striking resemblance between Formulas (3) and (6). In fact, if we view the means of the separate sets as individual values and the number of values in each set as the relevant weight, we see that (3) is merely a special case of (6).

Mode and Median

The *mode* of a set of data is the value that occurs most frequently. To determine the mode, we simply count the number of occurrences of each observed value. For example, suppose we have the set of data

$$\begin{aligned} X: x_1 &= 8; x_2 = 10; x_3 = 12; x_4 = 8; x_5 = 9; \\ x_6 &= 7; x_7 = 10; x_8 = 13; x_9 = 10; x_{10} = 6. \end{aligned} \qquad (7)$$

We note that the value 10 occurs three times, the value 8 occurs twice, and every other value occurs only once. Thus, 10 represents the mode of the data, since its frequency is greater than that for any other observed value. Since the role of the mode in elementary statistics is so minor, we shall not even discuss its characteristics.

Of somewhat more importance is the measure of location called the *median*. It is the value of the middle observation (or any value between the two middle observations) when the observations are arranged in ascending or descending order of magnitude. In other words, a median of a set of data is any value such that the number of observed values that are larger equals the number that are smaller. To determine a median, we arrange the data of concern into an ordered array and simply choose it by inspection.

In the case of an odd number of observations, the median is well defined and easily determined. For example, if we have the set of data

$$X: 12, 19, 23, 14, 38,$$

the median can be found by forming the array

$$X: 12, 14, 19, 23, 38$$

and noting that the middle item according to magnitude is 19. In general, when n is odd, the median of n numbers is the value of the $[(n + 1)/2]$th largest observation.

It should be noted that our definition refers to "a median" rather than "the median" —because when we have an even number of values, no "middle" value exists. In such case we have an *indeterminate median*; therefore, according to our definition, any value between the two middle observations may be considered a median. However, we usually take (as a matter of convention) the median to be the mean of the values between the two middle observations. For example, for the data in the ordered array

$$X: 12, 14, 19, 23, 26, 38,$$

any value between 19 and 23 represents a median. However, so that we may refer to a specific value, we ordinarily say that the median is $(19 + 23)/2 = 21$. In general, when n is even, the median of n numbers is taken to be the value midway between the $(n/2)$th and $[(n + 2)/2]$th largest observations.

In what kind of situation might we wish to use the median rather than the mean as a descriptive measure of a set of data? The fact that it requires no calculation is not itself reason enough, since this advantage is offset by the fact that determination of a median requires forming an array. However, if the set of data includes extreme values, the median may more clearly reflect the central tendency of the data than does the mean, because the median is not so greatly affected by extreme values as is the mean.

Problems

1. Computer A can execute a certain job in 0.50 minutes, while computer B can do it in 0.15 minutes.

   a. If 500 such jobs are executed on computer A and 2,000 on computer B, what is the average time per job?

   b. If computer A executes 1,000 such jobs and computer B executes 1,000, what is the average time per job?

   c. Suppose both computers operate for 30 minutes on such jobs. Calculate the average time per job by using as weights the relative numbers of jobs completed by the computers in any given time period.

2. Suppose we have the following data for the average amounts of sales for the last 4 weeks:

Week	*Average sale*	*Number of sales*
1	\$13.52	250
2	16.63	300
3	14.81	200
4	11.95	250
		1000

Calculate the amount of the average sale for the last 4 weeks as a whole, to the nearest cent.

3. Suppose we have the following data on the prices, contribution margins, and quantities sold of 3 products for the last month:

Product	*Price*	*Contribution margin*	*Quantity sold*
A	\$2.00	30%	3,500
B	5.00	25%	4,500
C	4.00	50%	2,000
		105%	10,000

Calculate to the nearest cent the aggregate average amount of contribution margin per unit sold for the three products as a whole.

4. Given $y_1 = 3, y_2 = 6, y_3 = 4, y_4 = 5, y_5 = 7$.

a. Compute $\bar{y} = \frac{1}{5} \sum_{i=1}^{5} y_i$. b. Show that $\sum_{i=1}^{5} (y_i - \bar{y}) = 0$.

c. Prove that

$$\sum_{i=1}^{n} (x_i - \bar{x}) = 0$$

for any set of values $x_i (i = 1, 2, \ldots, n)$. $\left(\textit{Hint:} \text{ Use } \bar{x} = \frac{1}{n} \sum_{i=1}^{n} x_i.\right)$

3.4 Measures of Dispersion for Ungrouped Data

The measure of location alone is seldom sufficient to describe a set of data. Some measure of dispersion is also usually required. Suppose that Company A and Company B each has eight salesmen who have been with their respective companies for ten years. Yearly salaries (in dollars) for the salesmen are

A: 9,000; 13,000; 35,000; 26,000; 39,000; 14,000; 21,000; 19,000.
B: 21,000; 22,000; 24,000; 26,000; 18,000; 20,000; 23,000; 22,000. (1)

Suppose further that an aspiring salesman has a choice of positions between these two companies and that the companies appear equal with regard to other important factors (insurance, retirement, etc.). Which sales force should he join? Naturally, he would probably be interested in the arithmetic mean of the salaries paid by the two companies. Appropriate computations would show, however, that this particular descriptive measure would be of very little help in the situation under consideration, since

$$\bar{x}_A = \frac{9{,}000 + 13{,}000 + \cdots + 19{,}000}{8} = \frac{176{,}000}{8} = 22{,}000 \tag{2}$$

and

$$\bar{x}_B = \frac{21{,}000 + 22{,}000 + \cdots + 22{,}000}{8} = \frac{176{,}000}{8} = 22{,}000. \tag{3}$$

However, careful scrutiny of the data reveals that, although the means are equal, the spread of the salaries of Company A around $\bar{x}_A$ is much greater than that of the salaries of Company B around $\bar{x}_B$. Before making a decision, a salesman might wish to investigate the reason for this difference in dispersion. Further examination might reveal that the reason for lack of spread in Company B salaries is that their sales procedures are so systematized that individual effort and ingenuity make very little difference in ultimate salary. In such case, a creative and innovative individual might not wish to be so constrained, and, thus, might choose Company A. Further, it is easy to see that other lines of reasoning could lead logically to different decisions.

There are several measures of dispersion; we shall mention only four: the *range*, the *mean deviation*, the *variance* and the *standard deviation*. Since the variance and the standard deviation are by far the most important for our purposes, we shall treat the range and mean deviation very lightly and then deal with the variance (and, thus, the standard deviation) rather extensively. Our discussion for the present will be limited to *ungrouped* data.

Range

The *range* of a set of data is defined as the difference between the largest and the smallest observed values. Its primary advantage is that it is quite easy to determine. Its disadvantages are so many that its direct application in business statistics is severely limited. Nevertheless, it does provide some (limited) measure of dispersion. Further, in some specific areas of application (e.g., quality control) the range, primarily because of its relation to the standard deviation, has great indirect utility.

Mean Deviation

You will frequently encounter terms using the word "deviation." We should agree what is meant by a deviation. Naturally, if a measurement is to deviate, there must be some frame of reference within which we measure the deviation. We must further qualify the description by indicating some entity or value from which the deviations occur. Therefore, with regard to a given set of data, "deviations from the mean," "deviations from the mode," and "deviations from the median" are all acceptable examples which imply differences between the individual values of observations and some fixed entity. However, the term "deviations from the (arithmetic) mean " occurs so often in statistics that it is customary to call this particular deviation simply "deviation." Whenever we need to speak of other types of deviations (e.g., deviation from the median), we shall make the meaning clear by the context.

Now, consider a set of values $x_1, x_2, \ldots, x_n$ with mean $\bar{x}$. The corresponding deviations of the values from the mean will be $x_1 - \bar{x}, x_2 - \bar{x}, \ldots, x_n - \bar{x}$. If all individual values are concentrated closely around the mean, it is obvious that the dispersion of the data is small. Conversely, the less closely concentrated the individual values are around their mean, the greater their dispersion. Thus, it seems intuitively clear that a good measure of dispersion might involve the deviations about the mean. A generally accepted measure of dispersion is the *mean deviation*, which is defined as the mean of the absolute values of the deviations from the mean. That is,

$$MD = \frac{1}{n} \sum_{i=1}^{n} |x_i - \bar{x}|, \tag{4}$$

where MD represents the mean deviation associated with n values, x_i is the ith value, and $\bar{x}$ is the mean of the n values. Consider an example. Let us find the mean deviation of the salaries of Company A in (1). To do so, we construct Table 3.1 using the following steps:

Table 3.1
Mean deviation calculations

x_i	$x_i - \bar{x}$	$\|x_i - \bar{x}\|$
9,000	−13,000	13,000
13,000	−9,000	9,000
35,000	13,000	13,000
26,000	4,000	4,000
39,000	17,000	17,000
14,000	−8,000	8,000
21,000	−1,000	1,000
19,000	−3,000	3,000
176,000	0	68,000

$$\bar{x} = \frac{176{,}000}{8} = 22{,}000.$$

$$MD = \frac{68{,}000}{8} = 8{,}500.$$

1. List the values of the observations (x_i).
2. Compute $\bar{x}$.
3. Compute the deviations of values from the mean ($x_i - \bar{x}$).
4. Take the absolute value of the deviations (Column 3).
5. Sum the absolute deviations and divide by the number of observations.

Since we found, on page 40, that $\bar{x}_A = 22{,}000$, we have

$$\begin{aligned} MD &= \frac{1}{8} \sum_{i=1}^{8} |x_i - \bar{x}_A| \\ &= \frac{1}{8}(|9{,}000 - 22{,}000| + |13{,}000 - 22{,}000| + \cdots + |19{,}000 - 22{,}000|) \\ &= \frac{68{,}000}{8} = 8{,}500. \end{aligned}$$

Similar calculations reveal that the mean deviation of the salaries of Company B is only 1,750. Thus, we see that the two values of the measure of dispersion immediately reflect the result we obtained through an intuitive argument. That is, the variation is greater for Company A than for Company B.

While the mean deviation does provide us with a measure that actually reflects dispersion, the absolute values involved in the formula make it mathematically difficult to use in further analysis. This being true, a similar measure, the variance, has become a much better known and more often used measure of disperison.

Variance

Definition 3.1

The ***variance*** *of a set of data is defined as the mean of the squared deviations of the observations from the mean.*

That is, by definition,

$$s_x^2 = \frac{1}{n} \sum_{i=1}^{n} (x_i - \bar{x})^2, \tag{5}$$

where s_x^2 represents the variance of n values of the variable X, x_i is the ith value, and $\bar{x}$ is the mean. Due to its tremendous utility in the theory of statistics, this is one of the most important definitions in the entire discipline. Thus, you should take extra care to assimilate both the verbal and notational definitions of the variance. Its use will arise repeatedly.

From a calculation standpoint, the definition of the variance is somewhat unwieldy. Fortunately, there exists a formula for the variance which is equivalent to Formula (5) and often much less demanding. The computational formula for the variance is

$$s_x^2 = \frac{\sum_{i=1}^{n} x_i^2 - \frac{(\sum_{i=1}^{n} x_i)^2}{n}}{n}, \tag{6}$$

where all symbols have the same meaning as in Relation (5). Note that $\sum_{i=1}^{n} x_i^2$ represents the "sum of the squared values," while $(\sum_{i=1}^{n} x_i)^2$ represents the "square of the sum of the values." This is a very important distinction.

For a demonstration that the definition of the variance (5) is equivalent to the computational Formula (6), see Supplement 3.A.

Standard Deviation

Note from the relations for the mean and variance that, while the value of the mean is always measured in the same unit of measure as the original observations, the variance, by definition, is given in terms of "squares" of the unit of measure associated with the original observations. However, it is often convenient to have dispersion in the same unit of measure as the mean. To obtain such a measure, we simply take the square root of the variance and call it the *standard deviation*. That is,

$$s_x = \sqrt{\frac{1}{n} \sum_{i=1}^{n} (x_i - \bar{x})^2},$$

where s_x is the standard deviation of n values of the variable X.

Problems

1. Suppose we have the following data on the number of milling machines produced each day last week:

$$x_1 = 41, \quad x_2 = 36, \quad x_3 = 39, \quad x_4 = 36, \quad x_5 = 33.$$

a. Find the range of units produced daily.

b. Calculate the mean deviation from the mean.

c. Compute the variance of the units produced daily by the *formula which defines* the variance. Round s_x^2 to one decimal place.

d. Calculate s_x^2 to one decimal place for the units produced daily, using the *computational formula* for the variance. Your answer should approximate your answer to c.

e. Compute the standard deviation of this distribution from your answer to d. Round s_x to one decimal place.

2. Suppose we have the following data on the number of milling tool sets produced each day last week:

$$y_1 = 26, \quad y_2 = 21, \quad y_3 = 24, \quad y_4 = 21, \quad y_5 = 18.$$

a. Calculate s_y^2 to one decimal place by the *computational formula* for the variance.

b. Compute s_y from your answer to Part a, rounding to one decimal place.

c. Note the relationship between x_i in Problem 1 and y_i in this problem. Compare your answers to Parts a and b of this problem with your answers to Parts d and e of Problem 1. Note any relationships between the answers and state what characteristics of the data cause these relationships.

3. Suppose we have the following data on the number of new accounts contacted by your company's salesmen each week for the last eight weeks:

$$x_1 = 14, \quad x_2 = 17, \quad x_3 = 8, \quad x_4 = 9, \quad x_5 = 17,$$

$$x_6 = 11, \quad x_7 = 16, \quad x_8 = 4.$$

a. Find the range of new accounts contacted weekly.

b. Calculate the mean deviation from the mean.

c. Find the variance of the new accounts contacted weekly by the *formula which defines* the variance. Round your answer to one decimal place.

d. Compute s_x^2 by the *computational formula*. Round your answer to one decimal place. Check that your answer approximates the s_x^2 found in Part c.

e. Calculate the standard deviation of this distribution from your answer to Part c, rounding to one decimal place.

4. The following data gives the number of new accounts contacted by a company's salesmen each week for the last eight weeks:

$$y_1 = 11, \quad y_2 = 14, \quad y_3 = 5, \quad y_4 = 6, \quad y_5 = 14,$$

$$y_6 = 8, \quad y_7 = 13, \quad y_8 = 1.$$

a. Compute the variance of the new accounts contacted weekly by the *formula which defines* s_y^2. Round your answer to one decimal place.

b. Calculate the standard deviation of this distribution from your answer to Part a.

c. Note the relationship between x_i in Problem 3 and y_i in this problem. Compare your answers to Parts a and b of this problem with your answers to Parts c and e of Problem 3. Note any relationships between the answers and state what characteristics of the data cause these relationships.

5. The monthly sales (in 1,000 units) of Company A and Company B were as follows:

	Jan.	*Feb.*	*Mar.*	*April*	*May*	*June*
Company A	18	18	18	18	18	18
Company B	10	24	12	23	13	26

The average monthly sales in units of the two companies are equal.

a. Does the mean adequately describe these two distributions? If the operations of the two companies were identical in all other respects, would the companies be equally safe in using the mean as a measure to determine optimum inventory levels?

b. Suggest another measure that would indicate the difference between these two distributions.

6. Let s_1^2 be the variance of the following distribution of X: 4, 5, 7, 9, 10.

 a. If three sevens are added to this distribution and the variance of the new distribution is called s_2^2, is $s_2^2 \leq s_1^2$? Explain your answer intuitively.

 b. If two fours and two tens are added to the original distribution and the variance of this new distribution is called s_3^2, is $s_3^2 \leq s_1^2$? Explain your answer intuitively.

7. Given a distribution X: $x_1, x_2, x_3, \ldots, x_n$, prove the relation

$$\frac{\sum_{i=1}^{n} x_i^2}{n} = s_x^2 + \bar{x}^2.$$

(*Hint*: Use the computational formula for s_x^2.)

8. Ten observations have a mean of 25 and a variance of 5. Find the sum of the squared observations. Round your answer to one decimal place.

3.5 *Transformations*

When we change an original set of data (a variable) to a new set by means of a well-defined arithmetic operation, we ordinarily say that "the original variable has been transformed to a new variable" or that "we have made a transformation on the variable of interest." For two important reasons, we need to understand what happens to the mean and variance of a set of data under a transformation. First, transformations often make it much easier to calculate descriptive measures such as the mean and variance. Much more important, however, is the fact that your understanding of what happens under a transformation enhances your ability to digest the theory underlying many of the most important statistical techniques to come. We shall restrict our discussion to *linear transformations* of data—that is, those in which a new variable is obtained through a linear function of a given variable. For example, the general mathematical form of a linear transformation on a variable X is

$$Y = c_1 + c_2 X, \tag{1}$$

where X is the variable to be transformed, c_1 and c_2 are any constants, and Y is the new variable obtained through the linear transformation. Since we are concerned only with the practical aspects of what happens under certain transformations, our approach will be elementary and straightforward, and results will be demonstrated only for sets of data involving a finite number of values. Nevertheless, except in very rare instances, corresponding results can be developed for continuous variables involving an infinite number of possible values.

First, what do we mean when we refer to a transformation (similar to the one represented in (1)) on a variable? We essentially mean that we wish to arrive at a new set of data by performing the indicated operation on every valuc in the set of data under consideration. Thus (if we restrict consideration to a finite number of values), under the transformation of (1), each of the n values in the original set will give rise to a value in the new set, with the ith value of Y being a linear function of the ith value of X—that is,

$$y_i = c_1 + c_2 x_i \quad (i = 1, 2, \ldots, n). \tag{2}$$

In the special case where c_2 has a value of one, we see that the ith value of Y is obtained simply by adding a constant to the ith value of X. We denote such a transformation as an *additive transformation.* Similarly, in the special case where c_1 has a value of zero, we see that the ith value of Y is obtained simply by multiplying the ith value of X by a constant. This type of transformation is called a *multiplicative transformation.* Thus, a linear transformation is comprised of a combination of an additive and a multiplicative transformation. We shall refer to c_1 as the *additive constant* and to c_2 as the *multiplicative constant.* The major question to be considered here is "Under a linear transformation, how are the mean and variance of the transformed variable related to the mean and variance, respectively, of the original variable?" Let us begin discussion of this question by an example designed simply to illustrate the "how" of transformations and related reasoning.

Example 1

Given the set of values X: 1, 3, 5, 7, 9, carry out the following steps:

(a) Calculate the mean and variance of X.
(b) Make the transformation $Y = 7 + 8X$.
(c) Calculate the mean and variance of Y.
(d) Note the relations between the respective means and variances.

Solution

(a) We have

$$\bar{x} = \frac{1}{n} \sum_{i=1}^{n} x_i = \frac{1}{5}(25) = 5$$

and

$$s_x^2 = \frac{1}{n} \sum_{i=1}^{n} (x_i - \bar{x})^2 = \frac{1}{5}[(1-5)^2 + (3-5)^2 + \cdots + (9-5)^2] = \frac{40}{5} = 8.$$

(b) Under the transformation, we have

$$y_1 = 7 + 8x_1 = 7 + 8(1) = 15, \qquad y_2 = 7 + 8x_2 = 7 + 8(3) = 31,$$
$$y_3 = 7 + 8x_3 = 7 + 8(5) = 47, \qquad y_4 = 7 + 8x_4 = 7 + 8(7) = 63,$$
$$y_5 = 7 + 8x_5 = 7 + 8(9) = 79.$$

Thus, the values of Y become Y: 15, 31, 47, 63, 79.

(c) We have

$$\bar{y} = \frac{1}{n}\sum_{i=1}^{n} y_i = \frac{1}{5}(15 + 31 + \cdots + 79) = \frac{235}{5} = 47$$

and

$$s_y^2 = \frac{1}{n}\sum_{i=1}^{n}(y_i - \bar{y})^2 = \frac{1}{5}[(15 - 47)^2 + (31 - 47)^2 + \cdots + (79 - 47)^2]$$

$$= \frac{1}{5}(1{,}024 + 256 + \cdots + 1{,}024) = \frac{2560}{5} = 512.$$

(d) Recall that the foregoing linear transformation involved an additive constant of 7 and a multiplicative constant of 8. Careful comparison reveals that the mean of Y can be written in terms of the mean of X as

$$\bar{y} = 7 + 8\bar{x} = 7 + 8(5) = 47.$$

Also, the variance of Y can be written in terms of the variance of X as

$$s_y^2 = (8)^2 s_x^2 = 64(8) = 512.$$

Note that the mean is affected by both the additive and multiplicative constants, while the variance is affected only by (the square of) the multiplicative constant. □

The facts demonstrated in Example 1 can be generalized as follows:

Rule 3.4

Under a linear transformation, the mean of the transformed variable is equal to the sum of the additive constant and the product of the multiplicative constant and the mean of the original variable.

Rule 3.5

Under a linear transformation, the variance of the transformed variable is equal to the product of the square of the multiplicative constant and the variance of the original variable.

We can prove these rules for any (finite) set of data by using only the definitions which we have developed previously. Let us consider the linear transformation of Equation (1), where

$$y_i = c_1 + c_2 x_i.$$

We see that

$$\bar{y} = \frac{1}{n}\sum_{i=1}^{n} y_i = \frac{1}{n}\sum_{i=1}^{n}(c_1 + c_2 x_i) = \frac{1}{n}\sum_{i=1}^{n} c_1 + \frac{1}{n}\sum_{i=1}^{n} c_2 x_i$$

$$= \frac{nc_1}{n} + c_2\left(\frac{1}{n}\sum_{i=1}^{n} x_i\right) = c_1 + c_2\bar{x}. \qquad (3)$$

Also,

$$
\begin{aligned}
s_y^2 &= \frac{1}{n}\sum_{i=1}^{n}(y_i - \bar{y})^2 = \frac{1}{n}\sum_{i=1}^{n}[(c_1 + c_2 x_i) - (c_1 + c_2 \bar{x})]^2 \\
&= \frac{1}{n}\sum_{i=1}^{n}(c_2 x_i - c_2 \bar{x})^2 = \frac{1}{n}\sum_{i=1}^{n} c_2^2 (x_i - \bar{x})^2 \\
&= c_2^2\left[\frac{1}{n}\sum_{i=1}^{n}(x_i - \bar{x})^2\right] = c_2^2 s_x^2. \qquad (4)
\end{aligned}
$$

As special cases of Rules 3.4 and 3.5, we have four additional rules.

Rule 3.6

Under an additive transformation, the mean of the transformed variable is equal to the sum of the additive constant and the mean of the original variable.

Rule 3.7

Under an additive transformation, the variance of the transformed variable is the same as the variance of the original variable.

Rule 3.8

Under a multiplicative transformation, the mean of the transformed variable is equal to the product of the multiplicative constant and the mean of the original variable.

Rule 3.9

Under a multiplicative transformation, the variance of the transformed variable is equal to the product of the square of the multiplicative constant and the variance of the original variable.

Rules 3.6–3.9 follow quite easily from rules 3.4 and 3.5. You are invited to prove them in a fashion similar to that of the proofs of Rules 3.4 and 3.5.

Example 2

Suppose we have a set of data (X) with mean $\bar{x}$ and variance s_x^2. Determine, in terms of the mean and variance of X, the mean and variance of the variables obtained through the following transformations:

(a) $Y = X - 8$; (b) $R = 7X$; (c) $Q = X/12$; and (d) $T = (X - 5)/7$.

Solution

(a) The additive constant is -8 and the multiplicative constant is 1. Thus, $\bar{y} = \bar{x} - 8$ and $s_y^2 = s_x^2$.

(b) The additive constant is 0 and the multiplicative constant is 7. Thus, $\bar{r} = 7\bar{x}$ and $s_r^2 = 49s_x^2$.

(c) The additive constant is 0 and the multiplicative constant is 1/12. Thus, $\bar{q} = \bar{x}/12$ and $s_q^2 = s_x^2/144$.

(d) The additive constant is $-5/7$ and the multiplicative constant is 1/7. Thus, $\bar{t} = \bar{x}/7 - 5/7 = (\bar{x} - 5)/7$ and $s_t^2 = s_x^2/49$. □

It will enhance your understanding to check each of the foregoing results by (1) selecting a specific set of data, (2) actually making the indicated transformations, (3) using the transformed values to calculate the means and variances through use of basic definitions, and (4) noting that the results of Example 2 hold.

Standardized Variables and Unit Distributions

Two linear transformations are of particular importance. Suppose we have a variable X, with mean $\bar{x}$ and variance s_x^2 (or, standard deviation s_x). When we make the transformation

$$Y = \frac{X}{s_x}, \tag{5}$$

we obtain what is known as a *standardized variable*. In other words, we can obtain a set of standardized values by dividing each value in the original set by the standard deviation of the data under consideration. Note that s_x is a particular constant. Thus, it follows quite easily from Rule 3.9 that the variance of a standardized variable is one. That is, under the special transformation of (5),

$$s_y^2 = (\text{square of multiplicative constant}) \cdot (\text{variance of } X) = \left(\frac{1}{s_x}\right)^2 \cdot (s_x^2) = 1.$$

A *unit distribution* is defined to be a distribution with mean equal to zero and variance (or standard deviation) equal to one. Unit distributions are so vital in statistics that we will use them repeatedly. Any distribution may be converted to a unit distribution by the proper transformation. Consider the transformation

$$Z = \frac{X - \bar{x}}{s_x}. \tag{6}$$

Since $\bar{x}$ and s_x are constants (for a given set of data), the additive constant in (6) is $-\bar{x}/s_x$ and the multiplicative constant is $1/s_x$. Thus, $c_1 = -\bar{x}/s_x$ and $c_2 = 1/s_x$. Hence,

$$\bar{z} = c_1 + c_2\bar{x} = \frac{-\bar{x}}{s_x} + \left(\frac{1}{s_x}\right)\bar{x} = 0 \quad \text{and} \quad s_z^2 = c_2^2 s_x^2 = \left(\frac{1}{s_x^2}\right)s_x^2 = 1.$$

Thus, the transformation (6) will *always* provide us a unit distribution.

Example 3

Given the following set of data for the variable X, X: 10, 14, 14, 14, 16, 16,

(a) find the mean and variance of X;
(b) determine the values which result from making the transformation $Z = (X - \bar{x})/s_x$;
(c) show numerically that the distribution of the transformed variable Z is a unit distribution. That is, show that $\bar{z} = 0$ and $s_z^2 = 1$.

Solution

(a) We have

$$\bar{x} = \frac{10 + 14 + \cdots + 16}{6} = 14,$$

$$s_x^2 = \frac{1}{n}\sum_{i=1}^{n}(x_i - \bar{x})^2 = \frac{1}{6}[(10 - 14)^2 + (14 - 14)^2 + \cdots + (16 - 14)^2]$$

$$= \frac{1}{6}(16 + 0 + \cdots + 4) = \frac{1}{6}(24) = 4,$$

and

$$s_x = 2.$$

(b) Under the transformation

$$Z = \frac{X - \bar{x}}{s_x} = \frac{X - 14}{2},$$

the values of Z become Z: $-2, 0, 0, 0, 1, 1$.

(c) Next,

$$\bar{z} = \frac{1}{n}\sum_{i=1}^{n} z_i = \frac{1}{6}(-2 + 0 + 0 + 0 + 1 + 1) = 0$$

and

$$s_z^2 = \frac{1}{n}\sum_{i=1}^{n}(z_i - \bar{z})^2 = \frac{1}{6}[(-2 - 0)^2 + (0 - 0)^2 + \cdots + (1 - 0)^2] = \frac{6}{6} = 1. \quad \square$$

The concepts used in the foregoing transformation lead to certain shortcuts for calculating the mean and variance of ungrouped data. For coverage of related materials, see Supplement 3.B.

Problems

1. Suppose we have the following distribution:

$$x_1 = 4, \quad x_2 = 3, \quad x_3 = 7, \quad x_4 = 11, \quad x_5 = 10.$$

   a. Standardize the variable by dividing each observation by s_x and calculate the variance of the standardized variable. Round your answer to two decimal places and check to see that it is approximately equal to one.

b. Transform the original variable by subtracting $\bar{x}$ from each observation and computing the mean of this new distribution. Round your answer to two decimal places and check to see that it is approximately equal to zero.

c. Transform the variable by the formula

$$z = \frac{x - \bar{x}}{s_x}$$

and calculate $\bar{z}$ and s_z. Round your answers to two decimal places. Check to see that $\bar{z}$ is approximately equal to zero and s_z is approximately equal to one.

2. Given the distribution

$$y_1 = 2, \quad y_2 = 4, \quad y_3 = 6, \quad y_4 = 8, \quad y_5 = 10:$$

a. Calculate the mean and variance of Y.

b. Make the transformation $R = Y + 3$ and calculate $\bar{r}$ and s_r^2. Note that $\bar{r} = \bar{y} + 3$ and $s_r^2 = s_y^2$.

c. Make the transformation $Q = 3Y$ and find $\bar{q}$ and s_q^2. Note that $\bar{q} = 3\bar{y}$ and $s_q^2 = (3)^2 s_y^2$.

d. Make the transformation $V = \dfrac{Y - \bar{y}}{s_y}$ and calculate $\bar{v}$ and s_v^2. Round all numbers to two decimal places. Note that V is a unit distribution.

3. Given the set of data

$$X: 3, 5, 6, 7, 9,$$

calculate $\bar{x}$ and s_x^2 and:

a. Make the transformation $Y = X - 8$ and calculate $\bar{y}$ and s_y^2 from the new set of data.

b. Make the transformation $R = 7X$ and calculate $\bar{r}$ and s_r^2 from the new set of data.

c. Make the transformation $Q = X/12$ and calculate $\bar{q}$ and s_q^2 from the new set of data.

d. Make the transformation $T = \dfrac{X - 5}{7}$ and calculate $\bar{t}$ and s_t^2 from the new set of data.

e. Compare your answers to Parts a, b, c, and d of this problem with the results of Example 2. Note the relationships between the mean and variance of each of the transformed variables and the mean and variance of the original variable.

3.6 Measures of Location for Grouped Data

Formulas for descriptive measures involving grouped data are important in two circumstances: when we may have available only a set of grouped data (that is, we may not know the value of each individual observation) and when we need to calculate a descriptive measure for a very large number of observed values. In the latter case, labor can be reduced substantially by grouping the data before determining the descriptive measure. Grouped-data formulas for the mean, the median, and the mode follow.

Mean—Grouped Data

The mean of a set of grouped data may be calculated by the formula

$$\bar{x} \cong \frac{\sum_{i=1}^{k} m_i f_i}{\sum_{i=1}^{k} f_i}, \tag{1}$$

where k is the number of classes into which the data are divided, f_i is the frequency for the ith class, and m_i is the midpoint of the ith class. It should be noted that the formula does not require equal class intervals. *However, it does not apply if one of the classes is an open class*, unless it is possible to determine the midpoint of such a class. Note also that, since we do not have the values of the individual observations, we make the implicit assumption that the mean of the values in each given class takes on a value equal to the midpoint of that class. (For this reason, Formula (1) provides only an approximation to the mean of the data.) Thus, $m_i f_i$ represents the sum of the values in each class, and when we sum such values over all the classes, we obtain an approximation to the sum of all the values. Further, the sum of the frequencies of the various classes represents the total number of observations. Thus, under the assumption that the midpoint of each class is representative of the mean of the values of such class, the computational formula of (1) is consistent with our previous definition of the arithmetic mean as the sum of the values divided by the number of values.

Example 1

Suppose we are required to find the mean of the grouped data of Table 2.2 (reproduced as the first two columns of Table 3.2).

Solution

We note that very little calculation is required. We simply determine and tabulate the midpoint of each class (column 3); compute the product of each respective midpoint and frequency (column 4); sum the products of step 2; sum the frequencies, and finally calculate the ratio of Formula (1). Necessary calculations are given in Table 3.2. Note that the result (54.75) is a very good approximation to the exact mean (55.0375), which we calculated in Example 2, page 36. □

From Rules 3.4 through 3.9 concerning the effects of linear transformations on the mean and the variance of a set of data, we obtain the more workable formula for the mean of grouped data (when class intervals are of equal length):

$$\bar{x} \cong A + c \cdot \frac{\sum_{i=1}^{k} f_i d_i}{\sum_{i=1}^{k} f_i}, \tag{2}$$

Table 3.2

Calculation of mean and variance of grouped data

(1) Class (minutes/ sales call)	(2) Frequency (f_i)	(3) Midpoint (m_i)	(4) $m_i f_i$	(5) $(m_i - \bar{x})$	(6) $(m_i - \bar{x})^2$	(7) $f_i(m_i - \bar{x})^2$
20–29	6	24.5	147.0	−30.25	915.0625	5,490.375
30–39	10	34.5	345.0	−20.25	410.0625	4,100.625
40–49	16	44.5	712.0	−10.25	105.0625	1,681.000
50–59	18	54.5	981.0	− 0.25	0.0625	1.125
60–69	12	64.5	774.0	+ 9.75	95.0625	1,140.750
70–79	10	74.5	745.0	+19.75	390.0625	3.900.625
80–89	8	84.5	676.0	+29.75	885.0625	7,080.500
	80		4,380.0			23,395.000

$$\bar{x} \cong \frac{\sum_{i=1}^{k} m_i f_i}{\sum_{i=1}^{k} f_i} = \frac{4,380}{80} = 54.75. \qquad s_x^2 \cong \frac{\sum_{i=1}^{k} f_i(m_i - \bar{x})^2}{n} = \frac{23,395}{80} = 292.4375.$$

where A is an approximated mean (ordinarily, the midpoint of one of the class intervals near the central tendency of the frequency distribution), k is the number of classes, c is the class interval length, f_i is the frequency of the ith class, and d_i is a term called the *class deviation.* The latter term requires clarification. We ordinarily define the class deviation of the class containing the approximated mean to be zero. Then, the class deviation of any other given class becomes the algebraic difference (measured in class intervals) between the midpoint of the given class and the approximated mean. However, the assignment of class deviations to the various classes is not so difficult as the definition might lead you to believe. Note that application of the definition simply results in an assignment of -1 as the class deviation of the class immediately smaller than the class containing the approximated mean, -2 as the class deviation of the next smaller class, etc., and assignment of $+1$ as the class deviation of the class immediately larger than the class containing the approximated mean, $+2$ as the class deviation of the next larger class, etc. Formula (2) can be derived directly from Formula (1).

Example 2

Utilize the data of the frequency distribution of Table 2.2 to illustrate simultaneously the application of Formula (2) and the equivalence of Formulas (1) and (2).

Solution

The frequency distribution is reproduced as the first two columns of Table 3.3. and the steps necessary for computing the mean by Formula (2) are given in the remainder of the table. Table 3.3 is constructed as follows: we choose an approximated mean equal to the midpoint (54.5) of the 50–59 class (note that this choice is arbitrary); as dictated by our approximation of step 1, assign zero as the value of the class deviation of the class 50–59; determine the class deviations for the other classes; perform the computational steps indicated in Formula (2). Note that the result is identical to the value obtained through use of Formula (1).

Table 3.3

Alternate method of calculating mean and variance of grouped data

(1) Class (minutes/ sales call) X	(2) Frequency (f_i)	(3) Class deviation (d_i)	(4) $f_i d_i$	(5) d_i^2	(6) $f_i d_i^2$
20–29	6	−3	−18	9	54
30–39	10	−2	−20	4	40
40–49	16	−1	−16	1	16
50–59	18	0	0	0	0
60–69	12	1	12	1	12
70–79	10	2	20	4	40
80–89	8	3	24	9	72
	80		2		234

Approximated mean $= A = 54.5$. Class interval length $= c = 10$.

$$\bar{x} \cong A + \frac{\sum_{i=1}^{k} f_i d_i}{\sum_{i=1}^{k} f_i} \cdot c = 54.5 + \frac{2}{80} \cdot 10 = 54.5 + 0.25 = 54.75.$$

$$s_x^2 \cong c^2 \left[\frac{\sum_{i=1}^{k} f_i d_i^2 - \frac{(\sum_{i=1}^{k} f_i d_i)^2}{\sum_{i=1}^{k} f_i}}{\sum_{i=1}^{k} f_i} \right] = 100 \left[\frac{234 - \frac{(2)^2}{80}}{80} \right]$$

$$= 100 \left[\frac{234 - 0.05}{80} \right] = 100 \left[\frac{233.95}{80} \right] = \frac{23{,}395}{80}$$

$$= 292.4375.$$

□

It should be noted that we can generalize Formula (2) to make it applicable when class intervals are not of equal length. The formula becomes

$$\bar{x} \cong A + c \cdot \frac{\sum_{i=i}^{k} f_i(m_i - A)/c}{\sum_{i=1}^{k} f_i}, \tag{3}$$

where A is any approximated mean, m_i is the midpoint of the ith class interval, c is any constant chosen to make data manipulation easy, and all other symbols are as given in conjunction with (2). However, when class intervals are not of equal length, it is often less complicated to simply use Formula (1).

You will gain further insight into the manipulation of grouped data if you will satisfy yourself of this fact. Under the assumption that A is chosen as the midpoint of one of the class intervals and c is the length of class intervals of equal length, straightforward calculation of the $(m_i - A)/c$ values in Formula (3) yields values identical to those arbitrarily assigned to the d_i's in Formula (2).

Mode and Median—Grouped Data

The mode of a set of grouped data is ordinarily taken to be the midpoint of the class having the highest frequency and thus can be determined by mere inspection of the frequency distribution. For example, simply by looking at the frequency distribution of Table 3.3 we see immediately that the class 50–59 has the highest frequency. Then, our approximation for the mode of the distribution becomes $(50 + 59)/2 = 54.5$. A set of data may have more than one mode. In such case, a graph of the frequency distribution will exhibit more than one peak. The maximum peak(s) is (are) termed the *primary* mode(s). Others are secondary modes. In this situation, we can approximate the modes (primary and secondary) by the midpoints of the class intervals in which the peaks appear.

Determination of the median for grouped data is slightly more complicated. We approximate the median of a frequency distribution by determining the median class (the class containing the median value) and approximating the median by a special sort of linear interpolation within that class. To determine the median class, we simply accumulate frequencies through the class interval containing the median value. Then, appropriate interpolation is accomplished by using the formula

$$x_{\text{med}} \cong L + \frac{\frac{n}{2} - (cf)_m}{f_m} c_m, \tag{4}$$

where x_{med} represents the median of n values of the variable X, L is the lower class boundary of the median class, n is the total number of observed values, $(cf)_m$ is the cumulative frequency up to the median class, f_m is the frequency of the median class, and c_m is the length (in units) of the median class. Careful study of the formula and related applications will reveal that the resulting interpolation is based on the assumption that the values in the median class are evenly distributed. The formula is applicable for either an even or odd number of observations. The procedure can best be illustrated by an example.

Example 3

Approximate the median of the data of Table 3.3.

Solution

Noting that the set is comprised of a total of 80 values, we know that, if we could list the data in ascending order, we would estimate the median to be the mean of the fortieth and the forty-first largest observations. However, since we have only grouped data, we are not able to do this. Nevertheless, we can find the class containing these values. If we consider cumulative frequencies, the frequency distribution tells us that 32 of the values are less than 50 and 50 of the values are less than 60. Thus, the fortieth and the forty-first largest values must lie somewhere in the class 50–59. When the data are grouped, this is all that we can say with certainty. Further information concerning the value of the median can be obtained only through an assumption about how the values are distributed within the class 50–59. Assuming that the values in the median class are evenly distributed, we arrive at the approximate median by use of (4); that is,

$$x_{\text{med}} \cong L + \frac{\frac{n}{2} - (cf)_m}{f_m} c_m = 49.5 + \frac{40 - 32}{18} \cdot 10 = 49.5 + \frac{80}{18}$$

$$= 49.5 + 4.44 = 53.94. \qquad \square$$

Problems

1. The dollar amounts of the sales per call on customers during the last month were classified as follows:

Amount of sale	*f*
\$ 0–99	200
100–199	75
200–299	40
300–399	25
400–499	10
500–999	5
1000–4999	15

Using this grouped *discrete* data:

a. Calculate the mean amount of sale per call to the nearest \$1.

b. Find the median amount of sale per call to the nearest \$1 by linear interpolation.

c. Which of these two figures seems most indicative of the size of the orders being obtained by salesmen on their calls? Why?

2. Suppose we have the following grouped *continuous* data for the tardiness of employees:

Tardiness in minutes X	*Number of employees* f
$0 \leq X < 5$	31
$5 \leq X < 10$	19
$10 \leq X < 15$	13
$15 \leq X < 30$	9
$30 \leq X < 60$	5
$60 \leq X < 120$	2
$120 \leq X < 240$	1
	80

a. Compute to the nearest minute the average time by which the employees were tardy.

b. Use linear interpolation to find the median time by which the employees were tardy. Round your answer to the nearest minute.

c. Compare the results of Parts a and b and state which measure is most indicative of the tardiness of employees. Why?

3.7 *Measures of Dispersion for Grouped Data*

The variance and standard deviation are the primary measures of dispersion ordinarily used when we deal with grouped data. However, since the standard deviation can always be obtained merely by taking the square root of the variance, we shall devote the section to computational formulas for the variance. We shall consider three formulas. The formula corresponding to the definition of the variance is

$$s_x^2 \cong \frac{\sum_{i=1}^{k} f_i(m_i - \bar{x})^2}{n}, \tag{1}$$

and the easier computational formula is

$$s_x^2 \cong c^2 \left[\frac{\sum_{i=1}^{k} f_i d_i^2 - \dfrac{(\sum_{i=1}^{k} f_i d_i)^2}{n}}{n} \right], \tag{2}$$

where $\bar{x}$ is the mean of the set of data under consideration, k is the number of classes, f_i is the frequency of the ith class, m_i is the midpoint of the ith class, c is the class interval length (once again, this assumes that the classes are of equal length), d_i is the class deviation of the ith class (see Formula (2), page 52), and n is the total number of observations (or total observed frequency, $\sum_{i=1}^{k} f_i$). Note that Formulas (1) and (2) are equivalent. Further, since it is obvious from comparing (1) with the original definition of the variance,

$$s_x^2 = \frac{1}{n} \sum_{i=1}^{n} (x_i - \bar{x})^2,$$

that we are again assuming that the mean of each given class can be represented by the appropriate midpoint, it immediately follows that both formulas represent only approximations to the variance.

Although it is nearly always preferable to use (2) rather than (1) in computing the variance of grouped data, we shall illustrate the use of both formulas. Refer to Table 3.2 for an example of the application of (1). We first calculate the mean of the data using columns 1 through 4. Next, we construct columns 5, 6, and 7. Finally, we compute the variance by extracting the appropriate values from the table and substituting them directly into (1). Required calculations are shown in the last line of the table.

Table 3.3 provides an example of the application of (2). Having already constructed columns 1 through 4 in calculating the mean of the data, we need only construct columns 5 and 6 and employ (2) directly. Note that the result given in the bottom line is identical to that obtained for the variance in Table 3.3 through use of (1).

As stated when we calculated the mean, the results will not be changed if a different midpoint is used as an approximated mean. You can ascertain this fact by choosing an approximated mean in the interval 40–49 and then calculating the mean and variance, respectively, by Formulas (2), page 52, and (1), above. Results should be identical to those of Table 3.3.

If the data are grouped into intervals of unequal length, Formula (2) can be modified to suit the occasion. However, it is often best to use another formula which can be derived directly from (1). That is, for the case of unequal class intervals, we should use

$$s_x^2 \cong \frac{\sum_{i=1}^k f_i m_i^2 - \frac{(\sum_{i=1}^k f_i m_i)^2}{n}}{n}, \tag{3}$$

where all symbols are as described in conjunction with (1) and (2).

Problems

1. The data of Problem 2, in Section 3.8 have been grouped as follows:

Man hours/unit X	*Number of units* f
$17 \le X < 18$	5
$18 \le X < 19$	4
$19 \le X < 20$	2
$20 \le X < 21$	1
$21 \le X < 22$	1
$22 \le X < 23$	1
$23 \le X < 24$	0
$24 \le X < 25$	1
	15

Round all answers to this problem to two decimal places.

a. Using the frequency table above, calculate the average number of man hours per unit ($\bar{y}$) by the grouped-data method, using Formula (1).

b. Compute the mean number of man hours per unit by the grouped-data method, but use an approximated mean of 20, using Formula (3).

c. Compare the $\bar{y}$'s found in Parts a and b and check to see that they are equal.

d. Calculate the median number of man hours per unit from the grouped data by linear interpolation.

e. Determine the modal class of the number of man hours per unit.

f. Compute the variance of the man hours per unit by (2), the *computational* formula for the variance for grouped data (not the formula which defines s_y^2).

2. Given n_x values of the variable X with mean $\bar{x}$ and variance s_x^2 and n_y values of the variable Y with mean $\bar{y}$ and variance s_y^2, suppose we combine them to form a new set of values of the variable Z. We now have n_z values of the variable Z with mean $\bar{z}$ and variance s_z^2. Prove that

$$s_z^2 = \frac{n_x s_x^2 + n_y s_y^2 + n_x(\bar{x} - \bar{z})^2 + n_y(\bar{y} - \bar{z})^2}{n_z}$$

(*Hint:* Start with $s_z^2 = \dfrac{\sum_{i=1}^{n_x}(x_i - \bar{z})^2 + \sum_{i=1}^{n_y}(y_i - \bar{z})^2}{n_z}$,

expand the expressions, substitute

$$\sum_{i=1}^{n_x} x_i = n_x \bar{x} \quad \text{and} \quad \sum_{i=1}^{n_y} y_i = n_y \bar{y},$$

and simplify.)

3. Suppose we have the following distributions:

$$X: 11, 13, 17, 19 \qquad Y: 9, 11, 13, 11$$

All symbols are defined as in Problem 2 and all answers are to be rounded to one decimal place.

a. Calculate $\bar{x}$, $\bar{y}$, and $\bar{z}$.

b. Compute s_x^2 and s_y^2.

c. Find s_z^2 by the ungrouped data *formula which defines* the variance.

d. Calculate s_z^2 by the formula proved in Problem 2 and compare your answer with the s_z^2 found in Part c of this problem. They should be approximately equal.

4. All symbols are defined as in Problem 2.

a. Prove the formula

$$s_z^2 = \frac{n_x s_x^2 + n_y s_y^2}{n_z} + \frac{n_x n_y (\bar{x} - \bar{y})^2}{n_z^2}$$

(*Hint:* Start with the formula proved in Problem 2.)

b. Calculate s_z^2 by the formula proved in Part a of this problem from the measures calculated in Problem 3 a and b. Round your answer to one decimal place. Compare your answer with the answers to c and d of Problem 3 and check that they are approximately equal.

5. A set of grouped data has a total of 15 observations. The mean of this data is 2 and the sum of the products of the frequency of each class and the square of the midpoint of that class is 105. Find the standard deviation of this set of grouped data. Round your answer to two decimal places.

6. A set of grouped data has a mean of 5 and a variance of 15. The sum of the products of the frequency of each class and the square of the midpoint of that class is 800. Find the total number of observations, to the nearest integer.

7. A set of grouped data has a total of 25 observations. The sum of the products of the class frequency of each class and the class deviation of that class is 20 and the sum of the products of the class frequency of each class and the square of the class deviation of that class is 116. If the standard deviation is 12 and class intervals are of equal length, what is the class interval length?

3.8 *Measure of Relative Dispersion (Coefficient of Variation)*

Given two sets of data, one representing values of a variable X the other representing values of a variable Y, suppose we want to know which set reflects greater variation relative to its mean. Since the standard deviation provides a measure of dispersion, we could compute it for each set and compare the results. However, since the standard deviation (or any measure of dispersion discussed so far) measures absolute dispersion, the results might not be meaningfully comparable. For example, suppose we want to determine which of two retail salesmen is the more consistent performer relative to his average sales, salesman X or salesman Y. Suppose further that historical records of ten weeks' activity (dollar sales per week) are:

X: 4,500, 5,300, 6,200, 4,000, 5,600, 5,100, 4,700, 3,900, 4,600, 5,800;

Y: 7,700, 7,100, 6,700, 5,900, 6,500, 8,000, 6,400, 7,200, 8,500, 5,800.

The respective standard deviations are (Table 3.4)

$$s_x = \$723.95 \quad \text{and} \quad s_y = \$844.75.$$

We might conclude that, since his measure of dispersion is smaller, salesman X is relatively more consistent. However, is it an inherent variation in salesman Y's performance that causes his standard deviation to be larger than that of salesman X, or should some other factor be considered? Inspection reveals that the magnitude of Y's sales is generally larger than that of X's. This is reflected by the fact that $\bar{y} = 6{,}980 > \bar{x} = 4{,}970$. If we convert each of the standard deviations to relative units by dividing each by the appropriate mean, we see that "the dispersion of X relative to the mean of X" is larger than "the dispersion of Y relative to the mean of Y." That is, while X has a smaller dispersion from an absolute standpoint, Y has the smaller dispersion from a relative standpoint.

Table 3.4

Calculating means and variances

x_i	$x_i - \bar{x}$	$(x_i - \bar{x})^2$	y_i	$(y_i - \bar{y})$	$(y_i - \bar{y})^2$
4,500	−470	220,900	7,700	+720	518,400
5,300	+330	108,900	7,100	+120	14,400
6,200	+1,230	1,512,900	6,700	−280	78,400
4,000	−970	940,900	5,900	−1,080	1,166,400
5,600	+630	396,900	6,500	−480	230,400
5,100	+130	16,900	8,000	+1,020	1,040,400
4,700	−270	72,900	6,400	−580	336,400
3,900	−1,070	1,144,900	7,200	+220	48,400
4,600	−370	136,900	8,500	+1,520	2,310,400
5,800	+830	688,900	5,800	−1,180	1,392,400
49,700		5,241,000	69,800		7,136,000

$$\bar{x} = \frac{\sum x}{n} = \frac{49{,}700}{10} = 4{,}970.$$

$$s_x^2 = \frac{\sum_{i=1}^{n} (x_i - \bar{x})^2}{n} = \frac{5{,}241{,}000}{10} = 524{,}100.$$

$$s_x = \sqrt{524{,}100} \cong 723.95.$$

$$\bar{y} = \frac{\sum y}{n} = \frac{69{,}800}{10} = 6{,}980.$$

$$s_y^2 = \frac{\sum_{i=1}^{n} (y_i - \bar{y})^2}{n} = \frac{7{,}136{,}000}{10} = 713{,}600.$$

$$s_y = \sqrt{713{,}600} \cong 844.75.$$

Since so many situations arise in which the variance associated with a phenomenon increases as the magnitude of the variable being measured, it is often more meaningful to compare the spread of two distributions in terms of their respective relative dispersions. To do this we compute a measure of relative dispersion called the *coefficient of variation* for each distribution. The coefficient of variation of a variable X is defined as

$$V_x = \frac{s_x}{\bar{x}}, \tag{1}$$

where V_x is the coefficient of variation of X. Thus, whether we are dealing with grouped or ungrouped data, we calculate the coefficient of variation of a variable X simply by taking the ratio of the standard deviation of X to the mean of X.

With regard to the above example, we have

$$V_x = \frac{s_x}{\bar{x}} = \frac{723.95}{4970} = 0.1457 \quad \text{and} \quad V_y = \frac{s_y}{\bar{y}} = \frac{844.75}{6980} = 0.1210.$$

Since $V_y < V_x$, we conclude that, when we consider the average magnitude of sales, salesman Y exhibits less relative variation than salesman X. Thus, we say that his sales performance is more consistent (if we measure consistency in relation to average sales).

Problems

1. The data of Problem 2, page 71, have been grouped as follows:

Sheets of steel used X	*Number of tanks* f
115–119	5
120–124	2
125–129	3
130–134	2
135–139	1
140–144	0
145–149	2
	15

a. Using the frequency table above, calculate the average number of sheets of steel used per unit by the grouped data method. Round your answer to one decimal place.

b. Calculate $\bar{x}$ by the grouped-data method, but use an approximated mean of 127. Round your answer to one decimal place.

c. Compute the mean of the sheets of steel used per unit by the grouped-data method, but use an approximated mean of 122. Round to one decimal place.

d. Check the results of Parts a, b, and c, to see that they are equal.

e. Compare the means found in a, b, and c of Problem 2, page 71, with those found in Parts a, b, and c of this problem and check to see that they are approximately equal. Explain intuitively why they are not exactly equal.

f. Calculate the median number of sheets of steel used per unit from the grouped data by linear interpolation. Round to one decimal place.

g. Determine the modal class of sheets of steel used per unit.

h. Compute s_x for the sheets of steel used per unit by the formula which defines the standard deviation for grouped data. Round to one decimal place.

i. From your earlier answers, calculate the coefficient of variation of the sheets of steel used per unit. Round V_x to two decimal places.

2. Suppose we have the following continuous data for the number of man hours used in completing tank assemblies.

$y_1 = 22.51$	$y_6 = 17.97$	$y_{11} = 18.41$
$y_2 = 17.63$	$y_7 = 20.21$	$y_{12} = 18.13$
$y_3 = 19.42$	$y_8 = 17.55$	$y_{13} = 18.41$
$y_4 = 24.16$	$y_9 = 19.36$	$y_{14} = 21.88$
$y_5 = 18.03$	$y_{10} = 17.55$	$y_{15} = 17.55$

Round all answers in this problem to two decimal places.

a. Calculate the arithmetic mean ($\bar{y}$) of the man hours worked per unit by the ungrouped-data method.

b. Compute $\bar{y}$ by the ungrouped-data method, but use an approximated mean of 18.00.

c. Calculate the arithmetic mean of the man hours worked per unit using an approximated mean of 16.00.

d. Check your answers to Parts a, b, and c, to see that they are equal.

e. Determine the median number of man hours worked per unit.

f. Determine the modal number of man hours worked per unit.

g. Compute the standard deviation of the number of man hours worked per unit by the *computational* formula.

h. Calculate the coefficient of variation (V_y) of the number of man hours worked per unit from your answers to earlier parts of this problem.

3. For a distribution with $V_x = 0.8$ and $\bar{x} = 50$, calculate the standard deviation to one decimal place.

4. Given the distribution of Y with $V_y = 0.5$ and $s_y^2 = 16.0$, calculate $\bar{y}$, rounding to one decimal place.

5. Suppose we have two distributions, A and B, with $V_a = V_b = 0.3, \bar{a} = 27$, and $\bar{b} = 87$. Calculate s_a and s_b to one decimal place and state which is larger.

6. Company G has average monthly sales of \$18,000 with a coefficient of variation of 0.6. Company H's average monthly sales are \$84,000 with a coefficient of variation of 0.3. Which company's monthly sales have the larger variance?

7. Given a distribution $X: x_1, x_2, x_3, \ldots, x_n$, prove that

$$V_x = \frac{s_x}{\bar{x}} = \sqrt{\frac{\sum_{i=1}^{n}\left[\frac{x_i - \bar{x}}{\bar{x}}\right]^2}{n}}.$$

3.9 Physical Meaning of Descriptive Measures

Suppose we have collected four sets of data by taking a very large number of observations from each of four variables (A, B, C, and D). Suppose further that for each set of data we have constructed a frequency polygon by using such narrow intervals that we obtain the frequency curves of Figure 3.1. What is the physical relationship between these frequency curves and the descriptive measures discussed in previous sections? If we could compute the various measures, we would note the following facts.

1. The *mean* of a distribution represents the center of gravity of that distribution. That is, if given a physical model of the distribution of A (Figure 3.1), for example, we would find that the mean is the point where the model balances perfectly on a fulcrum.
2. The *mode*(s) of a distribution is (are) the highest point(s) attained by the curve representing the distribution. (See curve A in Figure 3.1).
3. The *median* of a distribution is the value that separates the area under the appropriate curve into two equal parts. (See curve A in Figure 3.1). It follows that the median is the value on the horizontal scale of a cumulative distribution corresponding to 50 percent on the vertical axis (that is, 50 percent of the values are less than the median).

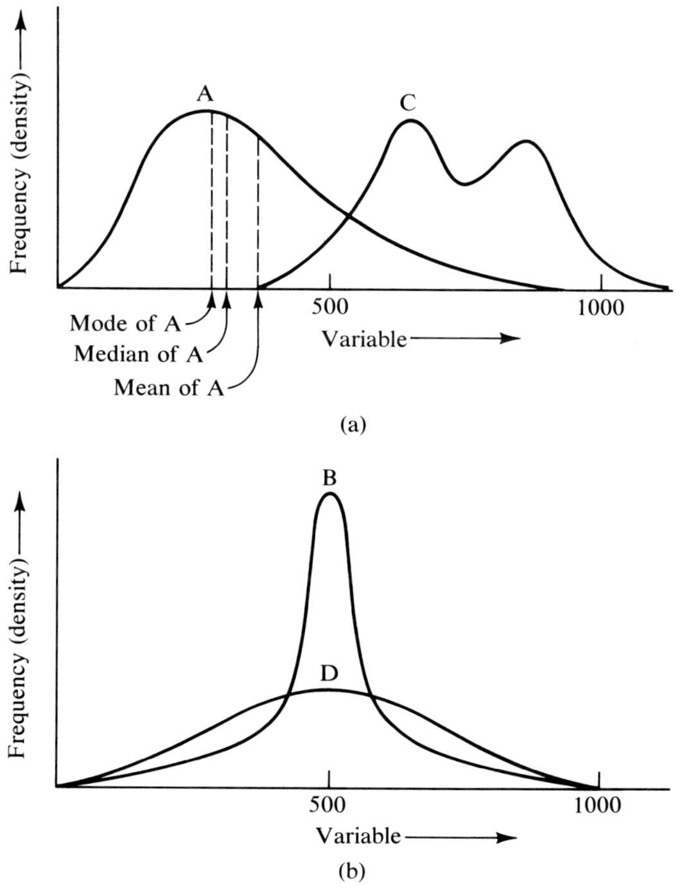

Figure 3.1

4. The physical meaning of the standard deviation (variance) will be discussed in subsequent chapters. Here we need only note that it gives an indication of how closely the values are concentrated around the mean. For example, since curves B and D have identical means (Figure 3.1), it is very easy to see that the distribution of B is concentrated near the mean, while that of D is more widely dispersed. Standard deviation computations would reflect this same fact; in other words, we would find $s_d > s_b$.

Other miscellaneous facts of interest concerning distributions are:

1. If the curve to the right of the mean for a distribution is the mirror image of the curve to the left of the mean, we say the curve (or distribution) is *symmetrical.* Curves B and D are symmetrical.
2. When a distribution is not symmetrical, it is said to be *skewed.* Curve A is skewed. When the curve has a long tail to the right (as curve A of Figure 3.1), we say the curve is skewed to the right (positively). If the tail is to the left, the curve is said to be skewed to the left (negatively).
3. A curve having only one peak (e.g., curve B, Figure 3.1) is described as *unimodal*; a curve having two peaks (curve C, Figure 3.1) is said to be *bimodal*; a curve having three peaks is *trimodal*; etc.
4. For unimodal distributions, the relations between the mean, the median, and the mode are as follows. If the distribution is symmetrical, the three measures are equal. If the distribution is skewed to the right, as curve A, we have

$$\text{mean} > \text{median} > \text{mode}.$$

On the other hand, if the distribution is skewed to the left,

$$\text{mean} < \text{median} < \text{mode}.$$

Problems

1. Suppose we have the following ungrouped discrete data on the hourly wages paid to employees in the production department:

Hourly wage X	*Number of employees* f
\$15.00	1
7.50	3
5.00	19
4.00	14
3.00	21
2.00	23
1.50	4
	75

a. Calculate the mean wage rate by the ungrouped-data method, accurate to the nearest cent.

b. Find the median wage rate.

c. Determine the modal wage rate.

d. Draw a rough line graph of this distribution and indicate the locations of the mean, median, and mode.

e. State whether this distribution is skewed in the positive or the negative direction.

2. Suppose we have the following grouped continuous data for the output of amplifiers produced by a given company.

Output in watts	f
$70 \le X < 72$	2
$72 \le X < 74$	10
$74 \le X < 76$	13
$76 \le X < 78$	8
$78 \le X < 80$	3
	36

a. Compute the mean output, accurate to the nearest watt.

b. Determine the median output by linear interpolation, to the nearest watt.

c. Find the modal output by selecting the midpoint of the modal class.

d. Draw an approximate frequency curve of this distribution and indicate the locations of the mean, median, and mode found in Parts a, b, and c.

3.10 Summary

Sets of data can be characterized (either partially or completely) by *descriptive measures*. The most important of such measures describe the location or the dispersion associated with a given set of data. The best-known measures of location are the *mean*, the *median*, and the *mode*, while the best-known measures of dispersion are the *variance*, the *standard deviation*, the *range*, and the *mean deviation*.

While all of these measures are important in descriptive statistics, the mean and the variance (or standard deviation) also play vital, almost dominant, roles in elementary statistical inference, especially inference about an entire population. It can be shown that the mean and the variance are more stable in a statistical sense than the other measures. Furthermore, they are mathematically and theoretically tractable. This is not to say, however, that they are to be considered as universally more important. There are, in fact, specific areas of both theoretical and applied statistics in which those measures which we have treated rather lightly (and even some we have not mentioned at all, e.g., quartiles) prove quite useful. It is obviously impractical to attempt to treat all such situations in a tcxt of this nature.

Table 3.5

Summary of Chapter 3 formulas

Type of measure	*Situation*	*Formula*	*Equation and section*
		Ungrouped data	
Mean	Definition	$\bar{x} = \frac{1}{n}\sum_{i=1}^{n} x_i$	(1), Sec. 3.3
Mean	Given means of k sets of data	$x = \frac{n_1\bar{x}_1 + n_2\bar{x}_2 + \cdots + n_k\bar{x}_k}{n_1 + n_2 + \cdots + n_k}$	(3), Sec. 3.3
Weighted mean	Standard	$\bar{x}_w = \frac{\sum_{i=1}^{k} x_i w_i}{\sum_{i=1}^{k} w_i}$	(6), Sec. 3.3
Mean	Shortcut	$\bar{x} = A + \frac{1}{n}\sum_{i=1}^{n} d_i'$	(3), Supplement 3.B
Mean deviation	Standard	$MD = \frac{1}{n}\sum_{i=1}^{n} \lvert x_i - \bar{x} \rvert$	(4), Sec. 3.4
Variance	Definition	$s_x^2 = \frac{1}{n}\sum_{i=1}^{n} (x_i - \bar{x})^2$	(5), Sec. 3.4
Variance	Computational formula	$s_x^2 = \frac{\sum_{i=1}^{n} x_i^2 - \frac{(\sum_{i=1}^{n} x_i)^2}{n}}{n}$	(6), Sec. 3.4
		Grouped data	
Mean	Definition	$\bar{x} \cong \frac{\sum_{i=1}^{k} m_i f_i}{\sum_{i=1}^{k} f_i}$	(1), Sec. 3.6
Mean	Computational formula (for class intervals of equal length)	$\bar{x} \cong A + c \cdot \frac{\sum_{i=1}^{k} f_i d_i}{\sum_{i=1}^{k} f_i}$	(2), Sec. 3.6
Mean	Computational formula (for class intervals not of equal length)	$\bar{x} \cong A + c \cdot \frac{\sum_{i=1}^{k} f_i (m_i - A)/c}{\sum_{i=1}^{k} f_i}$	(3), Sec. 3.6
Variance	Definition	$s_x^2 \cong \frac{\sum_{i=1}^{k} f_i (m_i - \bar{x})^2}{\sum_{i=1}^{k} f_i}$	(1), Sec. 3.7
Variance	Computational formula (for class intervals of equal length)	$s_x^2 \cong c^2 \left[\frac{\sum_{i=1}^{k} f_i d_i^2 - \frac{(\sum_{i=1}^{k} f_i d_i)^2}{\sum_{i=1}^{k} f_i}}{\sum_{i=1}^{k} f_i} \right]$	(2), Sec. 3.7
Variance	Computational formula (for class intervals not of equal length)	$s_x^2 \cong \frac{\sum_{i=1}^{k} f_i m_i^2 - \frac{(\sum_{i=1}^{k} f_i m_i)^2}{\sum_{i=1}^{k} f_i}}{\sum_{i=1}^{k} f_i}$	(3), Sec. 3.7

Supplement 3.A Equivalence of the Computational Formula and the Definition of Variance

Let us first demonstrate the equivalence of Formulas (5) and (6) of Section 3.4 by example and then prove the general equivalence. Suppose we have the values for ten observations on a variable X,

$$X: 7, 9, 12, 8, 6, 5, 7, 6, 8, 11,$$

and we want to calculate the variance of the data. Relevant information and calculations for finding the variance by means of both the definition and the computational formula are given in Table 3.6. Note that the results are equivalent, but that the computational formula involves less work. The labor is even less if a calculator is used.

Table 3.6

Calculation of variance by definition and by computational formula

x_i	$x_i - \bar{x}$	$(x_i - \bar{x})^2$	x_i^2
7	−0.9	0.81	49
9	1.1	1.21	81
12	4.1	16.81	144
8	0.1	0.01	64
6	−1.9	3.61	36
5	−2.9	8.41	25
7	−0.9	0.81	49
6	−1.9	3.61	36
8	0.1	0.01	64
11	3.1	9.61	121
79	0	44.90	669

$$\bar{x} = \frac{79}{10} = 7.9.$$

By definition,

$$s_x^2 = \frac{\sum(x_i - \bar{x})^2}{n} = \frac{44.90}{10} = 4.49.$$

By computational formula,

$$s_x^2 = \frac{\sum x_i^2 - \frac{(\sum x_i)^2}{n}}{n} = \frac{669 - \frac{(79)^2}{10}}{10}$$

$$= \frac{669 - \frac{6241}{10}}{10} = \frac{669 - 624.1}{10} = \frac{44.9}{10} = 4.49.$$

Since both Formulas (5) and (6) of Section 3.4 involve a denominator of n, we can show a general equivalence of the two by showing only that the numerators are equal. We prove the equivalence as follows. Simply squaring the term $(x_i - \bar{x})$, we obtain

$$\sum_{i=1}^{n} (x_i - \bar{x})^2 = \sum_{i=1}^{n} (x_i^2 - 2\bar{x}x_i + \bar{x}^2).$$

Then, using Rule 3.1, we can express the right side of the above equation as

$$\sum_{i=1}^{n} x_i^2 - \sum_{i=1}^{n} 2\bar{x}x_i + \sum_{i=1}^{n} \bar{x}^2.$$

Noting that 2, $\bar{x}$, and $\bar{x}^2$ are constants (no subscript), we now apply Rules 3.2 and 3.3 to the second and third terms, respectively, to obtain

$$\sum_{i=1}^{n} x_i^2 - 2\bar{x}\sum_{i=1}^{n} x_i + n\bar{x}^2.$$

However, replacing $\bar{x}$ by $\sum_{i=1}^{n} x_i/n$ in the second and third terms, gives

$$\sum_{i=1}^{n} x_i^2 - 2\frac{(\sum_{i=1}^{n} x_i)(\sum_{i=1}^{n} x_i)}{n} + \frac{n(\sum_{i=1}^{n} x_i)^2}{n^2}.$$

Finally, through simple algebraic manipulation, we obtain

$$\sum_{i=1}^{n} x_i^2 - 2\frac{(\sum_{i=1}^{n} x_i)^2}{n} + \frac{(\sum_{i=1}^{n} x_i)^2}{n} = \sum_{i=1}^{n} x_i^2 - \frac{(\sum_{i=1}^{n} x_i)^2}{n}.$$

Therefore, we have proved that

$$\sum_{i=1}^{n} (x_i - \bar{x})^2 = \sum_{i=1}^{n} x_i^2 - \frac{(\sum_{i=1}^{n} x_i)^2}{n}.$$

From this, we see that Formulas (5) and (6) of Section 3.4 are equivalent.

Supplement 3.B Shortcut Method for Finding the Mean and Variance of Ungrouped Data

---

The transformations of Section 3.5 make it much easier to calculate the mean and variance of ungrouped data—particularly when the values are relatively large or decimal values are involved. Suppose we want the mean and variance of the following values of a variable X:

$$X: 8{,}325;\ 8{,}330;\ 8{,}328;\ 8{,}340;\ 8{,}332.$$

From Table 3.7, we see that direct application of appropriate definitions would give a mean of

$$\bar{x} = \frac{1}{n}\sum_{i=1}^{n} x_i = \frac{41{,}655}{5} = 8{,}331 \qquad (1)$$

and a variance of

$$s_x^2 = \frac{1}{n}\sum_{i=1}^{n} (x_i - \bar{x})^2 = \frac{128}{5} = 25.6. \qquad (2)$$

However, our task will be much easier if we

1. make the linear transformation

$$Y = X - 8{,}325,$$

2. find the mean and variance of Y,
3. note that $X = Y + 8{,}325$, and
4. use Rules 3.6 and 3.7, respectively, to find the mean and variance of X.

Table 3.7

Mean and variance calculations

x_i	$x_i - \bar{x}$	$(x_i - \bar{x})^2$
8,325	−6	36
8,330	−1	1
8,328	−3	9
8,340	9	81
8,332	1	1
41,655		128

Under the transformation $Y = X - 8{,}325$, corresponding values of Y become

$$Y: 0, 5, 3, 15, 7.$$

Then,

$$\bar{y} = \frac{1}{n}\sum_{i=1}^{n} y_i = \frac{30}{5} = 6 \quad \text{and} \quad s_y^2 = \frac{1}{n}\sum_{i=1}^{n}(y_i - \bar{y})^2 = \frac{128}{5} = 25.6.$$

However, since $X = Y + 8{,}325$, we have

$$\bar{x} = \bar{y} + 8{,}325 = 6 + 8{,}325 = 8{,}331,$$

and

$$s_x^2 = s_y^2 = 25.6.$$

Note that these results are respectively identical to those obtained in (1) and (2) using the definitions. Although the time saved by the above coding procedure appears small here, much effort is saved when we have a large number of values. The general approach for finding the mean of data for which a simple additive transformation will provide numbers that are easy to manipulate is

$$\bar{x} = A + \frac{1}{n}\sum_{i=1}^{n} d_i', \tag{3}$$

where $\bar{x}$ is the mean of a set of data, A is any approximated mean (usually approximated by inspection), d_i' is the difference between the ith value of X and the approximated mean $(x_i - A)$, and n is the number of observations.

Also, since each d_i' is obtained simply through an additive transformation of the corresponding x_i value, it is immediately evident that the variance of the x values will be the same as the variance of the d' values. Thus, we determine the variance of the x values simply by calculating the variance of the d' values.

When the original measurements involve decimals, numbers which are easy to manipulate can be obtained by a linear transformation involving both additive and multiplicative constants. For example, suppose we wish to find the mean and the variance of X for the values

$$X: 1.56, 1.67, 1.48, 1.53, 1.61. \tag{4}$$

Direct application of the Formulas of (1) and (2) will give us

$$x = 1.57 \quad \text{and} \quad s_x^2 = 0.00428. \tag{5}$$

However, careful study of the data shows that we can make things easier by multiplying each number by 100 and subtracting 155 from the results; that is, by making the transformation

$$Y = -155 + 100X. \tag{6}$$

Such transformation provides us with the new set of data

$$Y: 1, 12, -7, -2, 6,$$

from which we immediately have

$$\bar{y} = \frac{1}{5}(1 + 12 - 7 - 2 + 6) = 2$$

and

$$s_y^2 = \frac{1}{5}[(1 - 2)^2 + (12 - 2)^2 + (-7 - 2)^2 + (-2 - 2)^2 + (6 - 2)^2]$$

$$= \frac{1}{5}(1 + 100 + 81 + 16 + 16) = \frac{214}{5} = 42.8.$$

However, by solving the transformation equation of (6) for X in terms of Y, we see that

$$X = 1.55 + \frac{Y}{100}.$$

Thus, the additive constant is 1.55, the multiplicative constant is 1/100, and by Rules 3.4 and 3.5, respectively, we have

$$\bar{x} = 1.55 + \frac{\bar{y}}{100} = 1.55 + 0.02 = 1.57$$

and

$$s_x^2 = \left(\frac{1}{100}\right)^2 s_y^2 = \frac{42.8}{10{,}000} = 0.00428.$$

Note once again that these results are identical with those obtained in (5) by direct application of the formulas for the mean and variance in (1) and (2).

Problems

1. Prove that $\bar{x} = A + \frac{1}{n}\sum_{i=1}^{n} d_i'$ is equivalent to $\bar{x} = \frac{1}{n}\sum_{i=1}^{n} x_i$.

$\left(\textit{Hint:}\ \text{Use } \sum_{i=1}^{n} d_i' = \sum_{i=1}^{n} (x_i - A).\right)$

2. Suppose we have the following discrete data for the number of sheets of steel used in manufacturing each tank assembly during the first week of operations on a new contract:

$x_1 = 131$	$x_4 = 128$	$x_7 = 119$	$x_{10} = 125$	$x_{13} = 116$
$x_2 = 146$	$x_5 = 123$	$x_8 = 132$	$x_{11} = 118$	$x_{14} = 118$
$x_3 = 115$	$x_6 = 138$	$x_9 = 149$	$x_{12} = 121$	$x_{15} = 129$

Round all answers to one decimal place.

a. Calculate the arithmetic mean of the sheets of steel used per unit using the ungrouped data method. (That is, calculate $\bar{x}$.)

b. Compute $\bar{x}$ using an approximated mean of 125.

c. Calculate the arithmetic mean of the sheets of steel used per unit using an approximated mean of 134.

d. Compare your answer to Parts a, b, and c, and check to see that they are equal.

e. Determine the median number of sheets of steel used per unit.

f. Determine the modal number of sheets of steel used per unit.

g. Compute the standard deviation of the number of sheets of steel used per unit by the formula which defines the standard deviation. Use the $\bar{x}$ found in Part a for your calculations. Calculate s_x accurate to two decimal places.

3. Given A is any constant, $D = \bar{x} - A$, and $R^2 = \dfrac{\sum_{i=1}^{n} (x_i - A)^2}{n}$, prove that $s_x^2 = R^2 - D^2$.

(*Hint:* Begin with $s_x^2 = \frac{1}{n}\sum_{i=1}^{n} x_i^2 - \left[\frac{1}{n}\sum_{i=1}^{n} x_i\right]^2$,

substitute $\frac{1}{n}\sum_{i=1}^{n} x_i = \bar{x}$,

substitute $\bar{x} = D + A$, expand and simplify.)

4. Given the distribution

$$X: 835, 840, 850, 860, 865,$$

calculate s_x^2 by the formula $s_x^2 = R^2 - D^2$, where R and D are defined as in Problem 3 and $A = 845$. Round your answer to one decimal place.

5. The distribution of net weights (in grams) of packages of a new drug was found to be

299.4, 300.1, 300.6, 299.8, 300.1.

Calculate the variance of this distribution by the method of Problem 4 with $A = 299.0$. Round the result to two decimal places.

Important Terms and Concepts

absolute deviation	location	relative dispersion
absolute value	measure of dispersion	sample
additive transformation	measure of location	standard deviation
class deviation	mean	standardized variable
coding	mean deviation	summation notation
coefficient of variation	median	transformation
descriptive measure	mode	ungrouped data
deviation	multiplicative transformation	unit distribution
dispersion	population	variance
grouped data	range	weighted mean
linear transformation		

Review Problems

1. Class A 1/4-inch bolts always have a diameter between 0.2499 inch and 0.2501 inch. Class D 1/4-inch bolts have diameters that vary from 0.24 inch to 0.26 inch.

   a. Would it be safe to characterize these two types of bolts as identical? Will Class D 1/4-inch bolts always fit in the 0.2495-inch holes in a product your company is producing?

   b. What type of measure in addition to a measure of location is useful in describing distributions of bolt diameters?

2. a. Suppose we have the distribution X: 4, 8, 7, 10, 6. Calculate

$$\sum_{i=1}^{n}(x_i - \bar{x})^2 \quad \text{and} \quad \sum_{i=1}^{n} x_i^2 - \frac{(\sum_{i=1}^{n} x_i)^2}{n}$$

   and note that they are equal.

   b. Prove generally that

$$\sum_{i=1}^{n}(x_i - \bar{x})^2 = \sum_{i=1}^{n} x_i^2 - \frac{(\sum_{i=1}^{n} x_i)^2}{n}.$$

   Hint: Expand the quadratic in the left-hand expression and substitute $\bar{x} = \dfrac{\sum_{i=1}^{n} x_i}{n}$.

   c. Transform

$$s_x^2 = \frac{\sum_{i=1}^{n}(x_i - \bar{x})^2}{n}$$

   into

$$s_x^2 = \frac{\sum_{i=1}^{n} x_i^2}{n} - \left(\frac{\sum_{i=1}^{n} x_i}{n}\right)^2$$

   by applying the relation proved in Part b.

3. Suppose we have the distribution X: 15, 16, 19, 22, 23. Round all answers to one decimal place.

   a. Calculate $\bar{x}$ and s_x^2 by the formulas that define each of these measures.

   b. Compute $s_x^2 + \bar{x}^2$.

   c. Calculate $\frac{1}{n}\sum_{i=1}^{n} x_i^2$ and check to see that it is approximately equal to the quantity calculated in Part b.

4. A distribution of size 25 has a variance of 16 and the sum of the squared observations is 625. Find the mean of this distribution. Round your answer to one decimal place.

5. Suppose we have the distribution $X: x_1, x_2, \ldots, x_n$. Make the transformation $Z = \dfrac{X - \bar{x}}{s_x}$ and prove that $\bar{z} = 0$ and $s_z^2 = 1$. (*Hint:* Convert to a linear transformation and use the appropriate rules of transformations.)

6. a. Prove that, under an additive transformation, the mean of the transformed variable is equal to the *sum* of the additive constant and the mean of the original variable. (That is, prove Rule 3.6.)

   b. Prove that, under an additive transformation, the variance of the transformed variable is the *same* as the variance of the original variable. (That is, prove Rule 3.7.)

   c. Prove that, under a multiplicative transformation, the variance of the transformed variable is equal to the *product* of the *square* of the multiplicative constant and the variance of the original variable. (That is, prove Rule 3.8.)

   d. Prove that, under a multiplicative transformation, the variance of the transformed variable is equal to the *product* of the *square* of the multiplicative constant and the variance of the original variable. (That is, prove Rule 3.9.)

7. Using the distribution

$$X: 40, 45, 50, 55, 60:$$

   a. Calculate $\bar{x}$ and s_x^2 by the formulas which *define* the mean and variance.

   b. Using $A = 45$, calculate $D = \bar{x} - A$ and $R^2 = \dfrac{\sum_{i=1}^{n}(x_i - A)^2}{n}$ and note that $s_x^2 = R^2 - D^2$.

8. Prove that

$$\bar{x} = A + c \cdot \frac{\sum_{i=1}^{k} f_i d_i}{\sum_{i=1}^{k} f_i}$$

is equivalent to

$$\bar{x} = \frac{\sum_{i=1}^{k} m_i f_i}{\sum_{i=1}^{k} f_i}$$

(*Hint:* $m_i = A + d_i c$.)

9. Prove the computational formula for the variance of grouped data,

$$s_x^2 = c^2 \left[\frac{\sum_{i=1}^{k} f_i d_i^2 - \dfrac{(\sum_{i=1}^{k} f_i d_i)^2}{n}}{n} \right].$$

(*Hint:* Start with $s_x^2 = \frac{1}{n}\sum_{i=1}^{k} f_i(m_i - \bar{x})^2$, substitute

$$m_i = A + d_i c \quad \text{and} \quad \bar{x} = A + c \cdot \left[\frac{\sum_{i=1}^{k} f_i d_i}{n}\right],$$

expand and simplify the expression.)

10. Suppose we have the distribution X: 8, 10, 11, 12, 14. Round all answers to two decimal places.

    a. Calculate $\bar{x}$, s_x, and V_x by the *formulas which define* those measures.

    b. Calculate V_x by the formula

$$V_x = \sqrt{\frac{\sum_{i=1}^{n} \left[\frac{x_i - \bar{x}}{\bar{x}}\right]^2}{n}}$$

    and check that your answer is approximately equal to the V_x found in Part a.

11. A distribution of size 10 has a coefficient of variation of 3 and the sum of the squared observations is 1090. Find the mean of this distribution accurate to two decimal places.

Four

Probability Theory

4.1 Introduction

In the first three chapters we have been concerned with the area of statistics generally known as descriptive statistics. In statistical inference we make inferences about a population by using information obtained from a statistical sample taken from that population. Statistical inference is based largely upon probability theory, primarily because probability theory provides a means of determining the reliability of inferences. In this chapter, we shall learn enough about the basic concepts of probability theory to understand the conclusions that result from the application of statistical techniques to data analysis, as well as the reasons behind the requirements for probability sampling in the collection of data. For further reading in probability, see Mosteller, et al (4.1).

Probability theory was developed in the seventeenth century to predict the long-term outcomes of games of chance—tossing coins, rolling dice, spinning roulette wheels, drawing cards, and so forth. In all of these games, the outcome for any specific trial is uncertain, but each possible outcome will appear with a certain long-term regularity. The predictability of this regularity is the concern of probability theory.

In many less frivolous areas—business, economics, engineering, biology, psychology, physics, chemistry and meteorology, to name only a few—we encounter similar uncertainties and long-term regularities. For example, a production manager for an electronics firm does not know which transistors from a production line will be defective, but he does know what percentage will be defective in the long run. An economist cannot predict which people will spend less than $500 for clothing during a given year, but he can predict with some degree of certainty how many people will spend less than that amount. A geneticist cannot be certain that a specific baby will be born with blue eyes, but he can predict what percentage of babies will have this characteristic. Such predictions are based on probability.

Statements such as "the probability of thundershowers is 40 percent," "the probability of team A winning the world series is 0.7," and "the probability of obtaining three heads on three tosses of a coin is 1/8" are commonplace. The average person has an intuitive idea (regardless of its validity) of what is meant by each of the statements. However, such statements present two major problems: accuracy of interpretation and calculation of the numerical value. After briefly considering the first problem from the viewpoint of various definitions of probability, we shall be concerned primarily with calculation of probabilities and related values.

Defining probability is difficult; and several different definitions have evolved. We shall begin by considering several approaches which lead to somewhat different interpretations of the meaning of probability. These include the *classical* (or *a priori*) *approach*, the *frequency* (or *a posteriori*) *approach*, and the *subjective approach*.

Classical, or A Priori, Probability

As we said before, the theory of probability originated as a means of predicting long-term outcomes of games of chance. For example, to find the probability that one toss of a fair coin will result in a head, we can argue as follows: since the coin is fair and there are only two ways it can fall, we would expect it to fall heads one-half of the time and tails one-half of the time. Hence, the probability of the event of a head is given the value 1/2. This type of reasoning leads to the classical definition of probability.

Definition 4.1

If an event can occur in n mutually exclusive and equally likely ways, and if m of these outcomes have an attribute A, then the ***probability of A*** *is the fraction m/n.*

Using the definition is easy enough in simple cases, but its use is not always obvious. Particular attention must be given to the terms "mutually exclusive" (to be defined and discussed later) and "equally likely." For example, suppose we wish to compute the probability of getting two heads on two tosses of a fair coin. We might argue that there are three possible outcomes: two heads, two tails, or one head and one tail. Since there are three outcomes and only one has the desired attribute (two heads), an assumption that the outcomes are equally likely would yield a probability of 1/3. However, the three outcomes are not equally likely. The outcome of one head and one tail can occur in two ways. That is, the tail may appear on the first toss and the head on the second, or the head may appear on the first toss and the tail on the second. Thus, there are four equally likely outcomes: HH, HT, TH, TT. The first has the desired attribute, while the others do not. Thus, the correct probability is 1/4. We would have the same result if two fair coins were tossed simultaneously.

Some rather serious drawbacks are associated with the classical definition of probability. It must be modified somewhat when the number of possible outcomes is infinite. An even more serious defect shows up when we have an event that can occur in n ways but the ways are not equally likely. Finally, the classical approach provides no means of answering such questions as "What is the probability that a rocket launched tomorrow will reach the moon?" or "What is the probability of the Dallas Cowboys winning a specific football game?" or "What is the probability that a specific automobile made by Ford Motor Company will get at least 24 miles per gallon of gasoline?" All these questions are of great interest. However, in order to answer them, we must modify our definition of probability.

Frequency, or A Posteriori, Probability

A coin that seems well balanced and symmetrical has been tossed 200 times. The outcomes are listed in Table 4.1. We note that the relative frequencies are quite close to long-run expected relative frequencies for a balanced coin, that is, 1/2. In fact, we might be willing to use the relative frequencies in Table 4.1 as approximations for the probabilities of heads and tails on some future toss of that particular coin.

Table 4.1

Results of tossing a coin 200 times

Outcome	*Frequency*	*Observed relative frequency*	*Long-run expected relative frequency*
H	95	0.475	0.500
T	105	0.525	0.500
Total	200	1.000	1.000

It seems reasonable to assume for this experiment that there exists a number, say π, which is the probability of a head. If the coin appears symmetrical and well balanced, we might use Definition 4.1 to determine that π is approximately equal to 1/2. This would only be an approximation, however, since we do not know for certain that the two cases, heads and tails, are exactly equally likely. As an alternative, we could toss the coin a large number of times, recording the results as in Table 4.1, and use the relative frequency of heads as an approximation of π. In such case, we would have applied what is often called the *frequency-theory approach based on experience.*

Whether we use relative frequency or Definition 4.1 for the probability, π, seems unimportant in the above case. On the other hand, suppose our coin is unbalanced, so that we feel sure that heads and tails are not equally likely. For this situation, we can postulate a value for π. However, the classical definition does not provide an appropriate means of approximating the value for π. Hence, we must necessarily use the frequency approach.

In many business situations, our observations have an element of uncertainty or unpredictability. In such situations, it is often reasonable to postulate that the probability of a specific event is equal to a number π, which is approximated by taking the relative frequency of the event (from a large number of trials). Although we term this the frequency-theory approach based on experience, the method is occasionally called *statistical probability.*

Let us assume that a series of observations (or experiments) can be made under quite uniform conditions. That is, an event can be repeated many times (the outcome being noted each time) under identical conditions, but there is an uncontrollable variation that is "random." Thus, each observation (or outcome) is individually unpredictable. In many cases, such observations fall into certain classes which have quite stable relative frequencies. The relative frequency of a particular outcome is then used to approximate π. How many times must the experiment be repeated to assure a stable relative frequency? Since the answer depends upon the stability of the entire series of experiments, it is difficult to give a rigorous answer. The best we can do is say that the experiment should be repeated until the relative frequencies that will be used as probabilities appear to change very little as the number of experiments gets larger. Thus, the frequency-theory approach based on experience will not suffice entirely. We shall extend it to include the concept of "the limit of the relative frequency of occurrence of an event as the number of trials increases."

In the *frequency-theory approach based on limits*, we are still concerned with the relative frequency m/n of an event. However, under this approach, we define and interpret the probability $P(A)$ of event A as that number which the relative frequency of event A approaches as the number of trials increases without bound. Thus, we say that $P(A)$ is the limit of the relative frequency of the event A as n tends to infinity.

Subjective Probability

The two approaches to probability discussed so far are not satisfactory for all situations. For example, they will not answer such questions as "What is the probability of an investment returning a net profit of at least \$10,000 during the first year?" This type of question involves an event that cannot be repeated under similar conditions with any reasonable certainty. For such a situation, we use an approach involving *subjective probability*. The subjective approach is relatively new, but in recent years much attention has been devoted to it in related literature. We do not attempt an extensive treatment, but we introduce the concept so that you can see how it is linked to the axiomatic development of probability theory. In the subjective approach, one interprets probability as a measure of personal confidence in a particular statement. Since the concept is difficult to treat in a general fashion, we merely exemplify it below.

L. J. Savage (4.2), one of the pioneers of subjective probability, appeals to intuition. He suggests that we associate the definition of probability with "the price we would pay for a particular contingency." For example, suppose we write a contract that promises a person \$100 if an investment returns \$10,000 during the first year. What would the person be willing to pay for the contract before he has seen what happens in the first year? If he would be willing to pay \$50, we take this to mean that the probability of the investment returning \$10,000 is 50/100 for him. For someone willing to pay \$75 for the contract, the probability of the event is 75/100. Under this approach, different people may assign different probabilities to the occurrence of the same event. However, the subjective approach assumes that those involved will be reasonable, not arbitrary, in their probability assignments.

In general, when probabilities are set forth in a situation that contains no objective (in the sense of the classical approach) nor empirical (in the sense of the frequency approach) basis for such probabilities, we say that the probabilities are subjective. In this sense, subjective probabilities are necessarily expressions of personal judgment. Thus, even though a subjective probability may be based on experience with events similar to the one in question, different individuals still may assign different subjective probabilities to the same event. Thus, even among experts in probability and statistics, there exists a great deal of controversy regarding the subjective approach.

Problems

1. Explain *statistical inference.*
2. What two conditions must be satisfied before the definition of *classical* or *a priori* probability can be applied?
3. Explain *statistical probability.*
4. Explain each of the *objective* approaches to probability.
5. Explain the *subjective* approach to probability.

4.2 Set–Theory Notation and Concepts

Now that we have a common understanding about the general meaning of probability, we shall pursue the axiomatic development of probability theory. First, however, we shall review the set-theory notation and concepts we shall need.

In order to develop probability from the set-theoretic viewpoint, we begin with a generalized definition of what constitutes a "set."

Definition 4.2

A ***set*** *is a collection of well-defined, distinct objects with common characteristics. The objects are called elements.*

In other words, a set is an aggregate of elements that have certain specified properties. The population of interest constitutes the universal set (S), or sample space, which is the totality of conceivable outcomes for each situation that we encounter. As this universal set includes all outcomes, the probability of the universal set is unity.

In order to use this approach to probability, we must first discuss the properties of sets and operations with sets that are relevant for our studies. To aid in visualizing the operations, it is convenient for us to geometrically represent sets using Venn diagrams as shown in Figure 4.1. The universal set is represented by a rectangle and regions within the rectangle represent subsets of the universal set.

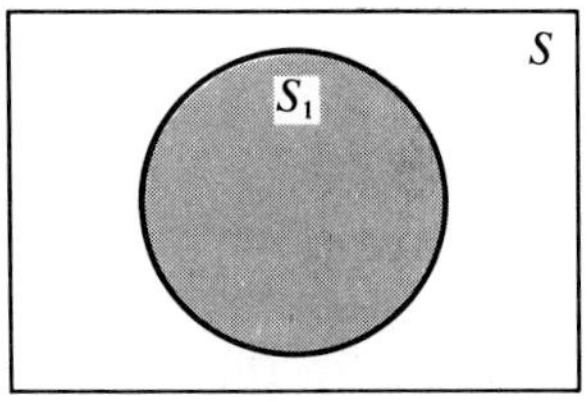

Figure 4.1
Geometric representation of a set

Definition 4.3

*A set S_1 is a **subset** of set S_2 if and only if every element of S_1 is also an element of S_2. We write this relationship as*

$$S_1 \subset S_2.$$

Among all the possible outcomes of an experiment, we may be interested in the outcomes having a particular attribute A. The aggregate of the events possessing the attribute A composes a set. Let us call it set A. The totality of outcomes is our universal set S, and A is a subset of S ($A \subset S$), because every element of A is also an element of S.

By definition, each set is a subset of itself. Also, the set containing no elements—the null, or empty, set—denoted $\varnothing$, must necessarily be a subset of every set. The importance of $\varnothing$ will be seen later in this chapter.

Set Operations

The fundamental operations in set theory are *union* and *intersection*. Other operations relevant to probability are the *complement*, *difference*, and *exclusive union* (Boolean sum).

Definition 4.4

*The **union** of two sets S_1 and S_2 is the set consisting of all elements belonging to S_1, to S_2, or to both. The union of sets S_1 and S_2 is denoted as $S_1 \cup S_2$.*

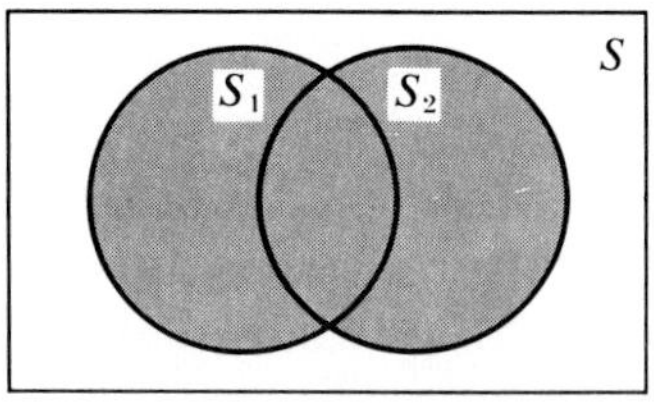

Figure 4.2

The union of sets

Pictorially, this union is the shaded area of Figure 4.2. As indicated in the figure, the union operation on two sets generates another set containing all the elements that are in either set or both sets.

Example 1

In a particular department made up of 20 individuals, 12 are men and 8 are women. Of the 20, 7 are salaried employees and 13 are hourly employees. Three of the men are salaried. Let S_1 represent the set of men and S_2 represent the set of salaried employees. Our interest is in the set of employees that consists of men or salaried employees or both. This set is the union $S_1 \cup S_2$. The set $S_1 \cup S_2$ consists of the 12 men and the 4 salaried women.

Example 2

Let
$$A = \{2, 3, 9\},$$
$$B = \{3, 9, 7, 5\},$$
$$C = \{4\}, \quad \text{where } \{\,\} \text{ enclose the elements of a set.}$$

Then,
$$A \cup B = \{2, 3, 9, 7, 5\},$$
$$B \cup C = \{3, 9, 7, 5, 4\},$$
$$A \cup C = \{2, 3, 9, 4\}.$$

Definition 4.5

The ***intersection*** *of sets S_1 and S_2 is the set of all elements that are contained in both S_1 and S_2 and is denoted by $S_1 \cap S_2$. This set is represented by the shaded area of Figure 4.3.*

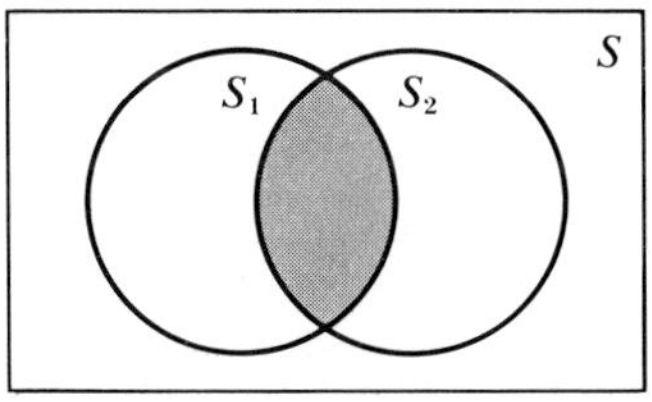

Figure 4.3
The intersection of sets

The intersection operation generates a set that contains the elements common to the sets on which we are operating. If the set generated by the intersection operation is the empty set (that is, the sets have no common elements), then the sets are said to be *mutually exclusive.*

Example 3

Referring to Example 1 concerning the department employees, the intersection set $(S_1 \cap S_2)$ is the set of male employees that are salaried. An element belongs to $S_1 \cap S_2$ if and only if the element belongs to S_1 *and* to S_2.

Example 4

Let $A = \{2, 3, 9\}$ and $B = \{3, 9, 7, 5\}$.
Then, $A \cap B = \{3, 9\}$ because these two elements are members of both sets.

Definition 4.6

*The **complement** of a set S_1 with respect to the universal set S is the set of all elements of S that are not elements of S_1. The complement of S_1 is written $\bar{S}_1$ and is represented in our Venn diagram as the shaded area in Figure 4.4.*

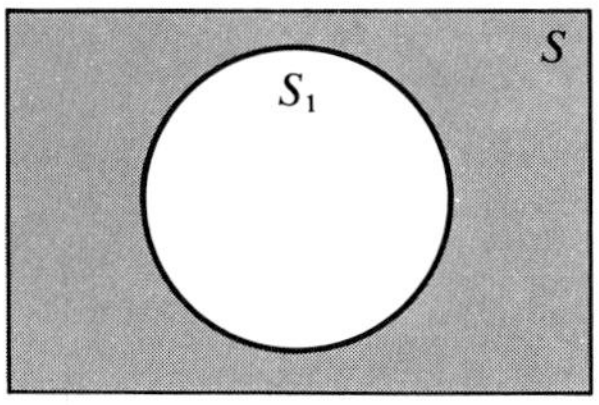

Figure 4.4
The complement of a set

The complement operation on a set, denoted by an overbar, generates the set containing all the elements in the universal set except those elements in the set itself.

Example 5

In order to contrast the operations, once again consider the department employees. Suppose the set of interest is the set consisting of employees who are not men. The set of men is represented by S_1; therefore, the set of women can be represented by $\bar{S}_1$. The eight women are elements of $\bar{S}_1$.

Definition 4.7

*The **difference** of two sets S_1 and S_2, denoted by $S_1 - S_2$, is the set of all elements in S_1 that are not elements of S_2.*

This operation generates a set containing all the elements unique to set S_1; thus, any element in $S_1 \cap S_2$ is excluded from the set $S_1 - S_2$. The difference $S_1 - S_2$ can be represented as the shaded area in Figure 4.5.

If sets S_1 and S_2 are mutually exclusive, then $S_1 - S_2$ is the set of all elements in S_1, or simply S_1. Similarly, $S_2 - S_1$ is the set S_2. Conversely, if $S_1 - S_2 = S_1$ or if $S_2 - S_1 = S_2$, then S_1 and S_2 are mutually exclusive.

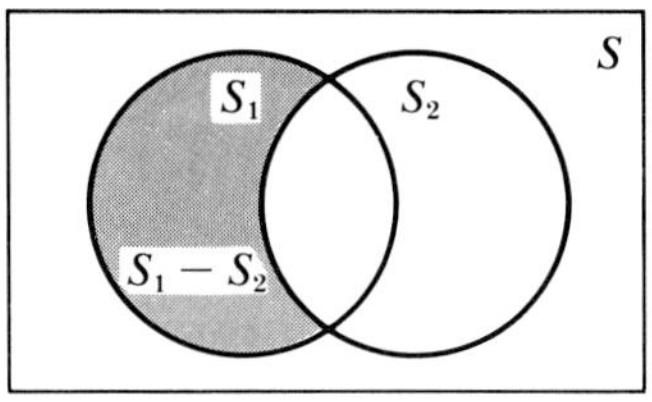

Figure 4.5
The difference of set S_1 and S_2:
$S_1 - S_2$

Example 6

Consider the two sets S_1 and S_2, where

$$S_1 = \{12, 13, 9\} \quad \text{and} \quad S_2 = \{13, 9, 17, 15\}.$$

The difference operation performed on S_1 and S_2 generates the following two sets:

$$S_1 - S_2 = \{12\},$$

because 12 is the only element of S_1 that is not an element of S_2, and

$$S_2 - S_1 = \{17, 15\}.$$

Note again that the difference set contains the elements of S_2 that are not in S_1.

Definition 4.8

*The **exclusive union** (Boolean sum) of two sets S_1 and S_2 is the set of all elements of S_1 and S_2 that are not common to S_1 and S_2. This set is designated as $S_1 \,\Delta\, S_2$.*

This operation is very similar to the union operation with the exception that $S_1 \,\Delta\, S_2$ does not include the elements of $S_1 \cap S_2$. The exclusive union is illustrated in Figure 4.6 by the shaded region.

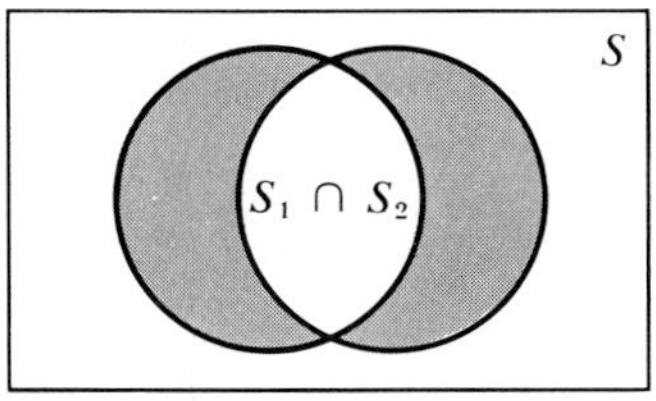

Figure 4.6
The exclusive union (Boolean sum) of sets

Example 7

The exclusive union of sets S_1 and S_2 in Example 6 consists of the elements that are members of S_1 or S_2 but not members of $S_1 \cap S_2$. Recalling that

$$S_1 = \{12, 13, 9\} \quad \text{and} \quad S_2 = \{13, 9, 17, 15\},$$

we have

$$S_1 \,\Delta\, S_2 = \{12, 17, 15\}.$$

Problems

1. Five midwestern brokerage firms are licensed to sell securities within certain states. The following sets represent the states in which the firms are licensed.

$$\begin{aligned} A &= \{\text{Illinois, Indiana, Missouri}\}, \\ B &= \{\text{Illinois, Iowa, Kansas, Nebraska}\}, \\ C &= \{\text{Illinois, Indiana, Missouri, Ohio}\}, \\ D &= \{\text{Iowa, Kansas, Missouri, Nebraska}\}, \\ E &= \{\text{Iowa, Kansas, Nebraska, Ohio}\}. \end{aligned}$$

Determine whether the following statements are true or false.

a. $A \subset C$, b. $E \subset D$,

c. $\bar{A} = E$, d. $\bar{C} = A$,

e. $A \cup C = C$, f. $A \cup E = S$,

g. $A \cup D = S$, h. $B \cup D = \{\text{Iowa, Kansas, Nebraska}\}$,

i. $A \cap B = \{\text{Illinois}\}$, j. $A \cap C = A$,

k. $A \cap S = \bar{E}$, l. $A \cap \bar{E} = \varnothing$,

m. $A \cap E = \varnothing$, n. $B \cap D = B \cap E$,

o. $A - C = \varnothing$, p. $B - D = \{\text{Illinois}\}$,

q. $A - E = A$, r. $A \,\Delta\, E = A \cup E$,

s. $A \,\Delta\, C = \varnothing$, t. $C \,\Delta\, E = A \,\Delta\, D$.

2. The following sets represent the products that four companies manufacture.

$$\begin{aligned} A &= \{\text{automobiles, radios, medicine, movies}\}, \\ B &= \{\text{copiers, radios, movies}\}, \\ C &= \{\text{detergents, medicine, movies}\}, \\ D &= \{\text{automobiles, copiers, movies}\}. \end{aligned}$$

List the products contained in each of the following sets.

a. $\bar{A}$, b. $\bar{B}$, c. $\bar{C}$, d. $A \cup D$,

e. $A \cup C$, f. $B \cap C$, g. $A \cap D$, h. $C \cap D$,

i. $A \cap B$, j. $(A \cap B) \cap C$, k. $\bar{A} \cap B$, l. $A \cap S$,

m. $A - D$, n. $B - C$, o. $C - D$, p. $A \,\Delta\, B$,

q. $A \,\Delta\, C$, r. $C \,\Delta\, D$.

3. For two sets A and B, draw Venn diagrams for each of the following sets.

a. $A \cup B$, b. $A \cap B$, c. $\bar{A}$, d. $\bar{B}$,

e. $A \cup \bar{B}$, f. $\bar{A} \cup B$, g. $\bar{A} \cup \bar{B}$, h. $\bar{A} \cap B$,

i. $A \cap \bar{B}$, j. $\bar{A} \cap \bar{B}$, k. $\overline{(A \cap B)}$, l. $\overline{(A \cup B)}$,

m. $A - B$, n. $B - A$, o. $\bar{A} - B$, p. $A \Delta B$,

q. $A \Delta \bar{B}$, r. $\bar{A} \Delta B$.

4. For the three sets A, B, and C, draw Venn diagrams for each of the following sets.

a. $(A \cap B) \cup C$, b. $(A \cap B) \cup \bar{C}$, c. $(A \cap \bar{B}) \cap C$,

d. $(A \cap \bar{B}) \cap \bar{C}$, e. $\overline{(A \cap B)} \cap C$, f. $\overline{(A \cap B)} \cap \bar{C}$,

g. $(A \cup B) \cap C$, h. $(A \cup B) \cap \bar{C}$, i. $(A \cup \bar{B}) \cap C$,

j. $(\bar{A} \cup \bar{B}) \cap C$, k. $(A - B) - C$, l. $(A - C) \cup B$,

m. $A \cup (B \Delta C)$, n. $(A \Delta B) - C$, o. $A - (B \Delta C)$.

5. Use Venn diagrams to verify that

a. $A \cap \varnothing = \varnothing$, b. $A \cup S = S$, c. $(A \cap B) \cup (A \cap \bar{B}) = A$,

d. $A \cup (A \cap B) = A$, e. $B \cup (A \cap B) = B$.

6. In a recent survey of 50 families, the numbers of current subscribers of various sports magazines were found to be: *Sports Illustrated* (SI), 25; *Golf Digest* (GD), 18; and *Field and Stream* (FS), 5. Of those who subscribed to SI, nine also subscribed to GD but not to FS; two subscribed to FS but not to GD; and one subscribed to FS and GD. Of the remaining families who subscribed to GD, only one subscribed to FS but not to SI.

How many families subscribed to:

a. *Sports Illustrated* only?

b. *Field and Stream* only?

c. *Golf Digest* only?

d. None of the magazines?

e. *Golf Digest* or *Sports Illustrated*?

f. *Golf Digest* or *Field and Stream*?

g. *Sports Illustrated* or *Field and Stream*?

h. *Sports Illustrated* or *Golf Digest* or *Field and Stream*?

7. An automobile manufacturer makes 30 different types of major parts for the bodies of three automobiles: Chaffy, Gord, and Rumbler. Of these 30 types of parts, four will fit only on a Chaffy, five will fit only on a Gord, two will fit only on a Rumbler, six will fit on a Chaffy and on a Gord but not on a Rumbler, one will fit on a Chaffy and on a Rumbler but not on a Gord, three will fit on a Gord and on a Rumbler but not on a Chaffy, and nine will fit on all three automobiles.

Draw a Venn diagram for the sets of parts for the respective automobiles and determine how many different types of parts are needed for each of the following combinations.

a. A Gord or a Chaffy, b. A Gord or a Rumbler, c. A Chaffy or a Rumbler,

d. A Gord and a Chaffy, e. A Gord and a Rumbler, f. A Chaffy and a Rumbler.

8. Of the 500 managers of a large corporation, 353 received a raise in salary, 224 received a bonus, and 100 received both. How many received neither a raise nor a bonus?

9. In a group of 175 employees, 130 signed up for hospitalization insurance, 80 signed up for disability insurance, and 25 did not sign up for either insurance.
   a. How many employees signed up for both insurances?
   b. How many signed up for hospitalization insurance only?
   c. How many signed up for disability insurance only?

4.3 Counting Techniques—Permutations and Combinations

In the axiomatic development of probability theory, it is impossible to divorce the theory completely from the classical and frequency definitions of probability. In particular, we shall use these earlier approaches to determine values for elementary probabilities. However, in both the classical and frequency approaches, we must count the possible outcomes—sometimes a very difficult task. Here, we shall consider certain useful counting techniques, including permutations and combinations.

Some counting techniques enable us to enumerate the elements of given sets. To obtain the probability of an event A as a subset of our universal sample space S, we must often count the elements of S and the elements of A. The ratio of the number of elements of A to the number of elements of S represents the probability of A. It is thus useful to present two rules for counting elements.

Rule 4.1

If sets S_1 and S_2 contain n_1 and n_2 elements, respectively, then the number of ways in which we can first select an element of S_1 and then an element of S_2 is $n_1 \cdot n_2$.

This rule can be generalized to apply to elements for more than two sets.

Rule 4.2

If sets $S_1, S_2, \ldots, S_k$ contain respectively $n_1, n_2, \ldots, n_k$ elements, then the number of ways in which we can first select an element of S_1, then an element of $S_2, \ldots,$ and finally an element of S_k is the product of their number of elements, namely, $n_1 \cdot n_2 \cdot n_3 \cdots n_k$.

To help visualize the counting situation, we construct a tree diagram with each branch representing a particular outcome. As an example, consider a young man contemplating summer employment. He may work for one of three companies A, B, or

C. The choice of being an hourly or a salaried worker and the choice of three work shifts—day, evening, or night—are available to him. How many different ways may he select a company, a pay basis, and a shift? Three sets are involved, specifically, the set of companies, the set of pay bases, and the set of work times, with three, two, and three elements, respectively. By Rule 4.2, the number of ways he may select a job is $3 \cdot 2 \cdot 3 = 18$. The tree diagram of Figure 4.7 illustrates all possible selections. The end point of the tree represents a particular job specifying the company, pay status, and work shift. The counting rule is seen to apply, as there are 18 distinct branches to the tree. Although we have a very simple situation here, we shall see in Section 4.9 that the tree-diagram approach can be used to simplify highly complicated situations.

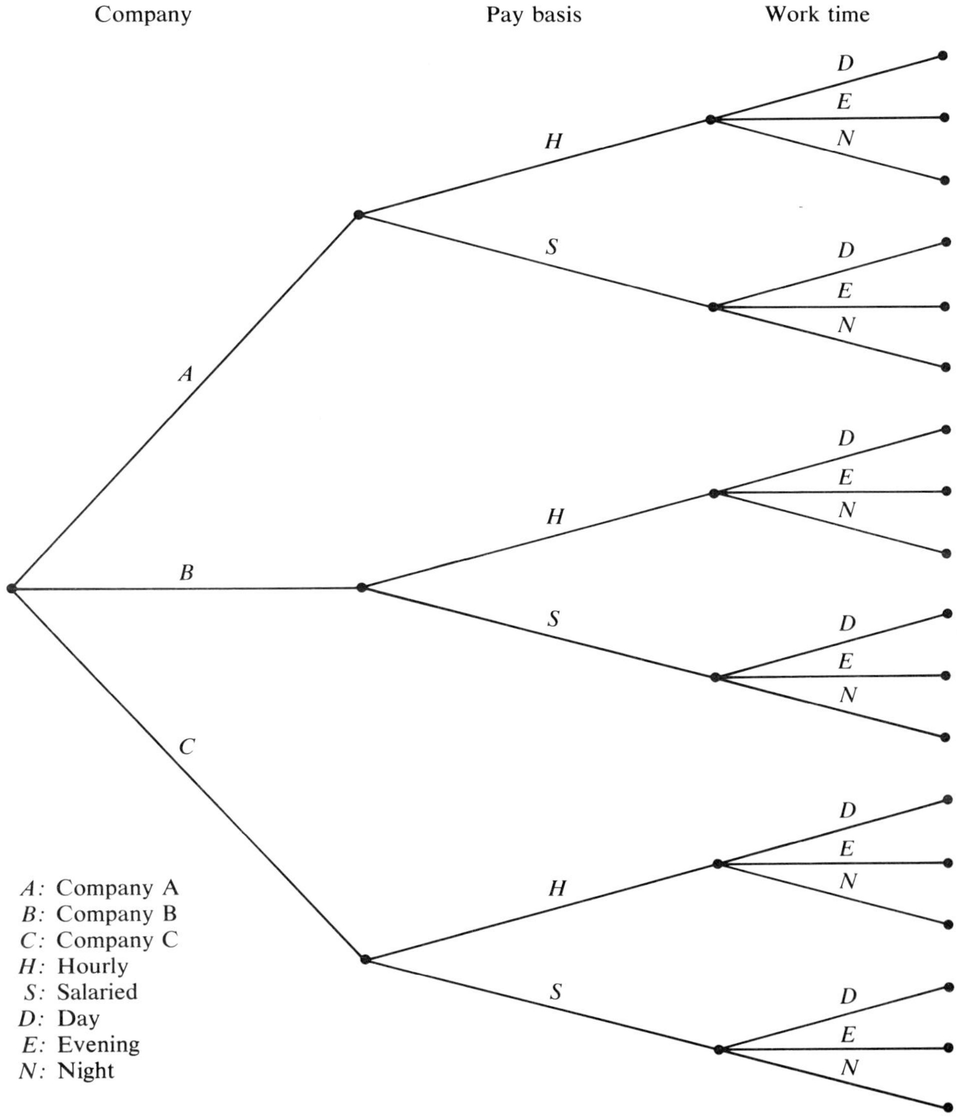

Figure 4.7

Tree diagram depicting choices of summer employment

Permutations and Combinations

Two other very important counting concepts involve *permutations* and *combinations*. Generally, permutations and combinations are expedient when we want to determine the number of ways in which a set of n items can be collected into groupings of size r. Groupings that are differentiated by both the items included in them and the order in which the items appear are called *permutations*. On the other hand, groupings that are differentiated only by the items included in them (and not by the order in which the items appear) are known as *combinations*. It follows that many permutations can be obtained from one combination. The general problem can be easily solved by using the following relationships.

Rule 4.3

For the general case where we have n distinguishable items and wish to determine the number of permutations which can be formed by grouping r of the n items at a time, we have

$$_nP_r = \frac{n!}{(n-r)!} = n \cdot (n-1) \cdot (n-2) \cdots (n-r+1),$$

where $_nP_r$ is the number of permutations of n things taken r at a time and $n!$, read "n factorial" is the product of all integers less than or equal to n.

An example using factorial notation is

$$6! = 6 \cdot 5 \cdot 4 \cdot 3 \cdot 2 \cdot 1 = 720.$$

Also, by definition, $0! = 1$.

As a special case of Rule 4.3, the number of permutations of n things taken n at a time is

$$_nP_n = \frac{n!}{(n-n)!} = \frac{n!}{0!} = \frac{n!}{1} = n!. \tag{1}$$

Rule 4.4

For the general case where we have n distinguishable items and wish to determine the number of combinations that can be formed by grouping r of the n items at a time, we have

$$_nC_r = \binom{n}{r} = \frac{n!}{r!(n-r)!},$$

where both $_nC_r$ and $\binom{n}{r}$ denote the number of combinations of n things taken r at a time.

Using the equations of Rules 4.3 and 4.4 and Equation (1), we note that

$$_nC_r = \frac{_nP_r}{r!}. \tag{2}$$

This relationship results from the fact that $r!$ permutations (permutations of r things taken r at a time) can be formed from each distinct combination.

Consider the problem of determining the number of combinations and permutations possible when the four letters A, B, C, and D are grouped three at a time. Without the foregoing formulas, we might simply list and enumerate all possibilities. This is done in the following table for both combinations and permutations.

Combinations:	*ABC*		*ABD*		*ACD*		*BCD*	
Corresponding permutations:	*ABC*	*ACB*	*ABD*	*ADB*	*ACD*	*ADC*	*BCD*	*BDC*
	BAC	*BCA*	*BAD*	*BDA*	*CAD*	*CDA*	*CBD*	*CDB*
	CAB	*CBA*	*DAB*	*DBA*	*DAC*	*DCA*	*DBC*	*DCB*

We count 4 combinations and 24 permutations. However, we also can solve the problem by using Rule 4.4 for combinations,

$$_nC_r = {_4C_3} = \frac{4!}{3!1!} = 4$$

and Rule 4.3 for permutations,

$$_nP_r = {_4P_3} = \frac{4!}{(4-3)!} = \frac{4!}{1!} = 4! = 24.$$

The table also shows that each combination can be permuted to obtain $r! = 3! = 6$ different permutations. For example, *ABC*, *ACB*, *BAC*, *BCA*, *CAB*, and *CBA* are all different permutations that represent the same combination.

Example 1

How many different sets of four officers can be obtained from a set of ten candidates for any of the offices of president, vice-president, secretary, and treasurer?

Solution

Obviously, order is important in this problem. Thus, since the four positions to be filled are different (distinguishable), we can apply Equation (1) of Rule 4.3 to determine how many different ways the four positions can be filled from the group of ten people. We have

$$_{10}P_4 = \frac{10!}{(10-4)!} = \frac{10!}{6!} = 10 \cdot 9 \cdot 8 \cdot 7 = 5{,}040. \qquad \square$$

Example 2

How many different grievance committees of five can be selected from a group of ten employees?

Solution

This is a combination problem, since any group of five employees will give only one combination regardless of how they are arranged. Thus, the number of different committees is

$$ {}_{10}C_5 = \frac{10!}{5!5!} = \frac{10 \cdot 9 \cdot 8 \cdot 7 \cdot 6}{5 \cdot 4 \cdot 3 \cdot 2 \cdot 1} = 252. \qquad \square $$

Example 3

Suppose six women are eligible for promotion to supervisor and four supervisor positions are open. How many different sets of promotions are possible if all available positions are considered equivalent?

Solution

Since positions are not distinguishable, we are concerned with combinations. Thus,

$$ {}_6C_4 = \frac{6!}{4!2!} = \frac{6 \cdot 5}{2 \cdot 1} = 15. \qquad \square $$

Example 4

How many different sets of promotions are possible in Example 3 if each of the supervisor positions is of a different grade?

Solution

This is a permutation problem. That is, if two employees exchange possible promotion positions, we have a different promotion set, even though the combination of employees is identical. Hence, the number of different sets is

$$ {}_6P_4 = \frac{6!}{(6-4)!} = 6 \cdot 5 \cdot 4 \cdot 3 = 360. $$

The items in the foregoing cases are all individually distinguishable. However, suppose we are interested in the number of different permutations obtainable from the group of letters A, A, B, and B. The four letters of this group are not all distinguishable from each other, since we have two groups of two letters each. The permutations become

$$\begin{array}{ccc} AABB, & ABAB, & ABBA, \\ BAAB, & BABA, & BBAA. \end{array}$$

These are all the visibly different permutations that can be obtained. Why? Because there is no distinguishable difference between either the two A's or the two B's. If there were such a difference, we would have ${}_4P_4 = 24$ permutations. For example, if instead of A, A, B, and B we had A_1, A_2, B_1, and B_2, then we would have four distinguishable items for which we could use Equation (1) of Rule 4.3. □

Fortunately, there is a well-known relationship for determining the number of different permutations in the foregoing problem. It gives us a means of determining the number of different permutations obtainable from a group of n items consisting of k subgroups of items, where the items within each subgroup are not distinguishable, but the items in one subgroup are distinguishable from those in another subgroup. The relationship is

$$ {}_nP_{r_1, r_2, \ldots, r_k} = \frac{n!}{r_1!\,r_2!\cdots r_k!}, \qquad (3)$$

where n is the total number of items under consideration, k is the number of distinguishable subgroups, $r_i (i = 1, 2, \ldots, k)$ is the number of nondistinguishable items in subgroup i, $r_1 + r_2 + \cdots + r_k = n$, and ${}_nP_{r_1, r_2, \ldots, r_k}$ is the number of permutations obtainable from n items consisting of the k distinguishable subgroups.

Example 5

How many different permutations can we obtain when arranging a group of nine numbers if the nine numbers are made up of three 4's, four 5's, and two 8's?

Solution

Using (3), we have

$$ {}_9P_{3,4,2} = \frac{9!}{3!4!2!} = 1{,}260. $$

We see that 1,260 different nine-digit numbers can be constructed from the nine numbers. □

Example 6

Five persons are being considered for promotion to five positions—two foremen, two supervisors, and one manager. How many sets of promotions are possible if we consider the sets to be different when persons receive different *types* of promotions.

Solution

We are merely concerned with the number of permutations obtainable from five persons when they are to be divided into three distinguishable groups of sizes 2, 2, and 1. The number is

$$_5P_{2,2,1} = \frac{5!}{2!2!1!} = 30. \qquad \square$$

Example 7

How many sets of promotions can we select from eight employees under the ground rules of the preceding example?

Solution

This problem is different from the earlier one, in that we have more potential employees than we do positions. However, if we view the state of "not being selected for a promotion" as a fourth type of position, we will simply be dividing the eight players into four distinguishable groups of sizes 2 (foremen), 2 (supervisors), 1 (managers), and 3 (no promotion). Thus, we have

$$_8P_{2,2,1,3} = \frac{8!}{2!2!1!3!} = \frac{8 \cdot 7 \cdot 6 \cdot 5 \cdot 4 \cdot 3 \cdot 2 \cdot 1}{2 \cdot 1 \cdot 2 \cdot 1 \cdot 1 \cdot 3 \cdot 2 \cdot 1} = 1{,}680.$$

Note that Rule 4.4 is a special case of (3). That is, since $r + (n - r) = n$, when we find the number of combinations of n things taken r at a time, we are essentially finding the number of permutations obtainable from n items when they are to be divided into two distinguishable groups of size r and $(n - r)$. Thus,

$$_nP_{r,n-r} = \frac{n!}{r!(n-r)!} = {_nC_r}. \tag{4}$$

Using this same reasoning, we note that the number of combinations of n things taken r at a time is equal to the number of combinations of n things taken $(n - r)$ at a time. That is,

$$_nC_r = \frac{n!}{r!(n-r)!} = {_nC_{n-r}}. \tag{5}$$

For example, the number of combinations of five items taken two at a time is 10:

$$_5C_2 = \frac{5!}{2!(5-2)!} = \frac{5!}{2!3!} = 10;$$

and the number of combinations of five items taken three at a time is also 10:

$$_5C_3 = \frac{5!}{3!(5-3)!} = \frac{5!}{3!2!} = 10. \qquad \square$$

Practical Distinction between Permutation and Combination Problems

It is usually easy to recognize that a given counting problem is *either* a permutation *or* a combination problem. Deciding which of the two counting techniques is appropriate is often a little harder. Recall that the key in this latter decision is whether *order* is to be considered in differentiating one combination from another. There is a general rule of thumb to help you decide (once you know that the problem involves either permutations or combinations) when to apply the counting techniques of permutations and combinations, respectively. The steps to be followed are:

1. Choose any single appropriate combination.
2. Choose any two permutations of the combination chosen.
3. If the two sets chosen in step 2 represent different groupings under the ground rules as established in the statement of the problem, then the problem should be solved by permutations. If not, then the problem should be solved by combinations.

To illustrate the rule of thumb, let us consider again the problem of Example 1. If we denote the positions of president, vice-president, secretary, and treasurer as P, V, S, and T, respectively, and the ten individuals as $A, B, C, D, \ldots, J$, we see that one possible permutation satisfying the type groupings we seek is

Position	P	V	S	T
Individual	A	B	C	D.

However, permuting this combination, we can obtain the arrangement

Position	P	V	S	T
Individual	D	B	C	A.

The question now merely becomes that of deciding whether, under the conditions of the problem, the two arrangements represent clearly distinguishable groupings. In this case, it is obvious that giving the presidency to D is a different assignment than giving it to A. Thus, *order* is important, and we should use the counting technique of permutations.

Conversely, suppose we simply want to form a four-member committee from the ten individuals. Any positioning in this situation would be purely arbitrary, and, by our rule of thumb, the two permutations $ABCD$ and $DBCA$ do not constitute clearly distinguishable groupings under the conditions of the problem. Thus, order is *not* important and we would use combinations as the applicable counting technique.

Problems

1. If there are 5 ways to get from Chicago to Pittsburgh and 4 ways to get from Pittsburgh to New York, how many different ways are there to get from Chicago to New York via Pittsburgh? Construct a tree diagram to illustrate the routes.

2. If an automobile manufacturer has available 4 different colors of paint, 3 different body styles, and 5 different interior colors, how many different types of automobiles can be manufactured? Construct a tree diagram to illustrate the different types of automobiles.

3. Explain the difference between *permutations* and *combinations*, and give an example involving the use of each.

4. Calculate:

   a. ${}_5P_3$, b. ${}_5C_3$, c. ${}_{10}P_6$,

   d. ${}_9C_5$, e. ${}_{11}P_{1,8,2}$, f. ${}_{12}P_{3,5,4}$.

5. How many distinct groups of 8 letters can be selected from the group *A, B, C, D, E, F, G, H, I, J, K, L*

   a. when order is *not* considered?

   b. when order *is* considered?

6. How many different 3-digit numbers can be constructed when zero is always considered and

   a. any digit may appear only once?

   b. any digit may not appear more than twice?

   c. any digit may appear up to three times?

7. How many different management teams (1 president, 2 vice-presidents, 2 plant managers, 2 production managers, 1 purchasing manager, 1 sales manager, 2 planning managers) can be selected from 11 employees when the team is considered different if the employees exchange *types* of positions?

8. How many different management teams (2 plant managers, 2 production managers, 2 quality control managers, 3 sales managers, 2 planning managers) can be selected from 15 employees when the team is considered different if the employees exchange *types* of positions?

9. How many different ways can a group of 20 employees be divided into

   a. 4 groups of 5 employees?

   b. 5 groups of 4 employees?

   c. 2 groups of 10 employees?

4.4 Elementary Probability Concepts

In order to understand the operating rules of probability, we must first consider certain fundamental concepts and definitions.

Definition 4.9

*A **random experiment** is a process or operation which results in one of a number of possible outcomes such that the outcome is determined only by chance and is, therefore, unpredictable.*

Example 1

Consider the process of tossing a fair die. This process is a random experiment, since the possible outcomes of 1, 2, 3, 4, 5, and 6 are determined strictly by chance and it is impossible to predict exactly which of the six values will occur on any given toss.

Example 2

Consider the process of selecting a card from a well-shuffled deck of ordinary playing cards. Since the cards are well shuffled, the selection of any particular card is strictly by chance. It is impossible to predict the exact card to be obtained on any given selection. Hence, this process is a random experiment.

Definition 4.10

*A **random event** is an event or occurrence generated by a random experiment.*

In Example 1 the occurrence of a 4 on the face of the die represents a random event if the die has been tossed without any restrictions. Likewise, in Example 2 the occurrence of the ace of spades represents a random event if both the selection is made from an ordinary (unmarked) deck of playing cards without the selector observing the faces of the cards and the cards have been thoroughly shuffled before selection.

Definition 4.11

*A **random variable** is a variable the values of which (1) are determined by the outcome of a random event and (2) can be ordered according to numerical value.*

Note that the second part of Definition 4.11 is rather restrictive. It represents the essential difference between the possible outcomes of a random experiment and the possible values of a random variable. For instance, in Example 1 the number appearing on the face of the die is the random variable of interest. The random variable receives its value as the result of a random event (our tossing the die and observing the number face up). Further, the possible values (1, 2, ... , 6) can be ordered according to numerical value: $6 > 2$, $3 > 1$, and so forth.

On the other hand, although the suit and rank of the card selected in Example 2 is the result of a random event, it cannot be construed strictly as a random variable in the sense of Definition 4.11. It makes no sense, for example, to speak of the four of clubs being larger or smaller than the four of hearts. However, we can create a random variable by arbitrarily assigning numerical values to the various possible combinations of suit and rank.

Definition 4.12

*A **simple event** is one that cannot be represented as a collection or set of two or more events.*

In Example 1 each of the six events that may occur—that is, a 1, 2, 3, 4, 5, or 6 on the face of the die—is a simple event. A simple event in Example 2 is, for instance, the event of drawing the ace of spades.

Definition 4.13

*A **compound event** is one that can be represented as a collection or set of two or more simple events.*

An exemplary compound event for Example 1 is the event of getting an even number on the face of the die as the result of only one toss. This particular event can be represented as the collection of the three simple events of getting a 2, 4, or 6 on the face of the die. That is, if we obtain one of the simple events 2, 4, or 6, the compound event "even number" will have occurred. For Example 2, an exemplary compound event is the event of drawing a card from the heart suit. This event can be represented by the collection of the thirteen simple events of getting the A, 2, 3, 4, 5, 6, 7, 8, 9, 10, J, Q, or K from the heart suit; for, if one of these simple events occurs, the compound event "heart" will necessarily have occurred.

Definition 4.14

*A **sample space** (or **population of events**) is the collection of all possible simple events that can occur as the result of a given random experiment.*

Definition 4.15

*A **sample** is a set of simple events (or observations) taken from a population of events.*

Definition 4.16

*A **simple random sample** is a sample selected by a random process such that any simple event remaining in the population at any given instant has the same chance of appearing in the sample as any other remaining simple event.*

This definition implies that there are two conditions required for simple random sampling. That is, (1) each simple event in the population has an equal chance of being selected, and (2) each possible sample of a specific size has the same chance of being selected as any other possible sample of the same size.

The population for Example 1 is the collection of the six simple events 1, 2, 3, 4, 5, and 6. A sample of size 2 from this population could be 1 and 3, while a sample of size 3 could be 1, 4, and 6. The population for Example 2 is the entire deck of 52 cards (i.e., the 52 simple events). A sample of size 3 from this population could be the ace of clubs, the king of hearts, and the seven of diamonds.

In Example 2, the sample obtained by randomly selecting a card from the well-shuffled deck of 52 cards is a simple random sample, since any given card will have the same opportunity of being selected as any other card. If the experiment of this example is such that more than one draw is required, we have the option of (1) replacing the card in the deck and reshuffling before the next draw or (2) keeping the selected card out of the deck. The first procedure is called *sampling with replacement*, and the second is called *sampling without replacement*. In either case, each remaining card has the same chance of being selected on the next draw. Hence, in this example, we obtain a simple random sample whether we sample with replacement or without replacement.

In Example 1 the sample obtained by tossing the die (if it is a fair die) is a simple random sample since each number has the same chance of appearing in the sample. Furthermore, any sample of size greater than 1 will be obtained by *essentially* replacing the event in the original population. This is due to the nature of the experiment of tossing a fair die (i.e., each simple event has the same chance of occurring on any repeated toss). Hence, the sampling procedure of this example is necessarily sampling with replacement.

Definition 4.17

*A **probability sample** is a sample obtained by a random process wherein each event has a known (not necessarily equal) probability of occurring.*

If in Example 1 the die were not fair (that is, the probabilities are known, but different for each possible value), the resulting sample would be a probability sample rather than a simple random sample. It should be noted that any simple random sample is also a probability sample, since the probabilities are equal and thus known. However, by the definition, a probability sample is not necessarily a simple random sample. In other words, a simple random sample is a more restrictive classification than a probability sample.

Definition 4.18

*Two events are **mutually exclusive** if they cannot occur simultaneously in one trial of the experiment of interest.*

The definition implies that the occurrence of one event excludes the occurrence of the other. It should also be noted that two mutually exclusive events have no simple events in common. Thus, any two simple events are mutually exclusive by definition.

Definition 4.19

A set of events is ***exhaustive*** *for a population if, when the events are viewed collectively, no simple event of the population is omitted from the set.*

Consider one toss of the die in Example 1. Let A be the event of getting a 1 or 2, B the event of getting a 3 or 4, C the event of getting a 2 or 3, and D the event of getting a 4, 5, or 6. That is, for example, event A will have occurred if either a 1 or 2 is obtained on a given toss. Since it is impossible for both to occur on one toss of the die, the events A and B are mutually exclusive events. On the other hand, events A and C are not mutually exclusive, since the occurrence of one does not exclude the occurrence of the other. That is, the occurrence of a 2 will mean that both A and C have occurred at the same time (on the same toss).

The events A, B, and D are exhaustive events since, when viewed collectively, they include (exhaust) all of the six possible simple events 1, 2, 3, 4, 5, 6. That is, since these six numbers are the only possible values for the variable, there is no simple event which is omitted from the collection of the three events.

Problems

1. Explain the relationships between a *random experiment*, a *random event*, and a *random variable*, and give an example of each.
2. Define *simple event* and *compound event*, and give an example of each.
3. Define *population* and *sample*, and give an example of each.
4. Define *simple random sampling*, and give an example.
5. What two conditions are required by the definition of *simple random sampling*?
6. Define *probability sampling*, and give an example which is not a simple random sample.
7. What two conditions are required by the definition of *probability sampling*?
8. a. Is a *probability sample* always a *simple random sample*? Why?

   b. Is a *simple random sample* always a *probability sample*? Why?
9. Define *mutually exclusive events*, and give an example.
10. Define a set of *exhaustive events*, and give an example.

4.5 Probability as a Set Function—Postulates of Probability

We are now ready to begin the axiomatic development of probability theory. In probability, as in mathematics, any set of operating rules is generally based on an explicitly stated set of assumptions. The assumptions, known as the *postulates of probability*, are as follows.

Let S be a sample space (or population of events) and let A represent any event in S. Then, if we represent the probability of the event A by the function $P(A)$, P is a probability function on the sample space S if the following three postulates are satisfied:

Postulate 1

$P(A)$ is a real number such that

$$P(A) \geq 0$$

for any event in S.

Postulate 2

If A_i and A_j are any mutually exclusive events in S, then

$$P(A_i \text{ or } A_j) = P(A_i) + P(A_j),$$

where the event "A_i or A_j" is satisfied if either event A_i or event A_j occurs.

Postulate 3

If $A_1, A_2, \ldots, A_n$ are mutually exclusive and exhaustive events in S, then

$$P(S) = P(A_1 \text{ or } A_2 \text{ or} \cdots \text{or } A_n) = P(A_1) + P(A_2) + \cdots + P(A_n) = 1.$$

Although these postulates are motivated by the classical and frequency approaches to probability, they provide a basis for a much more powerful and generally applicable approach. That is, they can be used to develop an idealized model of a real-world situation. However, it is important to note that, since the axiomatic approach does not always provide an expedient means for determining specific values for probabilities of mutually exclusive events, we often use one of the earlier approaches for determining such values. Further, even when the situation is such that use of the postulates will lead directly to specific probabilities for mutually exclusive events, it is often easier to use the basic counting principles associated with the other approaches.

Postulates 1 and 3 can be used to show that, for any event A,

$$0 \leq P(A) \leq 1. \tag{1}$$

This being true, Postulate 1 is often presented in such form. Also, as is implied in Postulate 3, Postulate 2 can be generalized to apply to any set of mutually exclusive events (say $A_1, A_2, \ldots, A_m$); that is,

$$P(A_1 \text{ or } A_2 \text{ or} \cdots \text{or } A_m) = P(A_1) + P(A_2) + \cdots + P(A_m), \tag{2}$$

where the event $(A_1 \text{ or } A_2 \text{ or} \cdots \text{or } A_m)$ is satisfied if either of the events $A_1, A_2, \ldots, A_m$ occurs.

In general, the logical operator "or" should be interpreted as "and/or." For example, the event "A_1 or A_2" is satisfied by the occurrence of event A_1; the occurrence of event A_2; or the simultaneous occurrence of both events A_1 and A_2. Obviously, the third occurrence cannot happen if events A_1 and A_2 are mutually exclusive. Thus, with regard to the postulates, the operator "or" does not have to be interpreted strictly. However, we shall later encounter many situations in which events of interest are not mutually exclusive. Thus, you should develop now the habit of interpreting "or" correctly.

Probability as a Set Function

We may regard or define a function as a correspondence (or mapping) between the elements of two sets S_1 and S_2 which assigns, maps, or coordinates a unique element of S_2 to each element of S_1 and coordinates at least one element of S_1 to each element of S_2. We shall designate S_1 as the domain of the function and S_2 as the range of the function. We are particularly concerned with the situation in which the elements of the domain are sets and the corresponding elements of the range are real numbers. When this situation exists we call the given function a *set* function.

An example of a set function that is most germane to our study is the function that assigns the number of elements in a subset S_1, $n(S_1)$, to each subset of the population. A useful property of this set function is the additive property. If sets S_1 and S_2 are disjoint, then $n(S_1 \cup S_2)$ equals $n(S_1) + n(S_2)$.

To continue the development of the set-theoretic approach to probability, we shall restate the earlier postulates of probability in terms of sets and set functions. We again denote the sample space by S, and the probability of the event A by $P(A)$ if the postulates are satisfied. However we now refer to the event A as a subset of the sample space S.

Postulate 1a

If A is any subset of S,

$$0 \leq P(A) \leq 1.$$

Postulate 2a

If we let S represent the entire sample space in an experiment,

$$P(S) = 1.$$

Postulate 3a

If A_i and A_j are any two disjoint (mutually exclusive) subsets of S, then

$$P(A_i \cup A_j) = P(A_i) + P(A_j).$$

Postulate 3a can be generalized as follows. If $A_1, \ldots, A_n$ constitute a sequence of disjoint (mutually exclusive) subsets of S, then,

$$P(A_1 \cup A_2 \cdots \cup A_n) = P(A_1) + P(A_2) + \cdots + P(A_n). \tag{3}$$

Hereafter, we shall use set-theory notation in all probability-related discussions. Further, when we refer to the "postulates of probability," we shall always mean those given in set-theory notation.

Example 1

Suppose we have ten different companies of which two manufacture only automobiles, three manufacture only airplanes, and five manufacture only electronic components. We wish to randomly select one company from this group such that each company has an equal chance of being selected. Hence, by the classical definition of probability, the probabilities of drawing a specific type of manufacturer are $P(\text{auto}) = 0.2$, $P(\text{plane}) = 0.3$, and $P(\text{component}) = 0.5$ (for example, of the ten possible outcomes, two have the attribute auto, so that $P(\text{auto}) = 2/10 = 0.2$).

For a better understanding of how these probabilities can be obtained by using the postulates, consider Table 4.2. Since there are ten companies and we are sampling

Table 4.2

Probability of simple and compound events

Simple event	*Probability*	*Compound event*	*Probability*
Auto_1	0.1	Automobile manufacturer	0.2
Auto_2	0.1		
Plane_1	0.1	Airplane manufacturer	0.3
Plane_2	0.1		
Plane_3	0.1		
Component_1	0.1	Electronic component manufacturer	0.5
Component_2	0.1		
Component_3	0.1		
Component_4	0.1		
Component_5	0.1		
	1.0		1.0

in such a way that each company has an equal chance of being selected, we can use Postulates 2a and 3a to solve for the common probability (equal chance) of the occurrence of each company. Thus, each manufacturer has a probability of 0.1 of being selected. Furthermore, since the ten simple events of selecting specific manufacturers are necessarily mutually exclusive (because each manufactures only one type of product), we may add the probabilities of individual events to obtain the probability of a compound event, the event of selecting an automobile manufacturer, for example.

Similarly, we can apply the postulates to determine the probabilities of two or more properly defined (mutually exclusive) compound events. For example, the probability of selecting a company that manufactures a transportation vehicle is, by Postulate 3a,

$$\begin{aligned} P(\text{transportation vehicle}) &= P(\text{auto} \cup \text{plane}) = P(\text{auto}) + P(\text{plane}) \\ &= 0.2 + 0.3 = 0.5. \end{aligned}$$

By the same token, the probability of selecting a company which manufactures a flight-related product is

$$\begin{aligned} P(\text{flight-related product}) &= P(\text{plane} \cup \text{component}) \\ &= P(\text{plane}) + P(\text{component}) = 0.3 + 0.5 = 0.8. \end{aligned}$$

Note that in both cases, we obtained the probability of one compound event by adding the probabilities of two other compound events—because (and only because), in each case, the two compound events involved in the addition are mutually exclusive. This leads us to a very important point. Postulate 3a does not apply in the case of compound events that are not mutually exclusive. Suppose we wish to find P(transportation vehicle or flight-related product). Using Postulate 3a, we would obtain

$$\begin{aligned} &P(\text{transportation vehicle} \cup \text{flight-related product}) \\ &\qquad = P(\text{transportation vehicle}) + P(\text{flight-related product}) \\ &\qquad = 0.5 + 0.8 = 1.3. \end{aligned}$$

However, Relation (1) tells us that this result is impossible, since, under our basic set of postulates, the probability of an event can never be greater than 1.0. We mention it here only to emphasize the vital role of the term "mutually exclusive" in postulates.

Let us develop one other point in this example. The probability of selecting a company which does not make automobiles can be obtained through use of Postulate 3a:

$$P(\text{not auto}) = P(\text{plane} \cup \text{component}) = P(\text{plane}) + P(\text{component}) = 0.3 + 0.5 = 0.8. \tag{4}$$

Further, the probability of selecting a company which manufactures either automobiles, airplanes, or components is, by Postulates 2a and 3a:

$$\begin{aligned} P(S) &= P(\text{plane} \cup \text{auto} \cup \text{component}) = P(\text{plane}) + P(\text{auto}) + P(\text{component}) \\ &= 0.3 + 0.2 + 0.5 = 1. \end{aligned}$$

This is necessarily true, since the events are both exhaustive and mutually exclusive. However, knowing that the three compound events are mutually exclusive and exhaustive, we can also find the probability of selecting a company which does not make automobiles by combining Postulates 2a and 3a and solving for the unknown sum; since

$$P(\text{plane}) + P(\text{component}) + P(\text{auto}) = 1,$$

we have

$$P(\text{plane}) + P(\text{component}) = 1 - P(\text{auto}).$$

Substituting from (4),

$$P(\text{not auto}) = 1 - P(\text{auto}) = 1 - 0.2 = 0.8.$$

The last result of Example 1 leads us to a very useful rule.

Rule 4.5

Let event $\bar{A}$ represent the failure of event A to occur. Then,

$$P(A) = 1 - P(\bar{A}).$$

Rule 4.5 follows directly from our set-theory approach. The complement of a set A, denoted $\bar{A}$, was defined by Definition 4.6 as the set containing all elements of S except for those contained in A. Since the intersection of A and $\bar{A}$ must be the null set, then A and $\bar{A}$ are disjoint, and

$$P(A) + P(\bar{A}) = P(S).$$

Thus, because all elements in S are either elements of A or elements of $\bar{A}$;

$$P(A) + P(\bar{A}) = 1 \quad \text{and} \quad P(A) = 1 - P(\bar{A}).$$

For Example 1, we can use this rule to calculate other probabilities. For example,

$$P(\text{auto} \cup \text{plane}) = 1 - P(\text{component}) = 1 - 0.5 = 0.5$$

and

$$P(\text{plane} \cup \text{component}) = 1 - P(\text{auto}) = 1 - 0.2 = 0.8.$$

Example 2

A large producer of cosmetics is interested in conducting a market research study concerning the effects of advertising in drug stores. For purposes of this study, a sample will be selected from the drugstores in a given geographical area. The distribution of drugstores in the area is given in Table 4.3, where X represents number of persons in a city.

The probabilities given in the table are those of choosing a drugstore in a city of the given size when randomly selecting a drugstore from the population of 10,000 drugstores in the geographical area of interest. Let X be the random variable which has

Table 4.3
Distribution of drugstores in cities of various size

Size of city	*Number of drugstores*	*Probability*
$X < 5{,}000$	424	0.0424
$5{,}000 \leq X < 15{,}000$	843	0.0843
$15{,}000 \leq X < 30{,}000$	1,427	0.1427
$30{,}000 \leq X < 50{,}000$	2,012	0.2012
$50{,}000 \leq X < 100{,}000$	2,384	0.2384
$X \geq 100{,}000$	2,910	0.2910
Total	10,000	1.0000

for its value the size of the city from which the selected drugstore comes. Then, by the classical approach, we obtain each of the following probabilities

$$\begin{aligned} P(X < 5{,}000) &= 0.0424, \\ P(5{,}000 \leq X < 15{,}000) &= 0.0843, \\ P(15{,}000 \leq X < 30{,}000) &= 0.1427, \\ P(30{,}000 \leq X < 50{,}000) &= 0.2012, \\ P(50{,}000 \leq X < 100{,}000) &= 0.2384, \\ P(X \geq 100{,}000) &= 0.2910. \end{aligned} \tag{5}$$

The events of (5) represent an arbitrarily defined set of mutually exclusive and exhaustive events. Thus, we can use the postulates to determine the probabilities of certain other compound events as follows. The probability of randomly selecting a drugstore from a city of 30,000 or larger is

$$\begin{aligned} P(X \geq 30{,}000) &= P(30{,}000 \leq X < 50{,}000 \cup 50{,}000 \leq X < 100{,}000 \cup X \geq 100{,}000) \\ &= P(30{,}000 \leq X < 50{,}000) + P(50{,}000 \leq X < 100{,}000) \\ &\quad + P(X \geq 100{,}000) \\ &= 0.2012 + 0.2384 + 0.2910 = 0.7306. \end{aligned}$$

This means that if we repeatedly sample with replacement from the population of drugstores, we expect an average of 7,306 drugstores per 10,000 to come from a city of at least 30,000 population.

The probability of randomly selecting a drugstore from a city having population between 15,000 and 100,000 is

$$\begin{aligned} P(15{,}000 \leq X < 100{,}000) &= P(15{,}000 \leq X < 30{,}000 \cup 30{,}000 \leq X < 50{,}000 \\ &\quad \cup 50{,}000 \leq X < 100{,}000) \\ &= P(15{,}000 \leq X < 30{,}000) + P(30{,}000 \leq X < 50{,}000) \\ &\quad + P(50{,}000 \leq X < 100{,}000) \\ &= 0.1427 + 0.2012 + 0.2384 = 0.5823. \end{aligned}$$

You should satisfy yourself, however, that we cannot obtain

$$P(X \leq 30{,}000 \cup 15{,}000 \leq X < 100{,}000)$$

simply by adding $P(X \leq 30{,}000)$ and $P(15{,}000 \leq X < 100{,}000)$. The two events of interest are not mutually exclusive.

Problems

1. State the three postulates of probability and the conditions necessary for them to be valid.
2. The results of a survey of 50 randomly selected retail outlets for a certain product classified 30 as small, 15 as medium, and 5 as large.
   a. Calculate the probabilities that a randomly selected retail outlet for this product will be small, medium, or large by applying the frequency-theory (*a posteriori*) approach based on experience.
   b. Use the results of Part a to illustrate each of the three postulates of probability.
3. A survey among 100 used car dealers was conducted to determine the number of cars they sold on a given day. The results were:

Number of cars	f
1	20
2	16
3	17
4	14
5	18
6	15

   a. Use the frequency-theory (*a posteriori*) approach based on experience to calculate the probabilities associated with each event.
   b. Use the results of Part a to illustrate each of the three postulates of probability.

4.6 *Joint, Marginal, and Conditional Probabilities*

The theory and related applications of joint, marginal, and conditional probabilities play an important role in statistics. We shall begin our discussion with an example designed specifically to illustrate the meaning of these terms.

Suppose we are concerned with a population of 1,000 individuals, each of whom is classified according to both educational level achieved and annual salary. Suppose further that the results of such classification are as given in Table 4.4. We are to choose one individual from the population by simple random sampling. As indicated in the table, the event A_1, for example, will occur if we select any one of the individuals who has at most eight years of education, regardless of what salary he makes. Similarly, B_3 will occur if we select an individual who earns more than \$15,000 per year, regardless of his education.

We may consider the events $A_1, A_2, A_3, B_1, B_2, B_3$ to be six subsets of the sample space S. In this context we may view joint, marginal, and conditional probabilities in terms of set operations. In considering an individual who has at most eight years

Table 4.4
Classification by educational level and salary

Level of education in years (Y)	*Event B_1* $0 < X \leq 5{,}000$	*Event B_2* $5{,}000 < X \leq 15{,}000$	*Event B_3* $X > 15{,}000$	*Total*
	Annual salary in dollars (X)			
Event A_1 $0 < Y \leq 8$	210	80	10	300
Event A_2 $8 < Y \leq 12$	400	150	50	600
Event A_3 $Y > 12$	30	50	20	100
Total	640	280	80	1,000

education *and* who earns more than \$15,000 per year, we are seeking the intersection of the events A_1 and B_3 (written $(A_1 \cap B_3)$). The number of people in set A_1 is 300 and the number in set B_3 is 80, but the number that are members of both A_1 and B_3 is only 10. Sets A_1, A_2, A_3 are mutually exclusive or disjoint sets, as are B_1, B_2, B_3. That is, $(A_i \cap A_j) = \varnothing$ for $i \neq j$ and $(B_i \cap B_j) = \varnothing$ for $i \neq j$.

Now, what if we draw an individual who has at most eight years education and who earns more than \$15,000 per year? In such case, both event A_1 and event B_3 will have occurred simultaneously (jointly). In such case, we call this a *joint event* and, as indicated above, denote it by the event "$A_1 \cap B_3$." In general, the operator $\cap$ is used to denote an event that corresponds to the simultaneous occurrence of the two events under consideration.

It is clear that, when there are n equally likely possible outcomes of a chance event and such outcomes may be classified according to two criteria (as in Table 4.4), the probability of a joint event may be obtained by the classical approach. For example, we obtain $P(A_1 \cap B_3)$ by taking the ratio of the number of people having both attributes of interest to the total number of individuals, that is,

$$P(A_1 \cap B_3) = \frac{10}{1{,}000}.$$

Now, in cross-classifying individuals according to education and annual salary, we have arbitrarily divided the 1,000 possible outcomes into nine mutually exclusive joint events. This is obvious, since one random selection will produce an individual having the attributes associated with only one cross-classification. It follows, then, that the sum of the probabilities of all the joint events will equal 1. That is,

$$\begin{aligned} P(A_1 \cap B_1) + P(A_1 \cap B_2) + \cdots + P(A_3 \cap B_3) &= \frac{210}{1{,}000} + \frac{80}{1{,}000} + \cdots + \frac{20}{1{,}000} \\ &= \frac{1{,}000}{1{,}000} = 1. \end{aligned} \tag{1}$$

Marginal Probabilities

The fact that the set $(A_i \cap A_j) = \emptyset$ for $i \neq j$ allows us to consider one classification, or subset, in relation to the entire sample space. The number of elements represents the size of a given subset. Then, the probability of the event A_1, neglecting all other events, is the ratio of the number of elements in the subset A_1 to the total number of elements in the sample space. This ratio is the *marginal probability* of A_1.

The fact that the joint events of interest are mutually exclusive allows another means of viewing this concept. We may be interested in only one of the criteria of classification, say education, while indifferent to the other, say annual salary. In such case, we may be interested, for example, in obtaining the probability that a randomly selected individual will have more than 12 years of education. We can obtain this immediately through the classical approach as

$$P(A_3) = \frac{100}{1{,}000}.$$

However, it is enlightening to apply Postulate 3a. Event A_3 is the same as the event "$(A_3 \cap B_1) \cup (A_3 \cap B_2) \cup (A_3 \cap B_3)$." Thus, since the joint events are mutually exclusive, we have

$$\begin{aligned} P(A_3) &= P[(A_3 \cap B_1) \cup (A_3 \cap B_2) \cup (A_3 \cap B_3)] \\ &= P(A_3 \cap B_1) + P(A_3 \cap B_2) + P(A_3 \cap B_3) \qquad (2) \\ &= 30/1000 + 50/1000 + 20/1000 = 100/1000. \end{aligned}$$

This explains why we call this type of probability a *marginal probability*. (Note that we obtained it above by dividing the number in the proper margin by the total possible outcomes.) The term "marginal" is used whenever one or more criteria of classification are ignored. In general, we can obtain the marginal probability of an event by summing the probabilities of all mutually exclusive joint events which satisfy the event in question.

Rule 4.6

If the event A can occur jointly with any subset of k mutually exclusive and exhaustive events $B_1, B_2, \ldots, B_k$, then the marginal probability of the event A is

$$P(A) = P(A \cap B_1) + P(A \cap B_2) + \cdots + P(A \cap B_k).$$

We often can determine the probability of an event occurring from information regarding the occurrence of another event. For example, let's say that a green ball has been drawn from an urn that originally contained six green balls and four blue balls. Nine balls remain, and we want to know the new probability of selecting a green ball. If we let the events of selecting colors green and blue be denoted by G and B, respectively, and the subscript of each correspond to the draw to which we have reference, the above conditional probability may be denoted by $P(G_2 | G_1)$. This probability statement is read as "the probability of selecting a green ball on the second draw given that a green ball was obtained on the first draw."

In general, for any events A and B, we shall have $P(A|B)$, "the probability of event A occurring given that event B has occurred." This is actually the probability of a conditional event. However, the probability of a conditional event is generally termed a *conditional probability*.

In the framework of set theory, the information regarding the occurrence of an event defines a reduced sample space. If we seek the probability of event S_2 occurring, given the knowledge that event S_1 has occurred, then the set S_1 represents our reduced sample space. The probability of S_2 now occurring depends upon the ratio of the number of sample points representing S_1 *and* S_2 to the total number of sample points in S_1. This is illustrated by the Venn diagram in Figure 4.8. The information

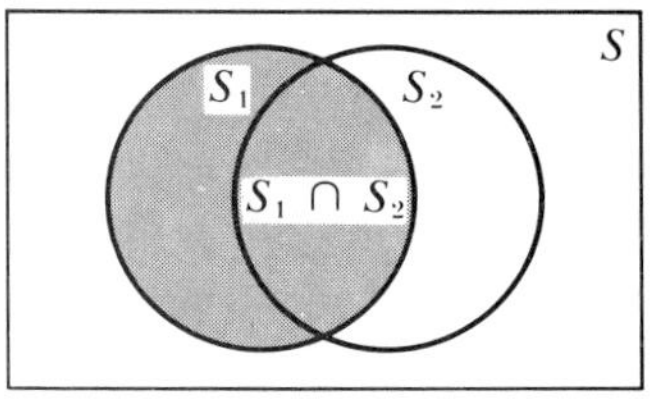

Figure 4.8

Conditional events

that S_1 has occurred reduces the sample space under consideration to the area shaded in the figure so that the desired probability is the ratio of the number of occurrences of $(S_1 \cap S_2)$ to the number of occurrences of S_1.

To determine conditional probabilities, we can construct the remaining population (perhaps only a subpopulation) from which we will be sampling under the given conditions. Then, we determine the probabilities in the usual manner of enumeration of possibilities. As a numerical example, suppose we have randomly selected an individual from the population given in Table 4.4 and have been told that he has no more than eight years education (event A_1 has occurred). Under this conditional information, how do we assess the probability that he earns more than \$15,000 per year? That is, what is $P(B_3|A_1)$? We are concerned only with the subpopulation of individuals having the attribute denoted by the event A_1. Thus, we limit consideration to these 300 individuals. Since we know that the randomly selected individual came from this subpopulation, we know the total number of possible outcomes has been reduced to 300. Of these 300, 10 have the attribute of interest, i.e., B_3. Thus, simply by limiting consideration to the proper subpopulation and enumerating outcomes (the classical approach), we obtain

$$P(B_3|A_1) = \frac{10}{300}. \tag{3}$$

This type of reasoning will lead us conveniently to a correct answer in many conditional probability situations. However, in many other situations it is very difficult, perhaps even impossible, to count cases (so to speak). For them, we require a more general approach for obtaining conditional probabilities. The following rules can be seen to hold graphically in Figure 4.8.

Rule 4.7

Let A and B be two events in a sample space S such that P(B) > 0. The conditional probability of event A given event B is

$$P(A \mid B) = \frac{P(A \cap B)}{P(B)}. \tag{a}$$

It also follows that, as long as P(A) > 0,

$$P(B \mid A) = \frac{P(A \cap B)}{P(A)}. \tag{b}$$

For the preceding case, using Table 4.4, we have

$$P(B_3 \cap A_1) = \frac{10}{1000} \quad \text{and} \quad P(A_1) = \frac{300}{1000}. \tag{4}$$

Thus, we can apply Rule 4.7(b) to obtain

$$P(B_3 \mid A_1) = \frac{P(B_3 \cap A_1)}{P(A_1)} = \frac{10/1000}{300/1000} = \frac{10}{300}. \tag{5}$$

The result is identical to that of (3), which we obtained by reducing consideration to the appropriate subpopulation and using the classical approach of counting cases.

In the foregoing example, the desired conditional probability is easy to obtain regardless of method. However, in some situations, the method involving direct enumeration in the appropriate subpopulation is much better suited, and in others the general formula of Rule 4.7(a) is most expedient. Unfortunately, it is not always obvious which method to use for determining conditional probabilities, and it is practically impossible to set forth general rules for choosing the proper method. Thus, you should learn to consider the complications involved in both methods before selecting one. The following examples are designed to enhance your ability to determine the best method.

Example 1

Suppose we are to conduct a survey among the 20 employees of Essex Shoe Company, who are categorized in the following table.

	M—Men	*W—Women*	*Total*
S—Salaried	4	3	7
H—Hourly	8	5	13
Total	12	8	20

The survey sample is to be chosen by simple random sampling without replacement. If we randomly select an individual and observe that we have selected a man, what is the probability that we have selected one who is salaried? In other words, what is $P(S \mid M)$?

Solution

Given the information of the table, it makes very little difference which method we choose for calculating the conditional probability. $P(S|M)$ can be obtained by considering only the population of men. Thus, the subpopulation of interest is the total of 12 men, 4 salaried and 8 hourly. Hence, we know immediately that $P(S|M) = 4/12$.

On the other hand, we can count cases to obtain the relevant joint and marginal probabilities to be

$$P(S \cap M) = 4/20 \quad \text{and} \quad P(M) = 12/20.$$

Thus, (a) of Rule 4.7 gives us

$$P(S|M) = \frac{P(S \cap M)}{P(M)} = \frac{4/20}{12/20} = \frac{4}{12}. \qquad \square$$

You may wish to verify that either method will yield the following probabilities for other conditional events of Example 1:

$$P(H|M) = 8/12; \quad P(S|W) = 3/8; \quad P(M|S) = 4/7; \quad P(M|H) = 8/13;$$
$$P(H|W) = 5/8; \quad P(W|S) = 3/7; \quad P(W|H) = 5/13.$$

Example 2

Suppose we have a set containing six good (G) light bulbs and four bad (B) light bulbs, from which we wish to select a simple random sample of size 2 without replacement. Determine the probability of drawing a bad bulb on the second draw given that a bad bulb has been drawn on the first draw. That is, find $P(B_2|B_1)$.

Solution

In this type of problem, selecting the appropriate method is critical. Let us illustrate by again calculating the desired probability with both methods. Note how very easy it is to count cases. To determine $P(B_2|B_1)$ by this method, we simply consider the subpopulation which remains after B_1 has occurred. Thus, under the condition of having selected a bad bulb on the first draw, the subpopulation from which we shall sample on the second draw consists of the original six good bulbs and the three remaining bad bulbs. Since we are selecting a simple random sample, the probability of any specific bulb being selected on the second draw is 1/9. There are three bad bulbs which constitute mutually exclusive events for the second draw. Thus, we may count cases (or apply Postulate 3a) to determine immediately

$$P(B_2|B_1) = 3/9 = 1/3. \tag{6}$$

Now, consider the excessive labor required if we choose to use Rule 4.7(a). This method requires that we solve

$$P(B_2|B_1) = \frac{P(B_1 \cap B_2)}{P(B_1)}.$$

Clearly, the marginal probability of event B_1 can be obtained easily (by counting cases) as

$$P(B_1) = 4/10.$$

However, it is very difficult to obtain the joint probability, $P(B_1 \cap B_2)$; for, in order to count cases, we must utilize permutations. There are ${}_{10}P_2$ ways of selecting two bulbs from the set. (Order is clearly important, and, even though some of the bulbs are identical, we must consider each as a distinguishable item if we are to include all possible draws.) Thus, our sample space consists of ${}_{10}P_2$ mutually exclusive events. Since we are using simple random sampling, each of these events has equal probability of occurring, or $1/{}_{10}P_2$. Thus, to obtain $P(B_1 \cap B_2)$, we must determine how many of these ${}_{10}P_2$ events satisfy the event $(B_1 \cap B_2)$. Since there are four bad bulbs available, ${}_4P_2$ events satisfy the event $(B_1 \cap B_2)$. Thus, by counting cases we have

$$P(B_1 \cap B_2) = {}_4P_2/{}_{10}P_2 = \frac{4!}{2!}\bigg/\frac{10!}{8!} = \frac{12}{90} = \frac{2}{15}.$$

Finally,

$$P(B_2 | B_1) = \frac{P(B_1 \cap B_2)}{P(B_1)} = \frac{2/15}{4/10} = \frac{20}{60} = \frac{1}{3}.$$

Although the answer is identical to (6), the solution is excessively laborious in this instance. However, the method should not be dismissed completely. It represents a more generally applicable method than that of counting cases, particularly for certain practical problems. □

Although it can hardly be called a rule, the calculations of Example 2 suggest it is often expedient to try initially to count cases in problems similar to the one involved. However, one also must be careful to avoid oversimplification. One reason we could count cases so easily in the foregoing problem is that we were considering the events in logical order. What if, for some reason, we needed to calculate $P(B_1 | B_2)$ in Example 2? In this case, it is apparently impractical to attempt to solve the problem by reducing consideration to a subpopulation and counting cases. Further, evaluation of the formula

$$P(B_1 | B_2) = \frac{P(B_1 \cap B_2)}{P(B_2)}$$

is even more difficult than before, since the calculation of $P(B_2)$ is not completely straightforward. Thus, here, neither method is expedient. However, the problem can be solved quite easily by another approach to be discussed later.

Example 3

The joint probabilities in the table below have been calculated from the information of the table in Example 1. Using these probabilities and Rules 4.6 and 4.7, find the following probabilities:

(a) $P(M)$	(b) $P(W)$	(c) $P(S)$	(d) $P(H)$
(e) $P(S \mid M)$	(f) $P(H \mid M)$	(g) $P(S \mid W)$	(h) $P(H \mid W)$
(i) $P(M \mid S)$	(j) $P(W \mid S)$	(k) $P(M \mid H)$	(l) $P(W \mid H)$

	M	W
S	4/20	3/20
H	8/20	5/20

Solution

(a) $P(M) = P(M \cap S) + P(M \cap H) = 4/20 + 8/20 = 12/20.$

(b) $P(W) = P(W \cap S) + P(W \cap H) = 3/20 + 5/20 = 8/20.$

(c) $P(S) = P(M \cap S) + P(W \cap S) = 4/20 + 3/20 = 7/20.$

(d) $P(H) = P(M \cap H) + P(W \cap H) = 8/20 + 5/20 = 13/20.$

(e) $P(S\,|\,M) = \dfrac{P(M \cap S)}{P(M)} = \dfrac{4/20}{12/20} = \dfrac{4}{12}.$ (f) $P(H\,|\,M) = \dfrac{P(M \cap H)}{P(M)} = \dfrac{8/20}{12/20} = \dfrac{8}{12}.$

(g) $P(S\,|\,W) = \dfrac{P(W \cap S)}{P(W)} = \dfrac{3/20}{8/20} = \dfrac{3}{8}.$ (h) $P(H\,|\,W) = \dfrac{P(W \cap H)}{P(W)} = \dfrac{5/20}{8/20} = \dfrac{5}{8}.$

(i) $P(M\,|\,S) = \dfrac{P(M \cap S)}{P(S)} = \dfrac{4/20}{7/20} = \dfrac{4}{7}.$ (j) $P(W\,|\,S) = \dfrac{P(W \cap S)}{P(S)} = \dfrac{3/20}{7/20} = \dfrac{3}{7}.$

(k) $P(M\,|\,H) = \dfrac{P(M \cap H)}{P(H)} = \dfrac{8/20}{13/20} = \dfrac{8}{13}.$ (l) $P(W\,|\,H) = \dfrac{P(W \cap H)}{P(H)} = \dfrac{5/20}{13/20} = \dfrac{5}{13}.$

□

Problems

1. Define *conditional probability*, and give an example.
2. Define *joint probability*, and give an example.
3. The following is a classification of the employees of a company:

Wage classification	*Division* A	B	C
$W1$	100	900	1000
$W2$	150	700	200
$W3$	250	400	200
$W4$	500	500	100

   a. State what the event $(B \cup W3)$ represents.
   b. Calculate $P(B \cup W3)$.
   c. State what the event $(A \cap W2)$ represents.
   d. Calculate $P(A \cap W2)$.
   e. State what the event $(C\,|\,W1)$ represents.
   f. Calculate $P(C\,|\,W1)$.
   g. State what $1 - P(B \cap W4)$ represents.
   h. Calculate $1 - P(B \cap W4)$.

4. Using the data in Problem 3, determine the marginal probabilities for each division and for each wage classification.

4.7 *Multiplication Rule*

We demonstrated in the last section how difficult it can be to compute joint probabilities through the classical approach. A rule known as the *multiplication rule for probabilities* provides another (often more expedient) means of computing joint probabilities in such situations.

Rule 4.8

For any events A and B

$$P(A \cap B) = P(A)P(B|A)$$

and

$$P(A \cap B) = P(B)P(A|B).$$

The first equality is obtained simply by solving (b) of Rule 4.7 for $P(A \cap B)$, while the second is obtained by solving (a) of Rule 4.7 for $P(A \cap B)$. Thus, (as we warned earlier) one must view these and the two equalities of Rule 4.8 not as formulas, but as relations involving joint, marginal, and conditional probabilities. In a sense, the relationship is circular. That is, for any given problem, we cannot use either relationship to solve for a desired probability until we have first determined values for other required probabilities.

Example 1

Suppose we have a set containing six good (G) transistors and four bad (B) transistors, from which we wish to select a simple random sample of size 2 without replacement. Use Rule 4.8 to find the following joint probabilities,

(a) $P(B_1 \cap B_2)$, (b) $P(B_1 \cap G_2)$, (c) $P(G_1 \cap B_2)$, (d) $P(G_1 \cap G_2)$,

where the subscripts 1 and 2 give the number of the draw.

Solution

We obtain the marginal probabilities $P(B_1)$ and $P(G_1)$ by considering the problem of selecting only one transistor from the original population. Thus,

$$P(B_1) = 4/10 = 2/5 \quad \text{and} \quad P(G_1) = 6/10 = 3/5.$$

We obtain relevant conditional probabilities by considering only the appropriate conditional populations and counting cases as in Example 2, Section 4.6. Hence,

$$P(B_2 | B_1) = 1/3, \qquad P(G_2 | B_1) = 2/3,$$
$$P(B_2 | G_1) = 4/9, \quad \text{and} \quad P(G_2 | G_1) = 5/9.$$

Using the above information, we find the required joint probabilities as

(a) $P(B_1 \cap B_2) = P(B_1)P(B_2 | B_1) = 2/5 \cdot 1/3 = 2/15.$
(b) $P(B_1 \cap G_2) = P(B_1)P(G_2 | B_1) = 2/5 \cdot 2/3 = 4/15.$
(c) $P(G_1 \cap B_2) = P(G_1)P(B_2 | G_1) = 3/5 \cdot 4/9 = 4/15.$
(d) $P(G_1 \cap G_2) = P(G_1)P(G_2 | G_1) = 3/5 \cdot 5/9 = 5/15.$

It can be noted that the four joint events are mutually exclusive and exhaustive. Thus, as we should expect, the sum of their probabilities is 1. □

Note that we could have determined $P(B_1 \cap B_2)$, for example, by

$$P(B_1 \cap B_2) = P(B_1 | B_2)P(B_2).$$

However, common sense and a little reasoning make it quite clear that this is not the relation to use in this instance. It requires us to obtain $P(B_1 | B_2)$ and $P(B_2)$, and neither task is easy in the problem at hand. As we pursue this area in a more general context, we shall summarize the operating rules which we have so far treated.

Let us consider all the marginal, joint, and conditional probabilities that might be of interest in Example 1 and see if we can develop some logical approach to deciding which methods and/or relations to use for each. The probabilities are

Marginal: $P(B_1), P(G_1).$ (1)

Conditional: $P(B_2 | B_1), P(B_2 | G_1), P(G_2 | B_1), P(G_2 | G_1).$ (2)

Joint: $P(B_1 \cap B_2), P(B_1 \cap G_2), P(G_1 \cap B_2). P(G_1 \cap G_2).$ (3)

Marginal: $P(B_2), P(G_2).$ (4)

Conditional: $P(B_1 | B_2), P(G_1 | B_2), P(B_1 | G_2), P(G_1 | G_2).$ (5)

The nature of the problem is such that the grouping we have used represents the logical order of solution. Why? Let us answer this by working backward through the grouping.

1. First, it is risky to rely on the intuition necessary to obtain the conditional probabilities of (5) by counting cases. (Make sure that you understand and agree with this statement.) Thus, we must use the relationship of Rule 4.7 for conditional probabilities. However, this requires that we know the respective marginal and joint probabilities of (3) and (4).
2. It is not practical (again, intuition is risky) to count cases to obtain the marginal probabilities of (4). Hence, we require the use of Rule 4.6, which necessarily entails the joint probabilities of (3).
3. While we can count cases (using permutations as in Example 2, page 110) to obtain the joint probabilities of (3), it is much easier to use Rule 4.8 for multiplication of probabilities. However, this requires the conditional and marginal probabilities of (2) and (1), respectively.

4. We can determine the conditional probabilities of (2) simply by reducing consideration to the appropriate subpopulations and by counting cases.
5. It is very easy to obtain the marginal probabilities of (4) by counting cases in the original population.

As an illustration, consider the case in which we desire $P(G_1 | B_2)$. We cannot count cases, so we use Rule 4.7 to obtain

$$P(G_1 | B_2) = \frac{P(G_1 \cap B_2)}{P(B_2)}. \tag{6}$$

But, by Rule 4.6,

$$P(B_2) = P(B_1 \cap B_2) + P(G_1 \cap B_2) \tag{7}$$

and, by Rule 4.8,

$$P(B_1 \cap B_2) = P(B_1)P(B_2 | B_1) \tag{8}$$

and

$$P(G_1 \cap B_2) = P(G_1)P(B_2 | G_1). \tag{9}$$

Combining steps (6) through (9), we have

$$\begin{aligned} P(G_1 | B_2) &= \frac{P(G_1 \cap B_2)}{P(B_2)} = \frac{P(G_1 \cap B_2)}{P(B_1 \cap B_2) + P(G_1 \cap B_2)} \\ &= \frac{P(G_1)P(B_2 | G_1)}{P(B_1)P(B_2 | B_1) + P(G_1)P(B_2 | G_1)}. \end{aligned} \tag{10}$$

Now, each of the probabilities in the last term of (10) can be found by counting cases. They are:

$$P(G_1) = \frac{6}{10}; \quad P(B_2 | G_1) = \frac{4}{9}; \quad P(B_1) = \frac{4}{10}; \quad P(B_2 | B_1) = \frac{3}{9}. \tag{11}$$

Substituting the values of (11) into (10), we finally obtain

$$P(G_1 | B_2) = \frac{6/10 \cdot 4/9}{(4/10)(3/9) + (6/10)(4/9)} = \frac{24/90}{36/90} = \frac{2}{3}. \tag{12}$$

It may be somewhat disconcerting to note that $P(G_1 | B_2)$ turns out to be exactly what we would have obtained by reasoning intuitively as follows: We have drawn one transistor without noting its condition, and the second draw has resulted in a bad transistor. This means that, even though one of the transistors is not in the set, we still have nine transistors the condition of which we do not know (a subpopulation). Six of them are good and three bad. Thus, intuitively, the one we have drawn on the first draw has six equally likely chances out of nine of being good. Thus, $P(G_1 | B_2) = 2/3$. However, even though the correct value for the probability of interest could have been obtained intuitively in this instance, the above reasoning process is valuable for at least two major reasons. First, it provides valid results in problems so complicated that intuition breaks down completely. Second, and perhaps even more important, it illustrates the logical approach to general probability problems involving formulas and rules that are circular.

Statistical Independence

The foregoing illustration shows that the probability of a specific event will vary as the conditional information regarding another event changes. For example,

$$P(B_2 | B_1) = 1/3 \neq P(B_2 | G_1) = 4/9.$$

In this case, the probabilities depend upon the given condition, or the events are dependent. When this is not the case, we have statistical independence of events. *Statistically independent events* are defined as follows.

Definition 4.20

Let A and B be two events in a sample space S. The events are independent if any one of the following conditions is satisfied:

$$P(A | B) = P(A), \qquad \text{(a)}$$

$$P(B | A) = P(B), \qquad \text{(b)}$$

or

$$P(A \cap B) = P(A) \cdot P(B). \qquad \text{(c)}$$

It may not be immediately obvious that the truth of any one of the conditions implies independence of the two events under consideration. However, it can be shown that if any one of the relationships in Definition 4.20 is satisfied, both of the other two will also be satisfied. Thus, if we wish to check whether two events are independent, we need only test for one of the three conditions. Conversely, if one of the three relationships is not satisfied, neither of the remaining two will be satisfied.

In addition to providing a means for checking the independence of two events, (c) of the definition allows a simpler solution for joint probabilities (when the events are independent) than does the general multiplication rule, Rule 4.8. However, given any doubt regarding dependence, Rule 4.8 should always be used. It will give the correct answer whether or not the events are independent.

Let us consider several examples that demonstrate the use of Definition 4.20.

Example 2

Suppose we have a set containing twelve good (G) transistors and eight bad (B) from which we wish to select a simple random sample of size 2 with replacement. Determine whether the two events B_1 and G_2 are statistically independent.

Solution

We need to test to see if $P(G_2) = P(G_2 | B_1)$. Knowing that we are sampling with replacement, we can determine the appropriate marginal and conditional probabilities by considering only the population from which we shall sample. Thus,

$$P(G_2) = 3/5 \quad \text{and} \quad P(G_2 | B_1) = 3/5.$$

3. Under what conditions is $P(A|B)$ not equal to $P(A \cap B)/P(B)$?

4. If $P(A|B) = \frac{1}{2}$ and $P(B) = \frac{2}{3}$, what is $P(A \cap B)$?

5. If $P(A \cap B) = \frac{1}{6}$ and $P(A|B) = \frac{1}{5}$ what is $P(B)$?

6. If $P(A) = \frac{1}{4}$, $P(B) = \frac{1}{3}$, and $P(B|A) = \frac{2}{3}$, what is $P(A|B)$?

7. A wine list contains seven red wines and eight white wines. Use Rule 4.8 to determine the following joint probabilities when selecting a simple random sample of size 2 *without replacement.*

   a. $P(R_1 \cap R_2)$, b. $P(R_1 \cap W_2)$, c. $P(W_1 \cap R_2)$, d. $P(W_1 \cap W_2)$.

8. For the wine list given in Problem 7, use the general multiplication rule to determine the following joint probabilities when selecting a sample of size 3 *without replacement.*

   a. $P(R_1 \cap R_2 \cap R_3)$, b. $P(R_1 \cap R_2 \cap W_3)$,
   c. $P(R_1 \cap W_2 \cap R_3)$, d. $P(R_1 \cap W_2 \cap W_3)$,
   e. $P(W_1 \cap R_2 \cap R_3)$, f. $P(W_1 \cap R_2 \cap W_3)$,
   g. $P(W_1 \cap W_2 \cap R_3)$, h. $P(W_1 \cap W_2 \cap W_3)$,
   i. $P(R_1 \cap R_2)$, j. $P(R_1 \cap W_2)$,
   k. $P(W_1 \cap R_2)$, l. $P(W_1 \cap W_2)$.

9. Rework Problem 7 for a simple random sample of size 2 *with replacement.*

10. Rework Problem 8 for a simple random sample of size 3 *with replacement.*

4.8 *Addition Rule*

Postulate 3a can be used to determine the probability of the event $A \cup B$ only if events A and B are mutually exclusive. In order to determine $P(A \cup B)$ when events A and B are not mutually exclusive, we use the more general *addition rule for probabilities.*

Rule 4.9

For any two events A and B

$$P(A \cup B) = P(A) + P(B) - P(A \cap B).$$

Postulate 3a is a special case of Rule 4.9, since $P(A \cap B) = 0$ if A and B are mutually exclusive events. It should be noted, on the other hand, that if we know that

$P(A \cap B) = 0$, we do not necessarily know that A and B are mutually exclusive events.

Example 1

The 1,000 current loans of a certain loan company are distributed as follows:

	I—Individual	*B—Business*
H—High risk	260	40
M—Medium risk	300	100
L—Low risk	140	160

Suppose that an auditor is sent in to examine the company records to determine whether or not his employer, a finance firm, should supply the loan company additional money to lend its clients. The auditor will base his decision on one loan record selected randomly from the files. If the loan is either a business loan or a low-risk individual loan, the company will obtain more money for lending purposes. Otherwise, it will get no more money. Find

(a) the probability that the loan company will receive more money, and

(b) the probability that the loan company will not receive more money.

Solution

Since the auditor is randomly selecting a record from the files, each loan record has an equal chance of being selected. Thus, from the table we can count cases to obtain the joint probabilities

$$P(H \cap I) = 0.260, \qquad P(H \cap B) = 0.040,$$
$$P(M \cap I) = 0.300, \qquad P(M \cap B) = 0.100,$$
$$P(L \cap I) = 0.140, \quad \text{and} \quad P(L \cap B) = 0.160.$$

(a) To solve the first problem, we can use Rule 4.9, which gives

$$P(\text{money}) = P(L \cup B) = P(L) + P(B) - P(L \cap B). \tag{1}$$

However, we must determine $P(L)$ and $P(B)$. We use Rule 4.6 to obtain

$$P(L) = P(L \cap B) + P(L \cap I) = 0.140 + 0.160 = 0.300 \tag{2}$$

and

$$\begin{aligned} P(B) &= P(H \cap B) + P(M \cap B) + P(L \cap B) \\ &= 0.040 + 0.100 + 0.160 = 0.300. \end{aligned} \tag{3}$$

Therefore, combining (1), (2), and (3), we have

$$P(\text{money}) = 0.300 + 0.300 - 0.160 = 0.440.$$

(b) We consider two alternatives for the solution of this problem. As the first alternative, we can apply Rule 4.5 to obtain

$$P(\text{no money}) = 1 - P(\text{money}) = 1 - 0.440 = 0.560.$$

As the second alternative, we can consider the event "no money" satisfied if we obtain either the event $(H \cap I)$ or the event $(M \cap I)$. These two events are mutually exclusive. Hence

$$\begin{aligned} P(\text{no money}) &= P[(H \cap I) \cup (M \cap I)] \\ &= P(H \cap I) + P(M \cap I) \\ &= 0.260 + 0.300 = 0.560. \end{aligned}$$ □

Since there are generally several correct ways to solve a given probability problem, it is easy to check a solution. If we do not arrive at the same solution through different approaches, we must conclude that our "logic" (or common sense) is in error for at least one of the methods. Let us consider another example for which at least two approaches are possible.

Example 2

Suppose we have a set containing six good diodes and four bad diodes from which we wish to select a simple random sample of size 2 with replacement. Determine the probability of selecting a good diode on the first draw or a bad diode on the second draw, that is, $P(G_1 \cup B_2)$.

Solution

For our first approach, we use Rules 4.8 and 4.9 to obtain

$$\begin{aligned} P(G_1 \cup B_2) &= P(G_1) + P(B_2) - P(G_1 \cap B_2) \\ &= P(G_1) + P(B_2) - P(G_1)P(B_2 | G_1) \\ &= P(G_1) + P(B_2) - P(G_1)P(B_2) \\ &= 3/5 + 2/5 - (3/5) \cdot (2/5) \\ &= 19/25. \end{aligned}$$

(We leave it to the student to show that G_1 and B_2 are independent.)

For our second approach, let us consider the four mutually exclusive and exhaustive joint events, $(B_1 \cap B_2)$, $(B_1 \cap G_2)$, $(G_1 \cap B_2)$, and $(G_1 \cap G_2)$. Since the event $(G_1 \cup B_2)$ is satisfied by the occurrence of any one of the events $(B_1 \cap B_2)$, $(G_1 \cap B_2)$, or $(G_1 \cap G_2)$, and since the latter three events are mutually exclusive, we may apply Postulate 3a to obtain

$$P(G_1 \cup B_2) = P(B_1 \cap B_2) + P(G_1 \cap B_2) + P(G_1 \cap G_2). \tag{4}$$

However, applying the multiplication rule (Rule 4.8) to each of the three terms of (4), we have

$$\begin{aligned} P(G_1 \cup B_2) &= P(B_1 \cap B_2) + P(G_1 \cap B_2) + P(G_1 \cap G_2) \\ &= P(B_1)P(B_2|B_1) + P(G_1)P(B_2|G_1) + P(G_1)P(G_2|G_1) \\ &= (2/5) \cdot (2/5) + (3/5) \cdot (2/5) + (3/5) \cdot (3/5) \\ &= 4/25 + 6/25 + 9/25 \\ &= 19/25. \end{aligned} \tag{5}$$

This second approach allows an even easier solution. Since the only mutually exclusive and exhaustive joint event that does not satisfy $(G_1 \cup B_2)$ is $(B_1 \cap G_2)$, we could apply Rule 4.5 to obtain

$$\begin{aligned} P(G_1 \cup B_2) &= 1 - P(B_1 \cap G_2) \\ &= 1 - P(B_1)P(G_2|B_1) \\ &= 1 - (2/5) \cdot (3/5) = 19/25. \end{aligned}$$ □

While it is not useful to generalize the addition rule to the extent that we did the multiplication rule, it is enlightening (particularly from a set-theory point of view) to extend Rule 4.9 slightly. Thus, let us consider the events A, B, and C, for which the addition rule becomes

$$\begin{aligned} P(A \cup B \cup C) = P(A) + P(B) + P(C) - P(A \cap B) - P(A \cap C) \\ - P(B \cap C) + P(A \cap B \cap C). \end{aligned} \tag{6}$$

These different subsets are graphically illustrated by the Venn diagram of Figure 4.13.

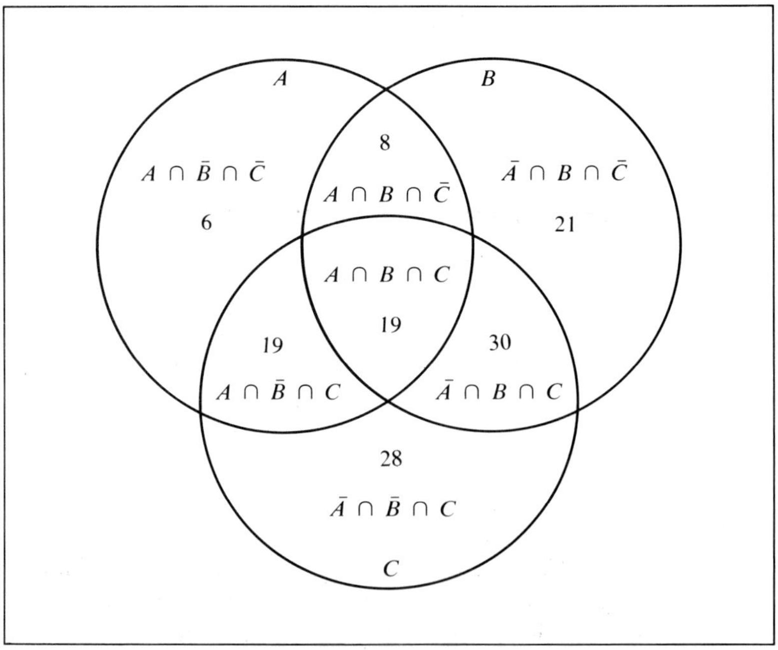

Figure 4.13
Generalization of the addition rule

If we evaluate the terms in (6) with respect to Figure 4.13, we can intuitively understand the formula. While adding the probabilities for A, B, and C, we accounted twice for each of the probabilities associated with sets $(A \cap B)$, $(A \cap C)$, and $(B \cap C)$. Thus, we must subtract the probability associated with each of these intersections, as shown in (6). However, in this subtraction, we inadvertently exclude the probability associated with the set common to all, that is, $(A \cap B \cap C)$. We must then add the probability associated with this set to complete the computation. Equation (6) provides a simple though rigorous manifestation of this intuition.

Example 3

A certain tire manufacturer conducted a survey of vehicles stopped at a toll booth in order to plan the orientation of his future advertising campaign. He observed 152 vehicles with the following results:

52 vehicles were convertibles;
78 vehicles had radial tires;
96 vehicles were produced by General Motors;
27 convertibles had radial tires;
38 convertibles were General Motors products;
49 General Motors products had radial tires;
19 vehicles were General Motors convertibles with radial tires.

Let us use the Venn diagram of Figure 4.13 to answer any additional questions about the survey. Let A, B, C represent convertibles, radial tires, and GM products, respectively.

As shown in the figure, it helps to denote each area of the Venn diagram in terms of complements and intersections. From this partitioning of the set, we can select any combination and determine the number in the appropriate classification. Since probability is expressed when possible in terms of counting, this approach serves equally well for measuring probability.

Problems

1. Events A, B, and C are defined as follows:

$$A = \{1, 3, 5, 6\}, \qquad B = \{2, 4, 6\}, \qquad C = \{1, 3, 7\}.$$

Find:

a. $(A \cup B)$, b. $(A \cap B)$, c. $(A \cup C)$,

d. $(A \cap C)$, e. $(B \cap C)$, f. $[A \cup (B \cup C)]$,

g. $[A \cap (B \cup C)]$, h. $[(A \cup C) \cap (A \cup B)]$.

i. Are A and B mutually exclusive? j. Are B and C mutually exclusive?

2. $P(A) = \frac{1}{2}, \quad P(B) = \frac{2}{3}, \quad P(C) = \frac{1}{3}.$

Events A, B, and C are statistically independent but not mutually exclusive. Find and plot in a Venn diagram:

a. $P(A \cap B \cap C)$, b. $P(A \cup C)$, c. $P(A \cup B \cup C)$,

d. $P(A \cap \bar{C})$, e. $P[(A \cap C) \cup \bar{B}]$,

f. the probability that at least one of the three events will occur.

3. Events E_1 and E_2 are mutually exclusive. Does this imply that

a. $P(E_2 | E_1) = 0$? b. $P(E_1 \cap E_2) = 0$? c. $P(E_2 \cup E_1) = P(E_2) + P(E_1)$?

d. $P(E_1 \cup E_2) = 0$? e. $P(E_2 \cap E_1) = P(E_2)P(E_1)$?

4. If a fair die is rolled, what is the probability of the outcome being

a. 3 or more? b. 5 or less?

c. greater than or equal to 4 but less than or equal to 2?

d. greater than 3 but less than 5?

5. If a pair of fair dice is tossed, what is the probability that the total number of spots facing upward is

a. 2? b. 3? c. 4?

d. 5? e. 6? f. 7?

g. 8? h. 9? i. 10?

j. 11? k. 12? l. 2 or 3 or 12?

m. 7 or 11? n. 4 or 10? o. 5 or 9?

p. 6 or 8? q. 2 or 3 or 7 or 11 or 12? r. 4 or 5 or 6 or 8 or 9 or 10?

The game defined above is commonly known as craps! The answers for Parts l, m, and r represent the probabilities for losing, winning, and continuing play, respectively, *on the first toss.*

6. Orders of a particular company have been classified as follows:

	Region			
Amount	*A*	B	C	*D*
I	50	100	50	10
J	100	50	75	25
K	75	30	100	75
L	75	20	25	140

a. State what $P(K|C)$ represents.

b. Calculate $P(K|C)$ by counting cases.

c. Compute $P(K|C)$ indirectly by using the appropriate probability theorems. Compare the result with your answer to Part b and check to see that they are equal.

d. State what $P(J \cup D)$ represents.

e. Find $P(J \cup D)$ by counting cases.

f. Compute $P(J \cup D)$ indirectly by using the appropriate probability rules. Compare this result with your answer to Part e and check to see that they are equal.

g. State what $P(L \cap A)$ represents.

h. Compute $P(L \cap A)$ by counting cases.

i. Calculate $P(L \cap A)$ indirectly by using the appropriate probability rules. Compare the result with your answer for Part h to see that they are equal.

4.9 *The Tree–Diagram Approach to Solving Probability Problems*

We introduced the tree diagram in Section 4.3 as a useful counting technique. Here we expand our discussion to demonstrate its power in solving probability problems. In our first example, we contrast its use with that of the general addition rule of probability.

Example 1

Consider a set containing seven batteries manufactured by Redico (R) and three batteries manufactured by Gocart (G). We are to select without replacement a sample of size $n = 3$ from the set. What is the probability that we will obtain a Redico battery on the first draw *or* a Gocart battery on the third draw—that is,

$$P(R_1 \cup G_3) = ? \tag{1}$$

Before attempting a solution, let us analyze the problem. Your first tendency probably is to apply the general addition rule of probability, by which

$$P(R_1 \cup G_3) = P(R_1) + P(G_3) - P(R_1 \cap G_3). \tag{2}$$

Note, however, that each of the three events on the right-hand side of this equation is itself a compound event, the probability of which is difficult to compute. We have previously applied the addition rule only to situations in which we seek the probability of one or the other or both of two events occurring on a single trial. Our current problem is not like this. We now seek the probability of one or the other or both of two events occurring on separate subparts of an experiment—*a very important distinction.*

Now, our intent is not to say that the general addition rule will not work in problems of this nature, but only that one must be extremely cautious to use it properly. More important, the tree diagram is a more powerful tool for this type of problem, as we shall demonstrate in the second approach to the solution of Example 1.

First approach to solution of Example 1

We have already noted, in (2), that

$$P(R_1 \cup G_3) = P(R_1) + P(G_3) - P(R_1 \cap G_3). \tag{3}$$

However, by a generalization of Equation (3), Section 4.8, we have

$$\begin{aligned} P(R_1) = P(R_1 \cap R_2 \cap R_3) + P(R_1 \cap R_2 \cap G_3) \\ + P(R_1 \cap G_2 \cap R_3) + P(R_1 \cap G_2 \cap G_3); \end{aligned} \tag{4}$$

$$\begin{aligned} P(G_3) = P(R_1 \cap R_2 \cap G_3) + P(R_1 \cap G_2 \cap G_3) \\ + P(G_1 \cap R_2 \cap G_3) + P(G_1 \cap G_2 \cap G_3); \end{aligned} \tag{5}$$

and

$$P(R_1 \cap G_3) = P(R_1 \cap R_2 \cap G_3) + P(R_1 \cap G_2 \cap G_3). \tag{6}$$

Substituting (4), (5), and (6) into (3), we obtain

$$\begin{aligned} P(R_1 \cup G_3) = P(R_1 \cap R_2 \cap R_3) + P(R_1 \cap R_2 \cap G_3) \\ + P(R_1 \cap G_2 \cap R_3) + P(R_1 \cap G_2 \cap G_3) \\ + P(R_1 \cap R_2 \cap G_3) + P(R_1 \cap G_2 \cap G_3) \\ + P(G_1 \cap R_2 \cap G_3) + P(G_1 \cap G_2 \cap G_3) \\ - P(R_1 \cap R_2 \cap G_3) - P(R_1 \cap G_2 \cap G_3). \end{aligned} \tag{7}$$

Simplifying, we obtain

$$\begin{aligned} P(R_1 \cup G_3) = P(R_1 \cap R_2 \cap R_3) + P(R_1 \cap R_2 \cap G_3) \\ + P(R_1 \cap G_2 \cap R_3) + P(R_1 \cap G_2 \cap G_3) \\ + P(G_1 \cap R_2 \cap G_3) + P(G_1 \cap G_2 \cap G_3). \end{aligned} \tag{8}$$

To solve (8), we must calculate each individual probability by using the general multiplication rule of probability. For example, by Equation (13), Sec. 4.7,

$$P(R_1 \cap R_2 \cap G_3) = P(R_1)P(R_2 | R_1)P(G_3 | R_1 \cap R_2). \tag{9}$$

Counting cases (by considcring appropriate subpopulations), we obtain

$$P(R_1) = \frac{7}{10}; \quad P(R_2 | R_1) = \frac{6}{9}; \quad P(G_3 | R_1 \cap R_2) = \frac{3}{8}. \tag{10}$$

Then, substituting from (10) into (9), we have

$$P(R_1 \cap R_2 \cap G_3) = \frac{7}{10} \cdot \frac{6}{9} \cdot \frac{3}{8} = \frac{126}{720} = \frac{7}{40}. \tag{11}$$

To solve the entire problem, we must repeat steps similar to (9) through (11) for each of the probabilities on the right-hand side of (8). □

While the foregoing approach is *correct*, it is difficult because of the many steps involved. Thus, we would welcome some "trick" with which to reduce the chance of error in both conceptualizing the problem and making the probability calculations. The tree diagram gives us just such a trick, as we shall see in our second approach to solving Example 1.

Note first that during our manipulations for the first solution, a very nice thing happened; we characterized the probability of $P(R_1 \cup G_3)$, the compound event of interest, as the sum of the probabilities of all the mutually exclusive joint events. In so doing, we reduced the problem to the simple situation in which we can apply Postulate 3a—a very important special case of the general addition rule of probability. A tree diagram lets us do this very easily.

Second approach to solution of Example 1

Our first task is to identify all the mutually exclusive and exhaustive joint events that can occur. A tree diagram serves this purpose nicely, as Figure 4.14 shows. We now can simply evaluate each of the mutually exclusive and exhaustive joint events to determine whether it represents the compound event of interest—that is, the event $R_1 \cup G_3$. In the tree diagram, a check ($\surd$) indicates each such event. Finally, to obtain the probability of the compound event $R_1 \cup G_3$, we simply calculate and add the

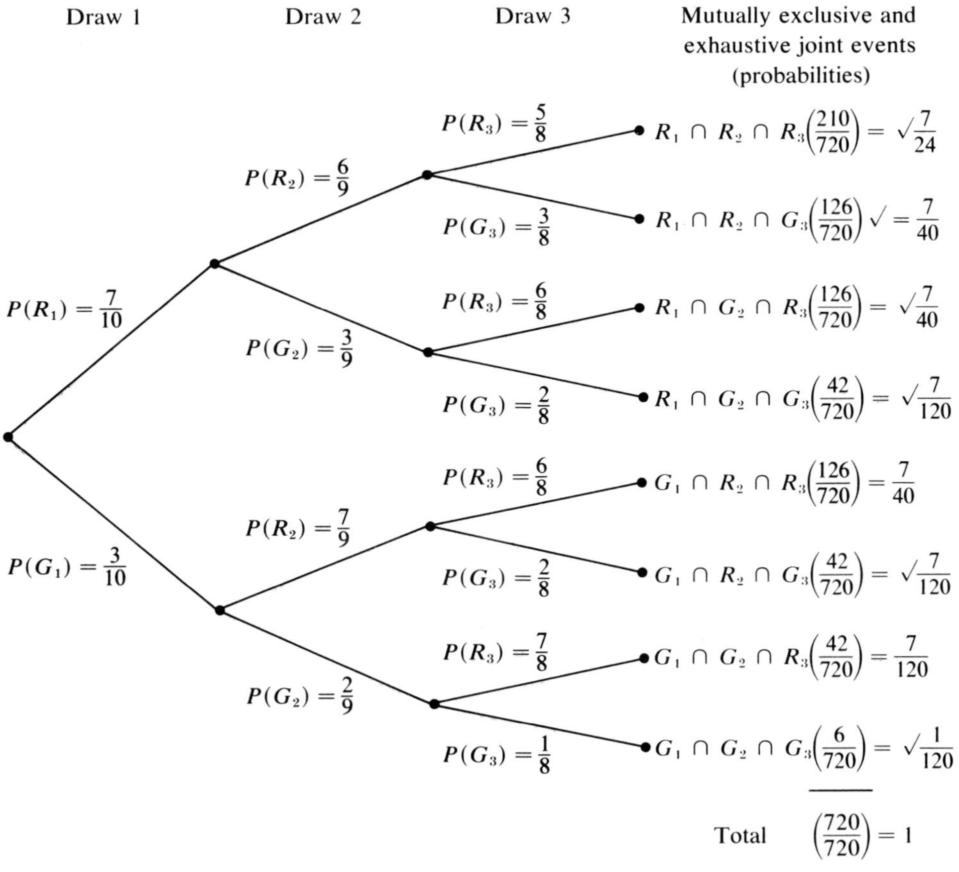

Figure 4.14
The mutually exclusive and exhaustive joint events in the experiment of Example 1

probabilities of the events checked. Thus,

$$\begin{aligned} P(R_1 \cup G_3) = P(R_1 \cap R_2 \cap R_3) + P(R_1 \cap R_2 \cap G_3) \\ + P(R_1 \cap G_2 \cap R_3) + P(R_1 \cap G_2 \cap G_3) \\ + P(G_1 \cap R_2 \cap G_3) + P(G_1 \cap G_2 \cap G_3). \end{aligned}$$

This result is identical to that obtained by applying the general addition rule of probability. □

Before going on, let us note the meaning of every facet of the tree of Figure 4.14. The tree is comprised of a sequence of "nodes" connected by "branches." The headings Draw 1, Draw 2, Draw 3 represent the subparts of the actions necessary to constitute one trial of the overall experiment. The labels on the branches represent the possible outcomes, or subevents, at any phase of the experiment, their subscripts representing the phase of the experiment in which each outcome, or subevent occurs. Each outcome is a "conditional" subevent—conditional on the occurrence of all subevents appearing on the branches leading from the first fork of the tree to the conditional subevent of interest. In Figure 4.14, for example, R_3, the subevent on branch 1 under "Draw 3," has a completely different meaning than R_3, the subevent on Branch 3 under "Draw 3." The former R_3 is interpreted as the outcome of a Redico battery on the third draw given that we have obtained a Redico battery on the first draw and the second draw—that is, given the occurrence of the event $R_1 \cap R_2$. The latter R_3 is interpreted as the outcome of a Redico battery on draw 3 given that we obtained a Redico battery on draw 1 and a Gocart battery on draw 2—that is, given the occurrence of the event $R_1 \cap G_2$.

The joint events listed on the right represent the mutually exclusive and exhaustive events that can occur. The numbers that follow represent the probabilities associated with each event. Note that the probability of any joint event is obtained by multiplying all the probabilities on the branches that lead from the first fork of the tree to the final branch that represents the event. *This type of calculation is actually an application of the general multiplication rule of probabilities, as will be demonstrated.*

Figure 4.15 demonstrates typical tree-diagram notation.

Calculations for Example 1

To calculate the probability of the relevant joint events, we shall evaluate $P(R_1 \cap R_2 \cap R_3)$. The general multiplication rule gives the same result we obtained by multiplying all probabilities on the branches, from the first fork of the tree to the branch on which the given event is represented. That is,

$$\begin{aligned} P(R_1 \cap R_2 \cap R_3) &= P(R_1)P(R_2|R_1)P(R_3|R_1 \cap R_2) \\ &= \frac{7}{10} \cdot \frac{6}{9} \cdot \frac{5}{8} = \frac{210}{720}. \end{aligned} \tag{12}$$

Performing similar operations for all other appropriate joint probabilities in the tree diagram and adding, we obtain the probability of the compound event of interest,

$$P(R_1 \cup G_3) = \frac{210}{720} + \frac{126}{720} + \frac{126}{720} + \frac{42}{720} + \frac{42}{720} + \frac{6}{720} = \frac{552}{720}. \tag{13}$$

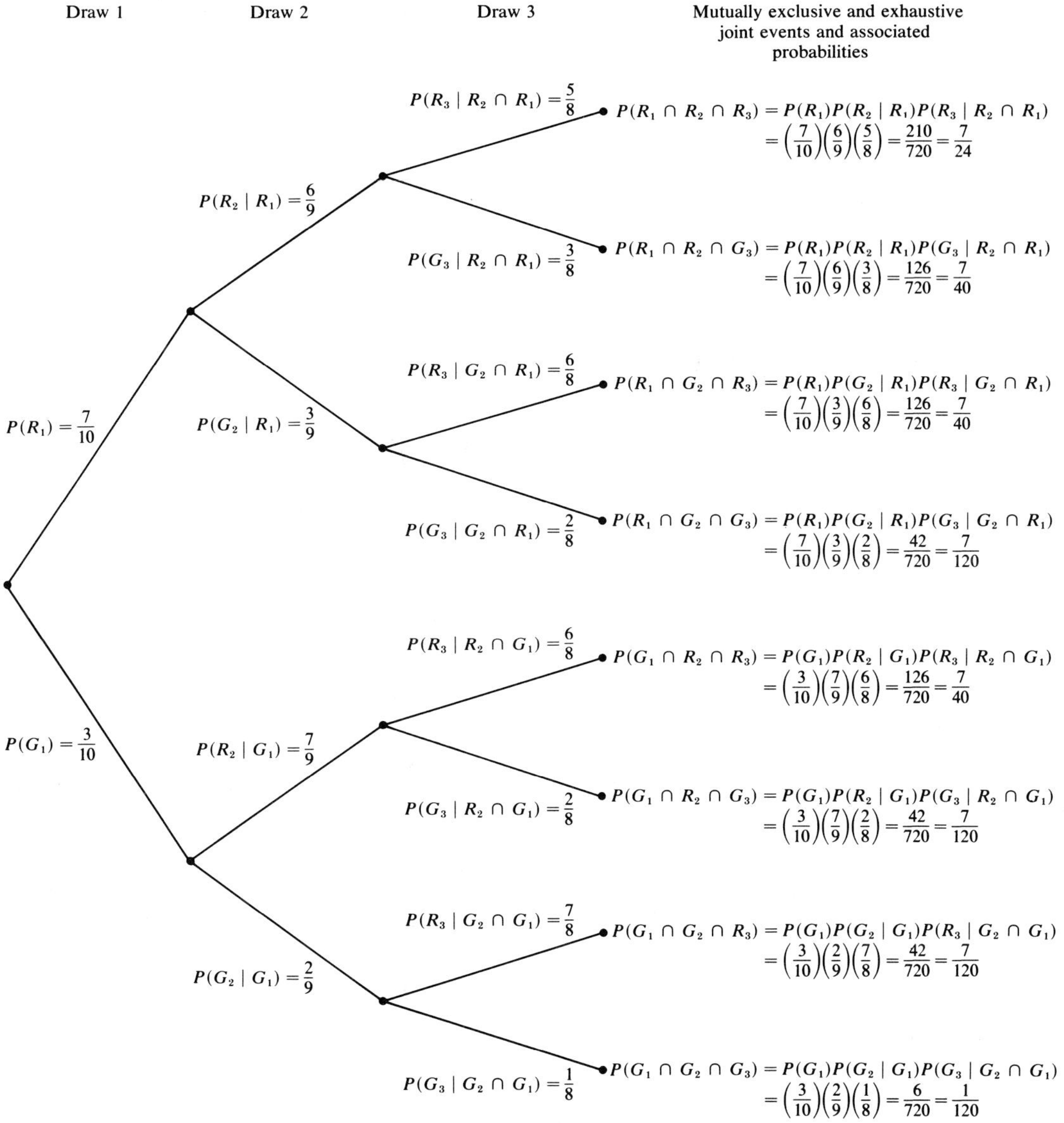

Figure 4.15
Detailed representation of Figure 4.14

Finally we note that (naturally) the sum of the mutually exclusive and exhaustive joint events is unity (one). Thus, in this particular case, it would have been easier to evaluate $P(R_1 \cup G_3)$ by subtracting from unity (one) the sum of the probabilities of the two mutually exclusive joint events that *do not* represent an occurrence of the event $(R_1 \cup G_2)$. That is, if either of the joint events $(G_1 \cap R_2 \cap R_3)$ or $(G_1 \cap G_2 \cap R_3)$ occurs, the compound event $(R_1 \cup G_3)$ will not be satisfied. Thus,

$$P(R_1 \cup G_3) = 1 - [P(G_1 \cap R_2 \cap R_3) + P(G_1 \cap G_2 \cap R_3)]. \qquad (14)$$

Generally, the tree-diagram method is applicable in the following situations:

1. The experiment is complicated, in that it consists of a series of subparts which may be conducted simultaneously or sequentially.
2. The probability sought involves one or more compound events so complicated that the solution is not easily visualized.
3. The nature of the problem is such that the compound event of interest can be represented as the union of a subset of the mutually exclusive and exhaustive events possible under the overall experiment.

The general steps to be followed in using the tree diagram are:

1. Identify and characterize the experiment.
2. State the compound event of interest.
3. Conceptualize a tree that identifies the mutually exclusive and exhaustive joint events which can occur under the experiment. In small-scale problems, it is feasible to "construct" the tree, but not in large-scale problems.
4. Select all of the mutually exclusive joint events whose union comprises the compound event of interest.
5. For each mutually exclusive joint event selected in step 4, obtain the probability by multiplying all of the probabilities appearing on the branches leading from the first fork of the tree to the event. (This is tantamount to applying the general rule of multiplication of probabilities to obtain the probability of each mutually exclusive joint event.)
6. Add the final probabilities of step 5 to obtain the probability of the compound event of interest.

Naturally, if it will simplify computations, one can replace steps 5 and 6 by subtracting from unity (one) the sum of the probabilities of all the events in the complement of the set of mutually exclusive joint events selected in step 4. Let us illustrate the technique in somewhat more complicated problems.

Example 2

We are to conduct the experiment of flipping a coin for which the probability of a head is 1/3 and the probability of a tail is 2/3. If we obtain a head, we are to roll a *fair* six-sided die. If we obtain a tail, we are to roll a *fair* four-sided die. What is the probability on one trial of the experiment of obtaining a tail or a 3. That is $P(T \cup 3) = ?$

Solution

Let us use in our tree diagram the following notation:

H = the occurrence of a head on the coin,

T = the occurrence of a tail on the coin,

$\boxed{n}$ = the occurrence of n on the six-sided die, where $n = 1, 2, \ldots, 6$,

$\triangle_n$ = the occurrence of n on the four-sided die, where $n = 1, 2, 3, 4$,

n = the occurrence of n on any die whatsoever, where $n = 1, 2, \ldots, 6$.

The tree diagram is illustrated in Figure 4.16. Inspecting it, we note that the union of the third mutually exclusive joint event and the last four mutually exclusive joint

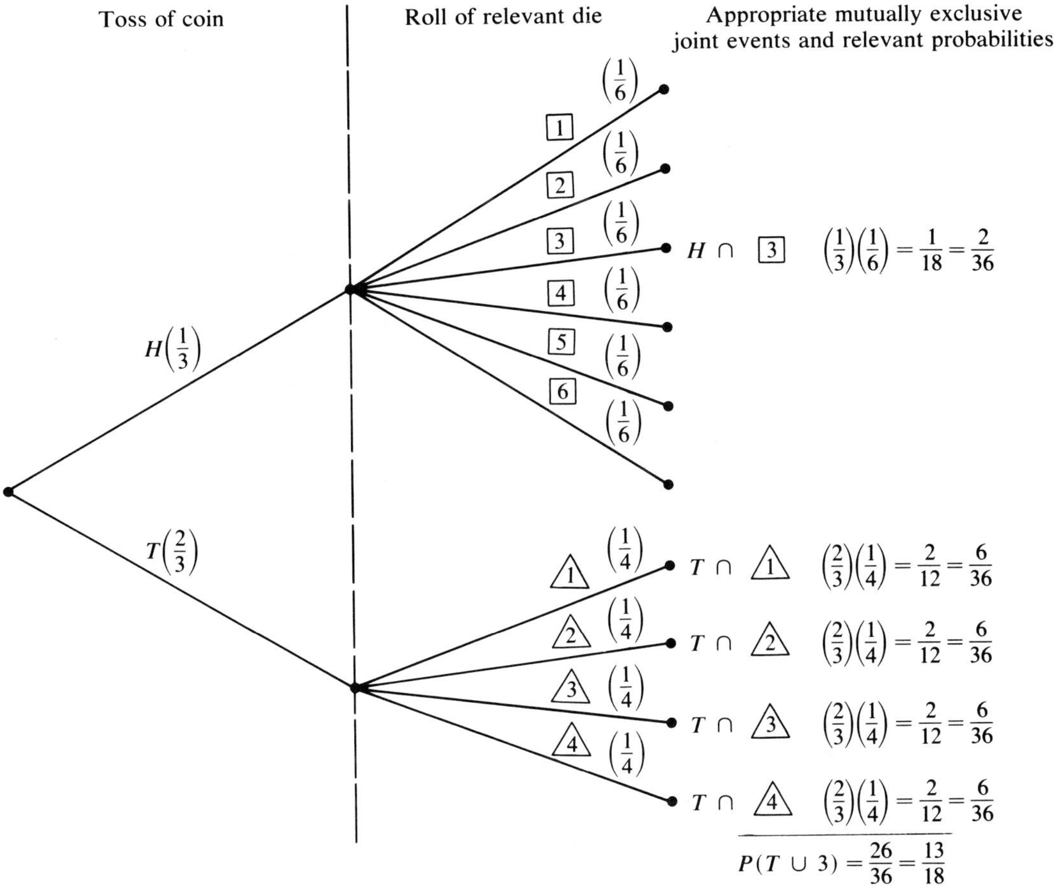

Figure 4.16
Tree diagram of Example 2

events constitute the compound event of interest, $T \cup 3$. Thus, multiplying and adding in our usual tree diagram fashion, we obtain

$$P(T \cup 3) = \frac{13}{18}.$$

In terms of the operating rules of probability, we have used the general rule of addition of probabilities,

$$P(T \cup 3) = P(T) + P(3) - P(T \cap 3), \tag{15}$$

by obtaining through the tree diagram

$$\begin{aligned}
P(T \cup 3) &= P(H \cap \boxed{3}) + P(T \cap \triangle_1) + P(T \cap \triangle_2) + P(T \cap \triangle_3) + P(T \cap \triangle_4) \\
&= P(H)P(\boxed{3}\,|\,H) + P(T)P(\triangle_1\,|\,T) + P(T)P(\triangle_2\,|\,T) \\
&\quad + P(T)P(\triangle_3\,|\,T) + P(T)P(\triangle_4\,|\,T) \\
&= \frac{1}{3}\cdot\frac{1}{6} + \frac{2}{3}\cdot\frac{1}{4} + \frac{2}{3}\cdot\frac{1}{4} + \frac{2}{3}\cdot\frac{1}{4} + \frac{2}{3}\cdot\frac{1}{4} \\
&= \frac{2}{36} + \frac{6}{36} + \frac{6}{36} + \frac{6}{36} + \frac{6}{36} = \frac{26}{36} = \frac{13}{18}.
\end{aligned}$$

□

Example 3

Suppose we have a set containing six starters manufactured by Realstart (R) and four starters manufactured by Goquick (G). We are to select without replacement a sample of size $n = 4$. What is the probability that we will obtain exactly one Realstart *or* exactly one Goquick on a given trial of the experiment. That is,

$$P(\text{exactly 1 Realstart} \cup \text{exactly 1 Goquick}) = ?$$

Solution

The tree diagram is shown in Figure 4.17. Note that we have inserted probabilities only on relevant branches. We see that satisfying eight of the sixteen mutually exclusive and exhaustive events of the experiment satisfies the compound event of interest. Thus, their union is equivalent to the compound event of interest. Finally, we simply multiply and add in the usual tree-diagram fashion to obtain

$$P(\text{exactly 1 Realstart} \cup \text{exactly 1 Goquick}) = \frac{2496}{5040}.$$

□

Additional Applications of Tree Diagrams

The tree diagram can be used for problems solved earlier in the chapter by less definitive techniques. In general, we may let the initial branches represent the first draw and the following branches represent the conditional events associated with successive draws. Let's take Example 1, page 113, where prior to the first draw the set contains ten transistors, four bad and six good. The drawing of a bad transistor comprises one branch and the drawing of a good transistor constitutes the second branch from the beginning node. The act of drawing one transistor and the possible outcomes can thus be represented as follows:

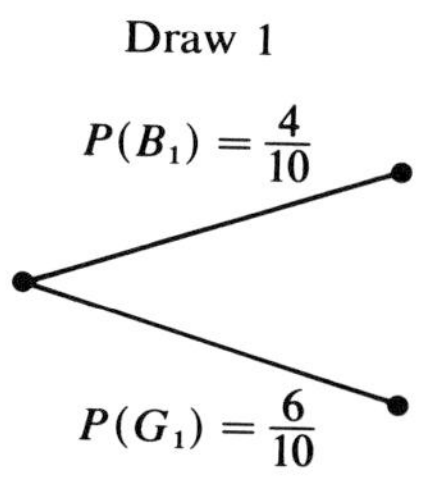

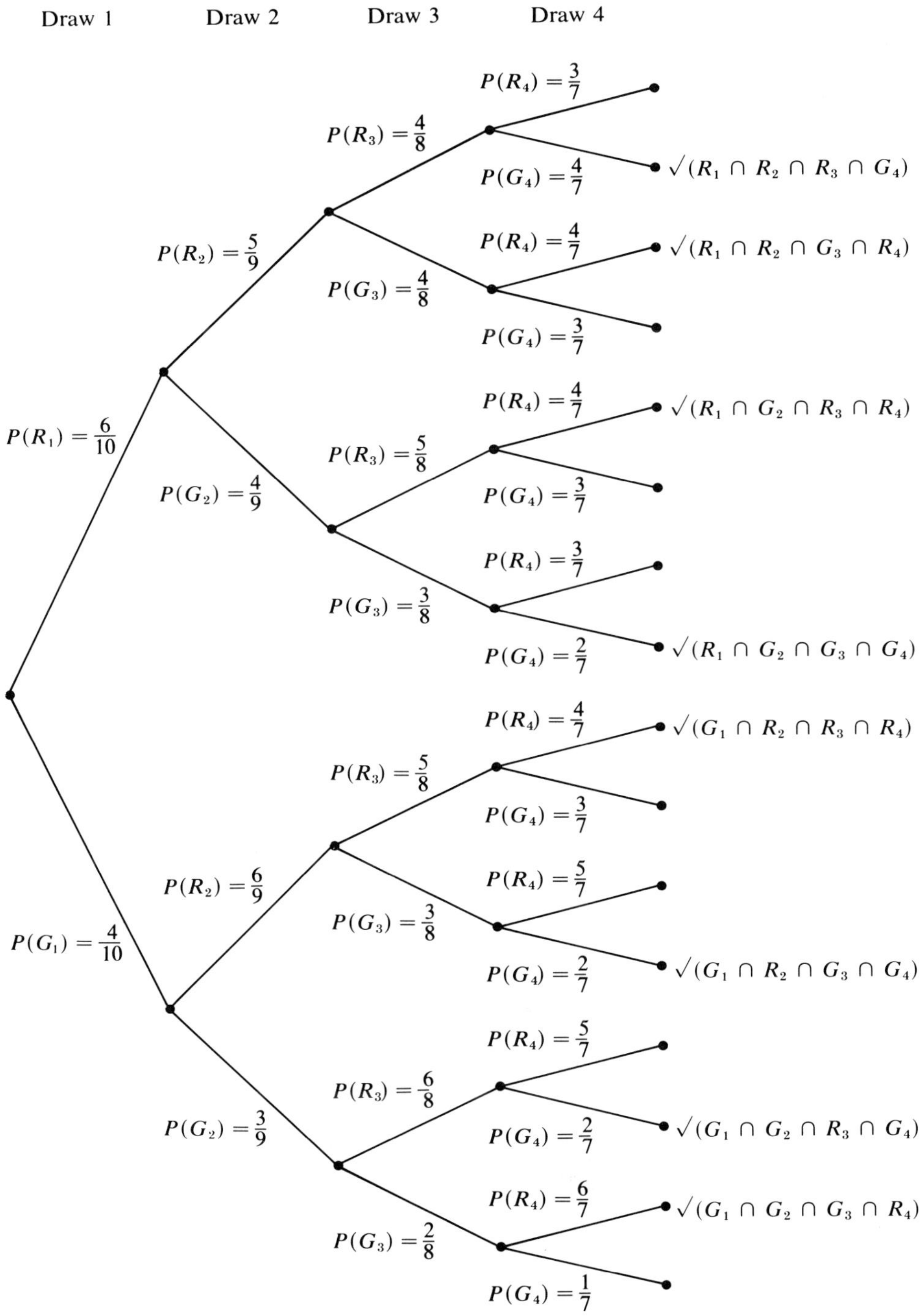

$P(\text{exactly 1 Realstart} \cup \text{exactly 1 Goquick}) = \frac{2496}{5040}$

Figure 4.17
Tree diagram of Example 3

After the first transistor has been drawn, the set contains nine transistors with the remaining distribution *dependent* upon the first draw. The second event is the drawing of the second transistor from the nine remaining. Using the knowledge from the first draw, the concluding branches may be constructed as:

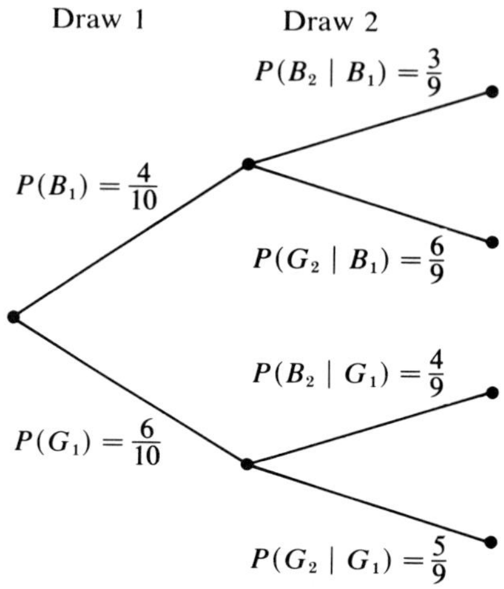

The joint probabilities we seek are the products of the probabilities associated with each particular branch of the decision tree as follows.

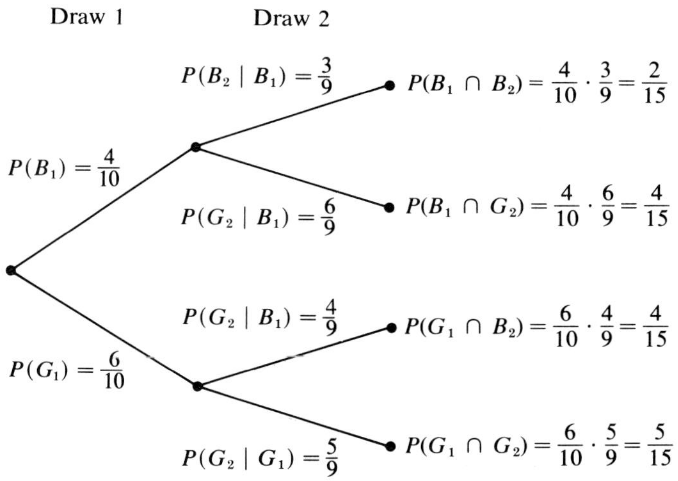

Problems

Use the tree diagram to solve each of the following problems.

1. Problem 8, Sec. 4.7.
2. Problem 10, Sec. 4.7.
3. Problem 2, Sec. 4.8.
4. Problem 4, Sec. 4.8.
5. Problem 5, Sec. 4.8.

4.10 Bayes' Theorem

An entire subset of statistics is labeled decision theory. In recent years, this facet of the discipline has been greatly influenced by what is called the Bayesian approach. Although we don't discuss decision theory and the Bayesian approach until Chapter 16, we lay the groundwork for it here. The following example introduces the central idea underlying Bayes' theorem.

Example 1

Suppose a set contains a number of parcels, each of which can be described by two attributes: source and destination. The possible sources are Rochester (R) and Boston (B), and the destinations are Albany (A) and New York (N). The parcels can be cross-classified into four categories with regard to source-destination combinations: $(R \cap A)$, $(R \cap N)$, $(B \cap A)$, and $(B \cap N)$. Suppose we do not know the total number of parcels. However, the relative numbers of parcels in each classification are as given in Table 4.5. Note that the classifications $(R \cap A)$, $(R \cap N)$, $(B \cap A)$,

Table 4.5
Probabilities of events of Example 1

	Source		
Destination	*Rochester* (R)	*Boston* (B)	*Destination*
Albany (A)	$P(R \cap A) = 2/10$	$P(B \cap A) = 4/10$	$P(A) = 6/10$
New York (N)	$P(R \cap N) = 1/10$	$P(B \cap N) = 3/10$	$P(N) = 4/10$
Source Totals	$P(R) = 3/10$	$P(B) = 7/10$	1.00

and $(B \cap N)$ may be viewed as joint events (e.g., $(R \cap A)$ represents the joint event of a parcel which has Rochester as its source and Albany as its destination). As such, they are mutually exclusive joint events. If we are concerned with source only, there are two mutually exclusive marginal classifications (events): R and B; similarly, with regard to destination there are two mutually exclusive marginal classifications (events): A and N.

Suppose we are to draw one parcel randomly from the set and we know certain conditional probabilities to be

$$P(R \mid A) = 1/3; \quad P(R \mid N) = 1/4; \quad P(B \mid A) = 2/3; \quad P(B \mid N) = 3/4. \tag{1}$$

Actually, these conditional probabilities can be obtained directly by applying the rule for calculating conditional probabilities (Rule 5, Sec. 4.6). Likewise, so can the conditional probabilities $P(A|R)$, $P(A|B)$, $P(N|R)$ and $P(N|B)$. However, for our purposes, we assume that the second set is not directly obtainable; this is precisely the kind of situation to which Bayes' theorem is applicable. Therefore, suppose we would like to be able, given only the source, to determine the probability that the parcel drawn will have a certain destination. How do we determine, for example, $P(A|R)$—that is, the probability that a certain parcel drawn is going to Albany given that it comes from Rochester. To solve for $P(A \cap R)$, we utilize the proper known information of (1) and Table 4.5 along with the appropriate operating laws of probability. (The situation is quite similar to the problem of the example associated with Relations (6)–(10) of Sec. 4.7.) Using Rule 4.7 for obtaining conditional probabilities, we have

$$P(A|R) = \frac{P(A \cap R)}{P(R)}. \tag{2}$$

However, since the denominator may be expressed as the sum of the probabilities of the two joint events $(A \cap R)$ and $(N \cap R)$, Formula (2) becomes

$$P(A|R) = \frac{P(A \cap R)}{P(A \cap R) + P(N \cap R)}. \tag{3}$$

Then, we simply use the marginal and conditional probabilities which we know and apply the multiplication rule (4.8) to all three terms of (3) in order to obtain as a final solution

$$P(A|R) = \frac{P(A)P(R|A)}{P(A)P(R|A) + P(N)P(R|N)} = \frac{(6/10)\cdot(1/3)}{(6/10)\cdot(1/3) + (4/10)\cdot(1/4)} = \frac{0.2}{0.3} = \frac{2}{3}. \tag{4}$$

Be certain that you can apply the reasoning of (2) through (4), above, in order to obtain other conditional probabilities as follows. The probability that a certain draw will yield a parcel that is going to New York given that it comes from Rochester is

$$P(N|R) = \frac{P(N)P(R|N)}{P(N)P(R|N) + P(A)P(R|A)} = \frac{(4/10)\cdot(1/4)}{(4/10)\cdot(1/4) + (6/10)\cdot(1/3)} = \frac{0.1}{0.3} = \frac{1}{3}. \tag{5}$$

The probability that a certain draw will yield a parcel that is going to Albany given that it comes from Boston is

$$P(A|B) = \frac{P(A)P(B|A)}{P(A)P(B|A) + P(N)P(B|N)} = \frac{(6/10)\cdot(2/3)}{(6/10)\cdot(2/3) + (4/10)\cdot(3/4)} = \frac{0.4}{0.7} = \frac{4}{7}. \tag{6}$$

Finally the probability that a certain parcel is going to New York given that it comes from Boston is

$$P(N|B) = \frac{P(N)P(B|N)}{P(N)P(B|N) + P(A)P(B|A)} = \frac{(4/10)\cdot(3/4)}{(4/10)\cdot(3/4) + (6/10)\cdot(2/3)} = \frac{0.3}{0.7} = \frac{3}{7}. \tag{7}$$

Note that the conditional probabilities of (4) and (5) total to unity (one). This is exactly what we should expect; for, given that the parcel is from Rochester, only two outcomes are possible: $A|R$ and $N|R$. Their probabilities must necessarily sum to one, since, under the information given, they are mutually exclusive and exhaustive. The same is true for the probabilities of (6) and (7).

Whether or not you realize it at this point, we have obtained the probabilities of (4) through (7) through a special application of the general theorem known as Bayes' theorem. Let us investigate the intuitive relationships manifested in (4) through (7) by viewing the probability distribution (with regard to distinction) associated with a random sample of size 1. Suppose we have selected one parcel. Given only the prior information that the makeup of the set is as shown in Table 4.5, we assess the probabilities of the event A (Albany) and the event N (New York) as 6/10 and 4/10, respectively. These are termed *prior probabilities.* However, suppose we decide to collect information on an accessible indicator which may possibly cause us to change our thinking about the original probability distribution of destination of the parcel drawn. Specifically, suppose we find the source of the parcel. Does this change our thinking? By all means! If the source (indicator) is Rochester, we now attach probabilities of 2/3 and 1/3 to the respective events A and N, because

$$P(A \mid R) = 2/3 \quad \text{and} \quad P(N \mid R) = 1/3.$$

On the other hand, if the source is Boston, we now attach probabilities of 4/7 and 3/7, respectively, to events A and N, because

$$P(A \mid B) = 4/7 \quad \text{and} \quad P(N \mid B) = 3/7.$$

We call such sets *posterior probabilities.* (The terms prior and posterior are related to the indicator in a given situation. Prior probabilities are based on information known before we have information on the indicator, and posterior probabilities are based on information that comes to us after information on the indicator.)

Although we have rather arbitrarily (even artificially) labeled one of the mutually exclusive sets of events as a set of states of nature (destination A or N) and the other as a set of possible outcomes of an indicator (source R or B), it is interesting to note that we can generalize even at this point. What we have done is change the probability distribution associated with a mutually exclusive set of states of nature by viewing a single outcome of an indicator. Since our particular indicator has only two possible outcomes, we can show the calculations for making the change simply by tabulation. Table 4.6 shows the outcome of the indicator for the event R and Table 4.7 the outcome

Table 4.6

Changing prior probabilities to posterior probabilities—source Rochester

Let indicator I_1 represent the outcome "parcel from Rochester"

(1) Events (states of nature) E_i	(2) Prior probabilities $P(E_i)$	(3) Joint probabilities $P(E_i \cap I_1) = P(E_i)P(I_1 \mid E_i)$	(4) Posterior probabilities $P(E_i \cap I_1)/\sum_i P(E_i \cap I_1)$
E_1 (destination Albany)	6/10	$\frac{6}{10} \cdot \frac{1}{3} = 0.2$	$\frac{0.2}{0.3} = \frac{2}{3}$
E_2 (destination New York)	4/10	$\frac{4}{10} \cdot \frac{1}{4} = 0.1$	$\frac{0.1}{0.3} = \frac{1}{3}$
		$\sum_i P(E_i \cap I_1) = 0.3$	1.0

Table 4.7

Changing prior probabilities to posterior probabilities—source Boston

Let indicator I_2 represent the outcome "parcel from Boston"

(1) Events (states of nature) E_i	(2) Prior probabilities $P(E_i)$	(3) Joint probabilities $P(E_i \cap I_2) = P(E_i)P(I_2\|E_i)$	(4) Posterior probabilities $P(E_i \cap I_2)/\sum_i P(E_i \cap I_2)$
E_1 (destination Albany)	6/10	$\frac{6}{10}\cdot\frac{2}{3} = 0.4$	$\frac{0.4}{0.7} = \frac{4}{7}$
E_2 (destination New York)	4/10	$\frac{4}{10}\cdot\frac{3}{4} = 0.3$	$\frac{0.3}{0.7} = \frac{3}{7}$
		$\sum_i P(E_i \cap I_2) = 0.7$	1.0

of the indicator for the event B. Each table provides the following information:

Column 1 describes the events of interest,

Column 2 presents probabilities of the possible events when no information is given regarding the indicator (i.e., prior probabilities),

Column 3 provides the probabilities of all joint events involving one specific outcome of the indicator (for example, column 3 of Table 4.7 gives the probabilities of all joint events involving the outcome of a parcel having Boston as its source).

Column 4 presents probabilities that should be attached to the events of interest in light of the relevant outcome of the indicator (i.e., posterior probabilities). Note, thus, that each row of calculations in both tables represents a specific application of Bayes' theorem. For example, the calculations of row 1 of Table 4.6 are identical to those of Equation (4).

Let us now proceed with a general statement of Bayes' Theorem.

Theorem 4.1 (Bayes' theorem)

Let $E_1, E_2, \ldots, E_n$ be a set of n mutually exclusive and exhaustive events and let $I_1, I_2, \ldots, I_m$ be another set of m mutually exclusive and exhaustive events (often referred to as outcomes of an indicator). Then, for any particular events E_r (the rth event) and I_t (the tth outcome of the indicator), we have

$$P(E_r|I_t) = \frac{P(E_r)P(I_t|E_r)}{\sum_{i=1}^{n} P(E_i)P(I_t|E_i)}.$$

We can see very easily that Bayes' theorem holds by viewing a specific event and a specific outcome of the indicator. Suppose we want to obtain the conditional probability of event E_4 given the occurrence of the outcome I_5 of the indicator, that is,

$P(E_4 | I_5)$. The conditional-probability formula (Rule 4.7) gives us

$$P(E_4 | I_5) = \frac{P(E_4 \cap I_5)}{P(I_5)}. \tag{8}$$

Applying the multiplication rule for probabilities, we find that the numerator can be written as

$$P(E_4 \cap I_5) = P(E_4)P(I_5 | E_4). \tag{9}$$

Further, the denominator of (8) can be written as the sum of all possible mutually exclusive joint events involving I_5; that is,

$$P(I_5) = P(E_1 \cap I_5) + P(E_2 \cap I_5) + \cdots + P(E_n \cap I_5). \tag{10}$$

Applying the multiplication rule for probabilities to each of these terms, we find

$$P(I_5) = P(E_1)P(I_5 | E_1) + P(E_2)P(I_5 | E_2) + \cdots + P(E_n)P(I_5 | E_n) \tag{11}$$

$$= \sum_{i=1}^{n} P(E_i)P(I_5 | E_i).$$

Substituting (9) and (11) into (8), we obtain

$$P(E_4 | I_5) = \frac{P(E_4)P(I_5 | E_4)}{\sum_{i=1}^{n} P(E_i)P(I_5 | E_i)}. \tag{12}$$

Note that this is a special case of the general statement of Bayes' theorem.

A More Practical Example

As emphasized before, we could have obtained all the conditional probabilities for Example 1 by direct application of the rule for obtaining conditional probabilities. Thus, our aim has been merely to exemplify the application of Bayes' theorem and verify that it yields results consistent with the usual operating laws of probability. Let us now examine an example wherein Bayes' theorem represents a more practical approach to certain probabilities than does the straightforward application of the rule for obtaining conditional probabilities.

Example 2

Each of eight delivery trucks for an automobile dealer holds ten cars, some with red (R) interiors and some with white (W) interiors. One (truck type A) has eight whites; three (truck type B) have six whites; and four (truck type C) have four whites. A truck is selected at random, and one car is selected at random from the truck. The interior is white. Viewing the color of the interior as an indicator, what do the posterior probabilities for the three types of trucks (A, B, and C) become as a result of the particular sample result obtained?

Solution

The three mutually exclusive events of interest are

1. the car came from truck type A (event A);
2. the car came from truck type B (event B); and
3. the car came from truck type C (event C).

The relevant outcome of the indicator is the obtaining of a white interior. Let us denote this outcome by W. Since the truck has been selected randomly, we assume each individual truck is equally likely to have been selected. Thus, prior to the time information is made available concerning the indicator, we assess the prior probabilities of the events A, B, and C as

$$P(A) = 1/8, \qquad P(B) = 3/8, \qquad P(C) = 4/8.$$

We wish to determine the posterior probabilities of these events in light of the specific outcome of the indicator: $P(A \mid W)$, $P(B \mid W)$, and $P(C \mid W)$. Straightforward application of the rule for conditional probabilities gives us, for example,

$$P(A \mid W) = \frac{P(A \cap W)}{P(W)}. \tag{13}$$

However, because of the nature of the problem, neither the numerator nor the denominator can be obtained simply by counting. Thus, we use Bayes' theorem to obtain

$$P(A \mid W) = \frac{P(A \cap W)}{P(W)} = \frac{P(A)P(W \mid A)}{P(A)P(W \mid A) + P(B)P(W \mid B) + P(C)P(W \mid C)}. \tag{14}$$

Knowing the makeup of each type of truck, we can obtain each of the probabilities of the right-hand side of (14) simply by counting. We then have

$$P(A \mid W) = \frac{(1/8) \cdot (4/5)}{(1/8) \cdot (4/5) + (3/8) \cdot (3/5) + (4/8) \cdot (2/5)} = \frac{4/40}{21/40} = \frac{4}{21}. \tag{15}$$

You might verify that the other posterior probabilities become

$$\begin{aligned} P(B \mid W) &= \frac{P(B)P(W \mid B)}{P(A)P(W \mid A) + P(B)P(W \mid B) + P(C)P(W \mid C)} \\ &= \frac{(3/8) \cdot (3/5)}{(1/8) \cdot (4/5) + (3/8) \cdot (3/5) + (4/8) \cdot (2/5)} = \frac{9/40}{21/40} = \frac{9}{21} \end{aligned} \tag{16}$$

and

$$\begin{aligned} P(C \mid W) &= \frac{P(C)P(W \mid C)}{P(A)P(W \mid A) + P(B)P(W \mid B) + P(C)P(W \mid C)} \\ &= \frac{(4/8) \cdot (2/5)}{(1/8) \cdot (4/5) + (3/8) \cdot (3/5) + (4/8) \cdot (2/5)} = \frac{8/40}{21/40} = \frac{8}{21}. \end{aligned} \tag{17}$$

Table 4.8

Calculations for changing prior probabilities to posterior probabilities—white interior

Let indicator I_1 represent the outcome "white interior"

Events (states of nature) E_i	Prior probabilities $P(E_i)$	Indicator probabilities $P(I_1 \mid E_i)$	Joint probabilities $P(E_i \cap I_1) = P(E_i)P(I_1 \mid E_i)$	Posterior probabilities $P(E_i \cap I_1)/\sum_i P(E_i \cap I_1)$
E_1 (truck type *A*)	1/8	4/5	$\frac{1}{8} \cdot \frac{4}{5} = \frac{4}{40}$	$\frac{4/40}{21/40} = \frac{4}{21}$
E_2 (truck type *B*)	3/8	3/5	$\frac{3}{8} \cdot \frac{3}{5} = \frac{9}{40}$	$\frac{9/40}{21/40} = \frac{9}{21}$
E_3 (truck type *C*)	4/8	2/5	$\frac{4}{8} \cdot \frac{2}{5} = \frac{8}{40}$	$\frac{8/40}{21/40} = \frac{8}{21}$
			$\sum_i P(E_i \cap I_1) = \frac{21}{40}$	1

You should also verify that the general tabular approach to calculating posterior probabilities, as exemplified by Table 4.8, gives results identical to (15)–(17).

Suppose the selected car has a red interior. The posterior probabilities resulting from this particular outcome of the indicator (color of interior) are as shown in Table 4.9. To make sure that you are familiar with both the application of Bayes' theorem and its relation to the tabular approach, you should apply Bayes' theorem to obtain results similar to (15) through (17) for the case in which the selected interior is red.

Table 4.9

Calculations for changing prior probabilities to posterior probabilities—red interior

Let indicator I_2 represent the outcome "red interior"

Events (states of nature) E_i	Prior probabilities $P(E_i)$	Indicator probabilities $P(I_2 \mid E_i)$	Joint probabilities $P(E_i \cap I_2) =$	Posterior probabilities $P(E_i \cap I_2)/\sum_i P(E_i \cap I_2)$
E_1 (truck type *A*)	1/8	1/5	$\frac{1}{8} \cdot \frac{1}{5} = \frac{1}{40}$	$\frac{1/40}{19/40} = \frac{1}{19}$
E_2 (truck type *B*)	3/8	2/5	$\frac{3}{8} \cdot \frac{2}{5} = \frac{6}{40}$	$\frac{6/40}{19/40} = \frac{6}{19}$
E_3 (truck type *C*)	4/8	3/5	$\frac{4}{8} \cdot \frac{3}{5} = \frac{12}{40}$	$\frac{12/40}{19/40} = \frac{12}{19}$
			$\sum_i P(E_i \cap I_2) = \frac{19}{40}$	1

□

Problems

1. A survey taken to determine the popularity of a certain political proposal among three economic segments of the population (low, middle, and upper income) yielded the results in the following table.

	Income			
	Low (L)	*Middle* (M)	*Upper* (U)	
For (F)	$P(F \cap L)$ 0.15	$P(F \cap M)$ 0.30	$P(F \cap U)$ 0.03	$P(F)$ 0.48
Against (A)	$P(A \cap L)$ 0.10	$P(A \cap M)$ 0.30	$P(A \cap U)$ 0.12	$P(A)$ 0.52
	0.25	0.60	0.15	1.00

   a. By use of Rule 4.7 for conditional probabilities, calculate:

$$P(L|F),\quad P(M|F),\quad P(U|F),\quad P(L|A),\quad P(M|A),\quad P(U|A).$$

   b. Using Bayes' theorem, calculate (show your work):

$$P(F|L),\quad P(F|M),\quad P(F|U),\quad P(A|L),\quad P(A|M),\quad P(A|U).$$

2. Three machines (X, Y, and Z) produce 30%, 45%, and 25%, respectively, of a factory's total output. These machines are known to produce 2%, 3%, and 4% defective items, respectively. An item selected at random is found to be defective. Find the probability that

   a. the defective item was produced by machine X,

   b. the defective item was produced by machine Y,

   c. the defective item was produced by machine Z.

3. Two stamping machines (A and B) produce defective parts, on the average, 2 times in 1000, and 15 times in 10,000 stampings, respectively. If you were to choose one of the machines at random, and obtained a nondefective part, what is the posterior probability that you chose machine A?

4. Each of 5 boxes contains 5 gears. Two (Type A) have 2 large gears and 3 small gears and the other three (Type B) have 4 large gears and 1 small gear. A box is chosen at random and two gears are drawn with replacement. Find the probability a Type-B box was selected if

   a. 2 large gears are observed,

   b. 1 large and 1 small gear are observed,

   c. 2 small gears are observed.

5. Three identical boxes contain 2 valves. One box contains 2 brass valves, another contains 2 aluminum valves, and the third a brass valve and an aluminum valve. A box is chosen at random and a valve is withdrawn. If the valve selected is aluminum, what is the probability that the remaining valve is also aluminum?

6. A credit department has found that 85 percent of those applying for credit are good credit risks. Sixty percent of the good credit risks are married, as are 30 percent of the bad credit risks.

   a. What is the probability that a married person is a poor credit risk?

   b. What is the probability that a single person is a good credit risk?

4.11 Probability Distributions

Recall from Chapter 2 that frequency distributions consist of the possible values of a given variable and the corresponding frequency for each value (or group of values). *Probability distributions* are similar, in that they consist of the possible values of a random variable and the corresponding probability for each value (or group of values). Actually, the relative frequency distribution of a variable is a probability distribution if the total frequency corresponds to the population size and the units are selected by simple random sampling.

Example 1

Consider the experiment of tossing a fair die. Let X be the random variable which has for its possible values the number of spots showing on the top side of the die. The probability distribution for the random variable X is shown on the right. Using this distribution and appropriate postulates, we can determine the probability of any event concerning the variable X. For example,

x	$P(X = x)$
1	1/6
2	1/6
3	1/6
4	1/6
5	1/6
6	1/6
	6/6 = 1.0

$$P(X \text{ is odd}) = P(X = 1 \text{ or } X = 3 \text{ or } X = 5) = \frac{1}{6} + \frac{1}{6} + \frac{1}{6} = \frac{1}{2},$$

$$P(X \text{ is even}) = P(X = 2 \text{ or } X = 4 \text{ or } X = 6) = \frac{1}{6} + \frac{1}{6} + \frac{1}{6} = \frac{1}{2},$$

$$P(X \geq 3) = P(X = 3 \text{ or } X = 4 \text{ or } X = 5 \text{ or } X = 6) = \frac{1}{6} + \frac{1}{6} + \frac{1}{6} + \frac{1}{6} = \frac{2}{3},$$

$$P(2 \leq X \leq 4) = P(X = 2 \text{ or } X = 3 \text{ or } X = 4) = \frac{1}{6} + \frac{1}{6} + \frac{1}{6} = \frac{1}{2},$$

$$P(X < 5) = P(X = 1 \text{ or } X = 2 \text{ or } X = 3 \text{ or } X = 4) = \frac{1}{6} + \frac{1}{6} + \frac{1}{6} + \frac{1}{6} = \frac{2}{3}.$$

Example 2

Let X be the number of typographical errors per page for a group of magazines. A probability distribution for X is shown on the right.

x	$P(X = x)$
0	0.26
1	0.04
2	0.30
3	0.10
4	0.14
5	0.16
	1.00

We can now use this distribution to make a probability statement about any random event from the population of events.

Example 3

Suppose we are sampling from a bin containing six good (G) capacitors and four bad (B) capacitors. If we select a sample of size 1, we have the probability distribution shown on the right, where X represents an arbitrary classification of the quality of the capacitor. However, if we select a sample of size 2 without replacement, the probability distribution becomes that given on the left below, where X now represents an arbitrary classification of the combined outcome of the two draws. For a sample of size 2 with replacement, the probability distribution is shown on the right.

Random event	x	$P(X = x)$
B	0	4/10
G	1	6/10
		10/10 = 1.0

Random event	x	$P(X = x)$
$B_1 \cap B_2$	0	2/15
$B_1 \cap G_2$	1	4/15
$G_1 \cap B_2$	2	4/15
$G_1 \cap G_2$	3	5/15
		15/15 = 1.0

Random event	x	$P(X = x)$
$B_1 \cap B_2$	0	4/25
$B_1 \cap G_2$	1	6/25
$G_1 \cap B_2$	2	6/25
$G_1 \cap G_2$	3	9/25
		25/25 = 1.0

Note that in the last two tables of the example we obtain different probability distributions for what appear on the surface to be identical situations. However, the two cases differ in the technique used for selecting the sample. Thus, as we cautioned earlier, the probabilities associated with various events necessarily depend on the experiment being conducted.

In the foregoing examples, we spoke of the tabular "association of the possible values of a random variable X with corresponding probabilities" as a probability distribution. We can often make such an association more conveniently by using a mathematical function. In such case, we represent a probability distribution notationally by an entity termed a *probability function* (in the discrete case) or a *probability density function* (in the continuous case). Consider an example. We have represented the probability distribution by a table in Example 1. However, we can also represent the distribution by a function, namely,

$$f(x) = 1/6 \quad (x = 1, 2, 3, 4, 5, 6), \tag{1}$$

where $f(x)$ is the probability function of the discrete random variable X and x represents any possible value that X may assume. Whenever we describe a probability function (or a probability density function) mathematically, we always include, as we have done in the right side of (1), the possible values that X may assume. In (1), the material $(x = 1, 2, 3, 4, 5, 6)$ to the right of the description of the probability function tells us that we are considering a discrete random variable, since it can assume only one of six possible values.

As another example, we can use a probability function to answer the question "What is the probability that x heads will appear if three coins are tossed?" Representing the answer in functional notation, we have

$$f(x) = \frac{\binom{3}{x}}{2^3} \quad (x = 0, 1, 2, 3). \tag{2}$$

However, we have the same results if we represent the probability distribution in tabular form, shown on the right.

x	$P(X = x)$
0	1/8
1	3/8
2	3/8
3	1/8
	8/8 = 1.0

We note that in each of the above cases, $f(x)$ corresponds to $P(X = x)$ for each possible value x. Furthermore, we note that the events represented by the possible values of X are mutually exclusive and exhaustive events. Hence, the properties of a probability function for a discrete random variable X are as follows:

1. The function is nowhere negative; that is, $f(x) \geq 0$ for any possible value of X.
2. The function $f(x)$ represents the probability that X assumes the possible value x.
3. $\sum f(x) = 1$, where the summation is over all possible values of X.

It should be noted that a function of a discrete random variable is a probability function if and only if the function satisfies all three of these requirements.

Note also that the variable X may actually describe the attribute of interest, as for the die-tossing in Example 1, or it may be simply a code as in Examples 2 and 3.

Another probability-associated concept is the *cumulative probability distribution function.* It is often necessary to evaluate probabilities such as $P(X < 8)$ and $P(3 < X \leq 7)$. Here, as well as in many other situations, it is convenient to define a new function known as the cumulative probability distribution function, or, more simply, the *cumulative distribution.* For a discrete probability function $f(x_i)$, $i = 1, 2, \ldots,$ the cumulative distribution $F(x)$ is defined by

$$F(x) = \sum f(x_i), \tag{3}$$

where the summation is over those values of i such that $x_i \leq x$.

We note that

$$F(x) = P(X \leq x),$$

where x is any possible value which X may assume;

$$P(a < X \leq b) = F(b) - F(a),$$

where a and b are any two possible values of X such that $a < b$; and

$$F(x) = 1,$$

where the appropriate summation is over all possible values of X.

We can see that the cumulative distribution and the probability distribution play roles analogous to the cumulative frequency distribution and frequency distribution, respectively.

Example 4

Suppose 50 employees are equally distributed in job grades 1 through 5. Using simple random sampling, find: (a) the probability distribution in both tabular and functional form, (b) the corresponding cumulative probability distribution, (c) $P(X \leq 4)$, and (d) $P(X > 3)$.

Solution

(a) The probability distribution in tabular form is shown on the right. In functional form, we have

$$f(x) = 1/5 \qquad (x = 1, 2, 3, 4, 5).$$

x	$P(X = x)$
1	0.2
2	0.2
3	0.2
4	0.2
5	0.2
	1.0

(b) The corresponding cumulative distribution is

x	$F(x)$
1	0.2
2	0.4
3	0.6
4	0.8
5	1.0

(c) $P(X \leq 4) = F(4) = 0.8.$

(d) $P(X > 3) = P(3 < X \leq 5) = F(5) - F(3) = 1.0 - 0.6 = 0.4.$ □

Problems

1. A pair of fair dice is rolled.
   a. List all possible outcomes and the probability of each.
   b. What is the probability of the outcome being less than five?
   c. What is the probability of the outcome being even?
   d. What is the probability of the outcome being both even and less than five?
   e. What is the probability of the outcome being even or less than five?
2. Four fair coins are tossed at one time.
   a. List all possible outcomes and the probability of each.
   b. What is the probability of the outcome being two or more heads?
   c. What is the probability of the outcome being an even number of heads? (For this problem, consider 0 as an even number.)
   d. What is the probability of the outcome being both an even number of heads and two or more heads?
   e. What is the probability of the outcome being an even number of heads or two or more heads?

3. Two boxes each contain five items numbered one through five. An item is to be drawn at random from each box and the sum of the numbers on the two items is the variable X.

   a. List each of the possible values of X and the probability associated with each value of X.

   b. What is $P(X \geq 6)$?

   c. What is $P(X \text{ is odd})$?

   d. What is $P(X \geq 6 \text{ and } X \text{ is odd})$?

   e. What is $P(X \geq 6 \text{ or } X \text{ is odd})$?

4. If a box contains 7 good and 6 defective parts and a simple random sample of size 2 is to be drawn, construct the probability distribution for the variable X, where X represents

   a. the number of good parts obtained and sampling is with replacement,

   b. the number of good parts obtained and sampling is without replacement,

   c. the number of defective parts obtained and sampling is with replacement,

   d. the number of defective parts obtained and sampling is without replacement.

5. The management team of a large manufacturing company consists of 6 production managers, 4 sales managers, and 2 quality control managers and a simple random sample of size 3 is to be drawn. Construct a probability distribution for the variable X, where X represents

   a. the number of production managers obtained and sampling is with replacement,

   b. the number of production managers obtained and sampling is without replacement,

   c. the number of sales managers obtained and sampling is with replacement,

   d. the number of sales managers obtained and sampling is without replacement,

   e. the number of quality control managers obtained and sampling is with replacement,

   f. the number of quality control managers obtained and sampling is without replacement,

   g. the number of managers who are in either sales or quality control and sampling is with replacement,

   h. the number of managers who are in either sales or quality control and sampling is without replacement.

6. Construct the probability function of the variable X which represents the total number of spots showing on the top surfaces for the experiment of tossing 3 fair dice.

7. What are the three properties of the probability function of a discrete random variable? Verify that the function obtained in Problem 6 is in fact a probability function.

8. Construct a cumulative distribution for a and b of Problem 5.

9. Construct a cumulative distribution for Problem 6 and find

   a. $P(X \geq 12)$, b. $P(X \leq 8)$.

4.12 Mathematical Expectation

In later chapters we shall use a statistical concept known as the mathematical expectation of a random variable X. It is simply a special weighted average of the possible values of X, the proper weights being the probabilities corresponding to the possible values (or groups of values) of X.

Definition 4.21

If $x_1, x_2, \ldots, x_n$ are the n possible values of the discrete random variable X, the ***mathematical expectation*** *of X (or the mean of the probability distribution of X) is*

$$E(X) = x_1 \cdot P(X = x_1) + x_2 \cdot P(X = x_2) + \cdots + x_n \cdot P(X = x_n)$$
$$= \sum_{i=1}^{n} x_i P(X = x_i),$$

where $P(X = x)$ is the probability that the random variable X assumes the value x.

The mathematical expectation is often referred to as an average or "expected" value. Note that, since $f(x_i) = P(X = x_i)$, we can also write $E(X) = \sum_{i=1}^{n} x_i f(x_i)$.

Example 1

Suppose Mr. Jones' steak house specializes in hamburgers and three types of steaks. The price (including beverage, dessert, and all trimmings) of each of the four items on the menu and the probability that a randomly selected customer will buy each are given in the table. What is the expected value of customer expenditure?

Item	*x (Price)*	$P(X = x)$
Sirloin	\$4.50	0.30
T-Bone	4.25	0.25
Rib Eye	3.95	0.35
Hamburger	1.45	0.10
		1.00

Solution

$$E(X) = \$4.50(0.30) + \$4.25(0.25) + \$3.95(0.35) + \$1.45(0.10) = \$3.94. \quad \square$$

Example 2

Mr. Jones, of Example 1, wishing to know the average (or expected) profit per customer, has determined his profit on each of the four items. The random variable now becomes profit per item rather than total price per item. What is the expected profit per customer?

Item	*x (Profit)*	$P(X = x)$
Sirloin	\$0.45	0.30
T-Bone	0.43	0.25
Rib Eye	0.47	0.35
Hamburger	0.25	0.10
		1.00

Solution

$$E(X) = \$0.45(0.30) + \$0.43(0.25) + \$0.47(0.35) + \$0.25(0.10) = \$0.4320.$$

Thus, if Mr. Jones has 100 customers during an evening, he expects to make $100 \times \$0.4320 = \$43.20$ profit for that evening. Even with the same number of customers, his profit will vary from one evening to the next as customer orders vary. In other words, the expected profit is an average. Hence, the \$43.20 is what he expects as his average profit for every 100 customers over a long period of time. □

The mathematical expectation, or expected value, is of considerable importance in decision making. For example, Mr. Jones could use expected values in determining the effects of price increases, the total profit increases due to various price changes, and the effects on profits of increases in costs (decreases in profits).

The calculation of mathematical expectations is often simplified by use of the following rules, which follow immediately from Definition 4.21.

Rule 4.10

If c is a constant, $E(c) = c.$

Rule 4.11

If c is a constant and g(X) is a function of X, $E[cg(X)] = cE[g(X)].$

Rule 4.12

If $g_i(X)$ is any function of X for $i = 1, 2, \ldots, n$, $E\left[\sum_{i=1}^{n} g_i(X)\right] = \sum_{i=1}^{n} E[g_i(X)].$

From the above three rules, we can obtain:

Rule 4.13

If a and b are constants, $E(a + bX) = a + bE(X).$

Example 3

Brown Lumber Company sells four grades of 1/4-inch den paneling. Suppose that the probability that a panel-purchasing customer will select a specific grade and corresponding prices are as given in the table. Suppose further that the company which supplies Brown Lumber Company is increasing its price

Grade	*Price* (\$) *per panel* (x)	$P(X = x)$
A	5.00	0.2
B	4.50	0.4
C	4.00	0.3
D	3.50	0.1
		1.0

by 10 percent, and the city decides to collect a tax of \$0.25 for each panel sold. Brown plans to increase the price of paneling accordingly, that is, by $0.25 + 1.1X$.

(a) Find the expected (average) price per panel prior to the price increase.

(b) Use Rule 4.13 to find the expected (average) price per panel after the price increase.

(c) Use the individual new prices to find the expected (average) price per panel after the price increase.

Assume in each case that relevant probabilities will be unchanged by the price increases.

Solution

(a) We use Definition 4.21 to find the average price per panel prior to the price increase

$$E(X) = 5.00(0.2) + 4.50(0.4) + 4.00(0.3) + 3.50(0.1) = 4.35.$$

(b) Using the result of (a) and Rule 4.13, we have

$$E(0.25 + 1.1X) = 0.25 + 1.1E(X) = 0.25 + 1.1(4.35) = 5.035.$$

(c) We first determine each of the individual prices after the increase to be:

Grade	*Price (\$) per panel* (x)	$P(X = x)$
A	$0.25 + 1.1(5.00) = 5.75$	0.2
B	$0.25 + 1.1(4.50) = 5.20$	0.4
C	$0.25 + 1.1(4.00) = 4.65$	0.3
D	$0.25 + 1.1(3.50) = 4.10$	0.1
		1.0

Then,

$$E(X) = 5.75(0.2) + 5.20(0.4) + 4.65(0.3) + 4.10(0.1) = 5.035.$$

Note that the results of (b) and (c) are identical. □

In Definition 4.21 we defined the mean of the probability distribution of X (or simply the mean of X). Considering the rules given in this section and our concept of the variance from Chapter 3, you can probably anticipate the definition of the variance of a probability distribution.

Definition 4.22

The variance of the probability distribution of X is given by

$$\text{Var}(X) = E[X - E(X)]^2.$$

We can see from Definition 4.21 and 4.22 that the variance of the probability distribution of X (or simply the variance of X) may be expressed as

$$\text{Var}(X) = \sum_{i=1}^{n} [x_i - E(X)]^2 \cdot P(X = x_i) = \sum_{i=1}^{n} [x_i - E(X)]^2 f(x_i). \tag{1}$$

Recall from Chapter 3 that the standard deviation of X is the square root of the variance. We shall frequently use mathematical expectations in the remainder of the text. For example, we shall use it in Chapter 5 to determine the mean and the variance of a random variable X, where the probability distribution of X is a specific function. More generally, we shall use the concept extensively in the study of sampling distributions and decision theory.

Problems

1. Define *mathematical expectation.*
2. From the following probability distribution of the discrete variable Y, determine $E(Y)$.

Y	1	2	3	4	5	6
$P(Y)$	$\frac{1}{20}$	$\frac{4}{20}$	$\frac{9}{20}$	$\frac{3}{20}$	$\frac{2}{20}$	$\frac{1}{20}$

3. The present values and probabilities of each of the possible outcomes of introducing a new product are as follows:

Outcome	*Present value*	*Probability*
A	\$ 105,000	0.75
B	1,400,000	0.10
C	−780,000	0.15

Determine the mathematical expectation of the present value of introducing the product.

4. The present values and probabilities of purchasing a new plant are as follows:

Outcome	*Present value*	*Probability*
A	\$ 850,000	0.85
B	450,000	0.10
C	−1,800,000	0.05

The present values and probabilities of buying a chain of stores are as follows:

Outcome	*Present value*	*Probability*
D	\$750,000	0.75
E	200,000	0.25

If a company must choose between these alternatives on the basis of expected monetary returns, should it buy the plant or the chain of stores?

5. An urn contains 7 black, 4 red, 3 white, and 1 blue ball. A player is to draw one ball. If it is black, he wins \$1; if it is red, he wins \$2; if it is white, he wins \$3; and if it is blue, he pays \$25. What is the mathematical expectation of this game? (In other words, if we let X represent "the amount won," what is $E(X)$?)
6. Given $E(X) = \$45.00$, find

a. $E(\$5 + 1.5X)$, b. $E(\$10 + 0.5X)$.

7. Given the probability distribution on the right, find

a. $E(X)$, b. $E(-2 + 6X)$,

c. $\mathrm{Var}(X)$ by the relation

$$\mathrm{Var}(X) = \sum_{i=1}^{n} [x_i - E(X)]^2 \cdot P(X = x_i).$$

x	$P(X = x)$
10	0.11
11	0.00
12	0.35
13	0.44
14	0.08
15	0.02

4.13 Continuous Probability Distributions

In Section 4.11 we discussed probability distributions for discrete variables and listed the properties that a function must satisfy in order to be classed as a probability function. The case wherein the random variable of interest is continuous can be treated somewhat similarly. However, we shall see a very important distinction in associated interpretations.

In many applications of probability, the random variable of interest can take on any real value within an interval (or intervals). For example, suppose we are interested in the probability that a typical light bulb made by Westinghouse will function properly for a period in excess of 2,000 hours of continuous use. Let the random variable X represent the length of useful life of a light bulb made by Westinghouse. In such case, we seek $P(X > 2{,}000)$. Theoretically, the variable X can take on any positive real value and is thus continuous.

Before we can compute probabilities for a continuous random variable, we must consider the properties its probability density function must satisfy. In general, we say that the function $f(x)$ is a *probability density function* for the continuous random variable X if and only if the following are satisfied:

1. The function is nowhere negative, i.e., $f(x) \geq 0$ for all possible values of X.
2. The probability, $P(a \leq X \leq b)$, where a and b are any two possible values of X such that $a \leq b$, is determined by the area under the curve $f(x)$ between the values a and b.
3. The total area under the curve $f(x)$ is equal to unity (one).

Example 1

Suppose we are firing a rifle at a 1-foot length of string which is stretched horizontally in a plane perpendicular to our line of fire. Further, suppose the probability density function of the random variable X, which is the distance our bullet hits from the left end of the string, is given by

$$f(x) = \begin{cases} 1, & 0 \leq x \leq 1, \\ 0, & \text{otherwise.} \end{cases}$$

Find $P(1/2 \leq X \leq 2/3)$.

Solution

We are given that $f(x)$ is a probability density function. Hence, $P(1/2 \leq X \leq 2/3)$ may be determined by finding the area between $f(x)$ and the x-axis and bounded by 1/2 on the left and 2/3 on the right. The area of interest is displayed by the shaded region in Figure 4.18. Since the region of interest is a rectangle, we can determine the appropriate area by use of plane geometry. Thus,

$$P(1/2 \leq X \leq 2/3) = \text{base} \cdot \text{height} = (2/3 - 1/2) \cdot 1 = 1/6.$$

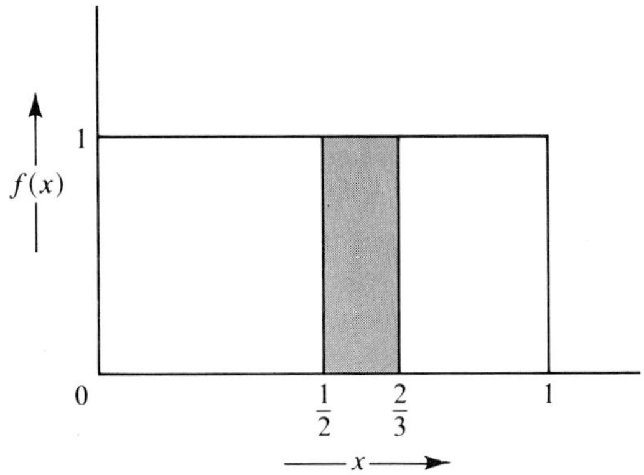

Figure 4.18
$P(1/2 \leq X \leq 2/3)$ □

The determination of $P(a \leq X \leq b)$ can be complicated in the case of continuous random variables. Finding the numerical value for such probability requires the use of integral calculus except for very simple functions such as that of Example 1. However, for several special probability density functions, the probabilities have been computed and recorded in tables, so that we will not be required to use integral calculus.

One further property of continuous random variables demands consideration. A probability density function, when evaluated at a given possible value of the continuous random variable X, does not provide the probability that the random variable assumes that particular value (as is the case for a discrete variable). The reasoning is as follows. In considering the properties of a probability density function for a continuous variable, we note that the probability $P(a \leq X \leq b)$ is determined by the area between the function and the x-axis and bounded by the values of a and b on the x-axis, where $a \leq b$. What is $P(X = c)$, where c is any possible value of X? To determine $P(X = c)$, we must find the area that lies between $f(x)$ and the x-axis and is bounded by the values c and c. Hence, this yields only a line which has no width and, therefore, zero area. Thus, if X is a continuous random variable, we have $P(X = c) = 0$, where c is any possible value of X. (Note that this is contrary to the results which were obtained for discrete random variables.) It does not mean that we cannot obtain a value of c for the variable, but, rather, that we have an extremely small chance. This does not appear unreasonable if we note that, for a continuous variable, we have an infinite (noncountable) number of possible values.

It is the foregoing difference that dictates separate names for the mathematical function that represents a discrete probability distribution (a probability function) and the one that represents a continuous probability distribution (a probability density function). We shall make such a distinction throughout all subsequent chapters.

Finally, the expected value and variance of a continuous distribution are defined in a manner parallel to that regarding corresponding definitions in Section 4.12 for the discrete case. However, since such definitions involve integral calculus, we do not include them.

Problems

1. What three conditions must be satisfied for a function to be the probability density function of a continuous random variable?

2. Given that the probability density function for the random variable X is

$$f(x) = \begin{cases} \frac{1}{2}, & 0 < x < 2, \\ 0, & \text{otherwise,} \end{cases}$$

determine

a. $P(X \leq 1)$, b. $P\left(\frac{1}{2} < X < \frac{3}{4}\right)$, c. $P\left(\frac{1}{2} \leq X < 2\right)$, d. $P(X \geq 1.5)$,

e. $P(X = 1)$, f. $P(X \geq 2.4)$, g. $P(X < 0)$,

h. whether the three conditions of Problem 1 are satisfied.

3. Given that the probability density function for the random variable X is

$$f(x) = \begin{cases} 2x, & 0 < x < 1, \\ 0, & \text{otherwise,} \end{cases}$$

determine

a. $P\left(X < \frac{1}{2}\right)$, b. $P\left(X \geq \frac{3}{4}\right)$, c. $P\left(\frac{1}{2} < X < \frac{3}{4}\right)$,

d. $P(X > 2)$, e. $P\left(X = \frac{1}{2}\right)$,

f. whether the three conditions of Problem 1 are satisfied.

4.14 Summary

The major objectives of this chapter have been accomplished if you now understand well the following formulas and related concepts:

1. *Permutations* $\quad {}_nP_r = \dfrac{n!}{(n-r)!}$.

2. *Combinations* $\quad {}_nC_r = \binom{n}{r} = \dfrac{n!}{r!(n-r)!}.$

3. *Permutations, k subgroups* $\quad {}_nP_{r_1, r_2, \ldots, r_k} = \dfrac{n!}{r_1!r_2!\ldots r_k!}.$

4. *Postulates of probability* — Let S be a sample space and let A represent any event in S. Then, if we represent the probability of event A by the function $P(A)$, we have that P is a probability function on the sample space S if the following three postulates are satisified.

   1a If A is any subset of S, $0 \le P(A) \le 1$.

   2a If we let S represent the entire sample space in an experiment, $P(S) = 1$.

   3a If A_i and A_j are any two disjoint subsets of S, then

$$P(A_i \cup A_j) = P(A_i) + P(A_j).$$

5. *Generalization of Postulate 3a* — If $A_1, A_2, \ldots, A_n$ constitute a sequence of disjoint subsets of S, then

$$P(A_1 \cup A_2 \cdots \cup A_n) = P(A_1) + P(A_2) + \cdots + P(A_n).$$

6. *Multiplication rule* $\quad P(A \cap B) = P(A)P(B\,|\,A) = P(B)P(A\,|\,B).$

7. *Addition rule* $\quad P(A \cup B) = P(A) + P(B) - P(A \cap B).$

8. *Conditional probability* $\quad P(A\,|\,B) = \dfrac{P(A \cap B)}{P(B)}$, where $P(B) \neq 0$.

9. *Statistical independence* — Events A and B are statistically independent if and only if any one of the following conditions is satisfied:

   (i) $P(A\,|\,B) = P(A)$.

   (ii) $P(B\,|\,A) = P(B)$.

   (iii) $P(A \cap B) = P(A) \cdot P(B)$.

10. *Mutually exclusive events* — If A and B are mutually exclusive events, $P(A \cap B) = 0$. However, $P(A \cap B) = 0$ does not imply that A and B are mutually exclusive.

11. *Probability of the complement of an event* $\quad P(A) = 1 - P(\bar{A}).$

12. *Bayes' theorem* $\quad P(E_r\,|\,I_t) = \dfrac{P(E_r)P(I_t\,|\,E_r)}{\sum_{i=1}^{n} P(E_i)P(I_t\,|\,E_i)}.$

13. *Mathematical expectation* $\quad E(X) = x_1 \cdot P(X = x_1) + x_2 \cdot P(X = x_2) + \cdots + x_n \cdot P(X = x_n).$

If c is constant,

$$E(c) = c.$$

If c is constant and $g(X)$ is a function of X,

$$E[cg(X)] = cE[g(X)].$$

If $g_i(X)$ is a function of X for each $i = 1, 2, \ldots, n$,

$$E\left[\sum_{i=1}^{n} g_i(X)\right] = \sum_{i=1}^{n} E[g_i(X)].$$

If a and b are constants,

$$E(a + bX) = a + bE(X).$$

14. *Variance of probability distribution* $\mathrm{Var}(X) = E[X - E(X)]^2 = \sum_{i=1}^{n} [x_i - E(X)]^2 \cdot P(X = x_i)$.

15. *Probability function* The function $f(x)$ is a probability function for the discrete random variable X if these three conditions are satisfied:
   (i) $f(x) \geq 0$ for all possible discrete values of x.
   (ii) $f(x) = P(X = x)$.
   (iii) $\sum f(x) = 1$, where the summation is over all possible values of X.

16. *Probability density function* The function $f(x)$ is a probability density function for the continuous random variable X if these three conditions are satisfied:
   (i) $f(x) \geq 0$ for every value x.
   (ii) $P(a \leq X \leq b)$ is the area under the curve $f(x)$ between the constant values a and b, where a and b are any two values of X such that $a \leq b$.
   (iii) The total area under the curve $f(x)$ is equal to unity (one).

Important Terms and Concepts

addition rule for probabilities
Bayes' theorem
branches
classical or *a priori* probability
combinations
complement of a set
complement of an event
compound event
conditional event
conditional probability
continuous probability distribution
cumulative probability distribution
cumulative probability distribution function
decision theory
decision tree
difference of sets
disjoint sets
exclusive union
exhaustive
expected value
frequency or *a posteriori* probability
indicator
intersection
joint event
joint probability
long-term regularity
marginal probability
mathematical expectation
multiplication rule for probabilities
mutually exclusive events
nodes
permutations
population
posterior probability
postulates of probability
prior probability
probability density function
probability distribution
probability distribution function
probability distribution mean
probability distribution variance
probability function
probability sample
random event

random experiment
random variable
sample
sample space
sampling with replacement
sampling without replacement
set
simple event
simple random sample
statistical independence
statistical inference
statistical probability
subjective probability
tree diagram
union

References

4.1 Frederick Mosteller, et al, *Probability with Statistical Applications*, 2d ed., Reading, Mass.: Addison-Wesley Publishing Company, Inc., 1970.

4.2 L. J. Savage, *Foundations of Statistics*, New York: John Wiley & Sons, Inc., 1954.

Review Problems

---

1. Explain with Venn diagrams the meaning of each of the following expressions.

   a. $P(A)$,
   b. $P(A \cup B)$,
   c. $P(\bar{B})$,
   d. $1 - P(A)$,
   e. $P(\bar{A} \cap B)$,
   f. $1 - P(\bar{B})$,
   g. $P[A \cap (\bar{B} \cup C)]$,
   h. $P[(\bar{A} \cap \bar{B}) \cup (B \cap C)]$,
   i. $P(A \cup \bar{B})$,
   j. $P[A \cup (B \cap C)]$,
   k. $P[(A \cup B) \cap (B \cup \bar{C})]$,
   l. $P(\bar{A} \cap \bar{B})$,
   m. $P([(A \cup \bar{B}) \cap (B \cap C)] \cup [\bar{A} \cup \bar{B}])$,
   n. $P(A \mid B)$,
   o. $P(B \mid \bar{A})$,
   p. $P[(A \cap C) \cup (B \mid \bar{A})]$,
   q. $P[B \mid (A \cup C)]$,
   r. $P([A \mid B] \cap [(C \mid B) \cup C])$.

2. A shipment contains 1 yellow, 3 red, and 6 blue sweaters. If we use simple random sampling *without* replacement and let the subscript indicate the number of the draw, find

   a. $P(R_2 \mid B_1)$,
   b. $P(B_2)$,
   c. $P(R_1 \mid B_2)$,
   d. $P(B_1 \mid R_2)$,
   e. $P(R_1 \cup B_2)$,
   f. $P(B_3)$,
   g. $P(B_1 \mid B_3)$,
   h. $P(B_1 \cup R_3)$.

3. Bin A contains 3 switches numbered 1 through 3, bin B contains 4 switches numbered 5 through 8, and bin C contains two switches numbered 2 and 4. One switch is to be drawn at random from each bin and X represents the sum of the numbers on the three switches.

a. List each possible value of X and the probability associated with each possible value.

b. What is $P(X < 12)$?

c. What is $P(X \text{ is even})$?

d. What is $P(X < 12 \cap X \text{ is odd})$?

e. What is $P(X \geq 12 \cup X \text{ is even})$?

4. A box contains 5 white (W), 3 black (B), and 2 red (R) ball-point pens. The pens are to be selected at random one at a time. Let the subscripts in the probability expressions indicate the number of the draw. Determine:

   a. $P(B_1)$,

   b. $P(W_2 | R_1)$ when sampling without replacement,

   c. $P(B_2)$ when sampling with replacement,

   d. $P(B_2)$ when sampling without replacement,

   e. $P(R_2 \cap W_1)$ when sampling without replacement,

   f. $P(R_2 \cup W_1)$ when sampling without replacement,

   g. $P(B_2 \cup W_2)$ when sampling without replacement,

   h. $P(B_3)$ when sampling with replacement,

   i. $P(B_3)$ when sampling without replacement.

5. Calculate the following probabilities for the simple random sampling of one card at a time from a standard deck of 52 playing cards.

   a. The probability of selecting a club on the first draw.

   b. The probability of selecting a nine or an eight on the second draw given that a two was selected on the first draw when sampling without replacement.

   c. The probability of selecting a five on the second draw given that a heart was selected on the first draw when sampling with replacement.

   d. The same probability calculated for Part c when sampling without replacement.

   e. The probability of selecting a king on the second draw and a ten on the first draw when sampling with replacement.

   f. The same probability calculated for Part c when sampling without replacement.

Five

Discrete Probability Distributions

5.1 Introduction

Having considered basic rules used with probability theory, we shall now investigate some special types of probability distributions that arise in both theory and applications. For most of these special distributions, we shall first develop the probability function for a specific example, and then derive the probability function for the general problem. For the general solution, we shall consider characteristics of the distribution such as the mean, variance, standard deviation, and cumulative probability function. Finally, we shall begin using appropriate tables, that make the task of determining probabilities for specific problems relatively simple.

In our investigations we shall consider a probability distribution as a mathematical model of a real-world situation. For each such model, we shall have a set of assumptions that characterize a large class of experiments. While the choice of a particular probability function as a model for a specific situation is somewhat arbitrary, it should not be completely so. Before using a specific model to describe a particular situation, we should justify the use of the model and make sure that our assumptions are realistic.

5.2 Binomial Distribution

It is important to understand the rationale underlying our treatment of the binomial distribution, which otherwise may seem unwieldy and prohibitively long. Therefore, the discussion that follows is of even greater import than those in succeeding sections of the chapter. Naturally, we first want to familiarize you with the binomial distribution.

However, we also wish to emphasize that both of the discrete probability distributions—the binomial and the hypergeometric—can be developed quite logically from elementary concepts of probability. From a philosophical standpoint, we cannot overemphasize this point.

The binomial distribution is used when repeated trials of a given event produce outcomes that are statistically independent. Further, for each trial, the event has only two mutually exclusive outcomes, such as heads or tails, good or bad, male or female, defective or nondefective, dead or alive, satisfied or dissatisfied, operative or inoperative, success or failure. Such a trial is known as a *Bernoulli trial* or, more specifically, a trial generated by a *Bernoulli process.*

Suppose we wish to determine the probability function of X, where X is the number of heads, for three tosses of a coin. Further, let us suppose that the coin is unfair, so that the probability of obtaining a head on a given toss is equal to 0.4 and, hence, the probability of obtaining a tail is 0.6. We shall begin by considering the probability distribution associated with one toss. Then we shall determine the probability distribution associated with three tosses.

The probability distribution in functional form for the experiment of one toss of our unfair coin is

$$f(x) = (0.4)^x(0.6)^{1-x} \quad (x = 0, 1), \tag{1}$$

where x represents the number of heads on a given toss. When we substitute the appropriate value of x into $f(x)$ we obtain the probability distribution shown on the right. The probability function given by (1) can be generalized for the case where π is the probability that X equals zero and $(1 - \pi)$ is the probability that X equals one. The more general probability function, known as a *Bernoulli probability function*, is

x	$P(X = x)$
0	0.6
1	0.4
	1.0

$$f(x) = \pi^x(1 - \pi)^{1-x} \quad (x = 0, 1). \tag{2}$$

Let us examine what occurs when we toss our coin three times. First, we note that there are only two possible mutually exclusive outcomes per toss, heads or tails. Second, since we are essentially sampling with replacement, the outcome of a given trial is statistically independent of the outcomes of any previous trials; in other words, the probabilities for a given toss are not changed by either knowledge or ignorance of outcomes of previous trials. Third, the probability of each possible outcome remains unchanged from trial to trial. Although we cannot prove these properties are correct, they appear to be reasonable for our problem. Therefore, we shall assume such properties.

To determine the probability function for three tosses of the coin, we must find the probability for each possible value of a random variable X, where X now represents the number of heads obtained on three tosses. In this example, the possible values of X are 0, 1, 2, 3. Let H and T represent the respective outcomes head and tail, and let the order of each represent the number of the toss. For example, HTH represents the event of heads on the first and third tosses and a tail on the second toss. With these assumptions, we determine the required probabilities to be

$$\begin{aligned}
P(X = 0) &= P(TTT) = P(T)P(T)P(T) = (0.6)(0.6)(0.6) = (0.6)^3 = 0.216, \\
P(X = 1) &= P(HTT \cup THT \cup TTH) = P(HTT) + P(THT) + P(TTH) \\
&= (0.4)(0.6)(0.6) + (0.6)(0.4)(0.6) + (0.6)(0.6)(0.4) \\
&= 3(0.4)(0.6)^2 = 0.432, \\
P(X = 2) &= P(HHT \cup HTH \cup THH) = P(HHT) + P(HTH) + P(THH) \\
&= (0.4)(0.4)(0.6) + (0.4)(0.6)(0.4) + (0.6)(0.4)(0.4) \\
&= 3(0.4)^2(0.6) = 0.288, \\
P(X = 3) &= P(HHH) = (0.4)(0.4)(0.4) = (0.4)^3 = 0.064.
\end{aligned} \tag{3}$$

The probability distribution is tabulated on the right.

x	$P(X = x)$
0	0.216
1	0.432
2	0.288
3	0.064
	1.000

Several points are so vital to the development of the functional form of the probability distribution that they should be spelled out in detail. First, the four compound events

$$X = 0, \qquad X = 1, \qquad X = 2, \qquad X = 3 \tag{4}$$

are mutually exclusive. For example, it is impossible on one repeat of the experiment of tossing three coins to obtain both "one head" ($X = 1$) and "two heads" ($X = 2$) simultaneously. The four events are also exhaustive. Thus, their probabilities will sum to unity.

Second, the joint events

$$HHH, HHT, HTH, THH, HTT, THT, TTH, TTT, \tag{5}$$

each of which satisfies one (and only one) of the events of (4), are obviously mutually exclusive. This being true, we can obtain the probability of any of the four compound events of (4) simply by adding the probabilities of all joint events of (5) which satisfy the compound event of interest. For example,

$$P(X = 2) = P(HHT \cup HTH \cup THH) = P(HHT) + P(HTH) + P(THH).$$

Third, the individual tosses of the coin are independent. Thus, we can compute the probability for each joint event of (5) simply by taking the product of the probabilities of the marginal events included in such joint event. However, in so doing, we see that if, when order is ignored, the overall make-up of the two joint events is the same (e.g., HHT and HTH), their probabilities of occurrence will be equal.

Having made these observations, we can determine the probability of a given compound event by first finding the probability for any specific joint event which satisfies the compound event of interest and then multiplying the probability of that joint event by the number of joint events which satisfy the compound event of interest. In other words, when we add the probabilities of the mutually exclusive joint events, we are actually adding equal numbers; hence, we may multiply. For example, to determine $P(X = 1)$, we can choose one of the joint events, say HTT, and compute

$$P(HTT) = P(H)P(T)P(T) = (0.4)(0.6)(0.6) = (0.4)(0.6)^2.$$

Next, we can determine how many such joint events satisfy the event $X = 1$. Since order is important, this may be done either by enumeration or by use of permutations. The use of permutations is much easier for complex problems. For our example, we note that we have two groups of outcomes with 2 in one group (tails) and 1 in the other group (heads). Thus (by Equation (3), page 91), we have

$$_3P_{1,2} = \frac{3!}{1!2!} = {_3C_1} = \binom{3}{1}.$$

Having thus obtained the number of joint events and the probability of each joint event which satisfies the compound event $X = 1$, we combine this information to obtain

$$P(X = 1) = \binom{3}{1}(0.4)(0.6)^2. \tag{6}$$

We could repeat the process for each of the possible values of X. However, we prefer to have a function of x such that we can determine $P(X = x)$ (the probability that the random variable X assumes the value x) simply by substituting x (the appropriate possible value of X) into the function. To determine the form of such function, let us reconsider our example for the more general case of $P(X = x)$, without x being specified. Thus, for the event $X = x$ to be satisfied, we must obtain x heads and $(3 - x)$ tails. Hence, we need consider only one specific joint event which yields exactly x heads. For simplicity, let us consider the event for which the first x outcomes are heads and the last $(3 - x)$ outcomes are tails. The probability for this joint event is $(0.4)^x(0.6)^{3-x}$. Now, to find the number of different orders for the outcomes that have exactly x heads and $(3 - x)$ tails, we determine

$$_3P_{x,3-x} = \frac{3!}{x!(3-x)!} = \binom{3}{x}.$$

Finally, we obtain the desired probability function by multiplying this result by the number of satisfying events. (Again, note that we are actually adding the equal probabilities of mutually exclusive events.) The desired probability function is

$$P(X = x) = f(x) = \binom{3}{x}(0.4)^x(0.6)^{3-x} \quad (x = 0, 1, 2, 3). \tag{7}$$

By substituting into $f(x)$ the possible values of X we obtain the same probabilities as we did in (3):

$$\begin{aligned} P(X = 0) = f(0) &= \binom{3}{0}(0.4)^0(0.6)^{3-0} = (1)(1)(0.216) = 0.216, \\ P(X = 1) = f(1) &= \binom{3}{1}(0.4)^1(0.6)^{3-1} = (3)(0.4)(0.36) = 0.432, \\ P(X = 2) = f(2) &= \binom{3}{2}(0.4)^2(0.6)^{3-2} = (3)(0.16)(0.6) = 0.288, \\ P(X = 3) = f(3) &= \binom{3}{3}(0.4)^3(0.6)^{3-3} = (1)(0.064)(1) = 0.064. \end{aligned} \tag{8}$$

Thus, we have generalized the solution to the initial problem of finding the probability function of X, where X is the number of heads that appear on three tosses of an unfair coin. This function is given as (7).

We usually speak of the two mutually exclusive events of interest as being either successes or failures. A *success* denotes the occurrence of the outcome of interest, while a *failure* denotes the nonoccurrence of the outcome of interest. Thus, in the foregoing problem, the event of obtaining a head is a success, while that of getting a tail is a failure. Here, success and failure have nothing to do with personal feelings. For example, if we wanted the probability of rolling three losing numbers in six tosses of a pair of dice, we would refer to the event of obtaining a losing number as a success—even though the outcome might be contrary to our personal desire.

We shall now consider the more general problem of determining the probability function of X, where X is the number of successes in n trials of an experiment for which there exist only two mutually exclusive outcomes, success and failure, with the probabilities of each being π and $(1 - \pi)$, respectively. We shall assume that the outcomes are statistically independent events and the probabilities of success and failure remain unchanged from trial to trial.

We first find the probability of a specific joint event which gives x successes and $(n - x)$ failures and multiply it by the number of events. This product gives us a general function for $P(X = x)$. Let us begin by considering that specific joint event for which the first x trials result in successes (S) and the last $(n - x)$ trials result in failures (F). We have,

$$P(\underbrace{S \cap S \cdots \cap S}_{x \text{ successes}} \cap \underbrace{F \cap F \cdots \cap F}_{(n-x) \text{ failures}}) = \underbrace{P(S)P(S) \cdots P(S)}_{x \text{ times}}\underbrace{P(F)P(F) \cdots P(F)}_{(n-x) \text{ times}}$$

$$= \underbrace{\pi \cdot \pi \cdots \pi}_{x \text{ values}} \cdot \underbrace{(1 - \pi)(\pi - \pi) \cdots (1 - \pi)}_{(n-x) \text{ values}} \tag{9}$$

$$= \pi^x(1 - \pi)^{n-x}.$$

Moreover, we have a group of n items (outcomes) of which there are x in one group (successes) and $(n - x)$ in the other group (failures). We wish to determine the number of different orders in which these outcomes may occur. Hence, again using Equation (3), page 91, we have

$$_nP_{x, n-x} = \frac{n!}{x!(n - x)!} = \binom{n}{x}. \tag{10}$$

Multiplying the results of (9) and (10) gives

$$P(X = x) = f(x) = \binom{n}{x}\pi^x(1 - \pi)^{n-x} \quad (x = 0, 1, 2, \ldots, n). \tag{11}$$

We need to check the function in (11) to determine whether it is actually a valid probability function according to the requirements given in Section 4.11:

1. $f(x) \geq 0$ for all x,
2. $f(x) = P(X = x)$, and
3. $\sum f(x) = 1$,

where the summation is over all possible values of X. Since each term in the product is nonnegative, the first requirement is immediately satisfied. Also, $f(x) = P(X = x)$, because we have derived the function for exactly that purpose. Finally, we need only determine whether $\sum f(x) = 1$. Thus,

$$\sum_{x=0}^{n} f(x) = \sum_{x=0}^{n} \binom{n}{x}\pi^x(1 - \pi)^{n-x} \tag{12}$$

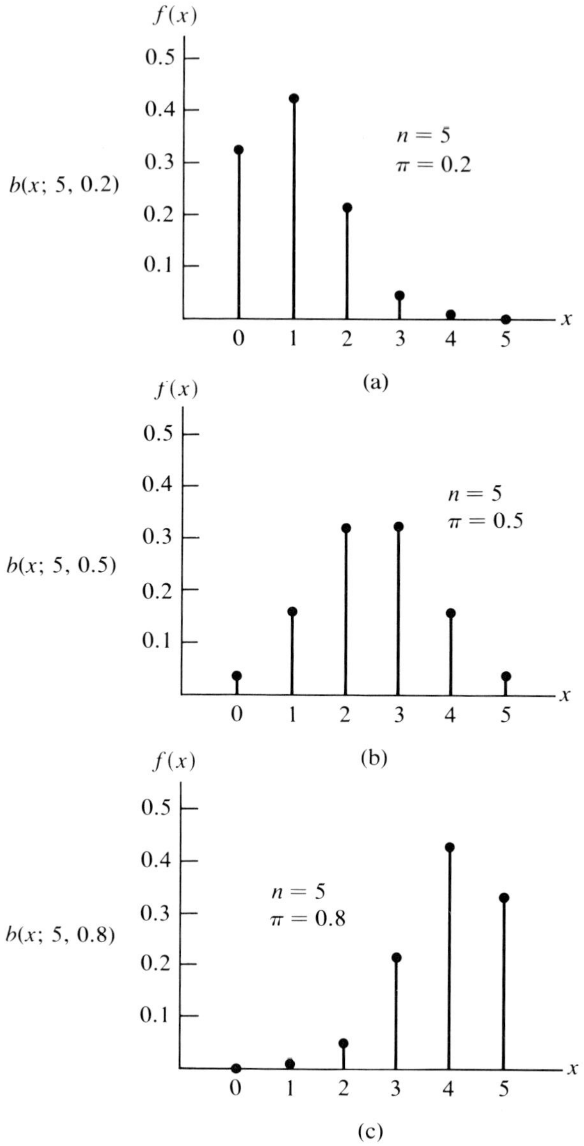

Figure 5.1

Shapes of various binomial distributions

must equal unity if $f(x)$ is to qualify as a valid probability function. We note that the right side of Equation (12) is the binomial expansion of

$$[\pi + (1 - \pi)]^n = 1^n = 1.$$

[Recall from algebra that

$$(a + b)^n = \sum_{x=0}^{n} \binom{n}{x} a^x b^{n-x}.]$$

Thus, $f(x)$ in (11) is in fact a valid probability function. Since the terms in such function are those of the *binomial expansion* of $[\pi + (1 - \pi)]^n$, the probability function in (11) is called the *binomial probability function*, while the corresponding probability distribution is simply called the *binomial distribution*. We shall denote the function by

$$b(x; n, \pi),$$

where x is any possible value which "number of successes" (X) can assume, n is the number of trials, and π is the probability of success on any specific trial. Functionally,

$$b(x; n, \pi) = \binom{n}{x}\pi^x(1 - \pi)^{n-x} \quad (x = 0, 1, \ldots, n), \tag{13}$$

where x is a variable, while n and π are parameters (constants for a given problem). The term $\binom{n}{x}$, known as the *binomial coefficient*, is always included in the binomial distribution. When referring to the binomial distribution, we shall always use the above notation. Since it is often difficult to see intuitively the reason for the term $\binom{n}{x}$, you may wish to refer to the development associated with (11) in which we have used the fact that

$$_nP_{x, n-x} = \binom{n}{x}.$$

In analyzing $b(x; n, \pi)$, we see that the distribution is symmetrical when $\pi = 0.5$. The shapes of binomial distributions for fixed n and different values of π are shown in Figure 5.1. There the probabilities are represented by the height of the lines above the respective x values, reflecting the fact that the variable of interest is a discrete random variable. To represent the probability distribution in the form of a relative frequency histogram (as in Chapter 2), we could determine the probabilities by computing the area under the histogram for a given class interval (that is, for a specific possible value of X). This is true because (and only because) each class interval has a length of one unit.

Example 1

Suppose we have a set containing three foremen and seven workers from which we wish to select with replacement a simple random sample of size $n = 3$. Find the probability of selecting (a) no workers, (b) one worker, (c) two workers, and (d) three workers.

Solution

Let X represent the number of workers (successes) in the sample. We have that $\pi = 7/10$, since this is the probability of success for one trial. Further, since we are sampling with replacement, the probabilities remain unchanged from trial to trial.

Finally, $n = 3$. Thus, we obtain answers to (a) through (d) simply by substituting appropriate values into the general formula for the binomial function, Equation (11):

$$\text{(a)}\quad P(X = 0) = b(0; 3, 0.7) = \binom{3}{0}(0.7)^0(0.3)^{3-0} = 0.027,$$

$$\text{(b)}\quad P(X = 1) = b(1; 3, 0.7) = \binom{3}{1}(0.7)^1(0.3)^{3-1} = 0.189,$$

$$\text{(c)}\quad P(X = 2) = b(2; 3, 0.7) = \binom{3}{2}(0.7)^2(0.3)^{3-2} = 0.441,$$

$$\text{(d)}\quad P(X = 3) = b(3; 3, 0.7) = \binom{3}{3}(0.7)^3(0.3)^{3-3} = 0.343.$$

Note that the four events are mutually exclusive and exhaustive. Hence, as we would expect, their probabilities sum to unity (one). □

Example 2

A shoe manufacturer knows from past history that 20 percent of his orders will be for more than 100 pairs of shoes. What is the probability that the next six orders he receives will contain three orders of more than 100 pairs of shoes, if, in fact, each order can be considered as the outcome of a Bernoulli process?

Solution

Since we are interested in orders of more than 100 pairs of shoes, we shall call such an order a success. Conversely, we shall denote by failure any order less than or equal to 100. We know that 20 percent of the orders are successes. Hence, $\pi = 0.2$. Further, $n = 6$ and $x = 3$. Thus,

$$P(X = 3) = h(3; 6, 0.2) - \binom{6}{3}(0.2)^3(0.8)^{6-3} = 0.08192. \qquad \square$$

Example 3

A bakery with five ovens finds that it can meet the demands of its customers with only four ovens. If the probability that an oven is inoperative on a given day is 0.1 and the failure of ovens can be considered as a Bernoulli process, determine the probability that the bakery will be able to meet its customer demands on a given day.

Solution

The bakery must have either four or five ovens operating to fulfill customer demands on a given day. Thus, let X represent the number of successes (operative ovens) on a given day. Then, $\pi = 0.9$ and $n = 5$, and

$$\begin{aligned} P(\text{success}) &= P(X = 4 \cup X = 5) \\ &= P(X = 4) + P(X = 5) \\ &= \binom{5}{4}(0.9)^4(0.1)^{5-4} + \binom{5}{5}(0.9)^5(0.1)^{5-5} \\ &= 0.32805 + 0.59049 = 0.91854. \end{aligned}$$ □

Cumulative Binomial Probabilities

Many practical situations require us to determine $P(X \leq x)$, where X is distributed according to the binomial distribution. From Example 3, we see that computing many of the values of $b(x; n, \pi)$ could be quite laborious, particularly for large values of n. For example, suppose we want to know the probability of obtaining 40 or fewer successes out of 90 trials of a Bernoulli process with $\pi = 0.3$. In order to obtain such a probability by direct application of the binomial probability function, we would have to evaluate and add 41 probability terms of the nature of (13); that is,

$$\begin{aligned} &P_b(X \leq 40 \mid n = 90, \pi = 0.3) \\ &\quad = P_b(X = 0 \mid n = 90, \pi = 0.3) + P_b(X = 1 \mid n = 90, \pi = 0.3) \\ &\qquad + \cdots + P_b(X = 40 \mid n = 90, \pi = 0.3) \\ &\quad = \binom{90}{0}(0.3)^0(0.7)^{90} + \binom{90}{1}(0.3)^1(0.7)^{89} + \cdots + \binom{90}{40}(0.3)^{40}(0.7)^{50}, \end{aligned}$$

where the subscript b denotes probabilities for the binomial distribution, thus distinguishing binomial probabilities from other probabilities of similar form. Obviously, we would like to avoid this difficult calculation. In order to circumvent it, we use the cumulative binomial distribution, denoted by $B(k; n, \pi)$. The relation between the cumulative binomial distribution and the binomial distribution is naturally

$$B(k; n, \pi) = P_p(X \leq k \mid n, \pi) = \sum_{x=0}^{k} b(x; n, \pi). \tag{14}$$

Information regarding the cumulative binomial distribution is given in Table A-6 in the Appendix for various values of x, n, and π. Before using this table, we must understand precisely what probabilities are given. For example, a table could present any one of the following types of probabilities:

$$P(X < x), \quad P(X \leq x), \quad P(X > x), \quad \text{or} \quad P(X \geq x).$$

Furthermore, from a table giving one of the types of probabilities listed, we can determine any of the other probabilities, as well as many more. Recall that, for any variable X,

$$P(a < X \leq b) = F(b) - F(a),$$

where $F(x)$ is the cumulative probability distribution of X. When we need to evaluate this kind of probability for the binomial distribution, we use Table A-6.

For the binomial distribution, $b(x; n, \pi)$, Table A-6 yields the probability

$$P_b(X \geq x \mid n, \pi)$$

for various possible combinations of x, n, and π. To find this in Table A-6, we must locate in the body of the table the probability that corresponds to appropriate values of x, n, and π. For example, to find cumulative binomial probabilities $B(k; n, \pi)$, we simply find

$$B(k; n, \pi) = P_b(X \leq k \mid n, \pi) = 1 - P_b(X \geq k + 1 \mid n, \pi). \tag{15}$$

Table A-6 is what we generally call a *greater-than-or-equal-to probability table*. It may be used to make any desired probability statement concerning events which follow a binomial distribution. For some examples, let us limit consideration to the binomial distribution with $n = 12$ and $\pi = 0.42$. We first locate the page which gives information for $n = 12$ and $\pi = 0.42$. Then we note that, regardless of our probability question, we must phrase it in terms of one or more probabilities of the form $P(X \geq x)$, since this is the type of information available in the table.

Example 4

Find:

(a) $P_b(X \geq 5 \mid n = 12, \pi = 0.42)$,
(b) $P_b(X > 5 \mid n = 12, \pi = 0.42)$,
(c) $P_b(X < 9 \mid n = 12, \pi = 0.42)$,
(d) $P_b(X \leq 4 \mid n = 12, \pi = 0.42)$,
(e) $P_b(X = 8 \mid n = 12, \pi = 0.42)$.

Solution

We are given that $n = 12$ and $\pi = 0.42$. First we must state each probability in the form $P(X \geq x)$, and then read the appropriate probability from Table A-6.

(a) $P_b(X \geq 5 \mid n = 12, \pi = 0.42) = 0.6175.$

(b) $P_b(X > 5 \mid n = 12, \pi = 0.42) = P_b(X \geq 6 \mid n = 12, \pi = 0.42) = 0.3889.$

(c) $P_b(X < 9 \mid n = 12, \pi = 0.42) = 1 - P_b(X \geq 9 \mid n = 12, \pi = 0.42)$
$= 1 - 0.0218 = 0.9782.$

(d) $P_b(X \leq 4 \mid n = 12, \pi = 0.42) = 1 - P_b(X \geq 5 \mid n = 12, \pi = 0.42)$
$= 1 - 0.6175 = 0.3825.$

(e) $P_b(X = 8 \mid n = 12, \pi = 0.42)$
$= P_b(X \geq 8 \mid n = 12, \pi = 0.42) - P_b(X \geq 9 \mid n = 12, \pi = 0.42)$
$= 0.0760 - 0.0218 = 0.0542.$ □

Although table A-6 does not provide probabilities for values of π larger than 0.5, we can use it to determine probabilities for binomial distributions where $\pi > 0.5$. To clarify the reasoning behind the proper procedure, let us consider the following example. Suppose we are to toss an unfair coin five times and we call the occurrence of a head a success, where the probability of a head on one toss is $\pi = 0.7$. Then, naturally, the occurrence of a tail becomes a failure, and the probability of a failure on one toss is

$$1 - \pi = 1 - 0.7 = 0.3.$$

Let X represent "number of successes" and Y represent "number of failures." Relevant probabilities for "number of successes" (heads) and "number of failures" (tails) are given in Table 5.1.

Table 5.1

Probabilities for tossing an unfair coin

(1) *Number of successes (Heads)* (x)	(2) $P(X = x)$	(3) *Number of failures (Tails)* (y)	(4) $P(Y = y)$
0	$\binom{5}{0}(0.7)^0(0.3)^5 = 0.0024$	5	$\binom{5}{5}(0.3)^5(0.7)^0 = 0.0024$
1	$\binom{5}{1}(0.7)^1(0.3)^4 = 0.0284$	4	$\binom{5}{4}(0.3)^4(0.7)^1 = 0.0284$
2	$\binom{5}{2}(0.7)^2(0.3)^3 = 0.1323$	3	$\binom{5}{3}(0.3)^3(0.7)^2 = 0.1323$
3	$\binom{5}{3}(0.7)^3(0.3)^2 = 0.3087$	2	$\binom{5}{2}(0.3)^2(0.7)^3 = 0.3087$
4	$\binom{5}{4}(0.7)^4(0.3)^1 = 0.3601$	1	$\binom{5}{1}(0.3)^1(0.7)^4 = 0.3601$
5	$\binom{5}{5}(0.7)^5(0.3)^0 = 0.1681$	0	$\binom{5}{0}(0.3)^0(0.7)^5 = 0.1681$

Each possible value of "number of heads" corresponds to one of the possible values of "number of tails." For example, every time we obtain three ($x = 3$) heads out of five tosses, we necessarily obtain two ($n - x = 2$) tails. Thus, as indicated in Table 5.1,

$$P(3 \text{ heads out of five trials} \mid P(H) = 0.7) = P(2 \text{ tails out of five trials} \mid P(T) = 0.3).$$

The equivalence also holds for probabilities of other corresponding events. The general result is as follows. Since we have designated heads as a success, we may seek, for example,

$$P_b(X = 3 \mid n = 5, \pi = 0.7),$$

where X is the number of successes. However, we can just as easily designate tails as a success. Then, since three heads with $P(H) = 0.7$ is the same event as two tails with

$P(T) = 0.3$, we need only find

$$P_b(X = 2|n = 5, \pi = 0.3).$$

In general, when we wish to find $P_b(X = x|n, \pi)$ for $\pi > 0.5$, we simply restate the problem in terms of $(n - x)$ successes with a probability of success of $(1 - \pi)$ on a given trial. That is,

$$P_b(X = x|n, \pi) = P_b(X = n - x|n, 1 - \pi). \tag{16}$$

Even more generally, we have for cumulative probabilities

$$P_b(X \geq x|n, \pi) = P_b(X \leq n - x|n, 1 - \pi). \tag{17}$$

Thus, to solve any binomial probability problem involving a value of $\pi > 0.5$, we merely interchange the roles of success with failure, π with $(1 - \pi)$, and x with $(n - x)$ and utilize the concept of (16). For example, suppose we wish to find

$$P_b(X \geq 5|n = 12, \pi = 0.58).$$

We see from (17) that

$$\begin{aligned} P_b(X \geq 5|n = 12, \pi = 0.58) &= P_b(X \leq 12 - 5|n = 12, \pi = 1 - 0.58) \\ &= P_b(X \leq 7|n = 12, \pi = 0.42). \end{aligned}$$

Then,

$$\begin{aligned} P_b(X \leq 7|n = 12, \pi = 0.42) &= 1 - P_b(X \geq 8|n = 12, \pi = 0.42) \\ &= 1 - 0.0760 = 0.9240. \end{aligned}$$

The general formula of (17) represents only one of many types of problems that need to be solved for the case of $\pi > 0.5$. In fact, there are so many such problems that it is fruitless to derive a general formula for each case. The wise reader, therefore, will rely on the line of reasoning leading to the statement of (17) rather than simply memorize the relation.

The Mean and the Variance of the Binomial Distribution

In many practical situations we deal with a value of n larger than any in Table A-6 or a value of π which is not in the table. In such situations, we generally can settle for an approximation of the probability. However, when extreme accuracy is mandatory, we can consult publications that include very extensive tables for the binomial distribution. More extensive tables are published by Harvard University Computation Laboratory (5.1), U.S. Army Ordnance Corps (5.2), National Bureau of Standards (5.3), Owen (5.4), and Romig (5.5). The Harvard and Ordnance Corps tables are the most extensive of these.

To determine approximate probabilities, we need to know the mean and the standard deviation for the binomial distribution. Even though we shall not discuss methods of approximation in this section, we shall find the mean and the standard deviation for the binomial distribution.

Recalling the definition of the mean of a probability distribution of a discrete random variable X (Definition 4.21), we can determine the mean of the binomial distribution as

$$E(X) = \sum_{x=0}^{n} xf(x) = \sum_{x=0}^{n} x\binom{n}{x}\pi^x(1-\pi)^{n-x}.$$

Let us begin with the special case $n = 3$. We have

$$\begin{aligned} E(X) &= \sum_{x=0}^{3} x\binom{3}{x}\pi^x(1-\pi)^{3-x} \\ &= 0 \cdot \binom{3}{0}\pi^0(1-\pi)^3 + 1 \cdot \binom{3}{1}\pi^1(1-\pi)^2 \\ &\quad + 2 \cdot \binom{3}{2}\pi^2(1-\pi)^1 + 3 \cdot \binom{3}{3}\pi^3(1-\pi)^0 \\ &= 0 + 3\pi(1-\pi)^2 + 6\pi^2(1-\pi) + 3\pi^3 \\ &= 3\pi[(1-\pi)^2 + 2\pi(1-\pi) + \pi^2] \\ &= 3\pi[(1-\pi) + \pi]^2 = 3\pi(1)^2 = 3\pi. \end{aligned}$$

For the general case, we obtain

$$E(X) = \sum_{x=0}^{n} x\binom{n}{x}\pi^x(1-\pi)^{n-x} = \sum_{x=0}^{n} x \cdot \frac{n!}{x!(n-x)!}\pi^x(1-\pi)^{n-x}. \tag{18}$$

However, since the first term yields 0 for $x = 0$, we may begin our summation with $x = 1$ and write (18) as

$$\begin{aligned} E(X) &= \sum_{x=1}^{n} x \frac{n!}{x!(n-x)!}\pi^x(1-\pi)^{n-x} \\ &= \sum_{x=1}^{n} \frac{n!}{(x-1)!(n-x)!}\pi^x(1-\pi)^{n-x}. \end{aligned}$$

Factoring out $n\pi$ from each term gives

$$E(X) = n\pi \sum_{x=1}^{n} \frac{(n-1)!}{(x-1)!(n-x)!}\pi^{x-1}(1-\pi)^{n-x}.$$

If we let $y = x - 1$ and indicate the true summation limits, the equation becomes

$$\begin{aligned} E(X) &= n\pi \sum_{y=0}^{n-1} \frac{(n-1)!}{y!(n-1-y)!}\pi^y(1-\pi)^{n-1-y} \\ &= n\pi \sum_{y=0}^{n-1} \binom{n-1}{y}\pi^y(1-\pi)^{n-1-y}. \end{aligned}$$

We recognize the summation of the last equation to be that for a binomial distribution with $(n - 1)$ trials. Thus, the sum is necessarily 1, and the mean for the binomial distribution is

$$E(X) = n\pi. \tag{19}$$

To determine the variance for the binomial distribution, let us again consider the case where $n = 3$. We have

$$\begin{aligned}
\operatorname{Var}(X) &= E[X - E(X)]^2 \\
&= \sum_{x=0}^{3} (x - 3\pi)^2 \binom{3}{x} \pi^x (1 - \pi)^{3-x} \\
&= 9\pi^2(1 - \pi)^3 + 3(1 - 3\pi)^2\pi(1 - \pi)^2 + 3(2 - 3\pi)^2\pi^2(1 - \pi) \\
&\quad + 9(1 - \pi)^2\pi^3 \\
&= 3\pi(1 - \pi)[3\pi(1 - \pi)^2 + (1 - 3\pi)^2(1 - \pi) + (2 - 3\pi)^2\pi + 3\pi^2(1 - \pi)] \\
&= 3\pi(1 - \pi)(3\pi - 6\pi^2 + 3\pi^3 + 1 - 6\pi + 9\pi^2 - \pi + 6\pi^2 - 9\pi^3 + 4\pi - 12\pi^2 \\
&\quad + 9\pi^3 + 3\pi^2 - 3\pi^3) \\
&= 3\pi(1 - \pi)(1) \\
&= 3\pi(1 - \pi).
\end{aligned}$$

While we omit details of the argument, the variance for any value of n may be obtained in a similar manner. Thus,

$$\operatorname{Var}(X) = n\pi(1 - \pi). \tag{20}$$

Problems

1. Define the *Bernoulli process.*
2. Define *success* and *failure.*
3. What is the *binomial coefficient*?
4. Suppose a car rental agency has available 6 station wagons, 9 sedans, and 5 compact cars from which we wish to select with replacement a simple random sample of size $n = 5$. Determine the probability of selecting

   a. three sedans, b. five sedans, c. two sedans,

   d. one sedan, e. zero sedans.
5. a. Using only the results of Problem 4, determine the probability of selecting four sedans.

   b. Use the binomial distribution to find the probability of selecting four sedans in Problem 4, and check to see that it equals the probability determined in Part a.

   c. Construct the probability distribution function of the experiment defined in Problem 4 in tabular form using the answers to that problem and Part b of this one.

   d. Construct a relative frequency histogram from the probability distribution constructed in Part c.
6. Suppose we have a shipment containing 13 radios and 12 stereos from which we wish to select with replacement a simple random sample of size $n = 15$. Determine the probability of selecting

   a. eight radios b. five stereos, c. six radios, d. twelve stereos.
7. The guidance system of a missile contains five redundant subsystems, three of which must fail before the system fails. If the probability of any one of the subsystems failing on any given flight is 0.1, what is the probability of the guidance system failing on a particular flight?

8. Suppose an assembly line has eight painting booths, at least six of which must be in operation in order to meet the production schedule on any given day. The probability that a mechanical failure will make a booth inoperative on a given day is 0.2, and the probability that the lack of an operator will make a booth inoperative on a given day is 0.1. If the failure of booths can be considered as a Bernoulli process and the causes of failure are independent, what is the probability that failures of painting booths will not prevent the assembly line from achieving scheduled production on a given day?

9. Use Table A-6 (Cumulative binomial distribution) to find

   a. $P_b(X \geq 7 | n = 11, \pi = 0.28)$,
   b. $P_b(X \geq 4 | n = 16, \pi = 0.37)$,
   c. $P_b(X < 9 | n = 13, \pi = 0.08)$,
   d. $P_b(X \leq 6 | n = 9, \pi = 0.48)$,
   e. $P_b(X > 12 | n = 18, \pi = 0.25)$,
   f. $P_b(X = 10 | n = 50, \pi = 0.13)$,
   g. $P_b(X > 3 | n = 8, \pi = 0.81)$,
   h. $P_b(X < 14 | n = 17, \pi = 0.64)$.

10. Use Table A-6 to find

   a. $P_b(X \geq 5 | n = 15, \pi = 0.06)$,
   b. $P_b(X \leq 16 | n = 19, \pi = 0.73)$,
   c. $P_b(X = 8 | n = 12, \pi = 0.95)$,
   d. $P_b(X > 7 | n = 14, \pi = 0.59)$,
   e. $P_b(8 \leq X \leq 13 | n = 13, \pi = 0.46)$,
   f. $P_b(3 < X < 8 | n = 10, \pi = 0.88)$,
   g. $P_b(9 \leq X \leq 11 | n = 18, \pi = 0.69)$,
   h. $P_b(12 \leq X \leq 15 | n = 50, \pi = 0.72)$.

5.3 Poisson Distribution

From the Bernoulli processes consisting of distinct trials, where the probability of success is constant from trial to trial, we turn to another situation wherein the probability of success is constant. However, we cannot view this process as a series of distinct trials. The Poisson distribution, named after the French mathematician Simeon Poisson, is generally used to compute probabilities for random variables which represent objects (or events) distributed over a fixed time or space. It is frequently a very good model to use for determining probabilities associated with such random variables as these:

1. The number of phone calls entering a switchboard during a fixed time.
2. The number of defects in a manufactured part.
3. The number of small organisms in a given volume of fluid.
4. The number of automobile deaths in a city during a fixed time.
5. The number of red cells in a blood specimen of a fixed volume.
6. The number of customers arriving at a service facility per fixed interval of time.
7. The number of sales of a particular product during a given period.
8. The number of typing errors per page.

When is the Poisson distribution a feasible tool? The process under consideration must satisfy certain assumptions before we can even consider using it as a model in a realistic situation. These assumptions are:

1. The number of occurrences of the phenomenon under consideration (hereafter termed a success) is independent from unit to unit.
2. The probability of a single occurrence in a very small unit is proportional to the size of the unit.
3. The probability of two or more occurrences in a very small unit is so small that it can be ignored.

Given these assumptions, one can derive the Poisson probability function for x, where x is the number of successes in a given space or time interval. The Poisson probability function is

$$p(x; \lambda) = \frac{e^{-\lambda}\lambda^x}{x!} \qquad (x = 0, 1, 2, \ldots), \tag{1}$$

where λ is the mean of the distribution (the average number of successes in a given space or time) and e is the constant 2.7182818.... As we can see from (1) the Poisson distribution for the number of occurrences x depends on (or is characterized by) the average number of occurrences per stated unit of space or time.

In many problems it is no easy task to be sure that the assumptions hold for a given situation. In fact, certainty often requires experience in the problem area or experimental evidence. For example, it is sometimes obvious that the independence requirement is not satisfied. If we doubt whether the Poisson distribution is appropriate for a specific situation, we may gather data from the process in question and compare the observed data with computed Poisson probabilities. Techniques to be used for such a comparison are presented later. First we must be able to determine Poisson probabilities from Table A-7 in the Appendix.

The cumulative probability distribution for the Poisson distribution is

$$P(k; \lambda) = \sum_{x=0}^{k} p(x; \lambda) = P_p(X \le k \mid \lambda), \tag{2}$$

where the subscript p denotes a Poisson probability. This is precisely what is given in Table A-7. That is, Table A-7 presents cumulative sums for $P_p(X \le k \mid \lambda)$, where X is the random variable representing the number of occurrences (successes) of the phenomenon of interest (in a given unit of time or space), k is any possible value of X, and λ is the average number of occurrences (successes) of the phenomenon of interest in the given unit of time or space. In order to determine $P_p(X \le k \mid \lambda)$ in Table A-7, we simply read the probability which corresponds to k and λ. Let us consider some examples.

Example 1

Suppose we have an automatic machine that produces many spools of wire per hour with an average of three defective spools per hour. Assuming a Poisson model, find the probability of the machine producing the following number of defective spools of wire in one hour:

(a) exactly three, (b) three or fewer, (c) fewer than three,
(d) three or more, (e) more than three.

Solution

We are given that $\lambda = 3$ (defects/hour). Therefore, we simply determine the appropriate probabilities from Table A-7:

(a) $P_p(X = 3|\lambda = 3) = P_p(X \leq 3|\lambda = 3) - P_p(X \leq 2|\lambda = 3)$
$= 0.647 - 0.423 = 0.224.$

(b) $P_p(X \leq 3|\lambda = 3) = 0.647.$

(c) $P_p(X < 3|\lambda = 3) = P_p(X \leq 2|\lambda = 3) = 0.423.$

(d) $P_p(X \geq 3|\lambda = 3) = 1 - P_p(X < 3|\lambda = 3)$
$= 1 - P_p(X \leq 2|\lambda = 3)$
$= 1 - 0.423 = 0.577.$

(e) $P_p(X > 3|\lambda = 3) = 1 - P_p(X \leq 3|\lambda = 3)$
$= 1 - 0.647 = 0.353.$ □

Example 2

What is the probability that the machine of Example 1 will produce seven defective spools in a period of two hours?

Solution

We recall that the Poisson distribution for the number of occurrences x depends on the average number of occurrences per stated unit of space or time. Thus, $\lambda = 6$ (defects/two hours) and the required probability is

$$P_p(X = 7|\lambda = 6) = P_p(X \leq 7|\lambda = 6) - P_p(X \leq 6|\lambda = 6)$$
$$= 0.744 - 0.606 = 0.138.$$ □

Example 3

Acme Electronics Company produces a special-purpose transistor for which it claims an operating life of 2,000 hours or more. The company knows from past history that about 4 of every 100 transistors will fail before 2,000 hours. Assuming the Poisson model, what is the probability that a large randomly selected group of Acme transistors will have zero failures in the first 2,000 hours of operation? One or more failures?

Solution

We have $\lambda = 0.04$ (failures per 2,000 hours). Hence, the probability of zero failures is

$$P_p(X = 0|\lambda = 0.04) = 0.961.$$

Further, the probability of one failure or more is

$$P_p(X \geq 1 | \lambda = 0.04) = 1 - P_p(X = 0 | \lambda = 0.04)$$
$$= 1 - 0.961 = 0.039. \qquad \square$$

These simple examples demonstrate the versatility of the Poisson distribution as a probability model. In addition to its direct utility, one primary use of the Poisson distribution is that of approximating probabilities for the binomial distribution when n is very large and π is very small. (It is generally considered feasible to use such approximation when $\pi \leq 0.1$ and $n\pi \leq 5$.) It is fortunate that the Poisson distribution closely approximates the binomial distribution for large n and small π, since tables of binomial probabilities generally omit probabilities for large values of n. To approximate a binomial probability, we use the mean of the binomial distribution ($n\pi$) as the mean of the Poisson distribution (λ) and determine the appropriate probability from the table of Poisson probabilities.

Example 4

A company buys parts from a certain vendor knowing that the vendor's manufacturing process produces 0.005 percent defective parts. What is the probability that the company will receive more than 3 defective parts from a lot of 20,000 parts?

Solution

A binomial model appears to fit this problem reasonably well. Thus,

$$P_b(X > 3 | 20{,}000, 0.00005) = 1 - P_b(X \leq 3 | n = 20{,}000, \pi = 0.00005)$$
$$= 1 - \sum_{x=0}^{3} b(x; 20{,}000, 0.00005).$$

This cannot be located in any available table, so we shall use the Poisson approximation with $\lambda = n\pi = 20{,}000(0.00005) = 1$. Thus,

$$\sum_{x=0}^{3} b(x; 20{,}000, 0.00005) \cong P_p(X \leq 3 | \lambda = 1) = 0.981,$$

where $\cong$ means "approximately equal to." Finally,

$$P_b(X > 3 | 20{,}000, 0.00005) \cong 1 - 0.981 = 0.019. \qquad \square$$

Recall from (1) that the Poisson distribution depends on only one parameter, λ, which we have called the mean of the distribution. Our reason for so labeling the parameter is that application of Definition 4.21 of the expected value of a distribution to the Poisson distribution yields

$$E(X) = \sum_{x=0}^{\infty} x \frac{e^{-\lambda}\lambda^x}{x!} = \lambda. \tag{3}$$

It can also be shown that the variance is

$$\operatorname{Var}(X) = E[X - E(X)]^2 = \sum_{x=0}^{\infty} (x - \lambda)^2 \frac{e^{-\lambda}\lambda^x}{x!} = \lambda. \tag{4}$$

Thus, we have the rather unusual feature that the Poisson distribution has equal mean and variance.

Problems

1. What three assumptions must be satisfied before we can consider applying the Poisson distribution to a given process?

2. On what single characteristic does the Poisson distribution depend?

3. An average of five customers per hour enter a small store. Assuming a Poisson model, determine
   a. the probability that six or fewer customers will enter the store during a one-hour period,
   b. the probability that exactly eleven customers will enter the store in a given one-hour period,
   c. the probability that at least twenty customers will enter the store in a given two-hour period,
   d. the probability that at least 15 but not more than 20 customers will enter the store in a given two-hour period.

4. Suppose the random access data file of a certain computer produces an average of one error in one hundred hours of operation. Assuming the Poisson model, what is the probability that
   a. the file will produce one error or more in a job which will require it to be in operation for 50 hours?
   b. the file will produce fewer than ten errors during 1,000 hours of operation?
   c. the file will produce more than seven but fewer than 13 errors during 500 hours of operation?

5. Suppose a high-speed loom produces an average of five flaws in each 100 yards of fabric woven. Assuming a Poisson model, what is the probability that
   a. the loom will produce more than ten flaws in a production run of 100 yards of fabric?
   b. the loom will produce fewer than four flaws in 50 yards of fabric?
   c. the loom will produce at least seven but not more than 12 flaws in 200 yards of fabric?

6. Under what specific conditions is it generally considered feasible to use the Poisson distribution to approximate probabilities for the binomial distribution?

7. Suppose that 0.5 percent of the nylon gears produced by a machine are defective. What is the probability that a lot of 600 of these gears will contain two or fewer defectives?

8. Determine the following binomial probabilities.
   a. $P_b(X \leq 4 | n = 5{,}000; \pi = 0.0004)$.
   b. $P_b(X > 1 | n = 100{,}000; \pi = 0.00003)$.
   c. $P_b(X < 3 | n = 50; \pi = 0.05)$.
   d. $P_b(2 \leq X \leq 5 | n = 500; \pi = 0.008)$.
   e. $P_b(X > 2 | n = 1{,}000; \pi = 0.003)$.

5.4 Hypergeometric Distribution

Recall that the development of the binomial distribution required us to consider a random process in which

1. each trial results in one of the only two mutually exclusive outcomes (termed success and failure),
2. the probability of success is constant from trial to trial,
3. each trial is independent of all others, and
4. the number of trials is fixed.

The binomial distribution was derived through repeated tosses of an unfair coin. Here, we were essentially sampling with replacement.

Let us consider the probability distribution for a similar situation, that of sampling without replacement from a finite population where each observation (trial) will result in one of only two mutually exclusive outcomes. As an example of such a situation, consider a set which contains ten batteries of which six are good and four bad. Suppose we wish to find the probability of obtaining four good batteries and two bad batteries when we select a simple random sample of size 6 without replacement. We shall use the classical approach and the properties of simple random sampling.

From simple random sampling, we know that, when order is disregarded, $\binom{10}{6}$ possible samples of size 6 can be selected from a population of size 10. Thus, the total number of possible outcomes for the experiment of interest is $\binom{10}{6}$. The next step is to determine how many outcomes satisfy the event of four good batteries and two bad batteries. Since four good batteries can be selected from six in $\binom{6}{4}$ ways and two bad batteries can be selected from four in $\binom{4}{2}$ ways, the total number of ways to obtain four good batteries and two bad batteries is

$$\binom{6}{4} \cdot \binom{4}{2}.$$

Thus, for sampling without replacement, we have from the classical definition of probability that the probability of selecting four good batteries and two bad batteries from a population of ten batteries, six good and four bad, is

$$\frac{\binom{6}{4}\binom{4}{2}}{\binom{10}{6}} = \frac{\dfrac{6!}{4!2!} \cdot \dfrac{4!}{2!2!}}{\dfrac{10!}{6!4!}} = \frac{3}{7}. \tag{1}$$

From the foregoing example, we see that first we can classify the outcomes for each trial into one of the only two mutually exclusive events (good and bad), second the probability of success (or failure) changes on each trial, third the trials are dependent upon outcomes of previous trials, and fourth the number of trials is fixed. The second and third properties become obvious if we use the conditional population approach to evaluate relevant conditional probabilities. For example,

$$P(G_2|B_1) = 6/9 \neq P(G_2|G_1) = 5/9,$$

which means that the probability for the second trial is dependent upon the outcome of the first trial. Further evidence that the probability of selecting a good battery changes from trial to trial is

$$P(G_2|B_1) = 6/9 \neq 6/10 = P(G_1).$$

The binomial distribution cannot be used as a model in this situation because of the second and third properties. Since these properties are a direct result of the fact that we are sampling *without* replacement, we note that it would be proper to use the binomial distribution as a model if we were sampling *with* replacement.

The conditions of the example may be generalized as follows:

1. The population from which we are sampling contains a items having a characteristic termed "success" and b items having a characteristic termed "failure." Thus, there are a total of $(a + b)$ items in the population.

2. We are to select *without* replacement a simple random sample of size n.

We wish to determine the probability of obtaining x successes in a sample of size n. Since we are not concerned with order, we can determine the total number of possible different samples of size n to be $\binom{a+b}{n}$. These samples are mutually exclusive and, because we are using simple random sampling, each is equally likely to occur. Using the classical definition of probability, we need now to find the number of samples which satisfy the event of x successes and $(n - x)$ failures. Again, since we are not concerned with order, we have $\binom{a}{x}$ different ways of obtaining x successes and $\binom{b}{n-x}$ different ways of obtaining $n - x$ failures. Moreover, for each of the different ways of obtaining x successes, we have $\binom{b}{n-x}$ ways of obtaining $(n - x)$ failures. Hence, we have a total of $\binom{a}{x}\binom{b}{n-x}$ different ways of obtaining x successes and $(n - x)$ failures. Therefore,

$$P_h(X = x|n, a, b) = \frac{\binom{a}{x}\binom{b}{n-x}}{\binom{a+b}{n}}, \tag{2}$$

where $P_h(X = x|n, a, b)$ is the probability of obtaining x successes out of n draws from a finite population containing two mutually exclusive groups of items, a of which are termed successes and b of which are termed failures. Relation (2) obviously holds only for the case of simple random sampling without replacement.

A random variable X having the probability function given by (2) is called a *hypergeometric random variable*, and the probability function is termed the *hypergeometric probability function*. This being the case, we employ the subscript h to denote a *hypergeometric* probability. For a more specific functional notation, we write (2) as

$$h(x; n, a, b) = \frac{\binom{a}{x}\binom{b}{n-x}}{\binom{a+b}{n}} \qquad (x = 0, 1, 2, \ldots, a). \tag{3}$$

There is a voluminous table [Lieberman and Owen (5.6)] that provides both cumulative probabilities and individual probabilities for the hypergeometric distribution. Nevertheless, individual values of $h(x; n, a, b)$ can be determined by substituting appropriate values for the binomial coefficients (combinations) into (3). Table A-8 in the Appendix gives the binomial coefficients for values of n up to 20. To determine $\binom{n}{r}$, we simply locate the value in the table which corresponds to n and r. For example, to evaluate $\binom{11}{5}$, simply read the value in Table A-8 which corresponds to $n = 11$ and $r = 5$, thus obtaining $\binom{11}{5} = 462$.

Example 1

Suppose we have received a shipment of 20 manufactured parts of which 8 are type A and 12 are type B. What is the probability of selecting without replacement a simple random sample of size 3 containing 2 parts of type A and 1 part of type B?

Solution

This problem satisfies the requirements for the hypergeometric model: the population can be classified into two mutually exclusive groups, and the sampling technique is that of simple random sampling without replacement. We have $a = 8$, $b = 12$, $x = 2$, and $n = 3$. Therefore, using (3), we have

$$P_h(X = 2 | n = 3, a = 8, b = 12) = h(2; 3, 8, 12) = \frac{\binom{8}{2}\binom{12}{1}}{\binom{20}{3}}.$$

From Table A-8, $\binom{8}{2} = 28$, $\binom{12}{1} = 12$, and $\binom{20}{3} = 1140$. Thus

$$P_h(X = 2 | n = 3, a = 8, b = 12) = \frac{28 \cdot 12}{1140} = \frac{28}{95} = 0.2947. \qquad \square$$

Example 2

Suppose we wish to conduct a market research study involving 27 electronic component manufacturers of which 12 are classified as "large" and 15 are classified as "small." If we select a simple random sample of size 4 without replacement from the group of 27 manufacturers, what is the probability of selecting 2 "large" and 2 "small" manufacturers?

Solution

Again, the problem satisfies the requirements for the hypergeometric model. Therefore, $x = 2$, $a = 12$, $b = 15$, and $n = 4$ gives

$$P_h(X = 2 \mid n = 4, a = 12, b = 15) = h(2; 4, 12, 15) = \frac{\binom{12}{2}\binom{15}{2}}{\binom{27}{4}}.$$

Using Table A-8, we obtain $\binom{12}{2} = 66$ and $\binom{15}{2} = 105$. We see that $\binom{27}{4}$ is not included in Table A-8. Thus, we manually compute $\binom{27}{4} = 17{,}550$, to obtain

$$P_h(X = 2 \mid n = 4, a = 12, b = 15) = \frac{66 \cdot 105}{17{,}550} = \frac{77}{195} = 0.3949. \qquad \square$$

A very useful feature of the hypergeometric distribution is that there exists a recursion formula which allows us, once we have computed $h(x; n, a, b)$ for a specific value x, to find probabilities for a value of $(x + 1)$ without employing (3). The recursion formula is

$$h(x + 1; n, a, b) = \frac{(n - x)(a - x)}{(x + 1)(b - n + x + 1)} h(x; n, a, b). \tag{4}$$

Example 3

Suppose Acme Electronics' inventory includes 18 boxes of transistors such that 7 boxes are of type A and 11 boxes are of type B. Let us consider the orders for the next 5 boxes of transistors to be a simple random sample of size 5 without replacement. Determine the probability that the orders contain either 2 or 3 boxes of type A. Use the recursion formula where applicable.

Solution

Since $n = 5$, $a = 7$ and $b = 11$, we obtain

$$\begin{aligned} P_h(X = 2 \cup X = 3 \mid n = 5, a = 7, b = 11) \\ &= P_h(X = 2 \mid n = 5, a = 7, b = 11) + P_h(X = 3 \mid n = 5, a = 7, b = 11) \\ &= h(2; 5, 7, 11) + h(3; 5, 7, 11). \end{aligned} \tag{5}$$

From Table A-8, we have

$$h(2; 5, 7, 11) = \frac{\binom{7}{2}\binom{11}{3}}{\binom{18}{5}} = \frac{21 \cdot 165}{8{,}568} = 0.4044. \tag{6}$$

Also, we have by (4) that

$$h(3; 5, 7, 11) = \frac{(5-2)(7-2)}{(3)(11-5+3)} \cdot h(2; 5, 7, 11) = \frac{15}{27}(0.4044) = 0.2247. \tag{7}$$

Finally, substituting the results of (6) and (7) into (5) gives

$$P_h(X = 2 \cup X = 3 \mid n = 5, a = 7, b = 11) = 0.4044 + 0.2247 = 0.6291. \quad \square$$

Binomial Approximations for Hypergeometric Probabilities

We have noted two reasons that the binomial model is not applicable in situations for which the hypergeometric probability function has been developed. First, the probability of success is not constant from trial to trial. Second, the outcomes of various trials are not independent. However, if the population size $(a + b)$ is large and $n/(a + b)$ is small, the probability of success changes very little from trial to trial and the outcomes are practically independent. Thus, if these conditions are satisfied, we would expect the binomial distribution to be a fairly good approximation of the hypergeometric distribution. It turns out that such is the case. For example, suppose a dealer has 5,000 items in inventory such that 2,000 are of type A and 3,000 are of type B. If we randomly select 5 items from the inventory without replacement, the probability that we have 3 of type A in our sample can be approximated by using the binomial distribution with $\pi = 2{,}000/5{,}000 = 2/5 = 0.4$ and $n = 5$. Thus,

$$\begin{aligned} P_h(X = 3 \mid n = 5, a = 2{,}000, b = 3{,}000) &= h(3; 5, 2{,}000, 3{,}000) = 0.23044 \\ &\cong P_b(X = 3 \mid n = 5, \pi = 0.4) \\ &= P_b(X \geq 3 \mid n = 5, \pi = 0.4) - P_b(X \geq 4 \mid n = 5, \pi = 0.4) \\ &= 0.3174 - 0.0870 = 0.2304. \end{aligned}$$

The Mean and Variance of the Hypergeometric Distribution

The mean and variance for the hypergeometric distribution can be determined in the same manner as for the binomial distribution. For the mean, we have

$$E(X) = \frac{na}{a + b}, \tag{8}$$

and the variance is given by

$$\mathrm{Var}(X) = \frac{nab(a + b - n)}{(a + b)^2(a + b - 1)}, \tag{9}$$

where n is the sample size, a is the number of successes in the population, and b is the number of failures in the population.

Problems

1. What is the essential difference between situations in which the hypergeometric distribution can be used as a model and situations in which the binomial distribution can be used as a model?

2. Suppose a bin of parts contains 11 usable and 4 unusable parts. What is the probability of selecting without replacement a simple random sample of size 4 containing
   a. exactly three usable parts?
   b. at least three usable parts?
   c. at most three usable parts?
   d. at least one but not more than three usable parts?
   e. exactly four usable parts?

3. Determine the following probabilities.
   a. $P_h(X = 3 | n = 5, a = 6, b = 9)$.
   b. $P_h(X = 0 | n = 7, a = 14, b = 3)$.
   c. $P_h(X = 9 | n = 10, a = 9, b = 10)$.
   d. $P_h(X = 4 | n = 7, a = 11, b = 14)$.
   e. $P_h(X = 2 | n = 6, a = 21, b = 4)$.

4. Suppose we wish to conduct an employee attitude survey in a small firm with 17 employees, 9 of whom are college graduates. If a simple random sample of size 5 is selected without replacement, determine the probabilities that the sample contains
   a. 2 college graduates,
   b. 3 college graduates,
   c. 2 or more college graduates,
   d. fewer than 3 college graduates.

5. Suppose we wish to conduct a survey of the size of the orders placed by the 19 customers of a firm of whom 9 are classified as "wholesale" and 10 "retail." Given a simple random sample of size 10 selected without replacement, determine the probabilities that
   a. exactly 3 retail customers are selected,
   b. more than 3 retail customers are selected,
   c. at most 3 retail customers are selected,
   d. at least 4 but not more than 6 retail customers are selected.

6. Under what circumstances may the binomial distribution be used as an approximation of the hypergeometric distribution? Why?

7. Recalculate Problem 4 for a large firm with 10,000 employees of whom 2,000 are college graduates.

8. Recalculate Problem 5 for a firm with 4,000 customers of whom 1,200 are classified as "wholesale" and 2,800 as "retail."

Important Terms and Concepts

Bernoulli probability function

Bernoulli process

Bernoulli trial

binomial probability function

cumulative binomial distribution

cumulative hypergeometric distribution

cumulative Poisson distribution

failure

hypergeometric probability function

Poisson probability function

success

References

5.1 Harvard University Computation Laboratory, *Tables of the Cumulative Binomial Probability Distribution*, Cambridge, Mass.: Harvard University Press, 1955.

5.2 U.S. Army Ordnance Corps, *Tables of Cumulative Binomial Probabilities*, Ordnance Corps Pamphlet ORDP 20-1, September, 1952.

5.3 National Bureau of Standards, *Tables of the Binomial Probability Distribution*, Applied Mathematics, Series 6, Washington, D.C.: U.S. Government Printing Office, 1950.

5.4 D. B. Owen, *Handbook of Statistical Tables*, Reading, Mass.: Addison-Wesley Publishing Company, Inc., 1962.

5.5 H. G. Romig, *50–100 Binomial Tables*, New York: John Wiley & Sons, Inc., 1953.

5.6 G. J. Lieberman and D. B. Owen, *Tables of the Hypergeometric Probability Distribution*, Stanford, California: Stanford University Press, 1961.

Review Problems

Sections from other chapters that you may wish to review before attempting to solve a particular problem are given after each part of each problem.

1. Suppose company A employs 3 managers and 7 salesmen and company B employs 6 managers and 4 salesmen. Letting X represent the number of managers selected from company A and Y represent the number of salesmen selected from company B, determine the following probabilities when selecting with replacement from company A a sample of size 5 and selecting with replacement from company B a sample of size 6. Assume that X and Y are independent.

   a. $P(X = 4 \cap Y = 3)$ (Section 4.7),

   b. $P(X \leq 2 \cap Y = 5)$ (Section 4.7),

   c. $P(X > 2 \cap Y \leq 2)$ (Section 4.7),

   d. $P(X = 4 \cup Y = 3)$ (Section 4.8),

   e. $P(X \leq 2 \cup Y = 5)$ (Section 4.8),

   f. $P(X > 2 \cup Y \leq 2)$ (Section 4.8).

2. A certain store sells widgets to in-store customers at the rate of eight per hour, a number that follows a Poisson distribution. In addition to selling widgets to the in-store customers, the salesmen conduct a telephone campaign. It is known that the probability of selling a widget via any one telephone call is 0.2. Letting X represent the number of widget sales per hour to in-store customers and Y represent the number of telephone sales, determine the following probabilities for a one-hour period in which 10 telephone calls were made. Assume the number of in-store sales and the number of telephone sales are independent.

   a. $P(X = 5 \cap Y = 4)$ (Section 4.7),

   b. $P(X \geq 7 \cap Y \geq 3)$ (Section 4.7),

   c. $P(X \geq 7 \cup Y \geq 3)$ (Section 4.8),

   d. $P(X = 5 \cup Y = 4)$ (Section 4.8),

   e. $P(X + T \leq 2)$ (Sections 4.7, 4.8).

3. Same as Problem 1 except that sampling from company B is without replacement. Compare the results of the two problems.

Six

The Normal Distribution

6.1 Introduction

It would be wise of you to review Sections 4.11 and 4.13 at this point, particularly the material on continuous probability distributions. Next to probability itself, the material associated with the normal distribution—the subject of this chapter—is perhaps the most basic body of knowledge in statistics. Hence, the importance of a keen understanding of related concepts cannot be overemphasized. Indeed, the concepts of the normal distribution indirectly contribute greatly to one's understanding of all statistics.

As you progress through this chapter, it may appear at first glance that many examples are repetitious. Not so. For the most part, each example illustrates a *different* principle. Elementary aspects of the normal distribution are given extensive treatment because of the importance of the normal distribution in its own right and because its concepts underlie other important continuous probability distributions. Hence, thorough coverage here will eliminate the necessity of reviewing the techniques each time we encounter a new probability distribution in subsequent chapters.

Unfortunately the real significance of the principles, techniques, and applications of the normal distribution is too often obscured by a mechanistic treatment. Although our aim is not mathematical sophistication, we cannot neglect the relation between the mechanics of problem solving involving the normal distribution and the reasoning behind these mechanics. Therefore, we shall cover some of the more elementary theoretical aspects of the normal distribution.

6.2 Background Information

We are concerned here with the specific continuous probability distribution known as the *normal distribution.* What does it mean to say that the variable X has a normal distribution? Simply that the probabilities associated with X can be described in a specific fashion, which is characterized by a form called a normal distribution. The fact can be expressed mathematically by the probability density function

$$f(x) = \frac{1}{b\sqrt{2\pi}} \exp\left[-\frac{1}{2}\left(\frac{x-a}{b}\right)^2\right] \quad (-\infty < x < \infty), \tag{1}$$

where exp represents e raised to the power indicated by the bracketed quantity that follows, x represents the possible value of X, π and e are constants (π being the ratio of the circumference of a circle to its diameter, $\cong 3.14159$, and e being the base of the Napierian logarithms, $\cong 2.71828$), and a and b are parameters associated with the probability distribution. As indicated, the range of possible values for the continuous variable having this distribution is theoretically $-\infty$ to $+\infty$.

If we assume values for a and b, we can plot $f(x)$ on a vertical axis and possible values of X on a horizontal axis, thus obtaining a graph of the probability distribution of the variable X. If $a = 50$ and $b = 10$, for example, we obtain for selected values of X the corresponding values for $f(x)$, as shown in Columns 1 and 7, respectively, of Table 6.1. Columns 2 through 6 show intermediate steps in the evaluation of $f(x)$ in

Table 6.1

Selected computed values of a normal curve, $a = 50$, $b = 10$

(1)	(2)	(3)	(4)	(5)	(6)	(7)
x	$\frac{1}{b\sqrt{2\pi}}$	$x - a$	$\frac{x-a}{b}$	$-\frac{1}{2}\left(\frac{x-a}{b}\right)^2$	$\exp\left[-\frac{1}{2}\left(\frac{x-a}{b}\right)^2\right]$	$f(x) = \frac{1}{b\sqrt{2\pi}} \times$ Column 6
20	0.03989	−30	−3	−4.5	0.011109	0.00044
30	0.03989	−20	−2	−2.0	0.135335	0.00540
40	0.03989	−10	−1	−0.5	0.606531	0.02420
50	0.03989	0	0	0	1.000000	0.03989
60	0.03989	10	1	−0.5	0.606531	0.02420
70	0.03989	20	2	−2.0	0.135335	0.00540
80	0.03989	30	3	−4.5	0.011109	0.00044

each particular case. The graph of the probability distribution that results when many such points are plotted is shown in Figure 6.1. (Recall from Section 4.13 that, since $f(x)$ is a continuous probability density function, the height of the curve at a given value of X cannot be viewed as the probability that X assumes such value.) The points indicated are those actually calculated in Table 6.1. The most obvious characteristics of the resulting curve are its bell shape and symmetry about the center point, $a = 50$.

If we were to repeat the process for other combinations of assumed values of a and b, we would obtain different curves but all would have the same general shape and characteristics. Thus, we see that the so-called normal distribution is really a *family of curves*. A specific normal distribution is obtained only when we fix the values of a and b. That is, the normal distribution is completely characterized by the two parameters a and b.

Two major points of interest occur immediately. First, although we have referred to the normal distribution as a probability distribution, we have not established that

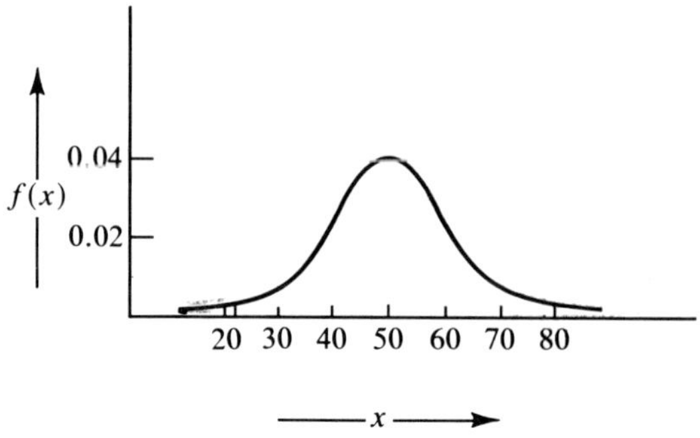

Figure 6.1
Graph of normal curve of Table 6.1

$f(x)$ of Equation (1) meets the requirements of a probability density function. Then, if it is in fact a probability density function, we would like to investigate the nature of the parameters a and b. As one might expect, the normal distribution does satisfy the general requirements for a probability distribution. Further, it can be shown that a represents the mean of the normal distribution, while b represents the standard deviation. Both of these facts are demonstrated by Mood and Graybill (6.1). We merely review here enough of the facts to crystallize the meaning. Recall from Chapter 4 the requirements that qualify a function as a probability density function:

1. It must be nowhere negative, that is, $f(x) \geq 0 \quad (-\infty < x < \infty)$.
2. The area under the curve given by the function must be equal to one.
3. The probability that the variable X will take on a value between any two possible values x_1 and x_2 is represented by the area under the curve between the two points x_1 and x_2.

Thus, the normal distribution is a genuine probability distribution, because all three requirements are satisfied by the function of Equation (1). Having emphasized earlier that the mean and the variance are important parameters associated with the probability distribution of a random variable, let us consider the mean and variance of the normal distribution. If we were to perform the mathematical operations necessary to determine the mean and the variance of the normal distribution, we would find that

$$E(X) = a \tag{2}$$

and

$$\text{Var}(X) = E[X - E(X)]^2 = b^2. \tag{3}$$

In other words, we would find that in Equation (1) a and b represent, respectively, the mean and the standard deviation of the normal variable X. (We often refer to a variable having a normal distribution as a *normal variable*.) Therefore, we write the probability density function of a normal variable X as

$$f(x) = \frac{1}{\sigma_x\sqrt{2\pi}} \exp -\frac{1}{2}\left(\frac{x - \mu_x}{\sigma_x}\right)^2 \quad (-\infty < x < \infty), \tag{4}$$

where μ_x and σ_x^2 represent the mean and variance of X, respectively. For brevity, we shall use the notation

$$X \sim \mathrm{N}(\mu_x, \sigma_x^2), \tag{5}$$

read as "X is distributed normally with mean μ_x and variance σ_x^2," to imply that the probability density function of X is as given by (4).

We have already seen what happens to the mean and the variance of any variable (say Y) when we make the transformation

$$Z = (Y - \mu_y)/\sigma_y;$$

we obtain a new variable Z with mean zero and variance one. This also holds true for a normal variable; in fact, we obtain an even better result by such a transformation, as follows:

Rule 6.1

If X is distributed normally with mean μ_x and variance σ_x^2, the transformation $Z = (X - \mu_x)/\sigma_x$ results in a variable Z which is normally distributed with mean zero and variance one; that is,

Given: $X \sim \mathrm{N}(\mu_x, \sigma_x^2)$.

Transformation: $Z = \dfrac{X - \mu_x}{\sigma_x}$.

Result: $Z \sim \mathrm{N}(0, 1)$.

This rule is not to be taken lightly. For many types of probability density functions, analogous results do not hold. However, its truth is vital for any variable having a normal distribution, as you will discover in subsequent developments.

In the foregoing case, the probability density function for the variable Z becomes

$$f(z) = \frac{1}{\sqrt{2\pi}} \exp\left(-\frac{z^2}{2}\right) \quad (-\infty < z < \infty). \tag{6}$$

Note that this exact form of the probability density function would be obtained if μ_x were replaced by 0 and σ_x by 1 in the probability density function given by (4). A variable having this specific probability distribution is commonly called a *standard normal variable*.

Problems

1. Construct a graph of a normal distribution with $a = 100$ and $b = 5$. Plot $f(x)$ on the vertical axis and X on the horizontal axis. First, construct a table in the form of Table 6.1 and calculate $f(x)$, accurate to 5 decimal places, for X equal to

   85, 90, 95, 100, 105, 110, and 115.

   Plot these points and connect them with a smooth curve.

2. a. Construct a table with the same format as Table 6.1 for a normal distribution with $a = 500$ and $b = 20$. Calculate $f(x)$ for values of X equal to

$$440, 460, 470, 480, 490, 500, 510, 520, 530, 540, \text{ and } 560,$$

rounding the results to 5 decimal places.

b. Construct a graph of the normal distribution in Part a by plotting the points and connecting them with a smooth curve.

3. Indicate the mean and standard deviation of each of the following variables.

a. $X \sim N(5, 100)$, b. $X \sim N(-100, 25)$, c. $X \sim N(800, 225)$,

d. Z where $X \sim N(75, 25)$ and $Z = \dfrac{X - 75}{5}$, e. a standard normal variable.

4. Given $X \sim N(500, 400)$, indicate the transformation required to transform X into a standard normal variable.

6.3 Properties of the Normal Curve

As we have just seen, the normal curve can be obtained simply by plotting, for various possible values of the normal variable X, values of the height of the curve associated with the corresponding probability density function. Thus, the normal curve is merely a graphical representation of the normal distribution. The nature of the normal probability density function endows the normal curve with certain properties that we state without proof:

1. The curve is bell-shaped and symmetrical about the mean μ_x.
2. Consequently, the mean bisects the area under the curve, and takes on the same value as the median and mode.
3. Theoretically, the range of the curve is from negative infinity to positive infinity; thus the curve comes closer and closer to the x-axis as it moves away from the mean in either direction.

Recall from Chapter 4 that, in any probabilistic situation, the sum of the probabilities of all possible mutually exclusive events is equal to 1. We saw in Section 6.2 that we actually represent graphically the probability of all possible outcomes of the normal variable X by the area under the normal curve (such area being equal to 1). Thus the probability of any subset of all possible outcomes can be obtained by finding the area under the normal curve which corresponds to that subset. From the foregoing properties of the normal curve we can derive a rule that applies to any normal distribution. Later you will see that the rule is simply a special case of the information obtainable from the area under the normal curve.

Rule 6.2

Given the normal curve for a normal variable with mean μ_x and variance σ_x^2 (standard deviation σ_x),

1. *the area under the curve between points located one standard deviation in each direction from the mean is* 0.6826,
2. *the area under the curve between the points located two standard deviations in each direction from the mean is* 0.9544,
3. *the area under the curve between the points located three standard deviations in each direction from the mean is* 0.9974.

Problems

Solve by using Rule 6.2 and logical reasoning.

1. Given $Z \sim N(0, 1)$, find the area under the curve between

   a. 1 and 2, b. -1 and 2, c. -3 and -1, d. -3 and 1.

2. Given $X \sim N(200, 100)$, find the area under the curve between

   a. 180 and 220, b. 200 and 230, c. 190 and 220, d. 180 and 190.

3. Given $X \sim N(19, 9)$, find the area under the curve between

   a. 10 and 16, b. 16 and 28, c. 22 and 28, d. 13 and 22.

6.4 Normal Area (Probability) Table

How do we determine the probability that a sample value chosen randomly from any normal population will come from a specific subset of that population? We shall begin by discussing the question for a specific normal population, that is, one with mean 0 and variance 1.

Suppose we have a normal variable Z such that $Z \sim N(0, 1)$. We should like to know the probability of a sample value being between any two given numbers (say z_1 and z_2) when the sample is selected randomly from the Z population. In other words, given $z_1 < z_2$, and $Z \sim N(0, 1)$,

$$P(z_1 < Z < z_2) = ? \tag{1}$$

We can answer this question by finding the area under the proper normal curve between the abscissa points z_1 and z_2—a difficult task, since it entails the use of integral calculus. Further, working from definitions, we would have to repeat the task to find the probability associated with any subset of the possible values of Z. Fortunately, there is a table, the normal area table, to help us to find desired probability information without integral calculus. For various values of Z, Table A-1 in the Appendix gives $P(Z > z)$, where $Z \sim N(0, 1)$ and z represents any possible positive value of the variable Z. The shaded area in Figure 6.2 corresponds to that which is plotted in Table A-1 for any positive value of Z.

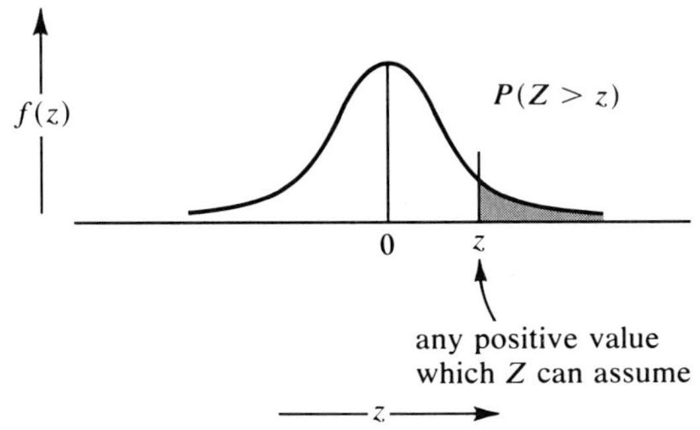

Figure 6.2
Illustration of information given in Table A-1

Values for the probability $P(Z > z)$ are in the body of Table A-1, while various possible values (z) are in the left-hand column and top row of the table. To find $P(Z > z)$, we perform the following steps:

1. Find the whole number and first decimal of z in the left column.
2. Find the second decimal of z in the top row.
3. Read $P(Z > z)$ at the point the column and row intersect.

Example 1

Given $Z \sim N(0, 1)$, find $P(Z > 1.38)$.

Solution

1. Find 1.3 in the left column.
2. Find 08 in the top row.
3. From the intersection of the row and the column read $P(Z > 1.38) = 0.0838$.

□

When it is necessary to find $P(Z > z)$ where z contains more than two decimal places, linear interpolation is ordinarily used. However, bear in mind that results are not completely accurate in this situation.

Example 2

Given $Z \sim N(0, 1)$, find $P(Z > 1.458)$.

Solution

Table A-1 shows $P(Z > 1.45) = 0.0735$ and $P(Z > 1.46) = 0.0721$. Hence, using linear interpolation, $P(Z > 1.458)$ must lie eight-tenths of the distance between 0.0735 and 0.0721. Thus, since

$$0.0735 - 0.0721 = 0.0014 \quad \text{and} \quad 0.8\,(0.0014) \cong 0.0011,$$

we have

$$P(Z > 1.458) = P(Z > 1.45) - 0.0011 = 0.0735 - 0.0011 = 0.0724. \tag{2}$$

□

Naturally, given $P(Z > z)$, one could find z by reversing the foregoing procedure.

Example 3

Suppose Z is a standard normal variable. Find the value of Z (say z^*) above which 8.85 percent of the values lie.

Solution

We find 0.0885 in the body of Table A-1. The whole number and first decimal (1.3) of z appear in column 1 in the same row as 0.0885, while the second decimal (0.05) of z appears in the top row in the same column as 0.0885. From this we infer $z^* = 1.35$. Thus, 1.35 is the value of Z above which 8.85 percent of the values lie. □

Obviously, not every probability problem associated with the standard normal variable can be directly solved in terms of $P(Z > z)$, the probability that the standard normal variable takes on a value greater than a positive number. Since the range of possible values for any normal variable is from $-\infty$ to $+\infty$, we might be interested in any of the following probabilities:

$$P(Z > z), \quad P(Z > -z), \quad P(Z < z), \quad \text{or} \quad P(Z < -z). \tag{3}$$

How can we use Table A-1 to find the last three probabilities?

First, to find $P(Z < z)$, the probability that the standard normal variable is less than some positive value, we note that, since the area under the curve is 1,

$$P(Z < z) = 1 - P(Z > z). \tag{4}$$

Next, to find $P(Z < -z)$, the probability that a standard normal variable is less than some negative value, we note from Figure 6.3 that, since the area under the curve is one and the curve is symmetrical about the value zero,

$$P(Z < -z) = P(Z > z). \tag{5}$$

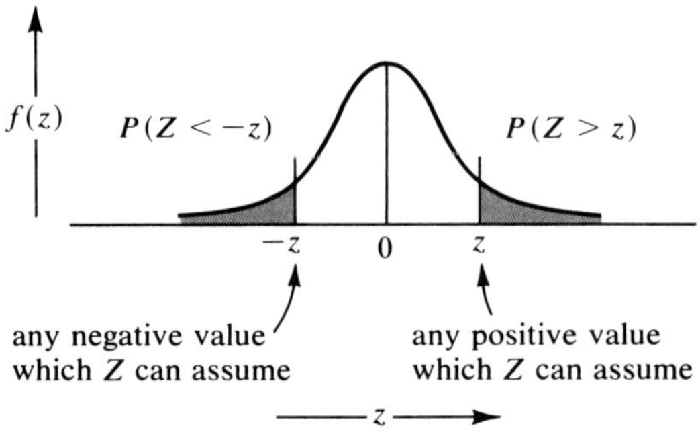

Figure 6.3
Graph of $P(Z < -z) = P(Z > z)$

Finally, to find $P(Z > -z)$, we note that, again since the area under the curve is 1,

$$P(Z > -z) = 1 - P(Z < -z),$$

and, by (5), we have

$$1 - P(Z < -z) = 1 - P(Z > z).$$

Hence,

$$P(Z > -z) = 1 - P(Z > z). \tag{6}$$

Table 6.2 summarizes the information related to Relations (4) through (6).

Table 6.2
How to determine various probabilities from Table A-1

Information desired	*Obtain from Table A-1*
$P(Z > +z)$	$P(Z > +z)$
$P(Z < +z)$	$1 - P(Z > +z)$
$P(Z > -z)$	$1 - P(Z > +z)$
$P(Z < -z)$	$P(Z > +z)$

In the examples that follow, you should in each case draw the standard normal distribution, shade the area required, and note its relation to the information from the table.

Example 4

Given $Z \sim N(0, 1)$, find $P(Z < 1.75)$.

Solution

$P(Z < 1.75) = 1 - P(Z > 1.75) = 1 - 0.0401 = 0.9599.$ □

Example 5

Given $Z \sim N(0, 1)$, find $P(Z < -2.76)$.

Solution

$P(Z < -2.76) = P(Z > 2.76) = 0.0029.$ □

Example 6

Given $Z \sim N(0, 1)$, find $P(Z > -1.15)$.

Solution

$P(Z > -1.15) = 1 - P(Z > 1.15) = 1 - 0.1251 = 0.8749.$ □

A wise student will make a habit of sketching the normal probability distribution and shading the area of interest when solving probability problems such as those of Examples 4–6. Now, how does all the foregoing relate to the general solution of the problem posed in Relation (1)? That is, how do we determine $P(z_1 < Z < z_2)$, where $z_1 < z_2$? Recall that we are merely looking for the area under the standard normal curve between the two points z_1 and z_2. Note in Figure 6.4 that when we subtract the

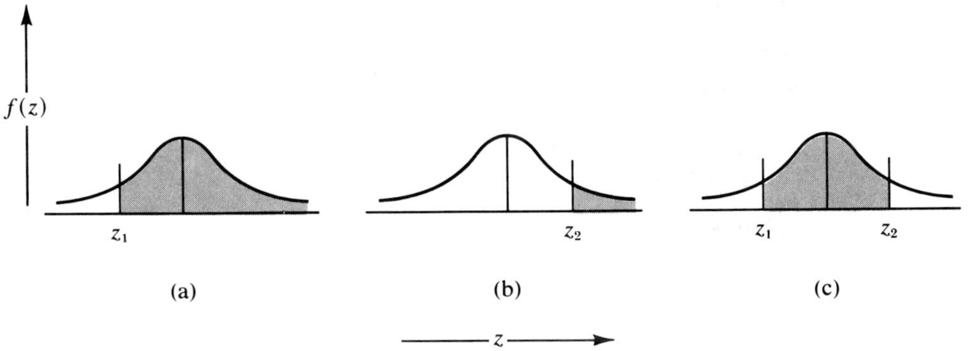

Figure 6.4
Graph of $P(z_1 < Z < z_2) = P(Z > z_1) - P(Z > z_2)$

value represented by the shaded area in (b) from the value represented by the shaded area in (a), we obtain the shaded area in (c). Such would be the case for any curve, as long as the same curve were represented by parts a, b, and c. However, it is obvious in our particular case that (a) of the figure represents $P(Z > z_1)$, (b) represents $P(Z > z_2)$, and (c) represents $P(z_1 < Z < z_2)$. Thus, intuitively, we have

$$P(z_1 < Z < z_2) = P(Z > z_1) - P(Z > z_2), \quad (7)$$

where z_1 and z_2 represent any possible values Z can assume, given $z_1 < z_2$.

Example 7

Given $Z \sim N(0, 1)$, find $P(0.78 < Z < 1.32)$.

Solution

Regardless of the algebraic signs of the two numbers, (7) tells us that, since $0.78 < 1.32$,

$$P(0.78 < Z < 1.32) = P(Z > 0.78) - P(Z > 1.32)$$
$$= 0.2177 - 0.0934 = 0.1243.$$

We obtain the shaded area $[P(0.78 < Z < 1.32)]$ depicted in (c) of Figure 6.5 by subtracting the shaded area in (b) from the shaded area in (a).

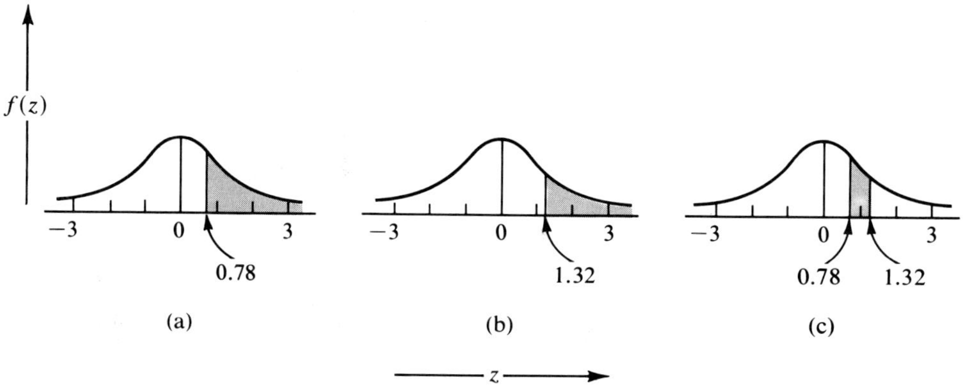

Figure 6.5
Graph of $P(0.78 < Z < 1.32) = P(Z > 0.78) - P(Z > 1.32)$ □

Example 8

Given $Z \sim N(0, 1)$, find $P(-1.18 < Z < 1.46)$.

Solution

By Relations (7) and (6)

$$P(-1.18 < Z < 1.46) = P(Z > -1.18) - P(Z > 1.46)$$
$$= 1 - P(Z > 1.18) - P(Z > 1.46)$$
$$= 1 - 0.1190 - 0.0721 \doteq 0.8089.$$

We obtain the shaded area $[P(-1.18 < Z < 1.46)]$ depicted in (c) of Figure 6.6 by subtracting the shaded area in (b) from the shaded area in (a).

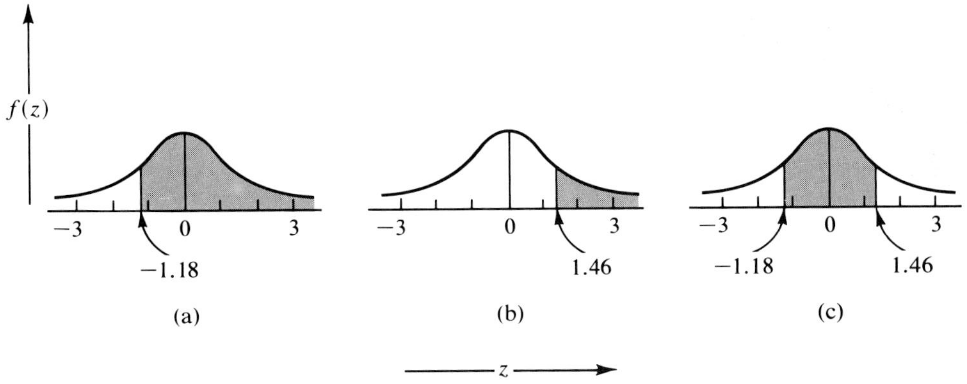

Figure 6.6
Graph of $P(-1.18 < Z < 1.46) = P(Z > -1.18) - P(Z > 1.46)$ □

Example 9

Given $Z \sim N(0, 1)$, find $P(-1.56 < Z < -0.68)$.

Solution

By Relations (7) and (6)

$$\begin{aligned} P(-1.56 < Z < -0.68) &= P(Z > -1.56) - P(Z > -0.68) \\ &= 1 - P(Z > 1.56) - [1 - P(Z > 0.68)] \\ &= P(Z > 0.68) - P(Z > 1.56) \\ &= 0.2483 - 0.0594 = 0.1889. \end{aligned}$$

The shaded area $[P(-1.56 < Z < -0.68)]$ in (c) of Figure 6.7 is obtained by subtracting the shaded area in (b) from the shaded area in (a).

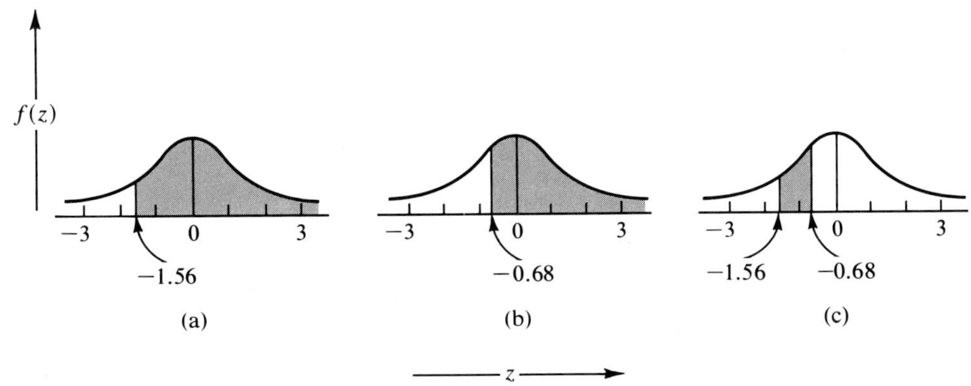

Figure 6.7
Graph of $P(-1.56 < Z < -0.68) = P(Z > -1.56) - P(Z > -0.68)$ □

Problems

1. Given $Z \sim N(0, 1)$, find
   a. $P(Z > 0.72)$, b. $P(Z > 2.19)$, c. $P(Z > 1.745)$, d. $P(Z > 0.193)$,
   e. the value of $Z(z^*)$ above which 13.35% of the values lie,
   f. the value of $Z(z^*)$ above which 3.759% of the values lie,
   g. the value of $Z(z^*)$ above which 41.02% of the values lie,
   h. the value of $Z(z^*)$ above which 0.416% of the values lie.
2. Given $Z \sim N(0, 1)$, find
   a. $P(Z > 1.68)$, b. $P(Z > 2.315)$, c. $P(Z > 0.916)$, d. $P(Z > 2.822)$,
   e. the value of $Z(z^*)$ above which 29.76% of the values lie,
   f. the value of $Z(z^*)$ above which 6.382% of the values lie,
   g. the value of $Z(z^*)$ above which 1.76% of the values lie,
   h. the value of $Z(z^*)$ above which 34.59% of the values lie.
3. Given $Z \sim N(0, 1)$, find
   a. $P(Z < 1.357)$, b. $P(Z < -2.826)$, c. $P(Z > -0.763)$,
   d. $P(Z < -1.988)$, e. $P(Z < 2.146)$, f. $P(Z > -0.342)$,
   g. the value of $Z(z^*)$ below which 31.04% of the values lie,
   h. the value of $Z(z^*)$ above which 5.01% of the values lie,
   i. the value of $Z(z^*)$ above which 89.07% of the values lie,
   j. the value of $Z(z^*)$ below which 67.659% of the values lie.
4. Given $Z \sim N(0, 1)$, find
   a. $P(Z < -2.431)$, b. $P(Z > -0.155)$, c. $P(Z > 1.969)$,
   d. $P(Z < 0.863)$, e. $P(Z > -2.374)$,
   f. the value of $Z(z^*)$ below which 56.36% of the values lie,
   g. the value of $Z(z^*)$ above which 89.52% of the values lie,
   h. the value of $Z(z^*)$ below which 46.87% of the values lie,
   i. the value of $Z(z^*)$ above which 15.27% of the values lie,
   j. the value of $Z(z^*)$ below which 94.81% of the values lie.
5. Given $Z \sim N(0, 1)$, find
   a. $P(1.82 < Z < 2.53)$, b. $P(-2.61 < Z < 0.14)$, c. $P(0.38 < Z < 0.77)$,
   d. $P(-2.25 < Z < -0.98)$, e. $P(-1.76 < Z < -1.43)$, f. $P(-1.87 < Z < 2.31)$.
6. Given $Z \sim N(0, 1)$, find
   a. $P(-2.79 < Z < 2.54)$, b. $P(-1.16 < Z < -0.63)$, c. $P(-2.89 < Z < -1.75)$,
   d. $P(2.07 < Z < 2.64)$, e. $P(0.08 < Z < 1.39)$, f. $P(-1.16 < Z < 0.50)$.

7. Given $Z \sim N(0, 1)$, find z^* such that

   a. $P(Z > z^*) = 0.0885$, b. $P(Z > z^*) = 0.7019$, c. $P(Z < z^*) = 0.0136$,

   d. $P(Z > z^*) = 0.9591$, e. $P(Z < z^*) = 0.4483$, f. $P(Z < z^*) = 0.8749$.

8. Given $Z \sim N(0, 1)$, find z^* such that

   a. $P(Z > z^*) = 0.1894$, b. $P(Z < z^*) = 0.0329$, c. $P(Z < z^*) = 0.9332$,

   d. $P(Z > z^*) = 0.6255$, e. $P(z^* < Z) = 0.0099$, f. $P(z^* > Z) = 0.9756$.

9. Given $Z \sim N(0, 1)$, find z^* such that

   a. $P(z^* < Z < -1.6) = 0.0441$, b. $P(z^* < Z < 0.71) = 0.2372$,

   c. $P(z^* < Z < 2.18) = 0.5490$, d. $P(-1.89 < Z < z^*) = 0.6334$,

   e. $P(-1.13 < Z < z^*) = 0.0914$, f. $P(0.29 < Z < z^*) = 0.3837$.

10. Given $Z \sim N(0, 1)$, find z^* such that

    a. $P(z^* < Z < 0.91) = 0.2160$, b. $P(-2.32 < Z < z^*) = 0.9497$,

    c. $P(z^* < Z < -1.08) = 0.1347$, d. $P(-0.31 < Z < z^*) = 0.0346$,

    e. $P(1.87 < Z < z^*) = 0.0168$, f. $P(z^* < Z < 2.59) = 0.9794$.

11. Suppose we had a table analogous to Table A-1 of the Appendix that gives $P(Z < -z)$, where Z is a standard normal variable, and $-z$ is any negative value Z can assume. Indicate the probabilities we would need to find in the table to solve the following probabilities.

    a. $P(1.54 < Z < 1.93)$, b. $P(-2.46 < Z < 0.78)$, c. $P(-1.34 < Z < -0.96)$.

12. Given a table, analogous to Table A-1, that gives $P(Z > -z)$, where Z is a standard normal variable and $-z$ is any negative value Z can assume, indicate the probabilities it should provide to solve

    a. $P(0.60 < Z < 1.54)$, b. $P(-2.93 < Z < 2.48)$, c. $P(-2.77 < Z < -1.85)$.

13. Solve Problem 1, page 193, using the methods of Section 6.4. Note that the answers are the same.

14. Solve Problem 2, page 193, using the methods of Section 6.4. Note that the answers are the same.

15. Solve Problem 3, page 193, using the methods of Section 6.4. Note that the answers are the same.

6.5 Key Property of the Normal Distribution—Probabilities Associated with General Normal Variables

We are now ready to go about finding the probability that a sample value chosen randomly from a general normal population will come from any given subset of that population. More specifically, we are concerned with finding

$$P(x_1 < X < x_2), \qquad x_1 < x_2, \tag{1}$$

given that $X \sim N(\mu_x, \sigma_x^2)$. Once again, we could use integral calculus to find the area under the proper normal curve between the two abscissa points. But the theoretical solution is even more difficult than in the case of a standard normal variable. To build tables similar to Table A-1 for all possible normal distributions would be an impossible task, since there are countless combinations of the mean μ_x and the variance σ_x^2. Fortunately, however, the property of the normal distribution given by Rule 6.1 enables us to circumvent the difficulty. You will recall that any normal distribution with a known mean μ_x and variance σ_x^2 can be transformed into a standard normal variable by the transformation $Z = (X - \mu_x)/\sigma_x$. Thus, any probability question concerning a general normal variable can be answered in terms of a corresponding probability question concerning a standard normal variable. For example, to find

$$P(x_1 < X < x_2), \tag{2}$$

we use the information

$$P(x_1 < X < x_2) = P(z_1 < Z < z_2), \tag{3}$$

where

$$z_1 = (x_1 - \mu_x)/\sigma_x \text{ and } z_2 = (x_2 - \mu_x)/\sigma_x.$$

Similarly,

$$P(X > x_1) = P\left(Z > \frac{x_1 - \mu_x}{\sigma_x}\right), \tag{4}$$

$$P(X < x_2) = P\left(Z < \frac{x_2 - \mu_x}{\sigma_x}\right), \tag{5}$$

and

$$P(x_3 < X < x_4) = P\left(\frac{x_3 - \mu_x}{\sigma_x} < Z < \frac{x_4 - \mu_x}{\sigma_x}\right), \tag{6}$$

where x_1, x_2, x_3, and x_4 are any possible values which X can assume, subject to the restriction that $x_3 < x_4$.

Example 1

Given $X \sim N(50, 100)$, find $P(55 < X < 65)$.

Solution

We can sketch the distribution of X (knowing that approximately 99.7 percent (Rule 6.2) of the values lie within three standard deviations of the mean) and label the scale as shown in the top part of Figure 6.8. Then, when we make the transformation $Z = (X - \mu_x)/\sigma_x$, we obtain the lower graph in Figure 6.8. Obviously, 60 (mean plus one standard deviation) on the x-scale corresponds to 1 on the z-scale. Further, 20 (mean minus three standard deviations) corresponds to -3 and 30 (mean minus two standard deviations) corresponds to -2, and so forth. Intuitively, it is obvious area. Thus, we use as an approximation,

$$P_b(29 \le X_D \le 46 \mid n = 100, \pi = 0.40) \cong P(28.5 \le X \le 46.5)$$

$$= P\left(\frac{28.5 - 40}{\sqrt{24}} < Z < \frac{46.5 - 40}{\sqrt{24}}\right)$$

$$= P(-2.347 < Z < 1.327) \tag{5}$$

$$= P(Z > -2.347) - P(Z > 1.327)$$

$$= 1 - P(Z > 2.347) - P(Z > 1.327)$$

$$= 1 - 0.0094 - 0.0923 = 0.8983.$$

Note that the approximation of Relation (5) is considerably better than that of Relation (4).

In general, when we wish to approximate binomial probabilities by the area under the appropriate normal curve, we add or subtract the continuity correction factor according to the rules illustrated in Example 1 of the preceding section. Proper use of the continuity correction factor will allow us to use the normal approximation even when we wish to determine the probability that X will assume one specific value. For example, if we wish to know

$$P_b(X = 35 \mid n = 100, \pi = 0.30),$$

we have

$$\mu_x = n\pi = 100(0.30) = 30 \quad \text{and} \quad \sigma_x^2 = n\pi(1 - \pi) = 100(0.3)(0.7) = 21.$$

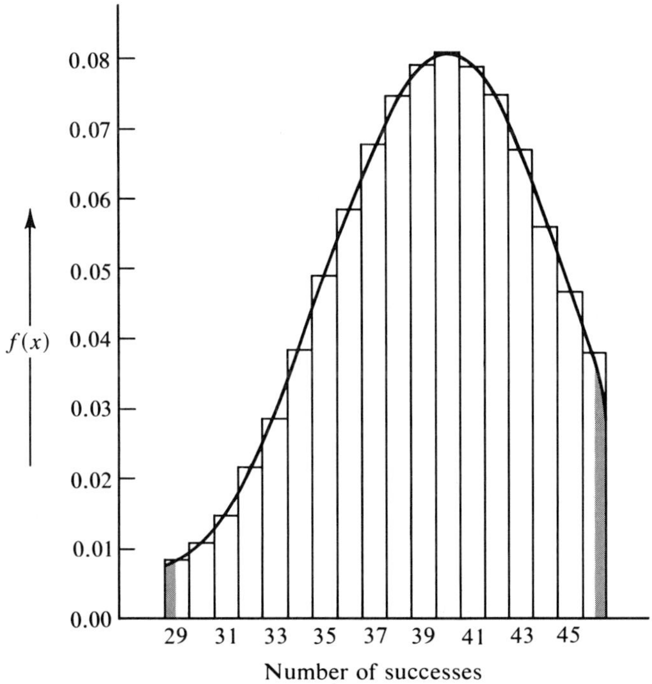

Figure 6.11

Normal approximation of $P_b(29 \le X \le 46 \mid n = 100, \pi = 0.40)$

Thus,

$$\begin{aligned}
P_b(X = 35 \mid n = 100, \pi = 0.30) &\cong P\left(\frac{34.5 - 30}{\sqrt{21}} < Z < \frac{35.5 - 30}{\sqrt{21}}\right) \\
&= P(0.98 < Z < 1.20) \\
&= P(Z > 0.98) - P(Z > 1.20) \\
&= 0.1635 - 0.1151 = 0.0484.
\end{aligned} \tag{6}$$

From Table A-6, the exact probability is 0.0468.

Problems

1. Use the normal distribution to approximate the following binomial probabilities.
   a. $P_b(X \geq 43 \mid n = 100, \pi = 0.50)$.
   b. $P_b(X < 79 \mid n = 100, \pi = 0.80)$.
   c. $P_b(1979 < X < 1988 \mid n = 10{,}000, \pi = 0.20)$.
   d. $P_b(X = 84 \mid n = 100, \pi = 0.90)$.
2. Use the normal distribution to approximate the following binomial probabilities.
   a. $P_b(X > 98 \mid n = 900, \pi = 0.10)$.
   b. $P_b(X \leq 351 \mid n = 1600, \pi = 0.20)$.
   c. $P_b(103 \leq X \leq 112 \mid n = 196, \pi = 0.5)$.
   d. $P_b(X = 67 \mid n = 144, \pi = 0.50)$.

6.9 *Ordinates of the Normal Distribution*

Recall that in Section 6.2, we developed a normal curve by assuming values for the parameters of the normal distribution and plotting the probability distribution function on the vertical axis versus various values of X on the horizontal axis. When we obtain such a graphical display of the probability distribution of the normal variable X, the height of the normal curve at any given value of X is called the ordinate of the corresponding normal curve. In the particular case of a normal variable with mean 0 and variance 1, the ordinate (height) of the normal curve can be obtained by plotting

$$f(z) = \frac{1}{\sqrt{2\pi}} \exp\left(-\frac{z^2}{2}\right)$$

versus various possible values of Z. Such information is given in Table A-2 of the Appendix for positive values of Z. To find the ordinate corresponding to any possible positive value of Z, say z, we carry out the following steps:

1. Find the whole number and first decimal of z in the left column.
2. Find the second decimal of z in the top row.
3. Read the desired ordinate of the standard normal curve from the point where the corresponding column and row intersect.

Since the curve is symmetrical, the ordinate for a negative value of Z will be the same as for the corresponding positive value of Z.

Example 1

Find the height, or ordinate, of the standard normal curve for $z = 1.68$.

Solution

In Table A-2, we find 1.6 in the left column, 0.08 in the top row, and read 0.0973, the proper ordinate, at the intersection of the row and column. □

As indicated in Section 6.2, it is possible to obtain a value for the ordinate of any normal curve simply by calculating, for a given value of the random variable X, the corresponding value given by the mathematical representation of the normal curve

$$f(x; \mu_x, \sigma_x^2) = \frac{1}{\sigma_x\sqrt{2\pi}} \exp\left[-\frac{1}{2}\left(\frac{x-\mu_x}{\sigma_x}\right)^2\right].$$

However, once again the nature of the normal distribution, spares us this arduous task. The table of ordinates of the standard normal curve provides sufficient information to determine ordinates associated with any normal curve. We can find the ordinate of the normal variable X corresponding to any given possible value of X (say x) by performing the following steps:

1. Compute $z = (x - \mu_x)/\sigma_x$.
2. Use Table A-2 to determine $f(z)$, the ordinate of the standard normal variable Z corresponding to the value z.
3. Calculate $f(x)$, the ordinate of the normal variable X corresponding to the value x by $f(x) = (1/\sigma_x)f(z)$.

Example 2

Use Table A-2 of the Appendix to verify the ordinates obtained in column 7 of Table 6.1 (page 189) for selected values of X, where $X \sim N(50, 100)$.

Solution

The solution is given in Table 6.3. Column 1 presents values of the variable X for which ordinates are desired, column 2 gives corresponding values of $z = (x - \mu_x)/\sigma_x$, column 3 gives corresponding ordinates (read from Table A-2 of the Appendix) of the standard normal variable, $f(z)$, and column 4 provides values of $f(x)$, the ordinates corresponding to the various X values. Note that the results of column 4 are the same as those obtained in column 7 of Table 6.1 through a direct calculation of $f(x)$.

Table 6.3
Selected values of a normal curve ($a = 50$, $b = 10$) determined from Table A-2

(1) x	(2) $z = \frac{x - \mu_x}{\sigma_x}$	(3) $f(z)$ (*from Table A-2*)	(4) $f(x) = \frac{1}{\sigma_x} f(z)$
20	−3	0.0044	0.00044
30	−2	0.0540	0.00540
40	−1	0.2420	0.02420
50	0	0.3989	0.03989
60	1	0.2420	0.02420
70	2	0.0540	0.00540
80	3	0.0044	0.00044

□

Problems

1. Find the height (ordinate) of the standard normal curve for each of the following Z values.
   a. $Z = 2.47$. b. $Z = 0.81$. c. $Z = 1.563$. d. $Z = 0.905$.
2. Find the height (ordinate) of the standard normal curve for each of the following Z values.
   a. $Z = 1.21$. b. $Z = -0.74$. c. $Z = 2.355$. d. $Z = -1.686$.
3. Use Table A-2 to verify the ordinates obtained in column 7 of the table prepared in Problem 1, page 191, for selected values of X, where $X \sim N(100, 25)$. Use the same format as that of Table 6.3.
4. Use Table A-2 to verify the ordinates obtained in column 7 of the table prepared in Problem 2, page 192, for selected values of X, where $X \sim N(500, 400)$. Use the same format as in Table 6.3.

6.10 Properties of Linear Combinations of Normal Variables

In many practical applications, it is necessary to determine the probability distribution of a random variable that results from a linear combination of other random variables. Properties associated with linear combinations of normal variables are particularly important to our further development of statistical theory. For now we shall merely state the properties without proof and exemplify their utility. While the examples are not completely realistic, they will give you facility in applying probabilistic reasoning to business-related problems.

If $X_1, X_2, \ldots, X_n$ are independent variables, the random variable Y obtained by adding such variables,

$$Y = X_1 + X_2 + \cdots + X_n, \tag{1}$$

has mean

$$\mu_y = \mu_1 + \mu_2 + \cdots + \mu_n \tag{2}$$

and variance

$$\sigma_y^2 = \sigma_1^2 + \sigma_2^2 + \cdots + \sigma_n^2. \tag{3}$$

However, when the n independent random variables are normally distributed, so that,

$$X_i \sim \mathrm{N}(\mu_i, \sigma_i^2) \qquad (i = 1, 2, \ldots, n),$$

the variable resulting from their addition is also normally distributed with mean and variance as given by (2) and (3), respectively.

Theorem 6.1

The sum of any number of independent normally distributed variables is itself normally distributed with mean equal to the sum of the means of the summed variables and variance equal to the sum of the variances of the summed variables.

Also, for any two independent normally distributed variables

$$X_1 \sim \mathrm{N}(\mu_1, \sigma_1^2) \quad \text{and} \quad X_2 \sim \mathrm{N}(\mu_2, \sigma_2^2),$$

the variable $W = X_1 - X_2$ has the properties

$$W \sim \mathrm{N}(\mu_1 - \mu_2, \sigma_1^2 + \sigma_2^2). \tag{4}$$

It can be shown through use of Relation (4) and Theorem 6.1 that the following more general theorem holds.

Theorem 6.2

A linear function of independent normally distributed variables

$$V = a_1X_1 + a_2X_2 + \cdots + a_nX_n + b$$

(where $a_1, a_2, \ldots, a_n$, and b are constants) is itself normal with mean

$$\mu_v = a_1\mu_1 + a_2\mu_2 + \cdots + a_n\mu_n + b$$

and variance

$$\sigma_v^2 = a_1^2\sigma_1^2 + a_2^2\sigma_2^2 + \cdots + a_n^2\sigma_n^2.$$

Example 1

Given that X_1, X_2, and X_3 are independent with

$$X_1 \sim \text{N}(5, 10), \quad X_2 \sim \text{N}(12, 20), \quad \text{and} \quad X_3 \sim \text{N}(15, 13);$$

find the mean, variance, and distribution of

$$T = 3X_1 - 5X_2 + 4X_3 + 8.$$

Solution

From Theorem 6.2

$$\begin{aligned} \mu_t &= 3\mu_1 - 5\mu_2 + 4\mu_3 + 8 \\ &= 3(5) - 5(12) + 4(15) + 8 = 23 \end{aligned}$$

and

$$\begin{aligned} \sigma_t^2 &= 9\sigma_1^2 + 25\sigma_2^2 + 16\sigma_3^2 \\ &= 9(10) + 25(20) + 16(13) = 798. \end{aligned}$$

Thus,

$$T \sim \text{N}(23, 798).$$

□

Example 2

Suppose a company markets three major products: W, which is sold only in South America, X, sold only in Australia, and Y, sold only in Europe. Historical data on monthly sales of the three products indicate that no trend (with regard to time) is evident and that sales for each product can be viewed as normally distributed. The means and standard deviations of sales (in thousands of dollars) of the products are

$$\begin{aligned} \mu_w &= 16.5, & \sigma_w &= 1.2 \\ \mu_x &= 12.7, & \sigma_x &= 1.6 \\ \mu_y &= 14.9, & \sigma_y &= 1.8. \end{aligned}$$

The company is negotiating for a substantial loan. Bank officials, having examined all important company records, have determined that the loan should be made if the probability that total company sales will exceed \$50,000 per month (based on past experience) is greater than 0.1. Using this criterion, will the bank make the loan?

Solution

Since the products are marketed on three separte continents, we may assume that the sales of the three products are independent. Thus, if we let T = total sales,

$$\mu_t = \mu_w + \mu_x + \mu_y = 16.5 + 12.7 + 14.9 = 44.1$$

and

$$\sigma_t^2 = \sigma_w^2 + \sigma_x^2 + \sigma_y^2 = 1.44 + 2.56 + 3.24 = 7.24.$$

That is, by Theorem 6.1 we have $T = \text{N}(44.1, 7.24)$. Then, the probability that total sales will exceed \$50,000 per month can be found by

$$P(T > 50) = P\left(Z > \frac{50 - 44.1}{\sqrt{7.24}}\right) = P(Z > 2.19) = 0.0143.$$

Since this is much less than 0.1, the bank will not make the loan. □

Problems

1. Given that X_1, X_2, and X_3 are independent with

$$X_1 \sim \text{N}(50, 25), \quad X_2 \sim \text{N}(65, 80), \quad \text{and} \quad X_3 \sim \text{N}(10, 100),$$

find the mean, variance, and distribution of

a. $R = 11X_1 + 9X_2 - 15X_3 - 75$,

b. $S = 2X_1 - 13X_2 - 3X_3 + 18$,

c. $T = 25X_3 - 14X_1 - 7$.

2. Given that X_1, X_2, and X_3 are independent with

$$X_1 \sim \text{N}(75, 15), \quad X_2 \sim \text{N}(-30, 43), \quad \text{and} \quad X_3 \sim \text{N}(-94, 8),$$

find the mean, variance, and distribution of

a. $T = 6X_1 + 3X_2 - 13X_3 + 4$,

b. $U = -5X_1 - 19X_2 + 7X_3 - 10$,

c. $V = -X_1 + 13X_2 - 6X_3 + 19$.

3. Given that X_1, X_2, and X_3 are independent with

$$X_1 \sim \text{N}(50, 100), \quad X_2 \sim \text{N}(75, 25), \quad \text{and} \quad X_3 \sim \text{N}(\mu_3, \sigma_3^2)$$

and

$$T \sim \text{N}(78, 22{,}365), \quad \text{where } T = 5X_1 - 3X_2 + 7X_3 - 17,$$

find μ_3 and σ_3^2.

4. Given that X_1, X_2, and X_3 are independent with

$$X_1 \sim \text{N}(-30, 36), \quad X_2 \sim \text{N}(\mu_2, \sigma_2^2), \quad X_3 \sim \text{N}(47, 19),$$

and

$$T \sim \text{N}(-311, 1{,}205), \quad \text{where } T = 2X_1 + 8X_2 - 5X_3 + 104,$$

find μ_2 and σ_2^2.

5. Suppose $X \sim \text{N}(75, 400)$. For $Y = 50 + 10X$, find $P(650 < Y < 850)$.

6. Suppose $X \sim \text{N}(130, 100)$. For $Y = 4X - 60$, find $P(482 < Y < 514)$.

Important Terms and Concepts

continuity correction factor

normal area table

normal curve

normal distribution

normal ordinate

normal variable

standard normal variable

References

6.1 A. M. Mood and F. A. Graybill, *Introduction to the Theory of Statistics*, New York: McGraw-Hill Book Company, 1963.

6.2 H. G. Romig, *50–100 Binomial Tables*, New York: John Wiley & Sons, Inc., 1953.

Review Problems

1. Suppose we conduct the experiment of drawing five independent values from a continuous infinite population which is *normally distributed* with mean 50 and variance 100. What is the probability that exactly 3 of the 5 values will be greater than 66.45? (*Hint*: This is a Bernoulli process—Section 5.2).

2. In Problem 1, what is the probability that from 10 to 17 *inclusive* of the values selected in 100 draws will be between 60.6 and 67.0?

3. Given $X \sim N(75, 25)$ and $Y \sim N(100, 225)$, X and Y independent, determine

   a. $P(69 < X < 82$ *and* $88 < Y < 97)$,

   b. $P(73 < X < 85$ *or* $76 < Y < 82)$,

   c. $P(63 < X < 71$ *and* $68 < X < 74)$,

   d. $P(106 < Y < 121$ *or* $115 < Y < 133)$.

   (Sections 4.6–4.8).

4. A company is competing for a subcontract to produce automobile engines. It has produced a trial run of 100 engines which were found to have horsepowers normally distributed with mean 260 and variance 25. The tooling up and production costs of the trial run were \$150,000. The automobile manufacturer will select an engine at random from the trial run and test its horsepower to determine its action on the subcontract. If the horsepower of the tested engine is over 256, a subcontract for 20,000 of the engines will be awarded at a profit of \$50 per engine for the subcontractor. If the horsepower is between 251 and 256, a subcontract for 10,000 of the engines at a profit of \$25 each will be awarded. If the horsepower is under 251, no contract will be awarded. The trial-run engines cannot be sold in any case and the company will not be reimbursed for them.

   What is the expected monetary value of competing for the subcontract?

   (Section 4.12)

5. This problem is the same as Problem 4 with the following modifications:

 a. The company has not yet committed the \$150,000 to seeking the contract and may instead invest in a proposal that guarantees a net return of \$75,000.

 b. While no trial-run engines have been produced, past experience indicates that the distribution of horsepowers will be as in Problem 4 if the trial run is made.

 c. The automobile manufacturer will randomly select 5 engines with replacement for testing and an engine will be classified as "good" if its horsepower is between 250.2 and 269.8.

 d. A subcontract for 10,000 engines at a profit of \$25 per engine will be awarded only if all 5 engines are "good." Otherwise, no contract will be awarded.

 Determine whether the company should begin the trial production run and seek the subcontract or adopt the guaranteed return proposal on the basis of expected monetary value.

 (Sections 4.12 and 5.2)

6. Given $X \sim N(120, 100)$ and $Y \sim N(180, 400)$, determine the value (say q) so that $P(X > q) = P(Y < q)$.

7. An aircraft manufacturer is seeking a supply of rivets with a minimum tensile breaking point of 4,100 lbs. Failure of any rivet to meet this specification could result in an aircraft disaster. Four different rivet manufacturers have a product which meets the basic requirements. Selection is to be based on minimizing the probability of failure.

 a. The mean breaking points of rivets produced by Companies *A*, *B*, *C*, and *D* are 4,200, 4,400, 4,600, and 5,000 lb, respectively. Should the rivets manufactured by Company *D* be chosen on the basis of this information only? Why?

 b. The standard deviations of the breaking points of the rivets of Companies *A*, *B*, *C*, and *D* are 200, 100, 200, and 400 lb, respectively. Is this information pertinent to the decision? Why?

 c. Assuming that the distributions are normal, what percentage of the rivets produced by each company fail to meet the specifications? Which supplier should be chosen on the basis of this criterion? Is it the best criterion for this decision? Why?

8. The scores of new management trainees (X) of a large corporation on the company aptitude test are normally distributed with mean 600 and standard deviation 100.

 The salaries of the trainees (Y) after 3 years assume the following distributions according to the test scores:

$$\begin{array}{cc} X < 550 & Y \sim N(7{,}000;\ 10{,}000) \\ 550 < X < 675 & Y \sim N(10{,}000;\ 1{,}000{,}000) \\ X > 675 & Y \sim N(15{,}000;\ 9{,}000{,}000) \end{array}$$

 Determine the probability that a trainee

 a. with a score of 580 will have a salary of over \$11,500 after 3 years,

 b. with a score under 675 will have a salary between \$8,000 and \$9,500 after 3 years,

 c. will have a salary under \$10,000 after 3 years.

 (Section 4.6)

Seven

Sampling Distributions

7.1 Introduction

Most people think of the word "statistics" only as the title of a discipline. Few realize that it has a very special and well-defined meaning all its own. In fact, most people mistakenly think that "statistics" and "numbers" are synonymous (thus, they view a statistician as, for example, the person who keeps track of the yardage gained by a team in the National Football League). It is true that numbers play a large role in the discipline of statistics and further, that numerical data of any kind are almost universally referred to as statistics. Analysts of all kinds work with "mileage statistics," "birth and death statistics," "consumer statistics," and so forth. However, to call this raw data "statistics" represents a very narrow interpretation of the term. Worse, it contributes greatly to peoples' misconceptions because it obscures even the elementary techniques associated with the discipline.

So far, we have discussed probability distributions, properties and characteristics of random variables, and the meaning of a probability distribution being defined on the possible values of a random variable. In fact, you should now be so conditioned that certain questions "pop to mind" at the mention of a variable "having a probability distribution"—questions about the shape of the distribution, the mean and variance associated with the variable, and even the exact mathematical function that will describe the distribution completely. Obviously variables play a vital role in statistics. However, just as the fact that "statistics involves numbers" fails to distinguish the statistician from accountants (and others), so does the fact that "statistics involves variables" fail to distinguish the statistician from the mathematician (and others). Then, what does set the statistician apart? We shall see that there are entities termed statistics which play this role. Further, we shall come to see even more clearly why statisticians are so interested in the properties of random variables and their distributions. It is a basic and important fact that these entities, which are functions of random variables (and, hence, are themselves special random variables), have the properties and characteristics of random variables. Thus, we shall learn, for example, that these entities called statistics have probability distributions, means, and variances, to name only a few of the properties. Finally, we shall see that when the "statistician deals with numbers" in a particular way, he is doing so because the body of theory associated with these entities called statistics in this discipline called statistics dictates that he do so.

7.2 Parameters and Statistics

In order to discuss parameters and statistics, we must refer repeatedly to populations and samples. Throughout the remainder of the text, therefore, we shall use the following symbols to refer to specific related calculations:

μ_x: population mean;	$\bar{x}$: sample mean;
σ_x^2: population variance;	s_x^2: sample variance;
N: population size;	n: sample size.

Note that, in denoting the population size and sample size, we have departed from our policy of having Greek symbols represent population characteristics and Roman symbols represent sample characteristics. Nevertheless, the convention of using N and n to represent the size of the population and the sample, respectively, has been so nearly universally adopted by statisticians that we feel the departure is warranted in this case.

Although we have mentioned the terms "parameter" and "statistic" (Sections 3.1 and 7.1), we have not defined them. Definitions are necessary to build the basic conceptual framework that promotes a facility in statistical inference.

Definition 7.1

*A **parameter** is a population descriptive measure that characterizes (either partially or completely) the probability distribution of a random variable.*

For example, when we fix the value of the entity λ in the Poisson probability function,

$$p(x; \lambda) = \frac{e^{-\lambda}\lambda^x}{x!} \quad (x = 0, 1, 2, \ldots), \tag{1}$$

we have completely characterized the probability distribution. Recall from Section 5.2 that in this function, λ is the mean of the distribution. The probability distribution in question is completely and uniquely determined simply through knowledge of the parameter λ. On the other hand, fixing the value of μ_x in the probability density function for a normal variable,

$$f(x; \mu_x, \sigma_x^2) = \frac{1}{\sigma_x\sqrt{2\pi}} \exp\left[-\frac{1}{2}\left(\frac{x - \mu_x}{\sigma_x}\right)^2\right] \quad (-\infty < x < \infty), \tag{2}$$

only partially characterizes the distribution, since one parameter (σ_x) may remain unspecified. Recall that we completely characterized a normal distribution in Section

6.2 when we specified $\mu_x = 50$ and $\sigma_x^2 = 100$. Thus, we plotted the normal curve and from it could extract any probabilistic type of information desired.

Two points regarding parameters are so important that you should fix them firmly in mind. First, a parameter is always associated with an entire population. In fact, it is not even necessary that the population have associated with it a mathematical function which represents the probability distribution. For example, a parameter may be viewed simply as any descriptive measure that can be calculated, counted, or noted through consideration of every possible observation in the entire population of interest. Second, although the parameter may be an unknown quantity, it is always viewed in the classical approach to statistics as a fixed entity (a constant). Among other things, these two points greatly facilitate the understanding of the distinction between a parameter and a statistic.

***Definition** 7.2*

*A **statistic** is any descriptive measure that can be calculated, counted, or noted through use of a random sample (a subset of the possible observations in a given population).*

Suppose we have the set of data

$$X: x_1 = 3; \quad x_2 = 5; \quad x_3 = 4; \quad x_4 = 8; \quad x_5 = 10.$$

Suppose further that we view the set of data as a population and seek the value of the population mean (parameter). Representing the population size by N, we obtain

$$\mu_x = \frac{1}{N}\sum_{i=1}^{N} x_i = \frac{1}{5}(30) = 6.$$

On the other hand, suppose we view the set of data as a sample of size 5 taken from a larger population. We find the value of the sample mean (a statistic) to be

$$\bar{x} = \frac{1}{n}\sum_{i=1}^{n} x_i = \frac{1}{5}(30) = 6.$$

Since the results are identical, the calculations shed no light whatsoever on the difference between a parameter and a statistic. However, the difference becomes clear when we view the manner in which the set of data was obtained in each instance. If we were to repeat the above experiment, we would always obtain a value of 6 in the calculation of the population mean, since the set of data will never change. The five observations represent all possible observations. However, when we view the five items as a sample, they do not represent all possible observations. They are simply the five observations which occurred in the one sample drawn from some larger population. The next repeat of the sampling procedure may well result in a different sample of size 5 and, thus, in a different value of the sample mean (a statistic).

Note that, in order to identify a given parameter, we need only define the population of interest and state the formula of (or method of obtaining) the given descriptive measure in terms of the observations of the population.

In order to identify a given statistic, we must not only define the population and state the formula of (or method of obtaining) the given descriptive measure in terms of the observations of a sample but also describe the sampling procedure to be used.

This one point is the essence of the difference between a parameter and a statistic. After identifying the parameter of interest, we can obtain the one and only value of the parameter simply by use of the appropriate formula or method. However, in general, there is no one specific value that a statistic will assume, no matter how explicitly it is identified. The value the statistic assumes in any instance depends on the particular sample selected. Thus, unlike a parameter, which is a constant, a statistic may assume any of various possible values according to the result of the given sampling process. Thus, we see that a statistic is a random variable.

Example 1

Suppose we have the population of values

$$X: x_1 = 3; \quad x_2 = 5; \quad x_3 = 11; \quad x_4 = 7; \quad x_5 = 9.$$

We wish to (a) compute the parameter called the population mean and (b) select a simple random sample of size 2 without replacement (order disregarded) and determine the value of the statistic called the sample mean. What is the difference between the possible outcomes of (a) and (b)?

Solution

There is only one possibility for the outcome of (a). We know that the result will be

$$\mu_x = \frac{1}{5}\sum_{i=1}^{5} x_i = \frac{35}{5} = 7.$$

However, the result obtained in (b) for the value of the statistic depends entirely on which particular sample of size 2 we randomly select. We may obtain any of the possible samples listed in Table 7.1. Thus, the statistic may assume a value equal to any of the possible sample means corresponding to the various possible samples.

Table 7.1

Possible sample means

Possible sample	*Resulting sample mean*	*Possible sample*	*Resulting sample mean*
$x_1 = 3, x_2 = 5$	$(3 + 5)/2 = 4$	$x_2 = 5, x_4 = 7$	$(5 + 7)/2 = 6$
$x_1 = 3, x_3 = 11$	$(3 + 11)/2 = 7$	$x_2 = 5, x_5 = 9$	$(5 + 9)/2 = 7$
$x_1 = 3, x_4 = 7$	$(3 + 7)/2 = 5$	$x_3 = 11, x_4 = 7$	$(11 + 7)/2 = 9$
$x_1 = 3, x_5 = 9$	$(3 + 9)/2 = 6$	$x_3 = 11, x_5 = 9$	$(11 + 9)/2 = 10$
$x_2 = 5, x_3 = 11$	$(5 + 11)/2 = 8$	$x_4 = 7, x_5 = 9$	$(7 + 9)/2 = 8$

Thus, the sample mean (statistic) may take on any one of the values 4, 5, 6, 7, 8, 9, 10, and we can at best attach probabilities to these possible values. □

We defer the procedure for determining exact probabilities for the example. The important point now is that a statistic, because of "chance" induced by sampling, must be viewed as having the same general characteristics as a random variable, while a parameter is a constant.

Problems

1. What parameter(s) *completely* characterize(s) the binomial distribution? If your answer contains symbols, describe their meaning completely.
2. What parameter(s) *completely* characterize(s) the hypergeometric distribution? Completely describe any symbols in your answer.
3. What parameter(s) *completely* characterize(s) the Poisson distribution? Symbols must be completely described.
4. What parameter(s) *completely* characterize(s) the normal distribution? Describe the symbols you use.
5. State briefly, but clearly, the *one essential* difference between a *parameter* and a *statistic*.
6. For each item in the following list, determine whether the term or symbol is or can best be associated with a parameter (p) or a statistic (s).

   a. μ_x, b. population, c. variable, d. N, e. s_x^2,
   f. sample, g. $\bar{X}$, h. n, i. constant, j. σ_x.

7. Suppose we have the population of values

   $$X: x_1 = 6;\quad x_2 = 8;\quad x_3 = 10;\quad x_4 = 14;\quad x_5 = 16.$$

   We wish to (a) compute the parameter called the *population mean* and (b) select a simple random sample of size 3 without replacement (order disregraded) and compute the statistic called the *sample mean*. What is the difference between the possible outcomes of (a) and (b)? Include a table listing all possible samples and the associated values of the statistic, accurate to one decimal place.

   Repeat the above steps for the population and sample medians and note the differences in the tables.

8. Suppose we have the population of values

   $$X: x_1 = 5;\quad x_2 = 8;\quad x_3 = 9;\quad x_4 = 11;\quad x_5 = 13;\quad x_6 = 14.$$

   We wish to (a) compute the parameter called population range, and (b) select a simple random sample of size 2 without replacement (order disregarded) and compute the statistic called the sample range. What is the difference between the possible outcomes of (a) and (b)? Include a table listing all possible samples and the associated values of the statistic.

   Repeat the above steps for the mean deviation from the mean, including a column for the sample mean ($\bar{x}$) in the table and rounding numbers to one decimal place.

7.3 *Sampling Distributions*

One of the most important concepts in statistical inference is the sampling distribution of a statistic.

Definition 7.3

*The **sampling distribution of a statistic** is the probability distribution which, for a given sampling scheme, associates probabilities with the possible values (or groups of possible values) of the statistic.*

This definition implies several requirements for the designation of a given sampling distribution. Generally speaking, the determination of the sampling distribution of a given statistic involves

1. defining the statistic of interest,
2. selecting a sampling scheme (size, procedure, with or without replacement, etc.), and type of sampling (simple random sampling, probability sampling, etc.),
3. designing the sample space of the statistic (all possible values the statistic can assume under the given sampling scheme), and
4. calculating the probabilities associated with various possible subsets of the designated sample space.

Example 1

Suppose we are to use simple random sampling with replacement (that is, where order is important) to select a sample of size $n = 2$ from the following population of size $N = 5$:

$$X: x_1 = 3; \quad x_2 = 7; \quad x_3 = 11; \quad x_4 = 15; \quad x_5 = 19.$$

Find and plot the sampling distribution of (a) the sample mean and (b) the sample range.

Solution

We can determine each sampling distribution by listing all possible samples, determining the probabilities associated with each, determining the value of the statistic which results from each, and, finally, associating probabilities with the various values the statistic can assume. Table 7.2 gives sufficient information for determining the appropriate sample spaces and building the required sampling distributions. Column 1 presents all possible samples; column 2 gives the probability of each; and columns 3 and 4 show values for the sample mean and sample range, respectively, that will result from any possible sample.

Table 7.2
Possible values of sample mean and sample range $X: x_1 = 3$; $x_2 = 7$; $x_3 = 11$; $x_4 = 15$; $x_5 = 19$

(1) Possible sample		(2)	(3)	(4)
Draw 1	*Draw 2*	*Probability of sample*	*Sample mean*	*Sample range*
$x_1 = 3$	$x_1 = 3$	1/25	3	0
$x_1 = 3$	$x_2 = 7$	1/25	5	4
$x_1 = 3$	$x_3 = 11$	1/25	7	8
$x_1 = 3$	$x_4 = 15$	1/25	9	12
$x_1 = 3$	$x_5 = 19$	1/25	11	16
$x_2 = 7$	$x_1 = 3$	1/25	5	4
$x_2 = 7$	$x_2 = 7$	1/25	7	0
$x_2 = 7$	$x_3 = 11$	1/25	9	4
$x_2 = 7$	$x_4 = 15$	1/25	11	8
$x_2 = 7$	$x_5 = 19$	1/25	13	12
$x_3 = 11$	$x_1 = 3$	1/25	7	8
$x_3 = 11$	$x_2 = 7$	1/25	9	4
$x_3 = 11$	$x_3 = 11$	1/25	11	0
$x_3 = 11$	$x_4 = 15$	1/25	13	4
$x_3 = 11$	$x_5 = 19$	1/25	15	8
$x_4 = 15$	$x_1 = 3$	1/25	9	12
$x_4 = 15$	$x_2 = 7$	1/25	11	8
$x_4 = 15$	$x_3 = 11$	1/25	13	4
$x_4 = 15$	$x_4 = 15$	1/25	15	0
$x_4 = 15$	$x_5 = 19$	1/25	17	4
$x_5 = 19$	$x_1 = 3$	1/25	11	16
$x_5 = 19$	$x_2 = 7$	1/25	13	12
$x_5 = 19$	$x_3 = 11$	1/25	15	8
$x_5 = 19$	$x_4 = 15$	1/25	17	4
$x_5 = 19$	$x_5 = 19$	1/25	19	0

We can determine the values in column 2 by either of two lines of reasoning. First, we know there are 25 possible samples. (That is, any of the five values may be selected on the first draw, and they may occur in any combination with the five values which can occur on the second draw; $5 \cdot 5 = 25$ possible samples.) Under simple random sampling, each possible sample has the same probability of being selected as any other sample. Thus, the probability associated with any sample is 1/25. The second line of reasoning is as follows. Since the sampling is with replacement, the probability of drawing any given value on a specific draw is 1/5. Thus, for example, by the independence of the draws,

$$P(x_1 \text{ on first draw} \cap x_1 \text{ on second draw}) = P(x_1 \text{ on first draw})\, P(x_1 \text{ on second draw})$$

$$= \frac{1}{5} \cdot \frac{1}{5} = \frac{1}{25}. \qquad (1)$$

Table 7.3 shows the sample spaces and sampling distributions associated with the two statistics of interest (the sample mean and sample range). We determine, for example, the sample space associated with the sample mean (column 1) by scanning column 3 of Table 7.2 and simply listing all the different possible values that can occur for the sample mean. Then, we determine the probability of any given possible value of the sample mean by adding the probabilities of the possible samples which result in that particular value. For example, since the possible samples are mutually exclusive events,

$$\begin{aligned} P(\text{sample mean} = 5) &= P(\text{“}x_1 \text{ on first draw} \cap x_2 \text{ on second draw”} \cup \text{“}x_2 \text{ on first draw} \cap x_1 \text{ on second draw”}) \\ &= P(x_1 \text{ on first draw} \cap x_2 \text{ on second draw}) \qquad (2) \\ &\quad + P(x_2 \text{ on first draw} \cap x_1 \text{ on second draw}) \\ &= 1/25 + 1/25 = 2/25. \end{aligned}$$

Other probabilities in either sampling distribution can be derived in a similar manner.

The sampling distributions corresponding to the data of Table 7.3 are graphed in Figure 7.1, where the height of the vertical line above a given possible value represents the probability associated with that value. This method of presenting a probability distribution, termed a *bar chart* is often used when the variable (or statistic) takes on only a small number of possible values.

Table 7.3

Sampling distributions of sample mean and sample range (Example 1)

Sampling distribution of sample mean		*Sampling distribution of sample range*	
(1) *Sample space*	(2) *Probability*	(3) *Sample space*	(4) *Probability*
3	1/25	0	5/25
5	2/25	4	8/25
7	3/25	8	6/25
9	4/25	12	4/25
11	5/25	16	2/25
13	4/25		
15	3/25		
17	2/25		
19	1/25		
Total	25/25 = 1.0	Total	25/25 = 1.0

□

The primary purpose of Example 1 is simply to acquaint you with the meaning of the sampling distribution of any statistic and to illustrate how a particular sampling distribution can be derived through use of only elementary probability concepts. However, note that, even though we may carry out the identical sampling scheme on exactly the same population, *the sampling distribution for one statistic will be different from that for another*. For example, it is easy to note several differences between the sampling distribution of the sample mean and the sampling distribution of the sample range simply by inspecting Figure 7.1: (1) the central tendencies are slightly different; (2) the dispersions are quite different; (3) one is skewed and the other is symmetrical. Also, it now becomes even more obvious that we may view (probability-wise) a statistic in the same manner as a random variable.

In the development of sampling distributions, we are concerned with three distinct types of sampling. If sampling is with replacement, order is important for

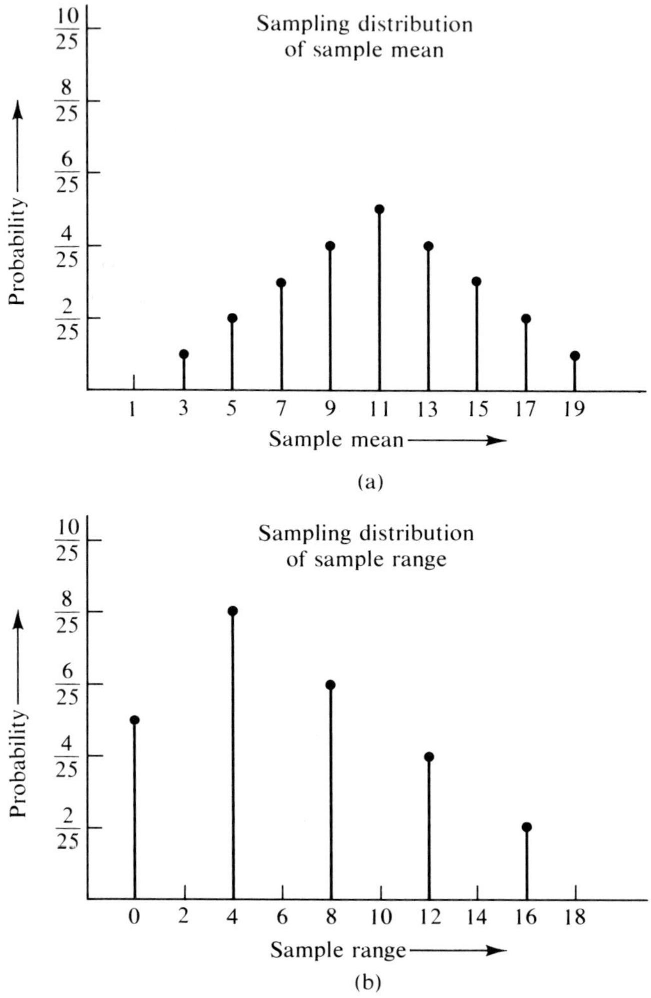

Figure 7.1
Sampling distributions of sample mean and sample range (Example 1)

probabilities to be properly associated with the resulting sample space. In sampling without replacement, order is important in some cases and unimportant in others. Nevertheless, we are more often concerned with the case where order is not important. Thus, unless otherwise stated, our policy will be to let *simple random sampling with replacement* imply that order is important and *simple random sampling without replacement* imply that order is not important.

In both cases, it often helps to determine quickly the number of possible samples that can occur under a given sampling scheme, as well as the probabilities associated with the occurrence of each. Let us consider these matters. Suppose we are to select with replacement a simple random sample of size n from a population of size N. The first element of the sample can be selected in N different ways, the second element in N different ways, ..., and the nth element in N different ways. Thus, the total number of possible samples becomes

$$\underbrace{N \cdot N \cdot N \cdot \cdots \cdot N}_{n \text{ times}} = N^n.$$

***Rule** 7.1*

The number of possible samples that can result from the selection with replacement (i.e., when order is important) of a simple random sample of size n from a population of size N is N^n.

Since we are considering simple random sampling, any of the N^n possible samples has the same probability of being selected.

***Rule** 7.2*

The probability associated with each possible sample when we select with replacement (i.e., when order is important) a simple random sample of size n from a population of size N is $1/N^n$.

Now, suppose we are to select without replacement a simple random sample of size n from a population of size N. The first element of the sample can be selected in N different ways, the second element in only $(N - 1)$ different ways, ..., and the nth element in $(N - n + 1)$ different ways. Thus, the total number of possible different samples, with order important, is

$$N \cdot (N - 1) \cdot (N - 2) \cdot \cdots \cdot (N - n + 1).$$

This is precisely $N!/(N - n)!$. However, each sample (which is a combination of n different elements) has $n!$ different orderings (permutations). Thus, the number of different ways we can select a sample of size n from a population of size N without replacement (when we disregard the order within each sample) is

$$\frac{N!}{n!(N - n)!}$$

This is simply the number of combinations of N things taken n at a time.

Rule 7.3

The number of possible samples which can result from the selection without replacement (order disregarded) of a simple random sample of size n from a population of size N is ${}_NC_n$.

Again, under simple random sampling, the samples are equally likely to occur.

Rule 7.4

The probability associated with each possible sample when we select without replacement (order disregarded) a simple random sample of size n from a population of size N is $1/{}_NC_n$.

Example 2

We are to use simple random sampling to select a sample of size $n = 2$ from the following population of size $N = 5$:

$$X: x_1 = 12; \quad x_2 = 6; \quad x_3 = 3; \quad x_4 = 15; \quad x_5 = 9.$$

(a) For the case of sampling with replacement (order important), determine the number of possible samples, list the possible samples, and determine the sample space and sampling distribution of the statistic known as the "sum of the sample observations."

(b) Repeat for the case of sampling without replacement (order disregarded).

Solution

(a) The number of possible samples is

$$N^n = 5^2 = 25.$$

For the case of sampling with replacement, Table 7.4 lists all possible samples (column 1), the probability of the occurrence of each possible sample (column 2), and the value of the statistic (sum of the sample observations) for each possible sample (column 3). Table 7.5 presents the resulting sample space (column 1) and corresponding probabilities (column 2) for sampling with replacement.

(b) The number of possible samples is

$${}_NC_n = {}_5C_2 = \frac{5!}{2!3!} = 10.$$

For the case of sampling without replacement, columns 4, 5, and 6 Table 7.4 and columns 3 and 4 of Table 7.5 present information similar to that listed above for the case of sampling with replacement. The sampling distributions for both of the above cases are shown in Figure 7.2.

Table 7.4

Possible values of the "sum of the sample observations" ($X: x_1 = 12;\ x_2 = 6;\ x_3 = 3;\ x_4 = 15;\ x_5 = 9$)

Sampling with replacement			*Sampling without replacement*		
(1) *Possible sample*	(2) *Probability of sample* $\left(\frac{1}{N^n}\right)$	(3) *Value of the statistic*	(4) *Possible sample*	(5) *Probability of sample* $\left(\frac{1}{{}_NC_n}\right)$	(6) *Value of the statistic*
12, 12	1/25	24			
12, 6	1/25	18	12, 6	1/10	18
12, 3	1/25	15	12, 3	1/10	15
12, 15	1/25	27	12, 15	1/10	27
12, 9	1/25	21	12, 9	1/10	21
6, 12	1/25	18			
6, 6	1/25	12			
6, 3	1/25	9	6, 3	1/10	9
6, 15	1/25	21	6, 15	1/10	21
6, 9	1/25	15	6, 9	1/10	15
3, 12	1/25	15			
3, 6	1/25	9			
3, 3	1/25	6			
3, 15	1/25	18	3, 15	1/10	18
3, 9	1/25	12	3, 9	1/10	12
15, 12	1/25	27			
15, 6	1/25	21			
15, 3	1/25	18			
15, 15	1/25	30			
15, 9	1/25	24	15, 9	1/10	24
9, 12	1/25	21			
9, 6	1/25	15			
9, 3	1/25	12			
9, 15	1/25	24			
9, 9	1/25	18			

Table 7.5

Sampling distribution of the "sum of the sample observations" (Example 2)

Sampling with replacement		*Sampling without replacement*	
(1) *Sample space (possible value of statistic)*	(2) *Probability*	(3) *Sample space (possible value of statistic)*	(4) *Probability*
6	1/25		
9	2/25	9	1/10
12	3/25	12	1/10
15	4/25	15	2/10
18	5/25	18	2/10
21	4/25	21	2/10
24	3/25	24	1/10
27	2/25	27	1/10
30	1/25		
Total	25/25 = 1.0	Total	10/10 = 1.0

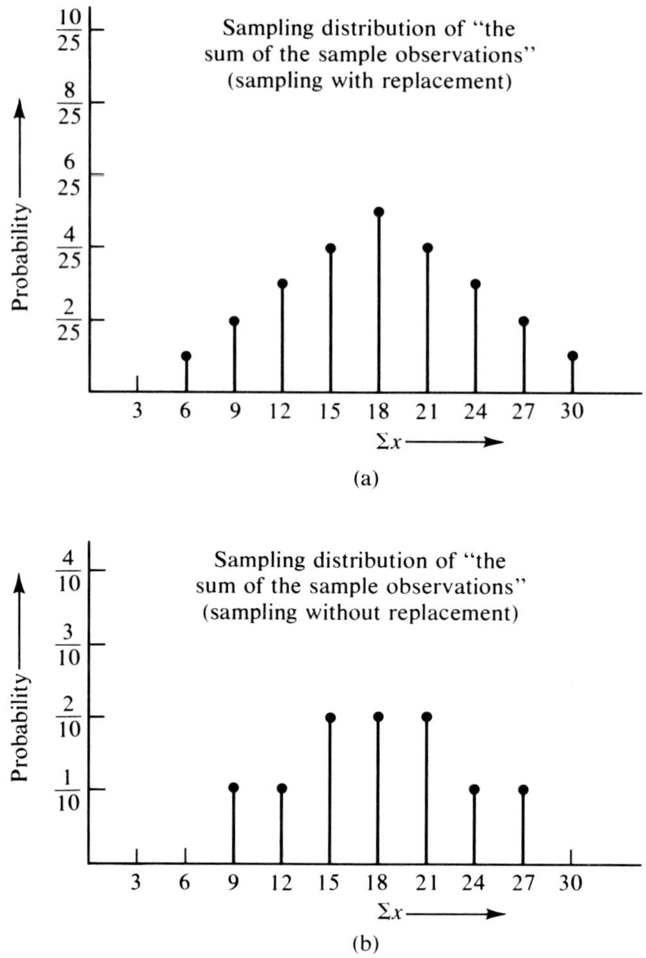

Figure 7.2
Sampling distribution of the "sum of the sample observations" (Example 2) □

Two points regarding the foregoing concepts and related examples are very important. First, we are now able to build, under any simple random sampling scheme (for a finite population), the sampling distribution for any statistic simply by following the steps illustrated. Second, even though direct application of the method for deriving sampling distributions is of limited utility, the method is invaluable to your understanding of more advanced statistical concepts.

Since statistics are similar to random variables and every statistic has associated with it a probability distribution, it follows that every statistic must have an expected value and a variance. How do we determine such measures? We proceed exactly as in the case of a variable. For example, when sampling from a finite population, we treat the possible values of the statistic under consideration exactly as we would the values of any set of data. Thus, if we are to consider a given statistic (say T), there will exist a finite (determinable by Rules 7.1 and 7.3) number of possible values T can assume under a given sampling scheme. If we are sampling with replacement, there will be N^n possible values, that is,

$$T: t_1, t_2, \ldots, t_{(N^n)}.$$

In this case, we obtain the expected value (mean) of T by

$$\mu_t \doteq E(T) = \frac{1}{N^n} \sum_{i=1}^{N^n} t_i \tag{3}$$

and the variance of T by

$$\sigma_t^2 \doteq \mathrm{Var}(T) \doteq E[T - E(T)]^2 = E(T - \mu_t)^2 = \frac{1}{N^n} \sum_{i=1}^{N^n} (t_i - \mu_t)^2, \tag{4}$$

where $\doteq$ should be read as "is defined to be equal to." In subsequent material, we shall often use the $\doteq$ symbol to introduce new notation. If we are sampling without replacement, there will be ${}_NC_n$ possible values of T, that is,

$$T: t_1, t_2, \ldots, t_{({}_NC_n)}.$$

and we can find the expected value (mean) and the variance of the statistic T by everywhere replacing N^n by ${}_NC_n$ in Equations (3) and (4), respectively.

Example 3

We are to use simple random sampling without replacement to select a sample of size $n = 3$ from the following population of size $N = 5$.

$$X: 3, 15, 12, 24, 9.$$

Find the expected value (mean) and the variance of each of the statistics called the "sample mean" and "sample median." Which sampling distribution has the greater dispersion?

Solution

The number of possible samples is

$${}_NC_n = {}_5C_3 = 10.$$

Table 7.6 presents possible samples (column 1) and corresponding values of possible sample means and sample medians (columns 2 and 5, respectively). Calculations (using the information of columns 3, 4, 6, and 7) for appropriate expected values and variances are given below the table. Note that $\sigma_{\bar{x}}^2 > \sigma_m^2$; that is, in this particular example, the variance of the sample mean is greater than the variance of the sample median. □

You must not infer too much from this last example. It cannot be taken for granted that the variance of the mean is in every situation greater than the variance of the median. The important point is that all statistics do have expected values and variances, and we are able to calculate them when dealing with finite populations.

It is not enough that we are able to build the sampling distribution and to calculate the mean and the variance of any statistic. The tasks are prohibitively laborious when

Table 7.6

Calculation of the mean and variance of the sample mean and sample variance ($X: x_1 = 3$; $x_2 = 15$; $x_3 = 12$; $x_4 = 24$; $x_5 = 9$), for Example 3

(1) Possible sample	(2) Sample mean ($\bar{x}_i$)	(3) $\bar{x}_i - \mu_{\bar{x}}$	(4) $(\bar{x}_i - \mu_{\bar{x}})^2$	(5) Sample median (m_i)	(6) $m_i - \mu_m$	(7) $(m_i - \mu_m)^2$
3, 15, 12	10	−2.6	6.76	12	0	0
3, 15, 24	14	1.4	1.96	15	3	9
3, 15, 9	9	−3.6	12.96	9	−3	9
3, 12, 24	13	0.4	0.16	12	0	0
3, 12, 9	8	−4.6	21.16	9	−3	9
3, 24, 9	12	−0.6	0.36	9	−3	9
15, 12, 24	17	4.4	19.36	15	3	9
15, 12, 9	12	−0.6	0.36	12	0	0
15, 24, 9	16	3.4	11.56	15	3	9
12, 24, 9	15	2.4	5.76	12	0	0
Total	126	0	80.40	120	0	54

Let $\bar{X}$ = sample mean; M = sample median; $\bar{x}_i$ = ith possible value of $\bar{X}$;
m_i = ith possible value of M; $\mu_{\bar{x}}$ = expected value of $\bar{X}$;
μ_m = expected value of M; $\sigma_{\bar{x}}^2$ = variance of $\bar{X}$;
σ_m^2 = variance of M.

$$\mu_{\bar{x}} = \frac{1}{{}_NC_n} \sum_{i=1}^{{}_NC_n} \bar{x}_i = \frac{126}{10} = 12.6, \qquad \mu_m = \frac{1}{{}_NC_n} \sum_{i=1}^{{}_NC_n} m_i = \frac{120}{10} = 12,$$

$$\sigma_{\bar{x}}^2 = \frac{1}{{}_NC_n} \sum_{i=1}^{{}_NC_n} (\bar{x}_i - \mu_{\bar{x}})^2 = \frac{1}{10}(80.40) = 8.04, \qquad \sigma_m^2 = \frac{1}{{}_NC_n} \sum_{i=1}^{{}_NC_n} (m_i - \mu_m)^2 = \frac{54}{10} = 5.4.$$

the size of the population being sampled is very large and impossible when the population is infinite. Finally, in order to approach statistical inference intelligently, we must be able to generalize results regarding the means, variances, and sampling distributions of various statistics. And the empirical approach does not provide a vehicle for drawing generalizations useful in statistical inference. Are we caught on the horns of dilemma? No, once again we can circumvent the problem by drawing from related statistical theory.

We devote the remainder of the chapter to discussing the general characteristics of two statistics called the sample mean and the sample proportion. In fact, much of the material of subsequent chapters will be based on such discussions. However, you should avoid concluding that these are the only statistics of importance in statistical inference. On the contrary, for some situations, other statistics are more useful and practical. For this reason, we have gone to great lengths to illustrate that other statistics can be treated just as the sample mean. We want to emphasize that the theory and applications associated with the sample mean and sample proportion represent only a small portion of that body of knowledge generally encompassed by the term "statistical inference."

On the other hand, associated theory is so straightforward, related results are so useful, and potential applications are so significant that a thorough treatment of the sample mean and sample proportion provides an excellent introduction to statistical inference. Thus, if you digest the concepts of this and subsequent chapters not only will you amass a useful kit of ideas and techniques but you also will take a giant stride toward understanding advanced statistical concepts.

Problems

1. Suppose we are to use simple random sampling with replacement (order important) to select a sample of size $n = 2$ from the following population of size $N = 5$:

$$X: x_1 = 21; \quad x_2 = 24; \quad x_3 = 27; \quad x_4 = 30; \quad x_5 = 33.$$

First construct a table similar to Table 7.2, rounding all numbers to one decimal place. Then tabulate and graph the sampling distribution of

a. the sample mean deviation from the mean,

b. the sample range.

2. Suppose we are to use simple random sampling without replacement (order disregarded) to select a sample of size $n = 2$ from the following population of size $N = 6$:

$$X: x_1 = 13; \quad x_2 = 17; \quad x_3 = 19; \quad x_4 = 25; \quad x_5 = 27; \quad x_6 = 31.$$

Find and plot the sampling distribution of the sample mean. Include a table similar to Table 7.2, accurate to one decimal place.

3. Indicate both the number of possible samples and the probability associated with each possible sample for simple random sampling in each of the following cases.

a. $n = 4$, $N = 10$, *with replacement*,

b. $n = 3$, $N = 11$, *without replacement*,

c. $n = 8$, $N = 15$, *without replacement*,

d. $n = 10$, $N = 20$, *with replacement*.

4. Indicate both the number of possible samples and the probability associated with each possible sample for simple random sampling in each of the following cases.

a. $n = 5$, $N = 12$, *without replacement*,

b. $n = 4$, $N = 15$, *with replacement*,

c. $n = 15$, $N = 100$, *with replacement*,

d. $n = 15$, $N = 20$, *without replacement*.

5. We are to select a simple random sample of size $n = 2$ from the following population of size $N = 5$:

$$X: x_1 = 15; \quad x_2 = 7; \quad x_3 = 19; \quad x_4 = 11; \quad x_5 = 23.$$

a. For the case of sampling *with* replacement (*order important*), determine the number of possible samples, list the possible samples, and determine the sample space and sampling distribution of the statistic known as the "sample range." Construct tables similar to Tables 7.4 and 7.5.

b. Repeat Part a for the case of sampling *without* replacement (*order disregarded*). Include the information in the same tables begun in Part a.

6. We are to select a simple random sample of size $n = 2$ from the following population of size $N = 6$:

$$X: x_1 = 31; \quad x_2 = 19; \quad x_3 = 13; \quad x_4 = 43; \quad x_5 = 37; \quad x_6 = 25.$$

a. For the case of sampling *with* replacement, determine the number of possible samples, list the possible samples, and determine the sample space and sampling distribution of the statistic known as the "sample mean." Construct tables similar to Tables 7.4 and 7.5.

b. Repeat Part a for the case of sampling *without* replacement. Include the information in the same tables begun in Part a.

7. Find the expected value (mean) and variance of the statistic "sample range" for the population and sampling scheme of Problem 5a. Construct a table including columns for the possible sample, R_i, $R_i - \mu_r$, and $(R_i - \mu_r)^2$. Round all numbers to two decimal places.

8. Find the expected value (mean) and variance of the statistic "sample mean" for the population and sampling scheme of Problem 6b. Construct a table including columns for the possible sample, $\bar{x}_i$, $\bar{x}_i - \mu_x$, and $(\bar{x}_i - \mu_x)^2$.

9. Under what condition(s) is it *impossible* to construct the sampling distribution of a statistic by the empirical method of Section 7.3?

7.4 The Sampling Distribution of the Sample Mean

Having developed sampling distributions that result for the sample mean under specific sampling schemes, we can now extend our capability, to answer probability questions about the sample mean for many different sampling situations. What we need is a general theory regarding the sampling distribution of the sample mean.

First, let us consider an example illustrating the relationship between the mean and variance of the sampling distribution of the sample mean and the respective mean and variance of the original distribution from which we are sampling.

Example 1

We are concerned with the sampling scheme of drawing without replacement (order disregarded) a simple random sample of size $n = 2$ from the following population of size $N = 8$:

$$X: 2, 3, 4, 5, 6, 7, 8, 9.$$

Show that

$$\mu_{\bar{x}} = \mu_x \quad \text{and} \quad \sigma_{\bar{x}}^2 = \frac{\sigma_x^2}{n} \cdot \frac{N-n}{N-1}, \tag{1}$$

where $\mu_{\bar{x}}$ and $\sigma_{\bar{x}}^2$ are the mean and the variance, respectively, of the sampling distribution of the sample mean and μ_x and σ_x^2 are the mean and the variance, respectively, of the original distribution (population) from which we are sampling. Plot the sampling distribution and note its shape and characteristics. (Note also that the distribution from which we are sampling is far from normal.)

Solution

We must

1. calculate the mean and the variance of the original population (μ_x and σ_x^2, respectively);
2. calculate
$$\frac{\sigma_x^2}{n} \cdot \frac{N-n}{N-1};$$
3. calculate $\mu_{\bar{x}}$ and $\sigma_{\bar{x}}^2$ by considering all possible sample means which can occur under the given sampling scheme; and
4. verify that the relations given by (1) are valid.

All relevant calculations are shown in Table 7.7. All possible samples are shown in column 1, while corresponding values of the sample mean are given in column 2. Columns 3 and 4 provide information necessary for calculation of the variance of the sample mean. Note the numerical relationship of the terms given by (1).

Table 7.7

Relations between mean and variance of the sample mean, and mean and variance of the population

Original population, X: 2, 3, 4, 5, 6, 7, 8, 9

(1) *Possible samples*	(2) *Sample mean* ($\bar{x}_i$)	(3) $\bar{x}_i - \mu_{\bar{x}}$	(4) $(\bar{x}_i - \mu_{\bar{x}})^2$
2, 3	2.5	−3.0	9.00
2, 4	3.0	−2.5	6.25
2, 5	3.5	−2.0	4.00
2, 6	4.0	−1.5	2.25
2, 7	4.5	−1.0	1.00
2, 8	5.0	−0.5	0.25
2, 9	5.5	0.0	0.00
3, 4	3.5	−2.0	4.00
3, 5	4.0	−1.5	2.25
3, 6	4.5	−1.0	1.00
3, 7	5.0	−0.5	0.25
3, 8	5.5	0.0	0.00
3, 9	6.0	0.5	0.25
4, 5	4.5	−1.0	1.00
4, 6	5.0	−0.5	0.25
4, 7	5.5	0.0	0.00
4, 8	6.0	0.5	0.25
4, 9	6.5	1.0	1.00
5, 6	5.5	0.0	0.00
5, 7	6.0	0.5	0.25
5, 8	6.5	1.0	1.00
5, 9	7.0	1.5	2.25
6, 7	6.5	1.0	1.00
6, 8	7.0	1.5	2.25
6, 9	7.5	2.0	4.00
7, 8	7.5	2.0	4.00
7, 9	8.0	2.5	6.25
8, 9	8.5	3.0	9.00
Total	154.0		63.00

$$\mu_x = \frac{1}{N}\sum_{i=1}^{N} x_i = \frac{2+3+\cdots+9}{8} = \frac{44}{8} = 5.5.$$

$$\sigma_x^2 = \frac{1}{N}\sum_{i=1}^{N}(x_i - \mu_x)^2$$
$$= \frac{(2-5.5)^2 + (3-5.5)^2 + \cdots + (9-5.5)^2}{8}$$
$$= \frac{42}{8} = 5.25.$$

$$\frac{\sigma_x^2}{n} \cdot \frac{N-n}{N-1} = \frac{5.25}{2} \cdot \frac{6}{7} = 2.25.$$

$${}_NC_n = \frac{8!}{2!6!} = \frac{8 \cdot 7}{2} = 28.$$

$$\mu_{\bar{x}} = \frac{1}{{}_NC_n}\sum_{i=1}^{{}_NC_n} \bar{x}_i = \frac{1}{28}\sum_{i=1}^{28} \bar{x}_i = \frac{154.0}{28} = 5.5.$$

$$\sigma_{\bar{x}}^2 = \frac{1}{{}_NC_n}\sum_{i=1}^{{}_NC_n}(\bar{x}_i - \mu_{\bar{x}})^2 = \frac{1}{28}\sum_{i=1}^{28}(\bar{x}_i - \mu_{\bar{x}})^2$$
$$= \frac{63.00}{28} = 2.25.$$

Note that

$$\mu_{\bar{x}} = \mu_x = 5.5 \text{ and } \sigma_{\bar{x}}^2 = \frac{\sigma_x^2}{n} \cdot \frac{N-n}{N-1} = 2.25.$$

Table 7.8 and Figure 7.3 present the sampling distribution of the sample mean. As usual, the sample space is obtained by listing (column 2 of Table 7.7) the possible values the sample mean can assume. The probability corresponding to each possible value is the ratio of the number of possible samples which result in that particular value for the sample mean to the total number of possible samples. For example, we will obtain a value of 5.5 for $\bar{x}$ if we select one of the samples (2, 9), (3, 8), (4, 7), or (5, 6). Thus, of 28 possible samples, 4 result in a value of $\bar{x} = 5.5$. Thus, $P(X = 5.5) = 4/28$. Other probabilities are obtained in a similar fashion.

Table 7.8
Sampling distribution of the sample mean

(1) *Sample space (possible value of sample mean)*	(2) *Probability*
2.5	1/28
3.0	1/28
3.5	2/28
4.0	2/28
4.5	3/28
5.0	3/28
5.5	4/28
6.0	3/28
6.5	3/28
7.0	2/28
7.5	2/28
8.0	1/28
8.5	1/28
Total	28/28 = 1.0

□

You should satisfy yourself that, when sampling is with replacement (order important),

$$\mu_{\bar{x}} = \mu_x \quad \text{and} \quad \sigma_{\bar{x}}^2 = \frac{\sigma_x^2}{n}. \tag{2}$$

In other words, we obtain the same results with the exception that the term $(N - n)/(N - 1)$ disappears from the variance of the sampling distribution of the sample mean. However, sampling with replacement is essentially equivalent to sampling from an infinite population. Thus, since the term $(N - n)/(N - 1)$ occurs when we are sampling from a finite population and fails to occur when we are sampling from an infinite population, we call it the *finite population correction factor*. We shall say more about this factor later.

The important thing is that Relations (1) and (2) always hold. However, we still need more general information on the sampling distribution of the sample mean to be able to answer related probability questions. We require information about the entire sampling distribution. Before going on however, let us consider typical problems that arise and point out the futility of the empirical approach (discussed in Section 7.3) to the development of required sampling distributions.

Asymptotically normal means that the sampling distribution of the sum will approach a normal distribution as the number (n) of random variables involved in the summation becomes large.

To determine how the general case of the central limit theorem provides us information concerning the sampling distribution of the sample mean, we must agree to view a sample in a specific fashion. Let us agree that when we select a random sample of size n from a population, we shall treat the n observations as coming from n independent random variables

$$X_1, X_2, \ldots, X_n,$$

all having the same distribution (an interpretation we shall follow consistently).

If the random sample is selected with replacement, the interpretation seems fairly natural, since the repeated draws will necessarily be independent. Further, if we are sampling from an infinite population, we are always essentially sampling with replacement. Strictly speaking, the observations cannot be viewed as independent if the population is finite and sampling is without replacement. Nevertheless, in most practical problems involving large finite populations and relatively small samples, even observations selected without replacement may be viewed as coming from n independent random variables.

Under our interpretations, the general case of the central limit theorem provides us with two very important special cases involving the sampling distribution of the sample mean: *the central limit theorem (infinite population)* and *the central limit theorem (finite population)*. However, in future references, both will be referred to as the central limit theorem. The specific theorem will ordinarily be clear from the context.

***Theorem** 7.2 (Central limit theorem—infinite population)*

When a simple random sample of size n is taken from an infinite population with mean μ_x and variance σ_x^2, then, as n increases, the sampling distribution of the sample mean approaches a normal distribution with mean μ_x and variance σ_x^2/n. That is, the sample mean is asymptotically normally distributed with mean (expected value)

$$\mu_{\bar{x}} \doteq E(\bar{X}) = \mu_x$$

and variance (a)

$$\sigma_{\bar{x}}^2 \doteq \mathrm{Var}(\bar{X}) = \frac{\sigma_x^2}{n};$$

thus, given an infinite population, a sufficiently large sample size, and

$$X \sim ?(\mu_x, \sigma_x^2),$$

then

$$\bar{X} \sim \mathrm{N}(\mu_x, \sigma_x^2/n). \tag{b}$$

***Theorem** 7.3 (Central limit theorem—finite population)*

When a simple random sample of size n is taken without replacement (order disregarded) from a finite population (of size N) with mean μ_x and variance σ_x^2, then, as n *increases, the sampling distribution of the sample mean approaches a normal distribution with mean μ_x and variance*

$$\frac{\sigma_x^2}{n} \cdot \frac{N - n}{N - 1}.$$

That is, the sample mean is asymptotically normally distributed with mean (expected value)

$$\mu_{\bar{x}} \doteq E(\bar{X}) = \mu_x$$

and variance (a)

$$\sigma_{\bar{x}}^2 \doteq \operatorname{Var}(\bar{X}) = \frac{\sigma_x^2}{n} \cdot \frac{N - n}{N - 1};$$

thus, given a finite population of size N, a sufficiently large sample size, and

$$X \sim ?(\mu_x, \sigma_x^2),$$

then

$$\bar{X} \sim \mathrm{N}\left(\mu_x, \frac{\sigma_x^2}{n} \cdot \frac{N - n}{N - 1}\right). \qquad \text{(b)}$$

One of the most astonishing facets of the central limit theorem is its disregard for the form of the distribution of the original variable. Whatever the original distribution, provided only that it have a finite variance, the sample mean will have approximately a normal distribution for large samples. Further, the condition that the variance be finite is not critically restrictive in applied statistics; it will be satisfied in almost any practical situation.

In two respects, the central limit theorem justifies the extensive effort devoted to the normal distribution in Chapter 6. First, the general theorem (Theorem 7.1) explains why many variables exhibit distributions that are approximately normal. Note that we do not even hint that the central limit theorem may be used to prove or even imply that a variable is normally distributed. It merely explains why one might expect to encounter quite often a variable that is approximately normally distributed. Thus, despite our emphasis in Chapter 6 that only a small portion of the utility of statistics is based on assumptions of normal variables, the central limit theorem makes us less uneasy about normal assumptions in some circumstances.

Second, and far more important, the special cases of the central limit theorem tell us that, even when assumptions of normal variables are suspect, the statistic known as the sample mean is, generally, approximately normally distributed regardless of the distribution of the original variable from which the sample is taken. Henceforth, we shall continually refer to the sample mean as being normally distributed; since we shall be drawing from the central limit theorem, the word "asymptotically" should be taken for granted.

Let us now view the central limit theorem practically. We have already said that, under a given sampling scheme, any statistic will have an associated sampling distribution, mean, and variance. For small populations, we can determine all such information from basic probability concepts. Now, the central limit theorem gives us a general method for obtaining such information for the specific statistic known as the sample mean. We can summarize our discussion about the sample mean as follows: Regardless of the type distribution (with rare exception) from which we are sampling the central limit theorem tells us that

1. the sampling distribution of the sample mean is asymptotically normal,
2. the mean (expected value) of the sample means is equal to the mean of the original population from which we are sampling, and
3. the variance of the sample mean is a well-defined function of the variance of the original population from which we are sampling. The function which describes the variance of the sample mean in terms of the variance of the original population depends on whether sampling is from a finite or infinite population. The proper relations are

$$\sigma_{\bar{x}}^2 = \frac{\sigma_x^2}{n} \text{ (infinite population)} \quad \text{and} \quad \sigma_{\bar{x}}^2 = \frac{\sigma_x^2}{n} \cdot \frac{N - n}{N - 1} \text{ (finite population).} \quad (3)$$

Note that 2 and 3 are simply restatements of (1) and (2). They are not a direct result of the central limit theorem. Also, note that the only difference between the two results of (3) is the finite population correction factor $(N - n)/(N - 1)$. There is nothing mystical about the factor. It just happens that this particular function of n and N occurs repeatedly in results of theoretical developments associated with finite populations, whereas it fails to appear in corresponding results associated with infinite populations.

Sampling Distributions and Infinite Repeated Sampling

As we have seen, we can generate the exact sampling distribution of any statistic as long as we are dealing with a rather small finite population. However, this is infeasible in the case of a large finite population and impossible in the case of an infinite population. Therefore, we resort to theoretical considerations to determine the form of sampling distributions of specific statistics for an infinite population.

A device that nicely bridges the chasm between the empirical construction of sampling distributions and the theoretical development of such distributions for large populations is that of considering the distribution that would result from carrying out an infinite set of independent repetitions of a given sampling scheme.

Let us begin by considering the experiment of rolling a fair die. We know that the probability of obtaining any given number is 1/6. However, suppose we did not know this or suppose the problem were so difficult that we could not know the probability by direct application of elementary probability concepts. We could (1) roll the die many times, (2) count the times each number occurs, and (3) convert such counts to relative frequencies. We know by the law of statistical regularity that the distribution of results would provide a relative frequency of approximately 1/6 for each possible outcome. In fact, if we could roll the die infinitely, we would obtain exactly 1/6 as the relative frequency (probability) of the various possible outcomes. (This type of reasoning manifests the relative-frequency-theory (based on limits) approach to probability.)

The same sort of reasoning is valid for any statistic. That is, under any well-defined sampling scheme from a specific population, we now know that there does in fact exist a sampling distribution which associates probabilities with the possible values (or groups of possible values) of the statistic. This distribution is precisely the one that would occur if we could perform the following infinite repeated sampling procedure:

1. Define the population and the statistic of interest (say T).
2. Select a sample consisting of n independent values from the population and calculate (or count or note) the statistic of interest—that is, determine an observed value of T by use of the n values.
3. Repeat the process of step 2 for an infinite number of independent samples. That is, determine t_i ($i = 1, 2, \ldots$), where t_i represents the value determined for the statistic T through use of the values selected in the ith independent sample.
4. Compile the probability distribution of T.

The resulting probability distribution would be the sampling distribution of the statistic T for the situation where the given sampling scheme is applied to the population of interest.

Now, it is obviously impossible to repeat step 2 an infinite number of times. Thus, you can simply consider the distribution for a very large number of repetitions.

Finally, the central limit theorem (infinite population) tells us that the sample mean $\overline{X}$ is asymptotically normally distributed. What does it mean in the context of our present discussion? Consider what would happen if we repeated the following sampling procedure:

1. Select from an infinite population having mean μ_x and variance σ_x^2 a sample consisting of n independent values and calculate

$$\bar{x} = \frac{1}{n} \sum_{i=1}^{n} x_i .$$

2. Repeat the process of step 1 for an infinite number of independent samples. That is, calculate $\bar{x}_i$ ($i = 1, 2, \ldots$), where $\bar{x}_i$ represents the sample mean for the ith independent sample.
3. Compile the probability distribution of $\overline{X}$.

The resulting probability distribution would be the sampling distribution of the sample mean. It would have mean $\mu_{\bar{x}} = \mu_x$ and variance $\sigma_{\bar{x}}^2 = \sigma_x^2/n$. Further, it would be approximately normal—the larger n, the more nearly normal.

We shall have repeated need in subsequent material to consider sampling distributions which result from various sampling schemes applied to populations having specific properties. In nearly every case, we will merely state important properties of the appropriate sampling distribution, indicating only that such properties can be shown to be true through theoretical proofs. *In such instances, you will find that use of a reasoning process involving the above infinite repeated sampling device will greatly enhance your understanding of the corresponding theoretical results under consideration.*

Problems

1. Rework Example 1, page 240, for sampling *with replacement* (*order important*) in order to show that

$$\mu_{\bar{x}} = \mu_x \quad \text{and} \quad \sigma_{\bar{x}}^2 = \frac{\sigma_x^2}{n}.$$

Round all numbers to two decimal places, except your final answers, which should be rounded to three decimal places.

2. We are concerned with the sampling scheme of drawing *without replacement* (*order disregarded*) a simple random sample of size $n = 2$ from the following population of size $N = 7$:

$$X: 7, 10, 9, 11, 13, 12, 15.$$

Show that

$$\mu_{\bar{x}} = \mu_x \quad \text{and} \quad \sigma_{\bar{x}}^2 = \frac{\sigma_x^2}{n} \frac{N - n}{N - 1}.$$

Plot the sampling distribution and note its shape and characteristics. Follow the format of Example 1, page 240.

7.5 *Typical Problems*

We saw in Chapter 6 that any normal variable can be transformed to a standard normal variable by the formula

$$Z = \frac{X - \mu_x}{\sigma_x},$$

where $Z \sim N(0, 1)$, X is the normal variable of interest, μ_x is the mean of that variable, and σ_x is the standard deviation. This formula holds for *any normal entity* (variable or a statistic). Thus, since the general statistic known as the sample mean has (by the central limit theorem) a normal distribution, we can obtain a standard normal distribution by the transformation

$$Z = \frac{\bar{X} - \mu_{\bar{x}}}{\sigma_{\bar{x}}}. \tag{1}$$

If the sampling scheme involves a sample of size n from an infinite population, we obtain a standard normal distribution by the transformation

$$Z = \frac{\bar{X} - \mu_x}{\sigma_x/\sqrt{n}}, \tag{2}$$

since, in the case of an infinite population, $\bar{X}$ is normally distributed with mean $\mu_{\bar{x}} = \mu_x$ and variance $\sigma_{\bar{x}}^2 = \sigma_x^2/n$, or

$$\bar{X} \sim N\left(\mu_x, \frac{\sigma_x^2}{n}\right).$$

On the other hand, if the sampling scheme involves a sample of size n from a finite population of size N, we obtain a standard normal distribution by the transformation

$$Z = \frac{\bar{X} - \mu_x}{\sqrt{\frac{\sigma_x^2}{n} \cdot \frac{N - n}{N - 1}}}, \tag{3}$$

since, in the case of a finite population of size N, $\bar{X}$ is normally distributed with mean $\mu_{\bar{x}} = \mu_x$ and variance

$$\sigma_{\bar{x}}^2 = \frac{\sigma_x^2}{n} \cdot \frac{N - n}{N - 1},$$

that is to say,

$$\bar{X} \sim N\left(\mu_x, \frac{\sigma_x^2}{n} \cdot \frac{N - n}{N - 1}\right).$$

Thus, (2) and (3) reveal a very simple means of answering a multitude of probability questions involving the sample mean.

Example 1

Suppose we are to take a simple random sample of size $n = 100$ from a population having mean 220 and variance 400. Assume that the variable of interest (X) is continuous, but we do not know the distribution of X.

(a) What is the probability that the sample mean will be greater than 225; that is, what is $P(\bar{X} > 225)$?
(b) What is $P(216 < \bar{X} < 223)$?
(c) What is the value, say $\bar{x}^*$, above which approximately 31.56 percent of all sample means chosen in the above manner will be? That is, find $\bar{x}^*$ so that $P(\bar{X} > \bar{x}^*) = 0.3156$.

Solution

The population is infinite. Thus, using the central limit theorem, 7.2(b), we simply apply Transformation (2) on both the general statistic $\bar{X}$ and the specific values of concern and then solve the resulting probability statements by the general approach developed in Chapter 6.

(a)
$$\begin{aligned} P(\bar{X} > 225) &= P\left(\frac{\bar{X} - \mu_x}{\sigma_x/\sqrt{n}} > \frac{225 - \mu_x}{\sigma_x/\sqrt{n}}\right) \\ &= P\left(Z > \frac{225 - 220}{20/10}\right) \\ &= P(Z > 2.5) = 0.0062. \end{aligned}$$

(b) $$P(216 < \bar{X} < 223) = P\left(\frac{216 - 220}{20/10} < Z < \frac{223 - 220}{20/10}\right)$$
$$= P(-2 < Z < 1.5)$$
$$= P(Z > -2) - P(Z > 1.5)$$
$$= 1 - P(Z > 2) - P(Z > 1.5)$$
$$= 1 - 0.0228 - 0.0668 = 1 - 0.0896 = 0.9104.$$

(c) $$P(\bar{X} > \bar{x}^*) = 0.3156 \Rightarrow P\left(Z > \frac{\bar{x}^* - \mu_x}{\sigma_x/\sqrt{n}}\right) = 0.3156$$
$$\Rightarrow P\left(Z > \frac{\bar{x}^* - 220}{20/10}\right) = 0.3156.$$

However, from Table A-1,
$$P(Z > 0.48) = 0.3156.$$
Thus,
$$\frac{\bar{x}^* - 220}{2} = 0.48.$$
Solving, we obtain
$$\bar{x}^* = 220.96.$$
□

Example 2

Suppose we are to take a simple random sample of size $n = 100$ from a finite population of size $N = 2{,}000$. Assume that the variable of interest (X) is continuous with mean 360 and variance 100 but that we do not know the distribution of X. What is the probability that the sample mean will be between 358 and 359?

Solution

Since the population size is finite, we use the central limit theorem 7.3(b), transformation (3), and appropriate methods from Chapter 6 to obtain

$$P(358 < \bar{X} < 359) = P\left(\frac{358 - \mu_x}{\sqrt{\frac{\sigma_x^2}{n} \cdot \frac{N - n}{N - 1}}} < Z < \frac{359 - \mu_x}{\sqrt{\frac{\sigma_x^2}{n} \cdot \frac{N - n}{N - 1}}}\right)$$
$$= P\left(\frac{358 - 360}{\sqrt{\frac{100}{100} \cdot \frac{1{,}900}{1{,}999}}} < Z < \frac{359 - 360}{\sqrt{\frac{100}{100} \cdot \frac{1{,}900}{1{,}999}}}\right)$$
$$= P(-2.051 < Z < -1.026)$$
$$= P(Z > -2.051) - P(Z > -1.026)$$
$$= 1 - P(Z > 2.051) - [1 - P(Z > 1.026)]$$
$$= P(Z > 1.026) - P(Z > 2.051)$$
$$= 0.1525 - 0.0201 = 0.1324.$$
□

If you had any difficulty whatsoever in following the steps of Examples 1 and 2, *you should review Section* 6.5 *very carefully.* In general, questions involving the sampling distribution of the sample mean can be solved by the same steps that would be followed in answering probability questions about any general normal distribution. The specific transformation [(2) or (3)] just happens to be slightly different from the usual one (Rule 6.1, page 191) because of the unique relationships between the mean and the variance of the sample mean and the respective mean and the variance of the original population from which we are sampling.

Let us now return to the problem raised in Example 3, of Sec. 7.4. It is easy to see that the central limit theorem provides the general approach needed to solve it.

Solution to Example 3 of Sec. 7.4

Let us assume that, although company officials are not willing to hazard a guess on the distribution of diameters of the parts to be produced if the company wins the preliminary subcontract, they have sufficient historical data to assume a population mean of 30.50 millimeters and a standard deviation of 4.00 millimeters for the diameters. Under this assumption, we can use the central limit theorem to find the probability that the follow-on contract will be awarded after the sample of size 300 is inspected. That is, since

$$\mu_x = 30.50, \quad \sigma_x = 4.00, \quad n = 300, \quad \text{and} \quad N = 20{,}000,$$

we have,

$$P(30.05 < \bar{X} < 30.95) = P\left(\frac{30.05 - \mu_x}{\sqrt{\frac{\sigma_x^2}{n} \cdot \frac{N-n}{N-1}}} < Z < \frac{30.95 - \mu_x}{\sqrt{\frac{\sigma_x^2}{n} \cdot \frac{N-n}{N-1}}}\right)$$

$$= P\left(\frac{30.05 - 30.50}{\sqrt{\frac{16.00}{300} \cdot \frac{19{,}700}{19{,}999}}} < Z < \frac{30.95 - 30.50}{\sqrt{\frac{16.00}{300} \cdot \frac{19{,}700}{19{,}999}}}\right)$$

$$= P\left(\frac{-0.45}{0.2292} < Z < \frac{0.45}{0.2292}\right) \cong P(-1.96 < Z < 1.96)$$

$$= P(Z > -1.96) - P(Z > 1.96)$$

$$= 1 - P(Z > 1.96) - P(Z > 1.96)$$

$$= 1 - 0.0250 - 0.0250 = 0.95.$$

Thus, if the company can produce parts with diameters having a mean of 30.50 millimeters and a standard deviation of 4.00 millimeters, the probability is 0.95 that it will be awarded the follow-on contract. Clearly, management must still exercise human judgment, but with the probability information available through the central limit theorem, it is now much better prepared to make a proper decision. □

Problems

1. Suppose we are to take a simple random sample of size $n = 50$ from an infinite population having mean 117 and variance 200. Assume that the variable of interest (X) is *continuous*, but that we *do not know the distribution* of X.

   a. What is $P(\bar{X} < 120.2)$?

   b. What is $P(115.4 < \bar{X} < 116.8)$?

   c. What is the value, say $\bar{x}^*$, above which approximately 91.31% of all sample means chosen in the above manner will be?

2. Suppose we are to take a simple random sample of size $n = 200$ from a population of size $N = 5{,}000$ having mean 3.81 and variance 2.63. Assume that the variable of interest (X) is *continuous* but that we *do not know the distribution* of X.

   a. What is $P(\bar{X} < 3.64)$?

   b. What is $P(3.51 < \bar{X} < 3.84)$?

   c. What is the value, say $\bar{x}^*$, such that approximately 44.97% of all sample means chosen lie between it and 40.5?

3. A firm has received an order for 5,000 capacitors from a large television manufacturer. If the order is accepted, the television manufacturer will select a simple random sample of 50 capacitors upon receipt of the shipment and will accept the shipment only if the average capacitance of the sample is between 1,900 and 2,120 microfarads (mfd). The capacitors could not easily be sold elsewhere, and the firm therefore wishes to know the probability that it would sustain a loss by accepting the order and producing the capacitors only to have the shipment rejected. The distribution of the capacitances of such capacitors is unknown but past experience indicates that the mean would be 2,000 and the variance 125,000. What is the probability that the shipment will be rejected if the order is accepted?

4. A large drug manufacturer is seeking a license to produce a new drug. The Food and Drug Administration will select a simple random sample of 100 capsules from the initial production run of 5,000 capsules and will issue the license if the capsules selected contain an average of between 17.98 and 18.02 milligrams of the drug. The distribution of the weights of the contents of the capsules is unknown, but experience indicates that the mean will be 18 milligrams and the standard deviation 0.89 milligrams. What is the probability that the company will receive a license to manufacture the new drug?

5. An automobile manufacturer is considering purchasing its tires from the Crosby Tire Co. It will select a simple random sample of 64 tires from the trial production run of 320 tires and test drive them to determine their average life in miles. If their average life is over 25,500 miles, it will purchase all its tires in the future from the Crosby Tire Co. The distribution of the tire mileages is unknown, but past experience indicates that the mean and standard deviation will be 26,000 miles and 2,000 miles, respectively. What is the probability that the automobile manufacturer will use the Crosby Tire Co. for all its tires?

6. The Underwriters Laboratories, Inc. must decide whether to certify an appliance sold in thousands of stores. After purchasing a simple random sample of 49 appliances from the essentially infinite population for testing, it will grant the certification if the average insulation breakdown voltage of the sample is over 350 volts. The distribution of the breakdown voltages is unknown, but experience indicates that the mean is 340 volts and the variance is 900 volts^2. What is the probability that the appliance will receive certification?

7.6 Continuity Correction Factor for Samples of Size *n*

We noted in Chapter 6 that a continuity correction factor is needed when we use a normal distribution to answer probability questions about a discrete variable. Must we follow this same policy in treating the sampling distribution of a statistic when the original distribution from which we are sampling is discrete? The answer is yes.

Up to now our discussions in this chapter have involved continuous variables. As we saw in Chapter 6, for a continuous variable, we essentially associate a probability of zero with any given possible value. It naturally follows that, when sampling from the population of a continuous variable, we must essentially associate a probability of zero with any given possible value of whatever statistic we may be considering. Thus, we have not been concerned with equalities in our probability statements, since, for example,

$$\begin{aligned} P(\bar{X} \geq \bar{x}) &= P(\bar{X} > \bar{x} \text{ or } \bar{X} = \bar{x}) = P(\bar{X} > \bar{x}) + P(\bar{X} = \bar{x}) \\ &= P(\bar{X} > \bar{x}) + 0 = P(\bar{X} > \bar{x}), \end{aligned} \tag{1}$$

where $\bar{x}$ represents any possible value $\bar{X}$ can assume when we are sampling from the population of a continuous variable.

When we are sampling from the population of a discrete variable, however, it often makes sense to associate a nonzero probability with a given possible value of the statistic under consideration. This fact must be taken into account when we use a continuous (normal) distribution to approximate the probabilities associated with the sampling distribution of a statistic (e.g., sample mean) if such distribution is generated by sampling from the population of a discrete variable. As in Chapter 6, we do this by using a continuity correction factor. However, since we are considering samples of size n (in Chapter 6 we were actually considering samples of size 1), the continuity correction factor to be used in this case is $1/2 \cdot 1/n$. As before, whether the continuity correction factor is to be added to or subtracted from the extreme points of the interval for which a probability is desired depends on whether the probability statement involves equalities, inequalities, or some combination of the two.

Example 1

Suppose we are to take a simple random sample of size $n = 49$ from an infinite population. Assume that the variable of interest (X) is discrete with mean 270 and variance 100, but that we do not know the distribution of X. What is the probability that the sample mean will be greater than or equal to 271 and less than or equal to 273?

Solution

Let $\overline{X}_D$ represent the sample mean for a discrete variable. Then, we wish to find $P(271 \leq \overline{X}_D \leq 173)$. We have

$$P(271 \leq \overline{X}_D \leq 273) = P\left(271 - \frac{1}{2}\cdot\frac{1}{n} < \overline{X} < 273 + \frac{1}{2}\cdot\frac{1}{n}\right)$$

$$= P\left(271 - \frac{1}{2}\cdot\frac{1}{49} < \overline{X} < 273 + \frac{1}{2}\cdot\frac{1}{49}\right) \cong P(270.99 < \overline{X} < 273.01)$$

$$= P\left(\frac{270.99 - 270}{10/7} < Z < \frac{273.01 - 270}{10/7}\right) = P(0.69 < Z < 2.11)$$

$$= P(Z > 0.69) - P(Z > 2.11) = 0.2451 - 0.0174 = 0.2277. \qquad \square$$

Now, let us return to the problem of Example 2 of Sec. 7.4.

Solution to Example 2 of Sec. 7.4

In this case, since records are available, we can go so far as to assume that the company has calculated the mean and standard deviation of "reading-speed increases" to be 850 words per minute and 400 words per minute, respectively. (Note that this assumption does not make it any easier for us to use the empirical approach of Section 7.3 in solving the problem.) We can now determine the probability that the sample mean of "reading-speed increases" of 200 persons selected randomly from company files will be less than 800 words per minute as

$$P(\overline{X}_D < 800) = P\left(\overline{X} < 800 - \frac{1}{2}\cdot\frac{1}{n}\right)$$

$$= P(\overline{X} < 799.9975) = P\left(Z < \frac{799.9975 - 850}{\sqrt{\frac{160{,}000}{200}\cdot\frac{9{,}800}{9{,}999}}}\right)$$

$$= P\left(Z < -\frac{50.0025}{28}\right) = P(Z < -1.786)$$

$$= P(Z > 1.786) = 0.0370.$$

Given such a small probability, the company may well decide to proceed with the proposal. At any rate, it now has relevant information. $\square$

It is obvious that the continuity correction factor has very little effect in the foregoing problem, mainly because the sample size is fairly large. We shall often ignore this factor in problems where it has such minute bearing.

Recall the term "asymptotically" used in the central limit theorem. Let us illustrate its importance by examining what can happen if we apply the central limit theorem when n is small. We shall use the information of Example 1 of Section 7.4. Note that we are sampling from a finite population of size 8—it is impossible for n to become large. Let's use the central limit theorem to approximate the probability that a sample of size 2 selected randomly from the population

$$X: 2, 3, 4, 5, 6, 7, 8, 9$$

will yield a sample mean of more than 7. Table 7.7 shows that $\mu_x = 5.5$ and $\sigma_x^2 = 5.25$. Thus,

$$\begin{aligned} P(\overline{X}_D > 7) &= P\left(\overline{X} > 7 + \frac{1}{2}\cdot\frac{1}{n}\right) = P\left(\overline{X} > 7 + \frac{1}{2}\cdot\frac{1}{2}\right) \\ &= P(\overline{X} > 7.25) = P\left(Z > \frac{7.25 - 5.5}{\sqrt{\frac{5.25}{2}\cdot\frac{6}{7}}}\right) \\ &= P\left(Z > \frac{1.75}{1.50}\right) = P(Z > 1.17) = 0.1210. \end{aligned}$$

From Table 7.8, we obtain the exact probability

$$\begin{aligned} P(\overline{X} > 7) &= P(\overline{X} = 7.5) + P(\overline{X} = 8) + P(\overline{X} = 8.5) \\ &= 2/28 + 1/28 + 1/28 = 4/28 = 0.1429. \end{aligned}$$

Even though we have used the continuity correction factor here, the normal approximation leads to an error of more than 15 percent. Obviously, then, we should be extremely cautious in applying the central limit theorem in situations involving small samples. In fact, the size of the error in our approximation indicates that we need some rule of thumb to guide us in deciding when it is appropriate to use the central limit theorem for approximating probabilities associated with the sampling distribution of the sample mean. For most cases, the approximation will be adequate whenever $n > 30$. However, the approximation will naturally improve as n increases.

Problems

1. Assume that the variable of interest (X) is *discrete*, that *the distribution* of X is unknown, and simple random sampling is to be used. Find the indicated probabilities.

   a. $X \sim ?(158, 225)$, $n = 25$, $N = \infty$. Find $P(\overline{X}_D \le 162)$.

   b. $X \sim ?(30, 200)$, $n = 50$, $N = \infty$. Find $P(28 \le \overline{X}_D \le 33)$.

   c. $X \sim ?(972, 12{,}000)$, $n = 30$, $N = 1{,}500$. Find $P(\overline{X}_D > 947)$.

   d. $X \sim ?(237, 3{,}240)$, $n = 40$, $N = 2{,}000$. Find $P(218 < \overline{X}_D \le 244)$.

2. Find the indicated probabilities under the same assumptions as in Problem 1.

   a. $X \sim ?(580, 450)$, $n = 50$, $N = \infty$. Find $P(\overline{X}_D > 574)$.

   b. $X \sim ?(73, 288)$, $n = 32$, $N = \infty$. Find $P(69 < \overline{X}_D < 74)$.

   c. $X \sim ?(902, 40{,}000)$, $n = 25$, $N = 5{,}000$. Find $P(\overline{X}_D \le 913)$.

   d. $X \sim ?(155, 750)$, $n = 30$, $N = 1{,}000$. Find $P(143 \le \overline{X}_D \le 159)$.

3. Assume that the variable of interest (X) is *continuous, the distribution* of X is unknown, and simple random sampling is to be used. Determine the value of $\bar{X}(\bar{x}^*)$ such that

   a. $X \sim ?(150, 400)$, $n = 225$, $N = \infty$, $P(\bar{X} < \bar{x}^*) = 0.1251$;

   b. $X \sim ?(1{,}200, 225)$, $n = 49$, $N = 674$, $P(\bar{X} > \bar{x}^*) = 0.6151$;

   c. $X \sim ?(30, 25)$, $n = 100$, $N = \infty$, $P(29 < \bar{X} < \bar{x}^*) = 0.5225$;

   d. $X \sim ?(230, 10{,}000)$, $n = 64$, $N = 388$, $P(\bar{X} < \bar{x}^*) = 0.8531$.

4. Assume that the variable of interest (X) is *discrete, the distribution* of X is unknown, and simple random sampling is to be employed. Determine the value of $\bar{X}(\bar{x}^*)$ such that

   a. $X \sim ?(600, 2{,}500)$, $n = 100$, $N = \infty$, $P(\bar{X} > \bar{x}^*) = 0.0808$;

   b. $X \sim ?(350, 100)$, $n = 400$, $N = \infty$, $P(\bar{X} < \bar{x}^*) = 0.9772$;

   c. $X \sim ?(870, 900)$, $n = 81$, $N = 405$, $P(\bar{X} > \bar{x}^*) = 0.6915$;

   d. $X \sim ?(195, 10{,}000)$, $n = 625$, $N = 10{,}625$, $P(\bar{x}^* < \bar{X} < 198) = 0.0722$.

5. A manufacturer has ordered a shipment of 3,000 boxes of small parts. When the shipment arrives, a simple random sample of 25 boxes will be taken and the number of parts in each box counted. If the average number of parts per box is less than 98, the shipment will be rejected. The distribution of the number of parts per box is unknown, but past experience indicates that the mean is 100 and the variance is 50. What is the probability that the shipment will be accepted?

7.7 *The Sampling Distribution of the Sample Proportion*

Another statistic that often brings up probability questions is the *sample proportion.* After stating the general theory regarding the sampling distribution of this statistic, we shall illustrate the significance of the theory.

Assume that we are sampling without replacement (order disregarded) from a population from which any draw will result in either a success or a failure. Let π represent the proportion of successes in the population. When the sample size n is large, the sampling distribution of the sample proportion, p, is approximately normal. For an infinite population, the mean of the sampling distribution of p is

$$\mu_p \doteq E(p) = \pi \tag{1}$$

and the variance of p is

$$\sigma_p^2 \doteq \text{Var}(p) = \frac{\pi(1 - \pi)}{n}. \tag{2}$$

That is, for large samples from an infinite population,

$$p \sim \text{N}\left(\pi, \frac{\pi(1 - \pi)}{n}\right).$$

In the case of a finite population of size N, the mean of the sampling distribution of p is the same as in (1), but the variance of p is

$$\sigma_p^2 \doteq \text{Var}(p) = \frac{\pi(1-\pi)}{n} \cdot \frac{N-n}{N-1}. \tag{3}$$

That is, for large samples from a finite population of size N,

$$p \sim \text{N}\left(\pi, \frac{\pi(1-\pi)}{n} \cdot \frac{N-n}{N-1}\right).$$

Note that the finite population correction factor occurs again in this case. To illustrate the truth of (1) and (3), suppose an urn contains five balls, three red and two blue. For identification, the three red balls are numbered 1, 2, and 3 and the two blue balls are numbered 4 and 5. We are to draw two balls without replacement (order disregarded). If we let the drawing of a red ball represent a success, the population proportion of successes (red balls) is $\pi = 3/5 = 0.6$. Let us demonstrate that $\mu_p = \pi$, and

$$\sigma_p^2 = \frac{\pi(1-\pi)}{n} \cdot \frac{N-n}{N-1},$$

Table 7.9

Relations between mean and variance of the sample proportion and the population proportion

(1) Possible samples	(2) Proportion of red balls in sample (p_i)	(3) $p_i - \mu_p$	(4) $(p_i - \mu_p)^2$
R_1R_2	1.0	0.4	0.16
R_1R_3	1.0	0.4	0.16
R_1B_4	0.5	−0.1	0.01
R_1B_5	0.5	−0.1	0.01
R_2R_3	1.0	0.4	0.16
R_2B_4	0.5	−0.1	0.01
R_2B_5	0.5	−0.1	0.01
R_3B_4	0.5	−0.1	0.01
R_3B_5	0.5	−0.1	0.01
B_4B_5	0.0	−0.6	0.36
Total	6.0		0.90

By definition, $\mu_p = \dfrac{1}{{}_NC_n} \sum_{i=1}^{{}_NC_n} p_i = \dfrac{1}{10} \sum_{i=1}^{10} p_i = \dfrac{6.0}{10} = 0.6.$

But, merely by counting, we have $\pi = 0.6$.

Thus, $\mu_p = \pi = 0.6$.

Again, by definition

$$\sigma_p^2 = \frac{1}{{}_NC_n} \sum_{i=1}^{{}_NC_n} (p_i - \mu_p)^2 = \frac{1}{10} \sum_{i=1}^{10} (p_i - \mu_p)^2 = \frac{1}{10}(0.90) = 0.09.$$

However, $\dfrac{\pi(1-\pi)}{n} \cdot \dfrac{N-n}{N-1} = \dfrac{(0.6)(0.4)}{2} \cdot \dfrac{3}{4} = \dfrac{0.72}{8} = 0.09.$

Hence, $\sigma_p^2 = \dfrac{\pi(1-\pi)}{n} \cdot \dfrac{N-n}{N-1} = 0.09.$

where p is the statistic known as the sample proportion of successes (red balls). Table 7.9 lists all possible samples (column 1) and the values of p that correspond to the various possible samples (column 2). Columns 3 and 4 give the necessary information to calculate σ_p^2. Pertinent calculations are given below the table. You may wish to verify for yourself that, if sampling is with replacement and order is considered important, the mean and variance of p are as given by (1) and (2) for the case of an infinite population. This is not surprising, since repeated sampling with replacement is equivalent to sampling from an infinite population.

Note that the only possible values p can assume under this scheme are 1.0, 0.5, and 0.0. Thus, the sample space is limited to three points. By counting (column 2, Table 7.9) the number of possible samples which result in the various possible values of p, we obtain the sampling distribution of Table 7.10. Now, let us return to the general case.

Table 7.10

Sampling distribution of the sample proportion

Mutually exclusive events (kinds of samples)	*Proportion of red balls in sample*	*Probability of event*
Two blue balls	0.0	1/10
One red ball and one blue ball	0.5	6/10
Two red balls	1.0	3/10

Using reasoning similar to that in Section 7.5, we can obtain a standard normal distribution by making the transformation

$$Z = \frac{p - \mu_p}{\sigma_p}. \tag{4}$$

If the sampling scheme involves a sample of size $n > 30$ from an infinite population, we obtain a standard normal distribution by making the transformation

$$Z = \frac{p - \pi}{\sqrt{\dfrac{\pi(1 - \pi)}{n}}}, \tag{5}$$

since, here, p is normally distributed with mean $\mu_p = \pi$ and variance $\sigma_p^2 = \pi(1 - \pi)/n$. On the other hand, if the sampling scheme involves a sample of size $n > 30$ from a finite population of size N, we obtain a standard normal distribution by making the transformation

$$Z = \frac{p - \pi}{\sqrt{\dfrac{\pi(1 - \pi)}{n} \cdot \dfrac{N - n}{N - 1}}}, \tag{6}$$

since, here, p is normally distributed with mean $\mu_p = \pi$ and variance

$$\sigma_p^2 = \frac{\pi(1 - \pi)}{n} \cdot \frac{N - n}{N - 1}.$$

Equations (5) and (6) can be used to solve a variety of problems involving the sampling distribution of the sample proportion. In addition to the restriction $n > 30$, we usually require that $n\pi(1 - \pi)$ be greater than 5.

Example 1

Suppose we are to select a simple random sample of size $n = 100$ from a population (essentially infinite) of transistors with population proportion defective $\pi = 0.20$. Sketch the sampling distribution of the sample proportion defective, p. Then, shade the area which represents the probability of $p < 0.14$. Next, solve for $P(p < 0.14)$. Finally, for clarity, relate the required area in the sampling distribution of p to the appropriate area under a standard normal curve. (Ignore the continuity correction factor throughout).

Solution

We know from (1) that p is normally distributed with mean $\mu_p = \pi = 0.2$ and from (2) we have the standard deviation

$$\sigma_p = \sqrt{\pi(1 - \pi)/n} = \sqrt{(0.2)(0.8)/100} = \sqrt{0.16/100} = 0.4/10 = 0.04.$$

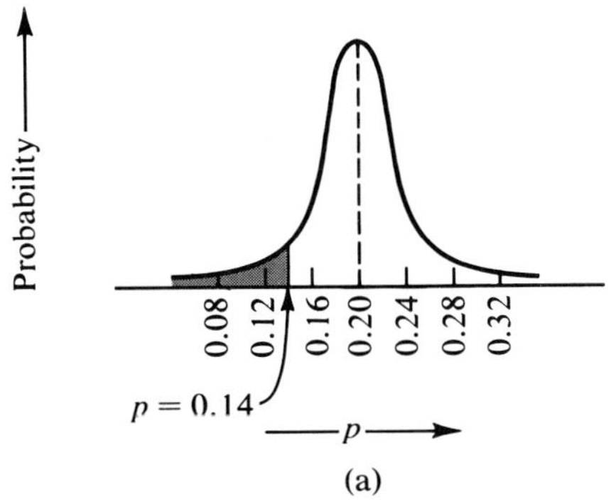

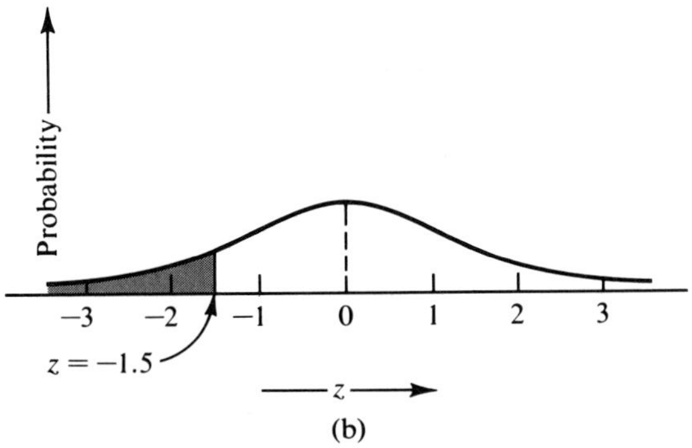

Figure 7.4
$P(p < 0.14)$

Thus, knowing that 99+ percent of the values of p will be between $\pi \pm 3(0.04)$, we can construct the curve of Figure 7.4a. Since we seek $P(p < 0.14)$, we shade the area to the left of $p = 0.14$. Using (1) and (2), we obtain

$$P(p < 0.14) = P\left(Z < \frac{0.14 - \mu_p}{\sigma_p}\right)$$

$$= P\left(Z < \frac{0.14 - 0.20}{\sqrt{\dfrac{(0.2)(0.8)}{100}}}\right) = P\left(Z < \frac{-0.06}{0.04}\right)$$

$$= P(Z < -1.5) = P(Z > 1.5) = 0.0668.$$

Finally, although the curve of Figure 7.4b is not drawn to the same scale as that of (a), we note that the area to the left of 0.14 in the sampling distribution of p corresponds to the area to the left of -1.5 for a standard normal variable. □

Example 2

Company A, which manufactures flashlight batteries, knows from past experience that its percentage defective is 30 percent. What is the probability of selecting a sample of size $n = 64$ with a sample proportion defective of 20 percent or greater when sampling from a lot of size $N = 1{,}000$? Use the continuity correction factor.

Solution

From (1) and (3),

$$\mu_p = \pi = 0.30$$

and

$$\sigma_p = \sqrt{\frac{\pi(1 - \pi)}{n} \cdot \frac{N - n}{N - 1}} = \sqrt{\frac{(0.30)(0.70)}{64} \cdot \frac{936}{999}} \cong 0.055. \tag{7}$$

Thus, the probability of selecting a sample of size $n = 64$ with a sample proportion defective of 20 percent or greater is

$$P(p_D \geq 0.20) = P\left(p > 0.20 - \frac{1}{2} \cdot \frac{1}{64}\right) = P(p > 0.1922). \tag{8}$$

However, using the values computed in (7) to make the proper transformation (4), we find that (8) becomes

$$P(p > 0.1922) = P\left(Z > \frac{0.1922 - \mu_p}{\sigma_p}\right)$$

$$= P\left(Z > \frac{0.1922 - 0.30}{0.055}\right) \cong P(Z > -1.96)$$

$$= 1 - P(Z > 1.96) = 1 - 0.0250 = 0.9750.$$

Thus, under the given sampling scheme, there is a very high probability of obtaining a sample proportion greater than or equal to 0.20. □

Example 3

If a fair die is tossed 100 times, what is the probability that at least 20 aces (ones) will appear.

Solution

This problem can be solved exactly through use of the binomial distribution as

$$P \text{ (at least 20 aces out of 100 tosses)} = P_b(X \geq 20 \mid n = 100, \pi = 1/6)$$

$$= \sum_{x=20}^{100} \binom{100}{x} (1/6)^x (5/6)^{100-x}.$$

However, without an extensive binomial table, one gets involved in a prohibitive amount of labor. Thus, we use the sampling distribution of the sample proportion to obtain a very good approximation as follows. "Obtaining at least 20 aces out of 100 tosses" is equivalent to "obtaining a sample proportion of at least 0.20 when taking a simple random sample of size $n = 100$ from an infinite population," because sampling is essentially with replacement in this case. Thus, for $n = 100$ and $\pi = 1/6$, we have

$$P\text{(at least 20 aces out of 100 tosses)} = P(p_D \geq 0.20)$$

$$\cong P\left(p > 0.20 - \frac{1}{2} \cdot \frac{1}{100}\right) = P(p > 0.195). \quad (9)$$

However,

$$\mu_p = \pi = 1/6 \cong 0.167 \quad \text{and} \quad \sigma_p = \sqrt{\frac{\pi(1-\pi)}{n}} = \sqrt{\left(\frac{1}{6}\right)\left(\frac{5}{6}\right) \cdot \frac{1}{100}} \cong 0.037.$$

Then, (9) becomes

$$P(p > 0.195) = P\left(Z > \frac{0.195 - \mu_p}{\sigma_p}\right)$$

$$= P\left(Z > \frac{0.195 - 0.167}{0.037}\right) \cong P(Z > 0.757) = 0.2245. \quad \square$$

Problems

1. Ignore the continuity correction factor. Suppose we are to select a simple random sample of size $n = 600$ from a population of parts with population proportion defective $\pi = 0.40$.

   a. Sketch the sampling distribution of the sample proportion defective p.

   b. Shade the area which represents the probability that the sample proportion defective will be greater than 0.385.

   c. Calculate $P(p < 0.37)$.

2. Assume simple random sampling and, without using the continuity correction factor, find the indicated probabilities.

   a. $\pi = 0.20$, $n = 200$, $N = \infty$. Find $P(p \leq 0.22)$.

   b. $\pi = 0.90$, $n = 100$, $N = \infty$. Find $P(0.88 < p < 0.91)$.

   c. $\pi = 0.30$, $n = 700$, $N = 8{,}000$. Find $P(p \geq 0.285)$.

   d. $\pi = 0.10$, $n = 100$, $N = 15{,}000$. Find p^* such that $P(0.07 < p < p^*) = 0.5721$.

3. A company that manufactures ball point pens knows from past experience that its percentage defective is 12%. What is the probability of selecting a simple random sample of size 144 with a sample proportion defective of over 14% when sampling from a lot of size $N = 2{,}000$? Use the continuity correction factor.

4. Assume simple random sampling and use the continuity correction factor in each case.

   a. $\pi = 0.3$, $n = 50$, $N = 500$. Find $P(p < 0.41)$.

   b. $\pi = 0.4$, $n = 35$, $N = 1{,}000$. Find $P(0.34 \leq p \leq 0.43)$.

   c. $\pi = 0.1$, $n = 60$, $N = \infty$. Find $P(0.03 < p < 0.08)$.

   d. $\pi = 0.2$, $n = 75$, $N = \infty$. Find p^* such that $P(p \leq p^*) = 0.9715$.

5. A pair of dice is tossed 1,000 times. What is the probability that a sum of 5 will appear on the top faces of the dice at least 100 times? Do not use the continuity correction factor. .8786

   a. Solve by using the normal approximation to the binomial distribution, where we are concerned with the number of successes.

   b. Solve by using the method of this chapter, where we are concerned with the proportion of successes.

   c. Check that your answers are approximately equal.

6. The experiment of drawing a card from a standard deck of cards and reshuffling the deck each time is repeated 150 times. What is the probability that fewer than 10 kings will appear? Use the continuity correction factor.

   a. Solve by using the normal approximation to the binomial distribution, where we are concerned with the number of successes.

   b. Solve by using the method of this chapter, where we are concerned with the proportion of successes.

   c. Check that your answers are approximately equal.

Table 7.11
Computational formulas for important parameters and statistics

Set of data	Variable	Distribution	Number of values in set	Computational formula: Mean	Computational formula: Variance
Population	X	Any kind	N	$\mu_x \doteq E(X) = \frac{1}{N}\sum_{i=1}^{N} x_i$	$\sigma_x^2 \doteq \mathrm{Var}(X) = \frac{1}{N}\sum_{i=1}^{N}(x_i - \mu_x)^2$
Sample	X	Any kind	n	$\overline{X} = \frac{1}{n}\sum_{i=1}^{n} x_i$	$s_x^2 = \frac{1}{n}\sum_{i=1}^{n}(x_i - \overline{X})^2$

Table 7.12
Important results regarding the sampling distribution of the sample mean and the sample proportion

Statistic	Relevant population parameters	Condition	Sampling distribution (type)	Mean	Variance
Sample mean ($\overline{X}$)	Population mean (μ_x) and population variance (σ_x^2)	Finite population (N) $n > 30$	Approximately normal	$\mu_{\bar{x}} \doteq E(\overline{X}) = \mu_x$	$\sigma_{\bar{x}}^2 \doteq \mathrm{Var}(\overline{X}) = \frac{\sigma_x^2}{n} \cdot \frac{N-n}{N-1}$
		Infinite population $n > 30$	Approximately normal	$\mu_{\bar{x}} \doteq E(\overline{X}) = \mu_x$	$\sigma_{\bar{x}}^2 \doteq \mathrm{Var}(\overline{X}) = \frac{\sigma_x^2}{n}$
Sample proportion (p)	Population proportion (π)	Finite population (N) $n > 30$ and $n\pi(1-\pi) > 5$	Approximately normal	$\mu_p \doteq E(p) = \pi$	$\sigma_p^2 \doteq \mathrm{Var}(p) = \frac{\pi(1-\pi)}{n} \cdot \frac{N-n}{N-1}$
		Infinite population $n > 30$ and $n\pi(1-\pi) > 5$	Approximately normal	$\mu_p \doteq E(p) = \pi$	$\sigma_p^2 \doteq \mathrm{Var}(p) = \frac{\pi(1-\pi)}{n}$

7.8 Summary

The terms "statistic" and "sampling distribution" represent two of the most important concepts of statistics. A *statistic* is a function of sample values taken from a population (that is, from the distribution of a random variable). Hence, as functions of random variables, they themselves may be viewed in the same fashion as random variables. Thus, any statistic has associated with it a mean, a variance, and, in fact, a probability distribution. The probability distribution of a statistic, called a *sampling distribution*, is derived by considering all possible values a statistic may assume under a given sampling scheme.

In this chapter, we dealt extensively with the sampling distributions of the sample mean and sample proportion, and saw that related theory helps to answer many practical probability questions. Tables 7.11 and 7.12 summarize the theory.

The importance of the sample mean and sample proportion notwithstanding, you should not overlook other statistics that play vital roles in the discipline. Given a specific sampling scheme, each such statistic has a mean, variance, and sampling distribution.

Important Terms and Concepts

asymptotically normal
bar chart
central limit theorem
finite population correction factor
mean of the statistic $T(\mu_t)$
mean of sample mean $(\mu_{\bar{x}})$
mean of sample proportion (μ_p)
parameter
population mean (μ_x)
population proportion (π)
population size (N)
population variance (σ_x^2)
sample mean $(\bar{X})$
sample proportion (p)
sample size (n)
sample space
sample variance (s_x^2)
sampling distribution
sampling procedure
sampling scheme
statistic
variance of the statistic $T(\sigma_t^2)$
variance of sample mean $(\sigma_{\bar{x}}^2)$
variance of sample proportion (σ_p^2)

References

7.1 H. Cramér, *Mathematical Methods of Statistics*, Princeton, N.J.: Princeton University Press, 1961.

7.2 A. M. Mood and F. A. Graybill, *Introduction to the Theory of Statistics*, New York: McGraw-Hill Book Company, 1963.

7.3 Frederick Mosteller, et al, *Probability with Statistical Applications*, 2d ed. Reading, Mass.: Addison-Wesley Publishing Company, Inc., 1970.

Review Problems

1. Here, we show by example that the binomial distribution is really a sampling distribution.

   Suppose we are to make 5 draws of one ball each with replacement (order important and draws independent) from an urn containing 25 red and 75 green balls. Letting a red ball be a success, use basic probability concepts to

   a. describe the sample space and graph the probability distribution of the Bernoulli process involved,

   b. list all possible samples in terms of the colors of the balls of the various draws,

   c. note the sample space associated with the number of red balls out of the five draws,

   d. graph the resulting sampling (binomial) distribution.

   (Section 5.2)

2. A company is seeking a government contract to supply nylon rope for the Army. A sample of 49 ropes will be selected at random from an essentially infinite population and tested for breaking strength. If the average breaking strength is less than 2,700 lb, no contract will be issued and the company will lose the \$10,000 it cost to submit its proposal. If it is between 2,700 and 2,850 lb, a contract will be received for 25,000 ropes at a profit of \$25,000, exclusive of proposal costs. If the average is over 2,850 lb, the contract will be for 100,000 ropes at a profit of \$150,000, exclusive of proposal costs. The distribution of the breaking strengths is not known, but the mean and variance are known to be 2,800 lb and 160,000 lb^2, respectively. What is the expected monetary value of seeking the contract?

   (Section 4.12)

3. Consider the following game. A die is tossed twice. If no aces are rolled, a sample of size 100 is randomly selected from the population $X_1 \sim ?\,(90; 40{,}000)$. If one ace is rolled, a sample of size 64 is randomly selected from the population $X_2 \sim ?\,(100, 625)$. If two aces are rolled, a sample of size 225 is randomly selected from the population $X_3 \sim ?\,(110; 10{,}000)$. If the sample mean ($\overline{X}$) is greater than 105, the player wins \$500. It costs \$100 to play. Would it be wise to play the game? Assume that the populations for X_1, X_2, and X_3 are infinite.

   (Sections 4.6, 4.12, and 5.2)

4. The number of sales which Mr. Browning makes per day is distributed as a Poisson distribution. On the average, he makes 4 sales per day. On days when he makes 3 sales, his earnings (X) average \$75 with a standard deviation of \$15. Find the probability that on a given day, he will make 5 sales *and* his earnings will be greater than \$140. He will obtain a bonus if, in the next 30 days, he accomplishes this feat more than twice. What is the probability that Mr. Browning will receive the bonus? Assume that daily earnings are proportional to the number of sales per day, while the standard deviation remains constant at \$15.

   (Sections 4.6, 5.2, and 5.3)

5. In Problem 3, page 253, suppose the cost of producing the capacitors is \$3,000 and the firm will receive \$5,000 for them if the shipment is accepted. If the shipment is rejected, there is a 10% chance it can be sold for \$3,000, a 25% chance it can be sold for \$1,000, and a 65% chance it cannot be sold at all. What is the expected monetary value of producing the order? What is the probability that a loss will be incurred?

   (Section 4.12)

Eight

Estimation

8.1 Introduction

Estimation involves using information from a sample statistic to estimate a population parameter. For example, we may use the mean of a sample to estimate the population mean, or we may use the proportion of successes in a sample to estimate the proportion of successes in a population. Here we shall study certain principles governing estimation and investigate specific methods of estimation. We begin by defining some terms.

Definition 8.1

*An **estimator** is a specific statistic used to estimate a population parameter.*

Definition 8.2

*An **estimate** is a single observed value of a given estimator.*

Definition 8.3

***A point estimate** produces a single-valued estimate of a parameter.*

Definition 8.4

***Interval estimation** is estimation involving the statement that a parameter lies within a given interval.*

Suppose we wish to estimate the mean of a given population by taking a sample of size $n = 50$ and computing the sample mean. In this case, the formula for the sample mean is the general formula for the estimation. Suppose further that we find the sample mean to be 28.5. Our estimate of the population mean also becomes 28.5. Note that this is a point estimate. Had we utilized a procedure to describe a range associated with the value of the population mean, we would have constructed an interval estimate.

The following problem faces us in estimation. We must estimate some unknown parameter associated with a certain probability distribution by using only sample information. It may be either impossible, under any circumstances whatsoever, to obtain the exact value of the parameter or it may be simply too difficult in a practical sense to obtain the exact value.

As an example of the first case, suppose a manufacturer of potato chips wishes to estimate the average number of ounces of potato chips packed into bags which are sold for a given price. It is impossible to weigh every bag. Further, the "true average" amount per bag may change from time to time because of changes in packaging conditions. In such case, it is impossible to define "the" population of interest. Nevertheless, from a profit standpoint alone, it is important that the manufacturer keep fairly close tabs on the parameter of interest (population mean).

As an example of the second case, suppose we want to determine the feasibility of introducing a new type of sleeping bag and thus want to know the percentage of families in a large city that camp out more than three times per year. It certainly is not practical to compute the exact percentage by questioning every family.

In either situation, we must answer several important questions before we can even begin to estimate the parameter of interest.

1. What estimator should we use?
2. What restraints (time, money, etc.) exist on the size of sample to be taken?
3. How large a discrepancy between the estimate and the true parameter value can we permit?
4. How confident must we be that the discrepancy is within set limits?

The next five sections are devoted to answering the first question. Proper attack of the last three necessarily requires the complete involvement of the person charged with the responsibility for a final decision. We cannot hope to discuss all of the ramifications of these questions.

Problems

1. Explain the difference between an *estimator* and an *estimate.*
2. Explain the difference between *point estimation* and *interval estimation.*
3. Name at least two conditions which can result in the need to estimate a parameter.
4. What four questions *must* be answered before we can intelligently begin the task of estimating a parameter?

8.2 *Point Estimation*

We shall begin our treatment of point estimation by viewing intuitively the concept of *good estimation.* After all, we cannot select a good estimator without some general agreement about what is meant by "good."

Suppose three people in our company have the job of estimating weekly sales one month in advance. Suppose further that we have kept excellent records on each individual by tabulating each forecast and actual weekly sales, calculating the appropriate differences between sales and forecasts, building frequency distributions of such differences, and converting the frequency distributions into the smoothed approximate probability distributions of Figure 8.1. Based on the curves shown,

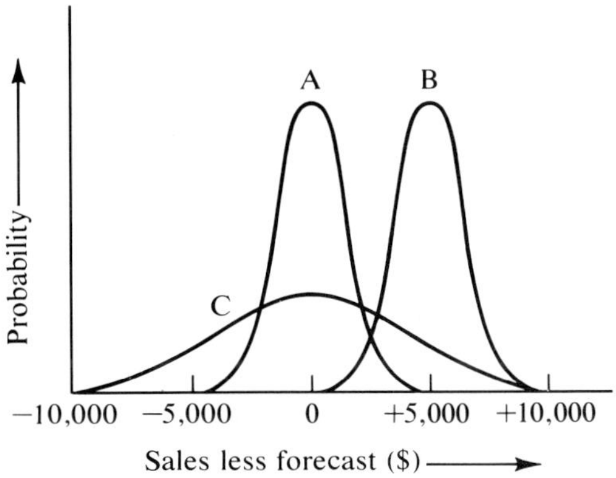

Figure 8.1
Approximate probability distributions for hypothetical sales forecasters

which of the three individuals, A, B, or C, would we intuitively classify as the best sales forecaster? There is little question that A appears to be the best. Why? We note that a perfect forecast will always result in a value of $0 for the variable "sales less forecast." This being the case, we note that B has done a poor job of forecasting, because his probability distribution is concentrated around +$5,000. Thus, even though the variance (dispersion) of B's values for "sales less forecast" is relatively small, he has a tendency to make pessimistic forecasts. When we contrast this with C's record, we note a different type of poor forecasting. While C appears to be optimistic just about as often as he is pessimistic, his variance (dispersion) for "sales less forecast" is very large. Thus, although his average value for "sales less forecast" is zero, he shows a tendency to err quite often in both directions. A has hit a happy medium, so to speak. His record exhibits two very good features: (1) his average value for "sales less forecast" is zero and (2) the variance (dispersion) of his values for "sales less forecast" is relatively small.

Statistical estimation and sales forecasting certainly are not identical, but the two share enough characteristics to give us insight on estimation by considering the main points of sales forecasting. Just as the value of zero for "sales less forecast" represented a perfect forecast above, there also exists some value—namely, the true value of the parameter we are trying to estimate—that represents perfection in the problem of estimation. Additionally, just as we could judge the ability of a sales forecaster by the relation of his historical probability distribution to the value representing perfection (zero), so can we judge the merit of an estimator (statistic) by the relation of its sampling distribution to the value representing perfection (the true value of the parameter which we are trying to estimate).

Since we repeatedly encounter in statistics the problem of estimating population parameters through use of an estimator (statistic) determined from a set of sample values, any discussion of specific estimation problems must begin with a systematic investigation of the general subject. Suppose that we have a sample from a population the distribution of which has a known mathematical form involving at least one unknown parameter and that we wish to select an estimator for a specific parameter. An infinite number of statistics (i.e., functions of the sample values) might be chosen. Two questions immediately arise. First, how do we best use the sample data to estimate the parameter? For example, it is usually conceded in general circumstances that we should use the sample mean to estimate the population mean. Why? Why not use the sample median? Both are measures of location. However, the sample mean is a better estimator of the population mean than is the sample median (or other statistics).

This leads immediately to the second major question. What do we mean by good estimates or, more generally, a good estimator? Obviously, if we limit consideration to a single sample, and thus to a single estimate, the best estimate is the one coming closest to the true value of the parameter to be estimated. However, we know that every estimator, and hence every estimate obtained through its use, is a function of sample values resulting from some sampling scheme. Thus, since an estimator is essentially a random variable, any estimate resulting from its use must be regarded as an observed value of that variable. This being the case, it is naturally impossible to predict any single value to be assumed by an estimate (obtained through use of a specific estimator) in a particular case. Thus, the goodness of an estimate cannot be judged from individual values, but only from the distribution of values which it can assume, that is, from the sampling distribution of the estimator which gives rise to the individual values (estimates). If the sampling distribution of a given estimator is closely concentrated around the true value of the parameter of interest, there is a high probability that any individual estimate will differ from such true value by only a small amount. This type of reasoning leads us immediately to an intuitive statement regarding a good estimator.

For an estimator of an unknown parameter to be classed as "good," we generally require that its sampling distribution be closely concentrated around the parameter to be estimated. Naturally then, all else being equal, we generally consider one estimator better than another if its sampling distribution has the greater concentration about the true parameter value. (Hence, a major reason for generally using the sample mean rather than the sample median—or other statistics—as an estimator of the population mean is that its sampling distribution is more closely concentrated around the population mean.) Thus, the question of how best to use sample data to arrive at an estimate of a parameter reduces to the question of how we would use our data in order to obtain estimates having maximum concentration, or, in terms more familiar to us, minimum dispersion, about the parameter of interest.

If we are to compare various estimators to determine which is better for a given parameter and if a criterion for goodness is the dispersion of the sampling distribution about the true value of the parameter, then it would seem that an effective decision process would require that we deal with some measure that reflects such dispersion. What measure should we use? Recall our discussion of similar problems in preceding chapters. In such instances, we were concerned with obtaining a reasonable measure of the dispersion of a distribution about its mean; we determined that the variance is a highly acceptable measure from both intuitive and theoretical standpoints. However, the variance of the sampling distribution under consideration is obviously not an acceptable measure; for, even though the variance of the distribution will give us a measure of the concentration about the mean of the distribution, it in no way follows generally that it will provide adequate information regarding the concentration of the distribution about some unknown parameter (unless, of course, we have the special situation where the parameter of interest and mean of the distribution are identical). Nevertheless, the verbal definition (Definition 3.1) of the variance of a set of data (or a distribution) gives us a guideline for obtaining a reasonable measure of the dispersion of a set of data (sampling distribution) about a parameter. That is, since the "expected value of the squared deviations about the mean (constant)" represents a reasonable measure of the dispersion of a set of data (or a distribution) about the mean, it stands to reason that the "expected value of the squared deviations about a parameter (constant)" should represent a reasonable measure of the dispersion of a set of data (or a distribution) about the parameter. This statement being theoretically justifiable [reference (8.1)], we shall adopt this type measure as a desirable criterion of goodness for an estimator.

Using this criterion in a situation where we know a parameter of interest, let us illustrate how it is possible to demonstrate which of two estimators would have been better to use for estimating such parameter had it not been known. *The example to be used involves two estimators which are measuring different characteristics, and, hence, is purely illustrative.* However, a thorough understanding of all the details associated with it will greatly enhance your understanding of concepts to be discussed subsequently.

Example 1

Given the finite population of size $N = 5$

$$X: x_1 = 3; \quad x_2 = 7; \quad x_3 = 11; \quad x_4 = 15; \quad x_5 = 19,$$

suppose we wish to select without replacement a simple random sample of size $n = 2$ in order to estimate the population mean. Further, we have decided to use either the sample mean or the sample range for our estimate. Which would be the better estimator?

Solution

Now, it is obvious that we can easily calculate the population mean to be

$$\mu_x = \frac{1}{n}\sum_{i=1}^{n} x_i = \frac{1}{5}\sum_{i=1}^{5} x_i = 11.$$

Let us use this fact in determining which of the two statistics would be the better estimator. For the sampling distribution of each, we must find the expected value of the squared deviations about the parameter of interest and then base our selection on a comparison of the resulting values. That is, we must calculate and compare

$$E(\bar{X} - \mu_x)^2 = \frac{1}{k}\sum_{i=1}^{k} (\bar{x}_i - \mu_x)^2 \tag{1}$$

and

$$E(R - \mu_x)^2 = \frac{1}{k}\sum_{i=1}^{k} (r_i - \mu_x)^2, \tag{2}$$

where $\bar{X}$ is the general statistic called the sample mean, R is the general statistic called the sample range, $\bar{x}_i$ and r_i are the ith possible values that the sample mean and sample range can assume, respectively, under the sampling scheme, μ_x is the mean of the population, and $k = {}_NC_n$ represents the number of possible samples. The following steps and Table 8.1 outline the procedure.

1. List all possible samples which can occur under the sampling scheme (column 1). Note that there are ${}_NC_n = \binom{5}{2} = 10$ equally likely possible samples.
2. Calculate the mean for each possible sample, that is, $\bar{x}_i$ for $i = 1, 2, \ldots, 10$ (column 2).
3. Calculate $\bar{x}_i - \mu_x$ and $(\bar{x}_i - \mu_x)^2$ for $i = 1, 2, \ldots, 10$ (columns 3 and 4).
4. Calculate $\frac{1}{10}\sum_{i=1}^{10} (\bar{x}_i - \mu_x)^2$. That is, average the values of column 4.
5. Calculate the range for each possible sample, that is, r_i for $i = 1, 2, \ldots, 10$ (column 5).
6. Calculate $(r_i - \mu_x)$ and $(r_i - \mu_x)^2$ for $i = 1, 2, \ldots, 10$ (columns 6 and 7).
7. Calculate $\frac{1}{10}\sum_{i=1}^{10} (r_i - \mu_x)^2$. That is, average the values of column 7.

It can be noted from Table 8.1 that the measure of dispersion about the parameter to be estimated ($\mu_x = 11$) is smaller for the sample mean than for the sample range. Obviously, from a probability standpoint, we are more likely to obtain a value near the true population mean by using the sample mean than by using the sample range. Thus, for this case, the sample mean is the better estimator for the population mean. This is no surprise since we have agreed that the range of a set of date (the sample in this case) is a measure of dispersion. We therefore would not expect it to be a good estimator for the population mean, a measure of location. Insignificant as this example may seem, the concepts used in it are important.

Example 1

Suppose we are concerned with drawing without replacement a simple random sample of size $n = 5$ from the following population of size $N = 7$:

$$X: 3, 6, 8, 9, 12, 15, 18.$$

Determine whether the sample mean, median, range, and variance are unbiased estimators for the population mean, median, range, and variance, respectively.

Solution

Since we are dealing with a finite population, we can answer this question empirically by (1) determining all values each of the estimators can assume, (2) calculating the expected value of each estimator, (3) calculating the parameters of interest, and, finally, (4) comparing the expected value of each estimator with the value of the appropriate parameter. Information pertaining to various parts of the question is given in Table 8.3. Let

$$\mu_x = \text{population mean}, \quad M = \text{population median},$$
$$R = \text{population range}, \quad \sigma_x^2 = \text{population variance},$$

and for the ith sample, let

$$\bar{x}_i = \text{sample mean}, \quad m_i = \text{sample median},$$
$$r_i = \text{sample range}, \quad s_i^2 = \text{sample variance};$$

and let

$$k = {}_N C_n.$$

Then,

$$\mu_x = \frac{1}{N}\sum_{i=1}^{N} x_i = \frac{3 + 6 + \cdots + 18}{7} = \frac{71}{7} \cong 10.143$$

and

$$E(\text{sample mean}) = \frac{1}{k}\sum_{i=1}^{k} \bar{x}_i = \frac{213.0}{21} \cong 10.143 = \mu_x \text{ (unbiased)};$$

$$M = 9 \quad \text{and} \quad E(\text{sample median}) = \frac{1}{k}\sum_{i=1}^{k} m_i = \frac{201}{21} \cong 9.57 \neq M \text{ (biased)};$$

$$R = 18 - 3 = 15 \quad \text{and} \quad E(\text{sample range}) = \frac{1}{k}\sum_{i=1}^{k} r_i = \frac{274}{21} \cong 13.05 \neq R \text{ (biased)};$$

$$\sigma_x^2 = \frac{1}{N}\sum_{i=1}^{N}(x_i - \mu_x)^2 = \frac{\sum_{i=1}^{N} x_i^2 - \dfrac{(\sum_{i=1}^{N} x_i)^2}{N}}{N} = \frac{883 - \dfrac{(71)^2}{7}}{7} \cong 23.27$$

and

$$E(\text{sample variance}) = \frac{1}{k}\sum_{i=1}^{k} s_i^2 = \frac{456}{21} \cong 21.71 \neq \sigma_x^2 \text{ (biased)}.$$

Table 8.3

Possible values for sample mean, median, range, and variance: simple random sampling ($N = 7,\ n = 5$)

Population X: 3, 6, 8, 9, 12, 15, 18 $\quad {}_NC_n = {}_7C_5 = 21$

Possible sample	Sample mean ($\bar{x}_i$)	Sample median (m_i)	Sample range (r_i)	Sample variance (s_i^2)	
3, 6, 8, 9, 12	7.6	8	9	$[(3-7.6)^2+(6-7.6)^2+(8-7.6)^2+(9-7.6)^2+(12-7.6)^2]/5$	= 9
3, 6, 8, 9, 15	8.2	8	12	$[(3-8.2)^2+(6-8.2)^2+(8-8.2)^2+(9-8.2)^2+(15-8.2)^2]/5$	= 15.76
3, 6, 8, 9, 18	8.8	8	15	$[(3-8.8)^2+(6-8.8)^2+(8-8.8)^2+(9-8.8)^2+(18-8.8)^2]/5$	= 25.36
3, 6, 8, 12, 15	8.8	8	12	$[(3-8.8)^2+(6-8.8)^2+(8-8.8)^2+(12-8.8)^2+(15-8.8)^2]/5$	= 18.16
3, 6, 8, 12, 18	9.4	8	15	$[(3-9.4)^2+(6-9.4)^2+(8-9.4)^2+(12-9.4)^2+(18-9.4)^2]/5$	= 27.04
3, 6, 8, 15, 18	10.0	8	15	$[(3-10)^2+(6-10)^2+(8-10)^2+(15-10)^2+(18-10)^2]/5$	= 31.60
3, 6, 9, 12, 15	9.0	9	12	$[(3-9)^2+(6-9)^2+(9-9)^2+(12-9)^2+(15-9)^2]/5$	= 18.00
3, 6, 9, 12, 18	9.6	9	15	$[(3-9.6)^2+(6-9.6)^2+(9-9.6)^2+(12-9.6)^2+(18-9.6)^2]/5$	= 26.64
3, 6, 9, 15, 18	10.2	9	15	$[(3-10.2)^2+(6-10.2)^2+(9-10.2)^2+(15-10.2)^2+(18-10.2)^2]/5$	= 30.96
3, 6, 12, 15, 18	10.8	12	15	$[(3-10.8)^2+(6-10.8)^2+(12-10.8)^2+(15-10.8)^2+(18-10.8)^2]/5$	= 30.96
3, 8, 9, 12, 15	9.4	9	12	$[(3-9.4)^2+(8-9.4)^2+(9-9.4)^2+(12-9.4)^2+(15-9.4)^2]/5$	= 16.24
3, 8, 9, 12, 18	10.0	9	15	$[(3-10)^2+(8-10)^2+(9-10)^2+(12-10)^2+(18-10)^2]/5$	= 24.40
3, 8, 9, 15, 18	10.6	9	15	$[(3-10.6)^2+(8-10.6)^2+(9-10.6)^2+(15-10.6)^2+(18-10.6)^2]/5$	= 28.24
3, 8, 12, 15, 18	11.2	12	15	$[(3-11.2)^2+(8-11.2)^2+(12-11.2)^2+(15-11.2)^2+(18-11.2)^2]/5$	= 27.76
3, 9, 12, 15, 18	11.4	12	15	$[(3-11.4)^2+(9-11.4)^2+(12-11.4)^2+(15-11.4)^2+(18-11.4)^2]/5$	= 26.64
6, 8, 9, 12, 15	10.0	9	9	$[(6-10)^2+(8-10)^2+(9-10)^2+(12-10)^2+(15-10)^2]/5$	= 10.00
6, 8, 9, 12, 18	10.6	9	12	$[(6-10.6)^2+(8-10.6)^2+(9-10.6)^2+(12-10.6)^2+(18-10.6)^2]/5$	= 17.44
6, 8, 9, 15, 18	11.2	9	12	$[(6-11.2)^2+(8-11.2)^2+(9-11.2)^2+(15-11.2)^2+(18-11.2)^2]/5$	= 20.56
6, 8, 12, 15, 18	11.8	12	12	$[(6-11.8)^2+(8-11.8)^2+(12-11.8)^2+(15-11.8)^2+(18-11.8)^2]/5$	= 19.36
6, 9, 12, 15, 18	12.0	12	12	$[(6-12)^2+(9-12)^2+(12-12)^2+(15-12)^2+(18-12)^2]/5$	= 18.00
8, 9, 12, 15, 18	12.4	12	10	$[(8-12.4)^2+(9-12.4)^2+(12-12.4)^2+(15-12.4)^2+(18-12.4)^2]/5$	= 13.84
Total	213.0	201	274		456.00

We see that the sample mean is unbiased for the population mean. On the other hand, we note that no other estimator is unbiased for its namesake parameter. Thus, we see that obtaining an unbiased estimator is not an automatic process. □

Consistent Estimators

It is also desirable that an estimator become better as the sample size increase. That is, we would like the sampling distribution of the estimator to become more closely concentrated around the parameter of interest as the sample size n increases. An estimator having this property is said to be *consistent.*

Example 2

Given the population of size $N = 9$

$$X: 3, 5, 8, 11, 15, 16, 19, 23, 25,$$

we are to select a simple random sample in order to estimate the population median. Demonstrate empirically that the sample median is consistent for the population median.

Solution

In order to demonstrate consistency we develop empirically the sampling distributions of the sample median for several different sample sizes, and then compare these distributions to see whether, as n increases, the distributions tend to exhibit closer concentration around the parameter of interest. (We note by inspection that the population median is 15.) Information for the empirical development of the distributions of the sample median for cases of $n = 3$, $n = 5$, and $n = 7$, respectively, is given in Tables 8.4 through 8.6. Each table lists all possible samples under the given sampling schemes along with the corresponding values of the sample median. Resulting sampling distributions are given in Table 8.7 and Figure 8.3. You should assure yourself that you can derive the distributions of Table 8.7. If not, you should review Section 7.3. Since the sampling distribution of the sample median does in fact become more closely concentrated around the true population median (15) as the sample size increases (as Figure 8.3 shows), the sample median is consistent for the population median. □

Even though the results of Example 2 tell us that the sample median is a consistent estimator for the population median, they do not allow a general statement. Let us consider a case for which the results can be generalized. Consider again the sample mean as an estimator of the population mean. Suppose we are sampling from an infinite population with mean μ_x and variance σ_x^2. We know that, for samples of size n, the distribution of the sample mean $\overline{X}$ will have mean μ_x and variance σ_x^2/n. Thus, we see that, regardless of the value of n, the sampling distribution of $\overline{X}$ will be centered at μ_x (the population parameter). However, as n increases, the variance of $\overline{X}$ (its dispersion about μ_x) will decrease. Thus, we find that $\overline{X}$ is consistent for μ_x; for, as the sample size increases, the sampling distribution of $\overline{X}$ becomes more closely concentrated around the population mean μ_x.

Table 8.4
Possible values for sample median: Simple random sampling ($N = 9, n = 3$)

Population X: 3, 5, 8, 11, 15, 16, 19, 23, 25 $\quad {}_NC_n = {}_9C_3 = 84$

Possible sample	*Sample median*	*Possible sample*	*Sample median*	*Possible sample*	*Sample median*	*Possible sample*	*Sample median*
3, 5, 8	5	3, 15, 25	15	5, 15, 25	15	8, 23, 25	23
3, 5, 11	5	3, 16, 19	16	5, 16, 19	16	11, 15, 16	15
3, 5, 15	5	3, 16, 23	16	5, 16, 23	16	11, 15, 19	15
3, 5, 16	5	3, 16, 25	16	5, 16, 25	16	11, 15, 23	15
3, 5, 19	5	3, 19, 23	19	5, 19, 23	19	11, 15, 25	15
3, 5, 23	5	3, 19, 25	19	5, 19, 25	19	11, 16, 19	16
3, 5, 25	5	3, 23, 25	23	5, 23, 25	23	11, 16, 23	16
3, 8, 11	8	5, 8, 11	8	8, 11, 15	11	11, 16, 25	16
3, 8, 15	8	5, 8, 15	8	8, 11, 16	11	11, 19, 23	19
3, 8, 16	8	5, 8, 16	8	8, 11, 19	11	11, 19, 25	19
3, 8, 19	8	5, 8, 19	8	8, 11, 23	11	11, 23, 25	23
3, 8, 23	8	5, 8, 23	8	8, 11, 25	11	15, 16, 19	16
3, 8, 25	8	5, 8, 25	8	8, 15, 16	15	15, 16, 23	16
3, 11, 15	11	5, 11, 15	11	8, 15, 19	15	15, 16, 25	16
3, 11, 16	11	5, 11, 16	11	8, 15, 23	15	15, 19, 23	19
3, 11, 19	11	5, 11, 19	11	8, 15, 25	15	15, 19, 25	19
3, 11, 23	11	5, 11, 23	11	8, 16, 19	16	15, 23, 25	23
3, 11, 25	11	5, 11, 25	11	8, 16, 23	16	16, 19, 23	19
3, 15, 16	15	5, 15, 16	15	8, 16, 25	16	16, 19, 25	19
3, 15, 19	15	5, 15, 19	15	8, 19, 23	19	16, 23, 25	23
3, 15, 23	15	5, 15, 23	15	8, 19, 25	19	19, 23, 25	23

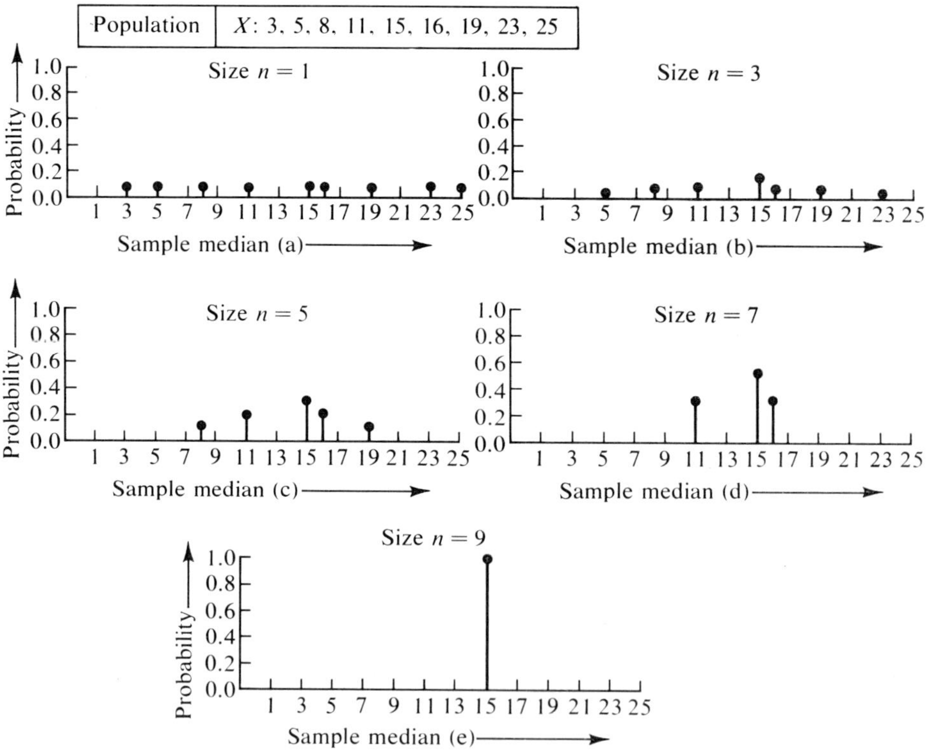

Figure 8.3
Relation between sampling distribution of sample median and sample size

Table 8.5

Possible values for sample median: Simple random sampling $(N = 9, n = 5)$

Population X: 3, 5, 8, 11, 15, 16, 19, 23, 25 $\quad {}_NC_n = {}_9C_5 = 126$

Possible sample	*Sample median*	*Possible sample*	*Sample median*	*Possible sample*	*Sample median*	*Possible sample*	*Sample median*
3, 5, 8, 11, 15	8	3, 5, 16, 19, 25	16	3, 11, 19, 23, 25	19	5, 11, 16, 19, 23	16
3, 5, 8, 11, 16	8	3, 5, 16, 23, 25	16	3, 15, 16, 19, 23	16	5, 11, 16, 19, 25	16
3, 5, 8, 11, 19	8	3, 5, 19, 23, 25	19	3, 15, 16, 19, 25	16	5, 11, 16, 23, 25	16
3, 5, 8, 11, 23	8	3, 8, 11, 15, 16	11	3, 15, 16, 23, 25	16	5, 11, 19, 23, 25	19
3, 5, 8, 11, 25	8	3, 8, 11, 15, 19	11	3, 15, 19, 23, 25	19	5, 15, 16, 19, 23	16
3, 5, 8, 15, 16	8	3, 8, 11, 15, 23	11	3, 16, 19, 23, 25	19	5, 15, 16, 19, 25	16
3, 5, 8, 15, 19	8	3, 8, 11, 15, 25	11	5, 8, 11, 15, 16	11	5, 15, 16, 23, 25	16
3, 5, 8, 15, 23	8	3, 8, 11, 16, 19	11	5, 8, 11, 15, 19	11	5, 15, 19, 23, 25	19
3, 5, 8, 15, 25	8	3, 8, 11, 16, 23	11	5, 8, 11, 15, 23	11	5, 16, 19, 23, 25	19
3, 5, 8, 16, 19	8	3, 8, 11, 16, 25	11	5, 8, 11, 15, 25	11	8, 11, 15, 16, 19	15
3, 5, 8, 16, 23	8	3, 8, 11, 19, 23	11	5, 8, 11, 16, 19	11	8, 11, 15, 16, 23	15
3, 5, 8, 16, 25	8	3, 8, 11, 19, 25	11	5, 8, 11, 16, 23	11	8, 11, 15, 16, 25	15
3, 5, 8, 19, 23	8	3, 8, 11, 23, 25	11	5, 8, 11, 16, 25	11	8, 11, 15, 19, 23	15
3, 5, 8, 19, 25	8	3, 8, 15, 16, 19	15	5, 8, 11, 19, 23	11	8, 11, 15, 19, 25	15
3, 5, 8, 23, 25	8	3, 8, 15, 16, 23	15	5, 8, 11, 19, 25	11	8, 11, 15, 23, 25	15
3, 5, 11, 15, 16	11	3, 8, 15, 16, 25	15	5, 8, 11, 23, 25	11	8, 11, 16, 19, 23	16
3, 5, 11, 15, 19	11	3, 8, 15, 19, 23	15	5, 8, 15, 16, 19	15	8, 11, 16, 19, 25	16
3, 5, 11, 15, 23	11	3, 8, 15, 19, 25	15	5, 8, 15, 16, 23	15	8, 11, 16, 23, 25	16
3, 5, 11, 15, 25	11	3, 8, 15, 23, 25	15	5, 8, 15, 16, 25	15	8, 11, 19, 23, 25	19
3, 5, 11, 16, 19	11	3, 8, 16, 19, 23	16	5, 8, 15, 19, 23	15	8, 15, 16, 19, 23	16
3, 5, 11, 16, 23	11	3, 8, 16, 19, 25	16	5, 8, 15, 19, 25	15	8, 15, 16, 19, 25	16
3, 5, 11, 16, 25	11	3, 8, 16, 23, 25	16	5, 8, 15, 23, 25	15	8, 15, 16, 23, 25	16
3, 5, 11, 19, 23	11	3, 8, 19, 23, 25	19	5, 8, 16, 19, 23	16	8, 15, 19, 23, 25	19
3, 5, 11, 19, 25	11	3, 11, 15, 16, 19	15	5, 8, 16, 19, 25	16	8, 16, 19, 23, 25	19
3, 5, 11, 23, 25	11	3, 11, 15, 16, 23	15	5, 8, 16, 23, 25	16	11, 15, 16, 19, 23	16
3, 5, 15, 16, 19	15	3, 11, 15, 16, 25	15	5, 8, 19, 23, 25	19	11, 15, 16, 19, 25	16
3, 5, 15, 16, 23	15	3, 11, 15, 19, 23	15	5, 11, 15, 16, 19	15	11, 15, 16, 23, 25	16
3, 5, 15, 16, 25	15	3, 11, 15, 19, 25	15	5, 11, 15, 16, 23	15	11, 15, 19, 23, 25	19
3, 5, 15, 19, 23	15	3, 11, 15, 23, 25	15	5, 11, 15, 16, 25	15	11, 16, 19, 23, 25	19
3, 5, 15, 19, 25	15	3, 11, 16, 19, 23	16	5, 11, 15, 19, 23	15	15, 16, 19, 23, 25	19
3, 5, 15, 23, 25	15	3, 11, 16, 19, 25	16	5, 11, 15, 19, 25	15		
3, 5, 16, 19, 23	16	3, 11, 16, 23, 25	16	5, 11, 15, 23, 25	15		

Table 8.6
Possible values for sample median: Simple random sampling ($N = 9, n = 7$)

Population X: 3, 5, 8, 11, 15, 16, 19, 23, 25 $\quad {}_NC_n = {}_9C_7 = 36$

Possible sample	*Sample median*	*Possible sample*	*Sample median*	*Possible sample*	*Sample median*
3, 5, 8, 11, 15, 16, 19	11	3, 5, 8, 15, 16, 23, 25	15	3, 8, 11, 15, 19, 23, 25	15
3, 5, 8, 11, 15, 16, 23	11	3, 5, 8, 15, 19, 23, 25	15	3, 8, 11, 16, 19, 23, 25	16
3, 5, 8, 11, 15, 16, 25	11	3, 5, 8, 16, 19, 23, 25	16	3, 8, 15, 16, 19, 23, 25	16
3, 5, 8, 11, 15, 19, 23	11	3, 5, 11, 15, 16, 19, 23	15	3, 11, 15, 16, 19, 23, 25	16
3, 5, 8, 11, 15, 19, 25	11	3, 5, 11, 15, 16, 19, 25	15	5, 8, 11, 15, 16, 19, 23	15
3, 5, 8, 11, 15, 23, 25	11	3, 5, 11, 15, 16, 23, 25	15	5, 8, 11, 15, 16, 19, 25	15
3, 5, 8, 11, 16, 19, 23	11	3, 5, 11, 15, 19, 23, 25	15	5, 8, 11, 15, 16, 23, 25	15
3, 5, 8, 11, 16, 19, 25	11	3, 5, 11, 16, 19, 23, 25	16	5, 8, 11, 15, 19, 23, 25	15
3, 5, 8, 11, 16, 23, 25	11	3, 5, 15, 16, 19, 23, 25	16	5, 8, 11, 16, 19, 23, 25	16
3, 5, 8, 11, 19, 23, 25	11	3, 8, 11, 15, 16, 19, 23	15	5, 8, 15, 16, 19, 23, 25	16
3, 5, 8, 15, 16, 19, 23	15	3, 8, 11, 15, 16, 19, 25	15	5, 11, 15, 16, 19, 23, 25	16
3, 5, 8, 15, 16, 19, 25	15	3, 8, 11, 15, 16, 23, 25	15	8, 11, 15, 16, 19, 23, 25	16

Table 8.7
Sampling distributions of sample median for increasing sample sizes

Population X: 3, 5, 8, 11, 15, 16, 19, 23, 25

$N = 9, n = 1$		$N = 9, n = 3$		$N = 9, n = 5$		$N = 9, n = 7$		$N = 9, n = 9$	
Sample space	*Probability*	*Sample space*	*Probability*	*Sample space*	*Probability*	*Sample space*	*Probability*	*Sample space*	*Probability*
3	1/9								
5	1/9	5	7/84						
8	1/9	8	12/84	8	15/126				
11	1/9	11	15/84	11	30/126	11	10/36		
15	1/9	15	16/84	15	36/126	15	16/36	15	1
16	1/9	16	15/84	16	30/126	16	10/36		
19	1/9	19	12/84	19	15/126				
23	1/9	23	7/84						
25	1/9								

Example 3

Suppose there exists a continuous variable X (infinite population) with mean 100 and variance 400. Show empirically that the sample mean $\bar{X}$ is a consistent estimator for the population mean μ_x.

Solution

Let us construct and compare the sampling distribution of the sample mean $\bar{X}$ for simple random samples of sizes 25, 100, and 400, respectively. We know that, for $n = 25$,

$$\mu_{\bar{x}} = \mu_x = 100 \quad \text{and} \quad \sigma_{\bar{x}}^2 = \sigma_x^2/n = \frac{400}{25} = 16;$$

that is, for $n = 25$,

$$\bar{X} \sim \text{N}(100, 16);$$

for $n = 100$,

$$\mu_{\bar{x}} = 100 \quad \text{and} \quad \sigma_{\bar{x}}^2 = \frac{400}{100} = 4,$$

which is to say, for $n = 100$,

$$\bar{X} \sim \text{N}(100, 4);$$

and for $n = 400$,

$$\mu_{\bar{x}} = 100 \quad \text{and} \quad \sigma_{\bar{x}}^2 = \frac{400}{400} = 1,$$

or for $n = 400$,

$$\bar{X} \sim \text{N}(100, 1).$$

The three sampling distributions are graphed in Figure 8.4, where it is clear that as sample size increases, the sampling distribution becomes more closely concentrated around the population mean $\mu_x = 100$.

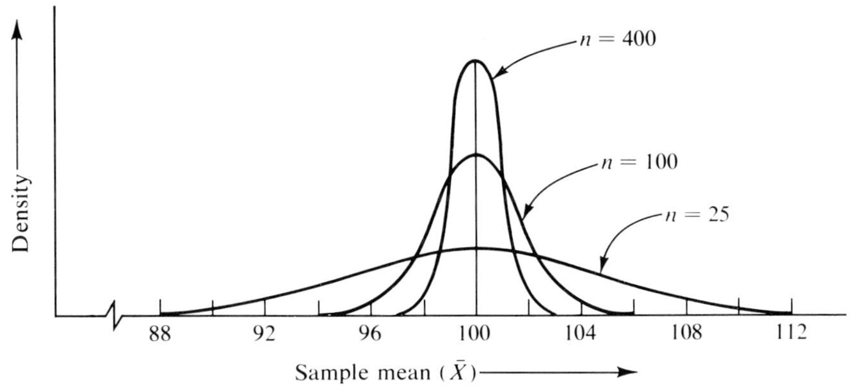

Figure 8.4

Sampling distribution of sample mean for increasing sample sizes

Note once again the importance of the relations between the mean and the variance of the sample mean and the mean and the variance, respectively, of the population being sampled. Also note that we could use an argument similar to that of Example 2 to show that the sample proportion is a consistent estimator of the population proportion. □

There are at least two types of consistency with which you should be familiar.

Definition 8.7

Let T be an estimator for the parameter θ and let $T_1, T_2, \ldots, T_n$ be a sequence of estimators representing T based respectively on samples of size 1, 2, . . . , n. We say that T is a ***squared-error consistent estimator*** *for θ if, for all possible values of θ,*

$$E[(T_n - \theta)^2]$$

approaches zero as n increases.

This simply says that if the *mean-squared error* of T approaches zero as n increases without bound in the case of an infinite population (or as n approaches N in the case of a finite population), the estimator T is squared-error consistent. Certainly, then, a squared-error consistent estimator will satisfy the rather loose definition we originally set forth for consistency. However, a much more useful criterion for a good estimator is that of *simple consistency.*

Definition 8.8

Let T be an estimator for the parameter θ, and let $T_1, T_2, \ldots, T_n$ be a sequence of estimators representing T based, respectively, on samples of size 1, 2, . . . , n. We say that T is a ***simple-consistent estimator*** *for θ if for every $\varepsilon > 0$ and for all possible values which θ can assume*

$$P(\theta - \varepsilon < T_n < \theta + \varepsilon)$$

approaches one as n increases.

Henceforth, we shall always mean simple consistency when we refer to "consistency." Practically speaking, Definition 8.8 says that an estimator is consistent if, for large samples, T_n tends to be close to θ.

Since $E[(T_n - \theta)^2] = \text{Var}(T_n) + [\theta - E(T_n)]^2$, Definition 8.7 states that squared-error consistency implies that both the variance and bias of T_n approach zero as n increases. However, it can be shown that if the variance and bias of an estimator approaches zero as n increases, the definition of simple consistency (8.8) is satisfied. Thus, a squared-error consistent estimator is always simple consistent, although the reverse does not hold.

Efficient Estimators

Even theoretical literature is sometimes contradictory in defining an *efficient estimator*. For the reader interested in a thorough treatment of the subject, we suggest Cramer (8.2). We only touch lightly on the subject.

8.4 The Maximum-Likelihood Method of Estimation

Just as it is generally impractical to use an empirical approach for finding a good estimator, it is much too laborious to start from scratch in each new problem to search for estimators that satisfy the criteria we have described. However, theoretical statisticians have set forth various general schemes of estimation that all but rarely lead directly to rather good estimators. The reader interested in digesting the theory should see Kendall (8.3). We shall limit consideration to only one such method of estimation—the *method of maximum likelihood.* This method represents a general approach to the problem of point estimation and results in estimators with such good characteristics that it deserves a more extensive treatment than we can give here. It is just too advanced for those without facility in calculus.

First let us discuss what the method automatically accomplishes. Recall that in Chapter 7 we saw how to determine the probability of obtaining a given sample result when we know certain characteristics of the population from which we are sampling. To understand the reasoning behind the method of maximum likelihood, we must view this process in reverse. In estimation, we are often in the situation of knowing the kind of distribution from which we are sampling but not knowing at least one parameter upon which the distribution depends. We thus draw a sample and use the information to estimate the parameter. A very interesting question, then, is "Of all the distributions (of a given type) from which we may be sampling, which one is most likely to have produced the given sample?" By this question, we mean "Based on the added (conditional) information that a given sample result has been obtained, what is the probability that the parameter of interest has a certain value?" However, classical statisticians do not like to view a parameter (a constant) as having associated with it a probability distribution. They prefer to ask, "Based on the added information that a random sample has given rise to a certain value of an estimator, what is the likelihood that the sample comes from the distribution for which the parameter has a certain value?"

As implied by its name, the method of maximum likelihood generally provides an estimator that maximizes the likelihood that we are sampling from a given distribution.

Properties of Maximum-Likelihood Estimators

What is so attractive about maximum-likelihood estimators? In any problem, there exist an untold number of candidate estimators for a particular unknown parameter. First, the maximum-likelihood estimator is consistent. Second, if a sufficient statistic exists for the parameter of interest, the maximum-likelihood estimator is sufficient. Further, if an efficient [in the sense discussed by Mood and Graybill (8.1)] estimator exists, the maximum-likelihood estimator is efficient.

The major drawback associated with the maximum-likelihood estimator is that it is often biased. (Thus, it is not efficient in the restricted sense of our definition of efficiency.) Hence, knowing only the probability distribution of the variable involved, we can, simply by applying the method of maximum likelihood, obtain an estimator that is highly desirable with regard to three of the criteria by which we measure goodness. Finally, even when the maximum-likelihood estimator is not attractive enough to warrant its use, it often points toward one that is sufficiently attractive.

First, let us clarify some terminology. *The sample proportion is the maximum-likelihood estimator of the population proportion.* What does this mean? It means that if we take a sample of size n from a population with an unknown proportion of successes and use the sample proportion to estimate the population proportion, the resulting value will always represent the proportion of successes in the population from which we are most likely to have obtained our particular sample. Let us use a case involving a finite population to see if this statement holds.

Example 1

Suppose we have a population consisting of four manufactured parts each of which can be classified as either good or bad. We do not know the proportion of good parts in the population but we have decided to take a simple random sample (with replacement) of size $n = 2$ and to use the sample proportion to estimate the population proportion. (We shall denote the proportion of good parts simply by "proportion.") Show that, whatever may be the true proportion (π) of the population we are actually sampling, using the sample proportion as an estimator is equivalent to choosing the population that is most likely to have given rise to our sample result. In other words, show empirically that the sample proportion is the maximum-likelihood estimator of the population proportion.

Solution

If we denote a good part by G and a bad part by B, we have the following possibilities for the population (and thus for the parameter π we are trying to estimate):

Arbitrary population identification number	1	2	3	4	5
Make-up of population	$4G, 0B$	$3G, 1B$	$2G, 2B$	$1G, 3B$	$0G, 4B$
Population proportion π	1.00	0.75	0.50	0.25	0.00

Now, what values can we possibly obtain for the sample proportion? We can obtain only one of the three following sample results and corresponding sample proportions (no order is implied):

Sample result	*Sample proportion* (p)
$2G, 0B$	1.00
$1G, 1B$	0.50
$0G, 2B$	0.00

Using p as an estimator of π, we will adopt the following policy:

1. If we obtain $p = 1.0$, our best estimate of π is 1.0; that is, we have sampled from population 1.
2. If we obtain $p = 0.5$, our best estimate of π is 0.5; we have sampled from population 3. (1)
3. If we obtain $p = 0.0$, our best estimate of π is 0.0; we have sampled from population 5.

To check the sample proportion as, in fact, the maximum-likelihood estimator of the population proportion, let us see if the possible sample proportions lead to an identical decision (estimation) policy when we empirically choose the population most likely to have given rise to each possible sample proportion. We begin by building the sampling distribution of the sample proportion p for each possible population. Since we are sampling with replacement, all draws are independent.

Population 1 ($\pi = 1.00$). All calculations are subject to the condition that we are sampling from the population containing four good and zero bad parts. Thus,

$$P(p = 0) = P(BB) = P(B)P(B) = \left(\frac{0}{4}\right)\left(\frac{0}{4}\right) = 0, \tag{2}$$

$$\begin{aligned} P(p = 0.5) &= P(GB \cup BG) = P(GB) + P(BG) \\ &= 2P(G)P(B) = 2\left(\frac{4}{4}\right)\left(\frac{0}{4}\right) = 0, \end{aligned} \tag{3}$$

and

$$P(p = 1.0) = P(GG) = P(G)P(G) = \left(\frac{4}{4}\right)\left(\frac{4}{4}\right) = 1.0. \tag{4}$$

The sampling distribution of p for samples of size $n = 2$ from population 1 is given in column 1 of Table 8.8 and in Figure 8.5a.

Population 2 ($\pi = 0.75$). We are sampling from the population in which there are three good parts and one bad part. Thus,

$$P(p = 0) = P(BB) = P(B)P(B) = \left(\frac{1}{4}\right)\left(\frac{1}{4}\right) = \frac{1}{16}, \tag{5}$$

$$\begin{aligned} P(p = 0.5) &= P(GB \cup BG) = P(GB) + P(BG) \\ &= 2P(G)P(B) = 2\left(\frac{3}{4}\right)\left(\frac{1}{4}\right) = \frac{6}{16}, \end{aligned} \tag{6}$$

and

$$P(p = 1.0) = P(GG) = P(G)P(G) = \left(\frac{3}{4}\right)\left(\frac{3}{4}\right) = \frac{9}{16}. \tag{7}$$

The sampling distribution of p for samples of size $n = 2$ from population 2 is given in column 2 of Table 8.8 and in Figure 8.5b.

You are encouraged to repeat the foregoing steps for Populations 3, 4, and 5 in order to verify that the sampling distribution of p for each are as given in Table 8.8 and in (c), (d), and (e), respectively, of Figure 8.5.

From Table 8.8, we see by scanning the second row, for example, that the probabilities of obtaining a sample proportion of $p = 0.5$ are 0, 6/16, 8/16, 6/16, and 0 for

Table 8.8

Sampling distributions of the sample proportion for various values of the population proportion

Possible values of p for n = 2	*Sampling distributions of p for n = 2*				
	Population 1 ($\pi = 1$)	*Population 2* ($\pi = 3/4$)	*Population 3* ($\pi = 1/2$)	*Population 4* ($\pi = 1/4$)	*Population 5* ($\pi = 0$)
0.00	0.0	1/16	4/16	9/16	1.0
0.50	0.0	6/16	8/16	6/16	0.0
1.00	1.0	9/16	4/16	1/16	0.0

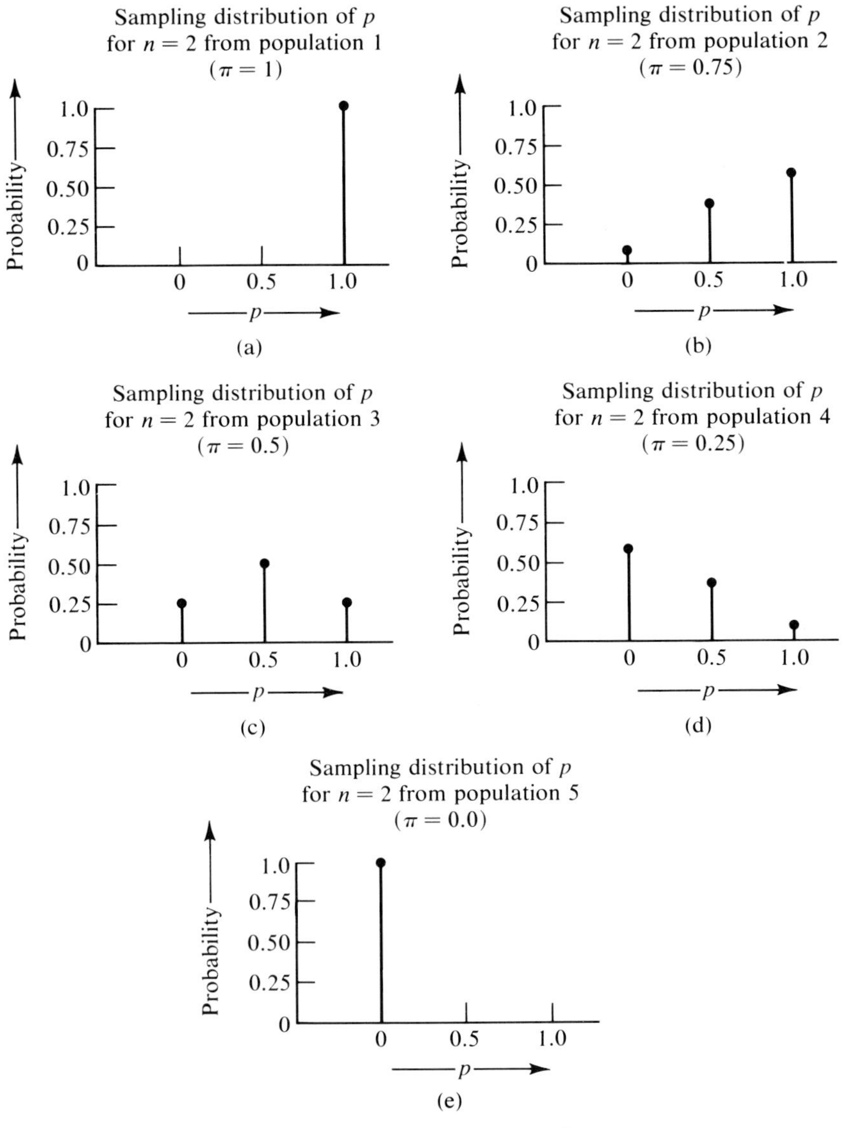

Figure 8.5

Sampling distributions of the sample proportion for various values of the population proportion (Example 1)

sampling from populations 1, 2, 3, 4, and 5, respectively. Thus, 8/16 being the greatest of the probabilities, we say that a sample proportion of $p = 0.5$ is more likely to have come from population 3 than from any other population. That is, $\pi = 0.5$ has maximum likelihood. Thus, if we obtain a value of $p = 0.5$, we estimate π to be 0.5.

Scanning the other rows we see that by always choosing the population which is most likely to have given rise to any particular value of p, we shall always estimate π to be the value that p takes on in the particular sample. Thus, our decision (estimation) policy based on an empirical maximization of the likelihood of the various possible populations is identical to the policy of (1) which resulted from our using directly the estimator p (a maximum-likelihood estimator of π). □

Note that this is only an illustrative example. It illustrates the essence of maximum-likelihood estimation because we were careful to select an example wherein the sample space of our estimator (p) contained no points that were not equal to one of our possible parameter values. Suppose we had considered samples of size $n = 3$. Then, the possible values for the estimator p would have been 0, 1/3, 2/3, and 1. Consider only the value 1/3. If we obtain a sample where $p = 1/3$, the fact that p is a maximum-likelihood estimator of π tells us that we should estimate π to be 1/3. On the other hand, if we use the empirical approach, we see that the probabilities of our obtaining a value of $p = 1/3$ from the possible populations are

Population 1 ($\pi = 1.0$)

$$P(p = 1/3) = P(GBB \cup BGB \cup BBG) = 3P(G)P(B)P(B) = 3(1)(0)(0) = 0;$$

Population 2 ($\pi = 0.75$)

$$P(p = 1/3) = 3P(G)P(B)P(B) = 3 \cdot \frac{3}{4} \cdot \frac{1}{4} \cdot \frac{1}{4} = \frac{9}{64};$$

Population 3 ($\pi = 0.50$)

$$P(p = 1/3) = 3P(G)P(B)P(B) = 3 \cdot \frac{1}{2} \cdot \frac{1}{2} \cdot \frac{1}{2} = \frac{3}{8} = \frac{24}{64};$$

Population 4 ($\pi = 0.25$)

$$P(p = 1/3) = 3P(G)P(B)P(B) = 3 \cdot \frac{1}{4} \cdot \frac{3}{4} \cdot \frac{3}{4} = \frac{27}{64};$$

Population 5 ($\pi = 0.0$)

$$P(p = 1/3) = 3P(G)P(B)P(B) = 3 \cdot 0 \cdot 1 \cdot 1 = 0.$$

Thus, by the empirical method, we would choose the population (population 4, $\pi = 0.25$) most likely to yield a sample for which $p = 1/3$, and would naturally estimate π to be 0.25. It is easy to see that the empirical approach breaks down here. However, much more important than the fact that the empirical approach breaks down is the reason for it. If we were sampling from an infinite population, π could conceivably assume any real value between zero and one, inclusive. Then, any value we might obtain for p through a sampling procedure would represent a possible value of π. Thus, in the limiting case, so to speak, the empirical reasoning would give us proper results.

8.5 Point Estimators for the Population Mean μ_x and the Population Proportion π

Let us now return to the two parameters we have discussed most extensively. What estimators should we use for μ_x and π in general situations? Our concern so far has been to find an estimator for an unknown parameter when the relevant probability distribution is known. However, the central limit theorem tells us that the sample mean is, for all practical purposes, a very good estimator for the population mean whether or not we know the form of the distribution involved. The fact is corroborated in specific instances by the method of maximum likelihood. For example, the maximum-likelihood estimator for the mean μ_x of a normal distribution is the sample mean $\bar{X}$. Analogous information on the sampling distribution of the sample proportion would indicate that the sample proportion is a very good estimator for the population proportion. However, because the two parameters μ_x and π are so important in practical statistics, it is not sufficient to determine good estimators for them. We must care about how good the estimators are and how individual estimates can be improved.

Regardless how good an estimator is, there will almost always be a discrepancy between a point estimate and the true value of the parameter being estimated. We call this discrepancy the error of the estimate.

Definition 8.11

Error *is the absolute value of the difference between the true value of a parameter being estimated and a point estimate obtained through use of a given estimator.*

Naturally, we would like the error of an estimate to be as small as possible as often as possible. And, since probability (sampling) distributions are associated with estimators, we can judge the relevant merits of a good estimator by measuring what is termed the risk associated with an estimator.

Definition 8.12

Risk *is the probability that the error of an estimate will be greater than a specified amount.*

Obviously, all else being fixed, risk will increase as the size of error decreases, and vice versa. However, we can, to some extent, control both by manipulation of the sample size.

Let us investigate the precise analytical relationships between error, risk, and sample size in the case where we wish to use the sample mean $\bar{X}$ as an estimator of the population mean μ_x. We shall take the case of an infinite population and a continuous variable, so that neither the finite population correction factor nor the continuity correction factor need be taken into account.

Estimation of Population Mean by Sample Mean—Relationships among Error, Risk, and Sample Size

Let e be a given value of error, α represent the risk associated with an error of size e, and n be the sample size. Now, suppose we are to take a simple random sample of size n and use $\bar{X}$ as an estimator for μ_x. (Note that we make no assumption about the form of the distribution from which we are sampling.) In this case, we can denote the error by

$$\text{error} = |\bar{X} - \mu_x| \tag{1}$$

and the risk associated with an error of size e as

$$\text{risk} \doteq \alpha = P(|\bar{X} - \mu_x| > e). \tag{2}$$

Note that (2) is the probability that $\bar{X}$ (the estimator) will differ in either direction from μ_x (the parameter of interest) by an amount greater than some value e. To determine relationships among error, risk, and sample size, we again call on the central limit theorem.

Suppose we are willing to tolerate an error of size e when we use $\bar{X}$ to estimate μ_x in a situation that the variance σ_x^2 of the population is known. In such case, the risk is by definition

$$\alpha = P(|\bar{X} - \mu_x| > e). \tag{3}$$

However, when we eliminate the absolute-value signs, (3) becomes

$$P(|\bar{X} - \mu_x| > e) = P(\bar{X} - \mu_x < -e \cup \bar{X} - \mu_x > e). \tag{4}$$

Since the events "$\bar{X} - \mu_x < -e$" and "$\bar{X} - \mu_x > e$" are mutually exclusive, (4) can be written as

$$P(\bar{X} - \mu_x < -e \cup \bar{X} - \mu_x > e) = P(\bar{X} - \mu_x < -e) + P(\bar{X} - \mu_x > e). \tag{5}$$

By the central limit theorem, we can transform the entity $(\bar{X} - \mu_x)$ into a standard normal variable (for n sufficiently large) simply by dividing by the standard deviation of the sample mean, i.e., by $\sigma_x/\sqrt{n}$. Making such division on both sides of each inequality and manipulating the probabilities by the method of Chapter 6, we obtain

$$\begin{aligned} P(\bar{X} - \mu_x < -e) + P(\bar{X} - \mu_x > e) &= P\left(Z < \frac{-e}{\sigma_x/\sqrt{n}}\right) + P\left(Z > \frac{e}{\sigma_x/\sqrt{n}}\right) \\ &= P\left(Z > \frac{e}{\sigma_x/\sqrt{n}}\right) + P\left(Z > \frac{e}{\sigma_x/\sqrt{n}}\right) \\ &= 2P\left(Z > \frac{e}{\sigma_x/\sqrt{n}}\right) = 2P\left(Z > \frac{e\sqrt{n}}{\sigma_x}\right). \end{aligned} \tag{6}$$

Combining (3) through (6), we have

$$\alpha = 2P\left(Z > \frac{e\sqrt{n}}{\sigma_x}\right). \tag{7}$$

(Note that we assumed σ_x^2 known.) Thus, we have immediately a relationship which will allow us to solve several types of problems:

1. We are to take a simple random sample of size n from an infinite population where σ_x^2 is known. We are to estimate μ_x by $\bar{X}$. We wish to find the risk associated with an error of size e. We have the solution directly from (7),

$$\alpha = 2P\left(Z > \frac{e\sqrt{n}}{\sigma_x}\right),$$

and need merely to find the probability in Table A-1 of the Appendix corresponding to a value of $z = e\sqrt{n}/\sigma_x$.

2. We are to take a simple random sample of size n from an infinite population where σ_x^2 is known. We are to estimate μ_x by $\bar{X}$. To find the size error associated with a risk of size α, we have from (7),

$$P\left(Z > \frac{e\sqrt{n}}{\sigma_x}\right) = \frac{\alpha}{2}. \tag{8}$$

However, this implies that

$$\frac{e\sqrt{n}}{\sigma_x} = z_{\alpha/2}, \tag{9}$$

where $z_{\alpha/2}$ is the value in Table A-1 such that

$$P(Z > z_{\alpha/2}) = \frac{\alpha}{2}. \tag{10}$$

Solving (9) gives

$$e = \frac{z_{\alpha/2}\sigma_x}{\sqrt{n}}. \tag{11}$$

3. We are to take a simple random sample of size n from an infinite population where σ_x^2 is known. We are willing to tolerate an error of size e. How large must n be to limit our risk to size α? Using (11), we merely solve for n:

$$\sqrt{n} = \frac{z_{\alpha/2}\sigma_x}{e}, \quad \text{or} \quad n = \frac{z_{\alpha/2}^2\sigma_x^2}{e^2}, \tag{12}$$

where, once again $z_{\alpha/2}$ is the value from Table A-1 such that

$$P(Z > z_{\alpha/2}) = \frac{\alpha}{2}. \tag{13}$$

4. We are to take a simple random sample of size n from an infinite population where σ_x^2 is known. We are willing to tolerate a risk of size α. How large must n be to limit our error to size e?

The solution is identical to that given by (12) and (13).

a. What risk is involved if he takes a sample of size $n = 225$ and estimates the population mean (μ_x) by the sample mean ($\bar{X}$)?

b. How large must n be to lower the risk to 0.05? Round your answer to the next larger whole number.

c. If $n = 64$, what is the size of the error associated with a risk of 0.10, accurate to two decimal places?

d. How large must n be in order to limit error to 20 hours with $\alpha = 0.10$? Round your answer to the next larger whole number.

4. A manufacturer of audio equipment knows that the variance of the power output of the stereo amplifiers he produces is 81 watts^2. He wishes to select a sample of size n to estimate the population mean by the sample mean.

   a. Determine the risk involved if $n = 144$ and the maximum acceptable error is one watt.

   b. How large must n be for the risk to be 0.01 with the sample size and error of Part a? Round your answer to the next larger whole number.

   c. If $n = 36$, what is the size of the error associated with a risk of 0.05, accurate to two decimal places?

   d. How large must n be to limit the error to 3 watts with $\alpha = 0.02$? Round your answer to the next larger whole number.

5. Suppose we are to take a simple random sample of size 400 from an infinite population known to have a variance of 25. What is the risk associated with an error of size 0.5?

6. Suppose we are to take a simple random sample of size 100 from an infinite population known to have a standard deviation of 15. What size error is associated with a risk of 0.01?

7. Develop formulas analogous to those of this section for the sample size, risk, and error, expressing each as a function of the other two for *finite* populations of size N.

8. Rework Problem 5 for a finite population of size $N = 1{,}500$.

9. Rework Problem 6 for a finite population of size $N = 800$.

10. Suppose we are to take a simple random sample from an infinite population known to have a variance of 144. How large must the sample be in order to limit to 0.05 the risk of making an error greater than 3 in estimating μ_x? Round your answer to the next larger whole number.

11. A company wishes to take a simple random sample to estimate the proportion of people in the United States watching television on a given evening. What size sample (rounded to the next larger whole number) should be taken to limit the error of the estimate to

    a. 0.01 with a risk of 5%?
    b. 0.10 with a risk of 5%?

8.6 Interval Estimation

In many situations we want to be more explicit about the uncertainty involved in estimating a parameter than we can by using point estimates. We turn then to *interval estimates.* For example, we feel better about saying that an unknown parameter lies between two numbers, say a and b, than about the statement that the parameter is equal to one given number. However, just as we know that it would be totally incorrect to say with certainty that an unknown parameter is equal to a given number, we know that it is incorrect to say with certainty that the unknown parameter lies between any two given numbers. We avoid this dilemma by constructing *confidence intervals.* Important as the concept of a confidence interval is, it is frequently misinterpreted in statistics.

We shall begin by developing a confidence interval (an interval estimate) for the mean of a population (variance known). Then, we shall interpret the meaning of confidence intervals. Next, we shall generalize the theoretical method for constructing confidence intervals and relate it to the construction of the confidence interval for the mean. Finally, we shall investigate the method for constructing a confidence interval for the difference between two population means when the variances of both populations are known. Throughout this section we shall be sampling from an infinite population.

Confidence Interval for the Population Mean (Variance Known)

Suppose we select a simple random sample of size $n = 50$ from a population having a variance (σ_x^2) of 200 and determine the sample mean to be $\bar{x} = 65$. Suppose further that our purpose is to estimate the population mean μ_x. If we were willing to live with a point estimate, we would estimate μ_x to be 65. If we wish to construct an *interval estimate* (confidence interval), how should we proceed? First, we must decide how "confident" we want to be with regard to the interval estimate. For this example, suppose we wish to find two numbers a and b such that we are "95 percent confident that the mean μ_x is between the two numbers a and b." In this case 0.95 represents a measure of our confidence, called a *confidence coefficient*. Having decided to establish a *confidence level* of 0.95, we can proceed directly to construct the confidence interval. We know that for any standard normal variable Z, we can make the probability statement

$$P(-1.96 < Z < 1.96) = 0.95. \tag{1}$$

That is, for any standard normal curve, 0.025 of the area under the curve is to the right of 1.96, and 0.025 of the area is to the left of -1.96. Hence, 0.95 of the area under the curve is between -1.96 and 1.96. If (1) holds for any standard normal variable, it certainly holds for our most famous standard normal variable to date. That is, since

$$\frac{\bar{X} - \mu_x}{\sigma_x/\sqrt{n}} \sim \text{N}(0, 1),$$

it follows from (1) that

$$P\left(-1.96 < \frac{\bar{X} - \mu_x}{\sigma_x/\sqrt{n}} < 1.96\right) = 0.95. \tag{2}$$

If we multiply all three terms of the inequality by $\sigma_x/\sqrt{n}$, Equation (2) becomes

$$P\left(-1.96\frac{\sigma_x}{\sqrt{n}} < \bar{X} - \mu_x < 1.96\frac{\sigma_x}{\sqrt{n}}\right) = 0.95. \tag{3}$$

Next, we manipulate the inequality of (3) to obtain

$$P\left(-1.96\frac{\sigma_x}{\sqrt{n}} < \bar{X} - \mu_x < 1.96\frac{\sigma_x}{\sqrt{n}}\right) = 0.95$$

$$\Rightarrow P\left(-\bar{X} - 1.96\frac{\sigma_x}{\sqrt{n}} < -\mu_x < -\bar{X} + 1.96\frac{\sigma_x}{\sqrt{n}}\right) = 0.95 \qquad (4)$$

$$\Rightarrow P\left(\bar{X} - 1.96\frac{\sigma_x}{\sqrt{n}} < \mu_x < \bar{X} + 1.96\frac{\sigma_x}{\sqrt{n}}\right) = 0.95.$$

Now, since it is a function of the general random variable $\bar{X}$, the relation of (4) represents a probability statement in the classical sense. Also, it leads us directly to a method for constructing confidence intervals. Suppose we consider a specific sample of size n which yields a specific sample mean $\bar{x}$. We then have

$$P\left(\bar{X} - 1.96\frac{\sigma_x}{\sqrt{n}} < \mu_x < \bar{X} + 1.96\frac{\sigma_x}{\sqrt{n}}\right) = 0.95$$

$$\Rightarrow P_{\mathrm{F}}\left(\bar{x} - 1.96\frac{\sigma_x}{\sqrt{n}} < \mu_x < \bar{x} + 1.96\frac{\sigma_x}{\sqrt{n}}\right) = 0.95. \qquad (5)$$

Note carefully the significance of P_{F} in Relation (5). That is, strictly speaking, the statement of (5) is not a probability statement in the same sense as (4) (i.e., in the classical sense). To understand this concept, compare statements (4) and (5) carefully. Note that (4) involves a statistic ($\bar{X}$). This being the case, it is permissible to write it as a 95 percent probability statement. On the other hand, (5) involves only the parameter μ_x and entities which represent numerical values associated with a specific sample result. That is, the statement of (5) involves only μ_x (unknown) and other known constants. Given any two numbers, the parameter μ_x, since it is a constant, is either between such numbers or not between such numbers. Thus, the probability that μ_x is between any two numbers is either one or zero; hence, it certainly cannot be 0.95, at least in the classical sense of probability. To avoid this sort of problem, we generally label this type of probability statement as a *fiducial probability statement* (one which involves an unknown parameter and the specific observed value(s) of the relevant random variable(s), not the random variable(s) per se). We call the interval

$$\left(\bar{x} - 1.96\frac{\sigma_x}{\sqrt{n}} < \mu_x < \bar{x} + 1.96\frac{\sigma_x}{\sqrt{n}}\right)$$

associated with the fiducial probability statement of (5) a 95 percent *confidence* interval.

Now, returning to the example, in which

$$\sigma_x^2 = 200, \quad n = 50, \quad \text{and} \quad \bar{x} = 65,$$

our 95 percent fiducial probability statement for the population parameter of interest (5) becomes

$$P_{\mathrm{F}}\left(65 - \frac{1.96\sqrt{200}}{\sqrt{50}} < \mu_x < 65 + \frac{1.96\sqrt{200}}{\sqrt{50}}\right) = 0.95$$

$$\Rightarrow P_{\mathrm{F}}(61.08 < \mu_x < 68.92) = 0.95. \qquad (6)$$

Thus, our 95 percent confidence interval is

$$61.08 < \mu_x < 68.92.$$

To interpret (6), let us first consider the incorrect interpretation that is often made.

Incorrect Interpretation of the Fiducial Probability Statement of (6): The fact that we have taken a sample of size $n = 50$ from a population having variance $\sigma_x^2 = 200$, obtained a sample mean of $\bar{x} = 65$, and used the formula of (5) to construct a 95 percent confidence interval for μ_x tells us that the probability is 0.95 that the population mean μ_x has a value between 61.08 and 68.92.

Why is this interpretation incorrect? Even though we do not know the value, there exists a single numerical value which represents the true value of the population mean. That value either is between 61.08 and 68.92, or it is not. Thus, it is completely incorrect to say that the probability is 0.95 that μ_x is between 61.08 and 68.92.

It is important to note that, unlike the classical probability statements we have encountered, the variable portion of the probability statement of (6) is essentially the interval itself, not the entity encompassed by the interval. That is, if we took a second sample of size $n = 50$, we would expect to obtain a value of $\bar{X}$ different from 65 (the sample mean of the first sample of size $n = 50$). The second sample would naturally give us a confidence interval with different end points. For example, suppose a second sample resulted in a sample mean of $\bar{x} = 66$. Then, our 95 percent fiducial probability statement for μ_x would be

$$P_{\text{F}}\left(66 - \frac{1.96\sqrt{200}}{\sqrt{50}} < \mu_x < 66 + \frac{1.96\sqrt{200}}{\sqrt{50}}\right) = 0.95$$

$$\Rightarrow P_{\text{F}}(62.08 < \mu_x < 69.92) = 0.95.$$

Obviously, other samples of size $n = 50$ could well result in different confidence intervals. Thus, on repeating the sampling process many times, we would discover a distribution of intervals which result from the method of construction of confidence intervals; and μ_x, as a constant, would be encompassed by some of these intervals and not by others.

Correct Interpretation of the Fiducial Probability Statement of (6): Having obtained a sample mean of $\bar{x} = 65$ for a sample of size $n = 50$ from a population having variance $\sigma_x^2 = 200$, and used the formula of (5) to construct a 95 percent confidence interval for μ_x, we know that for many repetitions of the process, 95 percent of the intervals so constructed would encompass the true value of μ_x.

Obviously, we are not always concerned with 95 percent confidence intervals. Thus, the formula of (5) needs to be generalized. To make our notation compatible with that used in our discussion of error and risk, let us henceforth refer to a $100(1 - \alpha)$ percent confidence interval (recall that α denotes risk), where α can take on any value between 0 and 1. (In our first example, $\alpha = 0.05$.) How do we construct such a confidence interval? Our reasoning is the same as that used in developing Relation (5). For any standard normal variable Z, we know that there exists some value $z_{\alpha/2}$ such that

$$P(-z_{\alpha/2} < Z < z_{\alpha/2}) = 1 - \alpha. \tag{7}$$

Naturally, $z_{\alpha/2}$ is the value in Table A-1 such that

$$P(Z > z_{\alpha/2}) = \alpha/2. \tag{8}$$

From (7) and the central limit theorem (if n is sufficiently large), we have

$$P\left(-z_{\alpha/2} < \frac{\bar{X} - \mu_x}{\sigma_x/\sqrt{n}} < z_{\alpha/2}\right) = 1 - \alpha. \tag{9}$$

Manipulating the inequalities of (9), we obtain the probability statement

$$P\left(\bar{X} - \frac{z_{\alpha/2}\sigma_x}{\sqrt{n}} < \mu_x < \bar{X} + \frac{z_{\alpha/2}\sigma_x}{\sqrt{n}}\right) = 1 - \alpha.$$

Finally, replacing the random variable $\bar{X}$ by a specific observed value $\bar{x}$, we obtain the general fiducial probability statement

$$P_F\left(\bar{x} - \frac{z_{\alpha/2}\sigma_x}{\sqrt{n}} < \mu_x < \bar{x} + \frac{z_{\alpha/2}\sigma_x}{\sqrt{n}}\right) = 1 - \alpha. \tag{10}$$

Thus, the $100(1 - \alpha)$ percent confidence interval for the population mean when the variance of the population is known is

$$\bar{x} \pm \frac{z_{\alpha/2}\sigma_x}{\sqrt{n}},$$

where $z_{\alpha/2}$ is as given in (8).

To construct a $100(1 - \alpha)$ percent confidence interval for the population mean (variance known), we perform the following steps:

1. Select a simple random sample of size n. If the population is normal, the sample can be of any size; otherwise, it should be greater than or equal to 30, roughly.
2. Compute the sample mean $\bar{x}$.
3. Determine the value $z_{\alpha/2}$ from Table A-1 such that $P(Z > z_{\alpha/2}) = \alpha/2$.
4. Construct the interval

$$\bar{x} \pm \frac{z_{\alpha/2}\sigma_x}{\sqrt{n}}. \tag{11}$$

Example 1

A simple random sample of $n = 50$ families in a large city (we may assume that the finite population correction factor is 1) is selected to estimate the average amount expended per month by a family for entertainment. The sample mean is found to be $\bar{x} = \$45$. The variance from past data is known to be 128. Assuming that income per month is normally distributed, find a 90 percent confidence interval for the average of all the families in the city.

Solution

Since our confidence coefficient is 0.90,

$$90 \text{ percent} = 100(1 - \alpha) \text{ percent}.$$

This implies that $\alpha/2 = 0.05$. Thus $z_{0.05} = 1.645$, since

$$P(Z > 1.645) = 0.05.$$

For $n = 50$, $\sigma_x^2 = 128$, and $\bar{x} = \$45$, we have, by (10),

$$P_F\left(\bar{x} - \frac{z_{0.05}\sigma_x}{\sqrt{n}} < \mu_x < \bar{x} + \frac{z_{0.05}\sigma_x}{\sqrt{n}}\right) = 0.90$$

$$\Rightarrow P_F\left(45 - \frac{1.645\sqrt{128}}{\sqrt{50}} < \mu_x < 45 + \frac{1.645\sqrt{128}}{\sqrt{50}}\right) = 0.90$$

$$\Rightarrow P_F(42.368 < \mu_x < 47.632) = 0.90.$$

□

Sample Size Required for a Confidence Interval of a Given Length

Sometimes we would like to specify both the confidence coefficient and the length of the confidence interval to be constructed. How do we achieve this? Careful investigation of (11) indicates that we must vary the sample size n. That is, we note that we obtain upper and lower extreme points of a $100(1 - \alpha)$ percent confidence interval by respectively adding $z_{\alpha/2}\sigma_x/\sqrt{n}$ to the observed sample mean $\bar{x}$ and subtracting $z_{\alpha/2}\sigma_x/\sqrt{n}$ from the observed sample mean $\bar{x}$. Thus, the length (say l) of a $100(1 - \alpha)$ percent confidence interval is necessarily

$$l = \frac{2z_{\alpha/2}\sigma_x}{\sqrt{n}}. \tag{12}$$

Now, once we fix the value of our confidence coefficient $(1 - \alpha)$, we may immediately obtain $z_{\alpha/2}$ from Table A-1. Further, we are assuming that σ_x^2 is known. It thus follows that the only entity left unspecified when we fix the confidence interval length l is the sample size n. Hence, in order to determine the sample size required for a specified confidence coefficient and confidence interval length, we merely solve (12) for n. That is, a sample of size

$$n = \frac{4z_{\alpha/2}^2\sigma_x^2}{l^2} \tag{13}$$

will provide a $100(1 - \alpha)$ percent confidence interval of length l.

Example 2

For the data of Example 1, determine the sample size required to permit construction of a 90 percent confidence interval of length 4 units ($).

Solution

For our confidence coefficient of 0.90, we have

$$90 \text{ percent} = 100(1 - \alpha) \text{ percent},$$

which implies $\alpha/2 = 0.05$; since $P(Z > 1.645) \cong 0.05$

$$z_{0.05} = 1.645.$$

Given $\sigma_x^2 = 128$ and $l = 4$, we use (13) to obtain

$$n = \frac{4z_{\alpha/2}^2 \sigma_x^2}{l^2} = \frac{4(1.645)^2 128}{16} = 86.59 \cong 87. \qquad \square$$

The Method of Pivotal Quantities—General Method for Constructing Confidence Intervals

Let us examine the approach, manifested in Relations 1 through 5, just used to develop the confidence interval for the population mean. A familiarity with the steps that follow will enhance your ability to comprehend tests of hypotheses in Chapter 9.

Problem

We are required to construct a $100(1 - \alpha)$ percent confidence interval for an unknown parameter θ. (Whether or not we know the mathematical form of the probability distribution of the variable of interest is unimportant.)

Steps in Solution

1. Find a statistic satisfying all of the following requirements:
   (a) The form of the sampling distribution of the statistic is known and independent of the value of the unknown parameter θ.
   (b) The statistic is a function of the unknown parameter θ.
   (c) The statistic is not a function of any other unknown—that is, its formula involves only constants, known parameters, and functions determinable through sample values taken from the population of interest.
2. Set up a probability statement involving the statistic of step 1.
3. Manipulate the probability statement of step 2 until the associated confidence interval involves only θ as its middle entity.
4. Select a random sample of size n from the population of interest and compute the observed value(s) of the random variable(s) (statistic) involved in the extreme entities of the derived probability statement of step 3, thus obtaining constants a and b such that

$$P_F(a < \theta < b) = 1 - \alpha.$$

A statistic satisfying the requirements of step 1 is sometimes called a *pivotal quantity for a given parameter* [see Huntsberger (8.4)]. A pivotal quantity for a given parameter is defined as a function of both the parameter and an estimator such that the sampling distribution of the function is known and is independent of the parameter of interest and any other unknown parameters. We often refer to the statistic satisfying the above three requirements simply as a pivotal quantity. [It is not always possible to find a satisfactory pivotal quantity. For alternatives, see Mood and Graybill (8.1).]

The four steps can be related to the development of the $100(1 - \alpha)$ percent confidence interval for the population mean μ_x (variance known) as follows. First, we find a statistic satisfying the requirements of step 1. If n is sufficiently large, we know (from the central limit theorem) that

$$\frac{\bar{X} - \mu_x}{\sigma_x/\sqrt{n}} \sim \mathrm{N}(0, 1). \tag{14}$$

Thus, Step 1a is satisfied. For b and c of step 1, note that the statistic is a function of the unknown parameter μ_x and the only other entities involved in the formula of the statistic are $\bar{X}$, σ_x, and n. An observed value of $\bar{X}$ can be calculated from sample values, σ_x represents a known parameter, and being the sample size, n is constant for any given problem.

Next, using (14), we set up the probability statement

$$P\left(-z_{\alpha/2} < \frac{\bar{X} - \mu_x}{\sigma_x/\sqrt{n}} < z_{\alpha/2}\right) = 1 - \alpha, \tag{15}$$

where $z_{\alpha/2}$ is the value such that

$$P(Z > z_{\alpha/2}) = \alpha/2.$$

From this, we obtain the fiducial probability statement

$$P_{\mathrm{F}}\left(\bar{x} - \frac{z_{\alpha/2}\sigma_x}{\sqrt{n}} < \mu_x < \bar{x} + \frac{z_{\alpha/2}\sigma_x}{\sqrt{n}}\right) = 1 - \alpha, \tag{16}$$

where $\bar{x}$ is a specific observed value of $\bar{X}$.

Whether or not we explicitly refer to them, the four steps will be integral to the development of every set of confidence intervals to be constructed in all subsequent chapters.

It is easy to see, once again, the very important role played by our original assumption of a known variance. For, without it, the statistic $(\bar{X} - \mu_x)/(\sigma_x/\sqrt{n})$ would not have satisfied requirements of step 1c and, hence, could not have been used for the desired confidence interval. □

Confidence Interval for the Difference between the Means of Two Populations (Variances of Both Populations Known)

Let us now estimate the difference between the means of two populations (variables X_1 and X_2) where the variances are known. The parameter of interest becomes $(\mu_1 - \mu_2)$, which we shall estimate by collecting sample information. The major question is exactly what sample information to collect. If we denote by $\bar{X}_1$ and $\bar{X}_2$ the sample means of two samples of size n_1 and n_2, respectively, we have, from the central limit theorem,

$$\bar{X}_1 \sim \mathrm{N}\left(\mu_1, \frac{\sigma_1^2}{n_1}\right) \quad \text{and} \quad \bar{X}_2 \sim \mathrm{N}\left(\mu_2, \frac{\sigma_2^2}{n_2}\right). \tag{17}$$

From the properties of linear combinations of normal variables (Section 6.10), it follows that the statistic $(\bar{X}_1 - \bar{X}_2)$ has the normal distribution

$$(\bar{X}_1 - \bar{X}_2) \sim \mathrm{N}\left(\mu_1 - \mu_2, \frac{\sigma_1^2}{n_1} + \frac{\sigma_2^2}{n_2}\right). \tag{18}$$

Once again, by transforming the normal statistic $(\bar{X}_1 - \bar{X}_2)$ (by subtracting its mean and dividing by its standard deviation), we obtain a standard normal variable:

$$\frac{(\bar{X}_1 - \bar{X}_2) - (\mu_1 - \mu_2)}{\sqrt{(\sigma_1^2/n_1) + (\sigma_2^2/n_2)}} \sim \mathrm{N}(0, 1). \tag{19}$$

Finally, the statistic of Relation (19) satisfies all the requirements to qualify it for use in constructing a confidence interval for the difference of two population means. To obtain the formula for a $100(1 - \alpha)$ percent confidence interval, we begin with

$$P\left(-z_{\alpha/2} < \frac{(\bar{X}_1 - \bar{X}_2) - (\mu_1 - \mu_2)}{\sqrt{(\sigma_1^2/n_1) + (\sigma_2^2/n_2)}} < z_{\alpha/2}\right) = 1 - \alpha, \tag{20}$$

where, again, $z_{\alpha/2}$ is the value such that

$$P(Z > z_{\alpha/2}) = \alpha/2. \tag{21}$$

By manipulating the inequality of (20), we obtain the fiducial probability statement

$$P_F\left[(\bar{x}_1 - \bar{x}_2) - z_{\alpha/2}\sqrt{\frac{\sigma_1^2}{n_1} + \frac{\sigma_2^2}{n_2}} < (\mu_1 - \mu_2) < (\bar{x}_1 - \bar{x}_2) + z_{\alpha/2}\sqrt{\frac{\sigma_1^2}{n_1} + \frac{\sigma_2^2}{n_2}}\right] = 1 - \alpha, \tag{22}$$

where $z_{\alpha/2}$ is as given by (21) and $\bar{x}_1$ and $\bar{x}_2$ are specific observations of the statistics $\bar{X}_1$ and $\bar{X}_2$.

Relation (22) tells us that we simply perform the following steps in order to set a $100(1 - \alpha)$ percent confidence interval.

1. Select simple random samples of size n_1 from Population 1 and of size n_2 from Population 2 (n_1 and n_2 both greater than 30, roughly).
2. Compute the sample means $\bar{x}_1$ and $\bar{x}_2$.
3. From Table A-1, determine the value $z_{\alpha/2}$ such that $P(Z > z_{\alpha/2}) = \alpha/2$.
4. Construct the interval

$$(\bar{x}_1 - \bar{x}_2) \pm z_{\alpha/2}\sqrt{\frac{\sigma_1^2}{n_1} + \frac{\sigma_2^2}{n_2}}. \tag{23}$$

Example 3

Mr. Fannin, wanting to purchase a data processing school, has narrowed his choice to two schools, A and B. They are priced the same. He decides to buy the school whose graduates have performed best on data processing tests administered by an independent source. He wants to construct a 95 percent confidence interval for the difference between the mean scores of students of School A and students of School B.

Records indicate that the variances of scores for students from the two schools are $\sigma_A^2 = \sigma_B^2 = 32$. From the independent source, Mr. Fannin randomly selects the records of 100 students from each school. Calculations result in $\bar{x}_A = 83$ and $\bar{x}_B = 75$. Construct a 95 percent confidence interval for $(\mu_A - \mu_B)$.

Solution

Since the desired confidence coefficient is 0.95,

$$95 \text{ percent} = 100(1 - \alpha) \text{ percent}.$$

This implies that $\alpha/2 = 0.025$, and, since $P(Z > 1.96) = 0.025$,

$$z_{0.025} = 1.96.$$

Substituting

$$n_A = n_B = 100, \sigma_A^2 = \sigma_B^2 = 32, \bar{x}_A = 83, \text{ and } \bar{x}_B = 75,$$

into (22) gives

$$P_F\left[(83 - 75) - 1.96\sqrt{\frac{32}{100} + \frac{32}{100}} < (\mu_A - \mu_B) < (83 - 75) + 1.96\sqrt{\frac{32}{100} + \frac{32}{100}}\right] = 95$$

$$\Rightarrow P_F[6.432 < (\mu_A - \mu_B) < 9.568] = 0.95.$$

The appropriate confidence interval becomes

$$6.432 < (\mu_A - \mu_B) < 9.568.$$

□

Classical Approach in Obtaining Confidence Intervals for the Population Proportion π

If we are sampling from an infinite population having only two possible outcomes, we can construct a $100(1 - \alpha)$ percent confidence interval for the population proportion π, where the sample size is sufficiently large, by using

$$P\left(-z_{\alpha/2} < \frac{p - \pi}{\sqrt{\frac{\pi(1 - \pi)}{n}}} < z_{\alpha/2}\right) = 1 - \alpha. \tag{24}$$

This is permissible because $(p - \pi)/\sqrt{\pi(1 - \pi)/n}$, which has a normal distribution with mean 0 and variance 1, meets the requirements of a statistic to be used in constructing a confidence interval. We can manipulate the probability statement of (24) until a somewhat complicated probability statement having π as the center term results. Therefore, we merely note here that such is a possibility. A more practical approach exists (Reference 8.4).

Problems

1. Explain the difference between *probability* and *confidence*.

2. Explain the difference between *strict probability* and *fiducial probability*.

3. Suppose a simple random sample of $n = 49$ people in the United States is selected to estimate the average amount an individual spends for recreation per year. The sample mean is \$500 and past experience indicates that the population standard deviation is \$90. Find the 95% confidence interval for the average annual recreation expenditure per person in the United States.

4. In Problem 3, find the 98% confidence interval if the sample size is $n = 625$.

5. In Problem 3, what sample size would be required to limit the length of the confidence interval to \$20?

6. In Problem 4, what sample size would be required to limit the length of the confidence interval to \$40?

7. What three characteristics must a statistic possess before it can be used in constructing a confidence interval for an unknown parameter? In other words, what constitutes a *pivotal quantity*?

8. A manufacturer wants to subcontract the manufacture of amplifier subassemblies to one of two subcontractors, *A* and *B*, on the basis of amplifier gain in the assemblies the competitors produce. He has decided to construct a 90% confidence interval for the difference between the mean gains of the amplifier subassemblies. Past experience indicates that the standard deviations of the gains of the units produced by the two subcontractors are $\sigma_A = 100$ and $\sigma_B = 80$. Simple random samples of 64 units from each of the two subcontractors result in sample means of $\bar{x}_A = 160$ and $\bar{x}_B = 175$. Construct a 90% confidence interval for $(\mu_A - \mu_B)$.

9. In Problem 8, construct an 80% confidence interval for $\mu_A - \mu_B$ with sample sizes of 225.

10. Suppose a canner wishes to compare the average weights in ounces of "one-pound" cans of fruit produced by his two machines, *A* and *B*. He knows from experience that the variances for the two machines are $\sigma_A^2 = 0.09 \text{ oz}^2$ and $\sigma_B^2 = 0.16 \text{ oz}^2$, respectively. The following random samples were selected:

Machine A (oz)				Machine B (oz)		
15.83	15.61	16.22	16.11	16.31	15.93	15.82
16.14	16.56	16.81	15.97	15.59	16.14	15.76
15.98	15.73	16.23	15.96	16.08	14.93	15.87
15.95	16.48	15.89	16.49	15.72	15.89	16.04
16.09	15.17	16.00	16.44	16.26	16.01	15.63
				16.13	15.92	15.55

a. Construct a 98% confidence interval for $\mu_A - \mu_B$.

b. Construct a 95% confidence interval for $\mu_A - \mu_B$.

11. Derive the confidence interval for the general case for μ_x when sampling from a finite population of size *N*.

12. Rework Problem 3 for a finite population of size 1,200 from a small town.

8.7 Summary

Generally, without investigating the entire population, we have no way of knowing the true value of a parameter of the probability distribution from which a given set of sample observations is taken. Thus, we face the often arduous task of determining what particular function (statistic) of the sample values to use to estimate the unknown parameter of interest. The sample statistics used for this purpose are called estimators. That is, the particular function of the sample values chosen to estimate a given unknown parameter is called an *estimator*. The specific numerical value obtained by evaluating an estimator in a given instance is called an *estimate*.

Since an estimator is a function of sample values, it has a sampling distribution which represents the theoretical distribution for the various possible values of the estimator corresponding to all possible samples of the same size that can be drawn from the population of interest. In order to select the best estimator for a given parameter, we must consider the characteristics of the sampling distributions of all candidates. In particular, we must view (1) the location of the distributions as they relate to the location (actual value) of the parameter of interest and (2) the dispersion of the distributions about the true value of the parameter of interest. Thus, in general, we say that the better of two estimators is the one whose sampling distribution is more closely concentrated around the true value of the parameter of interest.

Generally, an indication of whether the sampling distribution of an estimator is concentrated closely around a given parameter can be obtained by determining whether such estimator satisfies certain requirements, or criteria. Although they are not the only criteria by which estimators are judged, four of the most important are *unbiasedness*, *consistency*, *efficiency*, and *sufficiency*; and, of the various methods of estimation, the method of maximum likelihood produces attractive estimators as measured by these criteria.

Regardless of the estimator chosen, for a given parameter, there will nearly always be a discrepancy between an individual estimate (obtained through use of sample information) and the true value of the parameter. We call the absolute value of this discrepancy the *error* of an estimate and label the probability that the error will be larger than a specified size as *risk*. Despite our desire that both error and risk be small, for a fixed sample size n, the two vary inversely. Thus, for a given estimation scheme, we can control the size of both error and risk only by changing the sample size.

In particular, when the sample mean $\overline{X}$ is used as an estimator for the population mean μ_x (when variance is known), the relationship among error, risk, and sample size is given by

$$P\left(Z > \frac{e\sqrt{n}}{\sigma_x}\right) = \alpha/2, \tag{1}$$

where $Z \sim N(0, 1)$, e is a given error size, α is a given risk size, n is the sample size, and σ_x^2 is the known variance of the population. Using this relationship, we can,

sample mean of the charge-account balances, we may be able to convince ourselves of the validity (or nonvalidity) of the auditor's hypothesis. However, an inherent risk in this sampling procedure is that any conclusion we draw about the hypothesis may be wrong. This characteristic results directly from the properties associated with random variables.

The hypothesis to be tested, generally known as the *null hypothesis*, is denoted by H_0. Any time we test a hypothesis, we shall have an *alternative hypothesis*, which represents some form of negation of the null hypothesis. We shall denote the alternative hypothesis by H_1. Suppose that, in the example of the 50,000 charge accounts, we have two choices for the population mean, either \$55, as the auditor claims, or \$60; and we wish to check the auditor's claim. In this case, the null hypothesis is the statement that the population mean is equal to \$55, while the alternative hypothesis is the statement that the population mean is \$60. Such hypotheses are generally written as

$$H_0: \mu_x = \$55 \quad \text{and} \quad H_1: \mu_x = \$60, \tag{1}$$

where x represents the variable "balance of charge account."

The general steps for testing hypotheses are:

1. State the null hypothesis and the alternative hypothesis.
2. Determine the statistic to be used for testing the null hypothesis and its probability distribution.
3. Select a decision rule by which to judge an observed value of the statistic of step 2 and make a decision regarding the null hypothesis.
4. Perform the experiment required to obtain an observed value of the statistic of step 2.
5. Determine the action to be taken concerning the hypotheses of step 1 as a result of the decision rule of step 3 and the experiment of step 4.

Our primary purpose in the following pages will be to examine these steps in detail. However, we must first introduce relevant terminology, beginning with the types of errors that may occur in the testing of a given hypothesis.

Problems

1. Define *statistical hypothesis.*
2. What is the difference between the *null hypothesis* and the *alternative hypothesis*?
3. What are the general steps required for testing hypotheses?

9.2 Type I and Type II Errors

Whenever a statistical hypothesis is accepted or rejected on the basis of a sample, there is always the possibility of error. There are at least two hypotheses from which to choose and, by taking only sample information, we force ourselves to make a decision without access to full information on the population of interest.

The hypotheses from which we may choose are often referred to as *states of nature*. The decisions we make as a result of examining appropriate sample data are often called *actions*. Thus, we describe the possible states of nature as "H_0 True" and "H_0 False," and we describe the possible actions as "Accept H_0" and "Reject H_0." We relate the terminology to the null hypothesis H_0 rather than the alternative hypothesis H_1 because the alternative hypothesis often involves many possibilities of which at most one can be true, while the null hypothesis admits only one single possibility which is necessarily either true or false. Table 9.1 presents all combinations (as they relate to the null hypothesis) of the states of nature that may exist and the actions we may take in any specific test of a given null hypothesis.

Table 9.1

States of nature and actions in test of null hypothesis, H_0

	State of nature	
Action	*H_0 True*	*H_1 True*
Accept H_0	correct decision	Type II error
Reject H_0	Type I error	correct decision

Now, in order to test the null hypothesis of (1), page 317, versus the corresponding alternative hypothesis, suppose we select a random sample from the 50,000 charge accounts and compute the sample mean to be $\bar{x} = \$57$. Do we accept or reject the null hypothesis? Before we can make a decision, we must have a *decision rule* setting forth the conditions under which we accept or reject H_0.

As indicated in Table 9.1, a *Type I error* is defined to be the error of rejecting the null hypothesis when it is, in fact, true (the alternative hypothesis is false). A *Type II error* is defined to be the error of accepting the null hypothesis when it is actually false (the alternative hypothesis is true). It is obvious that the other two combinations of states of nature and actions will provide a correct decision. That is, if we either (1) accept H_0 when it is true or (2) reject H_0 when it is false, we will have made a correct decision. Obviously, the decision rule we select for any given test will be one of the primary factors which controls how often we make Type I or Type II errors.

Example 1

Suppose we plan to select a simple random sample of one ball from an urn containing six balls. Further, suppose we know that this urn contains either three red balls and three white balls or two red balls and four white balls. That is, we know that we are to sample from a population wherein the parameter "proportion of red balls" takes on a value of either $\pi = 3/6$ or $\pi = 2/6$. Suppose we are to decide the population by randomly selecting one ball. Let our decision rule be such that (1) if we select a red ball, we conclude that the urn contains three red balls and three white balls and $\pi = 3/6$; or (2) if we select a white ball, we conclude that the urn contains two red balls and four white balls and $\pi = 2/6$. Establish the null hypothesis that we are sampling from a population three red balls and three white balls; note the alternative hypothesis against which the null hypothesis is to be tested; describe the combinations of states of nature and actions which will comprise Type I and Type II errors, respectively; and determine the probabilities of making errors of Type I and Type II in testing the null hypothesis.

Solution

Let us denote the population of three red balls and three white balls as S_1 for the first state of nature and the population of two red balls and four white balls as S_2 for the second state of nature. Let us also denote the acts of obtaining a white ball by W and a red ball by R. We are to test the hypothesis that we are sampling from S_1. Hence, we shall state the null hypothesis in terms of parameters of S_1, namely, the proportion π of red balls in the population. Thus, we have H_0: $\pi = 1/2$ for our null hypothesis. Since the only alternative conclusion we can reach is that we are sampling from S_2, we shall state our alternative hypothesis in terms of the same parameter, so that $H_1 : \pi = 1/3$. Thus, we have, for our hypotheses

$$H_0 : \pi = 1/2 \quad \text{versus} \quad H_1 : \pi = 1/3.$$

Our decision rule requires us to accept H_0 if we select a red ball and thus reject H_0 if we select a white ball. The kinds of decisions which can result from all combinations of states of nature and actions are given below.

Action (based on noting attribute of sample of size n = 1)	*State of nature* H_0 *True* $S_1(\pi = 1/2)$	H_0 *False* $S_2(\pi = 1/3)$
Accept H_0 (obtain red ball)	correct decision	Type II error
Reject H_0 (obtain white ball)	Type I error	correct decision

Therefore, to determine the probabilities of making errors of Type I and Type II, we must determine $P(W \mid S_1)$ and $P(R \mid S_2)$, respectively. For S_1 we have

$$P(\text{Type I error}) = P(\text{rejecting } H_0 \mid H_0 \text{ true}) = P(W \mid S_1)$$
$$= 3/6 = 1/2.$$

Similarly, we have for S_2

$$P(\text{Type II error}) = P(\text{accepting } H_0 \mid H_0 \text{ false}) = P(R \mid S_2)$$
$$= 2/6 = 1/3.$$

We can also calculate the probabilities of making correct decisions for the different populations. We obtain

$$P(\text{correct decision} \mid S_1) = P(R \mid S_1) = 3/6 = 1/2,$$

and

$$P(\text{correct decision} \mid S_2) = P(W \mid S_2) = 4/6 = 2/3.$$

The actions, states of nature, and probabilities of taking such actions for each state of nature are given below.

	State of nature	
Action	H_0 *True* $S_1(\pi = 1/2)$	H_0 *False* $S_2(\pi = 1/3)$
Accept H_0	1/2	1/3
Reject H_0	1/2	2/3

□

Note from Example 1 that we are not sampling from two urns, but from only one urn, the exact contents of which are unknown. We know only that the urn contains either three red balls and three white balls or two red balls and four white balls.

The probability of making an error is called the *size* of the error. The probabilities of making Type I and Type II errors are generally denoted by α and β, respectively. That is, the size of Type I error is

$$\alpha = P(\text{Type I error}) = P(\text{rejecting } H_0 | H_0 \text{ true}),$$

and the size of Type II error is (1)

$$\beta = P(\text{Type II error}) = P(\text{accepting } H_0 | H_0 \text{ false})$$
$$= P(\text{accepting } H_0 | H_1 \text{ true}).$$

Occasionally, the Type I and Type II errors are referred to as the α *error* and β *error*, respectively. However, you are warned not to confuse α and β with α error and β error; α and β are probabilities, while α error and β error are two different kinds of mistakes.

Example 2

Suppose that we plan to select a simple random sample of size 1 from a normal population which has a known standard deviation of 20. Further, suppose that we know that the population mean is either 110 or 150 and we desire to test the null hypothesis that $\mu_x = 110$ against the alternative hypothesis that $\mu_x = 150$. That is, we wish to test

$$H_0: \mu_x = 110 \quad \text{versus} \quad H_1: \mu_x = 150.$$

For our decision rule let us conclude that (1) the population from which we are sampling has a mean of 110 if the sample value is less than 140 and (2) the population mean is 150 if the sample value is greater than or equal to 140. Determine the probabilities of making errors of Type I and Type II.

Solution

Our decision rule and the two populations under consideration are given in Figure 9.1. We note that we accept H_0 if x, the observed value of X, is less than 140 and reject

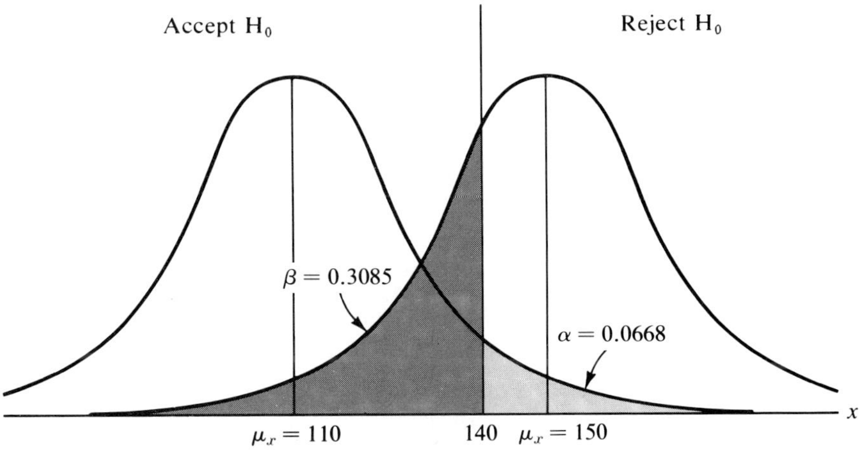

Figure 9.1

Decision rule, α, and β

H_0 if it is greater than or equal to 140. We determine the probability of a Type I error as

$$\alpha = P(\text{rejecting } H_0 | H_0 \text{ true}) = P(X \geq 140 | \mu_x = 110)$$
$$= P\left(Z \geq \frac{140 - 110}{20}\right) = P(Z \geq 1.50) = P(Z > 1.50) = 0.0668.$$

The probability of committing a Type II error is given by

$$\beta = P(\text{accepting } H_0 | H_1 \text{ true}) = P(X < 140 | \mu_x = 150)$$
$$= P\left(Z < \frac{140 - 150}{20}\right) = P(Z < -0.50) = P(Z > 0.50) = 0.3085. \quad \square$$

Problems

1. Explain the difference between *states of nature* and *actions.*
2. Explain the difference between a *Type I* error and *Type II* error.
3. Explain the difference between a Type I error and α.
4. Explain the difference between a Type II error and the *size* of a Type II error.
5. Supose that in a bin containing 10 parts we know that there are *either* 7 ball bearings and 3 small gears *or* 5 bearings and 5 gears. Further suppose that we are to decide the population from which we are sampling by randomly selecting one part. Let our decision rule be such that (1) if we select a ball bearing, we conclude that the bin contains 7 ball bearings and 3 gears—i.e., $\pi = 7/10$; and (2) if we select a gear, we conclude that the bin contains 5 ball bearings and 5 gears—i.e., $\pi = 5/10$.

a. Establish the null hypothesis that we are sampling from the population containing 7 ball bearings and 3 gears, and note the alternative hypothesis against which the null hypothesis is to be tested.

b. Describe the states of nature and actions which will comprise Type I and Type II errors, respectively.

c. Determine the probabilities of making errors of Type I and Type II if we use the above experiment and associated decision rule to test the null hypothesis.

6. Suppose there are 15 unlabeled cans in a shipment of canned goods, and we know that there are *either* 10 cans of tomatoes and 5 of beans among those unlabeled *or* 6 cans contain tomatoes and 9 contain beans. We are to decide from which population we are sampling by randomly selecting 2 cans *without replacement*. Let our decision rule be such that (1) if we select 2 cans of tomatoes we conclude that the population has 10 cans of tomatoes and 5 of beans, i.e., $\pi = 10/15$; and (2) if the sample does not have 2 cans of tomatoes, we conclude that there are 6 of tomatoes and 9 of beans—i.e., $\pi = 6/15$.

a. Establish the null hypothesis that we are sampling from the population that contains 10 cans of tomatoes and 5 cans of beans, and note the alternative hypothesis against which the null hypothesis is being tested.

b. Describe the states of nature and actions which will comprise Type I and Type II errors, respectively.

c. Determine the probabilities of making errors of Type I and Type II if we use the above experiment and associated decision rule to test the null hypothesis.

7. Suppose we plan to select a simple random sample of size 1 from a normal population known to have a standard deviation of 4. We know that the population mean is either 42 or 50, and we wish to test

$$H_0: \mu_x = 42 \quad \text{versus} \quad H_1: \mu_x = 50.$$

For our decision rule let us conclude that (1) $\mu_x = 42$ if the sample value is less than 43 and (2) $\mu_x = 50$ if the sample value is greater than or equal to 43.

a. For the given value (43) associated with the decision rule, determine the probabilities of making errors of Type I and Type II.

b. Determine the value associated with the decision rule that makes the probability of a Type I error equal 0.05 and, for this value, determine the probability of a Type II error.

c. Determine the value associated with the decision rule that makes the probability of a Type II error equal 0.10 and, for this value, determine the probability of a Type I error.

8. Suppose we plan to select a simple random sample of size 1 from a normal population known to have a standard deviation of 10. We know that the population mean is either 90 or 110, and we wish to test

$$H_0: \mu_x = 110 \quad \text{versus} \quad H_1: \mu_x = 90.$$

For our decision rule let us conclude that (1) $\mu_x = 110$ if the sample value is greater than 100 and (2) $\mu_x = 90$ if the sample value is less than or equal to 100.

a. For the given value (100) associated with the decision rule, determine the probabilities of making errors of Type I and Type II.

b. Determine the value associated with the decision rule that makes the probability of a Type I error equal 0.10 and, for this value, determine the probability of a Type II error.

c. Determine the value associated with the decision rule that makes the probability of a Type II error equal 0.05 and, for this value, determine the probability of a Type I error.

9.3 Testing Hypotheses Concerning μ_x When σ_x^2 is Known

Recall that a statistical hypothesis is a statement about the probability distribution for a specific population. In this section, we shall consider tests of hypotheses concerning the mean of a population with a known variance. Let us consider first the case of testing a hypothesis about the mean of a population which has as its probability distribution the normal distribution with a known variance.

Suppose that we are sampling from a normal population (one with a normal probability distribution) with a standard deviation equal to 10. Let us test the hypothesis that the population mean μ_x is equal to 35, with the only alternative value for μ_x being 25. Thus, following the 5 steps for testing a hypothesis, page 317, we have

$$H_0: \mu_x = 35 \quad \text{versus} \quad H_1: \mu_x = 25.$$

The variable of interest X is such that

$$X \sim N(\mu_x, 100).$$

From Section 7.5 we know that, if $X \sim N(\mu_x, \sigma_x^2)$, then the random variable $\bar{X}$ is such that $\bar{X} \sim N(\mu_x, \sigma_x^2/n)$ for a sample of any size n. Further, we know from Chapter 8 that $\bar{X}$ is an unbiased estimator of μ_x. Hence, we shall use $\bar{X}$ (step 2) as the statistic for testing our hypothesis. Let us use (arbitrarily, for the present) as our decision rule that of accepting H_0 if the observed value of $\bar{X}$ is greater than 29 and rejecting H_0 if it is less than or equal to 29. Thus, we have accomplished step 3. Suppose, as the experiment of step 4, we select a simple random sample of size $n = 16$ from the population under investigation and compute $\bar{x} = 30.1$. To determine what action we should take (step 5), we simply compare the observed value of $\bar{X}$ against the decision rule and note that $\bar{x} = 30.1$ is greater than 29. Thus, by our decision rule, we accept $H_0: \mu_x = 35$.

Let us next determine the probabilities of making errors of Type I and Type II, that is, find α and β. We first note that these probabilities could be determined without completing steps 4 and 5. That is, we note that the probabilities of Type I and Type II errors are not dependent upon the observed value of $\bar{X}$ (the statistic chosen for testing H_0) which is obtained on a given trial of an experiment. Thus, in addition to what is known about the original population, we determine α anb β by using only the information given by the null and alternative hypotheses, that related to the decision rule and general experiment, and facts about the statistic chosen for use in testing the null hypothesis. The possible combinations of states of nature, actions, and results of actions are given below.

Action (having taken sample of size $n = 16$ and calculating a value of the statistic $\bar{X}$)	*State of nature*	
	H_0 True (i.e., $\mu_x = 35$)	*H_0 False (i.e., $\mu_x = 25$)*
Accept H_0 (i.e., obtain $\bar{x} > 29$)	correct decision	Type II error
Reject H_0 (i.e., obtain $\bar{x} \leq 29$)	Type I error	correct decision

From $\sigma_x = 10$, $n = 16$, and $\overline{X} \sim N(\mu_x, \sigma_x^2/n)$, we obtain

$$\begin{aligned}\alpha &= P(\text{Type I error}) = P(\text{rejecting } H_0 | H_0 \text{ true})\\ &= P(\overline{X} \le 29 | \mu_x = 35).\end{aligned}$$

However, since we are sampling from a continuous distribution, we may drop the equality sign without affecting the probability. Thus,

$$\begin{aligned}\alpha &= P(\overline{X} \le 29 | \mu_x = 35) = P(\overline{X} < 29 | \mu_x = 35)\\ &= P\left(\frac{\overline{X} - \mu_x}{\sigma_x/\sqrt{n}} < \frac{29 - \mu_x}{\sigma_x/\sqrt{n}} \middle| \mu_x = 35\right)\\ &= P\left(Z < \frac{29 - 35}{10/4}\right) = P(Z < -2.4) = P(Z > 2.4)\\ &= 0.0082\end{aligned}$$

and

$$\begin{aligned}\beta &= P(\text{Type II error}) = P(\text{accepting } H_0 | H_0 \text{ false})\\ &\qquad\qquad\qquad\quad = P(\text{accepting } H_0 | H_1 \text{ true})\\ &= P(\overline{X} > 29 | \mu_x = 25) = P\left(Z > \frac{29 - 25}{10/4}\right)\\ &= P(Z > 1.6) = 0.0548.\end{aligned}$$

Figure 9.2 gives graphs of the above probabilities. If H_0 is true ($\mu_x = 35$), then $\overline{X} \sim N(35, 100/16)$, and the sampling distribution of $\overline{X}$ is given by the normal curve

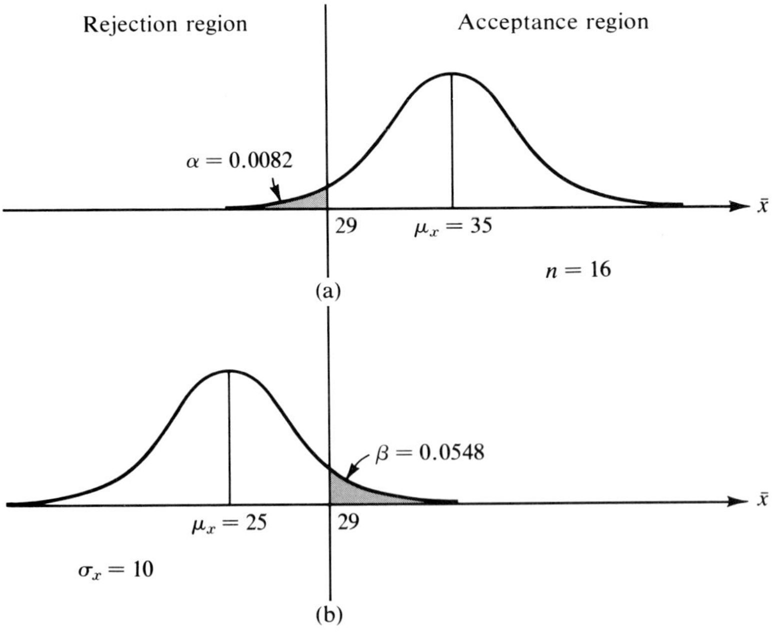

Figure 9.2

Decision rule, α, and β for sampling distributions of $\overline{X}$

centered at $\mu_x = 35$ (Figure 9.2a). H_0's being true is tantamount to our being in the situation of sampling from the normal distribution of the curve in Figure 9.2a. Since we decided to reject H_0 if we obtain a value of $\bar{X}$ less than or equal to 29, the probability that we reject H_0 when it is true is equal to the area under the curve to the left of 29. This is precisely what we calculated, that is, $\alpha = 0.0082$.

On the other hand, if H_1 is true ($\mu_x = 25$), then $\bar{X} \sim N(25, 100/16)$, and the sampling distribution of $\bar{X}$ is given by the normal curve centered at $\mu_x = 25$ in Figure 9.2b. That is, H_1's being true is tantamount to our sampling from the normal distribution of the curve in Figure 9.2b. Thus, according to our decision rule, the probability that we will accept H_0 when it is false (H_1 is true) is equal to the area under the curve to the right of 29. This is precisely what we calculated, that is, $\beta = 0.0548$.

One-Tail Tests

The foregoing example is somewhat unrealistic in the sense that there are only two possible values for μ_x. Most practical problems have many (sometimes an infinite number of) possible candidates for the true value of μ_x. However, in any specific situation, obviously only one value is the true value of μ_x.

Our decision rule used in the example was arbitrary. Generally it is determined by specifying the desired probability of a Type I error and finding the unique possible value of the test statistic which assures that the size of the Type I error is as specified. This manner of establishing the decision rule precludes our having more than one value of μ_x for our null hypothesis. The probability of a Type I error is often referred to as *risk*, or *level of significance*.

When we choose a decision rule where all of the rejection region is in one tail of the appropriate distribution, as shown in Figure 9.2a, we call this a *one-tail* test. To illustrate, consider again the situation of the foregoing example. Suppose we wish to test the hypothesis that $\mu_x = 35$ against the alternative hypothesis that $\mu_x < 35$. In such case, we have

$$H_0: \mu_x = 35 \quad \text{versus} \quad H_1: \mu_x < 35.$$

Suppose further that we specify the size of our Type I error as 0.05 ($\alpha = 0.05$). For a sample of size $n = 16$, we wish to find the unique possible value of $\bar{X}$, say $\bar{x}^*$, such that $\alpha = 0.05$.

If the observed value of $\bar{X}$ from our experiment (sample of size 16) is greater than $\bar{x}^*$, we will accept the null hypothesis as true. On the other hand, if our observed value of $\bar{X}$ is less than or equal to $\bar{x}^*$, we will reject the null hypothesis. In specifying that $\alpha = 0.05$, we are willing to employ a decision rule ($\bar{x}^*$) so that, even when sampling from a population having mean $\mu_x = 35$ (H_0 is true), we will obtain 5 percent of the time a sample mean which is less than the value $\bar{x}^*$. The value $\bar{x}^*$ is known as a *critical* value since we use it to decide whether to accept or reject the null hypothesis. To determine $\bar{x}^*$ for $\alpha = 0.05$, we proceed as follows. We know that

$$\alpha = P(\text{Type I error}) = P(\text{rejecting } H_0 \mid H_0 \text{ true}). \tag{1}$$

However, we have already agreed that obtaining a value of $\bar{X} \leq \bar{x}^*$ will cause us to reject H_0. Thus, in terms of our statistic and decision rule, the event $\bar{X} \leq \bar{x}^*$ is equivalent to the event rejecting H_0. Also, as we have defined our null hypothesis, the state of nature $\mu_x = 35$ is equivalent to the state of nature H_0 true. Thus, (1) becomes

$$P(\text{rejecting } H_0 \mid H_0 \text{ true}) = P(\bar{X} \leq \bar{x}^* \mid \mu_x = 35). \tag{2}$$

Transforming the general statistic $\bar{X}$ (and also the possible value $\bar{x}^*$) to a standard normal variable, we have

$$P(\bar{X} \leq \bar{x}^* \mid \mu_x = 35) = P\left(Z < \frac{\bar{x}^* - \mu_x}{\sigma_x/\sqrt{n}} \,\middle|\, \mu_x = 35\right). \tag{3}$$

Recalling that $\sigma_x = 10$ and $n = 16$, we have

$$P\left(Z < \frac{\bar{x}^* - \mu_x}{\sigma_x/\sqrt{n}} \,\middle|\, \mu_x = 35\right) = P\left(Z < \frac{\bar{x}^* - 35}{10/4}\right) = P\left(Z < \frac{\bar{x}^* - 35}{2.5}\right). \tag{4}$$

Combining (1) through (4) with the fact that we require $\alpha = 0.05$, we have

$$\alpha = P\left(Z < \frac{\bar{x}^* - 35}{2.5}\right) = 0.05. \tag{5}$$

From Table A-1, we see that

$$P(Z < -1.645) = 0.05. \tag{6}$$

Thus, (5) and (6) tell us that

$$\frac{\bar{x}^* - 35}{2.5} = -1.645, \tag{7}$$

since there is only one value of a standard normal variable with an area of 0.05 under the curve to its left. Solving (7), we obtain

$$\begin{aligned} \frac{\bar{x}^* - 35}{2.5} &= -1.645, \\ \Rightarrow \bar{x}^* &= 35 - (2.5)(1.645) \\ &= 35 - 4.1125 \\ &= 30.8875. \end{aligned} \tag{8}$$

Therefore, our decision rule (based on the specification that $\alpha = 0.05$) is that

1. if the observed value of $\bar{X}$ is greater than 30.8875, we shall accept H_0; and
2. if it is less than or equal to 30.8875, we shall reject H_0.

Note that the "equal to" portion of our decision rule may be included with either the greater-than portion or the less-than portion, but not both. The only remaining step in testing our hypothesis is to select a simple random sample of size $n = 16$, compute $\bar{x}$ from the sample, and compare the result against the decision rule in order to reach a conclusion. Another major difference between this example and the original one stems from the fact that the alternative hypothesis of the original example included only one possible value, whereas the alternative hypothesis of the present example includes all values less than 35. The difference is manifested in the determination of the value of β. We note that

$$\beta = P(\text{Type II error}) = P(\text{accepting } H_0 \mid H_1 \text{ true}).$$

However, in the present example, the alternative hypothesis is $\mu_x < 35$; hence,

$$P(\text{accepting } H_0 | H_1 \text{ true}) = P(\overline{X} > 30.8875 | \mu_x < 35).$$

Yet, we cannot determine an exact value for β without a specific value for μ_x. Thus, to determine a value for β, we see that β is actually a function of μ_x. That is,

$$\beta(\mu_x) = P(X > 30.8875 | \mu_x).$$

For example, if $\mu_x = 25$, we have

$$\beta(25) = P(\overline{X} > 30.8875 | \mu_x = 25) = P\left(Z > \frac{30.8875 - 25}{10/4}\right)$$
$$= P(Z > 2.36) = 0.0091.$$

On the other hand, if $\mu_x = 30$, we have

$$\beta(30) = P(\overline{X} > 30.8875 | \mu_x = 30) = P\left(Z > \frac{30.8875 - 30}{10/4}\right)$$
$$= P(Z > 0.36) = 0.3594.$$

Thus, under our decision rule, there is a much greater chance of making a Type II error if we are actually sampling from a distribution having mean $\mu_x = 30$ than if we are actually sampling from a distribution having $\mu_x = 25$.

In this chapter, we have considered only the situation in which we are sampling from a normal population. Recall from Chapter 7, that the central limit theorem can always be used to determine the probability distribution of the statistic $\overline{X}$ if our sample size is sufficiently large (generally if $n > 30$). *Thus, if our sample size is sufficiently large, we may use the normal distribution as the probability distribution for the statistic $\overline{X}$ regardless of the probability distribution of the population from which we are sampling.*

Let us consider several examples for a one-tail test when our sample is taken from a population which is not necessarily normal.

Example 1

A wire-manufacturing firm produces a special-purpose wire claimed to have an average breaking strength of 80 pounds and a standard deviation of 4 pounds. A customer purchases a carload of the wire and selects a sample of 64 spools of wire for testing purposes. The customer wants to know if the manufacturing firm is selling him wire of inferior quality, that is, wire with a breaking strength significantly less than 80 pounds. Assuming that the customer specifies a Type I error of 0.05 and that he has obtained a sample mean of $\bar{x} = 79$ pounds, test the appropriate hypothesis and determine the size of the Type II error for the situation where $\mu_x = 78$. (The term "significantly" has a specific meaning in relation to testing hypotheses. Generally, if we specify $\alpha = 0.05$, a sample result which causes us to reject the null hypothesis is said to exhibit a significant difference when compared to the value given by the null hypothesis for the unknown parameter of interest.)

Solution

Since the manufacturing firm claims that it is producing wire with an average breaking strength of 80 pounds, we shall use as our null hypothesis H_0: $\mu_x = 80$ pounds. Further, since the customer wishes to determine if he is getting wire of inferior quality, we shall make the alternative hypothesis H_1: $\mu_x < 80$ pounds. Now, if we accept H_0, we shall in essence conclude that the manufacturing firm is producing the quality product it claims. On the other hand, if we reject H_0 (when compared against H_1), we shall conclude that the product is inferior. Thus, the null and alternative hypotheses are

$$H_0: \mu_x = 80 \text{ pounds} \quad \text{versus} \quad H_1: \mu_x < 80 \text{ pounds.}$$

We shall see shortly that in establishing these hypotheses, the customer indicates that he wants to avoid the error of concluding that he is getting a poor product when he really is not. Ordinarily we construct hypotheses so as to avoid a Type I error.

Since we are testing a hypothesis about μ_x, we shall use $\bar{X}$ as our statistic. The sample size of $n = 64$ is sufficiently large to use the central limit theorem; so, we have

$$\bar{X} \sim \mathrm{N}(\mu_x, \sigma_x^2/n),$$

or, since $\sigma_x = 4$ and $n = 64$,

$$\bar{X} \sim \mathrm{N}(\mu_x, 0.25).$$

We can proceed by developing a decision rule—that is, finding the unique critical value of $\bar{x}^*$—so that (1) we will accept H_0 whenever an observed value of the statistic $\bar{X}$ is greater than $\bar{x}^*$ and (2) our probability of making a Type I error through use of such decision rule is 0.05. Thus, the critical value $\bar{x}^*$ must be chosen so that

$$P(\text{rejecting } H_0 | H_0 \text{ true}) = 0.05.$$

That is to say,

$$P(\bar{X} < \bar{x}^* | \mu_x = 80) = 0.05. \tag{9}$$

To determine $\bar{x}^*$, we proceed as before.

$$P(\bar{X} < \bar{x}^* | \mu_x = 80) = P\left(Z < \frac{\bar{x}^* - 80}{4/8}\right) = P\left(Z < \frac{\bar{x}^* - 80}{0.5}\right). \tag{10}$$

Combining (9) and (10), we have

$$P\left(Z < \frac{\bar{x}^* - 80}{0.5}\right) = 0.05.$$

However, from Table A-1, we obtain

$$P(Z < -1.645) = 0.05.$$

Therefore,

$$\frac{\bar{x}^* - 80}{0.5} = -1.645.$$

Solving for $\bar{x}^*$ gives

$$\bar{x}^* = 80 - (1.645)(0.5) = 79.1775.$$

Applying our decision rule to our specific sample result ($\bar{x} = 79$), we have $79 < 79.1775$. This means that we will reject H_0 and conclude that the customer is indeed receiving an inferior product, at least as compared with the manufacturer's claims.

Figure 9.3a shows a graph of our decision rule with $\alpha = 0.05$, along with the sampling distribution of $\bar{X}$ in the situation where μ_x is actually equal to 80 (the null hypothesis is true). In this example, the region to the right of our critical value, 79.1775, is called the *acceptance region*, while the region to the left of it is called the *rejection region*. Because we specified $\alpha = 0.05$, the area under the curve (which represents the sampling distribution of $\bar{X}$ when $\mu_x = 80$) to the left of the critical value is 0.05. Since the observed value of $\bar{X}$ ($\bar{x} = 79$) is in the rejection region, we reject H_0.

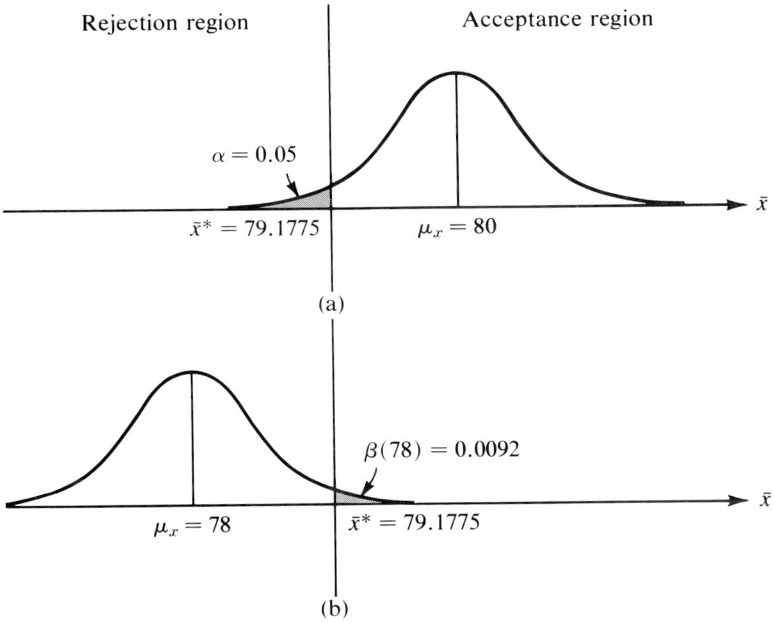

Figure 9.3

Decision rule, α, and β for one-tail test

In the second part of the problem, we are concerned with the specific alternative hypothesis H_1: $\mu_x = 78$. To determine the size of the Type II error for $\mu_x = 78$, we need to evaluate

$$\begin{aligned}\beta(78) &= P(\text{accepting } H_0 | H_1 \text{ true}) = P(\bar{X} > 79.1775 | \mu_x = 78) \\ &= P\left(Z > \frac{79.1775 - 78}{4/8}\right) = P(Z > 2.355) \\ &= 0.0092.\end{aligned}$$

We see that if, in fact, $\mu_x = 78$, there is a very small chance of committing a Type II error. Of course, $\alpha = 0.05$ is the probability of committing a Type I error. We fixed this in order to establish our decision rule. Figure 9.3b displays the sampling distribution of $\bar{X}$ when $\mu_x = 78$. Note that β is the area under the curve to the right of the critical value. □

Example 2

The manager of one branch of a large bank claims that his branch is making larger loans than the average-size loan for all branches combined. The average loan for the combined branches is \$5,000 and the standard deviation is \$2,000. An auditor randomly selects 100 loans from the manager's branch and obtains a sample mean of $\bar{x} = \$5{,}420$. Using $\alpha = 0.05$ and assuming there is no difference between the standard deviation for the branch and that for the combined branches, determine if the branch is making significantly larger loans than the \$5,000 average for all branches. Compute β for $\mu_x = \$5{,}500$.

Solution

The null and alternative hypotheses are

$$H_0\colon \mu_x = \$5{,}000 \quad \text{versus} \quad H_1\colon \mu_x > \$5{,}000.$$

We shall again use $\bar{X}$ as the test statistic. Here, by the central limit theorem, the statistic $\bar{X}$ has a normal distribution with a standard deviation of

$$\sigma_{\bar{x}} = \frac{\sigma_x}{\sqrt{n}} = 2{,}000/10 = 200.$$

To determine our decision rule for $\alpha = 0.05$, we proceed almost as before, the only difference being that we have a right-tail alternative hypothesis. Thus, we obtain $\bar{x}^*$ as follows. First we note that

$$\begin{aligned}\alpha &= P(\text{rejecting } H_0 \mid H_0 \text{ true}) = P(\bar{X} > \bar{x}^* \mid \mu_x = 5{,}000)\\ &= P\left(Z > \frac{\bar{x}^* - 5{,}000}{200}\right).\end{aligned} \tag{11}$$

However, since we have specified that $\alpha = 0.05$, (11) implies that

$$P\left(Z > \frac{\bar{x}^* - 5{,}000}{200}\right) = 0.05. \tag{12}$$

Again, however, we have from Table A-1 that $P(Z > 1.645) = 0.05$. Thus,

$$\begin{aligned}\frac{\bar{x}^* - 5{,}000}{200} &= 1.645,\\ \Rightarrow \bar{x}^* &= 5{,}000 + (200)(1.645)\\ &= 5{,}329.\end{aligned}$$

Our decision rule is demonstrated graphically in Figure 9.4a. It says that we should accept H_0 if our observed value of $\bar{X}$ is less than 5,329 and reject H_0 if our observed

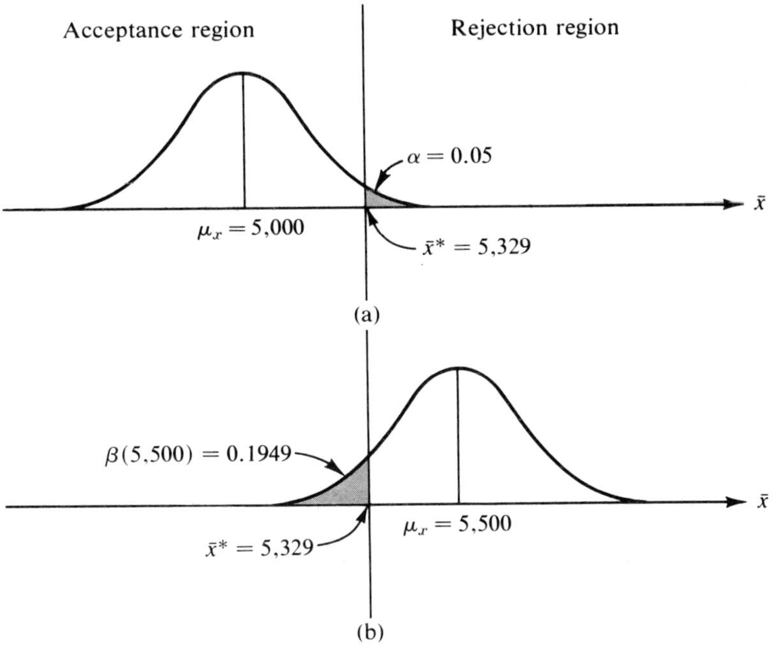

Figure 9.4

Decision rule, α, and β for one-tail test

value of $\bar{X}$ is greater than or equal to 5,329. Since the auditor has obtained a sample mean of $\bar{x}$ = \$5,420, we reject H_0. Thus, we conclude that the manager's claim that his branch is making larger loans is valid.

To compute β for the specific alternative hypothesis μ_x = \$5,500, we must evaluate the probability statement

$$\begin{aligned}\beta(5{,}500) &= P(\text{type II error})\\ &= P(\text{accepting } H_0 \mid H_1 \text{ true}) = P(\bar{X} < 5{,}329 \mid \mu_x = 5{,}500)\\ &= P\left(\bar{X} < \frac{5{,}329 - \mu_x}{\sigma_x/\sqrt{n}} \mid \mu_x = 5{,}500\right) = P\left(Z < \frac{5{,}329 - 5{,}500}{200}\right)\\ &= P(Z < -0.86) = P(Z > 0.86) = 0.1949.\end{aligned}$$

The probability of a Type II error is graphed in Figure 9.4b. □

You may wonder why, in the previous examples, we have calculated β (the probability of a Type II error) for rather arbitrarily selected values of the relevant alternative hypotheses. Our main purposes in so doing have been merely (1) to emphasize that the size of the Type II error can be calculated only for a specific value of the alternative hypothesis (i.e., that β is a function of the unknown parameter about which the test is being made) and (2) to demonstrate repeatedly the method for calculating β. These should not be construed as ultimate objectives in themselves but, rather, as a means to an end. Indeed, there is a much better reason for being able to calculate β

in any given instance. [Note that in many of the examples and problems of this and subsequent chapters, we shall include calculation of the probability (β) of Type II error for a single possible value which the parameter of interest may assume under the alternative hypothesis, to give you practice in these calculations. If a particular point is to be made regarding testing methodology, we will state it explicitly. Otherwise, you should avoid attaching particular significance to the specific parameter value used in the calculation of β.]

Although we have not discussed it explicitly, we have implied in our approach to tests of hypotheses that Type I error is the more critical of the two kinds of errors. That is, we have so far always fixed a value of α, determined a decision rule, and then checked to see what value β takes on for a fixed value of the parameter of interest. It turns out that this is the general approach in classical hypothesis testing. We are concerned with choosing the best possible test for a given hypothesis. What does this mean?

Let us consider it in terms of testing a null hypothesis about the population mean μ_x. Although we have used $\overline{X}$ as our test statistic, many statistics could serve our purpose. The major question is whether another statistic would give as good results in terms of α and β. For any given candidate statistic, then, we merely specify α to have the same value as was specified for the case of the statistic $\overline{X}$. Then, having made the two tests equivalent with regard to the probability of Type I error, we check to see which is better with regard to the probability of Type II error. However, it is by no means sufficient to compare β for $\overline{X}$ with β for the other statistic for only one possible value of μ_x. Rather, we must compare this type of information over the entire range of possible values μ_x can assume.

Even though it is unknown to us, there really is only one true value for μ_x. Thus, we construct a curve for β as a function of μ_x to get an idea of the probability of a Type II error for the different values μ_x can assume. This curve gives us a measure of the goodness of the test (decision rule based on $\overline{X}$) we are using. When we construct a similar curve for any other decision rule (test statistic), we immediately have a means of comparing the two decision rules.

We call this measure of goodness of a test (based on β) the *operating characteristic* of the test. Thus, the curve obtained by plotting $\beta(\mu_x)$ versus μ_x for a given decision rule (test) is known as the *operating characteristic curve* or simply, the *OC curve* of that decision rule (test), where $\beta(\mu_x)$ represents the value β assumes when the true value of the population mean is μ_x [see the discussion after Equation (8)]. Such curve for the test statistic ($\overline{X}$) of the example given earlier in this section for

$$H_0: \mu_x = 35 \quad \text{versus} \quad H_1: \mu_x < 35$$

is given in Figure 9.5. The real benefit of being able to calculate β is it provides means for constructing OC curves for determining which of the possible tests for a given hypothesis is the better test (i.e., which of several possible test statistics will provide the better decision rule, given that all decision rules have been constructed so as to yield a given probability of Type I error).

More generally, suppose we want to test a hypothesis about any unknown parameter θ and, hence, wish to select the better of several test statistics. As stated earlier, we generally specify the size of the Type I error. However, as seen above, the size of the Type II error depends upon the true value of θ, the parameter being tested (μ_x in our example). Since we cannot control both errors simultaneously for a given sample size, we generally choose, from all tests possessing the same size Type I error, one

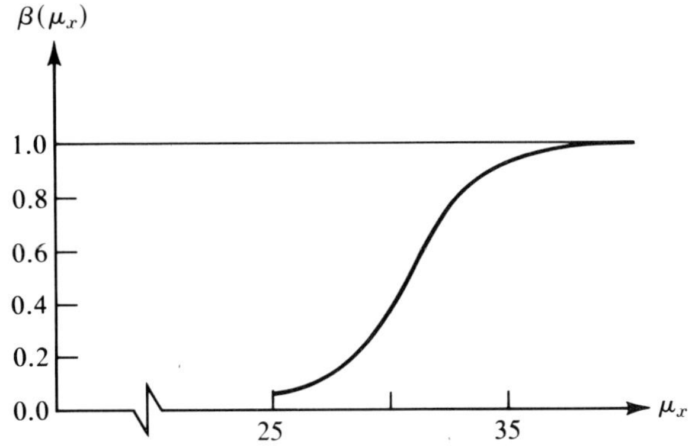

Figure 9.5

Operating characteristic curve

for which the Type II error is as small as possible. We try to do this by plotting an OC curve for each candidate statistic, that is, by plotting the function $\beta(\theta)$ versus various possible values of the unknown parameter θ.

Another function often used to determine the best test among several candidates is known as the *power function* $P(\theta)$. This function is directly related to $\beta(\theta)$ by the relationship

$$P(\theta) = 1 - \beta(\theta),$$

where θ is the parameter being tested. Referring to Table 9.1, we note that $P(\theta)$ represents the probability of making a correct decision when θ is a value such that H_1 is true. That is,

$$P(\text{correct decision} \mid H_1) = 1 - \beta(\theta).$$

In our example, the unknown parameter θ is the population mean μ_x. Thus,

$$P(\mu_x) = 1 - \beta(\mu_x).$$

The power function for $H_0: \mu_x = 35$ versus $H_1: \mu_x < 35$ is plotted in Figure 9.6.

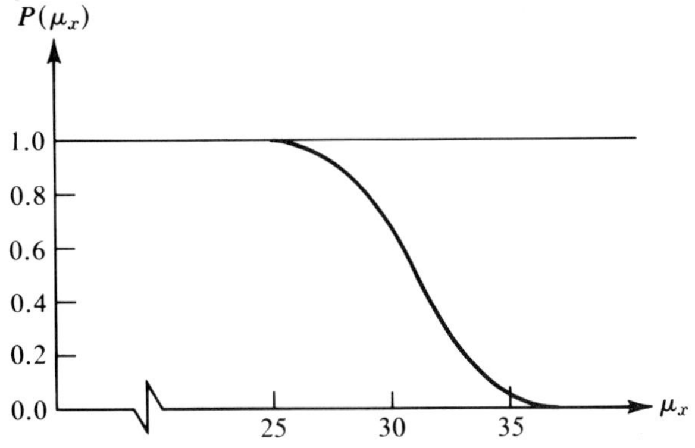

Figure 9.6

Power function

Example 3

Suppose, in a situation identical to that of Example 2, that we wish to consider a different decision rule. Let us consider accepting H_0 if our observed value of $\overline{X}$ is less than \$5,200 or greater than \$5,247 and rejecting H_0 if it is between \$5,200 and \$5,247. This decision rule is represented in Figure 9.7a. Note that it also has associated with it a value of $\alpha = 0.05$. Compute β for $\mu_x = \$5{,}500$ and compare it to the corresponding value obtained for $\beta(5{,}500)$ for the decision rule developed in Example 2.

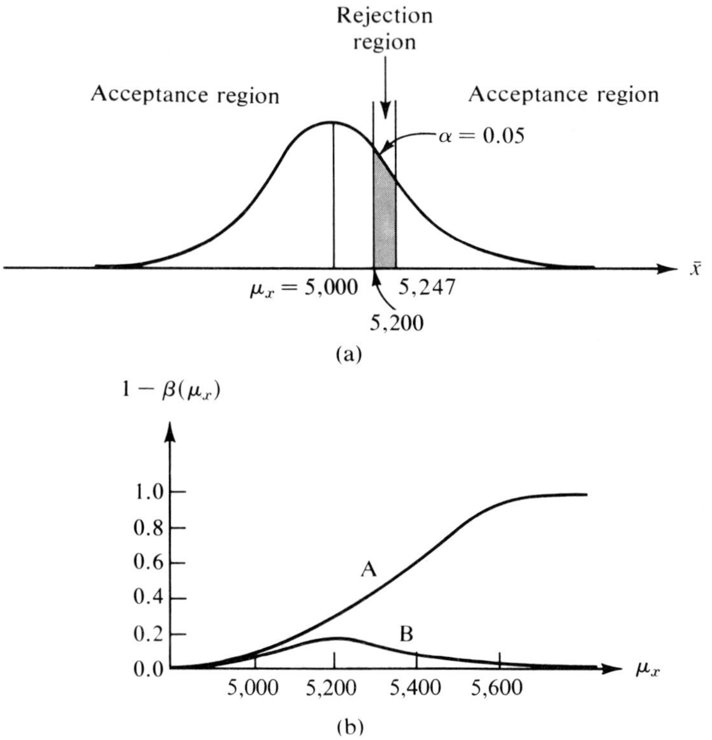

Figure 9.7

(a) Decision rule and α (Example 3) (b) Comparison of power functions for different values of μ_x (Examples 2 and 3)

Solution

To compute β for $\mu_x = \$5{,}500$, we must evaluate

$$\begin{aligned}
\beta(5{,}500) &= P(\text{accepting } H_0 \,|\, \mu_x = 5{,}500) \\
&= P(\overline{X} < 5{,}200 \text{ or } \overline{X} > 5{,}247 \,|\, \mu_x = 5{,}500) \\
&= P(\overline{X} < 5{,}200 \,|\, \mu_x = 5{,}500) + P(\overline{X} > 5{,}247 \,|\, \mu_x = 5{,}500) \\
&= P\left(Z < \frac{5{,}200 - 5{,}500}{200}\right) + P\left(Z > \frac{5{,}247 - 5{,}500}{200}\right) \\
&= P(Z < -1.50) + P(Z > -1.27) \\
&= 0.0668 + (1 - 0.1020) = 0.0668 + 0.8980 = 0.9648.
\end{aligned}$$

We note that the probability of a Type II error is considerably larger for this decision rule than it was for the decision rule of Example 2. Consequently, the value of the power function $P(\mu_x) = 1 - \beta(\mu_x)$ is considerably smaller here than Example 2, since

$$P(5{,}500) = 1 - \beta(5{,}500) = 1 - 0.9648 = 0.0352$$

for this example and

$$P(5{,}500) = 1 - 0.1949 = 0.8051$$

for Example 2. In other words, since the probability of a Type II error is larger, we have a smaller probability of making a correct decision when H_1 is true. Figure 9.7b gives a comparison of the power functions for different values of μ_x, where curve A represents the power function for the test given in Example 2 and curve B represents the power curve for the test of this example. □

In Example 3 we have obtained a different test than that of Example 2 simply by changing the decision rule. Another way to obtain a different test is to change the test statistic. For example, we may use the sample median rather than the sample mean in order to test a hypothesis on the population mean. However, to compare such a test with those of Examples 2 and 3 would require us to determine the probability distribution of the sample median for various values of μ_x. Although this problem is beyond the scope of this text, you should realize that many tests can be considered in the search for a best test.

Since our goal is to choose from all possible tests having the same size Type I error that one having Type II error as small as possible, we can achieve this goal by choosing that test whose power function is as large as possible. Recall again that the power function represents the probability of a correct decision when H_1 is true. For example, suppose we have two candidate tests A and B which yield the same size Type I error and their power functions are as given in Figure 9.7b. Since the power function of A is larger than the power function of B for all possible values of μ_x admitted under our alternative hypothesis, we would immediately choose test A as the better test.

In the example where the hypotheses were

$$H_0\colon \mu_x = 35 \quad \text{versus} \quad H_1\colon \mu_x < 35, \tag{13}$$

we began by determining a critical value $\bar{x}^*$ against which we could compare the observed value of $\bar{X}$. In this case, $\bar{x}^*$ represented the particular value in the sample space of $\bar{X}$ (under the assumption that the null hypothesis is true, i.e., $\mu_x = 35$) so that 5 percent ($\alpha = 0.05$) of the area of the sampling distribution of $\bar{X}$ would fall to the left of $\bar{x}^*$, in the critical region.

The general procedure (exemplified in previous examples) for making a one-tail 100α percent significance test concerning the mean μ_x of a population having known variance follows.

1. Set up the null hypothesis $H_0\colon \mu_x = \mu_0$, where μ_0 is any constant.
2. Specify the level of significance α.
3. Determine a decision rule (critical value) $\bar{x}^*$ which divides the sample space of $\bar{X}$ into a critical (rejection) region and an acceptance region. The critical value $\bar{x}^*$

should be such that, when the null hypothesis is true, 100α percent of the area under the curve representing the sampling distribution of $\bar{X}$ will be in the rejection region. That is, for a left-tail alternative hypothesis, $H_1: \mu_x < \mu_0$, find $\bar{x}^*$ such that

$$P(\bar{X} < \bar{x}^* \mid \mu_x = \mu_0) = \alpha;$$

for a right-tail alternative hypothesis, $H_1: \mu_x > \mu_0$, find $\bar{x}^*$ such that

$$P(\bar{X} > \bar{x}^* \mid \mu_x = \mu_0) = \alpha.$$

4. Take a sample of size n and compute the sample mean $\bar{x}$.
5. If $\bar{x}$ is the critical region, reject H_0. Otherwise, accept it.

Although this is only one way to approach the test of a null hypothesis of this nature, it is in some instances particularly expedient. If we must conduct an identical test repeatedly, it is particularly nice simply to determine a critical value $\bar{x}^*$ in the sample space of $\bar{X}$ against which we can compare the particular value our sample yields for $\bar{X}$ on any given occasion.

Equivalent Methods of Testing Hypotheses

There exist two other methods of testing a hypothesis concerning μ_x in which we do not directly compute $\bar{x}^*$. Even though the two methods are equivalent to that using $\bar{x}^*$ directly and yield identical decisions, you still should be familiar with them. In the first place, they are often less involved. Second, and more important, the concepts associated with these methods help one understand and apply the reasoning used in other hypothesis testing problems.

In determining $\bar{x}^*$, we used the fact that

$$\frac{\bar{X} - \mu_x}{\sigma_x/\sqrt{n}} \sim \mathrm{N}(0, 1) \tag{14}$$

and probabilities are preserved under such a transformation. In the first alternative method, we utilize this fact more explicitly. We test the null hypothesis by looking immediately at the probability distribution of the standard normal variable Z. The essential steps for making a one-tail 100α percent significance test concerning the mean μ_x of a population having known variance follow.

1. Set up the null hypothesis $H_0: \mu_x = \mu_0$, where μ_0 is any constant.
2. Specify the level of significance α.
3. Determine a critical value z^* which divides the distribution of Z into a critical (rejection) region and an acceptance region. The critical value z^* should be such that 100α percent of the area of the distribution of Z lies in the critical region. That is, for a left-tail alternative hypothesis, $H_1: \mu_x < \mu_0$, find z^* such that

$$P(Z < z^*) = \alpha;$$

for a right-tail alternative hypothesis, $H_1: \mu_x > \mu_0$, find z^* such that

$$P(Z > z^*) = \alpha.$$

4. Take a sample of size n, and compute the value

$$z = \frac{\bar{x} - \mu_0}{\sigma_x/\sqrt{n}},$$

where $\bar{x}$ is the observed sample mean, μ_0 is the value of μ_x given by the null hypothesis, and σ_x^2 is the known variance of the population from which we are sampling.

5. If z is in the critical region, reject H_0. Otherwise, accept H_0.

The second alternative method for effecting a one-tail 100α percent significance test concerning the mean μ_x of a population having known variance entails the following steps.

1. Set up the null hypothesis, $H_0: \mu_x = \mu_0$, where μ_0 is any constant.
2. Specify the level of significance α.
3. Take a sample of size n and compute the sample mean $\bar{x}$.
4. Determine the probability of obtaining from the population given by the null hypothesis a value equal to or "worse" than $\bar{x}$. By worse, we mean a value that has less tendency than $\bar{x}$ to support the null hypothesis. That is, for a left-tail alternative hypothesis, $H_1: \mu_x < \mu_0$, calculate

$$\alpha' = P(\bar{X} < \bar{x} \mid \mu_x = \mu_0),$$

or, for a right-tail alternative hypothesis, $H_1: \mu_x > \mu_0$, calculate

$$\alpha' = P(\bar{X} > \bar{x} \mid \mu_x = \mu_0).$$

5. If $\alpha' < \alpha$, reject H_0. Otherwise, accept H_0.

To see clearly the reasoning of this approach, consider that by specifying α, we mean that we do not want to make a Type I error more than 100α percent of the time. Now, note carefully the nature of α'. In each case above, α' represents the probability of obtaining from the population given by the null hypothesis a value equal to or worse than the value we have already obtained in one try. If α' is small enough (less than α), we say the sample is not likely to have come from the population given by the null hypothesis, and thus we reject H_0. On the other hand, if α' is not small enough (not less than α), we say there is not sufficient reason to believe that the sample came from a population other than that given by the null hypothesis. Hence, not having significant reason to reject H_0, we accept it.

To summarize the second method, we find the probability of obtaining (if we are really sampling from the distribution given by the null hypothesis) a value of $\bar{X}$ less than (for a left-tail H_1) or greater than (for a right-tail H_1) the observed value of $\bar{X}$ from our experiment and compare such value against α. If the calculated probability is smaller than α, we reject H_0, because this automatically tells us that the corresponding $\bar{x}$ would have been in the rejection region had we begun by determining a critical value $\bar{x}^*$. If the calculated probability is greater than or equal to α, we accept H_0, because this automatically tells us that the corresponding $\bar{x}$ would have been in the acceptance region had we begun by determining a critical value $\bar{x}^*$.

Figure 9.8 graphs, for the hypotheses of (13), the decision rules involving $\bar{x}^*$ and z^*. We can note from Figure 9.8 the equivalence of (1) comparing an observed value of $\bar{X}$ with the critical value $\bar{x}^*$, (2) comparing an observed value of Z with the critical

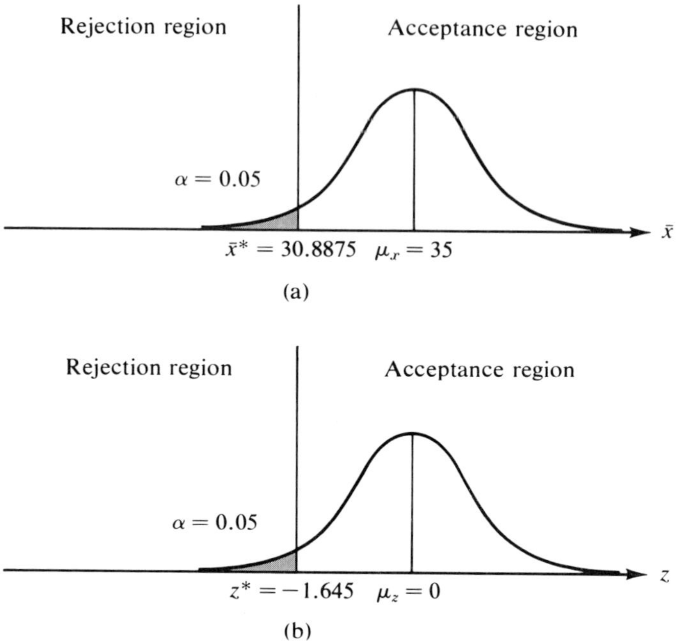

Figure 9.8

Decision rules in terms of critical values $\bar{x}^*$ and z^*

value z^*, and (3) comparing the probability of obtaining a value less than (or perhaps greater than) the observed value of $\bar{X}$ with $\alpha = 0.05$.

For example, suppose we obtained a value of $\bar{x} = 29$ for the sample mean of the 16 values from our experiment and wish to test the null hypothesis of (13). Don't be alarmed by the fact that we are shortly going to assume that $\bar{X}$ is normally distributed even though n is equal only to 16—unless you think we are relying on the central limit theorem. Recall that we are now referring to an example in which we have assumed that the original variable X is normally distributed. Let us now test the hypothesis of (13) by using each of our three methods given. Recall that $\sigma_x = 10$; hence $\sigma_{\bar{x}} = 10/\sqrt{16} = 2.5$.

By the decision rule for $\bar{x}$, we have $29 < 30.8875$, or $\bar{x} < \bar{x}^*$. Thus, we should reject H_0, as can be seen in Figure 9.8a. If we use the decision rule for z, we must compute

$$z = \frac{\bar{x} - \mu_0}{\sigma_{\bar{x}}} = \frac{29 - 35}{2.5} = -2.4.$$

Then, since $-2.4 < -1.645$, or $z < z^*$, we see from Figure 9.8b that we reject H_0. Using the method of comparing with $\alpha = 0.05$ the probability of obtaining from the population given by the null hypothesis a value of $\bar{X}$ less than the observed value of $\bar{x} = 29$, we have

$$P(\bar{X} < 29 \mid \mu_x = 35) = P\left(Z < \frac{29 - 35}{2.5}\right)$$
$$= P(Z < -2.4) = 0.0082.$$

We note that 0.0082 is less than 0.05. This automatically means that our observed $\bar{x}$ is less than $\bar{x}^*$ (even though we will not necessarily have calculated $\bar{x}^*$). Thus, we reject H_0. To satisfy ourselves that this is proper, we compare the area of $\alpha = 0.05$ with the area of 0.0082 in Figure 9.8a, and note that the $\bar{x}$ value associated with the area of 0.0082 must be to the left of the $\bar{x}^*$ associated with the area of $\alpha = 0.05$.

Example 4

A department-store manager has decided that the store can save money by installing a new billing system provided the average balance per account receivable is significantly greater than \$15. If the average balance is only \$15, the store will only break even and hence will not want to install the new system. The manager decides that he is willing to accept a 3 percent risk of making a Type I error. He selects a simple random sample of 81 accounts for which the sample mean is $\bar{x} = \$17$. From experience, he knows that the standard deviation is \$9. Determine whether the store should install the new billing system.

Solution

We wish to test the null hypothesis

$$H_0: \mu_x = \$15$$

versus the alternative hypothesis

$$H_1: \mu_x > \$15.$$

Using $\alpha = 0.03$ as the size of our Type I error, we find z^* such that

$$\alpha = P(Z > z^*) = 0.03.$$

Table A-1 shows $P(Z > 1.88) = 0.03$. Thus, $z^* = 1.88$, and our decision rule is as follows. If our calculated z under the null hypothesis is greater than 1.88, we reject H_0. Otherwise, we accept H_0. Since $\bar{x} = \$17$, we let μ_0 be the value of μ_x given by the null hypothesis and compute

$$z = \frac{\bar{x} - \mu_0}{\sigma_{\bar{x}}} = \frac{\bar{x} - \mu_0}{\sigma_x/\sqrt{n}} = \frac{17 - 15}{9/\sqrt{81}} = 2.00.$$

Since $2.00 > 1.88$, we reject the null hypothesis that $\mu_x = \$15$ and thus conclude that the store should install the new billing system.

In order to emphasize related reasoning, consider the method under which we compute α' and compare it to α. We have observed $\bar{x} = 17$. We find α' as

$$\begin{aligned}\alpha' &= P(\bar{X} > 17 \mid H_0 \text{ true}) = P(\bar{X} > 17 \mid \mu_x = 15)\\ &= P\left(Z > \frac{17 - 15}{9/\sqrt{81}}\right) = P(Z > 2.00) = 0.0228.\end{aligned}$$

Since α' is less than $\alpha = 0.03$ (the risk we are willing to take of making a Type I error), the probability is very small of our obtaining from a population with $\mu_x = 15$ the sample result which we have obtained. This being the case, we conclude that the sample came from another population—with mean greater than 15. □

Example 5

The manager of a supermarket knows from experience that he needs two express checkout lanes during peak periods if the average sale per customer is significantly less than \$5.00. Otherwise, he needs only one express lane. The manager selects a simple random sample of size $n = 49$ from the peak-period customers. He determines the sample mean to be $\bar{x} = \$4.65$. Allowing a 4 percent risk of a Type I error and using a historical standard deviation of \$2.00, determine whether the manager needs one or two express checkout lanes.

Solution

We wish to test

$$H_0: \mu_x = \$5.00 \quad \text{versus} \quad H_1: \mu_x < \$5.00.$$

We have $\sigma_{\bar{x}} = 2.00/\sqrt{49} = 0.29$. In order to assure $\alpha = 0.04$, we choose a critical value z^* such that

$$\alpha = P(Z < z^*) = 0.04.$$

Table A-1 shows

$$P(Z < -1.75) = 0.04.$$

Thus, $z^* = -1.75$. Our decision rule is thus to reject H_0 if our calculated z under the null hypothesis is less than -1.75 and accept H_0 otherwise. The observed value of Z is

$$z = \frac{\bar{x} - \mu_0}{\sigma_{\bar{x}}} = \frac{4.65 - 5.00}{0.29} = -1.21.$$

Since $-1.21 > -1.75$ (i.e., $z > z^*$), we lack sufficient reason to reject H_0. Thus, we accept the null hypothesis that $\mu_x = \$5.00$ and conclude that the manager does not need two express checkout lanes. □

Two-Tail Tests

So far we have considered only those tests that determine whether a population from which we are sampling has a mean significantly less than some specified value or a mean significantly greater. Often, however, we do not know which direction, if either, the mean of the population should be from the specified value. Then, we need to make what is generally called a two-tail test to determine whether the population mean is significantly different from the specified value of the null hypothesis.

The primary difference between a two-tail test and a one-tail test is as follows. Consider a given significance level α. For a *one-tail test*, we place all of the rejection region in one tail of the appropriate distribution (the single tail of the probability distribution curve corresponding to H_0 has an area of size α). For the *two-tail test*, we have two rejection regions with a total area of α. The two rejection regions are not necessarily of the same size. However, unless there is very definite reason for making them unequal, we shall make the risk the same size for each tail. We call such a test—a test with a risk of $\alpha/2$ in each tail—a *symmetrical test*.

Establishing our decision rule for a two-tail test is similar to the procedure for a one-tail test, except that there are two critical values to be determined. For these we use almost the same method we used for the single critical value, the only difference being that, we now use $\alpha/2$ as the area under each tail of the curve.

If this one change is incorporated, any of the three methods for making a one-tail test concerning the population mean can be used for making a two-tail (symmetrical) test concerning such parameter. Thus, if we wish to use directly the sampling distribution of $\bar{X}$, we find two critical values $\bar{x}_1^*$ and $\bar{x}_2^*$ such that

$$P(\bar{X} < \bar{x}_1^* \mid \mu_x = \mu_0) = \alpha/2 \quad \text{and} \quad P(\bar{X} > \bar{x}_2^* \mid \mu_x = \mu_0) = \alpha/2, \tag{15}$$

where μ_0 is again the value of μ_x given by the null hypothesis. The rejection region becomes any value in the sample space of $\bar{X}$ such that

$$\bar{X} < \bar{x}_1^* \quad \text{or} \quad \bar{X} > \bar{x}_2^*, \tag{16}$$

while the acceptance region becomes any value in the sample space of $\bar{X}$ such that

$$\bar{x}_1^* \leq \bar{X} \leq \bar{x}_2^*. \tag{17}$$

If we wish to effect the change by calculating

$$z = \frac{\bar{x} - \mu_0}{\sigma_x/\sqrt{n}} \tag{18}$$

and comparing it to critical values constructed directly from the distribution of Z, we begin by determining two critical values, z_1^* and z_2^*, such that

$$P(Z < z_1^*) = \alpha/2 \quad \text{and} \quad P(Z > z_2^*) = \alpha/2. \tag{19}$$

In this case, we reject H_0 if the calculated value of Z in (18) is less than z_1^* or greater than z_2^*, and accept H_0 otherwise. Note that in a symmetrical test, $z_1^* = -z_2^*$.

Finally, if we wish to use the procedure in which we calculate α' and compare it to a predetermined risk, we proceed exactly as in a one-tail test, except that we compare α' to $\alpha/2$ instead of α.

Example 6

A pharmaceutical firm has a machine which bottles a certain brand of nose drops. Since each bottle is supposed to contain 9 ounces, the machine is set to place 9 ounces in each bottle. However, even when the machine is operating properly, the contents of the filled bottles have a standard deviation of 1 ounce. The firm checks the machine periodically to determine whether it is operating properly, as follows. A foreman randomly selects a sample of size 16 and determines the average weight of the contents of the 16 bottles (the sample mean). Determine a decision rule for testing the null hypothesis that the machine is operating properly by both the method involving

$\bar{x}^*$ and the method involving z^*. Use a significance level of 5 percent and assume that the weights of the contents are normally distributed.

Solution

First of all, note the importance of the assumption that the weights of the contents are normally distributed. It is essential here. The sample size is so small ($n = 16$) that we are not at liberty to invoke the central limit theorem to get at the form of the distribution of $\bar{X}$. However, since the original population is normal, we know that $\bar{X}$ is normally distributed.

Since the machine may either overfill or underfill the bottles when it is out of control, we must use a two-tail test. Thus, we wish to test

$$H_0\colon \mu_x = 9 \text{ ounces} \quad \text{versus} \quad H_1\colon \mu_x \neq 9 \text{ ounces}.$$

To determine the decision rules, we use a risk of $\alpha/2 = 0.025$ in each tail, as indicated in Figure 9.9.

To determine the decision rule in terms of $\bar{x}^*$'s, we obtain

$$\sigma_{\bar{x}} = \frac{\sigma_x}{\sqrt{n}} = \frac{1}{\sqrt{16}} = 0.25$$

and proceed as follows.

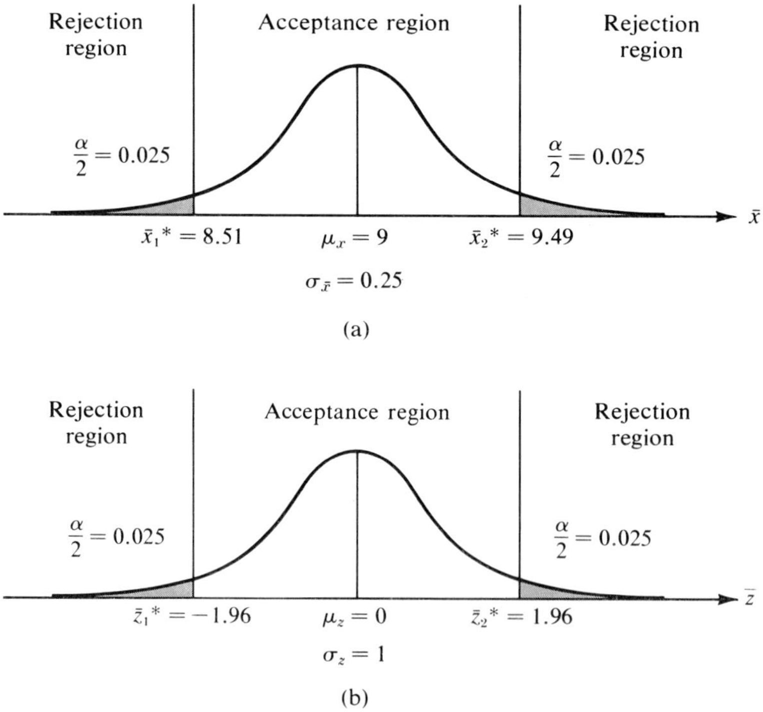

Figure 9.9

Decision rules in terms of $\bar{x}^*$ and z^* for two-tail test

For the left tail, we select $\bar{x}_1^*$ so that

$$P(\bar{X} < \bar{x}_1^* \mid \mu_x = 9) = \alpha/2 = 0.025. \tag{20}$$

However,

$$P(\bar{X} < \bar{x}_1^* \mid \mu_x = 9) = P\left(Z < \frac{\bar{x}_1^* - 9}{0.25}\right) = P(Z < -1.96). \tag{21}$$

For the right tail, we select $\bar{x}_2^*$ so that

$$P(\bar{X} > \bar{x}_2^* \mid \mu_x = 9) = \alpha/2 = 0.025. \tag{22}$$

However,

$$P(\bar{X} > \bar{x}_2^* \mid \mu_x = 9) = P\left(Z > \frac{\bar{x}_2^* - 9}{0.25}\right) = P(Z > 1.96). \tag{23}$$

Solving (20) and (21) for $\bar{x}_1^*$ and (22) and (23) for $\bar{x}_2^*$, we have

$$\frac{\bar{x}_1^* - 9}{0.25} = -1.96 \qquad\qquad \frac{\bar{x}_2^* - 9}{0.25} = 1.96$$

$$\Rightarrow \bar{x}_1^* = 9 - (0.25)(1.96) \qquad \text{and} \qquad \Rightarrow \bar{x}_2^* = 9 + (0.25)(1.96)$$

$$\Rightarrow \bar{x}_1^* = 8.51 \qquad\qquad \Rightarrow \bar{x}_2^* = 9.49.$$

Therefore, the decision rule in terms of $\bar{x}^*$'s is that of rejecting H_0 if the observed value of $\bar{X}$ is less than $\bar{x}_1^* = 8.51$ ounces or greater than $\bar{x}_2^* = 9.49$ ounces and accepting H_0 if the observed value of $\bar{X}$ is between $\bar{x}_1^* = 8.51$ ounces and $\bar{x}_2^* = 9.49$ ounces. This decision rule is graphed in Figure 9.9a.

To determine the decision rule in terms of z^*'s, we proceed as follows. For the left tail, we find z_1^* such that

$$P(Z < z_1^*) = \alpha/2 = 0.025,$$

and, for the right tail, we find z_2^* such that

$$P(Z > z_2^*) = \alpha/2 = 0.025.$$

From Table A-1, we obtain $z_1^* = -1.96$ and $z_2^* = 1.96$. Therefore, our decision rule in terms of z^* values is that of rejecting H_0 if the observed value of Z (the z value computed using the observed value of $\bar{X}$ and the value of μ_x given by the null hypothesis) is less than $z_1^* = -1.96$ or greater than $z_2^* = 1.96$ and accepting H_0 if the observed value of Z is between $z_1^* = -1.96$ and $z_2^* = 1.96$. This decision rule is graphed in Figure 9.9b. □

If we want to test the hypothesis in Example 6 by comparing the probability of obtaining a value of $\bar{X}$ which is greater (or smaller) than the observed value, we must compare the probability against $\alpha/2$ rather than α (as in the one-tail test). Of course, to make such a comparison, we must have an observed value of $\bar{X}$ just as we must before we can apply either of the two parts of the decision rule of Example 6 to test the null hypothesis of interest.

For problems such as Example 6, where the process of testing hypotheses is repetitive, it will save effort to establish $\bar{x}_1^*$ and $\bar{x}_2^*$ at the beginning and use them for our decision rule each time we select a sample of size 16. This precludes the necessity of determining a z value each time we obtain a sample of size 16. In many problems related to quality control (where the variance can be assumed), we wish to check periodically to determine whether a process has changed significantly. In such cases, one must repeatedly test a hypothesis, and thus it is advantageous to use the method involving $\bar{x}_1^*$ and $\bar{x}_2^*$.

For Example 6, the foreman can determine the observed value of $\bar{X}$ each time he selects a random sample of size 16 and determine whether the machine is out of control simply by comparing the observed value of $\bar{X}$ with $\bar{x}_1^* = 8.51$ and $\bar{x}_2^* = 9.49$. If the value is between 8.51 and 9.49, the foreman concludes that the machine is in control. This way he need only determine the sample mean, whereas, if he were to use either of the other method (the method involving z^* or the method involving α'), he would have to determine an observed value for Z or the probability α'. Either of these two methods obviously requires more computation than that of computing $\bar{x}$.

It should now be apparent that the decision rule can and should be established before we conduct the experiment required to obtain an observed value of the appropriate statistic. We should never use the observed value of the statistic to decide what the hypotheses should be. This is a common mistake, one we encourage you to avoid.

Example 7

The research and development department of the company that manufactures steel rods has discovered a new alloy which may possibly serve as a substitute for steel. The population of steel rods has an average breaking strength of 300 pounds. Departmental representatives have selected a random sample of 100 alloy rods and computed the sample mean to be 305 pounds. If the breaking strength of the population of rods made from the new alloy has a standard deviation of 20 pounds, determine whether the alloy rods have the same average breaking strength as the steel rods. Use a 5 percent significance level.

Solution

Since we wish to determine whether the alloy rods have the same average breaking strength as the steel rods, we must allow for the possibility of either a larger or smaller average by using a two-tail alternative hypothesis. Thus, we wish to test the null hypothesis

$$H_0: \mu_x = 300 \text{ pounds}$$

versus the two-tail alternative hypothesis

$$H_1: \mu_x \neq 300 \text{ pounds.}$$

Using Table A-1 and $\alpha/2$ area in each tail, we obtain $z_1^* = -1.96$ and $z_2^* = 1.96$ for our critical values. Our decision rule, then, becomes that of accepting H_0 if the observed value of Z is between -1.96 and 1.96 and rejecting H_0 if the observed value of Z is greater than 1.96 or less than -1.96. We can use a graph of the decision rule, such as Figure 9.9b, for reference.

Computing the value of $\sigma_{\bar{x}}$ and, finally, the observed value of Z, we obtain

$$\sigma_{\bar{x}} = \frac{\sigma_x}{\sqrt{n}} = \frac{20}{\sqrt{100}} = 2 \quad \text{and} \quad z = \frac{\bar{x} - \mu_0}{\sigma_x/\sqrt{n}} = \frac{305 - 300}{2} = 2.5.$$

Therefore, (since the observed value of Z is greater than 1.96), we reject H_0 and conclude that the average breaking strength of the alloy rods is not the same as for the steel rods. □

Example 8

An engine manufacturer uses an air-driven tool for tightening bolts when assembling engines. The tool is supposed to apply 200 foot-pounds of torque to the bolts. If much more torque is applied, the engine block may be damaged or the bolt may break. If much less is applied, the engine may experience excessive wear. An inspector randomly selects 81 bolts and manually determines the amount of torque that has been applied in order to determine whether the tool is operating properly. If the standard deviation for the torque delivered by the tool is 15 foot-pounds and the sample mean for the 81 selected bolts is 197 foot-pounds, determine whether the tool is operating properly. Since the loss incurred by removing the tool from operation for adjustment is rather high, make the test at a 1 percent significance level. □

Solution

Since the tool will need adjusting if the torque applied is either too much or not enough, we shall use a two-tail test. Thus, we shall test

$$H_0\colon \mu_x = 200 \text{ foot-pounds} \quad \text{versus} \quad H_1\colon \mu_x \neq 200 \text{ foot-pounds}.$$

Using $\alpha = 0.01$ ($\alpha/2 = 0.005$) and Table A-1, we obtain the decision rule to be that of accepting H_0 if the observed value of Z is between -2.575 and 2.575 and rejecting H_0 if it is not between such values. We determine $\sigma_{\bar{x}}$ and the observed value of Z as

$$\sigma_{\bar{x}} = \frac{\sigma_x}{\sqrt{n}} = \frac{15}{\sqrt{81}} = 1.67 \quad \text{and} \quad z = \frac{\bar{x} - \mu_0}{\sigma_x/\sqrt{n}} = \frac{197 - 200}{1.67} = -1.796.$$

The observed value of Z (-1.796) is between -2.575 and 2.575. We thus conclude that, at least at the 1 percent significance level, there is not sufficient reason to reject H_0. This being the case, we accept H_0 and conclude that the tool is operating properly. □

Example 9

Let us now consider a situation in which we wish to have more protection against an error in one direction than against an error in the opposite direction. That is, let us examine an example of a test that is not symmetrical. Suppose the manufacturer of Example 8 will accept a risk of 10 percent of removing the tool from operation when it is, in fact, operating properly. Suppose further, however, that the manufacturer wishes to construct a test so that, when the tool is operating properly, it will be removed from service because of excessive torque with a risk of only 2 percent. Using the sample information of Example 8, determine whether the tool is operating properly.

Solution

Since the manufacturer has specified $\alpha = 0.10$ and a risk of 2 percent in the right tail, we have $\alpha_1 = 0.08$ and $\alpha_2 = 0.02$. The situation is illustrated in Figure 9.10.

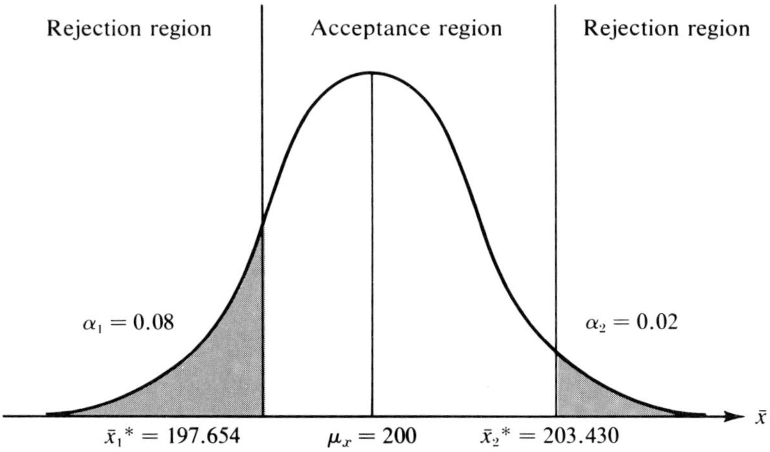

Figure 9.10

Decision rule for unsymmetrical two-tail test

We determine $\bar{x}_1^*$ and $\bar{x}_2^*$ as follows. For the left tail, we select $\bar{x}_1^*$ such that

$$P(\bar{X} < \bar{x}_1^* \,|\, \mu_x = 200) = \alpha_1 = 0.08. \tag{24}$$

However, since $\sigma_{\bar{x}} = 1.67$ (from Example 8),

$$P(\bar{X} < \bar{x}_1^* \,|\, \mu_x = 200) = P\left(Z < \frac{\bar{x}_1^* - 200}{1.67}\right)$$
$$= P(Z < -1.405). \tag{25}$$

For the right tail, we select $\bar{x}_2^*$ such that

$$P(\bar{X} > \bar{x}_2^* \,|\, \mu_x = 200) = \alpha_2 = 0.02. \tag{26}$$

However,

$$P(\bar{X} > \bar{x}_2^* \,|\, \mu_x = 200) = P\left(Z > \frac{\bar{x}_2^* - 200}{1.67}\right) = P(Z > 2.054). \tag{27}$$

Solving (24) and (25) for $\bar{x}_1^*$ and (26) and (27) for $\bar{x}_2^*$, we have

$$\frac{\bar{x}_1^* - 200}{1.67} = -1.405 \qquad\qquad \frac{\bar{x}_2^* - 200}{1.67} = 2.054$$

$$\Rightarrow \bar{x}_1^* = 200 - (1.67)(1.405) \qquad \text{and} \qquad \Rightarrow \bar{x}_2^* = 200 + (1.67)(2.054)$$

$$\Rightarrow \bar{x}_1^* = 197.654 \qquad\qquad \Rightarrow \bar{x}_2^* = 203.430.$$

Therefore, our decision rule in terms of $\bar{x}_1^*$ and $\bar{x}_2^*$ is to accept H_0 if the observed value of $\bar{X}$ is between $\bar{x}_1^* = 197.654$ and $\bar{x}_2^* = 203.430$ and reject H_0 if the observed value of $\bar{X}$ is not between these values.

Since the observed value of $\bar{X}$ is 197 foot-pounds (from Example 8), we reject H_0. Thus, we conclude that the tool is not operating properly. □

Problems

1. Suppose we wish to test the null hypothesis

$$H_0: \mu_x = 200$$

versus the alternative

$$H_1: \mu_x < 200.$$

That is, we wish to determine whether μ_x is "significantly" less than 200. The population from which we shall sample has a known standard deviation of 20. Suppose a simple random sample of size $n = 64$ is selected and a sample mean of $\bar{x} = 196$ is obtained. Use $\alpha = 0.05$ (a 5% significance level).

   a. Using the $\bar{x}^*$ method, sketch the sampling distribution of $\bar{X}$, indicate the acceptance and rejection regions, and determine whether μ_x is significantly less than 200.

   b. Repeat Part a using the z^* method and sketching the sampling distribution of Z.

   c. Using the α' method, state the associated decision rule and determine whether μ_x is significantly less than 200.

   d. Determine β for $\mu_x = 190$ [that is, determine $\beta(190)$].

   e. Assuming we wish to control the Type II error such that $\beta(190) = 0.05$, determine $\bar{x}^*$ and α.

   f. Why is the population from which the sample is taken not required to be normally distributed in order to make the above tests?

   g. Check to determine that the decisions reached in Parts a, b, and c are identical.

2. Use the data of Problem 1.

   a. Calculate β for $\mu_x = 187.5, 190.0, 192.5, 195.0, 197.5, 200.0, 202.5$, and 205.0.

   b. Plot an OC curve from the values calculated in Part a.

   c. Plot the power function from the values calculated in Part a on the same axes used for Part b and label the two curves.

3. A manufacturer of automobile batteries claims that a certain model has a capacity of 60 ampere-hours with a standard deviation of 5 ampere-hours. A large retail chain purchases several carloads of these batteries and selects a simple random sample of size $n = 100$ batteries in order to determine whether the batteries justify the claims made. Suppose that the customer specifies $\alpha = 0.10$ and that he obtains a sample mean of $\bar{x} = 58.3$ ampere-hours.

   a. Using the $\bar{x}^*$ method, sketch the sampling distribution of $\bar{X}$, indicate the acceptance and rejection regions, and determine whether the capacity of the batteries is significantly less than 60 ampere-hours.

   b. Repeat Part a using the z^* method and sketching the sampling distribution of Z.

   c. Using the α' method, state the associated decision rule and determine whether the capacity of the batteries is significantly less than 60 ampere-hours.

d. Determine $\beta(59)$.

e. Assuming we wish to control the Type II error so that $\beta(59) = 0.05$, determine $\bar{x}^*$ and α.

f. Why is the population from which the sample is taken not required to be normally distributed in order to make the above tests?

g. Check to determine that the decisions reached in Parts a, b, and c are identical, and state your conclusion in terms of the problem as well as whether you accept or reject the null hypothesis.

4. Rework Problem 3 with $\sigma_x^2 = 64$, $n = 49$, $\alpha = 0.01$, and $\bar{x} = 59.1$.

5. Suppose we wish to test

$$H_0: \mu_x = 80 \quad \text{versus} \quad H_1: \mu_x > 80.$$

(In other words, is μ_x "significantly" greater than 80?) The population from which we shall sample is normally distributed and has a known standard deviation of 20. Suppose a simple random sample of size $n = 16$ is selected and a sample mean of $\bar{x} = 89$ is obtained. We wish to operate at a 10% significance level (that is, $\alpha = 0.10$).

a. Using the $\bar{x}^*$ method, sketch the sampling distribution of $\bar{X}$, indicate the acceptance and rejection regions, and determine whether μ_x is significantly greater than 80.

b. Repeat Part a using the z^* method and sketching the sampling distribution of Z.

c. Using the α' method, state the decision rule and determine whether μ_x is significantly greater than 80.

d. Determine $\beta(86)$.

e. Assume we wish to control the Type II error such that $\beta(95) = 0.10$; determine $\bar{x}^*$ and α.

f. Why is the population from which the sample is taken required to be normally distributed in order to make the above tests?

g. Check to determine that the decisions reached in Parts a, b, and c are identical.

6. Use the data of Problem 5.

a. Calculate β for $\mu_x = 75, 80, 85, 90, 95, 100, 105$, and 110.

b. Plot an OC curve from the values calculated in Part a.

c. Plot the power function from the values calculated in Part a on the same axes used for Part b and label the two curves.

7. An electronic-components manufacturer claims that his top line of capacitors has a significantly higher average breakdown voltage than the average for his entire production. The average breakdown point for the entire production is known to be 600 volts, and the standard deviation for both the full line and the top line is known to be 100 volts. A customer decides to test the claim by selecting a simple random sample of 400 capacitors from the top line and obtains a mean breakdown of 608 volts. Use $\alpha = 0.05$.

a. Using the $\bar{x}^*$ method, sketch the sampling distribution of $\bar{X}$, indicate the acceptance and rejection regions, and determine whether μ_x is in fact significantly greater than 600 volts, as the manufacturer claims.

b. Repeat Part a using the z^* method and sketching the sampling distribution of Z.

c. Using the α' method, state the decision rule and determine whether μ_x is significantly greater than 600 volts.

d. Determine $\beta(615)$.

e. Assuming we wish to control the Type II error so that $\beta(615) = 0.05$, determine $\bar{x}^*$ and α.

f. Why is the population from which the sample is taken not required to be normally distributed?

g. Check to determine that the decisions reached in Parts a, b, and c are identical, and state your conclusion in terms of the problem as well as whether you accept or reject the null hypothesis.

8. Rework Problem 7 with $\sigma_x = 50$, $n = 225$, $\alpha = 0.01$, and $\bar{x} = 604$. In Part d calculate $\beta(602)$.

9. Use the data of Problem 7.

 a. Calculate β for $\mu_x = 590, 595, 598, 600, 602, 604, 606, 608, 610, 612, 615$, and 620.

 b. Plot an OC curve from the values calculated in Part a.

 c. Plot the power function from the values calculated in Part a on the same axes used for Part b and label the curves.

10. Use the data of Problem 7. However, rather than using a right-tail test, suppose the decision rule is such that we will accept H_0 if $601 \leq \bar{x} \leq 601.65$ and reject H_0 if $\bar{x}_1 < 601$ or $\bar{x} > 601.65$. (Let $\bar{x}_1^* = 601$ and $\bar{x}_2^* = 601.65$. Accept H_0 if $\bar{x}$ is between $\bar{x}_1^*$ and $\bar{x}_2^*$, and reject H_0 otherwise.)

 a. Determine α for the above test and compare its value against the significance level ($\alpha = 0.05$) given in Problem 7.

 b. For the above decision rule, determine β for $\mu_x = 590, 595, 598, 600, 602, 604, 606, 608, 610, 612, 615$, and 620. (Note that the values for μ_x are the same as in Problem 9a, but the β values are different.)

 c. Plot an OC curve from the values calculated in Part b.

 d. Plot the power function from the values calculated in Part b on the same axes used for Part c.

 e. Compare the results of Parts c and d with the respective results of Parts b and c of Problem 9 and explain the result.

11. The manager of a discount store knows from experience that the average sale when using a certain advertising procedure is \$4.00 and the standard deviation of the sales is known to be \$1.00. A representative of an advertising agency convinces the manager to use his agency by claiming that his agency's approach will significantly increase the average sale. A simple random sample of size $n = 100$ is selected from sales occurring at least five months after the advertising agency began servicing the discount-store account and a sample mean of $\bar{x} = \$4.15$ is obtained. Assume that the standard deviation remains unchanged and use $\alpha = 0.05$.

 a. State the appropriate null and alternative hypotheses to use in determining whether μ_x is significantly greater than the average sale associated with the old advertising procedure.

 b. Using any available test methods, sketch the sampling distribution of the appropriate statistic, indicate the acceptance and rejection regions, and determine whether the null hypothesis stated in Part a is true. Also, state your conclusion in terms of the problem as well as whether you accept or reject the null hypothesis.

12. A pharmaceutical firm produces a certain liquid drug for a corporation which operates a large chain of drug stores. The latter firm purchases large quantities of 6-ounce unlabeled bottles of this drug and labels each bottle with its own brand label. The corporation has a written agreement with the pharmaceutical firm stating that the average weight per bottle in any shipment must not be less than 6 ounces. The machine used to bottle the drug fills the bottles with a standard deviation of 0.5 ounces and so that the weights are *normally distributed.* An inspector for the pharmaceutical firm selects a simple random sample of size $n = 25$ from a large group of bottles which are ready for shipment and determines that the sample mean is $\bar{x} = 5.98$ ounces.

 a. State the appropriate null and alternative hypotheses to be used in determining whether the shipment should be sent to the corporation.

b. Using any available method for testing hypotheses and $\alpha = 0.05$, sketch the sampling distribution for the appropriate statistic, indicate the acceptance and rejection regions, and determine whether the null hypothesis stated in Part a is true. Also, state your conclusions in terms of the problem as well as whether you accept or reject the null hypothesis.

13. Suppose we wish to test

$$H_0: \mu_x = 120 \quad \text{versus} \quad H_1: \mu_x \neq 120.$$

(That is, we wish to determine whether μ_x is "significantly" different from 120.) The population from which we shall sample has a known standard deviation of 12. Suppose a simple random sample of size $n = 64$ is selected and a sample mean of $\bar{x} = 122.9$ is obtained. Use $\alpha = 0.05$.

a. Determine $\bar{x}_1^*$ and $\bar{x}_2^*$, sketch the sampling distribution of $\bar{X}$, indicate the acceptance and rejection regions, and using the $\bar{x}^*$ method, determine whether μ_x is significantly different from 120.

b. Determine z_1^* and z_2^*, sketch the sampling distribution of Z, indicate the acceptance and rejection regions, and using the z^* method, determine whether μ_x is significantly different from 120.

c. Using the α' method, state the decision rule and determine whether μ_x is significantly different from 120.

d. Calculate β for $\mu_x = 114.0, 115.5, 117.0, 118.5, 120.0, 121.5, 123.0, 124.5$, and 126.0.

e. Plot an OC curve from the values calculated in Part d.

f. Plot the power function from the values calculated in Part d on the same axes used for Part b and label the curves.

14. Use the data of Problem 13. Suppose that we wish to apply a nonsymmetrical test such that a risk of 1% exists in the left tail and a risk of 4% exists in the right tail ($\alpha_1 = 0.01$ and $\alpha_2 = 0.04$) rather than the usual symmetrical test ($\alpha_1 = \alpha_2 = \alpha/2 = 0.025$).

a. Determine $\bar{x}_1^*$ and $\bar{x}_2^*$, sketch the sampling distribution of $\bar{X}$, indicate the acceptance and rejection regions, and using the $\bar{x}^*$ method, determine whether μ_x is significantly different from 120.

b. Determine z_1^* and z_2^*, sketch the sampling distribution of Z, indicate the acceptance and rejection regions, and using the z^* method, determine whether μ_x is significantly different from 120.

c. Using the α' method, state the decision rule and determine whether μ_x is significantly different from 120.

d. Calculate β for $\mu_x = 114.0, 115.5, 117.0, 118.5, 120.0, 121.5, 123.0, 124.5$, and 126.0.

e. Plot an OC curve from the values calculated in Part d.

f. Plot the power function from the values calculated in Part d on the same axes used for Part b and label the curves.

g. Compare the results of Parts e and f with the results of e and f in Problem 13 and explain the results.

15. A company that manufactures steel pipes having an average diameter of 1.75 cm uses a process such that the standard deviation of the pipe diameters is 0.15 cm; the process is deemed out of control if the average diameter of the pipes is different from 1.75 cm. A simple random sample of size $n = 49$ is selected from a group of recently finished pipes, and the sample mean is determined to be $\bar{x} = 1.73$ cm.

a. State the null and alternative hypotheses to be used in determining whether the process is out of control.

b. Using any available method for testing hypotheses and $\alpha = 0.04$, sketch the sampling distribution for the appropriate statistic, indicate the acceptance and rejection regions, and determine whether or not the average diameter of the pipes which the process is currently producing is significantly different from 1.75 cm. Also, state your conclusions in terms of the problem as well as whether you accept or reject the null hypothesis.

c. Determine $\beta(1.85)$.

9.4 *The Method of Pivotal Quantities—A General Procedure for Testing Hypotheses Concerning a General Unknown Parameter* θ

Suppose we want to test a hypothesis about any general unknown parameter θ. How do we proceed? Obviously, our major concern is to select a statistic to be used for a decision rule for which the test yields a value of α for the probability of Type I error. Suppose we wish, for example, to test the null hypothesis

$$H_0: \theta = \theta_0$$

versus the one-tail alternative hypothesis

$$H_1: \theta > \theta_0,$$

where θ_0 is a given possible value θ can assume. We must find a statistic, say R, and determine a critical value r^*, one of the possible values in the sample space of R, such that

$$\begin{aligned} P(\text{Type I error}) &= P(\text{rejecting } H_0 \mid H_0 \text{ true}) \\ &= P(R > r^* \mid \theta = \theta_0) = \alpha. \end{aligned} \tag{1}$$

Then, we decide whether to accept or reject H_0 by taking a sample, computing an observed value of R (say r), and comparing r to r^*. If $r > r^*$, we reject H_0. Otherwise, we accept H_0.

This all sounds very nice. But, where in the world do we obtain R, and what is its nature? Strangely enough, the statistic R to be used for testing hypotheses concerning an unknown parameter θ must satisfy the same requirements as a statistic to be used for setting confidence intervals for an unknown parameter θ. In other words, it must be a pivotal quantity (Section 8.6). This being the case, R will be a function of the unknown parameter θ, say $R(\theta)$. The general steps in developing a one-tail, 100α percent significance test of a hypothesis concerning any unknown parameter θ, then, are as follows.

1. Set up the null hypothesis H_0: $\theta = \theta_0$, where θ_0 is any constant.
2. Specify the level of significance α.
3. Find a statistic (pivotal quantity) $R(\theta)$ satisfying the following requirements:
   a. The form of the sampling distribution of the statistic $R(\theta)$ is known and is independent of the value which θ may assume.
   b. The statistic is a function of the unknown parameter θ. Note that we imply this when we denote the statistic by $R(\theta)$.

c. The statistic is a function of no other unknowns except θ (i.e., the formula for the statistic involves only constants, known parameters, and functions that can be determined from sample values taken from the population of interest).

4. Determine a critical value r^* [one of the possible values in the sample space of the statistic $R(\theta)$] which divides the distribution of $R(\theta)$ into a critical (rejection) region and an acceptance region. The critical value r^* should be such that 100α percent of the area of the distribution of $R(\theta)$ lies in the critical region. That is, for a left-tail alternative hypothesis, $H_1: \theta < \theta_0$, find r^* such that

$$P[R(\theta) < r^*] = \alpha.$$

For a right-tail alternative hypothesis, $H_1: \theta > \theta_0$, find r^* such that

$$P[R(\theta) > r^*] = \alpha.$$

5. Take a sample of size n and compute the value

$$r = R(\theta_0), \tag{2}$$

that is, the value of $R(\theta)$ which results from the value of θ given by the null hypothesis.

6. If r is in the critical region, reject H_0. Otherwise, accept H_0.

A symmetrical two-tail 100α percent significance test can be accomplished similarly. The critical values become r_1^* and r_2^* such that

$$P[R(\theta) < r_1^*] = \alpha/2 \quad \text{and} \quad P[R(\theta) > r_2^*] = \alpha/2. \tag{3}$$

Thus, we test

$$H_0: \theta = \theta_0 \quad \text{versus} \quad H_1: \theta \neq \theta_0$$

by comparing r with r_1^* and r_2^*.

We realize that digesting the foregoing material is difficult for the beginner. However, repeated reference to the one-tail test (involving the distribution of Z) for μ_x (variance known) will help.

In making this test, why do we compute an observed value

$$z = \frac{\bar{x} - \mu_0}{\sigma_x/\sqrt{n}}$$

and compare it with a critical value z^*? Because under the assumption of a known variance, the statistic

$$Z = \frac{\bar{X} - \mu_x}{\sigma_x/\sqrt{n}} \sim \text{N}(0, 1) \tag{4}$$

and is thus a pivotal quantity for the unknown parameter μ_x. Note carefully that, when we do not assume known variance, the statistic of (4) is not a pivotal quantity. Then, a different statistic must be used and a different test is in order. We shall discuss this situation in a later chapter.

As with the general procedure for constructing confidence intervals, you will greatly benefit by referring to the above general procedure on each occasion when a new test of a hypothesis is introduced. Only careful study will allow you to understand and relate corresponding steps. The effort you make now will be greatly rewarded in the long run.

Problems

1. List briefly the general steps involving the method of pivotal quantities for developing a *one-tail, 100α percent significance test* of a hypothesis concerning any unknown parameter θ.
2. For the data of Problem 13, page 350, use the $\bar{x}^*$ method and illustrate each of the six general steps in developing the symmetrical two-tail, 5% significance test of the hypothesis concerning the unknown parameter μ_x used in Part a. Explain how the pivotal quantity you use satisfies the three requirements.
3. For the data of Problem 1, page 347, use the z^* method and illustrate each of the six general steps in developing the one-tail, 5% significance test of the hypothesis concerning the unknown parameter μ_x used in Part b. Explain how the pivotal quantity you use satisfies the three requirements.

9.5 Testing Hypotheses Concerning Population Proportions

We now turn to the testing of hypotheses concerning the proportion of successes (π) in a population where there are only two possible outcomes: successes and failures. The procedure for testing these hypotheses is, of course, a special case of the general procedure for testing hypotheses as given in (1), Sec. 9.1, and, even more specifically, a special case of the procedure of the last section. It is also quite similar to that of testing hypotheses concerning the population mean (Section 9.3). In fact, we shall be concerned with testing hypotheses about various population parameters throughout the text; and, in most cases, the tests will be very similar to those in Section 9.3. The primary difference between various tests will occur in step 2 of the general procedure. The statistic and its probability distribution will naturally change from one case to the next. However, having a good understanding of Sections 9.3 and 9.4, you will have little difficulty with testing hypotheses concerning any parameter.

Earlier we used the sample mean as a guide to the statistic to be used for the tests of hypotheses about the population mean. For tests of hypotheses about the population proportion π, we shall use the sample proportion p as a guide to the appropriate statistic. In Section 7.7, we discussed the sampling distribution of the sample proportion. We recall from Table 7.12 that

$$p \sim \mathrm{N}\left(\pi, \frac{\pi(1-\pi)}{n}\right) \tag{1}$$

or

$$p \sim \mathrm{N}\left(\pi, \frac{\pi(1-\pi)}{n} \cdot \frac{N-n}{N-1}\right), \tag{2}$$

provided, of course, that the sample size $n > 30$ and $n\pi(1-\pi) > 5$, where (1) is the proper distribution if we are sampling from an infinite population and (2) is appropriate for sampling from a finite population of size N. Thus, we have immediately that the relevant pivotal quantity for testing hypotheses concerning the population proportion of successes is

$$Z = \frac{p - \pi}{\sqrt{\dfrac{\pi(1 - \pi)}{n}}} \sim \mathrm{N}(0, 1) \tag{3}$$

for the case of sampling from an infinite population and

$$Z = \frac{p - \pi}{\sqrt{\dfrac{\pi(1 - \pi)}{n} \cdot \dfrac{N - n}{N - 1}}} \sim \mathrm{N}(0, 1) \tag{4}$$

for the case of sampling from a finite population of size N. Using (3), we see that the steps for making a one-tail, 100α percent significance test concerning the proportion of successes in an infinite population are as follows.

1. Set up the null hypothesis

$$H_0 : \pi = \pi_0,$$

where π_0 is any constant between 0 and 1.

2. Specify the level of significance α.
3. Determine a critical value [because of (3)] z^* such that 100α percent of the distribution of Z lies in the critical region. That is, for a left-tail alternative hypothesis, $H_1 : \pi < \pi_0$, find z^* such that

$$P(Z < z^*) = \alpha.$$

For a right-tail alternative hypothesis, $H_1 : \pi > \pi_0$, find z^* such that

$$P(Z > z^*) = \alpha.$$

4. Take a sample of size n and compute the value

$$z = \frac{p - \pi_0}{\sqrt{\dfrac{\pi_0(1 - \pi_0)}{n}}}, \tag{5}$$

where p is the sample proportion and π_0 is the value of the population proportion given by the null hypothesis.

5. If z is in the critical region, reject H_0. Otherwise, accept H_0.

The steps for a symmetrical two-tail test are again identical, except there are two critical values z_1^* and z_2^* such that

$$P(Z < z_1^*) = \alpha/2 \quad \text{and} \quad P(Z > z_2^*) = \alpha/2.$$

In a similar manner, we could use (4) to show that the steps for a 100α percent significance test concerning the proportion of successes in a finite population of size N should be the same except for step 4. For finite populations of size N, we compute

$$z = \frac{p - \pi_0}{\sqrt{\dfrac{\pi_0(1 - \pi_0)}{n} \cdot \dfrac{N - n}{N - 1}}}$$

instead of the observed value given by (5).

Of course, we may wish to view directly the sampling distribution of the sample proportion p. (We continue to include description of this method because of its extreme practicality in certain situations.) In this case, the critical value(s) for making a 100α percent test are accomplished by setting

$$P(\text{Type I error}) = P(p < p^* \mid \pi = \pi_0) = \alpha$$

for a left-tail test,

$$P(\text{Type I error}) = P(p > p^* \mid \pi = \pi_0) = \alpha$$

for a right-tail test, and

$$P(p < p_1^*) = \alpha/2 \quad \text{and} \quad P(p > p_2^*) = \alpha/2$$

for a symmetrical two-tail test. In either case, the value of p computed from a given sample is compared to the relevant critical value(s) in order to complete the test.

Example 1

An electronic-component manufacturer uses a process for making transistors from which 20 percent of the finished, untested transistors are defective. An engineer has developed a change in the process which he claims will significantly reduce the proportion of defective transistors. A manager for the company wishes to test the engineer's claim and hence randomly selects 100 transistors from a large group of finished, untested transistors made by the new process. The sample proportion of defective transistors is found to be $p = 0.14$. Assuming $\alpha = 0.05$, test the hypothesis that the new process is no better than the old. Also, for practice in calculating β, determine the value for β when $\pi = 0.15$.

Solution

Since we wish to know whether the proportion of defective transistors is 20 percent or significantly less than 20 percent, we decide to test

$$H_0: \pi = 0.20 \quad \text{versus} \quad H_1: \pi < 0.20.$$

Since we are sampling from a large population (for all practical purposes, it may be considered infinite), we shall use the variance as given in (1). We have a choice of (1) directly using the distribution of Z in order to determine a critical value z^* or (2) using the sampling distribution of p in order to determine a critical value p^*. Recall that the approach involving p^* is better when we must investigate the size of the β error. However, in order to make you as familiar as possible with the mechanics of both approaches, we shall again work the first part of the problem by both methods. Using the z^* approach, our specification that $\alpha = 0.05$ tells us that

$$P(Z < z^*) = \alpha = 0.05.$$

This implies immediately that $z^* = -1.645$. Thus, we shall reject H_0 if our calculated Z value is less than -1.645 and accept H_0 otherwise (Figure 9.11a). Using (5), we calculate z,

$$z = \frac{p - \pi_0}{\sqrt{\dfrac{\pi_0(1 - \pi_0)}{n}}} = \frac{0.14 - 0.20}{\sqrt{\dfrac{(0.2)(0.8)}{100}}} = -1.5.$$

Since our observed (calculated) value of Z, -1.5, is greater than our critical value, $z^* = -1.645$, we accept H_0 and conclude that the new process does not afford significant improvement.

Using the p^* approach, we determine a critical value p^* such that

$$P(\text{Type I error}) = \alpha = 0.05. \tag{6}$$

However,

$$\begin{aligned} P(\text{Type I error}) &= P(\text{rejecting } H_0 | H_0 \text{ true}) \\ &= P(p < p^* | \pi = 0.20) \\ &= P\left(Z < \frac{p^* - 0.20}{\sqrt{\dfrac{(0.2)(0.8)}{100}}}\right) = P\left(Z < \frac{p^* - 0.20}{0.04}\right). \end{aligned} \tag{7}$$

Combining (6) and (7), we have

$$P\left(Z < \frac{p^* - 0.20}{0.04}\right) = 0.05.$$

From Table A-1, we have

$$P(Z < -1.645) = 0.05.$$

Thus,

$$\frac{p^* - 0.20}{0.04} = -1.645. \tag{8}$$

Solving (8), we obtain as a critical value

$$p^* = 0.20 - 1.645(0.04) = 0.1342.$$

Our decision rule becomes that of rejecting H_0 if our observed value of p is less than $p^* = 0.1342$ and accepting H_0 otherwise. Using this decision rule, we would immediately accept H_0 because $0.14 > 0.1342$, that is, $p > p^*$.

To determine β when $\pi = 0.15$, we have

$$\begin{aligned} \beta(0.15) &= P(\text{type II error} | \pi = 0.15) \\ &= P(\text{accepting } H_0 | \pi = 0.15) \\ &= P(p > p^* | \pi = 0.15) \\ &= P(p > 0.1342 | \pi = 0.15). \end{aligned}$$

Recalling that σ_p depends upon the true value of π, we have, for $\pi = 0.15$,

$$\sigma_p = \sqrt{\frac{(0.15)(0.85)}{100}} = \sqrt{\frac{0.1275}{100}} = 0.036.$$

Thus,

$$\beta(0.15) = P\left(Z > \frac{0.1345 - 0.15}{0.036}\right)$$
$$= P(Z > -0.43) = 1 - P(Z < -0.43)$$
$$= 1 - 0.3336 = 0.6664.$$

The solution is demonstrated graphically in Figure 9.11b.

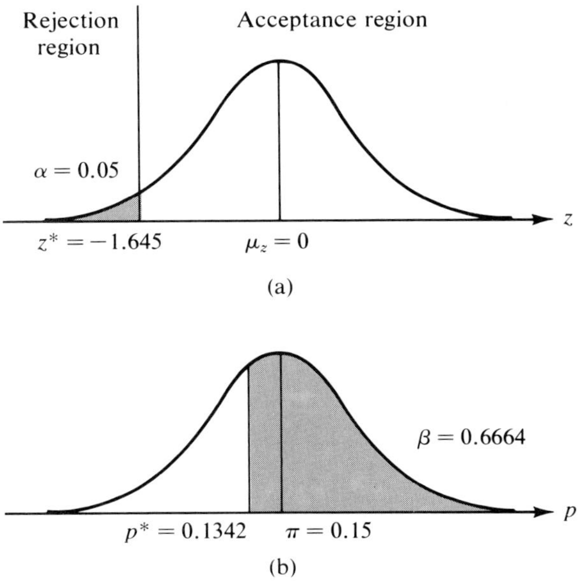

Figure 9.11

z^*, p^*, and β for left-tail test □

Example 2

Company A manufactures flashlight batteries. Past experience indicates that the percentage defective is 30 percent. A new step in the manufacturing process is introduced to reduce the percentage of defective batteries. A random sample of size $n = 64$ is selected from the only lot of size $N = 1{,}000$ which has been made using the new process. Fifteen batteries are found defective ($15/64 = 23.4$ percent defective). Has there been a significant decrease in the proportion of defective batteries? Use the 1 percent significance level.

Solution

Since we wish to know if there has been a significant decrease in the proportion of defective batteries, our hypotheses of interest concerning the new process are

$$H_0: \pi = 0.30 \quad \text{versus} \quad H_1: \pi < 0.30.$$

As before [except, since we are sampling from a finite population of size $N = 1{,}000$, we use the relationship of (2)] we have

$$z = \frac{p - \pi_0}{\sigma_p}$$

$$= \frac{p - \pi_0}{\sqrt{\dfrac{\pi_0(1 - \pi_0)}{n} \cdot \dfrac{N - n}{N - 1}}}$$

$$= \frac{0.234 - 0.30}{\sqrt{\dfrac{(0.3)(0.7)}{64} \cdot \dfrac{1{,}000 - 64}{1{,}000 - 1}}}$$

$$\cong \frac{0.234 - 0.30}{0.0554} \cong -1.19.$$

Using $\alpha = 0.01$, we find $z^* = -2.33$. Thus, our decision rule is to accept H_0 if the observed value of Z is greater than -2.33 and reject H_0 if the observed value of Z is less than -2.33. Hence, since $-1.19 > -2.33$, we accept H_0. That is, the data do not indicate a significant decrease in the proportion of defective batteries. □

Example 3

The executives of a large supermarket chain wish to determine whether to stock more bakery products. In the past, the stock level has been determined from the knowledge that 20 percent of all customers are purchasers of bakery goods. The executives will increase the amount of bakery products in stock if there has been a significant increase in the proportion of customers who purchase bakery goods. At numerous stores throughout the chain, a survey is conducted. A random sample of 100 customers is selected and, in each case, it is noted whether the customer buys bakery items. The result is that 28 percent of the customers in the sample purchase bakery items. Is the proportion of customers buying bakery items significantly greater than 20 percent? Use a significance level of 5 percent. Also, for practice, determine β for $\pi = 0.25$.

Solution

The hypotheses of interest are

$$H_0: \pi = 0.20 \quad \text{versus} \quad H_1: \pi > 0.20.$$

Since we need a critical value p^* in order to solve for β when $\pi = 0.25$, we shall perform the test by looking directly at the sampling distribution of the sample proportion p. In order to specify $\alpha = 0.05$, we require p^* such that

$$P(\text{Type I error}) = 0.05, \tag{9}$$

However,

$$\begin{aligned}
P(\text{Type I error}) &= P(\text{rejecting } H_0 \mid H_0 \text{ true}) \\
&= P(p > p^* \mid \pi = 0.20) \\
&= P\left(Z > \frac{p^* - \pi_0}{\sqrt{\dfrac{\pi_0(1 - \pi_0)}{n}}}\right) \\
&= P\left(Z > \frac{p^* - 0.20}{\sqrt{\dfrac{0.20(0.80)}{100}}}\right) = P\left(Z > \frac{p^* - 0.20}{0.04}\right). \qquad (10)
\end{aligned}$$

Combining (9) and (10), we have

$$P\left(Z > \frac{p^* - 0.20}{0.04}\right) = 0.05.$$

From Table A-1, we have

$$P(Z > 1.645) = 0.05.$$

Thus,

$$\frac{p^* - 0.20}{0.04} = 1.645.$$

Solving, we obtain

$$p^* = 0.20 + (0.04)(1.645) = 0.2658.$$

Our decision rule becomes that of rejecting H_0 if our observed value of p is greater than 0.2658 and accepting H_0 otherwise. Since our observed sample proportion is 0.28, we reject H_0 and conclude that there has been a significant increase in the proportion of customers purchasing bakery items.

Next, we use $p^* = 0.2658$ to determine the probability (under the decision rule p^*) of making a Type II error when $\pi = 0.25$. We thus determine $\beta(0.25)$ as follows:

$$\begin{aligned}
\beta(0.25) &= P(\text{accepting } H_0 \mid \pi = 0.25) \\
&= P(p < p^* \mid \pi = 0.25) = P(p < 0.2658 \mid \pi = 0.25) \\
&= P\left(Z < \frac{0.2658 - 0.25}{\sqrt{\dfrac{(0.25)(0.75)}{100}}}\right) = P\left(Z < \frac{0.2658 - 0.25}{0.0434}\right) \\
&= P(Z < 0.36) = 1 - P(Z > 0.36) \\
&= 1 - 0.3594 = 0.6406.
\end{aligned}$$

□

Example 4

A public health officer knows that five years ago 20 percent of the people in a certain large city had a specific disease. He wants to know if there has been a significant change in this proportion. He selects a random sample of $n = 100$ people within the city and finds that 27 percent of them have the disease. Let $\alpha = 0.05$. Do the data indicate that there has been a significant change in the proportion of people who have the disease? Determine, again for practice in such calculations, the value of β for $\pi = 0.15$.

Solution

The hypotheses of interest are

$$H_0: \pi = 0.20 \quad \text{versus} \quad H_1: \pi \neq 0.20.$$

The alternative hypothesis is two-sided, because we simply want to determine any significant change in the proportion of people having the disease—we have not specified the direction of the change.

Let us determine (in order to calculate β) critical values (p_1^* and p_2^*) with regard to the sampling distribution of p. To do so, we must find p_1^* and p_2^* such that

$$P(\text{Type I error}) = \alpha = 0.05,$$

that is, such that

$$P(p < p_1^* \mid \pi = 0.20) + P(p > p_2^* \mid \pi = 0.20) = 0.05.$$

We determine a symmetrical test by putting equal areas in each tail of the sampling distribution as shown in Figure 9.12a. Thus, we find p_1^* such that

$$P(p < p_1^* \mid \pi = 0.20) = 0.025 \qquad (11)$$

and p_2^* such that

$$P(p > p_2^* \mid \pi = 0.20) = 0.025. \qquad (12)$$

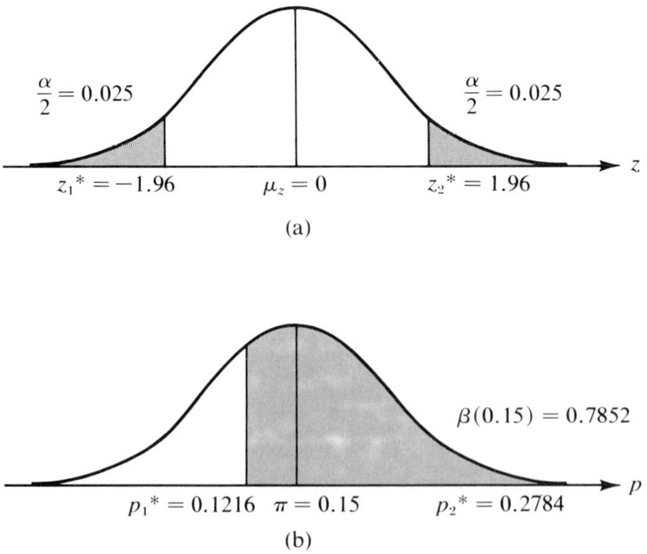

Figure 9.12

z^*, p^*, and β for symmetrical two-tail test

You should verify that solving (11) and (12) will give

$$p_1^* = 0.1216 \quad \text{and} \quad p_2^* = 0.2784.$$

Our decision rule becomes that of accepting H_0 if our observed value of p is between $p_1^* = 0.1216$ and $p_2^* = 0.2784$ and rejecting H_0 otherwise. In other words, if $p_1^* < p < p_2^*$, accept H_0, and if $p < p_1^*$ or $p > p_2^*$, reject H_0.

The observed value of p from our sample is 0.27. Thus, we accept H_0. That is, the sample data do not indicate a significant change in the proportion of people having the disease.

Now, to determine β for $\pi = 0.15$, we have

$$\begin{aligned}
\beta(0.15) &= P(\text{accepting } H_0 \mid H_1 \text{ true}) \\
&= P(p_1^* < p < p_2^* \mid \pi = 0.15) = P(0.1216 < p < 0.2784 \mid \pi = 0.15) \\
&= P\left(\frac{0.1216 - 0.15}{\sqrt{\dfrac{(0.15)(0.85)}{100}}} < Z < \frac{0.2784 - 0.15}{\sqrt{\dfrac{(0.15)(0.85)}{100}}}\right) \\
&= P\left(\frac{0.1216 - 0.15}{0.036} < Z < \frac{0.2784 - 0.15}{0.036}\right) \\
&= P(-0.79 < Z < 3.57) = 1 - P(Z > 0.79) - P(Z > 3.57) \\
&= 1 - 0.2148 - 0 = 0.7852.
\end{aligned}$$

This is represented graphically in Figure 9.12b. □

Note that the calculation for β in a two-tail test may look slightly different from similar calculations for one-tail tests, but it is made exactly according to our original definition of Type II error.

Problems

1. List briefly the five general steps for making a *one-tail, 100α percent significance test* concerning the proportion of successes in a *finite* population. Explain how the pivotal quantity you use satisfies the three necessary requirements.

2. Suppose we wish to test

$$H_0: \pi = 0.40 \quad \text{versus} \quad H_1: \pi < 0.40$$

(that is, we wish to determine whether π is "significantly" less than 0.40). Suppose a simple random sample of size $n = 100$ is selected and a sample proportion of $p = 0.32$ is obtained. Use $\alpha = 0.05$.

   a. Using the p^* method, sketch the sampling distribution of p, indicate the acceptance and rejection regions, and determine whether π is significantly less than 0.40.

   b. Repeat Part a using the z^* method and sketching the sampling distribution of Z.

   c. Using the α' method, state the decision rule and determine whether π is significantly less than 0.40.

d. Determine β for $\pi = 0.25$.

e. For the situation in which we wish to control the Type II error so that $\beta(0.25) = 0.05$, determine p^* and α.

f. Why are you permitted to use the normal distribution in making the above tests?

g. Use the p^* method and illustrate each of the five general steps involved in developing the one-tail, 5% significance test of the hypothesis concerning π in Part a. Explain how the pivotal quantity you use satisfies the three requirements.

3. The process for producing magnetic cores for critical applications in spacecraft guidance computers yields 80% rejects. A new manufacturing process is supposed to significantly reduce the proportion of defective cores. A simple random sample of 400 cores from a large batch produced by the new process is selected, and 296 are found to be defective. Use $\alpha = 0.05$.

a. Using the p^* method, sketch the sampling distribution of p, indicate the acceptance and rejection regions, and determine whether the proportion of defective cores produced by the new manufacturing process is significantly less than 80%.

b. Repeat Part a using the z^* method and sketching the sampling distribution of Z.

c. Using the α' method, state the decision rule and determine whether the proportion of defective cores produced by the new manufacturing process is significantly less than 80%.

d. Determine for $\pi = 0.76$.

e. Assume that we wish to control the Type II error such that $\beta(0.76) = 0.05$ and determine p^* and α.

4. Suppose we wish to test

$$H_0: \pi = 0.60 \quad \text{versus} \quad H_1: \pi > 0.60.$$

(In other words, we wish to determine whether π is "significantly" greater than 0.60.) Suppose a simple random sample of size $n = 96$ is selected and a sample proportion of $p = 0.69$ is determined. Use $\alpha = 0.01$.

a. Using the p^* method, sketch the sampling distribution of p, indicate the acceptance and rejection regions, and determine whether π is significantly greater than 0.60.

b. Repeat Part a using the z^* method and sketching the sampling distribution of Z.

c. Using the α' method, state the decision rule and determine whether π is significantly greater than 0.60.

d. Determine β for $\pi = 0.75$.

e. Assuming we wish to control the Type II error so that $\beta(0.75) = 0.05$, determine p^* and α.

f. Calculate β for $\pi = 0.60, 0.65, 0.70, 0.75, 0.80$, and 0.85.

g. Plot an OC curve from the values calculated in Part f.

h. Plot the power function from the values calculated in Part f on the same axes used for Part g and label the curves.

5. In a certain large city, the proportion of families who use a particular product has previously been 0.10. The manufacturer recently conducted an extensive advertising campaign in the city with the objective of significantly increasing this proportion. In order to evaluate its success, the firm selected a simple random sample of 64 families and found the proportion using the product to be 0.18. Use $\alpha = 0.01$.

a. Using any of the test methods, state the hypotheses, sketch the associated sampling distribution, indicate the acceptance and rejection regions, and determine whether there has been a significant increase in the proportion of families using the product. Explain your conclusion in terms of the problem as well as whether you accept or reject the null hypothesis.

b. Determine the value of β when $\pi = 0.20$.

c. Assuming we wish to control the Type II error so that $\beta(0.20) = 0.01$, determine p^* and π.

d. Use the z^* method and illustrate each of the five general steps involved in developing the one-tail, 1% significance test of the hypothesis concerning π used in Part a. Explain how the pivotal quantity you use satisfies the three requirements.

6. Rework Problem 5 for $n = 225$ and $\alpha = 0.10$.

7. Suppose we wish to test

$$H_0: \pi = 0.75 \quad \text{versus} \quad H_1: \pi \neq 0.75.$$

(That is, we wish to determine whether π is "significantly" different than 0.75.) A simple random sample of size $n = 121$ is selected and a sample proportion of $p = 0.82$ is obtained. Use $\alpha = 0.05$.

a. Using the p^* method, sketch the sampling distribution of p, indicate the acceptance and rejection regions, and determine whether π is significantly different than 0.75.

b. Repeat Part a using the z^* method and sketching the sampling distribution of Z.

c. Using the α' method, state the decision rule and determine whether π is significantly different than 0.75.

d. Determine β for $\pi = 0.62$.

8. Rework Problem 7 using a nonsymmetrical test such that $\alpha_1 = 0.04$ is the risk associated with the left tail and $\alpha_2 = 0.01$ is the risk associated with the right tail.

9.6 Effects on β of Changing Sample Size

In Sections 9.3 and 9.5 we determined decision rules for specified values of α and specified sample sizes. Our policy always assured us of a Type I error of size α, a risk we were able to control to any desired level. We have repeatedly emphasized that it is impossible to control both errors (Type I and Type II); that is, for a fixed sample size, one cannot decrease both errors simultaneously simply by changing the decision rule. Since we are often at liberty to specify (within cost and time constraints, of course) the sample size to be taken in any given experiment, it is interesting to consider the effect on Type II error of increasing the sample size.

Up to now, we have made no effort to control the probability of a Type II error although we have discussed the mechanics of calculating β in specific situations. Let us begin by considering various sample sizes for a specific example and observing the effect on β of the increasing sample sizes.

Suppose we wish to test, at the 5 percent significance level, the null hypothesis

$$H_0;\ \mu_x = 25$$

versus the one-sided alternative hypothesis

$$H_1: \mu_x > 25$$

and to determine β for $\mu_x = 27$. Note that by setting up the hypotheses in the above manner and specifying $\alpha = 0.05$, we are implying that the most critical error we can make is to say that $\mu_x > 25$ when it really is not. However, note also that our interest in β for the situation $\mu_x = 27$ may indicate that we also would like to control β for this specific situation. Suppose further that the standard deviation is known to be $\sigma_x = 6$. We shall make the test first with a sample size of 36, then with a sample of 100.

Sample size of $n = 36$ We wish to base our initial thoughts on a sample size of $n = 36$. We determine a critical value $\bar{x}^*$ such that, using $\bar{x}^*$ to arrive at a decision rule, we will have

$$P(\text{Type I error}) = \alpha = 0.05. \tag{1}$$

Now,

$$\begin{aligned} P(\text{Type I error}) &= P(\text{rejecting } H_0 \,|\, H_0 \text{ true}) \\ &= P(\bar{X} > \bar{x}^* \,|\, \mu_x = 25) \\ &= P\left(Z > \frac{\bar{x}^* - \mu_x}{\sigma_x/\sqrt{n}} \,\middle|\, \mu_x = 25\right) \\ &= P\left(Z > \frac{\bar{x}^* - 25}{6/\sqrt{36}}\right) = P\left(Z > \frac{\bar{x}^* - 25}{1}\right). \end{aligned} \tag{2}$$

Combining (1) and (2) gives

$$P\left(Z > \frac{\bar{x}^* - 25}{1}\right) = 0.05.$$

From Table A-1,

$$P(Z > 1.645) = 0.05.$$

Thus,

$$\frac{\bar{x}^* - 25}{1} = 1.645,$$

which implies

$$\bar{x}^* = 25 + 1.645 = 26.645.$$

In this case, then, the critical value $\bar{x}^* = 26.645$ is the value that assures $\alpha = 0.05$. Our decision rule is to accept H_0 if the observed value of $\bar{X}$ is less than 26.645 and reject H_0 if the observed value of $\bar{X}$ is greater than 26.645. We determine the value of β for the situation where $\mu_x = 27$ as

$$\begin{aligned} \beta(27) &= P(\text{accepting } H_0 \,|\, \mu_x = 27) \\ &= P(\bar{X} < 26.645 \,|\, \mu_x = 27) \\ &= P\left(Z < \frac{26.645 - \mu_x}{\sigma_x/\sqrt{n}} \,\middle|\, \mu_x = 27\right) \\ &= P\left(Z < \frac{26.645 - 27}{6/\sqrt{36}}\right) = P\left(Z < \frac{26.645 - 27}{1}\right) \\ &= P(Z < -0.36) = P(Z > 0.36) = 0.3594. \end{aligned}$$

Thus, for $n = 36$, the critical value $\bar{x}^* = 26.645$ yields $\alpha = 0.05$ and $\beta(27) = 0.3594$.

Sample Size of $n = 100$ For the sample size $n = 100$, we now determine a critical value $\bar{x}^*$ such that, using $\bar{x}^*$ to arrive at a decision rule, we will once again have

$$P(\text{Type I error}) = 0.05.$$

We determine $\bar{x}^*$ by following the same calculations as for (1) and (2), except that we everywhere replace $n = 36$ by $n = 100$. You should verify that these steps will yield a value of

$$\bar{x}^* = 25.987.$$

Our decision rule is now to accept H_0 if the observed value of $\bar{X}$ is less than 25.987 and reject H_0 if the value is greater than 25.987. We compute the value of β for $\mu_x = 27$ as

$$\begin{aligned}
\beta(27) &= P(\text{accepting } H_0 \mid \mu_x = 27) \\
&= P(\bar{X} < 25.987 \mid \mu_x = 27) \\
&= P\left(Z < \frac{25.987 - \mu_x}{\sigma_x/\sqrt{n}} \,\middle|\, \mu_x = 27\right) \\
&= P\left(Z < \frac{25.987 - 27}{6/\sqrt{100}}\right) = P\left(Z < \frac{25.987 - 27}{0.6}\right) \\
&= P(Z < -1.69) = P(Z > 1.69) = 0.0455.
\end{aligned}$$

Thus, for $n = 100$, the critical value $\bar{x}^* = 25.987$ yields $\alpha = 0.05$ and $\beta(27) = 0.0455$.

Comparing the values of β when $\mu_x = 27$ for sample sizes of 36 and 100, we have 0.3594 and 0.0455, respectively. We note that the value of β for $n = 36$ is considerably larger than for $n = 100$. From this example, we see that, for a fixed α and a fixed alternative hypothesis (fixed value of μ_x), the risk β of making a Type II error decreases as the sample size increases.

Determination of Sample Size

In the foregoing discussion we observed that the size of β is related to the sample size; all other conditions being fixed, the risk β of a Type II error decreases as the sample size increases. We would like, of course, to determine the relationship between the sample size and β for a specified value of μ_x, specifically any value that satisfies the alternative hypothesis. Such relationship will allow us to specify the desired value for β and then determine the required sample size.

To derive a general relationship, let us consider the case of a one-tail alternative hypothesis for a predetermined value of α. For example, let our hypotheses be

$$H_0\colon \mu_x = \mu_0 \quad \text{versus} \quad H_1\colon \mu_x > \mu_0, \tag{3}$$

where μ_0 is any possible value of μ_x. We shall use the (by now obviously) logical rule of accepting H_0 if the observed value of $\bar{X}$ is less than the critical value $\bar{x}^*$ and rejecting H_0 if the observed value of $\bar{X}$ is greater than $\bar{x}^*$. Ordinarily, of course, we would need to determine $\bar{x}^*$ in order to determine our decision rule. However, before we can establish such values, we must first specify sample size. Let us suppose further that we wish to have a probability of a Type II error equal to β when $\mu_x = \mu_1$, where μ_1 represents a specific value of μ_x satisfying the alternative hypothesis of (3). Moreover, we shall let the standard deviation be σ_0 when $\mu_x = \mu_0$ (i.e., under the null hypothesis) and σ_1 when $\mu_x = \mu_1$ (that is, under the fixed value of the alternative hypothesis). This does not preclude the possibility that $\sigma_0 = \sigma_1$. Assuming that in any given problem μ_0, μ_1, α, β, σ_0, and σ_1 will each have a specific (known) value, we proceed to determine their relationships to n and $\bar{x}^*$. For (3), we determine $\bar{x}^*$ such that

$$P(\text{Type I error}) = \alpha. \tag{4}$$

However, in terms of $\bar{x}^*$ and n,

$$\begin{aligned} P(\text{Type I error}) &= P(\text{rejecting } H_0 | H_0 \text{ true}) \\ &= P(\bar{X} > \bar{x}^* | \mu_x = \mu_0, \sigma_x = \sigma_0) \\ &= P\left(Z > \frac{\bar{x}^* - \mu_0}{\sigma_0/\sqrt{n}}\right). \end{aligned} \tag{5}$$

Combining (4) and (5) gives

$$P\left(Z > \frac{\bar{x}^* - \mu_0}{\sigma_0/\sqrt{n}}\right) = \alpha. \tag{6}$$

From Table A-1, we can determine the value z_α such that

$$P(Z > z_\alpha) = \alpha. \tag{7}$$

Since z_α is unique, we have

$$\frac{\bar{x}^* - \mu_0}{\sigma_0/\sqrt{n}} = z_\alpha. \tag{8}$$

We can manipulate (8) algebraically to obtain

$$\sqrt{n}(\bar{x}^* - \mu_0) = z_\alpha \sigma_0. \tag{9}$$

Now, recall that we have specified that, if $\mu_x = \mu_1$, we want a Type II error of size β. That is,

$$P(\text{Type II error}) = \beta. \tag{10}$$

But, in terms of $\bar{x}^*$ and n,

$$\begin{aligned} P(\text{Type II error}) &= P(\text{accepting } H_0 | \mu_x = \mu_1, \sigma_x = \sigma_1) \\ &= P(\bar{X} < \bar{x}^* | \mu_x = \mu_1, \sigma_x = \sigma_1) \\ &= P\left(Z < \frac{\bar{x}^* - \mu_1}{\sigma_1/\sqrt{n}}\right). \end{aligned} \tag{11}$$

Combining (10) and (11), we have

$$P\left(Z < \frac{\bar{x}^* - \mu_1}{\sigma_1/\sqrt{n}}\right) = \beta.$$

From Table A-1, we can determine the value $-z_\beta$ such that

$$P(Z < -z_\beta) = \beta. \tag{12}$$

And, since $-z_\beta$ is unique, we have

$$\frac{\bar{x}^* - \mu_1}{\sigma_1/\sqrt{n}} = -z_\beta. \tag{13}$$

Manipulating (13), we obtain

$$\sqrt{n}\,(\bar{x}^* - \mu_1) = -z_\beta \sigma_1. \tag{14}$$

Equations (9) and (14) constitute two equations in two unknowns, n and $\bar{x}^*$, whose values we seek. All other values in the equations have already been established. By subtracting (14) from (9), we eliminate the terms involving $\bar{x}^*$ and obtain

$$\sqrt{n}\,(\mu_1 - \mu_0) = z_\alpha \sigma_0 + z_\beta \sigma_1.$$

Solving for n, we get

$$n = \left(\frac{z_\alpha \sigma_0 + z_\beta \sigma_1}{\mu_1 - \mu_0}\right)^2. \tag{15}$$

Thus, to determine the sample size that assures a probability of a Type I error equal to α and a probability of a Type II error (when $\mu_x = \mu_1$) equal to β, we need only determine z_α and z_β from Table A-1, substitute their values into (15), and evaluate n. Once n has been determined, we can find $\bar{x}^*$ in the usual manner.

Equation (15) was derived for a right-tail alternative hypothesis. You should verify that the same equation will apply when we have a left-tail alternative hypothesis. Further, we can make (15) applicable for a two-tail test if we everywhere replace α by $\alpha/2$ and keep β less than 0.5. For the two-tail test, the sample size from (15) will not be exact for all cases. However, it will be sufficiently accurate for most practical situations.

You should also verify that a relation similar to (15) is applicable for testing hypotheses concerning proportions. Naturally, to develop such a relationship we must replace $\bar{X}$ by p, and μ_0 and μ_1 by π_0 and π_1, respectively. Such an analysis will show why two distinct standard deviations were considered in our general approach above. The general result for a one-tail test concerning proportions is

$$n = \left(\frac{z_\alpha \sqrt{\pi_0(1 - \pi_0)} + z_\beta \sqrt{\pi_1(1 - \pi_1)}}{\pi_1 - \pi_0}\right)^2, \tag{16}$$

where α is the value specified for probability of Type I error under the null hypothesis $H_0\colon \pi = \pi_0$, β is the value specified for probability of Type II error when $\pi = \pi_1$, z_α is such that $P(Z > z_\alpha) = \alpha$, z_β is such that $P(Z > z_\beta) = \beta$, and n is the sample size required to satisfy the constraints of the problem.

Example 1

Suppose we wish to test

$$H_0\colon \mu_x = 20 \quad \text{versus} \quad H_1\colon \mu_x > 20$$

at the 5 percent significance level. The standard deviation of the population is known to be 4, and we want a 10 percent risk of making a Type II error when $\mu_x = 22$. Find the sample size required to satisfy these constraints.

Solution

We have $\mu_0 = 20$ and $\mu_1 = 22$. We note that there is only one standard deviation given. However, as before, we assume that the standard deviation remains known and unchanged regardless of the value of μ_x. Hence, σ_0 and $\sigma_1 = 4$. From Table A-1, we find that $z_\alpha = z_{0.05} = 1.645$, since $P(Z > 1.645) = 0.05$, and $z_\beta = z_{0.10} = 1.28$, since $P(Z < -1.28) = 0.10$. Substituting into (15), we have

$$n = \left[\frac{(1.645)(4) + (1.28)(4)}{22 - 20}\right]^2 \cong (5.85)^2 \cong 34.$$

Thus, in order to satisfy the constraints of the problem, we should select a sample of size $n = 34$. □

Example 2

Suppose we wish to examine the following hypotheses concerning proportions.

$$H_0\colon \pi = 0.20 \quad \text{versus} \quad H_1\colon \pi < 0.20.$$

Suppose further that we wish $\alpha = 0.05$ and, for $\pi = 0.15$, we wish $\beta = 0.05$; that is, $\beta(0.15) = 0.05$. What sample size is required to assure the desired risks?

Solution

We have $\pi_0 = 0.20$ and $\pi_1 = 0.15$. From Table A-1 we obtain $z_\alpha = z_{0.05} = 1.645$ and $z_\beta = z_{0.05} = 1.645$. Substituting into (16), we obtain

$$\begin{aligned} n &= \left[\frac{(1.645)\sqrt{(0.2)(0.8)} + 1.645\sqrt{(0.15)(0.85)}}{0.15 - 0.20}\right]^2 \\ &= \left[\frac{(1.645)(0.40) + (1.645)(0.36)}{0.15 - 0.20}\right]^2 \\ &= \left(\frac{1.2502}{-0.05}\right)^2 \cong (25)^2 = 625. \end{aligned}$$

Therefore, a sample of size $n = 625$ will assure that our risk of a Type I error is 0.05 and the risk of a Type II error is 0.05 when $\pi = 0.15$. □

It is now apparent that we can control the risk of a Type II error by determining sample size before selecting the sample. Thus, in practice, we should determine sample size before conducting our experiment, with the selection being based upon the amount of risk we want for a specific value of the parameter under test. Of course, the decision maker must always consider the cost of conducting his experiment and, hence, temper his desire for low risk with the high cost generally associated with large sample sizes. In practice, a compromise must usually be made.

Problems

1. Prove that Equation (15) applies when we have a left-tail alternative hypothesis.
2. Determine the sample size in Problem 3 of Sec. 9.3, page 347, which will limit α to 0.05 and $\beta(59)$ to 0.20. Determine $\bar{x}^*$ for this sample size and value of α.
3. Determine the sample size required in Problem 7, page 348, that will limit α to 0.01 and $\beta(615)$ to 0.05. Determine $\bar{x}^*$ for this sample size and value of α.
4. Determine the sample size required in Problem 15, page 350, that will limit α to 0.05 and $\beta(1.85)$ to 0.10. Determine $\bar{x}_1^*$ and $\bar{x}_2^*$ for this sample size and value of α.
5. Determine the sample size required in Problem 3 of Sec. 9.5, page 362, which will limit α to 0.10 and $\beta(0.75)$ to 0.05. Determine p^* for this sample size and value of α.
6. Determine the sample size required in Problem 5 of Sec. 9.5, page 362, which will limit α to 0.05 and $\beta(0.20)$ to 0.05. Determine p^* for this sample size and value of α.

9.7 *Testing Hypotheses Concerning the Difference of Two Means (Variance Known)*

We often want to compare the means of two different populations rather than comparing a single population mean against a specific value. However, in order to test hypotheses, we must again follow the general steps for testing hypotheses as set forth in Sections 9.1 and 9.4. We then combine the theorems of Section 6.10 on linear combinations of normal variables with what we learned in Section 9.3 about testing hypotheses and establish tests concerning the difference of two means.

First we must determine the pivotal quantity applicable when testing hypotheses on the difference between the means of two populations. Recall from Section 6.10 that if we independently select samples of size n_1 (≥ 30) from a population (variable X_1) having mean μ_1 and variance σ_1^2 and size $n_2 (\geq 30)$ from a population (variable X_2) having mean μ_2 and variance σ_2^2, the statistic $(\bar{X}_1 - \bar{X}_2)$ has a normal distribution:

$$(\bar{X}_1 - \bar{X}_2) \sim \mathrm{N}\left(\mu_1 - \mu_2, \frac{\sigma_1^2}{n_1} + \frac{\sigma_2^2}{n_2}\right). \tag{1}$$

If we denote $\bar{X}_1 - \bar{X}_2$ as a single variable D, then we have

$$D = \bar{X}_1 - \bar{X}_2,$$

$$\mu_d = \mu_1 - \mu_2,$$

$$\sigma_d^2 = \frac{\sigma_1^2}{n_1} + \frac{\sigma_2^2}{n_2},$$

and

$$D \sim N(\mu_d, \sigma_d^2).$$

Note that the statistic D is conceptually no different from the sample mean $\bar{X}$ that we used for testing hypotheses about μ_x in Section 9.3. In both instances, we simply want a statistic to test hypotheses about a population mean. For the difference of two means, we must exert a little more effort to determine the observed value of the random variable and its mean and variance. Recalling, from Section 6.5, that any normal entity can be transformed to a standard normal variable, we have immediately the relevant pivotal quantity for testing hypotheses concerning the difference between the means $(\mu_1 - \mu_2)$ of two populations,

$$Z = \frac{(\bar{X}_1 - \bar{X}_2) - (\mu_1 - \mu_2)}{\sqrt{(\sigma_1^2/n_1) + (\sigma_2^2/n_2)}} \sim N(0,1). \tag{2}$$

Note that, if the original distributions (of X_1 and X_2) from which we are sampling are normal, (1) and (2) hold even if we relax the requirements that n_1 and n_2 both be large (≥ 30).

Thus, the statement of the hypotheses for the difference between the means of two populations will be stated in terms of $(\mu_1 - \mu_2)$ rather than a single parameter μ_x, and our observed value of Z will be computed using the observed values of $\bar{X}_1$ and $\bar{X}_2$ as

$$z = \frac{(\bar{x}_1 - \bar{x}_2) - (\mu_1 - \mu_2)_0}{\sqrt{(\sigma_1^2/n_1) + (\sigma_2^2/n_2)}}, \tag{3}$$

where $(\mu_1 - \mu_2)_0$ here represents the value of $(\mu_1 - \mu_2)$ as given by the null hypothesis.

Example 1

The officials of a steel company that has developed a new alloy for making steel rods believe that the new rods are stronger than their old standard steel rods. They decide to conduct an experiment to determine whether the rods made from the new alloy are significantly stronger. For the experiment a sample of size $n_1 = 49$ is selected from the new alloy rods and a sample of size $n_2 = 64$ from the steel rods. Breaking strengths of all rods are recorded. The standard deviations for the breaking strengths are known to be $\sigma_1 = 21$ pounds and $\sigma_2 = 24$ pounds for the alloy rods and steel rods, respectively. The results of the experiment reveal that the average breaking strengths for the two samples are $\bar{x}_1 = 318$ pounds and $\bar{x}_2 = 314$ pounds. Using the 5 percent significance level, determine whether the breaking strength of the alloy rods is significantly greater than that of the steel rods.

Solution

We wish to test

$$H_0: \mu_1 - \mu_2 = 0$$

versus the right-tail alternative

$$H_1: \mu_1 - \mu_2 > 0.$$

As always, we find a critical value z^* by setting

$$P(\text{Type I error}) = \alpha = 0.05.$$

That is, we find z^* such that

$$P(Z > z^*) = 0.05.$$

We find $z^* = 1.645$. Thus, our decision rule for a 5 percent significance level is that of accepting H_0 if the observed value of Z is less than 1.645 and rejecting H_0 if the observed value of Z is greater than 1.645.

We compute the observed value of Z by Formula (3). Thus the observed value of Z is

$$z = \frac{(\bar{x}_1 - \bar{x}_2) - (\mu_1 - \mu_2)_0}{\sqrt{(\sigma_1^2/n_1) + (\sigma_2^2/n_2)}} = \frac{(318 - 314) - 0}{\sqrt{[(21)^2/49] + [(24)^2/64]}}$$

$$\cong \frac{(318 - 314) - 0}{4.24} \cong 0.94.$$

Finally, since $z = 0.94$ is less than $z^* = 1.645$, we accept H_0 and conclude that the breaking strength of the alloy rods is not significantly stronger than that of the steel rods. □

Note that the alternative hypothesis in Example 1 would have to be changed to a left-tail alternative if either of two things is changed. If the problem is to determine whether the breaking strength of the alloy rods is significantly less than that of steel rods, then the appropriate alternative hypothesis, of course, has to be made a left-tail alternative. By the same token, if we reverse the population identifications in Example 1 and still wish to guard against the same type of error, it becomes necessary to state the alternative hypothesis as a left-tail alternative. The first change, in fact, alters the problem from that given in the example. In other words, although the only apparent change is in the alternative hypothesis, we are, under such a change, deciding to protect against a different kind of error.

Example 2

A company that considers building a new plant in one of two cities in Germany is studying the cost of rent for employees in each city. It wishes to know whether there is a significant difference between the average rents for the two cities. A random sample of size 100 is selected from each city, with the sample averages of rent being $\bar{x}_1 = \$86$ and $\bar{x}_2 = \$93$. The standard deviation of rent is known to be $\sigma_1 = \$10$ and $\sigma_2 = \$12$ for the respective cities. Using the 5 percent significance level, test to see whether there is a significant difference between the rental prices of employee housing for the two cities.

Solution

Since we are interested in determining only whether there is a significant difference, we test

$$H_0\colon \mu_1 - \mu_2 = 0$$

versus the two-tail alternative

$$H_1: \mu_1 - \mu_2 \neq 0.$$

In this case, we determine two critical values, z_1^* and z_2^*. Since we still require that

$$P(\text{Type I error}) = \alpha = 0.05,$$

we determine a symmetrical test by finding z_1^* and z_2^* such that

$$P(Z < z_1^*) = \alpha/2 = 0.025 \quad \text{and} \quad P(Z > z_2^*) = \alpha/2 = 0.025.$$

From Table A-1, we have $z_1^* = -1.96$ and $z_2^* = 1.96$. Thus, the decision rule for a 5 percent significance level is that of accepting H_0 if the observed value of Z is between -1.96 and 1.96 and rejecting H_0 if the observed value of Z is not between -1.96 and 1.96.

The observed value of Z is computed to be

$$z = \frac{(\bar{x}_1 - \bar{x}_2) - (\mu_1 - \mu_2)_0}{\sqrt{(\sigma_1^2/n_1) + (\sigma_2^2/n_2)}} = \frac{(86 - 93) - 0}{\sqrt{[(10)^2/100] + [(12)^2/100]}}$$

$$\cong \frac{(86 - 93) - 0}{1.56} \cong -4.49.$$

Thus, we reject H_0, since $z = -4.49$ is less than $z_1^* = -1.96$ and hence is not between -1.96 and 1.96. The company may conclude that there is a significant difference between the rental prices for the two cities. □

Example 3

A publisher of educational teaching aids has available two different types of aids for teaching statistics. The board of regents of a statewide university system wishes to know whether one is better than the other. The members know that both aids are currently in use at several large universities in the state. To determine if there is a significant difference between grades of students using the two teaching aids, a random sample is selected from each of the two groups of students, one using aid 1 and another using aid 2. From the group using aid 1, a sample of size $n_1 = 80$ is taken with an average exam grade of $\bar{x}_1 = 68$ points, and a sample of size $n_2 = 120$ is taken from the group using aid 2 with a resulting sample mean of $\bar{x}_2 = 65$ points. The standard deviations are known to be $\sigma_1 = 8$ points and $\sigma_2 = 12$ points. Determine, at the 2 percent significance level, whether there is a significant difference between the average grades for students using aid 1 and students using aid 2.

Solution

Since we wish to determine if there is a significant difference between the average grades for the two aids, we must test

$$H_0: \mu_1 - \mu_2 = 0$$

versus the two-tail alternative

$$H_1: \mu_1 - \mu_2 \neq 0.$$

Again we must determine two critical values z_1^* and z_2^*. However, for this problem, we require that

$$P(\text{Type I error}) = \alpha = 0.02.$$

Thus, we determine a symmetrical test by finding z_1^* and z_2^* such that

$$P(Z < z_1^*) = \alpha/2 = 0.01 \quad \text{and} \quad P(Z > z_2^*) = \alpha/2 = 0.01. \tag{4}$$

From Table A-1, we see that $z_1^* = -2.33$ and $z_2^* = 2.33$. Thus, our decision rule for the 2 percent signficant level is that of accepting H_0 if the observed value of Z is between -2.33 and 2.33 and rejecting H_0 if the observed value of Z is not between -2.33 and 2.33.

From the experiment, we obtain the observed value of Z as

$$z = \frac{(\bar{x}_1 - \bar{x}_2) - (\mu_1 - \mu_2)_0}{\sqrt{(\sigma_1^2/n_1) + (\sigma_2^2/n_2)}} = \frac{(68 - 65) - 0}{\sqrt{[(8)^2/80] + [(12)^2/120]}}$$

$$\cong \frac{(68 - 65) - 0}{1.41} \cong 2.13.$$

Since 2.13 is between -2.33 and 2.33, we accept H_0. The board of regents may conclude that, at the 2 percent significance level, there is no significant difference between the results from the two different teaching aids. Note that we have assumed in making this conclusion that the only factor influencing the results is that of teaching aid used. Obviously, there may well be other factors. Thus, we must be careful in drawing conclusions of the above nature. The only really absolute statement permitted by the above test is that there is no reason to believe that there is a significant difference between grades for the two groups of students. □

Problems

1. Suppose we wish to test

$$H_0: \mu_1 - \mu_2 = 0 \quad \text{versus} \quad H_1: \mu_1 - \mu_2 < 0.$$

The populations from which we shall sample are both normally distributed and have respective standard deviations of $\sigma_1 = 15$ and $\sigma_2 = 20$. Suppose simple random samples of sizes $n_1 = 25$ and $n_2 = 16$ are selected from the respective populations and sample means of $\bar{x}_1 = 107$ and $\bar{x}_2 = 118$ are obtained. We wish to operate at a 5% significance level.

a. Using the z^* method, sketch the sampling distribution of Z, indicate the acceptance and rejection regions, and determine whether μ_1 is significantly less than μ_2.

b. Using the α' method, state the decision rule and determine whether μ_1 is significantly less than μ_2.

c. Using the z^* method, illustrate each of the six general steps in developing the one-tail, 5% significance test of the hypothesis concerning $\mu_1 - \mu_2$ that was used in Part a. Explain how the pivotal quantity you use satisfies the three requirements.

d. Determine $\beta(-20)$.

e. For the situation in which we wish to control the Type II error so that $\beta(-20) = 0.10$, determine z^* and α.

f. Why is it necessary that the two populations each be normally distributed for this problem?

2. Two different processes are used to produce automobile tires. Process 1 is the standard method, while process 2 is a new method that is believed to produce tires having a longer average life. The standard deviations of the tire life for the two processes are known to be $\sigma_1 = 1{,}600$ miles and $\sigma_2 = 2{,}400$ miles. Simple random samples of sizes $n_1 = 64$ and $n_2 = 64$ are selected, and the sample means are determined to be $\bar{x}_1 = 18{,}250$ miles and $\bar{x}_2 = 19{,}400$ miles.

   a. Using any of the test methods and $\alpha = 0.01$, state the hypotheses, sketch the associated sampling distribution, indicate the acceptance and rejection regions, and determine whether the average life of the tires made by process 1 is significantly less than the average life of the tires made by process 2. Explain your answer in terms of the problem as well as whether you accept the null hypothesis.

   b. Why is it not necessary that each of the two populations be normally distributed for this problem?

   c. Determine $\beta(-2{,}000)$.

   d. Assuming we wish to control the Type II error so that $\beta(-2{,}000) = 0.10$, determine z^* and α.

3. Suppose we wish to test

$$H_0\colon \mu_1 - \mu_2 = 0 \quad \text{versus} \quad H_1\colon \mu_1 - \mu_2 > 0.$$

   The standard deviations of the two populations are known to be $\sigma_1 = 9$ and $\sigma_2 = 9$. Suppose simple random samples of sizes $n_1 = 81$ and $n_2 = 81$ are selected and sample means of $\bar{x}_1 = 75.2$ and $\bar{x}_2 = 74.1$ are obtained. We wish to operate at the 10% significance level.

   a. Using the z^* method, sketch the sampling distribution of Z, indicate the acceptance and rejection regions, and determine whether μ_1 is significantly greater than μ_2.

   b. Using the α' method, state the decision rule and determine whether μ_1 is significantly greater than μ_2.

   c. Determine $\beta(4.0)$.

   d. Assuming we wish to control the Type II error so that $\beta(4.0) = 0.05$, determine z^* and α.

   e. Why is it not necessary that the two populations be normally distributed for this problem?

4. Use the data of Problem 2, except interchange the identification of the two processes. Thus, the associated values become $n_1 = 64$, $n_2 = 64$, $\sigma_1 = 2{,}400$, $\sigma_2 = 1{,}600$, $\bar{x}_1 = 19{,}400$, and $\bar{x}_2 = 18{,}250$.

   a. State the hypotheses for determining whether μ_1 is significantly greater than μ_2.

   b. Using any of the test methods and $\alpha = 0.01$, sketch the associated sampling distribution, indicate the acceptance and rejection regions, and determine whether the average life of the tires made by (new) process 1 is significantly greater than the average life of the tires made by (new) process 2.

   c. Determine $\beta(2{,}000)$.

   d. Assuming we wish to control the Type II error so that $\beta(2{,}000) = 0.10$, determine z^* and α.

5. A new manufacturer of electronic resistors claims that his 5-watt resistors have a greater current-carrying capacity than those of the same rating made by his competitors. A large purchaser of resistors decides to test the claim that the new brand is *significantly* better than the brand he is presently using. He selects simple random samples of size $n_1 = 100$ from the new brand and $n_2 = 225$ from his present brand. The standard deviations are known to be $\sigma_1 = 0.5$ and $\sigma_2 = 0.8$, respectively, and each population is assumed to be normally distributed. The sample means were found to be $\bar{x}_1 = 5.3$ and $\bar{x}_2 = 5.0$. Use $\alpha = 0.10$.

a. State the hypotheses for determining whether the new brand has a significantly greater current-carrying capacity than the present brand.

b. Using the z^* method, sketch the sampling distribution of Z, indicate the acceptance and rejection regions, and determine whether the current-carrying capacity of the new brand is significantly greater than that of the present brand. Explain your conclusion in terms of the problem as well as whether you accept the null hypothesis.

c. Determine $\beta(0.30)$.

d. Assuming we wish to control the Type II error so that $\beta(0.30) = 0.05$, determine z^* and α.

e. Is the assumption that each population is normally distributed necessary? Why?

6. Suppose we wish to test

$$H_0: \mu_1 - \mu_2 = 0 \quad \text{versus} \quad H_1: \mu_1 - \mu_2 \neq 0.$$

The populations from which we shall sample have known standard deviations of $\sigma_1 = 20$ and $\sigma_2 = 32$. Suppose simple random samples of sizes $n_1 = 49$ and $n_2 = 64$ are selected, and sample means of $\bar{x}_1 = 494$ and $\bar{x}_2 = 480$ are obtained. We wish to operate at the 1% significance level.

a. Using the z^* method, sketch the sampling distribution of Z, indicate the acceptance and rejection regions, and determine whether μ_1 is significantly different than μ_2.

b. Using the α' method, state the decision rule and determine whether μ_1 is significantly different than μ_2.

c. Using the z^* method, illustrate each of the five general steps in developing the two-tail, 1% significance test of the hypothesis concerning $\mu_1 - \mu_2$ that was used in Part a. Explain how the pivotal quantity that you use satisfies the three requirements.

d. Determine $\beta(25)$.

e. Why is it not necessary that each population be normally distributed for this problem?

7. Use the data of Problem 5. However, change the objective to that of determining whether the current-carrying capacity of the new brand of resistors is significantly *different* than the current-carrying capacity of the present brand. Use $\alpha = 0.10$.

a. State the hypotheses.

b. Using the z^* method, sketch the sampling distribution of Z, indicate the acceptance and rejection regions, and determine whether the current-carrying capacity of the new brand is significantly different than that of the present brand. Explain your conclusions in terms of the problem as well as whether you accept the null hypothesis.

c. Determine $\beta(0.30)$.

9.8 Statement of Alternative Hypotheses for One-Tail Tests

Occasionally, a decision maker is interested in testing a null hypothesis against a one-tail alternative hypothesis, but is puzzled about which alternative to use. For example, suppose the management of an investment fund wishes to establish a sales force within a specific area. However, the management knows that the average monthly income per family within the area should be about \$800. In other words, the firm knows that the salesmen will be successful if the average income is significantly greater than \$800 and they will be unsuccessful if the average is significantly less than \$800. Let us consider two different situations. First, suppose the management is aggressive and wishes to establish a sales force unless strong evidence indicates an unsuccessful venture. It particularly wishes to avoid the error of not establishing a sales force when it should, or the error of saying $\mu_x <$ \$800 when μ_x is actually \$800 or greater. Since we always construct hypotheses so that the error we try to avoid corresponds to Type I error, we construct our hypotheses as

$$H_0: \mu_x = \$800 \quad \text{versus} \quad H_1: \mu_x < \$800.$$

Thus, the management will establish the sales force if it accepts H_0. This is exhibited in Figure 9.13a.

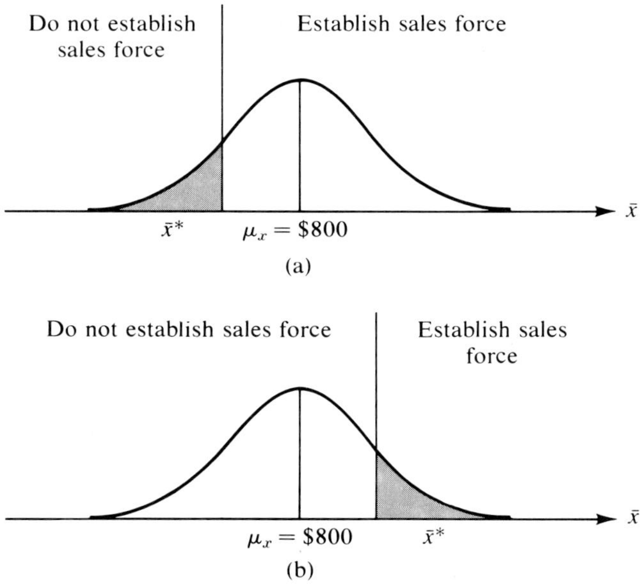

Figure 9.13

Decision rules for different alternative hypotheses

Next, suppose that the management is conservative and thus does not wish to establish a sales force in the area unless it is reasonably assured of a successful venture. That is, it wants to avoid the error of establishing the sales force when it should not, or it wishes to avoid saying that $\mu_x >$ \$800 when μ_x is actually \$800 or less. In this case, the hypotheses are constructed as

$$H_0: \mu_x = \$800 \quad \text{versus} \quad H_1: \mu_x > \$800.$$

The management will establish the sales force if it rejects H_0. The situation is shown in Figure 9.13b.

In both situations, management must ultimately make the decision regarding the appropriate alternative hypothesis for a specific problem. In general, this is simply a matter of (1) deciding which error in a given situation is the more important to avoid, and (2) constructing the hypotheses so that this specific error becomes Type I error.

Problems

1. Suppose a firm wishes to market its product in a certain area if over 20% of the families in the area express an interest in the product. The company will conduct a statistical survey of families from the area.

   a. Determine what hypotheses the firm should use if it wants to avoid the error of marketing the product when a satisfactory market does not exist.

   b. For the hypotheses in Part a, determine the decision rule, sketch the associated sampling distribution, and indicate the acceptance and rejection regions for $n = 100$ and $\alpha = 0.05$.

   c. Determine what hypotheses the firm should use in order to avoid the error of not marketing the product when a satisfactory market exists.

   d. For the hypotheses in Part c, determine the decision rule, sketch the associated sampling distribution, and indicate the acceptance and rejection regions for $n = 100$ and $\alpha = 0.10$.

   e. For the decision rule of Part b, determine β for $\pi = 0.04, 0.08, 0.12, 0.16, 0.20, 0.24, 0.28, 0.32$, and 0.36. Plot the OC curves for these values.

2. A national business magazine publisher will increase its charge for advertising if 25% or more business executives read the magazine, otherwise the charge will remain unchanged.

   a. Determine the hypotheses the publisher should use if he wishes to avoid the error of not increasing the charge when the proportion of readers is greater than 25%.

   b. For the hypotheses in Part a, determine the decision rule, sketch the associated sampling distribution, and indicate the acceptance and rejection regions for $n = 75$ and $\alpha = 0.05$.

   c. Determine the hypotheses the publisher should use if he wishes to avoid the error of increasing the charge when the proportion of readers is less than 25%.

   d. For the hypotheses in Part c, determine the decision rule, sketch the associated sampling distribution, and indicate the acceptance and rejection regions for $n = 75$ and $\alpha = 0.05$.

   e. For the decision rule of Part b, determine β for $\pi = 0.10$, 0.15, 0.20, 0.25, and 0.30. Plot the OC curves for these values.

3. A savings and loan company will establish a branch in a certain city if the average annual income per family is above \$8,000, but not if the average annual income per family is below \$8,000. It is known that the standard deviation of annual income per family is \$1,000.

   a. Determine what hypotheses the company should use in order to avoid the error of establishing the branch when the average income is below \$8,000.

   b. For the hypotheses determined in Part a, determine the decision rule, sketch the associated sampling distribution, and indicate the acceptance and rejection regions for $n = 100$ and $\alpha = 0.03$.

   c. Determine the hypotheses the company should use in order to avoid the error of not establishing the branch when the average income is above \$8,000.

   d. For the hypotheses determined in Part c, determine the decision rule, sketch the associated sampling distribution, and indicate the acceptance and rejection regions for $n = 100$ and $\alpha = 0.03$.

   e. For the decision rule of Part b, determine β for $\mu_x = \$7,600$; \$7,700; \$7,800; \$7,900; \$8,000; \$8,100; \$8,200; \$8,300; and \$8,400. Plot the OC curves for these values.

Important Terms and Concepts

acceptance region
accepting the null hypothesis
alternative hypothesis
critical region
critical value
decision rule
null hypothesis
one-tail test
operating characteristic curve
power function
probability of Type I error
probability of Type II error
rejecting the null hypothesis
rejection region
significance test
statistical hypothesis
symmetrical test
test statistic
tests of hypotheses
two-tail test
Type I error
Type II error

Review Problems

1. Suppose we select a simple random sample of size 4 from a *normal* population known to have a standard deviation of 20. We know that the population mean is either 63 or 80, and we wish to test

$$H_0: \mu_x = 80 \quad \text{versus} \quad H_1: \mu_x = 63.$$

For our decision rule let us conclude that (1) $\mu_x = 80$ if $\bar{x} > 70$ or (2) $\mu_x = 63$ if $\bar{x} \leq 70$.

a. Determine α and β.

b. Explain the significance of the assumption that the population is *normally* distributed.

c. Determine the value associated with the decision rule which makes $\alpha = 0.08$ and, for this value, determine β.

d. Determine the value associated with the decision rule which makes $\beta = 0.01$ and, for this value, determine α.

2. A machine automatically drills holes in parts at a preset angle. Regardless of the angle for which the machine is set, the standard deviation of the angle of the holes is 1°. Suppose the machine is set to drill holes at an angle of 15°. A simple random sample of 64 parts is selected periodically and the average angle of the holes is calculated to determine whether the machine is operating properly. The machine is considered to be out of control if the average angle is significantly different from 15°. On the most recent sample, a sample mean of $\bar{x} = 15.25°$ was obtained.

a. State the null and alternative hypotheses to be used in determining whether the machine is out of control.

b. Using any available method for testing hypotheses and $\alpha = 0.08$, sketch the sampling distribution of the appropriate statistic, indicate the acceptance and rejection regions, and determine whether the average angle of the holes which the machine is drilling is significantly different from 15°. Also, state your conclusions in terms of the problem as well as whether you accept or reject the null hypothesis.

3. Use the data of Problem 2. However, suppose it is more serious if the machine goes out of control on the high-angle side (angles greater than 15°) so that a 2% risk is acceptable for the low angles ($\alpha_1 = 0.02$ is the risk in the left tail) and a 6% risk is required for the high angles ($\alpha_2 = 0.06$ is the risk in the right tail).

a. State the appropriate null and alternative hypotheses to use in determining whether the machine is out of control.

b. Using any available method for testing hypotheses, $\alpha_1 = 0.02$, and $\alpha_2 = 0.06$, sketch the sampling distribution of the appropriate statistic, indicate the acceptance and rejection regions, and determine whether the average angle of the holes which the machine is drilling is significantly different from 15°. Also, state your conclusions in terms of the problem as well as whether you accept or reject the null hypothesis.

4. It is known that 10 years ago 90% of American families had an annual income within a certain range. It is desired to determine whether there has been a significant change in the proportion of families having annual income within the specified range. A sample of size $n = 625$ families is selected, and 525 families are found to have an annual income within the specified range. Use $\alpha = 0.05$.

a. Using any of the test methods, sketch the associated sampling distribution, indicate the acceptance and rejection regions, and determine whether there has been a significant change in the proportion of families having an annual income in the specified range. Explain your conclusion in terms of the problem as well as whether you accept or reject the null hypothesis.

b. Determine β for $\pi = 0.85$.

5. A chemist is considering two different adhesives for a certain manufacturing process. It is known that the standard deviations of the breaking strength of the adhesives are $\sigma_1 = 5$ lb and $\sigma_2 = 9$ lb. Simple random samples of sizes $n_1 = 50$ and $n_2 = 81$ are selected from the two groups of items made by the process using each adhesive. Sample means of $\bar{x}_1 = 102.1$ lb and $\bar{x}_2 = 104.4$ lb were obtained. Use $\alpha = 0.04$.

a. State the hypotheses for determining whether the average breaking strengths associated with the two adhesives are significantly different.

b. Using any of the test methods, sketch the associated sampling distribution, indicate the acceptance and rejection regions, and determine whether the average breaking strength of the items made with adhesive 1 is significantly different than the average breaking strength of those made with adhesive 2.

c. Determine $\beta(-5.0)$.

6. An insurance company is considering increasing the premiums for fire insurance policies within a certain city in which there have been 1,000 fires during the past year. The company wishes to increase the premiums if the average loss per fire is over \$1,500; however, it wants most to avoid the error of increasing the premiums when it shouldn't. Rather than evaluate the loss for each of the 1,000 fires, a simple random sample of size $n = 225$ is selected, and a sample mean of $\bar{x} = \$1,520$ is obtained. The standard deviation of the loss per fire for the 1,000 fires is known to be \$300. The company wishes to operate at the 10% significance level.

a. Determine the hypotheses the company should use to determine whether it should increase premiums.

b. Using the $\bar{x}^*$ method and the hypotheses of Part a, sketch the associated sampling distribution, indicate the acceptance and rejection regions, and determine whether the company should increase premiums.

c. Determine $\beta(\$1,600)$ for the hypotheses of Part a.

d. Develop a formula similar to the one given in Section 8.5 for determining the sample size required in order to estimate μ_x within an error of e with a risk of α when sampling from a *finite* population.

e. Use the formula developed in Part d to determine the sample size required for estimating μ_x within an error of \$40 with a risk of 5%.

f. Develop the formula for a $100\alpha\%$ confidence interval about μ_x when sampling from a *finite* population.

g. Determine the 95% confidence interval about μ_x using the formula obtained in Part f.

(Sections 4.6, 7.3, 7.4, 8.5, and 8.6)

7. A candidate for a political office of a certain city wishes to know whether the majority of voters favors a certain project. A simple random sample of size $n = 400$ is selected from the city's 2,000 registered voters, and 188 are found to be in favor of the project. Use $\alpha = 0.05$.

   a. Determine the hypotheses the candidate should use if he wishes most to avoid the error of supporting the project when he shouldn't.

   b. Using the p^* method and the hypotheses of Part a, sketch the associated sampling distribution, indicate the acceptance and rejection regions, and determine whether the candidate should support the project.

   c. Determine $\beta(0.55)$ for the hypotheses of Part a.

   d. Determine the hypotheses the candidate should use if he wishes most to avoid the error of not supporting the project when he should.

   e. Using the p^* method and the hypotheses of Part d, sketch the associated sampling distribution, indicate the acceptance and rejection regions, and determine whether the candidate should support the project.

   f. Determine $\beta(0.45)$ for the hypotheses of Part d.

   g. Develop a formula similar to the one given in Section 8.5 for determining the sample size required in order to estimate π within an error of e with a risk of α when sampling from a *finite* population.

   h. Use the formula developed in Part g to determine the sample size required for estimating π within an error of 0.05 and with a risk of 10 percent.

   i. Why are we unable to estimate π by using a confidence interval estimator similar to the one used for obtaining a confidence interval estimate for μ_x?

   (Sections 4.6, 7.3, 7.7, 8.5, and 8.6)

8. A machine that places sugar in a ten-pound bag is deemed in control if the average amount of sugar in the bags is 10 pounds, and out of control otherwise. The weights of sugar placed in the bags are normally distributed with a known standard deviation of 0.4 lb. In order to determine whether the machine is out of control, a sample of size $n = 16$ bags was selected and the weights are given below. Use $\alpha = 0.05$

10.9	10.3	9.8	10.6
9.7	9.8	10.4	10.4
9.9	9.6	10.3	10.2
10.1	9.9	10.2	10.3

   a. State the hypotheses to be used in determining whether the machine is out of control.

   b. Using the $\bar{x}^*$ method, sketch the associated sampling distribution, indicate the acceptance and rejection regions, and determine whether the machine is out of control.

   c. Determine the sample variance, sample standard deviation, and sample range.

   d. Determine the sample size required for estimating μ_x within an error of 0.05 lb with a risk of 1%.

   e. Determine $\beta(10.5)$.

   f. Determine the 95% confidence interval about μ_x.

   g. Determine the sample size required for a 95% confidence interval about μ_x of length 0.1.

   h. Is it legitimate to test the above hyptheses and determine the confidence interval for a sample of size 16? Why?

   (Sections 3.3, 3.4, 3.6, 6.10, 7.4, 8.5, and 8.6)

9. An inspector knows that he has a large group of firecrackers of which either 70% are defective and 30% are nondefective, or 40% are defective and 60% are nondefective. To determine whether a firecracker is defective requires a test which destroys it. The inspector wants to test the following hypotheses to

determine which of the two possible populations he has:

$$H_0: \pi = 0.7 \quad \text{versus} \quad H_1: \pi = 0.4.$$

His decision rule is to accept H_0 if 2 or more of the firecrackers from a sample of size 4 are defective and reject H_0 otherwise.

a. Using the binomial distribution, determine α and β for this decision rule.

b. Why is it inappropriate to use the normal distribution for this problem?

c. Would it be inappropriate to use the hypergeometric distribution for this problem? Why?

d. Why would you use the binomial distribution for this problem?

(Sections 5.2, 5.4, 6.8, and 7.7)

10. Suppose the inspector of Problem 9 has a group of 10 firecrackers of which either 7 are defective and 3 are nondefective, or 4 are defective and 6 are nondefective.

a. Using the hypergeometric distribution, determine α and β for the hypotheses and decision rule given in Problem 9.

b. Why is it inappropriate to use the normal distribution for this problem?

c. Would it be inappropriate to use the binomial distribution for this problem? Why?

d. Why would you use the hypergeometric distribution rather than the binomial distribution?

(Sections 4.7, 5.2, 5.4, 6.8, and 7.7)

11. A manufacturer of metal gears knows that the number of defects per gear follows a Poisson distribution. When the process producing the gears is in control, the average number of defects is 3. However, when the process is out of control, it usually produces gears with an average of 6 defects. To determine whether the process is in control, the manufacturer tests the following hypotheses:

$$H_0: \lambda = 3 \quad \text{versus} \quad H_1: \lambda = 6.$$

The decision rule the manufacturer uses is to accept H_0 if a randomly selected gear has 5 or less defects and reject H_0 otherwise.

a. Using the Poisson distribution, determine α and β.

b. Repeat Part a when the decision rule is to accept H_0 if 2 randomly selected gears have a total of 10 or less defects and reject H_0 otherwise,

(Section 5.3)

12. The relative frequency distribution for an infinite population from which we wish to sample is given below.

X	Relative frequency: f
60	0.30
61	0.25
62	0.15
63	0.10
64	0.10
65	0.10
	1.00

A simple random sample of size 40 is selected from a population believed to be the one corresponding to the above relative frequency distribution. The observed values for the sample are given below.

65	60	62	60
61	61	63	65
63	60	65	64
60	65	64	64
60	63	60	63
61	63	61	62
62	62	60	61
62	60	62	60
60	61	61	64
61	64	60	65

a. Determine the mean, variance, and standard deviation for the variable given in the relative frequency distribution.

b. Determine the mean, variance, and standard deviation of the values in the sample.

c. Determine $P(X > 62.4)$, $P(X < 63.8)$, and $P(60.2 < X < 64.1)$, where X is the variable given in the relative frequency distribution. Explain why you computed the probabilities in the manner that you did.

d. The variance for the population from which the sample of 40 values was selected is known to be the same as that for the given relative frequency distribution. However, it is suspected that the mean of this population is *different* from the mean of the relative frequency distribution. Using the $\bar{x}^*$ method and $\alpha = 0.03$, state the hypotheses to be used in determining whether the mean of the population from which the sample was taken is significantly different from the mean for the given relative frequency distribution, sketch the associated sampling distribution, indicate the acceptance and rejection regions, and determine whether the mean of the population is significantly different from the mean of the relative frequency distribution.

e. Determine the 95% confidence interval for the mean of the population from which the sample was taken. Note whether this interval contains the mean for the relative frequency distribution. Does this appear to agree with the conclusion reached in Part d? Why?

f. Determine $\beta(64)$.

(Sections 2.3, 3.7, 3.8, 4.8, 4.11, 7.4, and 8.6)

13. Two groups of assemblers perform the same task by different methods. The distributions of assembly time required for one unit are believed to be normal for each group, and the standard deviations are known to be $\sigma_1 = 3$ minutes and $\sigma_2 = 4$ minutes. The manufacturing manager wishes to determine whether there is a significant difference between the average times required for the two methods. Simple random samples of sizes $n_1 = 25$ and $n_2 = 25$ are selected from the two groups and the sample means are determined to be $\bar{x}_1 = 31$ minutes and $\bar{x}_2 = 33$ minutes. Use $\alpha = 0.01$.

a. State the appropriate hypotheses for determining whether there is a significant difference between the average times for the two methods.

b. Instead of using the z^* method of testing hypotheses, follow the same general steps used for the $\bar{x}^*$ method and develop an $(\bar{x}_1 - \bar{x}_2)^*$ method of testing hypotheses concerning the difference between two means, sketch the sampling distribution of $\bar{X}_1 - \bar{X}_2$, indicate the acceptance and rejection regions for this distribution, and determine whether there is a significant difference between the average times for the two methods.

c. Determine the 99% confidence interval for $\mu_1 - \mu_2$. Note whether the difference between the sample means is within the limits of the confidence interval. Does this appear to agree with the conclusion reached in Part b?

d. Determine β for $\mu_1 - \mu_2 = 3$.

(Sections 7.4 and 8.6)

Ten

The Chi-Square Distribution

10.1 Introduction

In the past four chapters, we have been primarily concerned with using the standard normal distribution—mainly because we needed to make probability statements about the sample mean, set confidence intervals, and test hypotheses about the sample mean when the variance of the population is known. Primarily because of the central limit theorem, we have used the sample mean as our basic sample statistic.

Now, many times, we wish to make probability statements about a statistic, construct confidence intervals, and test hypotheses concerning a parameter by using a statistic for which we must know the sampling distribution. Generally, when we must construct a confidence interval for or test a hypothesis about an unknown parameter we must find an appropriate pivotal quantity; and a primary requirement for such an entity is that we must know its distribution.

There are many reasons why the chi-square (χ^2) distribution has become one of the most important in statistics. One that we shall only mention in passing is the fact that, as with the normal distribution, the probability distributions of some phenomena simply have a tendency to exhibit characteristics similar to those of a chi-square distribution. More important to us are the applications of the chi-square distribution in

1. setting confidence intervals and testing hypotheses for the variance of a normal population,
2. conducting goodness-of-fit tests, and
3. conducting tests of independence.

First, let us investigate the meaning and properties of the chi-square distribution.

Definition 10.1

Given a sequence of k independent random variables $Z_1, Z_2, \ldots, Z_k$ such that each is normally distributed with mean zero and variance one, we define the chi-square *variable with k degrees of freedom as*

$$U = Z_1^2 + Z_2^2 + \cdots + Z_k^2,$$

and write

$$U \sim \chi_k^2,$$

where χ_k^2 denotes a chi-square variable with k degrees of freedom.

Although we refer to χ_k^2 as a variable, it is indeed a statistic, and the corresponding distribution is a sampling distribution. Note that $Z_1^2 + Z_2^2 + \cdots + Z_k^2$ is the sum of squares of k independent random variables, not the sum of squares of k sample observations. Since we are unaccustomed to the arithmetic manipulation of variables, it is difficult to interpret the physical meaning of the distribution of such an entity. Strictly speaking, the definition says that the distribution of U is obtained by finding the sampling distribution of the sum of squares of $Z_1, Z_2, \ldots, Z_k$. However, we can only describe the resulting distribution indirectly—by calling on the basic idea of infinite repeated sampling as it relates to the concept of a sampling distribution.

Suppose we have k distinct and independent standard normal distributions for $Z_1, Z_2, \ldots, Z_k$, and we repeatedly select a sample of k items, so that one is drawn from each independent distribution. Let z_{ij} represent a single observation drawn from the jth distribution in the ith sample. Conceptually, a chi-square distribution with k degrees of freedom would result if we were able to repeat the following process.

1. Take one observation from each of the k independent standard normal distributions: $z_{i1}, z_{i2}, \ldots, z_{ik}$.
2. Square each observation and compute a single observation from a chi-square distribution as

$$U_i = Z_{i1}^2 + Z_{i2}^2 + \cdots + Z_{ik}^2.$$

3. Repeat steps 1 and 2 for an infinite number of samples, that is, for $i = 1, 2, \ldots, \infty$.
4. Compile the probability distribution of the U_i's ($i = 1, 2, \ldots, \infty$). The result will be the probability distribution of U, a chi-square variable with k degrees of freedom.

In other words, a *chi-square variable with k degrees of freedom* can be defined as *the sum of squares of k independent standard normal variables.*

Now, what do we mean by *degrees of freedom*? A rather strict interpretation is that the number of degrees of freedom associated with a chi-square variable is the number of independent (standard normal) random variables that conceptually go into the makeup of the variable. For a more intuitive understanding of the term, let's compare two ways of estimating the variance of a population by taking a sample of size n—first when we know the value of the population mean μ_x, and then when we do not know μ_x.

In the first instance, we estimate the variance by $\sum(X_i - \mu_x)^2/n$. Here, the n terms $X_i - \mu_x$ are all independent; hence, each makes an independent contribution to the estimation of σ_x^2. Thus, we do not lose any degrees of freedom in estimating σ_x^2.

On the other hand, when we do not know μ_x, we must replace it by the sample mean $\bar{X}$ and estimate the variance by $\sum(X_i - \bar{X})^2/n$. Recall that $\sum(X_i - \bar{X}) = 0$. This automatically means that the n terms, $X_i - \bar{X}$, are not independent because, as soon as we know $n - 1$ of the terms, the value of the remaining term is fixed. This fact, resulting from our use of an estimate of μ_x rather than μ_x itself, causes us to lose a degree of freedom in our estimation of σ_x^2. Ultimately, we shall see that, in the general problem of estimation, we lose a degree of freedom for each parameter that is replaced by a sample estimate.

If we recall that any normal variable can be transformed into a standard normal variable, we see immediately that we do not necessarily have to be sampling from standard normal distributions in order to obtain a statistic having a chi-square distribution. That is, if we have a sequence of k independent normal variables X_1, $X_2, \ldots, X_k$, with means $\mu_1, \mu_2, \ldots, \mu_k$ and variances $\sigma_1^2, \sigma_2^2, \ldots, \sigma_k^2$, respectively, we have for $i = 1, 2, \ldots, k$,

$$\frac{X_i - \mu_i}{\sigma_i} \sim \mathrm{N}(0, 1),$$

and, hence, the statistic

$$\sum_{i=1}^{k} \left(\frac{X_i - \mu_i}{\sigma_i}\right)^2 \sim \chi_k^2. \qquad (1)$$

Further, recalling our discussion of Chapter 7 in which we concluded that repeated sampling from an infinite population is equivalent to the process of collecting a sample comprised of observations from independent populations, we have the following extremely important fact.

Rule 10.1

Suppose that from the population of a normal variable X with mean μ_x and variance σ_x^2 we select a sample of size n by simple random sampling. Since the values of such a sample may be viewed as observations from n independent random variables X_1, X_2, $\ldots$, X_n, we have that the statistic

$$\sum_{i=1}^{n} \left(\frac{X_i - \mu_x}{\sigma_x}\right)^2 \sim \chi_n^2.$$

Problems

1. What is meant by the number of degrees of freedom associated with a chi-square variable?
2. Explain why the following relation is strictly true only when we are sampling from an *infinite* population of a normal variable:

$$\sum_{i=1}^{n} \left(\frac{X_i - \mu_x}{\sigma_x}\right)^2 \sim \chi_n^2.$$

10.2 Properties of the Chi-Square Distribution

The probability density function of a variable U having the chi-square distribution with k degrees of freedom is

$$f(u; k) = \frac{1}{2^{k/2}\left(\frac{k}{2} - 1\right)!} u^{(k/2-1)} \exp(-u/2) \qquad (u > 0). \tag{1}$$

We note from (1) (and also from Definition 10.1) that a chi-square variable cannot assume a value less than zero. Figure 10.1 illustrates the graphical relationship of (1) for values of k of 3, 4, and 5. The derivation of the mathematical form of the chi-square distribution can be found in nearly any introductory mathematical statistics text, for example, Mood and Graybill (10.1).

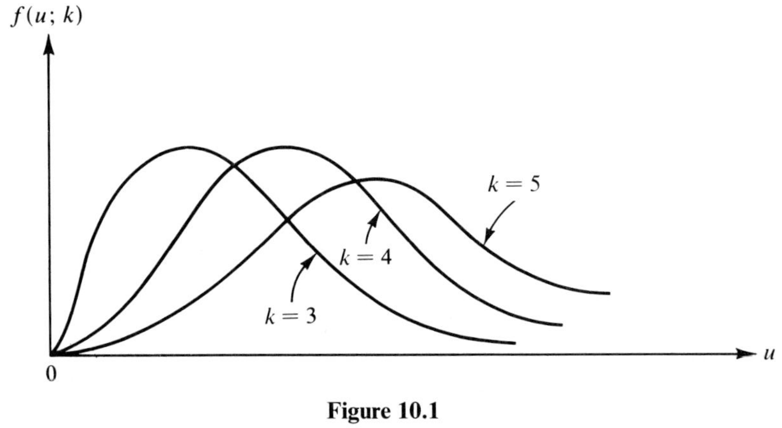

Figure 10.1

Chi-square distributions with various degrees of freedom

Recall that in the case in which we are required to determine probabilities involving any normal distribution, we are able to use the information of Table A-1 of the Appendix because of the fact that any normal variable can be transformed to a standard normal variable. Thus, one table of probabilities suffices for all problems involving normal probabilities. We are not so fortunate in the case of the chi-square distribution, for there exists no such transformation. Theoretically, there are an infinite number of chi-square distributions, one for each of the possible values which k (degrees of freedom) may assume. Practically speaking, however, we can circumvent this problem by (1) presenting less detailed probability information for the chi-square distribution than for the normal distribution (Table A-1) and (2) presenting probability information only for chi-square variables involving a not-too-large number of degrees of freedom.

Selected probability information for various chi-square distributions is given in Table A-3. Note that the table provides values for

$$P(\chi_k^2 > u_k),$$

where χ_k^2 represents the general chi-square variable with k degrees of freedom and u_k represents any value χ_k^2 can assume. Thus, we may need to solve two general types of problems involving chi-square probabilities. First, we may wish to find the probability (say α) that a chi-square variable with k degrees of freedom is greater than a specific value, say a. That is, given k and a, we may wish to find α such that

$$P(\chi_k^2 > a) = \alpha. \tag{2}$$

To answer such question, we:

1. Find k in the left column of Table A-3.
2. Find a in the same row of Table A-3 as k.
3. Read α [the probability of (2)] at the top of the column in which a appears.

Second, given k and α, we may wish to find the specific value (say a) so that

$$P(\chi_k^2 > a) = \alpha.$$

To answer this question, we:

1. Find k in the left column of Table A-3.
2. Find α in the top row of Table A-3.
3. Read a in the body of Table A-3 where the column in which α appears intersects the row in which k appears.

Example 1

Find the value so that the probability is 0.025 that a chi-square variable with $k = 11$ degrees of freedom is greater than such value; that is, find a such that

$$P(\chi_{11}^2 > a) = 0.025.$$

Solution

From the body of Table A-3, we see that $a = 21.9$. Thus,

$$P(\chi_{11}^2 > 21.9) = 0.025. \qquad \square$$

Example 2

Find the probability that a chi-square variable with $k = 15$ degrees of freedom is greater than 22.3.

Solution

From Table A-3, we have

$$P(\chi^2_{15} > 22.3) = 0.10. \quad \square$$

Example 3

Find

(a) $P(\chi^2_{16} > 9.31)$, (b) $P(\chi^2_{25} < 29.3)$, (c) $P(10.86 < \chi^2_{18} < 31.5)$,

(d) the value a such that $P(\chi^2_{22} > a) = 0.75$,

(e) the value a such that $P(\chi^2_{19} < a) = 0.025$,

(f) the value a such that $P(a < \chi^2_{26} < 41.9) = 0.725$.

Solution

Make sure you follow the reasoning involved in each solution.

(a) $P(\chi^2_{16} > 9.31) = 0.90$.

(b) $P(\chi^2_{25} < 29.3) = 1 - P(\chi^2_{25} > 29.3) = 1 - 0.25 = 0.75$.

(c) $P(10.86 < \chi^2_{18} < 31.5) = P(\chi^2_{18} > 10.86) - P(\chi^2_{18} > 31.5) = 0.90 - 0.025 = 0.875$.

(d) $P(\chi^2_{22} > a) = 0.75 \Rightarrow a = 17.25$.

(e) $P(\chi^2_{19} < a) = 0.025 \Rightarrow 1 - P(\chi^2_{19} > a) = 0.025$
$\Rightarrow P(\chi^2_{19} > a) = 0.975$
$\Rightarrow a = 8.91$.

(f) $P(a < \chi^2_{26} < 41.9) = 0.725 \Rightarrow P(\chi^2_{26} > a) - P(\chi^2_{26} > 41.9) = 0.725$
$\Rightarrow P(\chi^2_{26} > a) - 0.025 = 0.725$
$\Rightarrow P(\chi^2_{26} > a) = 0.750$
$\Rightarrow a = 20.8$. $\square$

Note that each row of Table A-3 represents a different probability distribution. Thus, it attempts to present as much information in one row as we have in the entire body of Table A-1. This being the case, it is obvious that in many situations we can only approximate probabilities associated with a given chi-square distribution. Consider an example.

Example 4

Find the probability α that a chi-square variable with $k = 12$ degrees of freedom is greater than the value 19.2; that is, find α such that

$$P(\chi^2_{12} > 19.2) = \alpha.$$

Solution

The value 19.2 does not appear in Table A-3 in the row corresponding to $k = 12$ degrees of freedom. However, 18.55 and 21.0 do appear and they encompass 19.2, the value of interest. Thus, since

$$P(\chi_{12}^2 > 21.0) = 0.05$$

and

$$P(\chi_{12}^2 > 18.55) = 0.10,$$

and since

$$18.55 < 19.2 < 21.0,$$

we obtain

$$P(\chi_{12}^2 > 21.0) < P(\chi_{12}^2 > 19.2) < P(\chi_{12}^2 > 18.55).$$

Thus,

$$0.05 < \alpha < 0.10. \qquad \square$$

Fortunately, for most of the problems involving chi-square probabilities, a bracketing of the required probability (as in Example 4) will suffice.

It is often necessary to distinguish notationally between a general chi-square variable and a single possible value of such variable. For this we use subscripts. That is, we denote by χ_k^2 the general chi-square variable with k degrees of freedom. On the other hand, we denote by $\chi_{\alpha,k}^2$ the single possible value of the general chi-square variable with k degrees of freedom such that

$$P(\chi_k^2 > \chi_{\alpha,k}^2) = \alpha. \tag{3}$$

For example, the value of the chi-square variable with 13 degrees of freedom which has 0.10 area to its right is 19.81, and such value is denoted by $\chi_{0.10,\,13}^2 = 19.81$. That is,

$$P(\chi_{13}^2 > \chi_{0.10,\,13}^2) = P(\chi_{13}^2 > 19.81) = 0.10.$$

Additive Property of Chi-Square Variables

For the normal distribution, we have seen that certain convenient properties result from the addition or subtraction of independent, normal variables. A similar property holds for chi-square variables. (We reiterate that for many kinds of variables such is not the case.) If we have two independent chi-square variables, say, U_1 and U_2, with k_1 and k_2 degrees of freedom, respectively, then

$$U_1 + U_2 \sim \chi_{k_1+k_2}^2. \tag{4}$$

For example, if U_1 and U_2 are independent, random variables with

$$U_1 \sim \chi_6^2 \quad \text{and} \quad U_2 \sim \chi_8^2,$$

then

$$(U_1 + U_2) \sim \chi^2_{14}.$$

This demonstrates the *additive property of chi-square variables*, which holds in general for m independent chi-square variables. In other words, if $U_1, U_2, \ldots, U_m$ are independent, random variables such that

$$U_i \sim \chi^2_{k_i}, \quad i = 1, 2, \ldots, m,$$

then

$$(U_1 + U_2 + \cdots + U_k) \sim \chi^2_{k_1 + k_2 + \cdots + k_m}. \tag{5}$$

We shall greatly exploit this property in subsequent chapters.

Problems

1. Determine

   a. $P(\chi^2_{24} > 33.2)$. b. $P(\chi^2_9 < 4.17)$. c. $P(10.28 < \chi^2_{21} < 16.34)$.

   d. the value a such that $P(\chi^2_{13} > a) = 0.975$.

   e. the value a such that $P(\chi^2_{40} < a) = 0.99$.

   f. the value a such that $P(21.6 < \chi^2_{18} < a) = 0.225$.

2. Determine

   a. the value a such that $P(\chi^2_{70} > a) = 0.10$.

   b. $P(\chi^2_4 < 1.923)$.

   c. the value a such that $P(a < \chi^2_{29} < 28.3) = 0.49$.

   d. $P(\chi^2_{11} > 13.70)$.

   e. $P(10.86 < \chi^2_{24} < 45.6)$

   f. the value a such that $P(\chi^2_{14} < a) = 0.025$.

3. Determine

   a. $P(\chi^2_{25} < 31.2)$. b. $P(7.03 < \chi^2_{16} < 15.34)$.

   c. the value a such that $P(\chi^2_7 > a) = 0.22$.

   d. the value a such that $P(a < \chi^2_{90} < 124.1) = 0.97$.

4. Determine

   a. the value a such that $P(13.56 < \chi^2_{28} < a) = 0.25$.

   b. $P(\chi^2_9 < 2.95)$.

   c. $P(14.04 < \chi^2_{22} < 32.1)$.

   d. the value a such that $P(6.02 < \chi^2_{15} < a) = 0.17$.

5. Suppose $U_1 \sim \chi^2_{12}$ and $U_2 \sim \chi^2_{15}$, with U_1 and U_2 *independent*. Find

$$P(U_1 + U_2 > 18.11).$$

10.3 Testing Hypotheses Concerning the Variance of a Normal Distribution

In Chapter 9, when we tested hypotheses concerning the mean we used the sample mean $\bar{X}$ as our statistic. In every case there, the sample mean $\bar{X}$ appeared in the statistic (pivotal quantity) meeting the requirements for use in testing a hypothesis concerning the population mean. For large sample situations, we were able to test a hypothesis concerning the population mean regardless of the form of the distribution from which sampling was accomplished by drawing on the central limit theorem. When sample sizes were too small to permit use of the central limit theorem, we assumed that sampling came from a normal distribution. In either case, we stressed the importance of the underlying assumptions.

In order to test a hypothesis concerning the variance of a population, we must begin exactly as we did in the case of the population mean. That is, we must find a suitable pivotal quantity. However, this is not easy for general populations. Thus, we shall begin with the basic assumption that we are sampling from a normal distribution with mean μ_x and variance σ_x^2. Then, if we select a simple random sample of size n from the normal population, the n random variables $X_1, \ldots, X_n$ are independent, and we can use the definition of a chi-square variable to write

$$\sum_{i=1}^{n} \left(\frac{X_i - \mu_x}{\sigma_x} \right)^2 \sim \chi_n^2. \tag{1}$$

Now, if we assume that we know μ_x, the statistic of (1) becomes suitable (a pivotal quantity) for testing a hypothesis concerning the variance σ_x^2 of a normal population.

However, we seldom know the true value of μ_x. Thus, for a practical test statistic, we must search for another pivotal quantity. What if we use the sample information to estimate μ_x through use of the unbiased estimator $\bar{X}$? It turns out that the statistic $\sum_{i=1}^{n}[(X_i - \bar{X})/\sigma_x]^2$ has a chi-square distribution with $(n - 1)$ degrees of freedom; in other words, for a sample of size n from a normal population,

$$\sum_{i=1}^{n} \left(\frac{X_i - \bar{X}}{\sigma_x} \right)^2 \sim \chi_{n-1}^2. \tag{2}$$

Note that our use of the capital letter X in Relation (2) implies a general conceptual statistic. As soon as a specific sample is selected, so that we can refer to specific observations of X, we would denote a single observed value of the statistic by

$$\sum_{i=1}^{n} \left(\frac{x_i - \bar{x}}{\sigma_x} \right)^2.$$

In light of (2), we immediately have a pivotal quantity to use in testing a hypothesis concerning the variance of a normal population. That is, the statistic

$$\sum_{i=1}^{n} \left(\frac{X_i - \bar{X}}{\sigma_x} \right)^2 = \frac{\sum_{i=1}^{n}(X_i - \bar{X})^2}{\sigma_x^2}$$

(1) has a known distribution which is independent of the unknown parameter σ_x^2,
(2) is a function of the unknown parameter σ_x^2, and
(3) is a function of no other unknowns.

Thus, the essential steps of the method for making a one-tail, 100α percent significance test concerning the variance σ_x^2 of a normal population are as follows. (You may wish to review the general procedure for testing hypotheses concerning a general unknown parameter θ in Section 9.4.)

1. Set up the null hypothesis

$$H_0: \sigma_x^2 = \sigma_0^2,$$

where σ_0^2 is any value the parameter σ_x^2 can assume.

2. Specify the level of significance α.
3. Determine the size (n) of the sample to be taken. (This fixes the number ($n - 1$) of degrees of freedom to be used in establishing the appropriate critical value.)
4. Determine a critical value χ_{n-1}^{2*} that divides the chi-square distribution with ($n - 1$) degrees of freedom into a critical (rejection) region and an acceptance region. This value should be such that 100α percent of the area of the chi-square distribution with ($n - 1$) degrees of freedom lies in the critical region. That is, for a left-tail alternative hypothesis, $H_1: \sigma_x^2 < \sigma_0^2$, find χ_{n-1}^{2*} such that

$$P(\chi_{n-1}^2 < \chi_{n-1}^{2*}) = \alpha. \tag{3}$$

We determine χ_{n-1}^{2*} as follows.

$$P(\chi_{n-1}^2 < \chi_{n-1}^{2*}) = \alpha \Rightarrow 1 - P(\chi_{n-1}^2 > \chi_{n-1}^{2*}) = \alpha$$
$$\Rightarrow P(\chi_{n-1}^2 > \chi_{n-1}^{2*}) = 1 - \alpha.$$

In our previous notation for tabular chi-square values, we see immediately that, for a left-tail alternative hypothesis,

$$\chi_{n-1}^{2*} = \chi_{1-\alpha,n-1}^2. \tag{4}$$

On the other hand, for a right-tail alternative hypothesis, $H_1: \sigma_x^2 > \sigma_0^2$, we find χ_{n-1}^{2*} such that

$$P(\chi_{n-1}^2 > \chi_{n-1}^{2*}) = \alpha. \tag{5}$$

This tells us immediately that, for a right-tail alternative hypothesis,

$$\chi_{n-1}^{2*} = \chi_{\alpha,n-1}^2. \tag{6}$$

5. From a sample of size n, obtain (under the assumption that the null hypothesis is true, or, $\sigma_x^2 = \sigma_0^2$) one observed value of the statistic of (2); that is, compute

$$\frac{\sum_{i=1}^{n}(x_i - \bar{x})^2}{\sigma_0^2}, \tag{7}$$

where x_i is the ith sample observation, $\bar{x}$ is the observed sample mean, and σ_0^2 is the value of σ_x^2 given by the null hypothesis.

6. If the value of (7) is in the critical region, reject H_0. Otherwise, accept H_0.

If we wish to make a symmetrical two-tail, 100α percent significance test concerning the variance of a normal population, the procedure is the same except that we must find two critical values. In this case, we determine $(\chi^2_{n-1})^*_1$ and $(\chi^2_{n-1})^*_2$ such that

$$P[\chi^2_{n-1} < (\chi^2_{n-1})^*_1] = \alpha/2 \quad \text{and} \quad P[\chi^2_{n-1} > (\chi^2_{n-1})^*_2] = \alpha/2. \tag{8}$$

Solving the relations of (8) for $(\chi^2_{n-1})^*_1$ and $(\chi^2_{n-1})^*_2$ yields as critical values for a symmetrical test

$$(\chi^2_{n-1})^*_1 = \chi^2_{1-\alpha/2,n-1} \quad \text{and} \quad (\chi^2_{n-1})^*_2 = \chi^2_{\alpha/2,n-1}. \tag{9}$$

To implement the two-tail test, we still calculate the value of (7). If it is between $\chi^2_{1-\alpha/2,n-1}$ and $\chi^2_{\alpha/2,n-1}$, we accept H_0. Otherwise, we reject H_0.

The approach for a nonsymmetrical, two-tail, 100α percent significance test is quite similar. We still force the two tail areas to total α, but relax the requirement that they be equal.

It is interesting to note that, for a given sample of size n,

$$\sum_{i=1}^{n} \left(\frac{x_i - \bar{x}}{\sigma_x}\right)^2 = \frac{\sum_{i=1}^{n}(x_i - \bar{x})^2}{\sigma_x^2} = \frac{ns_x^2}{\sigma_x^2} = \frac{(n-1)\hat{\sigma}_x^2}{\sigma_x^2}, \tag{10}$$

where s_x^2 is the sample variance,

$$s_x^2 = \frac{\sum_{i=1}^{n}(x_i - \bar{x})^2}{n}, \tag{11}$$

and $\hat{\sigma}_x^2$ is the unbiased estimate of the population variance,

$$\hat{\sigma}_x^2 = \frac{\sum_{i=1}^{n}(x_i - \bar{x})^2}{n-1}. \tag{12}$$

This being the case, it is often convenient to calculate and use either

$$\frac{ns_x^2}{\sigma_0^2} \tag{13}$$

or

$$\frac{(n-1)\hat{\sigma}_x^2}{\sigma_0^2} \tag{14}$$

instead of the value of (7) in conducting the test. Both are useful, with the choice depending on whether s_x^2 or $\hat{\sigma}_x^2$ has been computed from the sample data in a given problem.

Suppose that we wish to determine, at the 5 percent significance level, whether the variance of a specific normal distribution is significantly greater than 100. First, we determine the decision rule for testing

$$H_0: \sigma_x^2 = 100 \quad \text{versus} \quad H_1: \sigma_x^2 > 100.$$

Since we do not know μ_x, we shall estimate it by using the estimator $\bar{X}$. Then we use the test statistic as given in (2). Thus, we must determine the chi-square value with $(n - 1)$ degrees of freedom which has 0.05 area to its right (that is, $\chi^2_{0.05,n-1}$). Suppose we wish to select a sample of size $n = 16$ for our experiment. Then, we must find a chi-square value with $n - 1 = 15$ degrees of freedom. From Table A-3, we determine that $\chi^2_{0.05,15} = 25.0$. Therefore, as shown in Figure 10.2, we will accept H_0 if the observed value of our test statistic, $\sum_{i=1}^{n}(x_i - \bar{x})^2/\sigma_0^2$, is less than 25.0, and we will reject

H_0 if it is greater than 25.0. Suppose, after selecting our sample of size $n = 16$, we determine $s_x^2 = 195$. The observed value of our test statistic (7) becomes

$$\frac{\sum_{i=1}^{n}(x_i - \bar{x})^2}{\sigma_0^2} = \frac{ns_x^2}{\sigma_0^2} = \frac{16(195)}{100} = 31.20.$$

Therefore, since $31.20 > 25.0$, we reject H_0. The physical relationship between our observed value of the appropriate statistic and the critical region is shown in Figure 10.2.

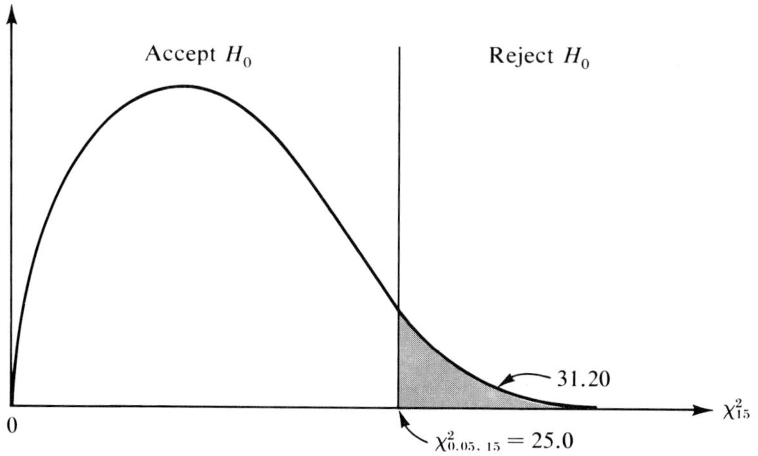

Figure 10.2
Decision rule for a chi-square test

Example 1

A process for manufacturing steel pipes has been changed. The variance of the diameters of the pipes from the previous process was 80 square millimeters. From the new products, a random sample of size $n = 27$ is selected to determine whether there has been a significant decrease in the variance of the diameters. The sample variance s_x^2 is found to be 71 square millimeters. Assuming the distribution of diameters is normal, has there been a significant decrease in the variance? Let $\alpha = 0.05$.

Solution

Since we are concerned with a significant decrease in the variance, we have as our hypotheses

$$H_0: \sigma_x^2 = 80 \text{ square millimeters} \quad \text{versus} \quad H_1: \sigma_x^2 < 80 \text{ square millimeters}.$$

We must determine the decision rule for (13). This simply means we must find the value for $\chi^2_{0.95,\,26}$, since the area under the left tail of the chi-square curve must be $\alpha = 0.05$ and the appropriate number of degrees of freedom is $n - 1 = 26$. From Table A-3, we find

$$\chi^2_{0.95,\,26} = 15.38.$$

Hence, our decision rule is that of accepting H_0 if the observed value of our test statistic (7) is greater than 15.38 and rejecting H_0 if it is less than 15.38. The observed value of our test statistic, using (7) and (13) is

$$\frac{\sum_{i=1}^{n}(x_i - \bar{x})^2}{\sigma_0^2} = \frac{ns_x^2}{\sigma_0^2} = \frac{27(71)}{80} = 23.96.$$

Therefore, we accept H_0, since $23.96 > 15.38$, as shown in Figure 10.3.

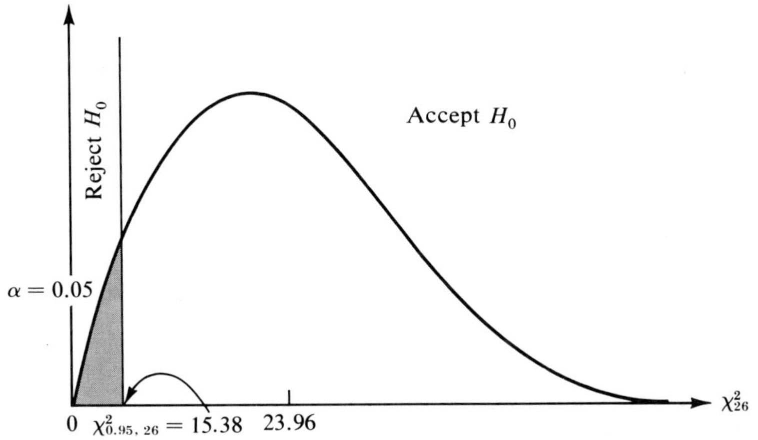

Figure 10.3
Decision rule

□

Example 2

A manufacturer of dietetic foods knows from experience that the variance of weights of women on normal diets is $\sigma_x^2 = 200$ square pounds. The manufacturer wishes to determine if there is a significant difference (at the 5 percent significance level) between the variance for the women on normal diets and the women who are on a special diet using one of its products. A group of 15 women on the special diet is randomly selected, and the unbiased estimate of the population variance is computed to be $\hat{\sigma}_x^2 = 170$ square pounds. Assuming the weights of the women to be normally distributed, is the variance of the weights of women on the special diet significantly different from 200 square pounds?

Solution

The hypotheses of interest are

$$H_0: \sigma_x^2 = 200 \text{ square pounds} \quad \text{versus} \quad H_1: \sigma_x^2 \neq 200 \text{ square pounds.}$$

Since we have a two-sided alternative hypothesis, we divide the risk equally between the two tails. Thus, we must determine our decision rule by finding the values for $\chi^2_{0.975,\,14}$ and $\chi^2_{0.025,\,14}$. From Table A-3, we find

$$\chi^2_{0.975,\,14} = 5.63 \quad \text{and} \quad \chi^2_{0.025,\,14} = 26.1.$$

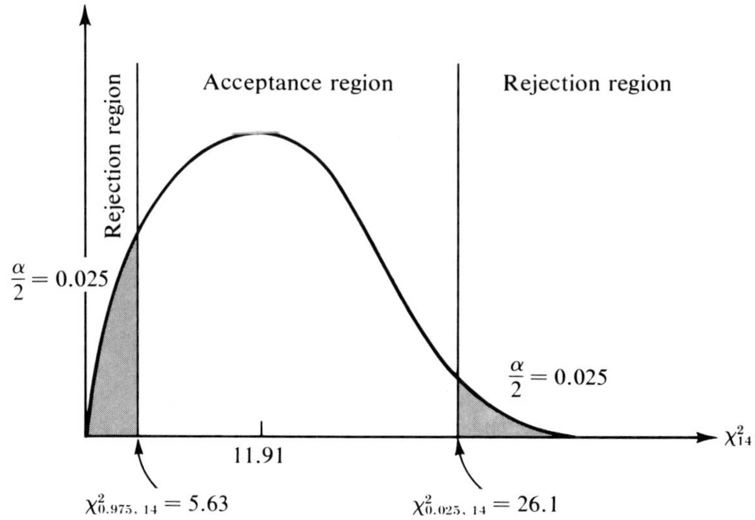

Figure 10.4
Decision rule

This immediately gives us the decision rule, shown graphically in Figure 10.4, that we accept H_0 if our test statistic is between 5.63 and 26.1 and reject H_0 if the test statistic is not between these values.

Calculating our test statistic [(7) and (14)], we have

$$\frac{\sum_{i=1}^{n}(x_i - \bar{x})^2}{\sigma_0^2} = \frac{(n-1)\hat{\sigma}_x^2}{\sigma_0^2} = \frac{14(170)}{200} = 11.90.$$

Since 11.90 is between 5.63 and 26.1, we accept H_0. Thus, we may conclude that there is not a significant difference (at the 5 percent level) between the variance of the weight of women on a normal diet and the variance of the weight of women on the special diet. □

Problems

1. List the essential steps of the method for making a symmetrical two-tail $100\alpha\%$ significance test concerning the variance σ_x^2 of a *normal* population.

2. Suppose we wish to test

$$H_0: \sigma_x^2 = 50 \quad \text{versus} \quad H_1: \sigma_x^2 > 50.$$

Given $n = 30$ and $s_x^2 = 63$, test the above hypotheses at the 1% significance level. Assume we are sampling from a normal population.

3. Suppose we wish to test

$$H_0: \sigma_x^2 = 200 \quad \text{versus} \quad H_1: \sigma_x^2 < 200.$$

Given $n = 25$ and $s_x^2 = 150$, test the above hypotheses at the 10% significance level. Assume we are sampling from a normal population.

4. Suppose we wish to test

$$H_0: \sigma_x^2 = 20 \quad \text{versus} \quad H_1: \sigma_x^2 \neq 20.$$

Given $n = 20$ and $\hat{\sigma}_x^2 = 22$, conduct a symmetrical two-tail test at the 5% significance level. Assume we are sampling from a normal population.

5. A process for manufacturing a special alloy has been changed. The variance of the weights of standard bars of the alloy with the old process was 0.63 lb^2. A random sample of 15 bars from the new process has a variance of 0.81 lb^2. *Assuming the distribution of weights is normal,* has there been a significant increase at the 10% significance level in the variance of the weights?

6. It is known that the variance of the daily sales in a small store is 10,000 dollars^2. During a test of a new advertising policy, a random sample of 30 days was selected and the unbiased estimate of the new population variance was determined to be 5,000 dollars^2. Assuming the daily sales to be *normally distributed,* was there a significant decrease at the 2% significance level in the variance of the daily sales?

7. The variance of the intelligence quotients of job applicants tested by a large corporation was 100 points^2 before a new screening procedure to be used prior to testing was introduced. A random sample of 24 test scores taken after the new procedure was introduced had a sample variance of 50 points^2. Assuming the intelligence quotient test scores are *normally distributed,* has there been a significant change at the 5% significance level in the variance of the test scores?

10.4 Confidence Intervals for the Variance of a Normal Distribution

We shall now consider the problem of constructing a $100(1 - \alpha)$ percent confidence interval for the variance of a normal population. This should be a very simple matter, because we can use the same pivotal quantity used in developing the test of a hypothesis concerning the variance of a normal population (Section 10.3). [Again, you may find it helpful to refer to the general procedure for constructing a $100(1 - \alpha)$ percent confidence interval for the unknown parameter θ (Section 8.6).] That is, we already know that, if we are sampling from a normal distribution, the statistic

$$\frac{\sum_{i=1}^{n}(X_i - \bar{X})^2}{\sigma_x^2} \sim \chi_{n-1}^2. \tag{1}$$

We also know that there are two values, $\chi_{1-\alpha/2,n-1}^2$ and $\chi_{\alpha/2,n-1}^2$, such that

$$P(\chi_{1-\alpha/2,n-1}^2 < \chi_{n-1}^2 < \chi_{\alpha/2,n-1}^2) = 1 - \alpha. \tag{2}$$

Since Relation (2) holds for any chi-square variable with $(n - 1)$ degrees of freedom, it must necessarily hold for the statistic of (1). Thus, we have

$$P\left(\chi_{1-\alpha/2,n-1}^2 < \frac{\sum_{i=1}^{n}(X_i - \bar{X})^2}{\sigma_x^2} < \chi_{\alpha/2,n-1}^2\right) = 1 - \alpha, \tag{3}$$

where, as usual, $\chi_{\alpha/2,n-1}^2$ and $\chi_{1-\alpha/2,n-1}^2$ are the chi-square values (from Table A-3) with $(n - 1)$ degrees of freedom and with $\alpha/2$ and $(1 - \alpha/2)$ area, respectively, to the right of them. Through algebraic manipulation of the inequalities in the probability statement of (3), we can obtain a set of inequalities in which σ_x^2 is bounded above and below. This probability statement will remain valid as long as we do not violate any of the basic rules of the algebra of inequalities. The first step is to divide each of the three terms of the set of inequalities by $\sum_{i=1}^n (X_i - \bar{X})^2$. This does not change the direction of the inequality signs since $\sum_{i=1}^n (X_i - \bar{X})^2$ is necessarily nonnegative. When we perform such division, (3) becomes

$$P\left[\frac{\chi^2_{1-\alpha/2,n-1}}{\sum_{i=1}^n (X_i - \bar{X})^2} < \frac{1}{\sigma_x^2} < \frac{\chi^2_{\alpha/2,n-1}}{\sum_{i=1}^n (X_i - \bar{X})^2}\right] = 1 - \alpha. \tag{4}$$

Next, taking the reciprocal of each term in the inequalities of (4), we change the direction of the two inequalities, thus obtaining

$$P\left[\frac{\sum_{i=1}^n (X_i - \bar{X})^2}{\chi^2_{1-\alpha/2,n-1}} > \sigma_x^2 > \frac{\sum_{i=1}^n (X_i - \bar{X})^2}{\chi^2_{\alpha/2,n-1}}\right] = 1 - \alpha. \tag{5}$$

Arranging the inequalities in the usual form, so that the smallest value appears on the left-hand side, we have the $(1 - \alpha)$ probability statement

$$P\left[\frac{\sum_{i=1}^n (X_i - \bar{X})^2}{\chi^2_{\alpha/2,n-1}} < \sigma_x^2 < \frac{\sum_{i=1}^n (X_i - \bar{X})^2}{\chi^2_{1-\alpha/2,n-1}}\right] = 1 - \alpha. \tag{6}$$

Once again, since (6) involves a general statistic (i.e., a variable), it represents a probability statement in the classical sense. However, as soon as we consider the results of a specific example, we will obtain the fiducial probability statement

$$P_F\left[\frac{\sum_{i=1}^n (x_i - \bar{x})^2}{\chi^2_{\alpha/2,n-1}} < \sigma_x^2 < \frac{\sum_{i=1}^n (x_i - \bar{x})^2}{\chi^2_{1-\alpha/2,n-1}}\right] = 1 - \alpha. \tag{7}$$

Similarly, because of the relations between $\sum_{i=1}^n (x_i - \bar{x})^2$, s_x^2, and $\hat{\sigma}_x^2$, we have, for a specific sample of size n,

$$P_F\left[\frac{ns_x^2}{\chi^2_{\alpha/2,n-1}} < \sigma_x^2 < \frac{ns_x^2}{\chi^2_{1-\alpha/2,n-1}}\right] = 1 - \alpha \tag{8}$$

and

$$P_F\left[\frac{(n-1)\hat{\sigma}_x^2}{\chi^2_{\alpha/2,n-1}} < \sigma_x^2 < \frac{(n-1)\hat{\sigma}_x^2}{\chi^2_{1-\alpha/2,n-1}}\right] = 1 - \alpha. \tag{9}$$

Thus, using a specific sample of size n from a normal population, we can construct a $100(1 - \alpha)$ percent confidence interval for the variance σ_x^2 by implementing either of the equivalent formulas given by (7), (8), or (9).

Example 1

We have taken a simple random sample of size $n = 21$ from a normal population and have determined $s_x^2 = 118$. Construct a 95 percent confidence interval for the population variance and give a strict interpretation of such confidence interval.

Solution

Since $1 - \alpha = 0.95$, we have $\alpha = 0.05$; thus, $\alpha/2 = 0.025$ and $1 - \alpha/2 = 0.975$. From Table A-3, we find

$$\chi^2_{0.025,\,20} = 34.2 \quad \text{and} \quad \chi^2_{0.975,\,20} = 9.59.$$

Since s_x^2 has already been calculated to be 118, we obtain a 95 percent confidence interval simply by substituting the appropriate values into (8). Thus,

$$P_F\left(\frac{(21)(118)}{34.2} < \sigma_x^2 < \frac{(21)(118)}{9.59}\right) = 0.95$$

Simplifying, we obtain the 95 percent confidence interval for σ_x^2

$$72.5 < \sigma_x^2 < 258.4. \tag{10}$$

If we were to construct a very large number of intervals in the above manner, 95 percent of these intervals would include the true value of the unknown parameter σ_x^2. □

Example 2

From a random sample of 18 automobiles, the sample variance of gas mileages is found to be $s_x^2 = 21$. Assuming that the gas mileage of automobiles is normally distributed, find an 80 percent confidence interval for the population variance.

Solution

Since $1 - \alpha = 0.80$, we have

$$\alpha = 0.20, \quad \alpha/2 = 0.10, \quad \text{and} \quad 1 - \alpha/2 = 0.90.$$

From Table A-3, we find

$$\chi^2_{0.10,\,17} = 24.8 \quad \text{and} \quad \chi^2_{0.90,\,17} = 10.09.$$

Substituting into (8) gives an 80 percent confidence interval for σ_x^2 as

$$\frac{(18)(21)}{24.8} < \sigma_x^2 < \frac{(18)(21)}{10.09} \Rightarrow 15.2 < \sigma_x^2 < 37.5.$$ □

Problems

1. Assuming we are sampling from a *normal population*, construct the indicated confidence interval for the population variance for each situation.

   a. A 98% confidence interval for $n = 18$ and $\hat{\sigma}_x^2 = 27$.

   b. A 90% confidence interval for $n = 30$ and $s_x = 13$.

c. A 99% confidence interval for $n = 12$ and $\sum_{i=1}^{n} (x_i - \bar{x})^2 = 208$.

2. Assuming we are sampling from a *normal population*, construct the indicated confidence interval for the population variance for each situation.

a. An 80% confidence interval for $n = 23$ and $s_x^2 = 183$.

b. A 95% confidence interval for $n = 8$ and $\sum_{i=1}^{n} (x_i - \bar{x})^2 = 87$.

c. A 90% confidence interval for $n = 15$ and $\hat{\sigma}_x = 53$.

3. A random sample of 27 lawn mower engines is selected, and the resulting unbiased estimate of the population variance of their horsepower is 0.85 hp^2. Assuming that the horsepower of such engines is *normally distributed*, find a 98% confidence interval for the population variance.

4. A random sample of 16 sales made by the salesmen of a certain company had a sample variance of the amount of each sale of 10,000 $dollars^2$. Assuming that the amount of sales is *normally distributed*, find a 90% confidence interval for the population variance.

10.5 Goodness-of-Fit Tests

So far, we have used the chi-square distribution for testing hypotheses and setting confidence intervals for the population variance when sampling from a normal distribution. We shall now consider another way to use the chi-square distribution. Frequently, a researcher is interested in the number (frequency) of items, subjects, or responses in a given population that fall into various mutually exclusive categories. For example, a group of college students may be classified according to the number of semester credit hours each has completed. Or, a group of consumers may be classified according to the amount each spends for certain items. Usually in such situations only sample information can be obtained. Generally, however, sample results per se are of very little interest to the researcher. Rather, he would like to use sample results to draw an inference concerning the entire population of interest. Thus, the researcher uses his sample information to test the agreement between a hypothetical distribution and the distribution obtained by classifying the sample values into mutually exclusive categories. Such a test is called a *goodness-of-fit test*, from the fact that it is designed to show when sample data fails to "fit" the hypothetical distribution of interest.

Suppose we wish to test the null hypothesis that a set of sample data comes from a given population versus the alternative hypothesis that the data did not come from the population. We randomly select a group of n observations such that

1. each observation can be classified into one and only one of k mutually exclusive classes of categories and
2. the outcome of each observation is independent of the outcomes of all other observations.

Next, suppose we represent the group of data as a frequency distribution (Chapter 2). We determine an observed frequency of f_i in each class such that $\sum_{i=1}^{k} f_i = n$. By our hypotheses, we want to determine whether the observed sample frequency distribution agrees with some hypothetical or expected frequency distribution, that is, whether the observed frequencies (f_i) agree with the frequencies (e_i) we would expect to obtain from the hypothetical distribution. The latter frequencies, termed expected frequencies, may be determined directly from the null hypothesis being tested, subject, of course, to the restriction that $\sum_{i=1}^{k} e_i = n$.

The test statistic we shall use is

$$\sum_{i=1}^{k} \frac{(f_i - e_i)^2}{e_i}. \tag{1}$$

Karl Pearson has shown that, when (and only when) the f_i's come from the same population as the e_i's, the sampling distribution of this statistic approaches a chi-square distribution as n becomes infinitely large. Thus, we immediately have a statistic to use for testing the hypothesis of interest. Intuitively, we would expect the statistic of (1) to take on small values when the f_i's come from the same population as the e_i's (i.e., when the null hypothesis is true) as opposed to large values when they do not come from the same population (i.e., when the null hypothesis is false). The major question now is "How large must the value of the statistic of (1) become before we are willing to reject the null hypothesis?" The fact that the distribution of the statistic

$$\sum_{i=1}^{k} \frac{(f_i - e_i)^2}{e_i}$$

approaches a chi-square distribution with $(k - 1)$ degrees of freedom only when the f_i's come from the same population as the e_i's gives us an answer. If we are willing to tolerate a probability of Type I error of α, then we simply test our observed value of the statistic against the critical value $(\chi^2_{k-1})^* = \chi^2_{\alpha, k-1}$. Thus, if we use this critical value in testing the null hypothesis, we will reject H_0 100α percent of the time when it is true. (Later we shall set a guideline to tell us when we can use the chi-square distribution. As a rule, we shall require $e_i \geq 5$ for each class.)

In order to compare a group of observed frequencies with a group of expected frequencies, we must, of course, be able to determine the expected frequencies. For this we use the null hypothesis. However, the form of the null hypothesis depends on the nature of the problem. It may state directly the proportion of observations that fall into each class under the hypothesized population. On the other hand, the null hypothesis may simply set forth the hypothesized probability distribution; then the distribution can be used to determine the proportion of observations expected to fall into each class.

How do we determine the number of degrees of freedom? When we tested hypotheses and set confidence intervals for σ_x^2, we used $(n - 1)$ as our number of degrees of freedom, where n is the sample size. However, for the goodness-of-fit tests using (1), the only place that our sample size n enters the problem is in the calculation of the expected frequencies. Thus, to determine the degrees of freedom, we consider the number of classes k. If all parameters of the probability distribution in H_0 are specified, the number of degrees of freedom is $(k - 1)$. However, if any parameters are replaced by estimates obtained from the sample data, then the number of degrees of freedom is $(k - r - 1)$, where r is the number of parameters for which estimates are obtained from the sample data.

In light of the foregoing reasoning, the general format of the goodness-of-fit test is as follows.

1. State the hypotheses of interest. Though they may be stated in various forms, essentially they are:

$$H_0\text{: Sample data come from a given population}$$

versus

$$H_1\text{: Sample data do not come from a given population.} \tag{2}$$

2. Specify the level of significance α.

3. Determine (or note) the number of categories (k) of interest and the number (r) of unspecified population parameters that must be estimated from sample data. (This fixes the number $(k - r - 1)$ of degrees of freedom to be used in establishing the appropriate critical value.)

4. Determine a critical value $(\chi^2_{k-r-1})^*$ which divides the chi-square distribution with $(k - r - 1)$ degrees of freedom into a critical (rejection) region and an acceptance region. This should always be such that 100α percent of the area of the chi-square distribution with $(k - r - 1)$ degrees of freedom lies to the right of the critical value. That is,

$$(\chi^2_{k-r-1})^* = \chi^2_{\alpha, k-r-1}.$$

5. Take a sample of size n observations from the population and classify them according to the k categories. Note the f_i's.

6. Use the null hypothesis to calculate the expected number of observations (e_i) for each category. Calculations should be made subject to the restriction that

$$\sum_{i=1}^{k} e_i = n = \sum_{i=1}^{k} f_i.$$

7. Calculate the observed value of the statistic (1)

$$\sum_{i=1}^{k} \frac{(f_i - e_i)^2}{e_i}.$$

8. If the value of step 7 is greater than $\chi^2_{\alpha, k-r-1}$, reject H_0. Otherwise, accept H_0.

Note that all tests involving the statistic of (1) are right-tail tests. Even though the alternative hypothesis appears to be a two-tail alternative, it is not. For example, suppose we have a sample such that the observed frequency for each class is identical to the expected frequency for that class. Then we have a perfect fit and

$$\sum_{i=1}^{k} \frac{(f_i - e_i)^2}{e_i} = 0.$$

Recalling that a chi-square variable cannot have a value less than zero, we see that we would reject H_0 when we have a perfect fit if we were to use a two-tail test. Also, the farther apart the observed and expected frequencies, the larger the value of the test statistic. Thus, when using the chi-square variable for a goodness-of-fit test, we always use a right-tail test.

Example 1

Horse-racing fans often argue that in a race around a circular track certain post positions are more advantageous than others. Post positions, assigned to horses by a random draw, are numbered so that the horse closest to the inner rail of the track is in position 1, the horse second from the inner rail in position 2, etc. We shall test the effect of post position by analyzing race results classified according to post position for a season at a particular race track. Suppose the results for all five-horse races are

Post position	1	2	3	4	5	Total
Number of wins	58	50	40	32	20	200

Determine at the 5 percent significance level whether there is a significant difference between the numbers of wins for the different post positions.

Solution

The null hypothesis is that the same proportion of winners occurs for each post position. Thus, the only variation we expect between the various f_i's is due to chance. The alternative hypothesis is that the proportion of winners is not the same for all post positions. If we let π_i represent the proportion of winners from the ith post position, we may state the hypotheses as follows:

$$H_0: \pi_1 = \pi_2 = \pi_3 = \pi_4 = \pi_5 \quad \text{versus} \quad H_1: \text{the } \pi_i\text{'s are not all equal.}$$

Since we are using $\alpha = 0.05$ and there are $k = 5$ classes, the appropriate number of degrees of freedom for the test is $k - 1 = 4$, and we find the critical value for our decision rule from Table A-3 to be $\chi^2_{0.05,4} = 9.49$. Thus, the appropriate decision rule is to reject H_0 if the observed value of the statistic (1) is greater than 9.49 and accept H_0 if it is less than 9.49. Now, to test H_0, we must compute the observed value for the test statistic of (1). Since the π_i are equal under H_0, we find that $\pi_i = 1/5\,(i = 1, 2, \ldots, 5)$ when H_0 is true. Thus, our expected frequency for each post position is

$$n\pi_i = (200)(1/5) = 40,$$

and we calculate $\sum_{i=1}^{5} [(f_i - e_i)^2/e_i]$ as shown in Table 10.1.

Table 10.1

Goodness-of-fit calculations

Post position	*Observed number of* ⇒ *wins* f_i	*Expected number of* ⇒ *wins* e_i	$(f_i - e_i)$	$(f_i - e_i)^2/e_i$
1	58	40	18	8.1
2	50	40	10	2.5
3	40	40	0	0.0
4	32	40	−8	1.6
5	20	40	−20	10.0
Total	200	200	0	$\sum_{i-1}^{5} \frac{(f_i - e_i)^2}{e_i} = 22.2$

Since the observed value of the statistic (22.2) is greater than the critical value $\chi^2_{0.05,4} = 9.49$, we reject H_0. Hence, we conclude that there are advantages to certain post positions. Obviously, the test does not indicate which position(s) is (are) most advantageous. □

Example 2

The marketing department of a company that publishes a financial magazine bases its sales strategy on the assumption that, on average, for every eight subscribers there will be three with stock portfolios under \$10,000, two with portfolios in the \$10,000–\$50,000 range, two with portfolios in the \$50,000–\$100,000 range, and one with a portfolio in excess of \$100,000. A random sample of 400 subscribers is selected with the results shown at the right. Determine at the 5 percent significance level whether the data support the 3:2:2:1 ratio of subscribers on which the marketing department has based its strategy.

Class	*Observed frequency* (f_i)
under \$10,000	120
\$10,000–\$50,000	130
\$50,000–\$100,000	80
over \$100,000	70
	400

Solution

The hypotheses of interest are

$$H_0: \pi_1:\pi_2:\pi_3:\pi_4 = 3:2:2:1 \quad \text{versus} \quad H_1: \pi_1:\pi_2:\pi_3:\pi_4 \neq 3:2:2:1.$$

Since there are 4 classes, three degrees of freedom is the appropriate number to be used in testing H_0. The critical value for our decision rule from Table A-3 is $\chi^2_{0.05,3} = 7.81$. Our decision rule, then, is that of accepting H_0 if the observed value of the statistic (1) is less than 7.81 and rejecting H_0 if it is greater. Calculations for determining the observed value of the statistic are given in Tablc 10.2.

Since $27.00 > 7.81$, we reject H_0 and conclude that the marketing department has based its strategy upon an erroneous assumption.

Table 10.2
Goodness-of-fit calculations

Class	f_i	e_i	$(f_i - e_i)$	$(f_i - e_i)^2/e_i$
under \$10,000	120	150	−30	$(-30)^2/150 = 6.00$
\$10,000–\$50,000	130	100	30	$(30)^2/100 = 9.00$
\$50,000–\$100,000	80	100	−20	$(-20)^2/100 = 4.00$
over \$100,000	70	50	20	$(20)^2/50 = 8.00$
Total	400	400		$\sum_{i=1}^{4} \frac{(f_i - e_i)^2}{e_i} = 27.00$

□

Example 3

Suppose that we believe that the number of shipments arriving daily at a certain warehouse is a random variable distributed according to the Poisson probability distribution. However, we do not know the value of λ and, consequently, must estimate it from sample data. A sample of 300 days is selected with the number of shipments per day (X) being noted. The sample data, along with calculations for the sample mean, are presented in Table 10.3. Our estimate of λ is 2.4, since $\overline{X}$ is commonly used for estimating the mean of the Poisson distribution. Determine at the

Table 10.3
Calculation of sample mean

Class i	*Shipments per day* x_i	*Number of days* f_i	$x_i f_i$
1	0	30	0
2	1	99	99
3	2	100	200
4	3	83	249
5	4	52	208
6	5	19	95
7	6	10	60
8	7	7	49
Total		400	960

$$\bar{x} = \frac{\sum_{i=1}^{k} f_i x_i}{\sum_{i=1}^{k} f_i} = \frac{960}{400} \cong 2.4.$$

5 percent significance level whether the number of shipments arriving at the warehouse is distributed according to a Poisson distribution with $\lambda = 2.4$.

Solution

The hypotheses of interest are

H_0: X is distributed as $p(x; 2.4)$ versus H_1: X is not distributed as $p(x; 2.4)$.

To determine the expected frequencies, we find the probability (under the null hypothesis) for each possible value of X and multiply the probability by $n = 400$. Since Table A-7 gives only cumulative probabilities, we can determine the individual probabilities by subtraction. For example,

$$\begin{aligned} P(X = 1 \mid \lambda = 2.4) &= P(X \le 1 \mid \lambda = 2.4) - P(X \le 0 \mid \lambda = 2.4) \\ &= 0.308 - 0.091 = 0.217, \\ P(X = 2 \mid \lambda = 2.4) &= P(X \le 2 \mid \lambda = 2.4) - P(X \le 1 \mid \lambda = 2.4) \\ &= 0.570 - 0.308 = 0.262, \end{aligned}$$

and so forth. All relevant probabilities are given in Table 10.4, along with observed and expected frequencies.

Table 10.4
Preliminary goodness-of-fit calculations

Class i	*Shipments per day* x_i	*Observed number of days* f_i	$P(X = x_i)$ p_i	*Expected number of days* $e_i = np_i$
1	0	30	0.091	36.4
2	1	99	0.217	86.8
3	2	100	0.262	104.8
4	3	83	0.209	83.6
5	4	52	0.125	50.0
6	5	19	0.060	24.0
7	6	10	0.024	9.6
8	7	7	0.009	3.6
Total		400		398.8

We note two things in the table: the sum of the expected frequencies is less than 400, and the class for $x = 7$ has less than 5 for its expected frequency. The sum of the expected frequencies is less than 400 simply because there are more possible values of X than what we have included in our analysis (i.e., 8, 9, . . .). However, each such possible value has a very small probability of occurring. We must pool the observed frequency for $x = 6$ so that we have no class with an expected frequency less than 5. If, however, after pooling the observed frequencies for $x = 6$ and $x = 7$, we denote the category as the interval $X \geq 6$, we can use

$$P(X \geq 6 \mid \lambda = 2.4)$$

for determining the corresponding expected frequency. This will cause the sum of the expected frequencies to equal 400.

We now have seven classes, but have estimated one parameter in order to obtain expected frequencies. Thus, our number of degrees of freedom is $k - r - 1 = 7 - 1 - 1 = 5$. The critical value for our decision rule is $\chi^2_{0.05,5} = 11.07$. Therefore (see Table 10.5), since the observed value of the appropriate statistic is less than $\chi^2_{0.05,5}$ (i.e., $4.65 < 11.07$), we accept H_0 and conclude that the number of shipments arriving at the warehouse is distributed according to the Poisson distribution with $\lambda = 2.4$.

Table 10.5
Goodness-of-fit calculations

Class i	x_i	f_i	$P(X = x_i)$	e_i	$f_i - e_i$	$(f_i - e_i)^2/e_i$
1	0	30	0.091	36.4	−6.4	$(-6.4)^2/36.4 = 1.13$
2	1	99	0.217	86.8	12.2	$(12.2)^2/86.8 = 1.71$
3	2	100	0.262	104.8	−4.8	$(-4.8)^2/104.8 = 0.22$
4	3	83	0.209	83.6	−0.6	$(-0.6)^2/83.6 = 0.00$
5	4	52	0.125	50.0	2.0	$(2.0)^2/50.0 = 0.08$
6	5	19	0.060	24.0	−5.0	$(-5.0)^2/24.0 = 1.04$
7	6 or more	17	0.036	14.4	2.6	$(2.6)^2/14.4 = 0.47$
Total		400		400.0		$\sum_{i=1}^{7} \frac{(f_i - e_i)^2}{e_i} = 4.65$

□

Yates' Correction for Continuity

As we did for the normal approximation for discrete variables, we are using the chi-square distribution, which is for a continuous variable, to approximate probabilities associated with discrete variables. Thus, in certain cases we must make a correction in the statistic of (1) to compensate for the fact that we tend to underestimate the true probabilities when the appropriate chi-square distribution has only one degree of freedom. We implement the correction for continuity simply by subtracting 1/2 from the absolute value of $f_i - e_i$ $(i = 1, 2)$ before squaring this difference. It should be noted that we use the correction only if $|f_i - e_i| > 1/2$. The correction is commonly referred to as *Yates' correction for continuity.*

Example 4

In tossing a fair coin we expect 50 percent heads. Thus, for 300 tosses, the expected frequencies are 150 heads and 150 tails. Suppose that in 300 tosses of a coin we obtain 130 heads and 170 tails. Determine at the 5 percent significance level whether this is a fair coin.

Solution

The hypotheses of interest are

$$H_0: \pi_1 = \pi_2 \quad \text{versus} \quad H_1: \pi_1 \neq \pi_2.$$

We find the critical value from Table A-3 for $k - 1 = 2 - 1 = 1$ degree of freedom and $\alpha = 0.05$. This value is $\chi^2_{0.05,1} = 3.84$. Therefore, we accept H_0 if the observed value of the statistic of (1) is less than 3.84 and reject H_0 if it is greater.

Noting that we have one degree of freedom, we see immediately that we must use Yates' continuity correction, unless, of course $|f_i - e_i| \leqq 1/2$. We calculate the observed value of the statistic

$$\sum_{i=1}^{2} \frac{(|f_i - e_i| - 1/2)^2}{e_i} \tag{3}$$

as shown in Table 10.6. Since our observed value of the corrected (for continuity) statistic (3) is greater than the critical value $\chi^2_{0.05,1} = 3.84$, we reject H_0 and conclude that the coin is not fair.

Table 10.6
Goodness-of-fit calculations

Class i	f_i	e_i	$f_i - e_i$	$(\lvert f_i - e_i\rvert - 1/2)^2$	$(\lvert f_i - e_i\rvert - 1/2)^2/e_i$
Head	130	150	−20	$(19.5)^2$	2.535
Tail	170	150	20	$(19.5)^2$	2.535
Total	300	300		$\sum_{i=1}^{2} \frac{(\lvert f_i - e_i\rvert - 1/2)^2}{e_i} = 5.070$	

□

Related Remarks

The chi-square test for goodness-of-fit is certainly not the only goodness-of-fit test. One that is often better than the chi-square test is the *Kolmogorov-Smirnov* test, which treats the individual observations separately, rather than grouped into a frequency distribution. Moreover, the Kolmogorov-Smirnov test is applicable for small sample sizes, while the chi-square test is not. For detailed discussions concerning the Kolmogorov-Smirnov test, the reader is referred to Birnbaum (10.2), Massey (10.3), Lindgren (10.4), and Siegel (10.5).

In order to avoid breaking the continuity of the treatment of the chi-square distribution, we provide coverage of the Kolmogorov-Smirnov test in supplement 10.A.

Problems

1. An advertising agency wishes to test its belief that the proportions of a certain group of people in a major city who subscribe to various newspapers are as follows:

Newspaper A	39%
Newspaper B	14%
Newspaper C	23%
Newspaper D	24%

A random sample of 500 people from the group produced the following results:

Class	*Frequency*
Newspaper A	213
Newspaper B	88
Newspaper C	102
Newspaper D	97

Assuming the classes are mutually exclusive, determine at the 10% significance level whether the data support the proportions assumed by the advertising agency.

2. Given the table on the right, test the hypothesis that the distribution given by the e_i's is valid at the 5% significance level.

Class	e_i	f_i
A	2	1
B	3	5
C	5	6
D	7	6
E	10	11
F	13	12
G	17	16
H	22	24
I	30	30
J	18	20
K	7	4
L	2	1

3. Suppose we believe that the distribution of the number of telephone calls entering a switchboard per minute is distributed according to the Poisson probability distribution with $\lambda = 3.8$. A random sample of 200 minutes yielded the following results:

Class	*Frequency*	*Class*	*Frequency*
0	6	7	12
1	12	8	5
2	26	9	4
3	42	10	3
4	41	11	1
5	29	12	0
6	18	13	1

Determine at the 2% significance level whether the data support the hypothesized distribution.

4. The probability that a salesman will make a sale on any given call is believed to be 0.28. A random sample of 50 calls made by a salesman reveals that he made sales on 11 of them. Do the data support at the 1% significance level the hypothesized proportion of sales?

10.6 The Chi-Square Test of Independence

Quite often in the sciences, business, economics, and other research-oriented areas, we need to test hypotheses about the independence of two different classifications. For example, we may wish to test the hypothesis that customer reaction (favorable or unfavorable) to an increase in price for telephone service is independent of the type of service (business, residential, etc.) received. As another example, we may wish to determine whether the salary of a college graduate with five years of experience is independent of his major subject in college. Or, we may wish to test whether the color of an individual's eyes is independent of his hair color. In each case, we note that the population can be classified into at least two different categories for each of two different characteristics.

We shall examine a general method of testing the hypothesis that the two different characteristics are independent—that is, that the probabilities associated with one characteristic should not be affected by the state of the other characteristic. For example, if "customer reaction" and "type of service" are independent, the proportion of business customers strongly opposed to a rate increase should be the same as the proportion of residential customers having the same reaction. Of course, the proportions to which we refer are for the population of interest, not a sample. However, in order to test a hypothesis that two population characteristics are independent, we simply examine sample proportions associated with such characteristics. If the sample proportions are significantly different, we reject the hypothesis of independence; if they are not significantly different, we simply say that the sample does not contradict the null hypothesis (the independence of the characteristics). The chi-square test of independence is a special application of the goodness-of-fit test just described.

As an initial step in testing hypotheses of independence, we select a sample of n observations from the population of interest, classify them according to one characteristic of interest, then cross-classify them according to the other characteristic. For example, we may select a random sample of telephone users from the population of interest and classify them as business or residential users. Next, we may cross-classify them, for each user type, according to their reaction to the price increase.

Suppose that we select a sample of 200 telephone users and find that 78 are business users and 122 are residential users. Moreover, we classify each of the user types by their reaction to the rate increase as either "opposed" or "strongly opposed." We find that of the business users, 50 are opposed and 28 are strongly opposed. Of the residential users, we find that 80 are opposed and 42 are strongly opposed. This cross-classification can be represented as in Table 10.7.

Table 10.7
Cross-classification of reactions of telephone users

	Reaction		
User type	*Opposed*	*Strongly opposed*	*Totals*
Business	50	28	78
Residential	80	42	122
Total	130	70	200

The test statistic is determined in the same manner as that for the chi-square goodness-of-fit test; again, the related test is approximate. However, now, the population proportion for each category must be estimated. Recall from probability that if events A and B are independent events then $P(A \cap B) = P(A) \cdot P(B)$. Using the subscript i to represent the row and the subscript j to represent the column of Table 10.7, we represent the proportion of the population of users who are of the type i and of reaction j by π_{ij} and obtain the null hypothesis of independence,

$$H_0\colon \pi_{ij} = \pi_{i\cdot} \cdot \pi_{\cdot j} \tag{1}$$

for $i = 1, 2$, and $j = 1, 2$, where the subscripted "dot" represents a summation over the subscript it replaces. That is,

$$\pi_{i\cdot} \doteq \sum_{j=1}^{2} \pi_{ij} \quad \text{and} \quad \pi_{\cdot j} \doteq \sum_{i=1}^{2} \pi_{ij}. \tag{2}$$

Thus, to estimate π_{ij} under the null hypothesis, we need only estimate $\pi_{i\cdot}$ and $\pi_{\cdot j}$ by using the marginal totals of Table 10.7. That is, we estimate relevant population proportions by corresponding sample proportions as

$$\begin{aligned} \hat{\pi}_{1\cdot} &= p_{1\cdot} = \frac{78}{200} = 0.39; & \hat{\pi}_{2\cdot} &= p_{2\cdot} = \frac{122}{200} = 0.61; \\ \hat{\pi}_{\cdot 1} &= p_{\cdot 1} = \frac{130}{200} = 0.65; & \hat{\pi}_{\cdot 2} &= p_{\cdot 2} = \frac{70}{200} = 0.35. \end{aligned} \tag{3}$$

Using the entities of Table 10.9, we obtain the expected value for each cross-classification as

$$e_{11} = \frac{n_{1\cdot} \cdot n_{\cdot 1}}{n} = \frac{250 \cdot 75}{500} = 37.50, \qquad e_{12} = \frac{n_{1\cdot} \cdot n_{\cdot 2}}{n} = \frac{250 \cdot 110}{500} = 55.00,$$

$$e_{13} = \frac{n_{1\cdot} \cdot n_{\cdot 3}}{n} = \frac{250 \cdot 165}{500} = 82.50, \qquad e_{14} = \frac{n_{1\cdot} \cdot n_{\cdot 4}}{n} = \frac{250 \cdot 150}{500} = 75.00,$$

$$e_{21} = \frac{n_{2\cdot} \cdot n_{\cdot 1}}{n} = \frac{135 \cdot 75}{500} = 20.25, \qquad e_{22} = \frac{n_{2\cdot} \cdot n_{\cdot 2}}{n} = \frac{135 \cdot 110}{500} = 29.70,$$

$$e_{23} = \frac{n_{2\cdot} \cdot n_{\cdot 3}}{n} = \frac{135 \cdot 165}{500} = 44.55, \qquad e_{24} = \frac{n_{2\cdot} \cdot n_{\cdot 4}}{n} = \frac{135 \cdot 150}{500} = 40.50,$$

$$e_{31} = \frac{n_{3\cdot} \cdot n_{\cdot 1}}{n} = \frac{115 \cdot 75}{500} = 17.25, \qquad e_{32} = \frac{n_{3\cdot} \cdot n_{\cdot 2}}{n} = \frac{115 \cdot 110}{500} = 25.30,$$

$$e_{33} = \frac{n_{3\cdot} \cdot n_{\cdot 3}}{n} = \frac{115 \cdot 165}{500} = 37.95, \qquad e_{34} = \frac{n_{3\cdot} \cdot n_{\cdot 4}}{n} = \frac{115 \cdot 150}{500} = 34.50.$$

The observed frequencies are given in Table 10.10, with the expected frequencies in parentheses.

Table 10.10

Observed and expected frequencies

	Type of work				
Reaction	*Production*	*Clerical*	*Administration*	*Non-administrative professional*	*Total*
Better than existing plan	40 (37.50)	55 (55.00)	75 (82.50)	80 (75.00)	250
Comparable to existing plan	15 (20.25)	30 (29.70)	50 (44.55)	40 (40.50)	135
Not so good as existing plan	20 (17.25)	25 (25.30)	40 (37.95)	30 (34.50)	115
Total	75	110	165	150	500

Using these, we compute the observed value of the test statistic of (5) to be

$$\sum_{i=1}^{3} \sum_{j=1}^{4} \frac{(f_{ij} - e_{ij})^2}{e_{ij}} = \frac{(40 - 37.50)^2}{37.50} + \frac{(55 - 55.00)^2}{55.00} + \frac{(75 - 82.50)^2}{82.50}$$
$$+ \frac{(80 - 75.00)^2}{75.00} + \frac{(15 - 20.25)^2}{20.25} + \frac{(30 - 29.70)^2}{29.70}$$
$$+ \frac{(50 - 44.55)^2}{44.55} + \frac{(40 - 40.50)^2}{40.50} + \frac{(20 - 17.25)^2}{17.25}$$
$$+ \frac{(25 - 25.30)^2}{25.30} + \frac{(40 - 37.95)^2}{37.95} + \frac{(30 - 34.50)^2}{34.50}$$
$$= 4.359.$$

Since the observed value (4.359) of the statistic of (5) is less than the critical value $\chi^2_{0.05,6} = 12.59$, we accept H_0 and conclude that the true proportions corresponding to the three responses are the same for all four types of employees. In other words, the response of a typical employee is independent of employee type. □

Problems

1. A large company has conducted a survey of its customers to determine the degree of their satisfaction with the sales service. A random sample of 200 customers was selected and cross-classified as follows:

Degree of satisfaction	*Size of account* Small	Medium	Large	*Total*
Completely unsatisfactory	14	25	5	44
Needs improvement	29	34	6	69
Satisfactory	11	33	18	62
Exceptional	6	8	11	25
Total	60	100	40	200

Determine at the 2% significance level whether the data support the hypothesis that the size of the account is independent of the degree of satisfaction with the sales service.

2. A company wishes to determine whether the salaries of its personnel are independent of their scores on the company aptitude test. A random sample of 1,000 employees was cross-classified as follows:

Annual salary	*Test scores* <300	300–449	450–549	550–700	>700	*Total*
<\$8,000	53	76	208	94	19	450
\$8,000 to \$15,000	36	51	149	67	47	350
>\$15,000	11	23	93	39	34	200
Total	100	150	450	200	100	1,000

Determine at the 5% significance level whether the data support the hypothesis that the salaries are independent of the scores on the aptitude test.

3. A firm wishes to determine whether the completion time of the computer memory modules it manufactures by hand is independent of whether or not the assemblies pass the rigorous final inspection. A random sample of 150 of the modules was cross-classified as follows:

Assembly time (hours)	*Final inspection* Accepted	Rejected	*Total*
<40	28	12	40
40 to 60	31	39	70
>60	26	14	40
Total	85	65	150

Determine at the 10% significance level whether the data support the hypothesis that the assembly times are independent of the results of the final inspections.

10.7 Summary

While the chi-square distribution is important in its own right, a primary aim in this chapter has been to develop the distribution from a definitional standpoint, so as to enhance your understanding of its use later. Thus, we have emphasized that the sampling distribution of a chi-square statistic with k degrees of freedom is essentially the sum of squares of k independent standard normal variables.

In illustrating some of the uses of the chi-square distribution, we have viewed the distribution as it applies in problems of setting confidence intervals and testing hypotheses for the variance of a normal population; conducting goodness-of-fit tests; and conducting tests of independence. The chi-square distribution has a wide variety of uses we have not attempted to cover. However, many are of the approximation type, and should be used with extreme care. In fact several tests involving the chi-square distribution are among the most misused tests in the field of statistics. Thus, before employing the chi-square distribution in any situation, you should be sure that you fully understand relevant assumptions and consequences.

Supplement 10.A Kolmogorov–Smirnov Test

When considering the chi-square test for goodness-of-fit in Section 10.5, we noted that the *Kolmogorov-Smirnov test* is often better than the chi-square test. This test treats the individual observations separately, rather than grouping them into a frequency distribution, and it is applicable for small sample sizes. Like any goodness-of-fit test, it is concerned with the degree of agreement between a set of sample values and a specified theoretical distribution.

The Kolmogorov–Smirnov test involves determining the cumulative frequency distribution for a hypothesized theoretical distribution and comparing it with the observed cumulative frequency distribution. The statistic of interest is the maximum difference D that occurs at the point the values of the theoretical cumulative distribution function reach their greatest divergence. The statistic D is then compared against a critical value, obtained from the sampling distribution of D for the situation where the theoretical distribution is the one from which the sample is taken. Although we shall not consider the specific characteristics of the sampling distribution for D, we do need the critical values associated with the distribution to compare against the observed difference between the two cumulative distribution functions. For these, we use Table A-11 in the Appendix.

The methodology associated with the Kolmogorov–Smirnov test requires that we completely specify the theoretical cumulative distribution function for the null hypothesis H_0. We denote this function by $F_0(x)$. Thus, for any value x, the value of $F_0(x)$ is the proportion of the population values less than or equal to x. Next, we let $S_n(x)$ represent the observed cumulative frequency distribution function of a random sample of n values. That is, for any possible sample value x, $S_n(x) = k/n$, where k is the number of sample values less than or equal to x.

Now, if the distribution from which we are sampling is indeed the one specified in H_0, we would expect $F_0(x)$ to be reasonably close to $S_n(x)$ for every value of x. Any deviations between the two values would be considered as random variation. The Kolmogorov–Smirnov test concentrates on the maximum absolute difference between $F_0(x)$ and $S_n(x)$. Thus, the maximum deviation is given by

$$D = \text{Maximum}|F_0(x) - S_n(x)|.$$

The sampling distribution for D, when H_0 is true, is known. The critical values of D are given in Table A.11. If the observed value of D is less than the table value for the corresponding sample size and significance level, we accept H_0; otherwise, we reject H_0. When we accept H_0, we conclude that the population from which the sample was taken has as its cumulative distribution function the theoretical function given in H_0.

Example 1

The following ten values are thought to be from a normal distribution with mean 28 and variance 4:

26.0, 26.4, 28.3, 28.4, 28.5, 28.7, 29.4, 29.9, 31.2, 32.0

Determine whether or not the data are from a normal distribution with mean 28 and variance 4. Use the 5 percent level of significance.

Solution

Since we wish to determine if the data are from a normal distribution with mean 28 and variance 4, we have

$$H_0: X \sim \text{N}(28, 4) \quad \text{versus} \quad H_1: X \text{ is not distributed N}(28, 4).$$

Next, we determine the values of the theoretical and observed cumulative distribution functions. These values and their difference for each observed value are given in Table 10.11. The $F_0(x)$ values are obtained by determining $P(X \leq x)$ for a normal variable X with mean 28 and variance 4. That is, the values are obtained from Table A-1, where $z = (x - 28)/2$. The values of $S_{10}(x)$ are obtained by $S_{10}(x) = k/10$, where k represents the number of observed values which are less than or equal to x.

Table 10.11 shows that the difference with the largest absolute value corresponds to x value of 28.3. Thus, $D = 0.2596$. Also, from Table A-11, we determine the critical value of D for a sample size of 10 and a significance level of 5 percent. This critical value,

Table 10.11
Goodness-of-fit computations

k	x	$F_0(x)$	$S_{10}(x)$	$F_0(x) - S_{10}(x)$
1	26.0	0.1587	0.1000	0.0587
2	26.4	0.2119	0.2000	0.0119
3	28.3	0.5596	0.3000	0.2596
4	28.4	0.5793	0.4000	0.1793
5	28.5	0.5987	0.5000	0.0987
6	28.7	0.6368	0.6000	0.0368
7	29.4	0.7580	0.7000	0.0580
8	29.9	0.8289	0.8000	0.0289
9	31.2	0.9452	0.9000	0.0452
10	32.0	0.9773	1.0000	−0.0227

denoted $D_{.05,10}$, is 0.409. Comparing the critical value of D against the observed value of D, we note that the observed value 0.2596 is less than the critical value of 0.409. Hence, we accept H_0 and conclude that the data may indeed come from a normal distribution with mean 28 and variance 4. □

Important Terms and Concepts

additive property for chi-square variables

confidence interval for σ_x^2

chi-square distribution

chi-square variable

degrees of freedom

expected frequency

goodness-of-fit test

Kolmogorov–Smirnov test

observed frequency

test of hypothesis about σ_x^2

test of independence

References

10.1 A. M. Mood and F. A. Graybill, *Introduction to the Theory of Statistics*, New York: McGraw-Hill Book Company, 1963.

10.2 Z. W. Birnbaum "Distribution-free Tests of Fit for Continuous Distribution Functions," *Annals of Mathematical Statistics*, vol. 24, pp. 1–8, 1953.

10.3 F. J. Massey, Jr., "The Kolmogorov–Smirnov Test for Goodness of Fit," *Journal of the American Statistical Association*, vol. 46, pp. 68–78, 1951.

10.4 B. W. Lindgren, *Statistical Theory*, New York: The Macmillan Company, 1962.

10.5 S. Siegel, *Nonparametric Statistics for the Behavioral Sciences*, New York: McGraw-Hill Book Company, 1956.

10.6 W. J. Conover, *Practical Nonparametric Statistics*, New York: John Wiley & Sons, Inc., 1971.

10.7 J. D. Gibbons, *Nonparametric Statistical Inference*, New York: McGraw-Hill Book Company, 1971.

Review Problems

1. Suppose we have $U_1 \sim \chi^2_{10}$, $U_2 \sim \chi^2_4$, and $U_3 \sim \chi^2_{11}$, with U_1, U_2, and U_3 *independent*. For $U_4 = U_1 + U_2 + U_3$, determine the value a such that $P(U_4 < a) = 0.02$.

2. The variance of the residual noise levels of audio tapes produced by a certain manufacturer was 9 db^2 before the manufacturing process was changed. A random sample of 18 tapes selected after the change produced an unbiased estimate of the population variance of 15 db^2. Assuming the *distribution of noise levels is normal*, has there been a significant change at the 5% significance level in the variance of the noise levels?

3. During a limited offering of a special purchase, each customer is allowed only 1 of the items. It is believed that the number of items purchased by each customer is distributed according to the binomial probability distribution with $n = 1$ and $\pi = 0.3$. A random sample of 100 customers consisted of 61 who bought none and 39 who bought one. At the 5% significance level, do the data support the hypothesized distribution?

4. Given that U_1, U_2, and U_3 are independent chi-square variables such that

$$U_1 \sim \chi^2_2, \quad U_2 \sim \chi^2_4, \quad \text{and} \quad U_3 \sim \chi^2_9,$$

determine the following probabilities.

a. $P(U_1 + U_2 + U_3 \leq 22.3)$.

b. $P(U_1 + U_2 \leq 12.59 \cap U_3 > 8.34)$.

c. $P(U_1 > 5.99 \cup U_2 + U_3 < 9.30)$.

d. $P(U_1 + U_3 > 11.00 \cap U_2 > 8.50)$.

e. $P(U_1 + U_3 < 11.00 \cup U_2 < 8.50)$.

(Sections 4.7 and 4.8)

5. If U_1 and U_2 are independent chi-square variables, does it seem conceivable that $U_1 - U_2$ could be distributed according to the chi-square distribution? Justify your answer with an intuitive argument.

(Sections 3.5, 4.11, 4.13, and 6.10)

6. The demand per day (a discrete variable) for a given product is believed to be normally distributed. The frequencies for 1,000 days are given below:

Demand/day	*Frequency*
61–65	25
66–70	50
71–75	110
76–80	350
81–85	320
86–90	95
91–95	40
96–100	10
Total	1,000

Determine at the 5% significance level whether daily demand is normally distributed.

(Sections 3.7, 3.8, 6.4, and 6.7)

Eleven

Student's t Distribution

11.1 Introduction

For the most part, our concern in the past three chapters has been to make probability statements, determine confidence intervals, and test hypotheses in the situation where the appropriate statistic is normally distributed. In all cases, we have assumed that the variance is known, and that either we are selecting a sample from a normal population or the sample size is sufficiently large to use the central limit theorem.

If either of the above assumptions is not satisfied, we must determine other ways of making probability statements, determining confidence intervals, and testing hypotheses. One way to obviate these assumptions is to construct the sampling distribution of the appropriate statistic by empirical methods, as was done in Chapter 7. However, as we pointed out there, these methods are too laborious and time consuming to be practical. Another way is to determine what occurs when one assumption is satisfied and the other is not. This is precisely what was done by W. S. Gossett, a statistician who wrote under the pseudonym "Student."

Gossett derived the exact distribution of the statistic

$$\frac{\overline{X} - \mu_x}{\hat{\sigma}_x/\sqrt{n}} = \frac{\overline{X} - \mu_x}{\sqrt{\dfrac{\sum_{i=1}^{n}(X_i - \overline{X})^2}{n(n-1)}}} \tag{1}$$

for the situation in which a sample of any size n is selected from a normal population having an unknown variance. This distribution (known as *Student's t distribution*, *Student's distribution*, or, simply, the *t distribution*) and its relationship to probability, confidence intervals, and statistical hypotheses will be our primary concern here. We shall refer to the distribution as the t distribution.

11.2 Properties of the t Distribution

Let us begin by defining the t distribution.

Definition 11.1

If Z and U are independent random variables such that Z is distributed normally with mean zero and variance one and U is distributed according to the chi-square distribution with k degrees of freedom, then the statistic

$$W = \frac{Z}{\sqrt{U/k}} \tag{1}$$

is distributed according to the t distribution with k degrees of freedom.

We may express Definition 11.1 symbolically as follows: If Z and U are independent random variables such that

$$Z \sim \mathrm{N}(0, 1) \tag{2}$$

and

$$U \sim \chi_k^2, \tag{3}$$

then

$$\frac{Z}{\sqrt{U/k}} \sim t_k. \tag{4}$$

Note, as with our definition of the chi-square distribution (Definition 10.1), the entities involved in the algebraic manipulation (Z and U) are variables, not sample observations. Strictly speaking, we mean by (1) that the variable W is obtained by taking the ratio of the variable Z (a standard normal variable) and the square root of the variable U (a chi-square variable with k degrees of freedom) divided by its number of degrees of freedom k. Again, we can best describe the resulting distribution indirectly by using the basic idea of infinite repeated sampling as it relates to the concept of a sampling distribution.

Suppose we have a standard normal distribution (Z) and a chi-square distribution (U) with k degrees of freedom such that the two distributions are independent, and we repeatedly select a sample of one item drawn from each of the independent distributions. Let z_i and u_i represent single observations selected in the ith sample from the distributions of Z and U, respectively. Conceptually, a t distribution with k degrees of freedom would result from the following process.

1. Take one observation (say z_i) from the distribution of Z and one observation (say u_i) from the distribution of U.
2. Compute a single observation from a t distribution with k degrees of freedom as

$$w_i = z_i/\sqrt{u_i/k}.$$

3. Repeat steps 1 and 2 an infinite number of times; that is, for $i = 1, 2, \ldots, \infty$.
4. Compile the probability distribution of the w_i's ($i = 1, 2, \ldots, \infty$). The result will be the probability distribution of W, a t distribution with k degrees of freedom.

Since it is impossible to repeat the sampling procedure an infinite number of times, you should visualize the distribution that would result from a very large number of repeated samplings. We see by Definition 11.1 that the number of degrees of freedom associated with a t variable is simply the number of degrees of freedom associated with the chi-square variable that conceptually goes to make up the variable. You may wish to refer to Section 10.1 for our original discussion of the meaning of the number of degrees of freedom associated with a chi-square variable.

Note that our use of the lowercase t for the general variable is contrary to our usual policy of using capital letters for general variables and lowercase letters for single possible values of a variable. We make this departure in notation because t is so widely used for this variable, and also because T is generally used in statistics for a different purpose. Notation for distinguishing the general t variable from possible values is discussed later.

The probability density function for a t variable with k degrees of freedom is given by

$$f(t; k) = \frac{[(k-1)/2]!}{\sqrt{k\pi}[(k-2)/2]!} \cdot \frac{1}{[1 + t^2/k]^{(k+1)/2}} \qquad (-\infty < t < \infty). \tag{5}$$

As one might expect from the definition, the range of the t variable is the same as that of a standard normal variable. Further, as with a standard normal variable, the mean of the probability distribution of a t variable is zero, and the distribution is symmetrical about zero. The variance of a t variable with k degrees of freedom can be shown to be $k/(k-2)$ for $k > 2$. That is,

$$E(t) = 0 \qquad \text{and} \qquad \text{Var}(t) = \frac{k}{k-2} \qquad \text{for } k > 2.$$

As in the case of the chi-square distribution, probabilities associated with the t distribution depend on the number of degrees of freedom. Thus, there are an infinite number of t distributions, and it is obviously impossible to tabulate probability information for all of them. We circumvent this problem by presenting less detailed probability information for the t distribution than for the normal distribution, and presenting probability information only for t variables involving a not-too-large number of degrees of freedom.

Selected probability information for various t distributions is given in Table A-4 of the Appendix. Note that the table provides values for

$$P(W_k > w_k), \tag{6}$$

where W_k represents the general t variable with k degrees of freedom and w_k represents any positive value W_k can assume. We emphasize the term "positive" because, in this sense, the probability information given for the t distributions in Table A-4 is similar to that provided for the standard normal variable in Table A-1. In other words, since the t distribution is symmetrical about zero.

$$P(W_k < -w_k) = P(W_k > w_k), \tag{7}$$

(that is, the area under the curve to the left of a negative t value is equal to the area under the curve to the right of the corresponding positive t value) and we need only present probability information regarding positive t values. The reasoning here is similar to that for the standard normal distribution, which is also symmetrical about zero.

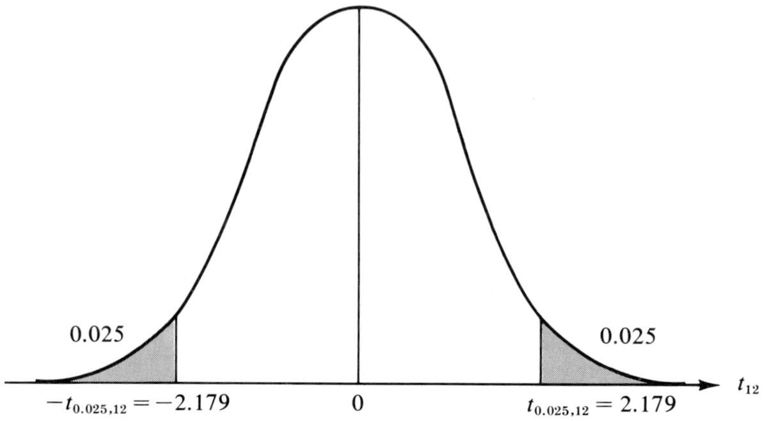

Figure 11.1

A t-distribution with 12 degrees of freedom

Figure 11.1 illustrates a t distribution with 12 degrees of freedom. Note that the area in each tail is 0.025. As usual, there are two general types of probability problems that can be solved. First, we may wish to find the probability (say α) that a t variable with k degrees of freedom is greater than a specific positive value, say b. That is, given k and b, what is α such that

$$P(t_k > b) = \alpha, \tag{8}$$

where t_k represents a t variable with k degrees of freedom? To answer such question, we

1. find k in the left column of Table A-4,
2. find b in the same row of Table A-4 as k,
3. read α [the probability of (8)] at the top of the column in which b appears.

Second, given k and α, we may wish to find the specific value (say b) so that

$$P(t_k > b) = \alpha.$$

To answer this question, we

1. find k in the left column of Table A-4,
2. find α in the top row of Table A-4,
3. read b in the body of Table A-4 where α's column intersects k's row.

Naturally, there are many variations of the foregoing problems. However, any such problems can be solved through relationships [similar to (7)] analogous to those associated with the standard normal distribution, which we treated in great detail in Chapter 6. That is

$$P(W_k < w_k) = 1 - P(W_k > w_k) \tag{9}$$

and

$$P(W_k > -w_k) = 1 - P(W_k > w_k), \tag{10}$$

where, again, W_k represents a general t variable with k degrees of freedom and w_k represents any positive value W_k can assume. Also,

$$P[(w_k)_1 < W_k < (w_k)_2] = P[W_k > (w_k)_1] - P[W_k > (w_k)_2], \tag{11}$$

where $(w_k)_1$ and $(w_k)_2$ are any values W_k can assume, subject only to the restriction that $(w_k)_1 < (w_k)_2$.

Example 1

Find the value so that the probability is 0.10 that a t variable with $k = 13$ degrees of freedom is greater than such value; that is, find b such that

$$P(t_{13} > b) = 0.10.$$

Solution

From the body of Table A-4, we see that $b = 1.350$. Thus,

$$P(t_{13} > 1.350) = 0.10. \qquad \square$$

Example 2

Find the probability that a t variable with $k = 17$ degrees of freedom is greater than 2.567.

Solution

From Table A-4, we have

$$P(t_{17} > 2.567) = 0.01. \qquad \square$$

Example 3

Find

(a) $P(t_{18} > 2.101)$. (b) $P(t_{27} > -1.314)$.
(c) $P(t_{23} < 0.858)$. (d) $P(t_{19} < -1.066)$.
(e) $P(0.687 < t_{20} < 2.528)$. (f) $P(-1.059 < t_{24} < 1.711)$.
(g) the value b such that $P(t_{24} > b) = 0.85$.
(h) the value b such that $P(t_{21} < b) = 0.025$.
(i) the value b such that $P(b < t_{28} < 1.701) = 0.75$.

Solution

Make sure you follow the reasoning involved in each solution.

(a) $P(t_{18} > 2.101) = 0.025$.

(b) $P(t_{27} > -1.314) = 1 - P(t_{27} > 1.314) = 1 - 0.10 = 0.90$.

(c) $P(t_{23} < 0.858) = 1 - P(t_{23} > 0.858) = 1 - 0.20 = 0.80$.

(d) $P(t_{19} < -1.066) = P(t_{19} > 1.066) = 0.15$.

(e) $$\begin{aligned} P(0.687 < t_{20} < 2.528) &= P(t_{20} > 0.687) - P(t_{20} > 2.528) \\ &= 0.25 - 0.01 = 0.24. \end{aligned}$$

(f) $$\begin{aligned} P(-1.059 < t_{24} < 1.711) &= P(t_{24} > -1.059) - P(t_{24} > 1.711) \\ &= 1 - P(t_{24} > 1.059) - P(t_{24} > 1.711) \\ &= 1 - 0.15 - 0.05 = 0.80. \end{aligned}$$

(g) $P(t_{24} > b) = 0.85$ implies that b is negative. Thus, let us denote the entity for which we desire a solution as $-c$, where $c > 0$, i.e., $b = -c$. Then,

$$\begin{aligned} P(t_{24} > b) = P(t_{24} > -c) = 0.85 &\Rightarrow 1 - P(t_{24} < -c) = 0.85 \\ &\Rightarrow P(t_{24} > c) = 0.15 \\ &\Rightarrow c = 1.059 \\ &\Rightarrow b = -c = -1.059. \end{aligned}$$

(h) $P(t_{21} < b) = 0.025$ implies that b is negative. Again, let $b = -c$. Then,

$$\begin{aligned} P(t_{21} < b) = P(t_{21} < -c) = 0.025 &\Rightarrow P(t_{21} > c) = 0.025 \\ &\Rightarrow c = 2.080 \\ &\Rightarrow b = -c = -2.080. \end{aligned}$$

(i) $$\begin{aligned} P(b < t_{28} < 1.701) = 0.75 &\Rightarrow P(t_{28} > b) - P(t_{28} > 1.701) = 0.75 \\ &\Rightarrow P(t_{28} > b) - 0.05 = 0.75 \\ &\Rightarrow P(t_{28} > b) = 0.80. \end{aligned}$$

Again, this implies that b is negative. Letting $b = -c$, we have

$$\begin{aligned} P(t_{28} > b) = P(t_{28} > -c) = 0.80 &\Rightarrow 1 - P(t_{28} > c) = 0.80 \\ &\Rightarrow P(t_{28} > c) = 0.20 \\ &\Rightarrow c = 0.855 \\ &\Rightarrow b = -c = -0.855. \end{aligned}$$ □

Note that, just as in the case of the chi-square distribution (Table A-3), each row of Table A-4 represents a different probability distribution. Thus, since we are again (as in Table A-3) attempting to present as much information in one row as we have in the entire body of Table A-1, it is obvious that in many situations we can only approximate probabilities. However, as with chi-square probabilities, a bracketing (see Example 4 of Sec. 10.2) of the required probability will suffice for most of the problems involving probabilities associated with a t distribution.

We use subscripts to distinguish a general t variable from a single possible value of such variable. Thus, we denote by t_k the general t variable with k degrees of freedom; and we denote by $t_{\alpha,k}$ the single possible value of the general t variable with k degrees of freedom such that

$$P(t_k > t_{\alpha,k}) = \alpha. \tag{12}$$

For example, the value of the t variable with 18 degrees of freedom which has 0.20 area to its right is 0.862, and it is denoted by $t_{0.20,18} = 0.862$. That is,

$$P(t_{18} > t_{0.20,18}) = P(t_{18} > 0.862) = 0.20.$$

We have repeatedly mentioned that the t distribution and the standard normal distribution possess similar properties. In Figure 11.2 we compare t distributions of

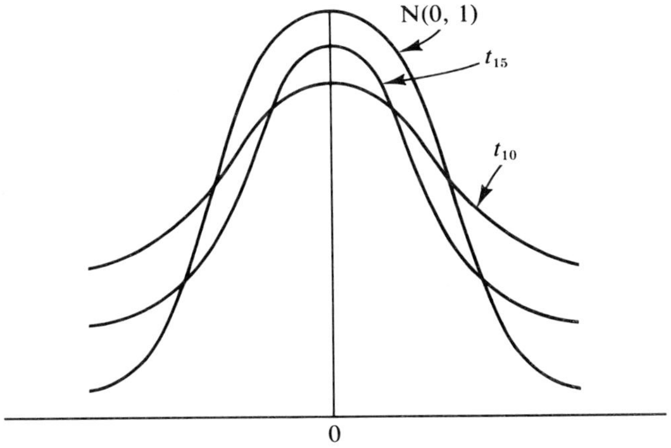

Figure 11.2

Comparison of the standard normal distribution and exemplary t-distributions

10 and 15 degrees of freedom. While the figure is not drawn to scale, it shows the general relationships between the two t distributions and the standard normal distribution. We see that the normal distribution is more concentrated around zero, while the two t distributions have more area under the tails. Further, we see that the t distribution with 15 degrees of freedom is more nearly like the normal distribution than the one with 10 degrees of freedom. This is true because, generally, the t distribution approaches the normal distribution as the number of degrees of freedom increases—as may be noted from the values given in Table A-4 for an infinite number of degrees of freedom, where the values are identical to those for corresponding probabilities from the standard normal distribution (Table A-1). The largest number of degrees of freedom for which t values and corresponding probabilities are given in Table A-4 is $k = 120$. Yet, note, for fixed probabilities, how close the t values for a t distribution with 120 degrees of freedom are to those with an infinite number of degrees of freedom (that is, values for a standard normal variable). This gives us an important license—for a large number of degrees of freedom, say greater than 120, we may use the normal distribution in lieu of the t distribution.

Problems

1. Define a t distribution with k degrees of freedom.
2. What is meant by the number of degrees of freedom for a t variable?
3. Determine

   a. $P(t_{10} > 2.764)$. b. $P(t_{15} > 2.947)$. c. $P(t_8 < 1.397)$.

   d. $P(t_{29} < 2.756)$. e. $P(-2.878 < t_{18} < -0.862)$.

   f. $P(-1.721 < t_{21} < 2.831)$. g. $P(0.690 < t_{16} < 2.921)$.

   h. the value a such that $P(t_{14} > a) = 0.10$.

   i. the value a such that $P(t_{22} < a) = 0.05$.

   j. the value a such that $P(-2.110 < t_{17} < a) = 0.97$.
4. Determine

   a. $P(t_{19} > 2.20)$. b. $P(t_3 < 2.00)$. c. $P(-1.440 < t_6 < 3.50)$.

   d. the value a such that $P(t_{11} > a) = 0.13$.

   e. the value a such that $P(t_{12} < a) = 0.23$.

   f. the value a such that $P(-1.372 < t_{10} < a) = 0.55$.

11.3 Testing Hypotheses Concerning the Mean of a Normal Population (Variance Unknown)

We observed in Chapter 10 that when sampling is from a normal distribution with mean μ_x and variance σ_x^2, the statistic

$$\sum_{i=1}^{n}\left(\frac{X_i - \overline{X}}{\sigma_x}\right)^2 \sim \chi^2_{n-1}. \tag{1}$$

Also, from Chapter 6, we know that if we are sampling from a normal distribution, the statistic

$$\frac{\overline{X} - \mu_x}{\sigma_x/\sqrt{n}} \sim \mathrm{N}(0, 1). \tag{2}$$

Further, even when the statistics of (1) and (2) are conceptually obtained through sampling from the same normal population, they are independent. This being true, if we take the ratio of the statistic of (2) and the square root of the statistic of (1) divided by the appropriate number of degrees of freedom $(n - 1)$, we will obtain (by Definition 11.1) a statistic having a t distribution with $(n - 1)$ degrees of freedom. Thus, the statistic

$$\frac{\dfrac{\bar{X} - \mu_x}{\sigma_x/\sqrt{n}}}{\sqrt{\dfrac{\sum_{i=1}^{n}\left(\dfrac{X_i - \bar{X}}{\sigma_x}\right)^2}{n-1}}} \sim t_{n-1}. \tag{3}$$

We can simplify (3) to obtain

$$\frac{\bar{X} - \mu_x}{\dfrac{\sigma_x}{\sqrt{n}} \cdot \dfrac{1}{\sigma_x}\sqrt{\dfrac{\sum_{i=1}^{n}(X_i - \bar{X})^2}{n-1}}} = \frac{\bar{X} - \mu_x}{\sqrt{\dfrac{\sum_{i=1}^{n}(X_i - \bar{X})^2}{n(n-1)}}} \sim t_{n-1}. \tag{4}$$

Note, once again, that our use of the capital letter X in Relation (4), implies a general conceptual statistic. As soon as a specific sample is selected, so that we can refer to specific observations of X, we would denote a single observed value of the statistic by

$$\frac{\bar{x} - \mu_x}{\sqrt{\dfrac{\sum_{i=1}^{n}(x_i - \bar{x})^2}{n(n-1)}}}.$$

Relation (4) provides us immediately with a pivotal quantity for problems involving a normal population with unknown variance. Thus, the essential steps for making a one-tail, 100α percent significance test concerning the mean μ_x of a normal population with unknown variance are as follows.

1. Set up the null hypothesis $H_0: \mu_x = \mu_0$, where μ_0 is any value μ_x can assume.
2. Specify the level of significance α.
3. Determine the size (n) of the sample to be taken. (This fixes the number $(n - 1)$ of degrees of freedom to be used in establishing the appropriate critical value.)
4. Determine a critical value t_{n-1}^* that divides the t distribution with $(n - 1)$ degrees of freedom into a critical (rejection) region and an acceptance region. This value should be such that 100α percent of the area of the t distribution with $(n - 1)$ degrees of freedom lies in the critical region. That is, for a left-tail alternative hypothesis,

$$H_1: \mu_x < \mu_0,$$

find t_{n-1}^* such that

$$P(t_{n-1} < t_{n-1}^*) = \alpha. \tag{5}$$

In our previous notation for tabular t values, we have from (5) that for a left-tail alternative hypothesis,

$$t_{n-1}^* = -t_{\alpha, n-1}. \tag{6}$$

On the other hand, for a right-tail alternative hypothesis,

$$H_1: \mu_x > \mu_0,$$

we determine t^*_{n-1} such that

$$P(t_{n-1} > t^*_{n-1}) = \alpha. \tag{7}$$

This tells us immediately that, for a right-tail alternative hypothesis,

$$t^*_{n-1} = t_{\alpha, n-1}. \tag{8}$$

5. From a sample of size n, obtain (under the assumption that the null hypothesis is true, or $\mu_x = \mu_0$) one observed value of the statistic of (4); that is, compute

$$\frac{\bar{x} - \mu_0}{\sqrt{\dfrac{\sum_{i=1}^{n} (x_i - \bar{x})^2}{n(n-1)}}}, \tag{9}$$

where x_i is the ith sample observation, $\bar{x}$ is the observed sample mean, and μ_0 is the value of μ_x given by the null hypothesis.
6. If the value of (9) is in the critical region, reject H_0. Otherwise, accept H_0.

The corresponding two-tail, 100α percent significance test is the same except that we must find two critical values. For a symmetrical test, they are

$$(t_{n-1})^*_1 = -t_{\alpha/2, n-1} \quad \text{and} \quad (t_{n-1})^*_2 = t_{\alpha/2, n-1}. \tag{10}$$

For a nonsymmetrical, two-tail, 100α percent test we still force the two tail areas to total α, but relax the requirement that they be equal.

Because

$$\frac{\sum_{i=1}^{n} (x_i - \bar{x})^2}{n(n-1)} = \frac{s_x^2}{n-1} = \frac{\hat{\sigma}_x^2}{n},$$

where s_x^2 is the sample variance and $\hat{\sigma}_x^2$ is the unbiased estimate of the population variance (Equations 11 and 12, Sec. 10.3), it is often convenient to calculate and use either

$$\frac{\bar{x} - \mu_0}{s_x/\sqrt{n-1}} \tag{11}$$

or

$$\frac{\bar{x} - \mu_0}{\hat{\sigma}_x/\sqrt{n}} \tag{12}$$

instead of (9) in conducting the t test.

Note carefully that now even the central limit theorem does not allow us to ignore the form of the distribution of the population from which we are sampling. This becomes apparent when we consider the derivation of the relevant t statistic [(1) through (4)]. For the variable of (1) to be a chi-square variable, the population from which we are sampling must have a normal distribution. Otherwise, Definition 10.1 of a chi-square variable will not be applicable. Thus, even though (2) is valid either when sampling is accomplished from a normal distribution or when the sample size is sufficiently large, the central limit theorem will apply only for the variable of (2) and not for the variable of (1). Thus, we can use the t test only when the distribution from which we are sampling is normal. Since the standard normal distribution and the t distribution are very similar, few problems should arise in testing hypotheses concerning μ_x when σ_x is unknown. The primary factors to remember when deciding which statistic to use are:

1. We use the standard normal statistic when we know the variance and either we take a sample of any size n from a normal population, or we take a sufficiently large sample from a population having practically any distribution.
2. We use the t statistic when we do not know the variance and take a sample of any size n from a normal population.

Example 1

A manufacturer of food products knows that a packaging process for dried beans is such that the weights of the packages are normally distributed. The mean of the weights is 18 ounces when the process is operating properly. However, the variance of the weights is unknown. The manufacturer selects a random sample of 17 packages from the process and determines the sample mean and sample variance to be $\bar{x} = 18.9$ ounces and $s_x^2 = 3.6$ square ounces. Determine at the 5 percent significance level whether the packaging process is out of control.

Solution

Since the process will be out of control if the average weight is either significantly greater or significantly less than 18 ounces, we use a two-tail test. Thus, our hypotheses are

$$H_0 : \mu_x = 18 \quad \text{versus} \quad H_1 : \mu_x \neq 18. \tag{13}$$

Since the population from which we are sampling is normal, we can test the null hypothesis, even though the variance is unknown, by using the t statistic. To determine the decision rule for our test, we note that our t statistic has $n - 1 = 16$ degrees of freedom. For the 5 percent significance level, we have $\alpha = 0.05$, or $\alpha/2 = 0.025$ for each tail. From Table A-4, we find

$$t_{0.025, 16} = 2.120,$$

which gives us a decision rule of accepting H_0 if the observed value of t is between -2.120 and 2.120 and rejecting H_0 if it is not between these values. This is shown graphically in Figure 11.3.

From (9) and (11) we compute the observed value of the t statistic,

$$\frac{\bar{x} - \mu_0}{\sqrt{\dfrac{\sum_{i=1}^{n} (x_i - \bar{x})^2}{n(n-1)}}} = \frac{\bar{x} - \mu_0}{s_x/\sqrt{n-1}} = \frac{18.9 - 18}{1.9/4} = 1.895.$$

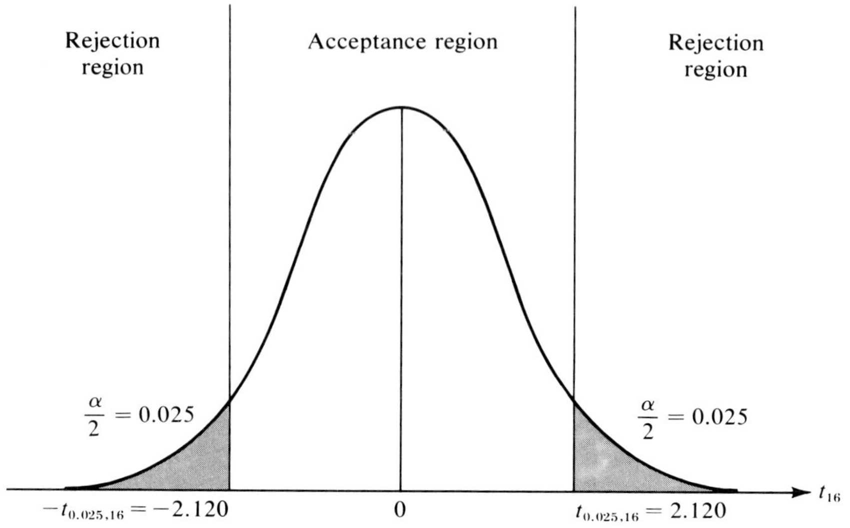

Figure 11.3

Critical regions for Example 1

Therefore, since 1.895 is between the two critical values −2.120 and 2.120, we accept H_0, and conclude that the process is not out of control. □

Let us review the reasoning behind a test of hypotheses of the nature of (13). We know that the statistic

$$\frac{\bar{X} - \mu_x}{\sqrt{\dfrac{\sum_{i=1}^{n} (X_i - \bar{X})^2}{n(n-1)}}} \tag{14}$$

has a t distribution with $(n - 1)$ degrees of freedom regardless of the true value of μ_x. Thus, when we calculate

$$\frac{\bar{x} - \mu_0}{\sqrt{\dfrac{\sum_{i=1}^{n} (x_i - \bar{x})^2}{n(n-1)}}}, \tag{15}$$

we have determined one observed value of the statistic of (14) if and only if μ_0 is the true value of μ_x. There are only two things that could cause the value of (15) to fall outside the limits −2.120 and 2.120. First, we may have inserted the wrong value of μ_x (i.e., μ_0) into the formula. Second, we may merely have obtained an unlikely observation. However, we do not choose to think that the latter is the reason, because we know this will happen to us only 5 percent of the time (and, by fixing $\alpha = 0.05$, we are willing to take the chance of rejecting the null hypothesis 5 percent of the time when it is true.) Thus, our only logical choice is to conclude that μ_0 is not the true value of μ_x.

Of course, in Example 1, the value of (15) did not fall outside the limits -2.120 and 2.120. This being the case, there is no significant indication that μ_0 is not the true value of μ_x.

Example 2

A steel company wishes to determine whether the breaking strength of steel rods made by a new method is significantly greater than the breaking strength for those made by the old method, for which it is 410 pounds. A random sample of size 28 is selected from the population of new rods, and the sample mean and unbiased estimate of the population variance are determined to be $\bar{x} = 418.7$ pounds and $\hat{\sigma}_x^2 = 74$ square pounds, respectively. Assuming that the breaking strength of the steel rods is distributed normally, determine at the 5 percent significance level whether the population mean of the new rods is significantly greater than 410 pounds.

Solution

The hypotheses of interest are

$$H_0: \mu_x = 410 \quad \text{versus} \quad H_1: \mu_x > 410.$$

The decision rule is to accept H_0 if the observed value of the test statistic (9) is less than the critical value $t_{0.05,\,27} = 1.703$ and reject H_0 if it is greater than 1.703.

Since $\hat{\sigma}_x^2$ has already been calculated from the data, our observed value of the test statistic is

$$\frac{\bar{x} - \mu_0}{\sqrt{\dfrac{\sum_{i=1}^{n}(x_i - \bar{x})^2}{n(n-1)}}} = \frac{\bar{x} - \mu_0}{\hat{\sigma}_x/\sqrt{n}} = \frac{418.7 - 410}{\sqrt{74}/\sqrt{28}} = \frac{8.7}{1.63} = 5.34.$$

Since 5.34 is greater than $t_{0.05,\,27} = 1.703$, we reject H_0. □

Problems

1. List the essential steps for making a symmetrical, two-tail, $100\alpha\%$ significance test concerning the mean of a *normal* population when the variation is unknown.
2. When testing hypotheses concerning the mean of a population with unknown variance, why is it necessary that the population be normally distributed rather than it being permissible to use a sample of sufficient size for $\bar{X}$ to be approximately normally distributed?
3. Suppose that we wish to test the hypotheses

$$H_0: \mu_x = 200 \quad \text{versus} \quad H_1: \mu_x < 200.$$

when sampling from a normally distributed population. A simple random sample of size 25 is selected and the sample mean and estimate of the population standard deviation are determined to be $\bar{x} = 195$ and $\hat{\sigma}_x = 20$. Determine at the 5% significance level whether μ_x is significantly less than 200.

4. A manufacturer of automobile batteries claims that a certain model has a capacity of 90 ampere-hours. An automobile manufacturer purchases a large quantity of these batteries and selects a simple random sample of 17 batteries to determine if the quality is inferior (that is, a capacity of less than 90 ampere-hours) to that claimed by the manufacturer. If the capacities are normally distributed and the sample is such that $\bar{x} = 86.2$ and $s_x = 8$, determine at the 5% significance level whether the batteries tested are inferior.

5. A soft-drink company wants its bottling machine to deposit ten ounces of soft drink into each bottle. From experience, it is known that the amounts deposited by the bottling machine are normally distributed. The company is concerned about depositing too little into the bottles; thus, they consider the bottling machine out of control if it is not depositing a sufficient amount of soft drink into the bottles. A simple random sample of size 25 is selected, and the values $\bar{x} = 9.5$ ounces and $\hat{\sigma}_x^2 = 0.0625$ ounces^2 are obtained. Determine at the 5% significance level whether the bottling machine is out of control.

6. The manager of a large supermarket knows from experience that when using a certain advertising procedure his average sale is \$7.50. The manager decides to try a new advertising procedure that he believes will significantly increase the average sale. After using the new procedure for some time, a simple random sample of size 15 is selected having the following dollar amounts of sale:

9.40	8.32	6.25	2.50	10.21
12.23	9.18	5.11	9.94	6.80
12.50	3.60	8.85	15.23	6.45

Assuming that the amount of sale is normally distributed, determine at the 10% significance level whether the average sale has been increased significantly.

7. A steel company produces precision-length steel bolts having an average length of 2.5 inches. The machine used can only cut the bolts longer than the required length if it goes out of control. A simple random sample of 26 is selected and the sample mean and variance are determined to be $\bar{x} = 2.62$ and $s_x^2 = 0.09$. If the length of the bolts is normally distributed, determine at the 1% significance level whether the machine is out of control.

8. Suppose that we wish to test the hypotheses

$$H_0\colon \mu_x = 300 \quad \text{versus} \quad H_1\colon \mu_x \neq 300$$

when sampling from a normally distributed population. A simple random sample of 16 units is selected and a sample mean of $\bar{x} = 312$ and an unbiased estimate of the population variance $\hat{\sigma}_x^2 = 121$ are obtained. Determine at the 5% significance level whether the population mean is significantly different from 300.

9. Using the data of Problem 5, suppose that the company is just as concerned about depositing too much soft drink into the bottles as about depositing too little.

10. Use the data of Problem 6. However, the manager now decides to try a new advertising procedure that is *less expensive* than the currently used procedure. Furthermore, he wishes to determine if there is a significant difference between the average sale for the currently used procedure and the average sale for the new procedure. Determine at the 10% significance level whether the average sale has been changed significantly.

11. Use the data of Problem 7 except assume that the machine can cut bolts that are either too long or too short when it is out of control. Determine at the 6% significance level whether the machine is out of control, using a non-symmetrical test with a risk of 1% in the left tail and 5% in the right tail.

11.4 *Confidence Intervals for the Mean of a Normal Population (Variance Unknown)*

In order to construct a $100(1-\alpha)$ percent confidence interval for the mean μ_x of a normal population when the variance is unknown, we simply use the pivotal quantity of Relation 4, Section 11.3, to apply the general procedure of Section 8.6. That is, knowing that

$$\frac{\bar{X}-\mu_x}{\sqrt{\dfrac{\sum_{i=1}^{n}(X_i-\bar{X})^2}{n(n-1)}}} \sim t_{n-1}, \tag{1}$$

we can determine that

$$P\left(\bar{X}-t_{\alpha/2,n-1}\sqrt{\frac{\sum(X_i-\bar{X})^2}{n(n-1)}} < \mu_x < \bar{X}+t_{\alpha/2,n-1}\sqrt{\frac{\sum(X_i-\bar{X})^2}{n(n-1)}}\right) = 1-\alpha. \tag{2}$$

(The mathematical development of (2) is given in Supplement 11.A)

Again, since (2) involves a general statistic, it represents a $(1-\alpha)$ probability statement in the classical sense—that is, a probability statement for an event having a probability of occurrence equal to $(1-\alpha)$. As usual, however, as soon as we consider the results of a specific sample, we obtain the fiducial probability statement

$$P_F\left(\bar{x}-t_{\alpha/2,n-1}\sqrt{\frac{\sum(x_i-\bar{x})^2}{n(n-1)}} < \mu_x < \bar{x}+t_{\alpha/2,n-1}\sqrt{\frac{\sum(x_i-\bar{x})^2}{n(n-1)}}\right) = 1-\alpha. \tag{3}$$

Since either $\hat{\sigma}_x$ or s_x may be computed for a set of sample data, we often express (3) in terms of $\hat{\sigma}_x$ or s_x. Hence, due to the relations between $\sum_{i=1}^{n}(x_i-\bar{x})^2$, s_x^2, and $\hat{\sigma}_x^2$, we have, for a specific sample of size n,

$$P_F(\bar{x}-t_{\alpha/2,n-1}\hat{\sigma}_x/\sqrt{n} < \mu_x < \bar{x}+t_{\alpha/2,n-1}\hat{\sigma}_x/\sqrt{n}) = 1-\alpha \tag{4}$$

and

$$P_F(\bar{x}-t_{\alpha/2,n-1}s_x/\sqrt{n-1} < \mu_x < \bar{x}+t_{\alpha/2,n-1}s_x/\sqrt{n-1}) = 1-\alpha. \tag{5}$$

The intervals associated with the fiducial probability statements of (3), (4), and (5) represent formulas for a $100(1-\alpha)$ percent confidence interval for the mean of a normal population when the variance is unknown. Thus, we can establish such an interval simply by implementing any one of these formulas.

Example 1

Determine the 95 percent confidence interval for μ_x when σ_x is unknown for the situation in which we select a simple random sample of size 18 from a normal population and obtain $\bar{x} = 24.1$ and $\hat{\sigma}_x = 3.2$ from the sample data.

Solution

Determining that the number of degrees of freedom is $n - 1 = 17$ and $\alpha = 0.05$, we then find

$$t_{\alpha/2, n-1} = t_{0.025, 17} = 2.11$$

from Table A-4 of the Appendix. Substituting the appropriate observed values into (4), we obtain the desired fiducial probability statement

$$\begin{aligned} P_F(\bar{x} - t_{0.025, 17}\hat{\sigma}_x/\sqrt{n} < \mu_x < \bar{x} + t_{0.025, 17}\hat{\sigma}_x/\sqrt{n}) \\ = P_F[24.1 - 2.11(3.2/\sqrt{18}) < \mu_x < 24.1 + 2.11(3.2/\sqrt{18})] \\ = P_F(24.1 - 1.59 < \mu_x < 24.1 + 1.59) \\ = P_F(22.51 < \mu_x < 25.69) = 0.05. \end{aligned}$$

Hence, the 95 percent confidence interval for μ_x is

$$22.51 < \mu_x < 25.69.$$

□

Example 2

A random sample of size 70 is selected from a normal population. The sample mean and sample standard deviation are determined to be 104.3 and 12, respectively. Determine the 99 percent confidence interval for μ_x.

Solution

Since σ_x is not given but estimated, and the population from which we are sampling is normal, we use the t distribution to establish the confidence interval for μ_x. The number of degrees of freedom is $n - 1 = 70 - 1 = 69$, and $\alpha = 0.01$. We obtain

$$t_{\alpha/2, n-1} = t_{0.005, 69} = 2.660$$

by using the closest available degrees of freedom (60) in Table A-4. (Note that we could interpolate to get a slightly better approximation of the t value for 69 degrees of freedom.) Substituting the values into (5), we obtain the 99 percent fiducial probability statement

$$\begin{aligned} P_F[104.3 - 2.66(12/\sqrt{69} < \mu_x < 104.3 + 2.66(12/\sqrt{69})] \\ = P_F(104.3 - 3.84 < \mu_x < 104.3 + 3.84) \\ = P_F(100.46 < \mu_x < 108.14) = 0.99. \end{aligned}$$

The 99 percent confidence interval is $100.46 < \mu_x < 108.14$. □

Note the similarity between the confidence interval for μ_x when σ_x is known and when σ_x is unknown. In comparing the two confidence intervals, we see that instead of σ_x and $z_{\alpha/2}$, as for the case where σ_x is known, we use $\hat{\sigma}_x$ and $t_{\alpha/2, n-1}$ for the case where σ_x is unknown. Otherwise, the confidence intervals are computed in the same manner. More importantly, they are also interpreted the same way.

Problems

Use the data of the specified problems in Section 11.3 and determine the confidence intervals for μ_x at the indicated confidence levels.

1. Problem 3; 90%.
2. Problem 4; 95%.
3. Problem 5; 98%.
4. Problem 6; 99%.
5. Problem 7; 99.9%.
6. Problem 8; 97%.

11.5 The Pooled Variance and Another Important Pivotal Quantity

In Chapter 8 we discussed the estimation of σ_x^2 through selection of one simple random sample of size n. We shall now consider the estimation of σ_x^2 through selection of two (or more) independent random samples from two (or more) populations having equal variances. Suppose we are sampling from two populations such that population 1 has mean μ_1 and variance σ_1^2 and population 2 has mean μ_2 and variance σ_2^2. Further, suppose that the values in the samples are represented by $x_{11}, x_{12}, \ldots, x_{1n_1}$ and $x_{21}, x_{22}, \ldots, x_{2n_2}$ for populations 1 and 2, respectively. That is, let the first subscript of the observed value represent the number of the population from which the sample is selected and the second subscript represent the number of the observation from that population. The sample means and variances can be determined by the usual formulas with only a slight adjustment in the subscript notation. Thus, for specific samples of size n_1 and n_2 from populations 1 and 2, respectively, we have

$$\bar{x}_1 = \frac{1}{n_1} \sum_{j=1}^{n_1} x_{1j}, \tag{1}$$

$$\bar{x}_2 = \frac{1}{n_2} \sum_{j=1}^{n_2} x_{2j}, \tag{2}$$

$$s_1^2 = \frac{1}{n_1} \sum_{j=1}^{n_1} (x_{1j} - \bar{x}_1)^2, \tag{3}$$

and

$$s_2^2 = \frac{1}{n_2} \sum_{j=1}^{n_2} (x_{2j} - \bar{x}_2)^2. \tag{4}$$

On the other hand, for unbiased estimates of σ_1^2 and σ_2^2, we use

$$\hat{\sigma}_1^2 = \frac{\sum_{j=1}^{n_1} (x_{1j} - \bar{x}_1)^2}{n_1 - 1} \tag{5}$$

and

$$\hat{\sigma}_2^2 = \frac{\sum_{j=1}^{n_2} (x_{2j} - \bar{x}_2)^2}{n_2 - 1}. \tag{6}$$

If the two populations from which we are sampling have equal variances, say both equal σ_x^2, then $\hat{\sigma}_1^2$ and $\hat{\sigma}_2^2$ are both unbiased estimates of σ_x^2. In this case, since we have two estimates of the same parameter, it seems reasonable to expect that we should get a better estimate of the parameter if we were to combine the two estimates. Thus, to obtain an estimate of σ_x^2 using $\hat{\sigma}_1^2$ and $\hat{\sigma}_2^2$, we simply take the average of (5) and (6). However, we take a weighted average, since we do not want the combined estimate to be influenced disproportionately by either of the estimates. We weight (5) and (6) by their respective degrees of freedom, since such a weighting scheme will give an unbiased estimate. Then our estimate of the variance is

$$\hat{\sigma}^2 = \frac{(n_1 - 1)\hat{\sigma}_1^2 + (n_2 - 1)\hat{\sigma}_2^2}{n_1 + n_2 - 2} \tag{7}$$

$$= \frac{\sum_{j=1}^{n_1} (x_{1j} - \bar{x}_1)^2 + \sum_{j=1}^{n_2} (x_{2j} - \bar{x}_2)^2}{n_1 + n_2 - 2}. \tag{8}$$

This estimate is generally termed the *pooled estimate* of the variance, or simply the *pooled variance.*

By comparing (8) with (3) and (4), we see that we can also represent the pooled variance as

$$\hat{\sigma}^2 = \frac{n_1 s_1^2 + n_2 s_2^2}{n_1 + n_2 - 2}. \tag{9}$$

You may wish to combine the properties for expected values, as given in Chapter 4, and the fact that $\hat{\sigma}_1^2$ and $\hat{\sigma}_2^2$ are unbiased estimates of σ_x^2 to show that $\hat{\sigma}^2$ is an unbiased estimate of σ_x^2.

The foregoing information holds regardless of the form of the distributions of the populations from which we are sampling. However, if the two populations from which we are sampling are both normal populations, then, combining this information with Relation (2), Section 10.3, we have immediately the distribution of two important statistics; that is,

$$\sum_{j=1}^{n_1} \frac{(X_{1j} - \bar{X}_1)^2}{\sigma_1^2} \sim \chi^2_{n_1 - 1} \tag{10}$$

and

$$\sum_{j=1}^{n_2} \frac{(X_{2j} - \bar{X}_2)^2}{\sigma_2^2} \sim \chi^2_{n_2 - 1}. \tag{11}$$

For two specific samples, the observed values for the statistic of (10) and (11) may be computed through use of sample data from each population and the respective relationships

$$\frac{\sum_{j=1}^{n_1}(x_{1j}-\bar{x}_1)^2}{\sigma_1^2}=\frac{(n_1-1)\hat{\sigma}_1^2}{\sigma_1^2}=\frac{n_1 s_1^2}{\sigma_1^2} \tag{12}$$

and

$$\frac{\sum_{j=1}^{n_2}(x_{2j}-\bar{x}_2)^2}{\sigma_2^2}=\frac{(n_2-1)\hat{\sigma}_2^2}{\sigma_2^2}=\frac{n_2 s_2^2}{\sigma_2^2}. \tag{13}$$

Moreover, due to the additive property for independent, chi-square variables, we can combine the variables of (10) and (11) to obtain

$$\frac{\sum_{j=1}^{n_1}(X_{1j}-\bar{X}_1)^2}{\sigma_1^2}+\frac{\sum_{j=1}^{n_2}(X_{2j}-\bar{X}_2)^2}{\sigma_2^2}\sim \chi^2_{n_1+n_2-2}. \tag{14}$$

The observed value of the statistic of (14) may be computed by combining (12) and (13). That is,

$$\begin{aligned}\frac{\sum_{j=1}^{n_1}(x_{1j}-\bar{x}_1)^2}{\sigma_1^2}+\frac{\sum_{j=1}^{n_2}(x_{2j}-\bar{x}_2)^2}{\sigma_2^2}&=\frac{(n_1-1)\hat{\sigma}_1^2}{\sigma_1^2}+\frac{(n_2-1)\hat{\sigma}_2^2}{\sigma_2^2}\\ &=\frac{n_1 s_1^2}{\sigma_1^2}+\frac{n_2 s_2^2}{\sigma_2^2}.\end{aligned} \tag{15}$$

Finally, for the situation where we have selected k specific samples of size $n_1, n_2, \ldots, n_k$ from k independent populations with means $\mu_1, \mu_2, \ldots, \mu_k$ and common variance σ^2, the formula for the pooled variance becomes

$$\hat{\sigma}^2=\frac{(n_1-1)\hat{\sigma}_1^2+(n_2-1)\hat{\sigma}_2^2+\cdots+(n_k-1)\hat{\sigma}_k^2}{n_1+n_2+\cdots+n_k-k} \tag{16}$$

$$=\frac{\sum_{j=1}^{k}(n_j-1)\hat{\sigma}_j^2}{(\sum_{j=1}^{k}n_j)-k}$$

$$=\frac{n_1 s_1^2+n_2 s_2^2+\cdots+n_k s_k^2}{n_1+n_2+\cdots+n_k-k}=\frac{\sum_{j=1}^{k}n_j s_j^2}{(\sum_{j=1}^{k}n_j)-k}, \tag{17}$$

where $\hat{\sigma}_j^2$ is the unbiased estimate of the variance computed from the jth sample and s_j^2 is the sample variance computed from the jth sample.

Now, if the k distributions are also normal, we have

$$\sum_{i=1}^{k}\left(\frac{\sum_{j=1}^{n_i}(X_{ij}-\bar{X}_j)^2}{\sigma_i^2}\right)\sim \chi_{(\sum_{i=1}^{k}n_i)-k}. \tag{18}$$

The observed value of the statistic in (18) may be determined by calculating either of the terms

$$\sum_{i=1}^{k}\frac{(n_i-1)\hat{\sigma}_i^2}{\sigma_i^2} \tag{19}$$

or

$$\sum_{i=1}^{k}\frac{n_i\cdot s_i^2}{\sigma_i^2}. \tag{20}$$

Problems

1. Suppose that $\sigma_1 = \sigma_2 = \sigma$ for the two populations from which we have sampled and that we have determined the sample variances $s_1^2 = 48$ and $s_2^2 = 60$ for two samples of size $n_1 = 16$ and $n_2 = 12$, respectively. Determine the pooled estimate of the variance.
2. Suppose that $\sigma_1 = \sigma_2 = \sigma$ for the two populations from which we have sampled and that we have determined unbiased estimates of the population variances $\hat{\sigma}_1^2 = 135$ and $\hat{\sigma}_2^2 = 180$ for two samples of size $n_1 = 15$ and $n_2 = 20$, respectively. Determine the pooled estimate of the variance.
3. Suppose that we have selected simple random samples as shown below from each of two populations having equal variances. Determine the pooled estimate of the variance.

Population	*Sample values*
1	12, 11, 9, 15, 10, 13, 11, 10, 9, 7, 15, 12
2	5, 12, 9, 13, 10, 8, 13, 14, 11

4. Suppose we have selected simple random samples from each of three populations having equal variances as shown below. Determine the pooled estimate of the variance.

Population	*Sample values*
1	18, 16, 12, 10, 9
2	11, 17, 21, 24, 10, 9
3	15, 18, 17, 9, 20

11.6 Testing Hypotheses Concerning the Difference between the Means of Two Normal Populations (Variances Unknown)

In many practical situations we wish to test the hypothesis that the difference between two population means is equal to a specific value. If the population variances are known for both populations and either both populations are normal or both relevant sample sizes are sufficiently large, then we may proceed as in Chapter 9 by using the test statistic (pivotal quantity)

$$Z = \frac{(\bar{X}_1 - \bar{X}_2) - (\mu_1 - \mu_2)}{\sqrt{(\sigma_1^2/n_1) + (\sigma_2^2/n_2)}}. \tag{1}$$

However, if the variances are unknown, as is usual in practical situations, we must use some other statistic for testing the relevant null hypothesis. The statistic we use is a t statistic. We shall now set forth procedures for testing hypotheses concerning the difference between the means of two normal populations when the unknown variances are either equal or unequal.

Case in Which the Variances of the Two Populations Are Equal ($\sigma_1^2 = \sigma_2^2$)

When we assume that the variances of the two populations are equal ($\sigma_1^2 = \sigma_2^2 = \sigma^2$), we can derive the t statistic

$$\frac{(\bar{X}_1 - \bar{X}_2) - (\mu_1 - \mu_2)}{\sqrt{\dfrac{\sum_{j=1}^{n_1} (X_{1j} - \bar{X}_1)^2 + \sum_{j=1}^{n_2} (X_{2j} - \bar{X}_2)^2}{n_1 + n_2 - 2}} \sqrt{\dfrac{1}{n_1} + \dfrac{1}{n_2}}} \sim t_{n_1+n_2-2}. \tag{2}$$

from relations given in Sections 11.5 and Definition 11.1. (The details of the mathematical development of (2) are given in Supplement 11.B). Relation (2) gives us a pivotal quantity to be used in testing a hypothesis concerning the difference of the means ($\mu_1 - \mu_2$) of two normal populations when the variances σ_1^2 and σ_2^2 are unknown but equal. That is, the statistic of (2) has a t distribution with $n_1 + n_2 - 2$ degrees of freedom regardless of the true value of the unknown parameter ($\mu_1 - \mu_2$). The procedure for making a 100α percent significance test for this situation becomes the following.

1. Set up the null hypothesis

$$H_0: (\mu_1 - \mu_2) = (\mu_1 - \mu_2)_0,$$

where $(\mu_1 - \mu_2)_0$ is any value the parameter ($\mu_1 - \mu_2$) can assume. [The most common hypothesis of this nature is that the means of the two populations are equal, that is, $(\mu_1 - \mu_2) = 0$.]

2. Specify the level of significance α.
3. Determine the sizes of the samples (n_1 and n_2) to be taken. (This fixes the number ($n_1 + n_2 - 2$) of degrees of freedom to be used in establishing the appropriate critical value.)
4. Determine a critical value $t^*_{n_1+n_2-2} = t_{\alpha,n_1+n_2-2}$. (Note that we can always arrange the null hypothesis in a one-tail test so that the alternative hypothesis can be right-sided. That is, if necessary, we can simply reverse the roles of μ_1 and μ_2.)
5. From a sample of size n_1 from population 1 and a sample of size n_2 from population 2, obtain [under the assumption that the null hypothesis is true, $(\mu_1 - \mu_2) = (\mu_1 - \mu_2)_0$] one observed value of the statistic of (2); that is, compute

$$\frac{(\bar{x}_1 - \bar{x}_2) - (\mu_1 - \mu_2)_0}{\sqrt{\dfrac{\sum_{j=1}^{n_1} (x_{1j} - \bar{x}_1)^2 + \sum_{j=1}^{n_2} (x_{2j} - \bar{x}_2)^2}{n_1 + n_2 - 2}} \sqrt{\dfrac{1}{n_1} + \dfrac{1}{n_2}}} \tag{3}$$

where x_{1j} is the jth sample observation from population 1, x_{2j} is the jth sample observation from population 2, $\bar{x}_1$ is the observed sample mean from population 1, $\bar{x}_2$ is the observed sample mean from population 2, and $(\mu_1 - \mu_2)_0$ is the value of the unknown parameter ($\mu_1 - \mu_2$) given by the null hypothesis.

6. If the value of (3) is in the critical region, reject H_0. Otherwise, accept H_0.

As always, the corresponding two-tail, 100α percent significance test is identical except that we must employ two critical values, $-t_{\alpha/2, n_1+n_2-2}$ and $t_{\alpha/2, n_1+n_2-2}$, for a symmetrical test. For a nonsymmetrical, two-tail, 100α percent test we still force the two tail areas to total α, but relax the requirement that they be equal.

Since, we know from Section 11.5 that for two specific samples,

$$\hat{\sigma} = \sqrt{\frac{\sum_{j=1}^{n_1}(x_{1j} - \bar{x}_1)^2 + \sum_{j=1}^{n_2}(x_{2j} - \bar{x}_2)^2}{n_1 + n_2 - 2}} = \sqrt{\frac{n_1 s_1^2 + n_2 s_2^2}{n_1 + n_2 - 2}},$$

it is often convenient to calculate and use

$$\frac{(\bar{x}_1 - \bar{x}_2) - (\mu_1 - \mu_2)_0}{\hat{\sigma}\sqrt{\frac{1}{n_1} + \frac{1}{n_2}}} \tag{4}$$

as the observed value of the pivotal quantity instead of (3) in conducting this test.

Example 1

Company A and Company B both manufacture automobile tires. The fleet manager of a car-rental agency plans to purchase new, high-quality tires for his fleet of automobiles and he wishes to determine if there is a significant difference between the average useful mileage of the best tires made by each company. The manager knows that the useful mileage of automobile tires follows a normal distribution. However, the means of these distributions depend on the manufacturer and the quality of the tires, while the variances are assumed to be the same, regardless of the manufacturer and quality of the tires. The manager selects a random sample of size $n_1 = 10$ from the best tires of Company A and a random sample of size $n_2 = 7$ from the best tires of Company B. He determines the sample means to be $\bar{x}_1 = 15{,}725$ miles and $\bar{x}_2 = 14{,}320$ miles, and the corresponding sample variances to be $s_1^2 = 210{,}000$ square miles and $s_2^2 = 225{,}000$ square miles. Using $\alpha = 0.05$, determine whether there is a significant difference between the average useful milcage of the best tires made by the two companies.

Solution

Since we assume the variance of the useful mileage of automobile tires is the same for different manufacturers and different quality tires, we can use the foregoing test. First, we use Equation (9, Sec. 11.5) to obtain the pooled variance as

$$\begin{aligned}\hat{\sigma}^2 &= \frac{n_1 s_1^2 + n_2 s_2^2}{n_1 + n_2 - 2} \\ &= \frac{(10)(210{,}000) + (7)(225{,}000)}{10 + 7 - 2} = \frac{2{,}100{,}000 + 1{,}575{,}000}{15} = 245{,}000.\end{aligned}$$

Thus

$$\hat{\sigma} = \sqrt{245{,}000} = 495.$$

Since we wish to determine if there is a significant difference between the means of the two populations, we simply test the null hypothesis that the means are equal

$$H_0: \mu_1 - \mu_2 = 0$$

versus the two-sided alternative hypothesis

$$H_1: \mu_1 - \mu_2 \neq 0.$$

Substituting the observed values into (4), we obtain the observed value of the pivotal quantity (under the assumption that H_0 is true)

$$t = \frac{(\bar{x}_1 - \bar{x}_2) - (\mu_1 - \mu_2)_0}{\hat{\sigma}\sqrt{\dfrac{1}{n_1} + \dfrac{1}{n_2}}}$$

$$= \frac{(15{,}725 - 14{,}320) - 0}{495\sqrt{\dfrac{1}{10} + \dfrac{1}{7}}} = \frac{1{,}405}{243.94} = 5.76.$$

Since the statistic of (2) is distributed according to the t distribution with $(n_1 + n_2 - 2)$ degrees of freedom and H_1 is a two-sided alternative with $\alpha = 0.05$, we compare the value of $t = 5.76$ against the critical values of the t distribution with $n_1 + n_2 - 2 = 10 + 7 - 2 = 15$ degrees of freedom. These critical values, $-t_{\alpha/2,15}$ and $t_{\alpha/2,15}$, are determined from Table A-4 such that

$$P(t_{15} < -t_{\alpha/2,15}) = \alpha/2 = 0.025$$

and

$$P(t_{15} > t_{\alpha/2,15}) = \alpha/2 = 0.025.$$

Thus, from Table A-4, we have

$$P(t_{15} < -2.131) = 0.025 \quad \text{and} \quad P(t_{15} > 2.131) = 0.025.$$

Therefore, $t_{0.025,15} = 2.131$, which means that we reject H_0—because the observed value of the pivotal quantity is not between -2.131 and 2.131. Our conclusion, then, is that there is a significant difference, at least at the 5 percent significance level, between the average useful mileage of automobile tires made by Company A and Company B. □

Example 2

Excellent Pharmaceutical Company and Simple Drugs, Incorporated, produce aspirin tablets. Excellent claims that its aspirin tablets (containing a secret, patented additive) will relieve headaches faster than any other aspirin on the market. To determine whether Excellent's claim is valid an independent research firm selects random samples of size $n_1 = 15$ and $n_2 = 15$ from aspirin made by Excellent and Simple, respectively. The research firm gives an aspirin to each of 30 randomly selected persons suffering from headaches and records the time required for each to recover from his headache. The sample results are $\bar{x}_1 = 8.4$ minutes and $s_1^2 = 4.2$

square minutes for Excellent's aspirin, and $\bar{x}_2 = 8.9$ minutes and $s_2^2 = 4.6$ square minutes for Simple's aspirin. Assuming that the time required to recover from a headache is normally distributed and the population variances are the same for the two aspirin, determine at the 5 percent significance level whether Excellent's aspirin cures headaches significantly faster than Simple's aspirin.

Solution

Since the population variances are assumed equal, we pool the sample variances to obtain the pooled variance (9, Sec. 11.5) as

$$\hat{\sigma}^2 = \frac{n_1 s_1^2 + n_2 s_2^2}{n_1 + n_2 - 2}$$

$$= \frac{(15)(4.2) + (15)(4.6)}{15 + 15 - 2} = \frac{132}{28} = 4.71.$$

Thus,

$$\hat{\sigma} = \sqrt{4.71} = 2.17.$$

We wish to determine if Excellent's aspirin cures headaches significantly faster than Simple's aspirin. Hence, we wish to test

$$H_0\colon \mu_1 - \mu_2 = 0$$

versus the one-sided alternative

$$H_1\colon \mu_1 - \mu_2 < 0.$$

The observed value (4) of our pivotal quantity (under the assumption that H_0 is true) is

$$t = \frac{(\bar{x}_1 - \bar{x}_2) - (\mu_1 - \mu_2)_0}{\hat{\sigma}\sqrt{\dfrac{1}{n_1} + \dfrac{1}{n_2}}}$$

$$= \frac{(8.4 - 8.9) - 0}{2.17\sqrt{\dfrac{1}{15} + \dfrac{1}{15}}}$$

$$= \frac{-0.5}{0.792} = -0.631.$$

Since we are comparing H_0 against a one-sided alternative, we use as our critical value $t_{\alpha, n_1 + n_2 - 2} = t_{0.95, 28} = -1.701$ (Table A-4); that is,

$$P(t_{28} < t_{0.95, 28}) = P(t_{28} < -1.701) = 0.05.$$

Since our observed $t = -0.631$ is greater than $t_{0.95, 28} = -1.701$, we conclude that the aspirin tablets made by Excellent do not relieve headaches significantly faster than those made by Simple. Hence, we accept H_0. □

Case in Which the Variances of the Two Populations Are Not Equal $(\sigma_1^2 \neq \sigma_2^2)$

How do we test a hypothesis concerning the difference between the means of two normal populations with unknown variances when we cannot justify the assumption that $\sigma_1^2 = \sigma_2^2$? It turns out that there does not exist an exact test in this situation. However, it can be shown that if sampling is accomplished from two independent normal populations, the statistic that results from modifying the statistic of (1) by replacing σ_1^2 and σ_2^2 by their respective unbiased estimators is approximately distributed as a t distribution. That is, the statistic

$$\frac{(\bar{X}_1 - \bar{X}_2) - (\mu_1 - \mu_2)}{\sqrt{\dfrac{\sum_{j=1}^{n_1}(X_{1j} - \bar{X}_1)^2}{n_1(n_1 - 1)} + \dfrac{\sum_{j=1}^{n_2}(X_{2j} - \bar{X}_2)^2}{n_2(n_2 - 1)}}} \tag{5}$$

is distributed approximately as a t distribution with degrees of freedom that must be determined from sample data. Based on this information, the procedure for making an approximate 100α percent significance test for this situation is as follows.

1. State the null hypothesis

$$H_0 : (\mu_1 - \mu_2) = (\mu_1 - \mu_2)_0,$$

   where $(\mu_1 - \mu_2)_0$ is any value that the parameter $(\mu_1 - \mu_2)$ can assume.
2. Specify the level of significance α.
3. From a sample of size n_1 from population 1 and a sample of size n_2 from population 2, obtain [under the assumption that the null hypothesis is true, $(\mu_1 - \mu_2) = (\mu_1 - \mu_2)_0$] one observed value of the statistic of (5). Note that for specific samples, we can obtain the observed value by computing

$$\frac{(\bar{x}_1 - \bar{x}_2) - (\mu_1 - \mu_2)_0}{\sqrt{(\hat{\sigma}_1^2/n_1) + (\hat{\sigma}_2^2/n_2)}}, \tag{6}$$

   where $\bar{x}_1$ and $\bar{x}_2$ are the means of the samples from populations 1 and 2, respectively, $\hat{\sigma}_1^2$ and $\hat{\sigma}_2^2$ are the unbiased estimates of σ_1^2 and σ_2^2, respectively, and $(\mu_1 - \mu_2)_0$ is the value of the unknown parameter $(\mu_1 - \mu_2)$ given by the null hypothesis.
4. Determine the number (k) of degrees of freedom to be used [that is, the number of degrees of freedom associated with the approximate t statistic of (5)] in the test, where k is the integer closest to

$$k' = \frac{[(\hat{\sigma}_1^2/n_1) + (\hat{\sigma}_2^2/n_2)]^2}{\dfrac{(\hat{\sigma}_1^2/n_1)^2}{n_1} + \dfrac{(\hat{\sigma}_2^2/n_2)^2}{n_2}}, \tag{7}$$

   where $\hat{\sigma}_1$, $\hat{\sigma}_2$, n_1, and n_2 are as given in conjunction with (6). Note that k' will usually assume a noninteger value. However, in most practical problems sufficient accuracy can be realized by using the nearest integer value as the number of degrees of freedom.

5. Determine a critical value $t_k^* = t_{\alpha,k}$. (Note once again that, by reversing the roles of μ_1 and μ_2 when necessary, we can always arrange the null hypothesis in a one-tail test so that the alternative hypothesis can be right-sided.)
6. If the value of (6) is in the critical region, reject H_0. Otherwise, accept H_0.

Corresponding two-tail tests are identical except that two critical values must be chosen so that the tail areas (which total α) are in the desired proportion.

Example 3

Suppose that in Example 1 it is unreasonable to assume that the variances of the two populations are equal. Use the same data, without the assumption of equal variances, to determine at the 5 percent significance level whether there is a significant difference between the average useful mileage of the tires made by Company A and those made by Company B.

Solution

Since we are sampling from normal distributions for which the variances are unknown and not necessarily equal, we simply calculate the observed value of the appropriate statistic (6) and use the value of (7) as a guide to our number of degrees of freedom. We have the same hypotheses as in Example 1. That is, we wish to test

$$H_0: \mu_1 - \mu_2 = 0$$

versus the two-sided alternative

$$H_1: \mu_1 - \mu_2 \neq 0.$$

Note that we must compute the observed value (6) of the statistic. We begin with

$$\frac{(\bar{x}_1 - \bar{x}_2) - (\mu_1 - \mu_2)_0}{\sqrt{(\hat{\sigma}_1^2/n_1) + (\hat{\sigma}_2^2/n_2)}}, \tag{8}$$

and recalling that

$$\hat{\sigma}_x^2 = \frac{ns_x^2}{n-1},$$

we convert the sample variances to unbiased estimates as

$$\hat{\sigma}_1^2 = \frac{10(210{,}000)}{9} = 233{,}333 \quad \text{and} \quad \hat{\sigma}_2^2 = \frac{7(225{,}000)}{6} = 262{,}500.$$

Now, substituting the appropriate values into (8), we have

$$\begin{aligned}\frac{(\bar{x}_1 - \bar{x}_2) - (\mu_1 - \mu_2)_0}{\sqrt{(\hat{\sigma}_1^2/n_1) + (\sigma_2^2/n_2)}} &= \frac{(15{,}725 - 14{,}320) - 0}{\sqrt{\dfrac{233{,}333}{10} + \dfrac{262{,}500}{7}}}\\ &= \frac{1{,}405}{\sqrt{60{,}833}}\\ &= \frac{1{,}405}{247} = 5.70.\end{aligned}$$

In order to obtain the critical values for the test, we must first use (7) to determine the degrees of freedom for our variable. We obtain

$$k' = \frac{[(\hat{\sigma}_1^2/n_1) + (\hat{\sigma}_2^2/n_2)]^2}{\frac{(\hat{\sigma}_1^2/n_1)^2}{n_1} + \frac{(\hat{\sigma}_2^2/n_2)^2}{n_2}}$$

$$= \frac{\left[\frac{233{,}333}{10} + \frac{262{,}500}{7}\right]^2}{\frac{\left(\frac{233{,}333}{10}\right)^2}{10} + \frac{\left(\frac{262{,}500}{7}\right)^2}{7}} = \frac{(60{,}833)^2}{54{,}444{,}288.89 + 200{,}892{,}857.1}$$

$$= \frac{3{,}700{,}653{,}889}{255{,}337{,}145.9} = 14.49 \cong 14.$$

We find the critical values (Table A-4) to be $-t_{\alpha/2,\,14} = -2.145$ and $t_{\alpha/2,\,14} = 2.145$. Hence, our decision rule is to accept H_0 if the observed value of (6) is between -2.145 and 2.145 and reject H_0 otherwise.

Since our observed value (5.70) is not between -2.145 and 2.145, we reject H_0. Thus, we conclude that there is a significant difference between the average mileage of tires made by the two companies. □

Problems

1. Suppose we wish to test the null hypothesis

$$H_0: \mu_1 - \mu_2 = 0 \quad \text{versus} \quad H_1: \mu_1 - \mu_2 > 0.$$

Suppose simple random samples of sizes $n_1 = 25$ and $n_2 = 15$ are selected from two normally distributed populations having equal variances. Sample means of $\bar{x}_1 = 83$ and $\bar{x}_2 = 80$, and sample variances of $s_1^2 = 31$ and $s_2^2 = 28$ are obtained. Determine at the 5% significance level whether the mean of population 1 is significantly greater than the mean of population 2.

2. A large textile manufacturer has two groups of employees performing the same tasks by different procedures. The times to perform the tasks are normally distributed for each group, and the variances of the groups are equal. Simple random samples of sizes $n_1 = 10$ and $n_2 = 15$ are selected. The sample means were found to be $\bar{x}_1 = 82$ minutes and $\bar{x}_2 = 70$ minutes, and the unbiased estimates of the individual variances were found to be $\hat{\sigma}_1^2 = 100$ minutes^2 and $\hat{\sigma}_2^2 = 144$ minutes^2. Determine at the 1% significance level whether the average time required by group 1 is significantly greater than the average time required by group 2.

3. The management of a supermarket chain wants to rearrange the layout of their stores according to one of two possible schemes that should favorably influence the average amount spent per customer. Two representative stores were selected and rearranged according to each possible scheme. After the schemes have been in use for some time, simple random samples of sizes $n_1 = 12$ and $n_2 = 15$ are selected from amounts spent by customers at stores using schemes 1 and 2, respectively. Sample means of $\bar{x}_1 = \$7.30$ and $\bar{x}_2 = \$9.10$ and sample standard deviations of $s_1 = \$3.00$ and $s_2 = \$3.00$ are determined. If the amount spent per customer is normally distributed and the population variances are equal, determine

at the 10% significance level whether the average amount spent per customer at the store arranged according to scheme 2 is significantly greater than that for scheme 1.

4. An educational testing service has developed two different multiple choice tests designed to test the same capability. Test 1 has five choices per question, while test 2 has four choices. Some individuals believe that, on the average, test 2 should yield a higher score because there are fewer wrong answers from which to choose. The test scores are normally distributed, and there is believed to be no difference between the variances of the scores for the two tests. Simple random samples of sizes $n_1 = 23$ and $n_2 = 19$ are selected from students' scores on test 1 and test 2, respectively. Sample means of $\bar{x}_1 = 64$ and $\bar{x}_2 = 67$ and standard deviation estimates of $\hat{\sigma}_1 = 4$ and $\hat{\sigma}_2 = 3$ are obtained. Determine at the 1% significance level whether test 2 yields a significantly greater average score than test 1.
5. Two groups of chickens were fed different diets for six weeks in order to measure the weight gains. Simple random samples of sizes $n_1 = 10$ and $n_2 = 20$ are selected from those chickens receiving diet 1 and diet 2, respectively. Sample means of $\bar{x}_1 = 18.0$ ounces and $\bar{x}_2 = 21.6$ ounces and sample standard deviations of $s_1 = 4$ ounces and $s_2 = 5$ ounces are determined. The weight gains are normally distributed, and the variances of the weight gains associated with the two diets are believed to be equal. Determine at the 5% significance level whether there is a significant difference between the average weight gains for the two diets.
6. Use the data of Problem 3 to determine at the 10% significance level whether the average amount spent per customer is significantly different for the two schemes.
7. Use the data of Problem 4 to determine at the 1% significance level whether the average scores for the two tests are significantly different.
8. Work Problem 2 for the situation where we cannot assume $\sigma_1 = \sigma_2$.
9. Work Problem 3 for the situation where we cannot assume $\sigma_1 = \sigma_2$.
10. Work Problem 5 for the situation where we cannot assume $\sigma_1 = \sigma_2$.

11.7 Confidence Intervals for the Difference between the Means of Two Normal Populations (Variances Unknown)

We shall now apply the general method of Section 8.6 to develop a $100(1 - \alpha)$ percent confidence interval for the difference between the means of two normal populations when the variances are unknown. Again, we must consider two situations: the case in which the population variances are equal, and the case in which they are not equal. For the first case, we use the pivotal quantity of (2, Sec. 11.6), while for the second case, we use the approximate t statistic of (5, Sec. 11.6).

Case in Which the Variances of the Two Populations Are Equal ($\sigma_1^2 = \sigma_2^2$)

For this case, we begin with the $(1 - \alpha)$ probability statement,

$$P(-t_{\alpha/2,\,n_1+n_2-2} < t_{n_1+n_2-2} < t_{\alpha/2,\,n_1+n_2-2}) = 1 - \alpha. \tag{1}$$

Review Problems

1. If U_1, U_2, and U_3 are independent random variables such that $U_1 \sim t_{13}$, $U_2 \sim \chi^2_7$, and $U_3 \sim N(0, 1)$, determine the following probabilities.

   a. $P(U_1 \leq 1.350 \cap U_2 \geq 6.35)$.
   b. $P(U_1 \geq -2.160 \cap U_3 \leq 1.84)$.
   c. $P(U_1 \leq 1.350 \cup U_2 \geq 6.35)$.
   d. $P(U_1 \geq 2.160 \cup U_3 \leq 1.84)$.

   (Sections 4.7, 4.8, 6.4, and 10.2)

2. Using the pooled estimate of the variance, derive the formula for a $100(1 - \alpha)\%$ confidence interval for σ^2 when selecting simple random samples from two normal populations with equal variances.

   (Sections 8.6, 10.2, and 10.4)

3. Using the formula derived in Problem 2 and the data of Problem 5, Section 11.6, determine the 95% confidence interval for σ^2.

4. Show that $\hat{\sigma}^2$ is an unbiased estimate of σ^2 when $\sigma_1 = \sigma_2$.

   (Section 4.12)

5. Determine the appropriate pivotal quantity for testing hypotheses concerning σ^2 when pooling the data from two normal populations with equal variances. Use the data of Problem 2, Section 11.6, to determine at the 5% significance level whether the variance is significantly greater than 81.

   (Sections 9.4, 10.2, and 10.3)

Twelve

Survey Sampling

12.1 Introduction

When we considered several measures of location and dispersion and their associated computational methods in Chapter 3, we were not concerned with how the values of the variables were obtained. In all cases except Chapter 13, probability sampling (Definition 4.17, page 97) was used to obtain the sample values. In Chapters 4 and 5, we considered probability and several special probability distributions; in Chapter 7, the development of sampling distributions; finally, in Chapters 8 and 9, hypothesis testing. In all of these, the type of probability sampling used was simple random sampling, where each element in the population is equally likely to appear in the sample.

We are now ready to consider some other sampling methods, several of which are particularly valuable in the areas of *survey sampling* and *experimental design.* In this chapter, we shall be mainly concerned with methods associated with survey sampling, and in Chapter 13 we shall consider some methods for experimental design. For both chapters, the discussion is only elementary and introductory. A thorough discussion of either topic would fill several volumes.

From a computational viewpoint, the discussions here are similar to those in Chapter 3, in that we appear to be primarily concerned with the descriptive measurements of the population parameters rather than with making inferences. However, wherever we can determine the probability distributions of the original population or apply results such as the central limit theorem, we can make statistical inferences. Nevertheless, the practical applications to survey sampling of most hypothesis tests are quite limited and the complexities associated with them are beyond the scope of this text.

Despite our concern with descriptive measures, we shall continue to select samples via probability sampling plans. The main reason for designing probability sampling plans is to minimize sampling and nonsampling errors. The primary requirement for a "good" sample is that it be representative of the population we are trying to describe. Of course, there are numerous ways of selecting "poor" samples, an obvious one being that of sampling from the wrong population. A classic example of such an error is the 1936 presidential poll conducted by *Literary Digest.* On the basis of results obtained from a sample of over two million names selected from automobile registrations and telephone directories, *Literary Digest* predicted that Landon would win an overwhelming victory over Roosevelt in the election that year. However, Roosevelt won by a considerable margin. The fallacy associated with the interpretation was that the sample was taken from a population having a greater proportion of high-income individuals than the entire voting population. Furthermore, Roosevelt's greatest support was among the low-income voters, whose opinions were hardly represented in the poll. This factor was not considered in the interpretation of the results. The resulting blunder is an excellent illustration of the importance of carefully identifying the population of interest and sampling from that particular population instead of from other, perhaps related, populations.

Our emphasis on probability sampling plans in this chapter does not mean that nonprobability samples are not important. As a matter of fact, nonprobability samples often give excellent results. Two types of nonprobability sampling are *quota* sampling, where the experimenter selects elements of the population until he satisfies a predetermined quota of respondents, and *judgment* sampling, where the experimenter intentionally selects, on the basis of his personal judgment, the specific elements of the population to be in the sample. Some advantages of these sampling plans are that they are generally simpler and less expensive than probability sampling plans. However, some very serious disadvantages are possible in nonprobability sampling plans. One such disadvantage is that the sampler may unknowingly select elements that deviate in the same direction, thus creating *biases.* A second disadvantage is that the range of variation as observed in such samples does not give a good measure of the variability in the population.

12.2 Simple Random Sampling

Recall from Chapter 4 that a simple random sample is a sample selected by a random process such that any simple event in the population at any given instant has the same chance of appearing in the sample as any other remaining simple event. Furthermore, in Chapter 7, we determined that, when selecting without replacement a sample of size n from a population of size N having mean μ_x and variance σ_x^2,

$$\mu_{\bar{x}} = \mu_x \quad \text{and} \quad \sigma_{\bar{x}}^2 = \frac{\sigma_x^2}{n} \cdot \frac{N-n}{N-1}.$$

Thus, we use $\bar{x}$ as the estimator of μ_x and $\hat{\sigma}_{\bar{x}}^2$ as the estimator of $\sigma_{\bar{x}}^2$. That is,

$$\hat{\mu}_x = \bar{x} \quad \text{and} \quad \hat{\sigma}_{\bar{x}}^2 = \frac{\hat{\sigma}_x^2}{n} \cdot \frac{N-n}{N-1}.$$

Moreover, recall that the term $(N - n)/(N - 1)$ is called the finite population correction factor, or simply fpc. The fpc is equal to one if we sample with replacement or sample from an infinite population. However, sampling with replacement is seldom beneficial in most practical situations. For example, in a voter preference survey, we generally would not be interested in interviewing the same voter twice. Similarly, a market researcher would not want to conduct a lengthy interview twice with the same consumer, and a quality control inspector would not test twice the same item from a new shipment. Observing the fpc, we note that for small values of n, it is near one and makes little difference in the estimate of $\sigma_{\bar{x}}^2$. A rule of thumb is to ignore the fpc if the sample size is less than 5 percent of the population size N.

The sampling plan for a simple random sample consists of selecting the elements of the population in a manner such that each element in the population has an equal chance of appearing in the sample. The sample may be selected through using a random number table or some other similar method.

By now you may wonder why simple random sampling is not always used and why we would even consider other sampling plans. There are two major factors that must be considered when determining a sampling plan—cost and precision. In practice, we want the plan to be as inexpensive as possible and, yet, we want to be as close to the expected value as possible—that is, we want to have as much precision as possible. Ideally, we would like to sample the entire population and obtain perfect estimates. Note that the fpc equals zero if we select the entire population and the sample mean equals the population mean. Unfortunately, it is generally impossible to sample the entire population because it is either too expensive or too time consuming, or both. Usually the sample will consist of only a small portion of the population unless the experimenter must have very precise estimates. Generally, an increase in precision results in an increase in cost.

Thus, one good reason for using sampling plans other than simple random sampling is that the same precision can be obtained at less cost. Another reason is that simple random sampling plans are not always as easy to implement as they might first appear. This is especially true when the population is large. For example, a simple random sample of 500 families from a city of 20,000 is not too easily obtained, and the cost of going from house to house in different neighborhoods is generally high. In subsequent sections, we shall consider two other sampling plans that are more advantageous than simple random sampling plans in certain situations. We will describe the conditions for which the different sampling plans are better. First, let us consider an example using a simple random sampling plan.

Example 1

A simple random sample of fifty blocks is selected from a city containing 500 blocks. The following data show the number of persons living on each block.

17	17	65	22	45	14	23	17	41	45
29	32	20	26	36	19	45	13	13	17
30	26	16	23	31	23	35	28	36	23
22	24	32	27	47	28	31	47	21	20
19	35	51	18	11	29	14	32	43	32

Determine the sample mean and the estimate of its precision—that is, $\hat{\sigma}_{\bar{x}}^2$.

Solution

For the sample mean and unbiased estimate of the population variance, we obtain

$$\bar{x} = \frac{1{,}410}{50} = 28.2 \quad \text{and} \quad \hat{\sigma}_x^2 = \frac{\sum_{i=1}^{n} (x_i - \bar{x})^2}{n - 1} = \frac{6{,}588}{49} = 134.45.$$

Thus, the estimate of the population mean μ_x is $\bar{x} = 28.2$ and its estimated precision is

$$\hat{\sigma}_{\bar{x}}^2 = \frac{134.45}{50} \cdot \frac{500 - 50}{500 - 1} = 2.42. \qquad \square$$

The same techniques for testing hypotheses and setting confidence intervals described in Chapters 9 through 11 can be applied in Example 1. Since we have already considered how to do this, we shall not review it at this point. You should note that we can make appropriate statistical inferences only if the sampling plan is carried out via probability sampling procedures of which simple random sampling is only one such procedure. Another advantage of probability sampling is that it helps eliminate unknown biases.

Problems

1. In a large eastern city there are 169 industrial firms employing 20 or more persons. The following are the employment figures based on a simple random sample of 34 firms. Determine the estimate of the average number of employees per firm and the precision of the estimate.

49	31	62	50	50	23	88	37
121	28	24	53	23	468	35	163
81	331	38	22	34	22	139	36
156	37	25	24	27	237	25	26
80	35						

2. A simple random sample of the expense accounts of 25 executives was selected from a group of 300 executives of a large aerospace firm. The following amounts were spent on entertainment during a particular week by each executive in the sample. Determine the sample variance and the estimate of its precision.

23	30	29	40	43
18	9	10	21	15
25	20	32	11	31
17	48	45	34	18
11	16	20	39	20

3. What is the difference between probability sampling plans and nonprobability sampling plans? Give two examples of each type.
4. Give an example where you would prefer to use probability sampling rather than nonprobability sampling.
5. Give an example where you would prefer to use nonprobability sampling rather than probability sampling.

12.3 Stratified Sampling

When the population is divided into a number of *strata*, or subpopulations, and a sample is taken from each stratum, the sampling process is called *stratified sampling*.

More specifically, if the sample for each stratum is selected via a simple random sampling plan, the process is known as *stratified random sampling.* There are many ways in which the sample may be selected from each population; however, we shall concern ourselves only with simple random sampling for each stratum unless we specify differently. Consequently, we shall condense the name to stratified sampling.

Some situations for which stratified sampling is applicable include those where we wish to classify the population into strata by age, sex, income level, geographical location, educational level, religious affiliation, political affiliation, or any other classification. There are several advantages of using stratified sampling plans in addition to obtaining estimates for the population. One is that we have estimates for each stratum. For example, we can estimate the proportion of voters in each precinct who favor a certain issue or candidate and then combine these estimates to obtain an estimate of the proportion of voters in the entire population who favor the issue or candidate.

A second advantage is that it may be simpler to stratify. For example, it would be easier to select a separate random sample from each precinct than to select a simple random sample from the entire population.

Also, stratification permits us to guarantee a "representative" sample. For example, if a simple random sampling plan were used to select voters of a population, it is possible to get a sample that is not very representative because it may contain a disproportionate number of voters affiliated with one party or a disproportionate number of white-collar workers, and so forth. Stratification permits us to select each group from which we wish to pick a small sample, thus guaranteeing a representative sample. Furthermore, stratification may result in greater precision for the estimates than simple random sampling. The conditions under which increased precision may be obtained will be discussed later.

As in the case of simple random sampling, the population size is N and the sample size is n: however, the population will now be divided into L strata of size $N_1, N_2, \ldots, N_L$, where

$$N = N_1 + N_2 + \cdots + N_L = \sum_{s=1}^{L} N_s.$$

A sample size of n_s is taken from the sth stratum, and the total sample size is

$$n = n_1 + n_2 + \cdots + n_L = \sum_{s=1}^{L} n_s.$$

The population mean for the sth stratum is μ_s and the n_s sample values are $x_{s1}, x_{s2}, \ldots, x_{sn_s}$. Because we are selecting the sample within each stratum via simple random sampling, we can utilize the results for simple random sampling. Thus, for the sth stratum, the estimate of μ_s is the sample mean $\bar{x}_s$ and the estimate of the variance $\sigma_{\bar{x}}^2$ of the estimator $\bar{x}_s$ is given by $\hat{\sigma}_{\bar{x}_s}^2$. The estimates are computed by the following equations:

$$\bar{x}_s = \frac{\sum_{i=1}^{n_s} x_{si}}{n_s} \tag{1}$$

and

$$\hat{\sigma}_{\bar{x}_s}^2 = \frac{\hat{\sigma}_s^2}{n_s} \cdot \frac{N_s - n_s}{N_s - 1}, \tag{2}$$

where $\hat{\sigma}_s^2$ is the unbiased estimate of the variance of the x values in the sth stratum; that is,

$$\hat{\sigma}_s^2 = \frac{\sum_{i=1}^{n_s} (x_{si} - \bar{x}_s)^2}{n - 1}.$$

Note that the only difference between Equations (1) and (2) and those equations used previously for computing a sample mean and the estimate of its variance is that the subscript s has been added to each term associated with the sth stratum. Hence, the estimates for the sth stratum are computed in exactly the same manner as the corresponding estimates for a simple random sample.

In order to estimate the population mean μ_x, we simply compute the weighted average of the stratum means, where each weight is the respective ratio of the stratum size N_s to the total population size N. Thus, the estimate of the population mean is given by

$$\hat{\mu}_x = \bar{x} = \frac{\sum_{s=1}^{L} N_s \bar{x}_s}{N}. \tag{3}$$

Recalling Rules 3.6 through 3.9 from Chapter 3 and using the independence property associated with the simple random sampling procedure, the variance of the estimate $\bar{x}$ of the population mean is given by

$$\sigma_{\bar{x}}^2 = \sum_{s=1}^{L} \left(\frac{N_s}{N}\right)^2 \sigma_{\bar{x}_s}^2.$$

Substituting the respective estimates for the variances of the individual strata, we get the estimate of the variance of $\bar{x}$,

$$\hat{\sigma}_{\bar{x}}^2 = \sum_{s=1}^{L} \left(\frac{N_s}{N}\right)^2 \cdot \frac{\hat{\sigma}_s^2}{n_s} \cdot \frac{N_s - n_s}{N_s - 1}. \tag{4}$$

It should be carefully noted that each weight is the proportion of the total population size each stratum represents, and it is determined as a ratio of *population* values instead of *sample* values.

Example 1

A market researcher studied a 100 block area of a large city to estimate the number of families living on a block. The 100 blocks were classified according to population density based on three types of housing: exclusive areas with low-density population, medium-density areas containing mainly one-family dwelling units, and high-density areas containing numerous apartments and multiple-family dwelling units. The numbers of blocks classified in the three strata were 25, 40, and 35, respectively. A simple random sample was selected from each stratum with samples of 5, 8, and 6, respectively. The data are shown in the following table.

Population density	*Families per block*
Low	11, 13, 13, 17, 16
Medium	21, 31, 32, 35, 26, 23, 22, 26
High	65, 45, 46, 35, 44, 41

Determine the sample mean and the estimate of its variance for each stratum. Also, determine the estimate of the population mean and the estimate of its variance.

Solution

The stratum sizes are $N_1 = 25$, $N_2 = 40$, and $N_3 = 35$. The sample sizes selected from the respective strata are $n_1 = 5$, $n_2 = 8$, and $n_3 = 6$. The sample means for the strata and the estimates of their variances are

$$\bar{x}_1 = \frac{70}{5} = 14, \qquad \hat{\sigma}^2_{\bar{x}_1} = \frac{6}{5}\left(\frac{25-5}{25-1}\right) = 1;$$

$$\bar{x}_2 = \frac{216}{8} = 27, \qquad \hat{\sigma}^2_{\bar{x}_2} = \frac{26.3}{8}\left(\frac{40-8}{40-1}\right) = 2.7;$$

$$\bar{x}_3 = \frac{276}{6} = 46, \qquad \hat{\sigma}^2_{\bar{x}_3} = \frac{102.4}{6}\left(\frac{35-6}{35-1}\right) = 14.6.$$

Using Equation (3), we obtain the estimate of the population mean.

$$\bar{x} = \frac{25}{100}(14) + \frac{40}{100}(27) + \frac{35}{100}(46) = 30.4.$$

Using Equation (4), we determine the estimate of the variance of $\bar{x}$.

$$\hat{\sigma}^2_{\bar{x}} = \left(\frac{25}{100}\right)^2(1) + \left(\frac{40}{100}\right)^2(2.7) + \left(\frac{35}{100}\right)^2(14.6) = 2.28. \qquad \square$$

Observing the stratification of the data of Example 1, we see that the data *within* each stratum are quite similar but quite dissimilar to the data in the other strata. In other words, there is less variability among the data within a stratum than between data in different strata. If we can stratify a population so that the variability within a stratum is less than that between strata, we can achieve greater precision in our estimates. Or, we can achieve the same precision using smaller sample sizes. Thus, if a heterogeneous population can be divided into homogeneous groups, we can obtain greater precision in the estimates based on the same total sample size.

Problems

1. Briefly discuss how stratified sampling could be used in designing a sample survey to determine voting preferences in the entire United States. Outline three different sampling plans for conducting the survey.
2. Explain why you might prefer stratified sampling over simple random sampling for the survey in Problem 1.
3. Discuss some of the problems that might arise in designing and conducting the sample survey in Problem 1.
4. A researcher surveyed a small eastern university to determine the amount of money the students actually received from their parents during a certain month. The enrollment

includes 4,000 males and 2,500 females. A stratified sampling plan was used to obtain the following data:

Females	150, 105, 130, 40, 75, 180, 110
Males	35, 105, 28, 15, 65, 25, 45, 25, 75, 63

Determine the sample mean and the estimate of its variance for each stratum. Also, determine the estimate of the population mean and the variance of the estimate.

5. All grocery stores in a certain area are assigned to strata according to certain sales criteria. A random sample is selected from each stratum and the number in stock of a certain product is determined. The results are given in the following table.

Stratum s	N_s	n_s	$\bar{x}_s$	$\hat{\sigma}_s^2$
1	19,850	3,000	4.1	34.8
2	3,250	600	13.0	92.2
3	1,007	340	25.0	174.2
4	606	230	38.2	320.4

Estimate the average number of items in stock for the entire population and the variance of the estimate.

12.4 Cluster Sampling

In stratified sampling we divided the population into strata, and selected a sample from each stratum. Occasionally, it is advantageous to divide the population into a large number of groups, or clusters, and to select a sample of clusters. For example, we might wish to divide a city into several hundred blocks and then select a sample of 50 blocks and measure the characteristics of interest for each block. Such a sampling design, called *cluster sampling,* is frequently used in surveys to reduce the cost of gathering data. Considering the expense of locating and interviewing an individual in the population, we see that cluster sampling could substantially reduce the expense of conducting the survey. The objective is to select clusters that are as *heterogeneous* as possible but small enough to reduce survey costs for such things as listing all members of the population and traveling from one interview to another.

In cluster sampling the cluster, or group, of individuals is the unit selected. In a simple cluster design, a random selection of clusters might be made first, and then each individual within those selected clusters included in the sample. Such a sampling design is known as a single-stage cluster design because sampling occurs only once in the process. On the other hand, multistage sampling would create a more complicated design. For example, we could take a random sample of the blocks within a city and then take a random sample of dwellings within each block. Finally, the interviewer could be instructed to select every third dwelling unit on the block. Thus, we see that sampling procedures can enter the process at several stages. Of course, in order to avoid some unknown biases, we assume some sort of probability sampling will be used.

We omit the formulas for estimating the various population parameters in situations utilizing cluster sampling because they would carry us far beyond our depth. These formulas depend on such things as how many clusters are chosen, whether the entire cluster is chosen or just a small portion is sampled, and whether or not the clusters are of equal size.

Problems

1. Give an example where cluster sampling could be utilized.
2. Give an example where cluster sampling would be more advantageous than simple random sampling or stratified sampling.
3. Give an example where stratified sampling would be more advantageous than cluster sampling or simple random sampling.
4. When would one prefer cluster sampling over simple random sampling or stratified sampling?

12.5 Summary

The relative advantages of the three types of sampling plans discussed in this chapter should be clear by now. Stratified sampling plans may provide the greatest precision for a fixed cost or the least cost for a specified precision when the population can be divided into homogeneous groups. Homogeneity of the groups permits us to obtain precise estimates based on small sample sizes from each group. Furthermore, stratified sampling permits us to obtain separate estimates for each group in addition to estimates for the entire population.

Cluster sampling is most useful when we can divide the population into similar groups that are individually heterogeneous and representative of the entire population. Thus, we can take large samples from a few of the groups at a cost substantially less than the cost of taking a simple random sample from the entire population.

Simple random sampling is most useful when the requirements for stratified or cluster sampling cannot be satisfied. It also may be the best plan when too little is known about the population to divide it into strata or clusters, or when the expense for simple random sampling is approximately the same as for the other plans.

The ideal sampling plan often involves a combination of two or more of the three sampling plans discussed in this chapter. However, these three sampling plans generally form the basis for most complex sampling plans. For further discussion of survey sampling designs see Cochran (12.1), Hansen, et al (12.2), or Raj (12.3 and 12.4).

Important Terms and Concepts

Cluster sampling
Multistage sampling
Nonprobability sampling
Probability sampling
Simple random sampling
Stratified sampling

References

12.1 W. G. Cochran, *Sampling Techniques*, New York: John Wiley & Sons, Inc., 1963.

12.2 M. H. Hansen, et al, *Sample Survey Methods and Theory*, Vol. 1, New York: John Wiley & Sons, Inc., 1953.

12.3 D. Raj, *Sampling Theory*, New York: McGraw-Hill Book Company, 1968.

12.4 D. Raj, *The Design of Sample Surveys*, New York: McGraw-Hill Book Company, 1972.

Thirteen

Analysis of Variance

13.1 Introduction

We have discussed ways to compare the difference between two population means and to test hypotheses about the equalities of several proportions. Since we often are interested in evaluating not just two but several different processes, methods, products, designs, and so forth, a general procedure for simultaneously testing hypotheses about the equalities of several means is important in statistical analysis.

The *analysis-of-variance* procedure was developed especially for the purpose of comparing several sample means. Introduced by Sir Ronald A. Fisher, a British statistician, the technique is essentially an arithmetic process for partitioning the total variation of a set of data into components associated with recognized sources of variation. This technique, one of the most powerful tools in the field of statistics, has been used to tremendous advantage in all fields of research in which data are measured quantitatively. Here, our major objective is to introduce the analysis-of-variance procedure; more extensive treatments can be found in Guenther (13.1) and Kirk (13.2).

In this chapter, we are concerned with testing hypotheses concerning the equality of several means and also testing hypotheses concerning the equality of two variances. To accomplish either task, we must use a new probability distribution. It represents the sampling distribution obtained by sampling from two independent chi-square populations and constructing a particular ratio. The distribution is named the *F distribution* in honor of Sir Ronald A. Fisher.

Definition 13.1

If U and V are independently distributed chi-square variables with m and n degrees of freedom, respectively, then the ratio

$$W = \frac{U/m}{V/n}$$

is distributed according to the F distribution with m and n degrees of freedom.

We may state Definition 13.1 symbolically as follows. If U and F are independent random variables such that

$$U \sim \chi_m^2 \quad \text{and} \quad V \sim \chi_n^2,$$

then

$$W = \frac{U/m}{V/n} \sim F_{m,n}. \tag{1}$$

Note, just as in the definitions of the chi-square distribution (Definition 10.1) and the t distribution (Definition 11.1) the entities (U and V) are variables, not sample observations. Strictly speaking, we mean by Definition 13.1 that the variable W is the ratio of the variable U (a chi-square variable with m degrees of freedom) divided by its number of degrees of freedom to the variable V (a chi-square variable with n degrees of freedom) divided by its number of degrees of freedom. We can again best describe a single resulting distribution indirectly by using the basic idea of infinite repeated sampling as it relates to the concept of a sampling distribution.

Suppose we have a chi-square distribution (U) with m degrees of freedom and another chi-square distribution (V) with n degrees of freedom such that the two distributions are independent and we repeatedly select a sample of one item drawn from each of the independent distributions. Let u_i and v_i represent single observations selected in the ith sample from the distributions of U and V, respectively. Conceptually, an F distribution with m and n degrees of freedom would result if we were able to repeat the following process.

1. Take one observation (say u_i) from the distribution of U and one observation (say v_i) from the distribution of V.
2. Compute a single observation from an F distribution with m and n degrees of freedom as

$$w_i = \frac{u_i/m}{v_i/n}.$$

3. Repeat steps 1 and 2 for an infinite number of samples, that is, for $i = 1, 2, \ldots, \infty$.
4. Compile the probability distribution of the w_i's ($i = 1, 2, \ldots, \infty$). The result is the probability distribution of W, an F distribution with m and n degrees of freedom.

Again, since it is obviously impossible to repeat the sampling procedure an infinite number of times, you should visualize the distribution that would result if the sampling process were repeated a very large number of times.

We can answer the question about the meaning of the numbers of degrees of freedom associated with an F distribution simply by referring to Definition 13.1. In other words, the degrees of freedom associated with the F variable are simply the numbers of degrees associated with the chi-square variables that conceptually go into the makeup of the variable. More specifically, the degrees of freedom m and n associated with an F variable refer to the number (m) of degrees of freedom associated with the chi-square variable in the numerator and the number (n) of degrees of freedom associated with the chi-square variable in the denominator. It is conventional to list the number of degrees of freedom for the numerator first and those for the denominator second. Thus, we refer to an F variable with m and n degrees of freedom as $F(m, n)$. You may again wish to refer to Section 10.1 for our original discussion of the meaning of the number of degrees of freedom associated with a chi-square variable.

The probability density function of the random variable (statistic) W having an F distribution with m and n degrees of freedom (the parameters of the distribution) is

$$f(w; m, n) \frac{\left(\dfrac{m+n-2}{2}\right)!}{\left(\dfrac{m-2}{2}\right)!\left(\dfrac{n-2}{2}\right)!} \left(\frac{m}{n}\right)^{m/2} \frac{w^{(m-2)/2}}{\left(1+\dfrac{mw}{n}\right)^{(m+n)/2}} \qquad (w > 0). \tag{2}$$

We see from (2) that the F distribution represents a two-parameter family of distributions and that the variable (statistic) having this distribution can assume only positive values.

Again, as in the case of the chi-square distribution and the t distribution, probabilities associated with the F distribution depend upon the numbers of degrees of freedom. Because the F distribution has associated with it two different numbers of degrees of freedom, there are an infinite number of F distributions, and we present only selected probability information for F variables in Table A-5 in the Appendix. Note that the table provides values for

$$P(W_{m,n} > w_{m,n}), \tag{3}$$

where $W_{m,n}$ represents the general F variable with m degrees of freedom in the numerator and n degrees of freedom in the denominator and $w_{m,n}$ represents any value $W_{m,n}$ can assume. Recall from (2) that $w_{m,n}$ must always be positive and, at least in this sense, is somewhat similar to the chi-square variable. The techniques for obtaining probabilities from Table A-5 for F variables are similar to those for obtaining chi-square variables from Table A-3 except for one major difference which we shall explain shortly.

Probabilities for the F distribution are given in Table A-5 for the percentage points 1, 2.5, 5, and 10 for selected combinations of the numerator (m) and denominator (n) degrees of freedom. Note that the distribution for each combination of m and n represents a different probability distribution. Thus, the amount of information provided for any given distribution is even less detailed than for chi-square distributions (Table A-3). As usual, there are two general types of probability problems that can be solved. First, we may wish to find the probability (say α) that an F variable with m and n degrees of freedom is greater than a specific value, say b. That is, given m, n, and b, we may wish to find (approximately, since we have available only selected percentage points) α such that

$$P(F_{m,n} > b) = \alpha, \tag{4}$$

where $F_{m,n}$ represents a general F variable with m and n degrees of freedom. To answer such question, we proceed as follows.

1. Find m in the top row of Table A-5.
2. Find n in the left column of Table A-5.
3. Locate the set of values at the intersection of the m column and the n row.
4. If b appears in the set found in step 3, read α as the value in the α column appearing in the same row as b.
5. If b does not appear in the set found in step 3, find the two numbers in the set between which b lies and call them b_1^* and b_2^* (that is, if two such numbers exist). Then, either

a. approximate α by linear interpolation between b_1^* and b_2^* and the two values in the α column corresponding to them, respectively, or

b. approximate α by the value in the α column corresponding to either b_1^* or b_2^*, whichever is nearer b.

The second type of problem is as follows. Given m, n, and α, we may wish to determine the specified value (say b) such that

$$P(F_{m,n} > b) = \alpha.$$

To answer this question, we perform the following steps.

1. Find n in the left column of Table A-5.
2. Locate the set of α values in the α column associated with n.
3. If α is in the set of α values found in step 2, read b as the value appearing in the same row as α and the same column as m.
4. If α is not in the set of values found in step 2, find the two numbers in the set between which α lies and call them α_1^* and α_2^* (that is, if two such numbers exist). Then, either

   a. approximate b by linear interpolation between α_1^* and α_2^* and the two values corresponding to them in the column labeled m, or

   b. approximate b by the value in the column labeled m corresponding to either α_1^* or α_2^*, whichever is nearer α.

Naturally, there are many variations of these two types of problems. However, for our purposes, the second type will be of greater concern to us than the first.

Example 1

Find the value so that the probability is 0.10 that an F variable with $m = 7$ and $n = 4$ degrees of freedom is greater than such value; that is, find b such that

$$P(F_{7,4} > b) = 0.10.$$

Solution

From the body of Table A-5, we see that $b = 3.98$. Thus,

$$P(F_{7,4} > 3.98) = 0.10.$$ □

Example 2

Find the probability that an F variable with $m = 9$ and $n = 10$ degrees of freedom is greater than 3.78.

Solution

From Table A-5, we have

$$P(F_{9,10} > 3.78) = 0.025. \qquad \square$$

Example 3

Find
(a) $P(F_{15,23} > 1.80)$. (b) $P(F_{4,9} > 4.72)$. (c) $P(F_{10,40} > 2.07)$.
(d) $P(F_{16,16} > 3.37)$. (e) $P(F_{12,4} < 5.91)$.
(f) the value b such that $P(F_{8,9} > b) = 0.01$.
(g) the value b such that $P(F_{10,25} < b) = 0.975$.
(h) the value b such that $P(F_{6,3} > b) = 0.05$.
(i) the value b such that $P(F_{4,21} > b) = 0.10$.

Solution

Make sure you follow the reasoning involved in each solution.

(a) $P(F_{15,23} > 1.80) = 0.10$ (b) $P(F_{4,9} > 4.72) = 0.025$.

(c) $P(F_{10,40} > 2.07) = 0.05$. (d) $P(F_{16,16} > 3.37) = 0.01$.

(e) $P(F_{12,4} < 5.91) = 1 - P(F_{12,4} > 5.91) = 1 - 0.05 = 0.95$.

(f) $P(F_{8,9} > 5.47) = 0.01 \Rightarrow b = 5.47$.

(g) $P(F_{10,25} < b) = 0.975 \Rightarrow P(F_{10,25} > b) = 1 - 0.975 = 0.025$.
$P(F_{10,25} > 2.61) = 0.025 \Rightarrow b = 2.61$.

(h) $P(F_{6,3} > 8.94) = 0.05 \Rightarrow b = 8.94$.

(i) $P(F_{4,21} > 2.23) = 0.10 \Rightarrow b = 2.23$. □

Note that there are at most four values in Table A-5 for a given probability distribution—the 1, 2.5, 5, and 10 percentage points for the F distribution corresponding to specific values of m and n. Since we attempt to present as much information in four values as we have in the entire normal table (Table A-1), in many situations we can only approximate probabilities. However, just as in the case of chi-square probabilities, a bracketing (see Example 4, Sec. 10.2) of the required probability will be adequate for most problems associated with an F distribution.

We again use subscripts to distinguish notationally between a general F variable and a single possible value of such variable. That is, throughout the text we shall denote by $F_{m,n}$ the general F variable with m and n degrees of freedom. On the other hand, we shall denote by $F_{\alpha,m,n}$ the single possible value of the general F variable with m and n degrees of freedom such that

$$P(F_{m,n} > F_{\alpha,m,n}) = \alpha. \tag{5}$$

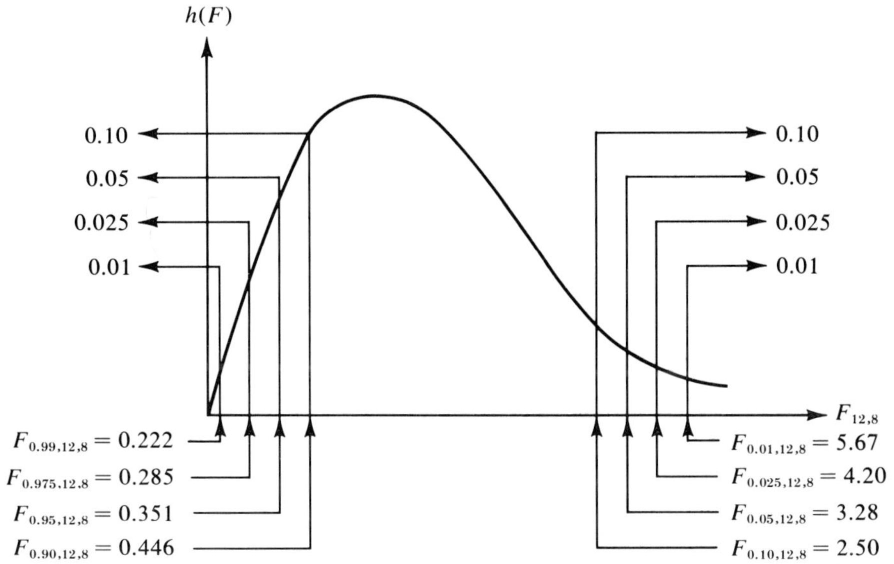

Figure 13.1

F values for the F distribution with 12 and 8 degrees of freedom

---

Problems

1. What do we mean by the numbers of degrees of freedom associated with an F variable?
2. In relationship to the parameters, how is the F distribution similar to the normal distribution but different from the t and chi-square distributions?
3. Determine

   a. $P(F_{8,6} > 8.10)$, b. $P(F_{3,7} > 4.35)$,
   c. $P(F_{10,9} > 2.42)$, d. $P(F_{4,6} < 4.53)$,
   e. $P(F_{6,4} < 6.16)$, f. $P(F_{10,14} < 3.15)$,
   g. the value b such that $P(F_{8,12} > b) = 0.05$,
   h. the value b such that $P(F_{20,3} > b) = 0.10$,
   i. the value b such that $P(F_{10,10} > b) = 0.025$,
   j. $F_{0.05,8,12}$, k. $F_{0.10,20,3}$, l. $F_{0.025,10,10}$.

4. Determine

   a. $P(F_{9,11} > 3.10)$, b. $P(F_{8,16} > 2.85)$, c. $P(F_{9,20} < 3.00)$,
   d. the value b such that $P(F_{9,15} > b) = 0.06$,
   e. the value b such that $P(F_{15,10} > b) = 0.02$,
   f. the value b such that $P(F_{8,3} < b) = 0.96$,
   g. $F_{0.06,9,15}$, h. $F_{0.02,15,10}$, i. $F_{0.96,8,3}$,

j. the value α such that $F_{\alpha, 8, 7} = 3.08$,

k. the value α such that $F_{1-\alpha, 12, 4} = 10.00$.

5. Determine the value of $F_{\alpha, m, n}$ for

a. $m - 8, \quad n = 6, \quad \alpha = 0.99$;

b. $m = 8, \quad n = 6, \quad \alpha = 0.975$;

c. $m = 8, \quad n = 6, \quad \alpha = 0.95$;

d. $m = 8, \quad n = 6, \quad \alpha = 0.90$;

e. $m = 12, \quad n = 15, \quad \alpha = 0.99$;

f. $m = 12, \quad n = 15, \quad \alpha = 0.975$;

g. $m = 12, \quad n = 15, \quad \alpha = 0.95$;

h. $m = 12, \quad n = 15, \quad \alpha = 0.90$.

13.2 Randomization

When we use the analysis-of-variance technique to compare the relationships of several means, we generally perform an experiment specifically designed to eliminate bias from known or unknown factors which may influence the results. The process of *randomization* is used primarily to eliminate the bias due to such factors.

Suppose an assembler can wire an electronic system by four different methods, and we want to study the effects of the methods on the time required to complete the wiring. If we have 20 assemblers available for the study, we should randomly assign each method to an equal number of assemblers—5 assemblers per method. That is, we should use randomization to allow for any variation in the ability of the assemblers. After each assembler is assigned a method, he assembles the system at a normal working pace, and his time to completion is recorded. To eliminate other effects that might cause biased results, we should randomize the location of an assembler within the assembly area; all assemblers will not be subjected to identical working conditions, and an assembler who is in a warmer environment might work slower than one in a cooler environment. If it happens that all assemblers who are assigned the same method are subjected to a warm environment, the average times for assembly will reflect the effects of both the given method and the working conditions (temperature).

Now, it is true that a specific arrangement obtained by randomization may result in one that favors a particular method. However, this is taken into consideration in the methods we use for testing hypotheses and determining confidence intervals in this type of situation. The essence of this type of reasoning can be summarized in the following important statement. Proper randomization validates the assumption of independent observations and renders a basis for measuring probabilities and confidence levels associated with the inference we make from our experimental results.

Problems

1. What is the primary purpose of randomization?
2. What benefits are gained from proper randomization?

13.3 Test for Equality of Variances

In Chapter 11, we considered the problem of pooling the estimates of variance from two different populations when we assumed that the variances of the two populations were equal. We also were required to make this assumption to obtain an exact test for the difference of two means when the variances were unknown. In this section, we develop a test for the equality of two variances. Naturally, such a test will not guarantee that we do or do not have equal variances. However, it will give us an indication of whether our assumption of equal variances is valid.

In order for Definition 13.1 to hold, the variables U and V must be independently distributed chi-square variables. If we are selecting samples of size n_1 and n_2 from two independent normal populations with mean μ_1 and μ_2 and variances σ_1^2 and σ_2^2, respectively, then, from Chapter 10, we know that

$$U = \frac{\sum_{j=1}^{n_1} (X_{1j} - \overline{X}_1)^2}{\sigma_1^2} \sim \chi^2_{n_1 - 1} \tag{1}$$

and

$$V = \frac{\sum_{j=1}^{n_2} (X_{2j} - \overline{X}_2)^2}{\sigma_2^2} \sim \chi^2_{n_2 - 1},$$

where X_{ij} represents the variable corresponding to the jth observation selected from the ith population and $\overline{X}_i$ represents the sample mean of the n_i observations selected from the ith population. Further, as we have indicated in conjunction with (14, Sec. 11.5), U and V are always independently distributed in this sampling situation. Thus, from Definition 13.1, with $m = n_1 - 1$ and $n = n_2 - 1$, we have

$$\frac{U/m}{V/n} = \frac{\dfrac{\sum_{j=1}^{n_1} (X_{1j} - \overline{X}_1)^2}{\sigma_1^2} \Big/ (n_1 - 1)}{\dfrac{\sum_{j=1}^{n_2} (X_{2j} - \overline{X}_2)^2}{\sigma_2^2} \Big/ (n_2 - 1)} \sim F_{n_1 - 1,\, n_2 - 1}. \tag{2}$$

Rearranging the right-hand term in the equality of (2) and substituting the sample values for two specific samples, we obtain the formula for computing an observed value of the statistic of (2). That is,

$$\frac{\dfrac{\sum_{j=1}^{n_1} (x_{1j} - \bar{x}_1)^2}{\sigma_1^2(n_1 - 1)}}{\dfrac{\sum_{j=1}^{n_2} (x_{2j} - \bar{x}_2)^2}{\sigma_2^2(n_2 - 1)}} = \frac{\hat{\sigma}_1^2/\sigma_1^2}{\hat{\sigma}_2^2/\sigma_2^2}, \tag{3}$$

where $\hat{\sigma}_1^2$ and $\hat{\sigma}_2^2$ are the unbiased estimates of the population variances for populations 1 and 2, respectively. Note that (3) represents a single observation on a statistic that satisfies all the requirements of a pivotal quantity to use for testing a hypothesis concerning the two parameters σ_1^2 and σ_2^2. That is,

1. the statistic of (2) is a function of σ_1^2 and σ_2^2 (the unknown parameters),
2. the distribution of the statistic is known and it holds regardless of the true values of σ_1^2 and σ_2^2 (that is, the form of the distribution is independent of σ_1^2 and σ_2^2), and
3. the statistic is a function of no other unknowns.

Thus, under the unique condition (and only under such condition) that $\sigma_1^2 = \sigma_2^2$ (say σ^2), (3) can be written as

$$F = \frac{\hat{\sigma}_1^2/\sigma_1^2}{\hat{\sigma}_2^2/\sigma_2^2} = \frac{\hat{\sigma}_1^2/\sigma^2}{\hat{\sigma}_2^2/\sigma^2} = \frac{\hat{\sigma}_1^2}{\hat{\sigma}_2^2}.$$

Thus, we immediately have a means of testing the hypothesis that the variances of two independent normal populations are equal.

The essential steps for making a 100α percent significance test concerning the equality of the variances of two independent normal populations are as follows.

1. Set up the null hypothesis
$$H_0: \sigma_1^2 = \sigma_2^2.$$
2. Specify the level of significance α.
3. Determine the sizes of the samples (n_1 and n_2) to be taken. (This fixes the numbers of degrees of freedom to be used in establishing the appropriate critical value for the relevant F distribution.)
4. Determine a critical value $F_{m,n}^* = F_{\alpha,m,n}$ (where m and n are appropriately determined from the sample sizes of step 3). Note that we can always arrange the null hypothesis in a one-tail test so that the alternative hypothesis can be right-sided. (For a two-tail test, we can also construct our test so that it is equivalent to a right-sided alternative hypothesis with a significance level of $100(\alpha/2)$ percent.) That is, if necessary, we can simply reverse the roles of σ_1^2 and σ_2^2.
5. Take a sample of size n_1 from population 1 and a sample of size n_2 from population 2 and obtain (under the assumption that the null hypothesis is true, or $\sigma_1^2 = \sigma_2^2$) one observed value of the statistic of (2); that is, compute
$$F = \frac{\hat{\sigma}_1^2}{\hat{\sigma}_2^2}. \tag{4}$$
6. If the value of (4) is in the critical region, reject H_0. Otherwise, accept H_0.

Note the major role played by the assumption that sampling is accomplished from two independently distributed normal populations. If this assumption is not satisfied, we cannot compare (4), the observed value of our pivotal quantity, with appropriate F values, since the statistic of (2) does not have an F distribution for such circumstances.

Example 1

We wish to determine if the variance of money spent per family for entertainment in city 1 is significantly greater than it is in city 2. For the two populations, the amounts of money spent are independent, normally distributed random variables. To conduct our study, we select independent random samples of size $n_1 = 15$ and $n_2 = 20$ from the two populations and determine the unbiased estimates of the variance to be $\hat{\sigma}_1^2 = 186$ and $\hat{\sigma}_2^2 = 81$. Is the variance for city 1 significantly greater at the 5 percent significance level than the variance for city 2?

Solution

The null hypothesis is

$$H_0: \sigma_1^2 = \sigma_2^2$$

versus the one-sided alternative hypothesis

$$H_1: \sigma_1^2 > \sigma_2^2.$$

We determine $F_{\alpha, m, n}$ from Table A-5 as our critical value for testing H_0 against H_1. We have $\alpha = 0.05$ and, from (2), $m = n_1 - 1 = 15 - 1 = 14$ and $n = n_2 - 1 = 20 - 1 = 19$. Thus, we obtain $F_{0.05, 14, 19} = 2.26$ as our critical value. Our decision rule, then, is that of accepting H_0 if the observed value of $F = \hat{\sigma}_1^2/\hat{\sigma}_2^2$ is less than 2.26 and rejecting H_0 if it is greater than 2.26.

Computing the observed value of F from (4), we obtain

$$F = \frac{\hat{\sigma}_1^2}{\hat{\sigma}_2^2} = \frac{186}{81} = 2.30.$$

Comparing 2.30 against our critical value of 2.26, we reject H_0 since $2.30 > 2.26$. Hence, our conclusion is that σ_1^2 is significantly greater than σ_2^2, or the variance of the amount of money spent per family in city 1 is significantly greater than it is in city 2. □

The situation considered in Example 1 is for a right-tail test. As discussed in Section 13.1, we must perform certain computations to obtain the F values associated with a left-tail test. This is not an insurmountable task, but a simpler way of performing a left-tail test is to rename the populations and state the problem in terms of a right-tail test.

Suppose we wish to test the null hypothesis

$$H_0: \sigma_1^2 = \sigma_2^2$$

versus the one-sided alternative hypothesis

$$H_1: \sigma_1^2 < \sigma_2^2. \tag{5}$$

One approach to making this left-tail test is to compute the value of the appropriate pivotal quantity,

$$F = \frac{\hat{\sigma}_1^2}{\hat{\sigma}_2^2}, \tag{6}$$

and compare it with the critical value $F_{1-\alpha, m, n}$. Obviously, this critical value must be determined from Equation (11) of Section 13.1.

On the other hand, we can make the test by simply renaming the populations. That is, if we let $\sigma_A^2 = \sigma_2^2$ and $\sigma_B^2 = \sigma_1^2$, we can test the hypotheses of (5) by testing

$$H_0: \sigma_A^2 = \sigma_B^2 \quad \text{versus} \quad H_1: \sigma_A^2 > \sigma_B^2. \tag{7}$$

In this case, we use (4) to compute

$$F = \frac{\hat{\sigma}_A^2}{\hat{\sigma}_B^2} \tag{8}$$

and compare its value against the critical value $F_{\alpha, n, m}$. (Note that the degrees of freedom have been interchanged.) For example, upon substitution of the original values into (8), we obtain

$$F = \frac{\hat{\sigma}_2^2}{\hat{\sigma}_1^2}, \tag{9}$$

which is the reciprocal of the value given in (6). However, the degrees of freedom for the F value compared against (9) must be reversed from those for the F value compared against (6). This is due to the nature of our definition of the F distribution (that is, degrees of freedom associated with the numerator and denominator), and also the fact that we have interchanged the numerator and denominator for this particular test.

Example 2

A food manufacturer is considering two different types of packaging machines. The salesman for machine A claims that his machine can achieve a smaller variance among the weights of the cans than can machine B. To conduct a study, the food manufacturer selects random samples of size $n_1 = 21$ and $n_2 = 15$ from the groups of cans filled by machines A and B, respectively. The unbiased estimates of the population variances are determined to be $\hat{\sigma}_1^2 = 10$ square ounces and $\hat{\sigma}_2^2 = 19$ square ounces. The weights of the cans are known to be normally distributed. Determine at the 5 percent significance level whether the variance of the weights of cans packaged by machine A is significantly less than the variance of those packaged by machine B.

Solution

The null hypothesis is

$$H_0: \sigma_A^2 = \sigma_B^2$$

versus the one-sided alternative hypothesis

$$H_1: \sigma_A^2 < \sigma_B^2.$$

Since this is a left-tail test, we may convert it to a right-tail test by simply stating our hypotheses as

$$H_0: \sigma_B^2 = \sigma_A^2 \quad \text{versus} \quad H_1: \sigma_B^2 > \sigma_A^2. \tag{10}$$

Thus, we compute (9) as

$$F = \frac{\hat{\sigma}_B^2}{\hat{\sigma}_A^2} = \frac{\hat{\sigma}_2^2}{\hat{\sigma}_1^2} = \frac{19}{10} = 1.90.$$

For our critical value, we determine the F value from Table A-5 with $n_2 - 1 = 15 - 1 = 14$ and $n_1 - 1 = 21 - 1 = 20$ degrees of freedom which has 0.05 area to its right. That is, $F_{0.05, 14, 20} = 2.23$. Our decision rule is that of accepting H_0 [when tested against the alternative hypothesis of (10)] if the observed value of our test statistic is less than the critical value 2.23 and rejecting H_0 [in favor of the alternative hypothesis as stated in (10)] if it is greater than 2.23.

Comparing the observed value (1.90) of the pivotal quantity against the critical value (2.23) and applying our decision rule, we accept H_0 and conclude that the variance of the weights of cans packaged by machine A is not significantly less than the variance of those packaged by machine B. That is, the variance for machine B is not significantly greater than the variance for machine A. □

When converting a left-tail test to a right-tail test as in Example 2, we should recall the vital point made in Chapter 9 that we should always state the hypotheses of interest prior to considering the sample data. If we erroneously allowed the sample data to determine the alternative hypothesis, we would never need a two-sided alternative hypothesis. Note particularly, then, that for Example 2, the alternative hypothesis was changed only to simplify the computations, not because of the sample data.

When we wish to make a two-tail test concerning the equality of two population variances, we use (when necessary) the same basic logic as that for converting a left-tail test to a right-tail test. However, because of the difference in the tail areas employed, it may appear that sample data are used to establish the alternative hypothesis. This is certainly not the case. That is, if the alternative hypothesis is supposed to be a two-sided alternative with a significance level of α, then the proper one-tail critical value (if one is used) must be for a significance level of $\alpha/2$ (assuming, of course, a symmetrical test). If our alternative hypothesis is, in fact, a two-sided alternative, then we can (when allowing the data to determine which tail to use for ease of computations only) perform the computations as though we were making a one-tail test with a significance level equal to one-half that for the associated two-tail test.

Example 3

A manufacturer of industrial thermostats is considering two methods for producing thermostats having as small a variation of temperature control as possible. To determine whether there is any difference between the variability in temperature control for the two methods, the manufacturer selects random samples of sizes $n_1 = 16$ and $n_2 = 13$ from the populations of thermostats made by method 1 and method 2, respectively. The unbiased estimates of the population variances are determined to be $\hat{\sigma}_1^2 = 8$ square degrees and $\hat{\sigma}_2^2 = 16$ square degrees. Using a significance level of 5 percent and assuming that the temperatures at which the thermostats open are normally distributed, determine whether there is a significant difference between the variability of temperature control for the two methods.

Solution

The hypotheses of interest are

$$H_0: \sigma_1^2 = \sigma_2^2 \quad \text{versus} \quad H_1: \sigma_1^2 \neq \sigma_2^2. \tag{11}$$

We could determine the two critical values $F_{0.975, 15, 12}$ and $F_{0.025, 15, 12}$ through use of Table A-5 and (5, Sec. 13.1) and compare the observed value of our pivotal quantity

$$F = \frac{\hat{\sigma}_1^2}{\hat{\sigma}_2^2} \tag{12}$$

with these values. Since, for our problem, this value will be less than one (that is, the value will be in the left tail of the F distribution), we would have to determine $F_{0.975, 15, 12}$ to implement the test in this manner. However, we shall simply determine one value from Table A-5 against which we can compare the value

$$F = \frac{\hat{\sigma}_2^2}{\hat{\sigma}_1^2}. \tag{13}$$

That is, this value will be greater than one and hence comparable to a right-tail critical value. Since we have reversed the numerator and denominator of (12), we must reverse the degrees of freedom. Thus, we now compare the value of (13) against $F_{0.025, 12, 15}$ and $F_{0.975, 12, 15}$. But, since (13) will be in the right tail of the F distribution for our problem, we need determine only one of the critical values; that is, $F_{0.025, 12, 15} = 2.96$.

Computing the value of (13), we obtain

$$F = \frac{\hat{\sigma}_2^2}{\hat{\sigma}_1^2} = \frac{16}{8} = 2.00.$$

Then, comparing the observed value (2.00) of our pivotal quantity against the critical value for the right tail ($F_{0.025, 12, 15} = 2.96$), we see that the observed value is less than the critical value. Thus, we accept H_0 and conclude that there is no significant difference between the variability of temperature control for the two methods. □

From Examples 1, 2, and 3, we see that each of the three types of problems (left-tail, right-tail, and two-tail) involving testing hypotheses concerning the equality of variances can be treated as a right-tail test. Thus, we require in each case only one critical value that can be read directly from Table A-5. You should be certain that you fully understand the reasoning behind this approach.

Problems

1. Suppose that simple random samples of size $n_1 = 10$ and $n_2 = 20$ are selected from two normal populations and that the estimates $\hat{\sigma}_1^2 = 25$ and $\hat{\sigma}_2^2 = 10$ are obtained. Determine at the 5% significance level whether σ_1^2 is significantly greater than σ_2^2.

2. Suppose that simple random samples of size $n_1 = 16$ and $n_2 = 21$ are selected from two normal populations and that the estimates $\hat{\sigma}_1^2 = 50$ and $\hat{\sigma}_2^2 = 112$ are obtained. Determine at the 5% significance level whether σ_1^2 is significantly less than σ_2^2.

Use the data of the indicated problems to determine whether σ_1^2 is significantly different than σ_2^2 at the given percent significance levels.

3. Problem 2, Section 11.5; 2%.
4. Problem 1, Section 11.5; 10%.
5. Problem 2, Section 11.6; 10%.
6. Problem 1, Section 11.6; 5%.

13.4 Completely Randomized, One-Variable Classification Model

In Chapters 7 through 11, as well as in the first three sections of this chapter, we have been concerned primarily with problems associated with sampling from either a single population or, at most, two separate populations. In the remainder of this chapter, we shall be concerned with problems associated with sampling from k populations. We begin by considering the situation in which the individual values are subject to a single criterion of classification, namely, the population to which they belong. As in the case for sampling from two separate populations, we can again completely identify any observed value with two subscripts. Hence, we represent any observed value by x_{ij}, where the first subscript ($i = 1, 2, \ldots, k$) represents the population from which the value comes and the second subscript ($j = 1, 2, \ldots, n_i$, where n_i is the sample size from population i) represents the number of the individual observed value from the ith population. For example, x_{35} represents the fifth observed value from the third population.

In Section 8.5, we discussed the error involved with point estimators. A specific error discussed was that for the sample mean ($\bar{X}$) when we use it as an estimator for the population mean (μ_x). This error was measured as $e = \bar{X} - \mu_x$. As an alternative method, we could have used each sample value as an estimator of the mean (such an estimator would at least be unbiased) and determined the error for, say, the ith sample value as $e_i = X_i - \mu_x$. Assuming, of course, that sampling is from a normal population, it is possible to show that this error is normally distributed with mean zero and variance σ_x^2.

As mentioned above, the variables of interest in this section now have two subscripts, Hence, when we use a single value (X_{ij}) as an estimator of the relevant (ith) population mean (μ_i), the error can be written as $e_{ij} = X_{ij} - \mu_i$, where μ_i is the mean for the ith population and e_{ij} is the error (or deviation from the population mean) associated with using X_{ij} as an estimator of μ_i. Thus, if each population from which we select a random sample is a normal population with mean μ_i and variance σ_x^2 (that is, all populations have the same variance) then the populations of $e_{ij}(i = 1, 2, \ldots, k;$

$j = 1, 2, \ldots, n_i$) are normally and independently distributed with means zero and variances σ_x^2. (Henceforth, to denote the variance of the e_{ij}'s, we shall use σ_e^2.) Now, letting

$$\mu = \frac{1}{k} \sum_{i=1}^{k} \mu_i \tag{1}$$

and

$$\tau_i = \mu_i - \mu, \tag{2}$$

we have

$$e_{ij} = X_{ij} - \mu_i = X_{ij} - \mu - \tau_i. \tag{3}$$

By solving the right and left sides of (3) for X_{ij} we can represent it as

$$X_{ij} = \mu + \tau_i + e_{ij} \qquad (i = 1, 2, \ldots, k; j = 1, 2, \ldots, n_i). \tag{4}$$

Relation (4) is a model representation of the jth observation taken from the ith population. Since τ_i represents the difference between the mean (μ_i) of one group (the ith population) and the overall mean (μ), it is generally termed the ith *group effect* or *treatment effect*. It can be seen from (2) that $\sum_{i=1}^{k} \tau_i = 0$.

The model of (4) is generally called the *completely randomized, one-variable classification model*. For this model, we shall assume that

1. μ is an unknown parameter,
2. the e_{ij}'s are normally and independently distributed with mean zero and variance σ_e^2, and (5a)
3. the τ_i's are unknown constants or parameters.

When the assumptions of (5a) are satisfied, (4) is called a *model for fixed effects*. That is, the group effects are assumed constant (not variable). This assumption is appropriate when the population from which we are sampling constitute the totality of all populations under consideration. However, when the populations from which we are sampling represent only a subset of a larger group of populations that might be considered, the τ_i's are assumed to be random variables that are normally and independently distributed with mean zero and variance σ_τ^2. In this case, (4) is termed a *model for random effects*, and the relevant assumptions may be stated as

1. μ is an unknown parameter,
2. the e_{ij}'s are normally and independently distributed with mean zero and variance σ_e^2, and (5b)
3. the τ_i's are normally and independently distributed with mean zero and variance σ_τ^2.

In the analysis of variance that we shall consider, the major difference between the model for fixed effects and the model for random effects is in the interpretation of the results. Relevant computations are identical for both models, and so are the mechanics for testing hypotheses for the two cases. However, the statements of the hypotheses for the two cases differ in that they involve fixed effects in one case and random effects in the other.

As stated at the beginning of the chapter, the analysis of variance is an arithmetic process for partitioning the total variation of a set of data into components associated with recognized sources of variation. This process, irrespective of the type of model assumed, is accomplished by partitioning the total sum of squares of the observed sample values into distinct parts so that the extent each recognized source of variation has contributed to the sum of squares can be determined.

We shall begin our study of the analysis of variance by relating the technique to the completely randomized, one-variable classification model. Let us consider, for our present model, a general set of observed values from k populations (represented notationally in Table 13.1), where, for example, n_i values have been randomly selected

Table 13.1

Data for one-way classification model

Population	*Observed values*					*Sum*	*Mean*
1	x_{11}	x_{12}	x_{13}	$\cdots$	x_{1n_1}	$x_{1\cdot}$	$\bar{x}_{1\cdot}$
2	x_{21}	x_{22}	x_{23}	$\cdots$	x_{2n_2}	$x_{2\cdot}$	$\bar{x}_{2\cdot}$
3	x_{31}	x_{32}	x_{33}	$\cdots$	x_{3n_3}	$x_{3\cdot}$	$\bar{x}_{3\cdot}$
$\vdots$	$\vdots$	$\vdots$	$\vdots$		$\vdots$	$\vdots$	$\vdots$
k	x_{k1}	x_{k2}	x_{k3}	$\cdots$	x_{kn_k}	$x_{k\cdot}$	$\bar{x}_{k\cdot}$
Total						$x_{\cdot\cdot}$	$\bar{x}_{\cdot\cdot}$

from the ith population. We shall refer to the n_i observed values from the ith population as the ith *group*. (Note that we can have either the same or a different number of values from each of the k populations.) In Table 13.1, the group totals and means, as well as the grand total (overall total of all observed values) and grand mean (overall mean or all observed values), are represented by the dot notation introduced in Chapter 10, page 410. Namely, the group totals are represented by

$$x_{i\cdot} = \sum_{j=1}^{n_i} x_{ij}, \tag{6}$$

the group means by

$$\bar{x}_{i\cdot} = \frac{x_{i\cdot}}{n_i} = \frac{1}{n_i}\sum_{j=1}^{n_i} x_{ij}, \tag{7}$$

the grand total by

$$x_{\cdot\cdot} = \sum_{i=1}^{k} x_{i\cdot} = \sum_{i=1}^{k}\sum_{j=1}^{n_i} x_{ij}, \tag{8}$$

and the grand mean by

$$\bar{x}_{\cdot\cdot} = \frac{1}{N}\sum_{i=1}^{k} x_{i\cdot} = \frac{1}{N}\sum_{i=1}^{k}\sum_{j=1}^{n_i} x_{ij}, \tag{9}$$

where

$$N = \sum_{i=1}^{k} n_i \tag{10}$$

is the total sample size. Note that N takes on a different meaning here than it has previously.

To construct the analysis-of-variance table for the selected model, we must consider what sources of variation are present and how we are to partition the total variation into portions corresponding to these sources. As usual, the total variation is determined by the total sum of squares of deviations of the observed values from the grand mean. (Hereafter, we shall refer to this entity as the *total sum of squares.*) Hence, this is the sum of squares that we must partition. One of the sources of variation is the differences among the group means, and it is determined by the sum of squares of deviations of the group means from the grand mean. The only other source of variation under our present model is that among the observed values within the groups (that is, within the samples taken from the various populations). This is clearly the variation within each group of the individual values about the sample mean for their group. This variation is determined by the sum of the squares of deviations of the individual values from the sample mean for the group to which they belong.

In order to understand the computations and the identities involved in the analysis-of-variance table, consider the total sum of squares (generally abbreviated as SS_{total}). We have

$$SS_{\text{total}} = \sum_{i=1}^{k} \sum_{j=1}^{n_i} (x_{ij} - \bar{x}_{..})^2. \tag{11}$$

If we insert within the parentheses of (11) a judiciously selected term equal to zero (namely, $-\bar{x}_{i.} + \bar{x}_{i.}$), we have

$$SS_{\text{total}} = \sum_{i=1}^{k} \sum_{j=1}^{n_i} (x_{ij} - \bar{x}_{i.} + \bar{x}_{i.} - \bar{x}_{..})^2. \tag{12}$$

(The reason for inserting this term may not be immediately obvious; nevertheless, it allows us an easy means of demonstrating an important relationship.) If we now expand (12) as follows, we obtain

$$\begin{aligned} SS_{\text{total}} &= \sum_{i=1}^{k} \sum_{j=1}^{n_i} (x_{ij} - \bar{x}_{..})^2 \\ &= \sum_{i=1}^{k} \sum_{j=1}^{n_i} [(x_{ij} - \bar{x}_{i.}) + (\bar{x}_{i.} - \bar{x}_{..})]^2 \\ &= \sum_{i=1}^{k} \sum_{j=1}^{n_i} (x_{ij} - \bar{x}_{i.})^2 + 2\sum_{i=1}^{k} \sum_{j=1}^{n_i} (x_{ij} - \bar{x}_{i.})(\bar{x}_{i.} - \bar{x}_{..}) + \sum_{i=1}^{k} \sum_{j=1}^{n_i} (\bar{x}_{i.} - \bar{x}_{..})^2. \end{aligned} \tag{13}$$

Now, the middle term on the right-hand side can be written as

$$2\sum_{i=1}^{k} \left[(\bar{x}_{i.} - \bar{x}_{..}) \sum_{j=1}^{n_i} (x_{ij} - \bar{x}_{i.}) \right],$$

and, from Chapter 3, we have

$$\sum_{j=1}^{n_i} (x_{ij} - \bar{x}_{i.}) = 0$$

for all i. This being true, the entire middle term on the right-hand side of (13) is equal to zero. Thus, (13) can be written as

$$\begin{aligned} \sum_{i=1}^{k} \sum_{j=1}^{n_i} (x_{ij} - \bar{x}_{..})^2 &= \sum_{i=1}^{k} \sum_{j=1}^{n_i} (x_{ij} - \bar{x}_{i.})^2 + \sum_{i=1}^{k} \sum_{j=1}^{n_i} (\bar{x}_{i.} - \bar{x}_{..})^2 \\ &= \sum_{i=1}^{k} \sum_{j=1}^{n_i} (x_{ij} - \bar{x}_{i.})^2 + \sum_{i=1}^{k} n_i(\bar{x}_{i.} - \bar{x}_{..})^2. \end{aligned} \tag{14}$$

The first term to the right of the equality of (14) is the sum of squares of the deviations of the individual values from the sample mean for the group to which they belong—that is, the variation among the values within the groups or populations. This term is known as the *sum of squares within groups* and is generally abbreviated as SS_{within}.

The second term to the right of the equality of (14) is the sum of squares of the deviations of group means from the grand mean. It is multiplied by n_i for each group in order to place the term on a per-observed-value basis. This term is called the *sum of squares of the deviations among group means* and is abbreviated as SS_{among}.

Using the foregoing abbreviations and (14), we have

$$SS_{\text{total}} = SS_{\text{within}} + SS_{\text{among}}. \tag{15}$$

This represents a highly useful partitioning of the total sum of squares. We will note shortly that it is particularly appropriate for our present model. The partitioning of the sum of squares is presented in notational form in Table 13.2 in conjunction with other related information.

Since we are sampling from k populations that are normally and independently distributed with a common variance σ_x^2, it can be shown that the sum of squares among groups (SS_{among}), when divided by the appropriate variance (say, σ_1^2, where σ_1^2 represents a variance to be explained in detail later), is an observed value of a statistic which is distributed according to the chi-square distribution with $(k - 1)$ degrees of freedom. Similarly, the sum of squares within groups (SS_{within}), when divided by the appropriate variance (say, σ_2^2, where σ_2^2 represents a variance to be explained in detail later), is an observed value of a statistic which is distributed according to the chi-square distribution with $(N - k)$ degrees of freedom and, moreover, is independent of the statistic for SS_{among}. However, mathematical verification of these statements is beyond the scope of this text and you are referred to Graybill (13.1) for further information.

The importance of the foregoing statements is their connection with testing a hypothesis concerning the group (or treatment) effects that we shall discuss later. First, however, let us become familiar with appropriate computational notation. The computational mechanics of the relevant test are given in Table 13.2. Note that the entities labeled *mean square* (MS_{among} and MS_{within}) are the observed values (excluding division by appropriate variances) for the two chi-square variables mentioned previously and are represented respectively by $SS_{\text{among}}/(k - 1)$ and $SS_{\text{within}}/(N - k)$.

The sums of squares given in Equation (14) are in terms of deviations. In Chapter 3, we discussed a computational formula for computing the sum of squares of deviations from a mean. Comparable formulas are available for the analogous sets of sums of squares in the analysis of variance as shown in Table 13.2. Thus, we can compute total sum of squares by the formula

$$SS_{\text{total}} = \sum_{i=1}^{k} \sum_{j=1}^{n_i} (x_{ij} - \bar{x}_{..})^2 = \sum_{i=1}^{k} \sum_{j=1}^{n_i} x_{ij}^2 - \frac{x_{..}^2}{N}, \tag{16}$$

among-groups sum of squares by the formula

$$SS_{\text{among}} = \sum_{i=1}^{k} n_i(\bar{x}_{i\cdot} - \bar{x}_{..})^2 = \sum_{i=1}^{k} \frac{x_{i\cdot}^2}{n_i} - \frac{x_{..}^2}{N}, \tag{17}$$

Table 13.2

Analysis-of-variance table for one-variable classification model

Source of variation	Degrees of freedom	Sum of squares	Mean square	F ratio
Among groups	$k-1$	$SS_{\text{among}} = \sum_{i=1}^{k} n_i(\bar{x}_{i\cdot} - \bar{x}_{\cdot\cdot})^2 = \sum_{i=1}^{k} \frac{x_{i\cdot}^2}{n_i} - \frac{x_{\cdot\cdot}^2}{N}$	$MS_{\text{among}} = \frac{SS_{\text{among}}}{(k-1)}$	$F = \frac{MS_{\text{among}}}{MS_{\text{within}}}$
Within groups	$N-k$	$SS_{\text{within}} = \sum_{i=1}^{k}\sum_{j=1}^{n_i}(x_{ij} - \bar{x}_{i\cdot})^2 = \sum_{i=1}^{k}\sum_{j=1}^{n_i} x_{ij}^2 - \sum_{i=1}^{k}\frac{x_{i\cdot}^2}{n_i}$	$MS_{\text{within}} = \frac{SS_{\text{within}}}{(N-k)}$	
Total	$N-1$	$SS_{\text{total}} = \sum_{i=1}^{k}\sum_{j=1}^{n_i}(x_{ij} - \bar{x}_{\cdot\cdot})^2 = \sum_{i=1}^{k}\sum_{j=1}^{n_i} x_{ij}^2 - \frac{x_{\cdot\cdot}^2}{N}$		

and within-groups sum of squares by the formula

$$SS_{\text{within}} = \sum_{i=1}^{k}\sum_{j=1}^{n_i}(x_{ij} - \bar{x}_{i\cdot})^2 = \sum_{i=1}^{k}\sum_{j=1}^{n_i} x_{ij}^2 - \sum_{i=1}^{k}\frac{x_{i\cdot}^2}{n_i}. \tag{18}$$

As stated earlier, the computations for the analysis of variance associated with our present model (4) are identical regardless of which assumptions [(5a) or (5b)] are valid. However, the statement of hypotheses and the interpretation of the results differ from the two cases because (5a) is the set of assumptions for fixed effects and (5b) is the set of assumptions for random effects. Having discussed the computational mechanics for the analysis of variance associated with model (4), let us now turn our attention to the statement and interpretation of hypotheses for the two different cases.

Table 13.3 summarizes Table 13.2 and includes two new columns representing the

Table 13.3

Analysis-of-variance table for one-variable classification model

Source of variation	Degrees of freedom	Mean square	Expected mean square	
			Fixed effects	Random effects
Among groups	$k-1$	MS_{among}	$\sigma_e^2 + \frac{1}{k-1}\sum_{i=1}^{k} n_i(\tau_i - \bar{\tau}_\cdot)^2$	$\sigma_e^2 + \frac{N^2 - \sum_{i=1}^{k} n_i^2}{N(k-1)}\sigma_\tau^2$
Within groups	$N-k$	MS_{within}	σ_e^2	σ_e^2
Total	$N-1$			

expected values of the variables (statistics) associated with corresponding entries in the mean-square column of Table 13.2. The columns labeled "Fixed effects" and "Random effects" correspond to the expected values of the variables associated with the entries in the mean-square column when the assumptions of (5a) and (5b), respectively, are valid. That is, the columns under the "Expected mean square" heading give those functions of the population parameters that are estimated unbiasedly by the statistics associated with the entries of the mean-square column for the two different cases. Thus, given the results of an experiment and the corresponding analysis-of-variance table, the numerical entry in the mean-square column for among-groups represents an estimate obtained through use of an unbiased estimator of

$$\sigma_e^2 + \frac{1}{k-1}\sum_{i=1}^{k} n_i(\tau_i - \bar{\tau}.)^2, \tag{19}$$

provided, of course, that the assumptions of (5a) are valid. If the assumptions of (5b) are valid, then it represents an estimate obtained through use of an unbiased estimator of

$$\sigma_e^2 + \frac{N^2 - \sum_{i=1}^{k} n_i^2}{N(k-1)}\sigma_\tau^2. \tag{20}$$

In both cases, the numerical entry in the mean-square column for within-groups represents an estimate obtained through use of an unbiased estimator of the variance of the error terms of model (4). For this reason, the within-groups source of variation is occasionally referred to as that due to experimental error, and, consequently, MS_{within} is often denoted by MS_{error}.

We noted earlier that the mean-square entries for among-groups and within-groups, when divided by the appropriate variances (σ_1^2 and σ_2^2, respectively), represent observed values for specific chi-square variables which have been divided by their respective degrees of freedom.

It can be shown that the appropriate variances are as follows. When the assumptions of (5a) are valid,

$$\sigma_1^2 = \sigma_e^2 + \frac{1}{k-1}\sum_{i=1}^{k} n_i(\tau_i - \bar{\tau}.)^2, \tag{21}$$

where σ_e^2 represents the variance of the error terms and τ_i represents the treatment effects of model (5a). On the other hand, when the assumptions of (5b) are valid,

$$\sigma_1^2 = \sigma_e^2 + \frac{N^2 - \sum_{i=1}^{k} n_i^2}{N(k-1)}\sigma_\tau^2, \tag{22}$$

where σ_τ^2 represents the variance of the random effects of model (5b). Further, for both cases

$$\sigma_2^2 = \sigma_e^2. \tag{23}$$

We note that the entities in the "Expected mean square" columns are actually variances [that is, the values in (21) through (23) are variances]. Therefore, the corresponding values in the "Mean square" column are unbiased estimates of the variances given in the "Expected mean square" columns, and, of course, the variances they estimate unbiasedly depend upon which assumption [(5a) or (5b)] is valid.

Moreover, we immediately have a pivotal quantity to use in testing appropriately selected hypotheses concerning the means of the various groups (treatments) when we assume a completely randomized, one-variable classification model. For, whether we assume fixed or random effects, the ratio

$$\frac{MS_{\text{among}}/\sigma_1^2}{MS_{\text{within}}/\sigma_2^2}$$

represents an observed value of an F variable with $(k - 1)$ and $(N - k)$ degrees of freedom, provided, of course, that σ_1^2 and σ_2^2 are chosen to suit the model assumptions—that is, provided they are chosen properly from among Equations (21) through (23). In effect, then, the general statistic represented by the single observation of the above provides us a way of obtaining a pivotal quantity for testing hypotheses concerning group (treatment) means.

Fixed-Effects Model

In the fixed-effect situation, the value

$$\frac{MS_{\text{among}}\Big/\left[\sigma_e^2 + \frac{1}{k-1}\sum_{i=1}^{k} n_i(\tau_i - \bar{\tau}.)^2\right]}{MS_{\text{within}}/\sigma_e^2} \tag{24}$$

represents an observed value of a statistic that

1. has a known $(F_{k-1,\,N-k})$ distribution the form of which is independent of the unknown parameters $\tau_i (i = 1, 2, \ldots, k)$,
2. is a function of the unknown parameters τ_i, and
3. can be shown to be, under an appropriately selected null hypothesis, a function of no other unknowns.

Suppose we are sampling from k populations that include all populations of interest and our model is given by (4); that is,

$$X_{ij} = \mu + \tau_i + e_{ij} \qquad (i = 1, 2, \ldots, k;\ j = 1, 2, \ldots, n_i).$$

Suppose further that we can assume a fixed-effects situation where the assumptions of (5a) are valid, and that our objective is to determine whether there are significant differences among the means of the various groups (treatments). Notationally [see (2)], the problem is that of determining whether any of the τ_i's are significantly different from zero.

What does this mean? Let us consider an example where we wish to determine whether there are any significant differences among the average useful lives of three types of automobile tires. The population averages of useful life may be represented by μ_1, μ_2, and μ_3 for the three types of tires. We could test the null hypothesis that the three population means are equal $(H_0\colon \mu_1 = \mu_2 = \mu_3)$. However, using (1), we can represent the average of the three population means by $\mu = (1/3)\sum_{i=1}^{3} \mu_i$. And using (2), we can represent the treatment (type-of-tire) effects by

$$\tau_1 = \mu_1 - \mu, \quad \tau_2 = \mu_2 - \mu, \quad \text{and} \quad \tau_3 = \mu_3 - \mu.$$

Now, if the population means (μ_i) are all equal, the treatment effect (τ_i) for each treatment (type of tire) is equal to zero. If this is the situation, we can express the null

represents the variance for both populations. In the fixed-effects model, we assume that the variances of all populations involved are equal and have labeled them as σ_e^2 for notational convenience. Thus, for any two populations (groups) involved in our fixed-effects model, the statistic

$$\frac{(\bar{X}_1 - \bar{X}_2) - (\mu_1 - \mu_2)}{\sigma_e\sqrt{\frac{1}{n_1} + \frac{1}{n_2}}} \sim \mathrm{N}(0, 1). \tag{29}$$

Further, for the fixed-effects model, we have from our original discussion of Table 13.2 and the material related to (23) that

$$\frac{MS_{\text{within}}}{\sigma_e^2} \tag{30}$$

represents a single observed value of a chi-square statistic with $(N - k)$ degrees of freedom which has been divided by its degrees of freedom, where N is the total sample size of the k samples involved. However, we already have labeled MS_{within} as the unbiased estimate of σ_e^2—that is, we have defined

$$MS_{\text{within}} \doteq \hat{\sigma}_e^2. \tag{31}$$

Thus, by replacing MS_{within} by $\hat{\sigma}_e^2$ in (30), taking the ratio of a single observed value of the statistic of (29) to the square root of the entity of (30), and cancelling the σ_e's, we have that

$$\frac{(\bar{x}_1 - \bar{x}_2) - (\mu_1 - \mu_2)}{\hat{\sigma}_e\sqrt{\frac{1}{n_1} + \frac{1}{n_2}}} \tag{32}$$

represents an observed value of a t statistic with $(N - k)$ degrees of freedom.

For an *LSD* test, we wish to determine the smallest difference between $\bar{x}_1$ and $\bar{x}_2$ for which our means will be considered significantly different. Thus, for a 100α percent significance level, we wish to determine the value of $(\bar{x}_1 - \bar{x}_2)$ that represents the *LSD* value. We denote the *LSD* value associated with a 100α percent significance level as $LSD(\alpha)$. That is,

$$LSD(\alpha) = \mathrm{Min}(\bar{x}_1 - \bar{x}_2),$$

where $\mathrm{Min}(\bar{x}_1 - \bar{x}_2)$ means "the smallest possible value $(\bar{x}_1 - \bar{x}_2)$ can assume such that the difference is considered significant." Since the value of $LSD(\alpha)$ represents the value of $(\bar{x}_1 - \bar{x}_2)$ for which (32) is equal to the critical value $t_{\alpha/2, N-k}$, we solve for $LSD(\alpha)$ simply by replacing the $(\bar{x}_1 - \bar{x}_2)$ term in (32) by $LSD(\alpha)$ and setting the resulting relation equal to the appropriate critical value. Upon substitution, we obtain

$$t_{\alpha/2, N-k} = \frac{LSD(\alpha) - (\mu_1 - \mu_2)}{\hat{\sigma}_e\sqrt{\frac{1}{n_1} + \frac{1}{n_2}}}. \tag{33}$$

Note that the sample sizes (n_1 and n_2) of (33) now represent the sample sizes associated with the two means whose differences we are comparing against $LSD(\alpha)$. Solving (33) for $LSD(\alpha)$, we obtain

$$LSD(\alpha) = (\mu_1 - \mu_2) + t_{\alpha/2, N-k} \cdot \hat{\sigma}_e\sqrt{\frac{1}{n_1} + \frac{1}{n_2}}. \tag{34}$$

However, since we are concerned with whether the group means are equal, the $(\mu_1 - \mu_2)$ term will be zero. [Recall that this is essentially what we are saying under our null hypothesis (25).] Equation (34) then becomes

$$LSD(\alpha) = t_{\alpha/2, N-k} \cdot \hat{\sigma}_e \sqrt{\frac{1}{n_1} + \frac{1}{n_2}}, \tag{35}$$

where α is the significance level for our test, $N = \sum_{i=1}^{k} n_i$, k is the number of groups, $\hat{\sigma}_e$ is obtained by using (31), and n_1 and n_2 represent the sample sizes associated with the two sample means being compared.

To determine which means in Example 1 are significantly different, we refer to the appropriate computations of Table 13.6. We note that the *LSD*(0.05) values are different for some comparisons because of the differences in sample sizes for the two means involved in the relevant absolute difference. We see that only the absolute differences involving group 4 means are greater than the corresponding *LSD*(0.05) value. Thus, the mean of group 4 is significantly different from the means of the other groups.

Table 13.6

LSD test for Example 1

Absolute difference between means	n_1 *and* n_2	*LSD*(0.05)	*Conclusion*
$\lvert\bar{x}_{1\cdot} - \bar{x}_{2\cdot}\rvert = \lvert 8.25\text{–}6.50\rvert = 1.75$	4, 4	3.587	Not significant
$\lvert\bar{x}_{1\cdot} - \bar{x}_{3\cdot}\rvert = \lvert 8.25\text{–}6.00\rvert = 2.25$	4, 3	3.877	Not significant
$\lvert\bar{x}_{1\cdot} - \bar{x}_{4\cdot}\rvert = \lvert 8.25\text{–}14.00\rvert = 5.75$	4, 4	3.587	Significant
$\lvert\bar{x}_{1\cdot} - \bar{x}_{5\cdot}\rvert = \lvert 8.25\text{–}10.00\rvert = 1.75$	4, 3	3.877	Not significant
$\lvert\bar{x}_{2\cdot} - \bar{x}_{3\cdot}\rvert = \lvert 6.50\text{–}6.00\rvert = 0.50$	4, 3	3.877	Not significant
$\lvert\bar{x}_{2\cdot} = \bar{x}_{4\cdot}\rvert = \lvert 6.50\text{–}14.00\rvert = 7.50$	4, 4	3.587	Significant
$\lvert\bar{x}_{2\cdot} - \bar{x}_{5\cdot}\rvert = \lvert 6.50\text{–}10.00\rvert = 3.50$	4, 3	3.877	Not significant
$\lvert\bar{x}_{3\cdot} - \bar{x}_{4\cdot}\rvert = \lvert 6.00\text{–}14.00\rvert = 8.00$	3, 4	3.877	Significant
$\lvert\bar{x}_{3\cdot} - \bar{x}_{5\cdot}\rvert = \lvert 6.00\text{–}10.00\rvert = 4.00$	3, 3	4.140	Not significant
$\lvert\bar{x}_{4\cdot} - \bar{x}_{5\cdot}\rvert = \lvert 14.00\text{–}10.00\rvert = 4.00$	4, 3	3.877	Significant

$$LSD(0.05) = t_{0.025, 13} \cdot \hat{\sigma}_e \sqrt{\frac{1}{n_1} + \frac{1}{n_2}} = (2.160)(2.349)\sqrt{\frac{1}{n_1} + \frac{1}{n_2}} = 5.074\sqrt{\frac{1}{n_1} + \frac{1}{n_2}}$$

The *LSD* test should be applied only when the null hypothesis has been rejected—that is, when there is a significant difference among group effects. Otherwise, we could possibly find a significant difference between two population means when the null hypothesis concerning the group effects from a larger number of groups has already been accepted as being true. These differences can be attributed to the random fluctuations that occur when random samples are selected from the same population or from different populations having the same mean.

The preceding discussion of *LSD* tests is intended merely to introduce the concept and by no means is intended as an exhaustive coverage of the subject.

Equal Sample Sizes

The computations we have considered are valid whether or not the samples taken from the various populations are of equal size. However, the computations simplify slightly when selecting random samples of equal size from each of the populations. For example, if we are selecting samples of size n from each of the k populations, we have

$$n_i = n \qquad (i = 1, 2, \ldots, k). \tag{36}$$

Thus, N becomes

$$N = \sum_{i=1}^{k} n_i = \sum_{i=1}^{k} n = nk. \tag{37}$$

We obtain the sums of squares and degrees of freedom associated with the analysis-of-variance table given in Table 13.7 by merely substituting equal sample sizes (n)

Table 13.7

Analysis-of-variance table for one-variable classification model (equal sample size)

Source of variation	Degrees of freedom	Sum of squares	Expected mean square: Fixed effects	Expected mean square: Random effects
Among groups	$k-1$	$SS_{\text{among}} = n\sum_{i=1}^{k}(\bar{x}_{i\cdot} - \bar{x}_{\cdot\cdot})^2 = \frac{\sum_{i=1}^{k} x_{i\cdot}^2}{n} - \frac{x_{\cdot\cdot}^2}{nk}$	$\sigma_e^2 + \frac{n}{k-1}\sum_{i=1}^{k}(\tau_i - \bar{\tau}_{\cdot})^2$	$\sigma_e^2 + n\sigma_\tau^2$
Within groups	$nk-k$	$SS_{\text{within}} = \sum_{i=1}^{k}\sum_{j=1}^{n}(x_{ij} - \bar{x}_{i\cdot})^2 = \sum_{i=1}^{k}\sum_{j=1}^{n} x_{ij}^2 - \frac{\sum_{i=1}^{k} x_{i\cdot}^2}{n}$	σ_e^2	σ_e^2
Total	$nk-1$	$SS_{\text{total}} = \sum_{i=1}^{k}\sum_{j=1}^{n}(x_{ij} - \bar{x}_{\cdot\cdot})^2 = \sum_{i=1}^{k}\sum_{j=1}^{n} x_{ij}^2 - \frac{x_{\cdot\cdot}^2}{nk}$		

into Table 13.2 in place of $n_i (i = 1, 2, \ldots, k)$. Also, (19) and (20) become (38) and (39), respectively; that is,

$$\sigma_e^2 + \frac{n}{k-1}\sum_{i=1}^{k}(\tau_i - \bar{\tau}_{\cdot})^2 \tag{38}$$

and

$$\sigma_e^2 + n\sigma_\tau^2. \tag{39}$$

Furthermore, (35) becomes

$$LSD(\alpha) = t_{\alpha/2, k(n-1)}\hat{\sigma}_e\sqrt{\frac{2}{n}} \tag{40}$$

Note that this formula provides only one value against which to compare all differences, thus reducing related work.

Let us consider an example for the case of equal sample sizes selected from each group.

Example 2

The White River Electronics Company has four different methods to be used by an assembler in wiring an electronic system. For a new system, they wish to determine which (if any) of the four methods is fastest. An experiment in which 20 assemblers are assigned to the four methods (5 to each method) is conducted with results measured in minutes to completion as given in Table 13.8. Determine at the 5 percent significance level whether there is a difference among the four methods.

Solution

There are only four methods available to the company. Thus, we use the assumptions of (5a), and our hypotheses of interest become

$$H_0 : \tau_i = 0 \qquad (i = 1, 2, 3, 4)$$

versus

$$H_1 : \tau_i \neq 0 \qquad (\text{for at least one } i;\ i = 1, 2, 3, 4).$$

From Table 13.8, we determine that $k = 4$ and $n = 5$. Further, for a 5 percent significance level, we determine our critical value from Table A-5 as $F_{\alpha, k-1, k(n-1)} = F_{0.05, 3, 16} = 3.24$. Our decision rule is to accept H_0 if the calculated F ratio from

Table 13.8

Minutes to completion by various assembly methods

Method	*Minutes to completion*					*Sum*	*n*	*Mean*
1	5	12	8	10	4	49	5	9.80
2	11	9	8	6	13	47	5	9.40
3	16	10	14	12	9	61	5	12.20
4	5	8	10	11	7	41	5	8.20
Total						198	20	9.90

the associated analysis-of-variance table is less than 3.24 and reject H_0 otherwise. Constructing the analysis-of-variance table, we obtain Table 13.9, where

$$SS_{\text{among}} = \frac{\sum_{i=1}^{k} x_{i\cdot}^2}{n} - \frac{x_{\cdot\cdot}^2}{nk} = \frac{1}{5}[(49)^2 + (47)^2 + (61)^2 + (41)^2] - \frac{(198)^2}{(5)(4)}$$

$$= 2{,}002.4 - 1{,}960.2 = 42.2,$$

$$SS_{\text{within}} = \sum_{i=1}^{k} \sum_{j=1}^{n} x_{ij}^2 - \frac{\sum_{i=1}^{k} x_{i\cdot}}{n} = [(5)^2 + (12)^2 + (8)^2 + (10)^2 + (14)^2$$
$$+ (11)^2 + (9)^2 + (8)^2 + (6)^2 + (13)^2 + (16)^2 + (10)^2$$
$$+ (14)^2 + (12)^2 + (9)^2 + (5)^2 + (8)^2 + (10)^2 + (11)^2 + (7)^2]$$
$$- \tfrac{1}{5}[(49)^2 + (47)^2 + (61)^2 + (41)^2]$$
$$= 2{,}136.0 - 2.002.4 = 133.6,$$

and

$$SS_{\text{total}} = \sum_{i=1}^{k} \sum_{j=1}^{n} x_{ij}^2 - \frac{x_{\cdot\cdot}^2}{nk} = [(5)^2 + (12)^2 + (8)^2 + (10)^2 + (14)^2 + (11)^2 + (9)^2$$
$$+ (8)^2 + (6)^2 + (13)^2 + (16)^2 + (10)^2 + (14)^2 + (12)^2 + (9)^2 + (5)^2$$
$$+ (8)^2 + (10)^2 + (11)^2 + (7)^2] - \frac{(198)}{(5)(4)}$$
$$= 2{,}136.0 - 1{,}960.2 = 175.8.$$

We note that the entities in the "Sum of squares" column of Table 13.9 satisfy Relation (15). That is, $SS_{\text{total}} = SS_{\text{within}} + SS_{\text{among}}$, or $175.8 = 133.6 + 42.2$.

Table 13.9

Analysis-of-variance table for data in Table 13.8

Source of variation	*Degrees of freedom*	*Sum of squares*	*Mean square*	*F ratio*
Among groups	3	42.2	14.07	1.685
Within groups	16	133.6	8.35	
Total	19	175.8		

From Table 13.9, we see that the F ratio (1.685) is less than the critical value ($F_{0.05,\,3,\,16} = 3.24$). Thus, we accept H_0 and conclude that there is not a significant difference among the four assembly methods. That is, because we do not have significant reason to do otherwise, we conclude that the group sample means in Table 13.8 are different due only to random chance. Thus, the company should employ some other criterion for selecting which assembly method to use. □

Random-Effects Model

Suppose that we are sampling from k populations that do not constitute the entire group of populations in which we are interested and, consequently, the k populations represent only a sample from the totality of populations of interest. Further, suppose that our model is given by (4) and satisfies the assumptions of (5b) for random effects. In this situation, we have from Table 13.3 that the among-groups mean square has an expected value of

$$\sigma_e^2 + \frac{N^2 - \sum_{i=1}^k n_i^2}{N(k-1)} \sigma_\tau^2. \tag{41}$$

We could test the hypothesis that $\sigma_\tau^2 = 0$ against the alternative hypothesis that $\sigma_\tau^2 > 0$. This test is identical to the test of equal group means in the case of a fixed-effects model. While we shall not demonstrate the truth of this, you may wish to verify it by taking steps similar to those in our development of the test of equal group means.

However, we are generally more interested in estimating σ_τ^2 than in testing hypotheses about it. That is, since we are concerned with a random sample from a large population of τ's and are thus not in a position to limit our interest to a specific set, we are interested in estimating the variability of the τ's. Our model for this case is sometimes referred to as a *components-of-variance model*. This is due to the τ's being random variables and, as such, contributing to the total variation associated with the X_{ij} variables. Thus, when we estimate σ_τ^2, we are actually estimating one of the components of the total variance (that is, the component due to the true differences in group means). The estimate of σ_e^2 is also a component of the total variance, since it estimates the variability of the error terms of model (4). Thus, the group effects and error effects are components of variance. The estimates of the components of variance may be used to estimate the variance of the variable X as given by model (4). This may be done as a result of Rule 4.12, page 151, and Relation (3, Sec. 6.10). The resulting relation is

$$\operatorname{Var}(X) = \sigma_\tau^2 + \sigma_e^2. \tag{42}$$

Making further direct application of Rule 4.12 concerning expected values, we can obtain an unbiased estimate of σ_τ^2 by simple algebraic manipulation of the mean-square terms from the analysis-of-variance table. That is, since the expected values of the variables, say A and B, for which MS_{among} and MS_{within} are the respective observed values, are given by

$$E(A) = \sigma_e^2 + \frac{N^2 - \sum_{i=1}^k n_i^2}{N(k-1)} \sigma_\tau^2 \tag{43}$$

and

$$E(B) = \sigma_e^2, \tag{44}$$

respectively, for the random-effects model of (4), we may subtract (44) from (43) and divide by the coefficient of σ_τ^2 to obtain

$$\sigma_\tau^2 = \frac{E(A) - E(B)}{\dfrac{N^2 - \sum_{i=1}^k n_i^2}{N(k-1)}}. \tag{45}$$

Then, we may simply substitute the unbiased estimates of $E(A)$ and $E(B)$, as given in the analysis-of-variance table (Table 13.2), into (45) in place of $E(A)$ and $E(B)$, respectively, to obtain the unbiased estimate of σ_τ^2. That is,

$$\hat{\sigma}_\tau^2 = \frac{MS_{\text{among}} - MS_{\text{within}}}{\dfrac{N^2 - \sum_{i=1}^k n_i^2}{N(k-1)}}. \tag{46}$$

If we have equal sample sizes for each of the k populations from which we are sampling, (46) becomes

$$\hat{\sigma}_\tau^2 = \frac{MS_{\text{among}} - MS_{\text{within}}}{n}. \tag{47}$$

Furthermore, the unbiased estimate of σ_x^2 (that is, the variance of X_{ij}, where $X_{ij} = \mu + \tau_i + e_{ij}$ as given by (4), and the assumptions of (5b) are valid) is given by

$$\hat{\sigma}_x^2 = \hat{\sigma}_\tau^2 + \hat{\sigma}_e^2, \tag{48}$$

where $\hat{\sigma}_\tau^2$ is obtained by using (47) and the corresponding values from Table 13.2 and $\hat{\sigma}_e^2$ is obtained by using (31) and the MS_{within} term from Table 13.2.

Example 3

Referring to the data of Example 1, suppose that there are numerous (actually, an infinite population is assumed) liquid-diet substitutes available commercially and that five products are randomly selected to estimate the total variation of weight losses due to the diet substitutes. Determine the unbiased estimate of the component of variance due to the differences in weight losses for the different diet substitutes. Also, determine the unbiased estimate of the variance of the variable X—that is, the variable given by (4) when the assumptions of (5b) are valid.

Solution

From Table 13.5, we have

$$k = 5, \quad N = 18, \quad MS_{\text{among}} = 39.30, \quad \text{and} \quad MS_{\text{within}} = 5.52.$$

We compute

$$\sum_{i=1}^{5} n_i^2 = (4)^2 + (4)^2 + (3)^2 + (4)^2 + (3)^2 = 66.$$

Substituting these values into (46), we obtain the unbiased estimate of σ_τ^2, namely,

$$\hat{\sigma}_\tau^2 = \frac{MS_{\text{among}} - MS_{\text{within}}}{\dfrac{N^2 - \sum_{i=1}^{k} n_i^2}{N(k-1)}} = \frac{39.30 - 5.52}{\dfrac{(18)^2 - 66}{18(4)}} = 9.43.$$

Substituting the unbiased estimates of σ_τ^2 (9.43) and σ_e^2 (5.52) into (48), we obtain the unbiased estimate of σ_x^2; that is,

$$\hat{\sigma}_x^2 = \hat{\sigma}_\tau^2 + \hat{\sigma}_e^2 = 9.43 + 5.52 = 14.95.$$ □

Example 4

Suppose the four different methods of assembling electronic systems that White River Electronics Company selects in Example 2 are from a large population of methods, so that model (4) is valid and the assumptions of (5b) are satisfied. Using the data for Example 2 (Table 13.8), determine the unbiased estimate of the component of variance due to the differences in methods. Also, determine the unbiased estimate of the variance of the variable X; that is, $\hat{\sigma}_x^2 = \hat{\sigma}_\tau^2 + \hat{\sigma}_e^2$.

Solution

In Example 2, the sample sizes ($n = 5$) are equal. Thus, we can use (47) to obtain the unbiased estimate of σ_τ^2. From the analysis-of-variance table (Table 13.9), we obtain $MS_{\text{among}} = 14.07$ and $MS_{\text{within}} = 8.35$. Substituting these values into (47), we obtain

$$\hat{\sigma}_\tau^2 = \frac{MS_{\text{among}} - MS_{\text{within}}}{n} = \frac{14.07 - 8.35}{5} = 1.14.$$

Thus, the estimate of the amount of variance attributable to the differences among methods is 1.14.

The unbiased estimate of σ_x^2 is obtained through substitution into (48); that is,

$$\hat{\sigma}_x^2 = \hat{\sigma}_\tau^2 + \hat{\sigma}_e^2 = 1.14 + 8.35 = 9.49. \qquad \square$$

In Example 4, the unbiased estimate of σ_x^2 was obtained by adding the unbiased estimates of σ_τ^2 and σ_e^2. You may wonder why we couldn't estimate σ_x^2 by the usual method which involves determining the mean square value for the total. That is,

$$MS_{\text{total}} = \frac{SS_{\text{total}}}{(N - 1)} = \frac{175.8}{19} = 9.25.$$

It can be shown by using the "Expected mean square" terms in Table 13.3 for the random effects that this yields a biased estimate which always underestimates the variance of X when the assumptions of (5b) are satisfied. The verification of this is left as an exercise.

Problems

1. State the assumptions for the completely randomized, one-way classification model for fixed effects.
2. State the assumptions for the completely randomized, one-way classification model for random effects.
3. Show algebraically that

$$SS_{\text{total}} = SS_{\text{within}} + SS_{\text{among}}.$$

4. Certain employees of a large manufacturing firm are given an advanced training course. They are divided into four groups, and each group is taught by a different instructor. Upon completion of the course, each employee is given an achievement test. Simple random samples of size 5 are selected from each group of test scores as follows.

Group	*Test scores*				
1	77	68	81	73	80
2	65	69	78	72	70
3	86	81	74	78	82
4	66	78	69	71	70

Assume that the test scores are normally distributed.

a. Determine at the 10% significance level whether there is a significant difference among instructor effects.

b. Regardless of the results of Part a, perform the *LSD* test at the 10% significance level.

c. In light of the results of Part a, explain the results of Part b.

5. A petroleum company wishes to determine the effects on gasoline mileage of five different blends of gasoline. Automobiles of identical make and style are randomly selected, and each automobile is assigned a particular blend of gasoline. The results in miles per gallon are as follows.

Blend	*Miles per gallon*				
1	15	17	14		
2	18	20	21	19	17
3	17	21	18	16	19
4	16	14	14	17	
5	19	22	18		

a. Assume that gasoline mileage is normally distributed, and determine at the 5% significance level whether there is significant difference among the average mileage for the five blends.

b. Regardless of the results of Part a, perform the *LSD* test at the 5% significance level.

c. In light of the results of Part a, explain the results of Part b.

6. Suppose that the employees of Problem 4 are divided into many groups and the data given are for four groups randomly selected from the numerous groups.

a. Assume that the test scores are normally distributed and determine the unbiased estimates of σ_e^2, σ_τ^2, and σ_x^2.

b. Explain what each of the components of Part a is estimating.

7. Suppose that the blends of Problem 5 are a sample from a larger population of blends and that the data given are for five blends randomly selected from the larger group of blends.

a. Assume that the gasoline mileages are normally distributed and determine the unbiased estimates of σ_e^2, σ_τ^2, and σ_x^2.

b. Explain what each of the components of Part a is estimating.

8. For the random effects model, use the "Expected mean square" values in Table 13.3 to determine the expected value of the term $SS_{\text{total}}/(N - 1)$. Note the bias.

13.5 Completely Randomized, Two-Variable Classification Model

In Section 12.4 we discussed how individual values can be classified by a single criterion (or variable), namely, the population from which the value was selected. We shall now consider the situation in which the individual values are subject to two criteria of classification. For example, a student may be classified according to his major subject and his academic rank (freshman, sophomore, etc.); an automobile may be classified according to its manufacturer and its weight; an individual may be classified according to his income and his educational level; or a garment may be classified according to its type of material and its pattern. In each case, we can identify each value with only two subscripts, unless, of course, we have more than one observed value per two-variable classification. In order to allow for more than one observed value per two-variable classification, we shall use three subscripts. Thus, we represent any observed value by x_{ijm} where i ($i = 1, 2, \ldots, k$) represents the first variable of classification, j ($j = 1, 2, \ldots, n$) represents the second variable of classification, and m ($m = 1, 2, \ldots, r$) represents the number of the individually observed value from the ith category of the first variable of classification and the jth category of the second variable of classification. For example, x_{312} is the second observed value from the third category of the first variable of classification and the first category of the second variable of classification.

A general set of observed values for the two-variable classification problem is given in Table 13.10, where A represents the first variable of classification and B represents the second variable of classification. (Note that we now have r observations for each combination of A and B classifications. Thus, we have equal sample sizes for each combination.)

Table 13.10

Observed values for two-variable classification model

First variable of classification	*Second variable of classification* B_1	B_2	$\cdots$	B_n
A_1	x_{111} x_{112} $\cdot$ x_{11r}	x_{121} x_{122} $\cdot$ x_{12r}	$\cdots$ $\cdots$ $\cdots$ $\cdots$	x_{1n1} x_{1n2} $\cdot$ x_{1nr}
A_2	x_{211} x_{212} $\cdot$ x_{21r}	x_{221} x_{222} $\cdot$ x_{22r}	$\cdots$ $\cdots$ $\cdots$ $\cdots$	x_{2n1} x_{2n2} $\cdot$ x_{2nr}
$\cdot$ $\cdot$ $\cdot$	$\cdot$ $\cdot$ $\cdot$	$\cdot$ $\cdot$ $\cdot$	$\cdots$ $\cdots$ $\cdots$	$\cdot$ $\cdot$ $\cdot$
A_k	x_{k11} x_{k12} $\cdot$ x_{k1r}	x_{k21} x_{k22} $\cdot$ x_{k2r}	$\cdots$ $\cdots$ $\cdots$ $\cdots$	x_{kn1} x_{kn2} $\cdot$ x_{knr}

The mean for the population from which the observed values in cell ij (that is, the group of values corresponding to the A_i and B_j classifications) are selected can be written as

$$\mu_{ij} = \mu + \alpha_i + \beta_j + (\alpha\beta)_{ij}, \qquad (1)$$

where μ is the average of the individual population means, α_i represents the effect due to the ith category of classification A, β_j represents the effect due to the jth category of classification B, and $(\alpha\beta)_{ij}$ represents the effect due to the ith and jth categories of the two classifications occurring together. The $(\alpha\beta)_{ij}$ term is known as the *interaction effect.*

To better understand the interaction effect, let us consider an example. Suppose an experiment is conducted to determine gasoline mileage for automobiles, and the variables of interest are four different types of gasoline and five different driving conditions. For this situation, we are interested not only in the separate effects of the driving conditions and the types of gasoline but their joint effects as well. For instance, one of the gasolines may be best for a certain driving condition while another type of gasoline may be best for a different driving condition. This differential effect is known as *interaction.* The interaction effect between classifications A and B is a measure of the failure of classification A to behave consistently from one category of classification B to another category of classification B. Thus, for our example, we have interaction present if the gasoline mileage for each type of gasoline is not the same in a relative sense for each different driving condition. For example, if the rankings of the average gasoline mileage for each type of gasoline differ considerably (more than can be attributed to chance) from one driving condition to the next, we take this as an indication that interaction is present.

If we assume that each of the populations from which we are sampling is normal, and the population means are as given by (1), then we may represent the variable X_{ijm} by the model

$$\begin{aligned} X_{ijm} &= \mu_{ij} + e_{ijm} \\ &= \mu + \alpha_i + \beta_j + (\alpha\beta)_{ij} + e_{ijm} \end{aligned} \left\{ \begin{aligned} i &= 1, 2, \ldots, k; \\ j &= 1, 2, \ldots, n; \\ m &= 1, 2, \ldots, r; \end{aligned} \right. \qquad (2)$$

where the error terms e_{ijm} are independent and normally distributed variables with means zero and variances σ_e^2. The α_i, β_j, and $(\alpha\beta)_{ij}$ are generally termed A effects, B effects, and AB effects, respectively. The usual model restrictions imposed on the entities of (2) are such that

$$\sum_{i=1}^{k} \alpha_i = 0, \quad \sum_{j=1}^{n} \beta_j = 0, \quad \text{and} \quad \sum_{i=1}^{k} \sum_{j=1}^{n} (\alpha\beta)_{ij} = 0.$$

Under the preceding model assumptions and restrictions, the major purpose of conducting an analysis of variance in the completely randomized, two-variable classification situation is generally to test the hypotheses that the A effects are zero, the B effects are zero, and/or the AB effects are zero. We shall justify and illustrate in what follows the appropriate analysis of variance for a fixed-effects model.

A summary of the data in Table 13.10 is given in Table 13.11, where

$$x_{ij\cdot} = \sum_{m=1}^{r} x_{ijm},$$

$$x_{i\cdot\cdot} = \sum_{j=1}^{n} x_{ij\cdot} = \sum_{j=1}^{n} \sum_{m=1}^{r} x_{ijm},$$

$$x_{\cdot j\cdot} = \sum_{i=1}^{k} x_{ij\cdot} = \sum_{i=1}^{k} \sum_{m=1}^{r} x_{ijm},$$

and

$$x_{\cdots} = \sum_{i=1}^{k} x_{i\cdot\cdot} = \sum_{j=1}^{n} x_{\cdot j\cdot} = \sum_{i=1}^{k} \sum_{j=1}^{n} x_{ij\cdot} = \sum_{i=1}^{k} \sum_{j=1}^{n} \sum_{m=1}^{r} x_{ijm}.$$

Table 13.11

Summary of data in Table 13.10

First variable of classification	*Second variable of classification* B_1	B_2	$\cdots$	B_n	*Total*
A_1	$x_{11\cdot}$	$x_{12\cdot}$	$\cdots$	$x_{1n\cdot}$	$x_{1\cdot\cdot}$
A_2	$x_{21\cdot}$	$x_{22\cdot}$	$\cdots$	$x_{2n\cdot}$	$x_{2\cdot\cdot}$
$\vdots$	$\vdots$	$\vdots$	$\vdots$	$\vdots$	$\vdots$
A_k	$x_{k1\cdot}$	$x_{k2\cdot}$	$\cdots$	$x_{kn\cdot}$	$x_{k\cdot\cdot}$
Total	$x_{\cdot 1\cdot}$	$x_{\cdot 2\cdot}$	$\cdots$	$x_{\cdot n\cdot}$	$x_{\cdots}$

The analysis-of-variance table associated with model (2) is given in Table 13.12, where the partitioning of the total variation is done in accordance with the sources of variation given in model (2). Appropriate computational formulas are

$$\begin{aligned}
SS_{\text{total}} &= \sum_{i=1}^{k}\sum_{j=1}^{n}\sum_{m=1}^{r}(x_{ijm}-\bar{x}_{\cdots})^2 = \sum_{i=1}^{k}\sum_{j=1}^{n}\sum_{m=1}^{r}x_{ijm}^2 - \frac{x_{\cdots}^2}{knr};\\
SS_A &= \sum_{i=1}^{k}\sum_{j=1}^{n}\sum_{m=1}^{r}(\bar{x}_{i\cdot\cdot}-\bar{x}_{\cdots})^2 = \frac{1}{nr}\sum_{i=1}^{k}x_{i\cdot\cdot}^2 - \frac{x_{\cdots}^2}{knr};\\
SS_B &= \sum_{i=1}^{k}\sum_{j=1}^{n}\sum_{m=1}^{r}(\bar{x}_{\cdot j\cdot}-\bar{x}_{\cdots})^2 = \frac{1}{kr}\sum_{j=1}^{n}x_{\cdot j\cdot}^2 - \frac{x_{\cdots}^2}{knr};\\
SS_{AB} &= \sum_{i=1}^{k}\sum_{j=1}^{n}\sum_{m=1}^{r}(\bar{x}_{ij\cdot}-\bar{x}_{i\cdot\cdot}-\bar{x}_{\cdot j\cdot}+\bar{x}_{\cdots})^2\\
&= \frac{1}{r}\sum_{i=1}^{k}\sum_{j=1}^{n}x_{ij\cdot}^2 - \frac{1}{nr}\sum_{i=1}^{k}x_{i\cdot\cdot}^2 - \frac{1}{kr}\sum_{j=1}^{n}x_{\cdot j\cdot}^2 + \frac{x_{\cdots}^2}{knr};
\end{aligned} \tag{3}$$

Table 13.12

Analysis-of-variance table of data in Table 13.10

Source of variation	*Degrees of freedom*	*Sum of squares*	*Mean square*	*F ratio*
A	$k-1$	SS_A	$MS_A = SS_A/(k-1)$	MS_A/MS_{error}
B	$n-1$	SS_B	$MS_B = SS_B/(n-1)$	MS_B/MS_{error}
AB	$(k-1)(n-1)$	SS_{AB}	$MS_{AB} = SS_{AB}/(k-1)(n-1)$	$MS_{AB}/MS_{\text{error}}$
Error	$kn(r-1)$	SS_{error}	$MS_{\text{error}} = SS_{\text{error}}/kn(r-1)$	
Total	$knr-1$	SS_{total}		

and

$$SS_{\text{error}} = \sum_{i=1}^{k} \sum_{j=1}^{n} \sum_{m=1}^{r} (x_{ijm} - \bar{x}_{ij\cdot})^2 = \sum_{i=1}^{k} \sum_{j=1}^{n} \sum_{m=1}^{r} x_{ijm}^2 - \frac{1}{r} \sum_{i=1}^{k} \sum_{j=1}^{n} x_{ij\cdot}^2.$$

Using the last term of each equation in (3), we note the identity

$$SS_{\text{total}} = SS_A + SS_B + SS_{AB} + SS_{\text{error}}. \tag{4}$$

This identity may be used as a check on the computations in the analysis-of-variance table. We further note that the degrees of freedom associated with the total sum of squares is equal to the sum of the degrees of freedom for the other sources of variation.

The expected mean squares for the variables associated with the mean-square column of Table 13.12 are given in Table 13.13. As was the case in Section 13.4, the

Table 13.13

Analysis-of-variance table for two-variable classification

Source of variation	*Degrees of freedom*	*Mean square*	*Expected mean square*
A	$k-1$	MS_A	$\sigma_e^2 + \frac{nr}{k-1}\sum_{i=1}^{k}\alpha_i^2$
B	$n-1$	MS_B	$\sigma_e^2 + \frac{kr}{n-1}\sum_{j=1}^{n}\beta_j^2$
AB	$(k-1)(n-1)$	MS_{AB}	$\sigma_e^2 + \frac{r}{(k-1)(n-1)}\sum_{i=1}^{k}\sum_{j=1}^{n}(\alpha\beta)_{ij}^2$
Error	$kn(r-1)$	MS_{error}	σ_e^2
Total	$knr-1$		

expected-mean-square values are the appropriate variances of the variables associated with the respective mean-square values. In order to obtain required chi-square statistics in each case, we must divide by the appropriate variance. (However, we note that the mean-square values are determined by dividing the sum-of-squares values by their respective degrees of freedom.) This is to say that

$$\frac{MS_A}{\sigma_e^2 + \dfrac{nr}{k-1}\displaystyle\sum_{i=1}^{k}\alpha_i^2}, \tag{5}$$

$$\frac{MS_B}{\sigma_e^2 + \dfrac{kr}{n-1}\displaystyle\sum_{j=1}^{n}\beta_j^2}, \tag{6}$$

$$\frac{MS_{AB}}{\sigma_e^2 + \dfrac{r}{(k-1)(n-1)}\displaystyle\sum_{i=1}^{k}\sum_{j=1}^{n}(\alpha\beta)_{ij}^2}, \tag{7}$$

and

$$\frac{MS_{\text{error}}}{\sigma_e^2} \tag{8}$$

represent observed values of chi-square statistics with $(k - 1)$, $[(k - 1)(n - 1)]$, and $[kn(r - 1)]$ degrees of freedom, respectively, which have been divided by their respective degrees of freedom. [Note that it can be shown that the values in the denominators of (5) through (8) are the "appropriate variances."] Furthermore, the variables associated with (5) through (8) can be shown to be independent.

With this information, we may take the ratio of any two of these observed values to obtain the corresponding observed value of an F variable. However, since our ultimate purpose is to test hypotheses about α_i, β_j, and $(\alpha\beta)_{ij}$, there are only certain ratios of interest to us. These are obtained by dividing each of (5), (6), and (7) by (8). That is, we have by Definition 13.1 that

$$\frac{\dfrac{MS_A}{\sigma_e^2 + \dfrac{nr}{k-1}\displaystyle\sum_{i=1}^{k} \alpha_i^2}}{\dfrac{MS_{\text{error}}}{\sigma_e^2}}, \tag{9}$$

$$\frac{\dfrac{MS_B}{\sigma_e^2 + \dfrac{kr}{n-1}\displaystyle\sum_{j=1}^{n} \beta_j^2}}{\dfrac{MS_{\text{error}}}{\sigma_e^2}}, \tag{10}$$

and

$$\frac{\dfrac{MS_{AB}}{\sigma_e^2 + \dfrac{r}{(k-1)(n-1)}\displaystyle\sum_{i=1}^{k}\sum_{j=1}^{n} (\alpha\beta)_{ij}^2}}{\dfrac{MS_{\text{error}}}{\sigma_e^2}} \tag{11}$$

are the observed values of an F variable with $(k - 1)$ and $[kn(r - 1)]$, $(n - 1)$ and $[kn(r - 1)]$, and $[(k - 1)(n - 1)]$ and $[kn(r - 1)]$ degrees of freedom, respectively. However, these observed values for F variables are of very little help to us in their present form because we do not know the values of σ_e^2 and the other terms involving α_i, β_j, and $(\alpha\beta)_{ij}$. They will, however, serve very nicely as the observed values of the pivotal quantities for tests of certain hypotheses. For example, suppose we wish to test the null hypothesis

$$H_0\colon \alpha_i = 0 \qquad (i = 1, 2, \ldots, k).$$

If the null hypothesis is true, the observed value of (9) becomes

$$\frac{\dfrac{MS_A}{\sigma_e^2 + \dfrac{nr}{k-1}\displaystyle\sum_{i=1}^{n} \alpha_i^2}}{\dfrac{MS_{\text{error}}}{\sigma_e^2}} = \frac{\dfrac{MS_A}{\sigma_e^2 + 0}}{\dfrac{MS_{\text{error}}}{\sigma_e^2}} = \frac{MS_A}{MS_{\text{error}}}.$$

Thus, if, and only if, the null hypothesis is true,

$$F_A \doteq \frac{MS_A}{MS_{\text{error}}} \tag{12}$$

represents an observed value of an F statistic with $(k - 1)$ and $[kn(r - 1)]$ degrees of freedom.

Suppose we wish to test the null hypothesis

$$H_0: \beta_j = 0 \qquad (j = 1, 2, \ldots, n).$$

If the null hypothesis is true in this case, the observed value of (10) becomes

$$\frac{\dfrac{MS_B}{\sigma_e^2 + \dfrac{kr}{n-1}\sum_{j=1}^{n}\beta_j^2}}{\dfrac{MS_{\text{error}}}{\sigma_e^2}} = \frac{\dfrac{MS_B}{\sigma_e^2 + 0}}{\dfrac{MS_{\text{error}}}{\sigma_e^2}} = \frac{MS_B}{MS_{\text{error}}}.$$

Thus, if, and only if, the null hypothesis is true,

$$F_B \doteq \frac{MS_B}{MS_{\text{error}}} \tag{13}$$

represents an observed value of an F statistic with $(n - 1)$ and $[kn(r - 1)]$ degrees of freedom.

Finally, suppose we wish to test the null hypothesis

$$H_0: (\alpha\beta)_{ij} = 0 \qquad (i = 1, 2, \ldots, k; j = 1, 2, \ldots, n).$$

If the null hypothesis is true, the observed value of (11) becomes

$$\frac{\dfrac{MS_{AB}}{\sigma_e^2 + \dfrac{r}{(k-1)(n-1)}\sum_{i=1}^{k}\sum_{j=1}^{n}(\alpha\beta)_{ij}^2}}{\dfrac{MS_{\text{error}}}{\sigma_e^2}} = \frac{\dfrac{MS_{AB}}{\sigma_e^2 + 0}}{\dfrac{MS_{\text{error}}}{\sigma_e^2}} = \frac{MS_{AB}}{MS_{\text{error}}}.$$

Thus, if, and only if, the null hypothesis is true,

$$F_{AB} \doteq \frac{MS_{AB}}{MS_{\text{error}}} \tag{14}$$

represents an observed value of an F statistic with $[(k - 1)(n - 1)]$ and $[kn(r - 1)]$ degrees of freedom.

Now, as in all past tests of hypotheses, it is only logical that if we calculate from sample data any of the observed values of (12) through (14) and obtain a value that is not at all likely to have come from the relevant F distribution, we should reject the corresponding null hypothesis. As in Section 13.4, each of the foregoing hypotheses will generally be stated in conjunction with two-sided alternative hypotheses. However, due to reasoning similar to that in Section 13.4, we make the tests by considering only the right tail of the appropriate F distribution.

All the foregoing reasoning leads us to the following essential steps for making 100α percent significance tests concerning the A effects, B effects, and AB effects.

1. Set up the null hypothesis and the corresponding alternative hypothesis for each case; that is, for A effects,

$$H_0\colon \alpha_i = 0 \qquad (i = 1, 2, \ldots, k)$$

versus (15)

$$H_1\colon \alpha_i \neq 0 \qquad (\text{for at least one } i;\ i = 1, 2, \ldots, k);$$

for B effects,

$$H_0\colon \beta_j = 0 \qquad (j = 1, 2, \ldots, n)$$

versus (16)

$$H_1\colon \beta_j \neq 0 \qquad (\text{for at least one } j;\ j = 1, 2, \ldots, n);$$

and, for AB effects,

$$H_0\colon (\alpha\beta)_{ij} = 0 \qquad (i = 1, 2, \ldots, k;\ j = 1, 2, \ldots, n)$$

versus (17)

$$H_1\colon (\alpha\beta)_{ij} \neq 0 \qquad (\text{for at least one combination of } i \text{ and } j;\ i = 1, 2, \ldots, k;\ j = 1, 2, \ldots, n).$$

2. Specify the level of significance α. (We generally use the same level for all three tests.)

3. Determine the values for k, n, and r. (This fixes the numbers of degrees of freedom to be used in establishing the appropriate critical values.)

4. Determine the critical values $F^*_{k-1, kn(r-1)}$, $F^*_{n-1, kn(r-1)}$, and $F^*_{(k-1)(n-1), kn(r-1)}$ (where k, n, and r are determined in step 3) that divide their respective F distributions into critical (rejection) regions and acceptance regions. This value for each test should always be such that 100α percent of the area of the corresponding F distribution lies to the right of the critical value. That is,

$$F^*_{k-1, kn(r-1)} = F_{\alpha, k-1, kn(r-1)}, \tag{18}$$

$$F^*_{n-1, kn(r-1)} = F_{\alpha, n-1, kn(r-1)}, \tag{19}$$

and

$$F^*_{(k-1)(n-1), kn(r-1)} = F_{\alpha, (k-1)(n-1), kn(r-1)} \tag{20}$$

for the hypotheses of (15), (16), and (17), respectively.

5. Take a sample of size r from each population corresponding to an $ij(i = 1, 2, \ldots, k; j = 1, 2, \ldots, n)$ category of classification and perform an analysis of variance by carrying out the calculations illustrated in Table 13.12. Thus, we obtain [under the assumption that H_0 is true in each of the cases (15), (16), and (17)] the observed values of three different F statistics as given in the F ratio column of Table 13.12. That is, from the analysis-of-variance table, compute the observed values as given in (12) through (14).

a. Calculate the values for the entries in the appropriate analysis-of-variance table.

b. Determine the unbiased estimates of σ_e^2, σ_τ^2, and σ_x^2.

5. Recall from Section 13.4, that regardless of whether the completely randomized, one-variable classification model is for fixed effects or random effects, the ratio

$$\frac{MS_{\text{among}}/\sigma_1^2}{MS_{\text{within}}/\sigma_2^2}$$

represents an observed value of an F variable with $(k - 1)$ and $(N - k)$ degrees of freedom, provided σ_1^2 and σ_2^2 are chosen properly to suit the model assumptions. Moreover, for the fixed-effects model, the variances σ_1^2 and σ_2^2 are selected from the expected-mean-square column for fixed effects, and the hypothesis tested is $H_0: \tau_i = 0$ for $i = 1, 2, \ldots, k$. Similarly, the variances σ_1^2 and σ_2^2 may be selected from the expected-mean-square column for random effects, and a hypothesis concerning σ_τ^2 may be tested.

a. What is the appropriate hypothesis concerning σ_τ^2? What is the appropriate alternative hypothesis?

b. Test the hypotheses of Part a at the 5% significance level using the data of Problem 5 in Section 13.4 (page 503).

c. Explain in terms of variance components the meaning of rejecting the null hypothesis of Part a.

(Section 9.4)

6. A petroleum company wishes to determine the effects of three different grades of motor oil on engine life of four different type engines. The company research team is convinced that there is a negligible interaction effect. Thus, to study the effects of the engine types and the grades of motor oil, an experiment is performed in which three engines of each type are randomly selected and each grade of motor oil is randomly assigned to one of the three engines. Each engine is operated until it needs a major overhaul. The number of operating hours are recorded for each engine and are given below in thousands of hours. The number of operating hours per engine is normally distributed.

Grade of oil	*Engine type*			
	1	2	3	4
A	15	17	14	19
B	17	18	16	21
C	18	19	18	23

a. Assuming that the company research team is correct in saying there is no interaction effect, compute the appropriate analysis-of-variance table.

b. Determine at the 5% significance level whether there is a significant engine-type effect and whether there is a significant oil-grade effect.

c. In light of the results of Part b, which engine type would you recommend for yielding the longest life? Which grade of oil would you recommend for that engine?

d. Why is it necessary to have more than one observation per treatment combination if we wish to determine the presence of an interaction effect?

7. After performing an analysis of variance, it is often desirable to determine a confidence interval for a treatment mean μ_i. Thus, if the assumptions for the fixed-effects model are satisfied, the estimator $\bar{X}_i$ is normally distributed with mean μ_i and variance σ_e^2/n_i.

a. Derive the general form for the confidence interval about μ_i when σ_e^2 is unknown.

b. Using the data from Problem 5 Section 13.4, determine a 95% confidence interval for the average mileage for each blend of gasoline.

(Sections 6.10, 8.6, 10.4, 11.7)

Fourteen

Linear Regression Analysis

14.1 Introduction

We have been concerned so far only with situations in which a single observation is made on each element in the sample. We are now ready to consider the case where two observations are made on each element in the sample. Then, in Chapter 15, we shall extend our consideration to the case of several observations per element. It is important that you fully understand the concepts presented in this chapter before proceeding to Chapter 15.

We term the case in which two characteristics are observed for each element in the sample a *bivariate* situation. The sample space of a bivariate experiment can be represented by a set of ordered pairs of measurements. For example, if we take a sample of students and let each student's score on an achievement examination be represented by x and his grade-point average by y, we obtain from the ith student the pair of numbers (x_i, y_i). For n students in the sample, we have a sample of size n consisting of the n ordered pairs

$$(x_1, y_1), (x_2, y_2), \ldots, (x_n, y_n).$$

In many subjects, the first basic idea introduced to a student is that of relationships between variables. For example, the quantity demanded of some commodity in a given market is a function of its price; the breaking strength of a steel rod is a function of the temperature to which it has been subjected; the food consumption of an individual is a function of his weight; and the force a specific object exerts on a surface is a function of its acceleration at the time of impact. Each of these examples represents a two-variable relation. Of course, more practical situations may require that these relations include several variables, but we limit our present discussion to the two-variable situation.

When a change in one variable accompanies a rather systematic change in another variable, we can often determine or predict the effect that a specific change in one variable may have upon the other. In other words, we are interested in the relationship between the two variables. One way to study this relationship is through *regression analysis*—that is, the analysis of results obtained by regressing data to a hypothetical, mathematical model. By regression, we mean working backward in the sense

Assumptions for Case B

1. The entities β_0 and β_1 are unknown population parameters to be estimated.
2. Y is a dependent observable variable inherently related to x in a linear fashion (that is, the mean of the variable is a linear function of the x values); but, because of random errors in observations on Y, observed values of Y will deviate from the line in a random manner.
3. The variable x is an independent mathematical (nonrandom) variable whose value can be determined without error.
4. The error variable e is a random variable that is not observable and accounts for the failure of the deterministic linear model $y = \beta_0 + \beta_1 x$ to hold for all paired observations on Y and x.
5. The key characteristics of e are that it is *normally distributed* with mean zero and variance σ_e^2 and, for any two values of Y (say Y_i and Y_j), the associated errors (e_i and e_j) are statistically independent.

In what follows, we shall refer to the situation in which no assumption is made concerning the form of the distribution of the error variable as *Case A*, and to the situation involving the assumption of a normal error variable as *Case B*, or the normal-theory case.

Figure 14.2 illustrates the basic model assumption of simple linear regression analysis for the normal-theory case. Note that the population means lie on the straight line $\mu_{y|x} = \beta_0 + \beta_1 x$, the variances of the distributions ($\sigma_{y|x}^2 = \sigma_e^2$) are the same regardless of the values assumed by the independent variable x, and the error distributions are normal. In other words, the probability distribution of Y, for a given value of x (that is, the conditional probability distribution of Y given x), is normal with mean $\beta_0 + \beta_1 x$ and variance σ_e^2. An illustration of a specific example of Case A would be similar except for the shapes of the error distributions.

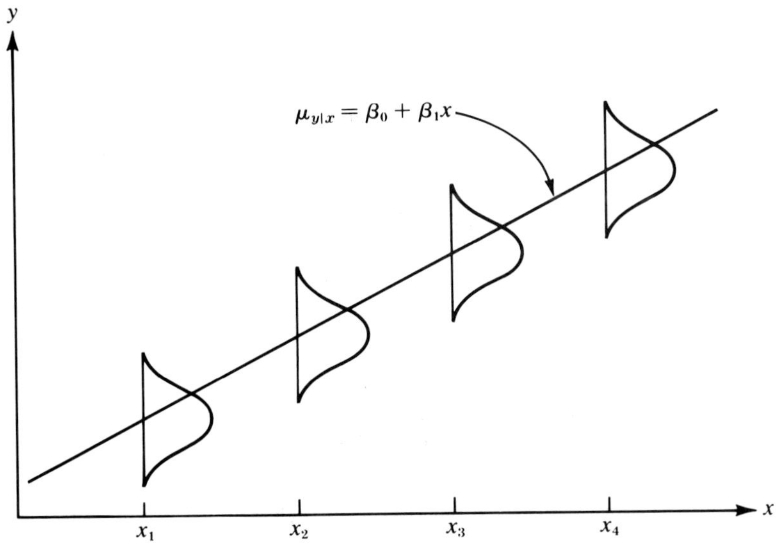

Figure 14.2 Linear probabilistic mathematical model: normal-theory case

Problems

1. What is the first major step in regression analysis?
2. What are the assumptions associated with the linear regression model given by
$$Y = \beta_0 + \beta_1 x + e?$$
3. What is the difference between Case A and Case B?
4. How is the use of Case B justified for many situations?

14.3 Estimators and Their Properties

Just as in Chapter 8, we are again confronted with the problem of determining estimators for population parameters. However, a major difference in our present situation is that from a single set of observations, we want to estimate the means ($\mu_{y|x}$) of numerous populations where these means are related to the independent variable x by the regression equation $\mu_{y|x} = \beta_0 + \beta_1 x$. Of course, we also face the problem of estimating the variance for each population. However, one of our basic assumptions is that these variances are equal. (Note that this assumption is not necessary; however, treatment of situations in which it does not hold is too advanced for this text.) Thus, this particular part of the problem is analogous to the case considered in Section 11.5 where we "pooled" estimates of a variance. To obtain estimates for the means and variance associated with the regression equation let us consider the situations created by the assumptions of both Case A and Case B.

Case A (Distributions of Error Variables Unknown)

Let us first consider the method of estimation known as the *method of least squares.* When we have selected a sample of n units and measured the two characteristics represented by Y and x, respectively, we have available n ordered pairs. If we graph in the usual fashion the values of Y corresponding to specific values of x, we obtain what is generally called a *scatter diagram* illustrated in Figure 14.3. (It should be emphasized here that one of the first steps in linear regression analysis should be to plot a scatter diagram of the observed data so that grossly inappropriate model assumptions can be ruled out.) How do we "fit" a line to the data points in a given scatter diagram? Obviously, there are an infinite number of lines that could be drawn on a given scatter

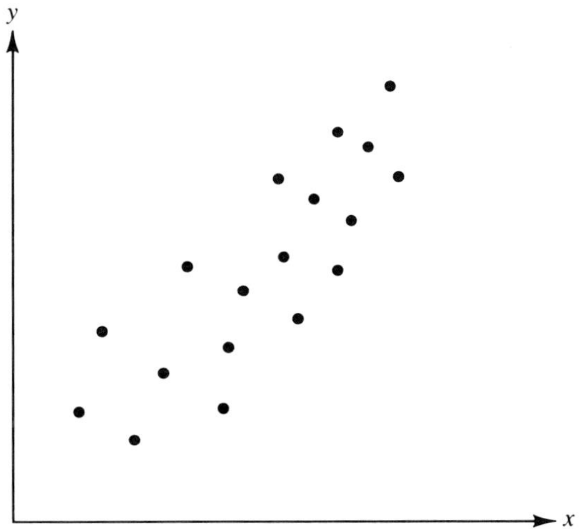

Figure 14.3 Scatter diagram

diagram and some of these lines would "fit the data quite well," while others would not. How, then, should we choose the line to be used for estimating $E(Y)$ as a function of x so that the line fits the sample data well?

A quite commonly used method for selecting such a line is the method of least squares. This method is essentially a formalization of the procedure whereby we visually fit a line to a set of points by moving a ruler around until it appears that the deviations of the points from the straight line are "small." The term *least squares* is used because, under the method of least squares, the sum of squares of the vertical deviations of the points from the least-squares line is smaller than the corresponding sum of squares of deviations from any other straight line.

Suppose we have observed n values of the nonrandom variable x (x_i, where $i = 1, 2, \ldots, n$) and the n corresponding values of the random variable Y (y_i, where $i = 1, 2, \ldots, n$). The resulting observations consist of the ordered pairs (x_i, y_i) ($i = 1, 2, \ldots, n$). Thus, we say that we have selected n sample units and recorded the values of two separate characteristics (the x characteristic and the y characteristic) for each sample unit. Suppose further that, under our assumptions for Case A, we wish to estimate the mean of the population of Y values for each x_i ($i = 1, 2, \ldots, n$)—that is, $\mu_{y|x_i}$ for $i = 1, 2, \ldots, n$. We have from the regression equation that

$$\mu_{y|x_i} = \beta_0 + \beta_1 x_i.$$

Thus, we must obtain estimates of β_0 and β_1 to estimate $\mu_{y|x_i}$. Letting $\hat{\mu}_{y|x_i}$, $\hat{\beta}_0$, and $\hat{\beta}_1$ represent least-squares estimates of $\mu_{y|x_i}$, β_0, and β_1, respectively, we can estimate $\mu_{y|x_i}$ by first estimating β_0 and β_1 and then using the relation

$$\hat{\mu}_{y|x_i} = \hat{\beta}_0 + \hat{\beta}_1 x_i. \tag{1}$$

Under this notation, the deviation of the ith observed value of Y from the estimate of the corresponding population mean [that is, from the straight line given by (1)] can be written as

$$y_i - \hat{\mu}_{y|x_i}. \tag{2}$$

A quite reasonable objective in fitting a straight line to a set of data points y_i ($i = 1, 2, \ldots, n$) is to determine the unique straight line for which the sum of squares of the deviations of (2) is minimized. Thus, our objective is to estimate the true regression line $\mu_{y|x_i}$ so that we minimize the sum of squares of deviations of the observed Y values from the corresponding estimated values. Relation (1) represents the result of such minimization; consequently, the minimum value of the sum of squares is given by

$$\sum_{i=1}^{n} (y_i - \hat{\mu}_{y|x_i})^2. \tag{3}$$

Hereafter, we shall label this entity as SS_{error}.

Substituting the right-hand term of (1) in place of $\hat{\mu}_{y|x_i}$ in (3), we obtain

$$SS_{\text{error}} = \sum_{i=1}^{n} [y_i - (\hat{\beta}_0 + \hat{\beta}_1 x_i)]^2. \tag{4}$$

Thus, the ith deviation is given by $y_i - (\hat{\beta}_0 + \hat{\beta}_1 x_i)$—the vertical distance between the observed value, y_i, and the estimated value, $\hat{\mu}_{y|x_i}$ (that is, $\hat{\beta}_0 + \hat{\beta}_1 x_i$), for a given value of x, namely x_i. Such deviations are illustrated in Figure 14.4.

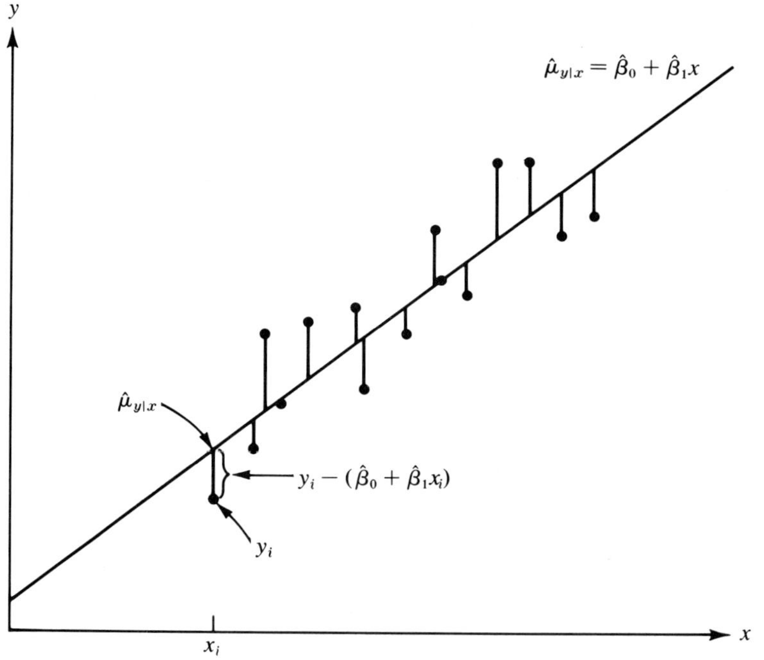

Figure 14.4 Deviations between y_i and $\hat{\mu}_{y|x}$

Under our stated objective, our problem is to find the estimates $\hat{\beta}_0$ and $\hat{\beta}_1$ for which (4) is a minimum. We use the method of least squares, to find these estimates. Because the derivation of these estimates is beyond the scope of this text, we simply state without proof the following important fact. It can be shown theoretically that we can obtain the estimates $\hat{\beta}_0$ and $\hat{\beta}_1$ for which the sum of squares of deviations of

(4) is a minimum by solving the simultaneous equations

$$n\hat{\beta}_0 + \hat{\beta}_1 \sum_{i=1}^{n} x_i = \sum_{i=1}^{n} y_i \tag{5}$$

and

$$\hat{\beta}_0 \sum_{i=1}^{n} x_i + \hat{\beta}_1 \sum_{i=1}^{n} x_i^2 = \sum_{i=1}^{n} x_i y_i \tag{6}$$

for $\hat{\beta}_0$ and $\hat{\beta}_1$. That is, if we solve Equations (5) and (6) for $\hat{\beta}_0$ and $\hat{\beta}_1$, we obtain the so-called *least-squares estimates* for β_0 and β_1 in the linear regression model. Equations (5) and (6) are known as the *normal equations* associated with the linear regression model. Solving these simultaneous equations for $\hat{\beta}_0$ and $\hat{\beta}_1$, we obtain

$$\hat{\beta}_1 = \frac{n \sum_{i=1}^{n} x_i y_i - (\sum_{i=1}^{n} x_i)(\sum_{i=1}^{n} y_i)}{n \sum_{i=1}^{n} x_i^2 - (\sum_{i=1}^{n} x_i)^2} = \frac{\sum_{i=1}^{n} (x_i - \bar{x})(y_i - \bar{y})}{\sum_{i=1}^{n} (x_i - \bar{x})^2} \tag{7}$$

and

$$\hat{\beta}_0 = \bar{y} - \hat{\beta}_1 \bar{x}. \tag{8}$$

Equations (7) and (8) are known as the *least-squares estimates*.

An in-depth treatment of the theory concerning either Case A or Case B is given in Graybill (14.1).

It can also be shown that the *unbiased least-squares* estimate of σ_e^2 is given by

$$\hat{\sigma}_e^2 = \frac{SS_{\text{error}}}{n-2} = \frac{\sum_{i=1}^{n} (y_i - \hat{\mu}_{y|x_i})^2}{n-2}. \tag{9}$$

To compute the value of $\hat{\sigma}_e^2$ directly from (9), we must compute $\hat{\mu}_{y|x_i}$ for each x_i, take the difference between the estimate and the observed value, square the difference, and, finally, add the squares prior to dividing by $(n-2)$. Fortunately, we can simplify this task by expressing $\hat{\sigma}_e^2$ in terms of values that must be computed in obtaining the least-squares estimates $\hat{\beta}_0$ and $\hat{\beta}_1$ and other values used in Chapter 3. That is,

$$SS_{\text{error}} = \sum_{i=1}^{n} (y_i - \hat{\mu}_{y|x_i})^2 = \sum_{i=1}^{n} (y_i - \bar{y})^2 - \frac{\hat{\beta}_1}{n}\left[n \sum_{i=1}^{n} x_i y_i - \left(\sum_{i=1}^{n} x_i\right)\left(\sum_{i=1}^{n} y_i\right)\right] \tag{10}$$

The term in the brackets of (10) is determined during the calculation of $\hat{\beta}_1$ using Equation (7). Expressing $\sum_{i=1}^{n}(y_i - \bar{y})^2$ in terms of its computational formula and substituting (10) into (9), we obtain

$$\hat{\sigma}_e^2 = \frac{SS_{\text{error}}}{n-2} = \frac{1}{n-2}\left\{\sum_{i=1}^{n} y_i^2 - \frac{(\sum_{i=1}^{n} y_i)^2}{n} - \frac{\hat{\beta}_1}{n}\left[n \sum_{i=1}^{n} x_i y_i - \left(\sum_{i=1}^{n} x_i\right)\left(\sum_{i=1}^{n} y_i\right)\right]\right\}. \tag{11}$$

It can be shown that (11) is an unbiased estimate of σ_e^2. We generally refer to the square root of $\hat{\sigma}_e^2$ as the *standard error*.

Case B (Normal-Theory Case)

When the assumptions of Case B are valid (that is, when the error variable is normally distributed), we can obtain estimates of β_0, β_1, and σ_e^2 using the method of maximum likelihood (Section 8.4). We are able to use the maximum-likelihood method in this instance only because we assume a specific probability distribution for the error variable. For this situation, it can be shown that the maximum-likelihood estimates for β_0 and β_1 are the same as the least-squares estimates of (7) and (8), respectively. Now, since maximum-likelihood estimators are consistent, sufficient, and efficient [at least in the sense as given in Mood and Graybill (14.2)] and the least-squares estimates $\hat{\beta}_0$ and $\hat{\beta}_1$ are unbiased, we have immediately that, under the normal-theory case, these estimates satisfy the four criteria mentioned in Chapter 8—that is, they are unbiased, consistent, sufficient, and efficient. The maximum-likelihood estimate of σ_e^2 is biased. However, when the assumptions for the normal-theory case are valid, the least-squares estimate of σ_e^2 satisfies all four criteria. Hence, we shall use (11) as our unbiased estimate of σ_e^2.

The important advantage of Case B over Case A is that, under the normal-theory case, each of the estimators has associated with it a known probability distribution (or *sampling distribution*). That is, the estimate of β_0, as given in (8), represents an observed value of a statistic having a normal distribution with mean

$$\mu_{\hat{\beta}_0} = \beta_0 \tag{12}$$

and variance

$$\sigma_{\hat{\beta}_0}^2 = \frac{\sigma_e^2 \sum_{i=1}^n x_i^2}{n \sum_{i=1}^n (x_i - \bar{x})^2}. \tag{13}$$

Further, the estimate of β_1, as given in (8), represents an observed value of a statistic having a normal distribution with mean

$$\mu_{\hat{\beta}_1} = \beta_1 \tag{14}$$

and variance

$$\sigma_{\hat{\beta}_1}^2 = \frac{\sigma_e^2}{\sum_{i=1}^n (x_i - \bar{x})^2}. \tag{15}$$

Similarly, the estimate of σ_e^2, as given in (11), represents an observed value of a statistic which, when divided by the variance σ_e^2 and multiplied by $(n - 2)$, has a chi-square distribution with $(n - 2)$ degrees of freedom. You are not expected to see the reasoning related to these important facts, but you should accept their truth. The reader interested in their development should see Graybill (14.1).

The most important thing to note is that the added assumption of normally distributed errors for Case B allows us to determine the probability (sampling) distributions for the estimators of β_0, β_1 and σ_e^2. This in turn lets us make probability statements about the estimators and construct pivotal quantities from which we determine confidence intervals and test hypotheses concerning the parameters β_0, β_1, and σ_e^2.

Example 1

The Speciality Toy Company manufactures special-purpose toys for customers on an exclusive basis. When a customer places his order, the company quotes a price

where $\hat{\sigma}_{\hat{\mu}_{y|x0}}$ is obtained by simply substituting $\hat{\sigma}_e^2$ into (15) in place of σ_e^2 and taking the square root of the result. The corresponding limits to the confidence interval are

$$\hat{\beta}_0 + \hat{\beta}_1 x_0 \pm t_{\alpha/2, n-2} \hat{\sigma}_e \sqrt{\frac{1}{n} + \frac{(x_0 - \bar{x})^2}{\sum_{i=1}^n (x_i - \bar{x})^2}}. \tag{17}$$

You should verify the results of (16) and (17).

Example 1

Suppose that the assumptions for Case B are valid. Using the data of Example 1, Section 14.3, determine the 90 percent confidence intervals for β_0, β_1, and $\mu_{y|x}$ for $x = 20$.

Solution

We have already computed the values for $\hat{\beta}_0$, $\hat{\beta}_1$, and $\hat{\sigma}_e^2$ in Example 1, Section 14.3. In fact, Table 14.2 yields all the results necessary for solving this problem except for the proper tabular t value. Thus, we simply need to determine $t_{\alpha/2, n-2}$ from Table A-4. Hence, including prior calculations from the example, we have

$$n = 12, \qquad \alpha = 0.10,$$
$$t_{\alpha/2, n-2} = t_{0.05, 10} = 1.812,$$
$$\hat{\beta}_0 = 64.247, \qquad \hat{\beta}_1 = -1.013,$$
$$\hat{\sigma}_e = \sqrt{\hat{\sigma}_e^2} = \sqrt{27.384} = 5.233,$$
$$\sum_{i=1}^n x_i^2 = 5{,}256,$$

and

$$\sum_{i=1}^n (x_i - \bar{x})^2 = \sum_{i=1}^n x_i^2 - \frac{(\sum_{i=1}^n x_i)^2}{n} = 5{,}256 - \frac{(228)^2}{12} = 924.$$

The 90 percent confidence intervals for β_0, from (10), is given by

$$\hat{\beta}_0 \pm t_{\alpha/2, n-2} \hat{\sigma}_e \sqrt{\frac{\sum_{i=1}^n x_i^2}{n \sum_{i=1}^n (x_i - \bar{x})^2}} \Rightarrow 64.247 \pm (1.812)(5.233)\sqrt{\frac{5{,}256}{12(924)}}$$
$$\Rightarrow 64.247 \pm 6.528$$
$$\Rightarrow 57.719 < \beta_0 < 70.775.$$

The 90 percent confidence interval for β_1, from (12), is given by

$$\hat{\beta}_1 \pm t_{\alpha/2, n-2} \frac{\hat{\sigma}_e}{\sqrt{\sum_{i=1}^n (x_i - \bar{x})^2}} \Rightarrow -1.013 \pm (1.812)\frac{5.233}{\sqrt{924}}$$
$$\Rightarrow -1.013 \pm 0.312$$
$$\Rightarrow -1.325 < \beta_1 < -0.701.$$

To determine the 90 percent confidence interval about $\beta_0 + \beta_1 x$ for $x = 20$, we need to compute only one more value from the data of Table 14.1—that is, $\bar{x} = 228/12 = 19.0$. The 90 percent confidence interval for $\beta_0 + \beta_1 x_0$, from (17), is given by

$$\hat{\beta}_0 + \hat{\beta}_1 x_0 \pm t_{\alpha/2, n-2}\hat{\sigma}_e \sqrt{\frac{1}{n} + \frac{(x_0 - \bar{x})^2}{\sum_{i=1}^n (x_i - \bar{x})^2}}.$$

For $x_0 = 20$, we obtain

$$\hat{\beta}_0 + \hat{\beta}_1 x_0 = 64.247 - 1.013(20) = 43.987.$$

Substituting 43.987 and the other relevant values into the formula, we obtain as the 90 percent confidence interval about $\mu_{y|20}$ [that is, $\beta_0 + \beta_1(20)$]

$$43.987 \pm (1.812)(5.233)\sqrt{\frac{1}{12} + \frac{(20 - 19)^2}{924}} \Rightarrow 43.987 \pm 2.755$$

$$\Rightarrow 41.232 < \beta_0 + \beta_1(20) < 46.742. \quad \square$$

Problems

1. Why are we unable to determine confidence intervals for β_0, β_1, and $\mu_{y|x}$ when the assumptions for Case A are satisfied?
2. Develop the fiducial probability statement given in Equation (11).
3. Develop the fiducial probability statement given in Equation (16).
4. Using the data of Problem 3, Section 14.3, determine the 95% confidence intervals for β_0, β_1, and $\mu_{y|x=2.0}$.
5. Using the data of Problem 4, Section 14.3, determine the 90% confidence intervals for β_0, β_1, and $\mu_{y|x=3.5}$.
6. Using the data of Problem 5, Section 14.3, determine the 98% confidence intervals for β_0, β_1, and $\mu_{y|x=17.5}$.

14.6 Confidence Interval for an Observation on *Y*

One of the characteristics of regression analysis that makes it so useful is the prediction capability it provides. For example, we may wish to predict a Y value which is yet to be selected for a specific value of x, say x_0. In order to predict the value of such observation, we may choose to use the estimate of the mean of the population of Y values associated with the x value (x_0) as our predicted value. In other words, we may choose to use

$$\hat{\mu}_{y|x_0} = \hat{\beta}_0 + \hat{\beta}_1 x_0 \tag{1}$$

as our prediction for an individual Y value. We generally denote the predicted value of a future Y value by $\hat{y}(x_0)$. In other words, we use the prediction equation

$$\hat{y}(x_0) = \hat{\beta}_0 + \hat{\beta}_1 x_0. \tag{2}$$

When we use the value given in (1) as our prediction for a Y value, we know that the Y value will vary from the predicted value in much the same manner as the observed values have deviated from the regression line. Thus, because we generally want an interval estimate of the Y value in conjunction with a point estimate, we shall determine a confidence interval about the predicted value $\hat{y}(x_0)$.

While the expected value of Y for a particular value of x (say x_0) is of interest to us, we are primarily interested in using our prediction equation (2), which is based upon observed data, to predict the value of a future observation. Intuitively, we see that the error of prediction (the deviation between the predicted and actual values) is composed of two factors. That is, since the actual value of Y at $x = x_0$ can be written as

$$Y(x_0) = \beta_0 + \beta_1 x_0 + e, \tag{3}$$

the deviation $[\hat{Y}(x_0) - Y(x_0)]$ will include the deviation between $\hat{Y}(x_0)$ and the expected value of $Y(x_0)$ plus the random amount e which represents the deviation of the predicted Y value from the expected value. With this additional amount (e) included in the deviation, we see that the variation for predicting a single future value of Y is greater than the variation for predicting the expected value of Y.

It can be shown that, when we predict a value of Y for $x = x_0$, the variance of the prediction error [that is, the variance of $\hat{Y}(x_0) - Y(x_0)$] is given by

$$\operatorname{Var}[\hat{Y}(x_0) - Y(x_0)] = \sigma_e^2\left[1 + \frac{1}{n} + \frac{(x_0 - \bar{x})^2}{\sum_{i=1}^{n}(x_i - \bar{x})^2}\right]. \tag{4}$$

Furthermore, the deviation $[\hat{Y}(x_0) - Y(x_0)]$ is normally distributed with mean zero. This being true, it can also be shown that

$$t = \frac{\hat{y}(x_0) - y(x_0)}{\hat{\sigma}_e\sqrt{1 + \dfrac{1}{n} + \dfrac{(x_0 - \bar{x})^2}{\sum_{i=1}^{n}(x_i - \bar{x})^2}}} \tag{5}$$

represents an observed value of a statistic that is distributed according to the t distribution with $(n - 2)$ degrees of freedom. Since this is the case, we may again construct the $100(1 - \alpha)$ percent fiducial probability statement

$$P_F\left[-t_{\alpha/2,n-2} < \frac{\hat{y}(x_0) - y(x_0)}{\hat{\sigma}_e\sqrt{1 + \dfrac{1}{n} + \dfrac{(x_0 - \bar{x})^2}{\sum_{i=1}^{n}(x_i - \bar{x})^2}}} < t_{\alpha/2,n-2}\right] = 1 - \alpha \tag{6}$$

and perform algebraic manipulations similar to those used in Section 14.5 to obtain a fiducial probability statement concerning $y(x_0)$. That is, we obtain

$$P_F\left[\hat{y}(x_0) - t_{\alpha/2,n-2}\hat{\sigma}_e\sqrt{1 + \frac{1}{n} + \frac{(x_0 - \bar{x})^2}{\sum_{i=1}^n (x_i - \bar{x})^2}} < y(x_0) < \hat{y}(x_0) + t_{\alpha/2,n-2}\hat{\sigma}_e\sqrt{1 + \frac{1}{n} + \frac{(x_0 - \bar{x})^2}{\sum_{i=1}^n (x_i - \bar{x})^2}}\right] = 1 - \alpha. \tag{7}$$

The corresponding $100(1 - \alpha)$ percent confidence interval is obtained by extracting the interval from the fiducial probability statement of (7). Thus, we obtain the following limits for the $100(1 - \alpha)$ percent confidence interval about $y(x_0)$, a future observed value for $x = x_0$:

$$\hat{y}(x_0) \pm t_{\alpha/2,n-2}\hat{\sigma}_e\sqrt{1 + \frac{1}{n} + \frac{(x_0 - \bar{x})^2}{\sum_{i=1}^n (x_i - \bar{x})^2}}. \tag{8}$$

Example 1

Suppose once again that the data of Example 1, Section 14.3, satisfy the assumptions of Case B. Determine the 95 percent confidence interval about a future observation for $x = 25$.

Solution

We first list the necessary values from Example 1, Section 14.3, for easy reference. Thus, we have

$$n = 12, \qquad \alpha = 0.05,$$
$$t_{\alpha/2,n-2} = t_{0.025,10} = 2.228,$$
$$\hat{\beta}_0 = 64.247, \qquad \hat{\beta}_1 = -1.013,$$
$$\hat{\sigma}_e = \sqrt{\hat{\sigma}_e^2} = \sqrt{27.384} = 5.233,$$
$$\sum_{i=1}^n x_i^2 = 5{,}256, \qquad \sum_{i=1}^n x_i = 228, \qquad \bar{x} = 19.0,$$

and

$$\sum_{i=1}^n (x_i - \bar{x})^2 = \sum_{i=1}^n x_i^2 - \frac{(\sum_{i=1}^n x_i)^2}{n} = 5{,}256 - \frac{(228)^2}{12} = 924.$$

From (8), the limits for the 95 percent confidence interval for a future observation at $x = x_0 = 25$ [that is, $y(25)$] is given by

$$\hat{y}(x_0) \pm t_{\alpha/2,n-2}\hat{\sigma}_e\sqrt{1 + \frac{1}{n} + \frac{(x_0 - \bar{x})^2}{\sum_{i=1}^n (x_i - \bar{x})^2}}.$$

Computing the value of $\hat{y}(x_0)$ when $x_0 = 25$, we obtain

$$\hat{y}(25) = \hat{\beta}_0 + \hat{\beta}_1(25) = 64.247 - 1.013(25) = 38.922.$$

Substituting $\hat{y}(25)$ and other relevant values into the formula for the limits of the confidence interval, we obtain

$$38.922 \pm (2.228)(5.233)\sqrt{1 + \frac{1}{12} + \frac{(25 - 19.0)^2}{924}}$$

$$\Rightarrow 38.922 - 12.351 < y(25) < 38.922 + 12.351$$

$$\Rightarrow 26.571 < y(25) < 51.273. \qquad \square$$

The 95 percent confidence interval of Example 1 represents an interval estimate of only a single future observation (that is, for $x = 25$). However, we could determine confidence intervals for as many values of x as we wish. Figure 14.6 represents an approximate graph of the confidence intervals for the x values throughout the range of values given in Example 1, Section 14.3. When the interval estimates are connected for the various x values, we generally refer to the resulting band as a *confidence belt*.

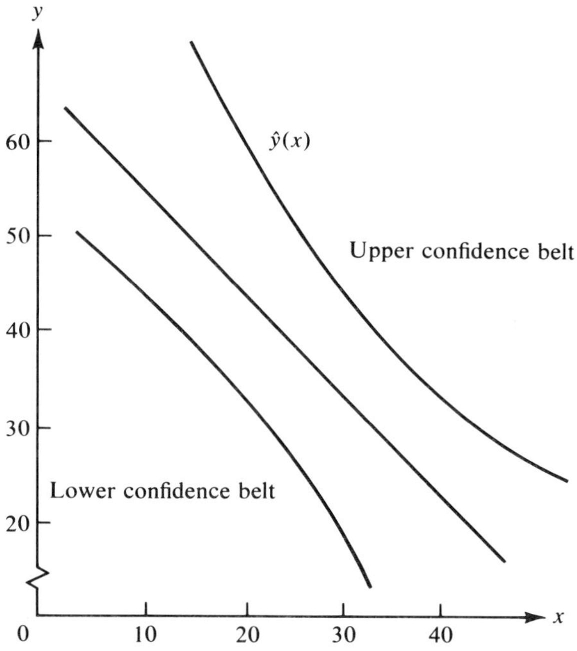

Figure 14.6 Confidence intervals for Example 1, Section 14.3

The interpretation of the confidence interval obtained in Example 1 is that we are 95 *percent confident* the future observation will lie between 26.571 and 51.273. That is, the value of the future observation will be between 26.571 and 51.273, unless we have been unfortunate enough to obtain as the observed value of our *confidence interval* one of the interval estimates among the 5 percent "bad" estimates in the population of all possible *interval estimates* that could be constructed by use of (8) for $x = 25$. However, we can never know whether we have been fortunate or not; thus, we simply rely on the *measure of goodness* for our estimate and say that we are 95 percent confident that $y(25)$ is between the stated limits.

Problems

1. Using the data of Problem 3, Section 14.3, determine the 95% confidence interval for a future observation on sales income when advertising cost is \$2,000.
2. Using the data of Problem 4, Section 14.3, determine the 90% confidence interval for a future observation on enrollment when the number of inquiries is 3,500.
3. Using the data of Problem 5, Section 14.3, determine the 98% confidence interval for a future observation when the nine-month estimated manpower level is 1,750.

14.7 Correlation Analysis

In the preceding sections, we have been concerned with analyzing sample data in order to estimate the coefficients of an assumed linear model and to make inferences concerning the population from which the sample is selected. More specifically, we have been concerned with various facets of treating an assumed linear relationship between two variables, one an independent mathematical (nonrandom) variable and the other a random variable assumed to be dependent upon the mathematical variable. In fact, we have constrained our independent variable to be nonrandom by assuming that the values of this variable are known or observed without error—that is, by assuming such values to be selected and controlled by the experimenter.

When two variables vary jointly, we call the probability distribution that associates probabilities with various combinations of their possible values a *joint probability distribution.* In this section, we shall consider a situation involving this type of distribution where the two variables of concern are both random variables. Specifically, we shall discuss methods that are appropriate when X and Y are random variables having a joint distribution and we wish to make inferences, rather than estimate a regression line, concerning the degree of linear relationship existing between X and Y.

If the joint probability distribution of X and Y is a *bivariate normal distribution* (roughly speaking, the conditional probability distribution of each variable is normal and the two variables are not necessarily independent), one of the parameters required for complete description of the distribution is the correlation coefficient ρ, or, more specifically, the *product moment correlation coefficient.* This correlation coefficient is a measure of the linear relationship between X and Y and is so defined that the higher its absolute value the stronger the linear relationship between X and Y. The value of ρ is in no way dependent upon the units in which X and Y are measured—it is a dimensionless quantity which can assume any value between -1 and 1. Furthermore, if ρ is either -1 or 1, the linear relationship is considered perfect and all of the points in the population lie on a straight line. Conversely, if ρ is equal to zero, there exists no linear relationship between X and Y, and we say that the variables are uncorrelated. A *positive correlation* means that as one variable increases the other also increases. Conversely, a *negative correlation* means that as one variable increases the other decreases. Note that ρ is a population parameter. It measures the degree of linear correlation that exists in the entire population of X and Y values.

You should keep in mind that the correlation coefficient measures only the linear relationship between two variables. It is possible for two variables to be related by a nonlinear relationship and, yet, for the correlation coefficient ρ to be equal to zero. This being the case, you should avoid assuming that a small correlation coefficient reflects no relationship whatsoever between two variables.

In correlation analysis, we wish to make inferences concerning the degree of linear relationship between X and Y. That is, since ρ is the population parameter that measures the linear relationship, we wish to make inferences concerning the value of ρ. For a sample size n, we estimate ρ by the formula

$$r = \frac{\sum_{i=1}^{n} (x_i - \bar{x})(y_i - \bar{y})}{\sqrt{\sum_{i=1}^{n} (x_i - \bar{x})^2 \sum_{i=1}^{n} (y_i - \bar{y})^2}}, \tag{1}$$

where r represents the statistic commonly referred to as the *sample correlation coefficient* and the x and y values are obtained from the sample of n ordered pairs (x_1, y_1), $(x_2, y_2), \ldots, (x_n, y_n)$.

Estimates of ρ (that is, values of r) are extremely variable for small sample sizes, particularly for ρ at or near zero. Unless the sample size is reasonably large, a single pair of values (one observation) can make a vast difference in the value of r. This makes it particularly difficult to detect values of ρ near zero. However, since small values of ρ are generally of little importance, this is not a very serious problem. Nevertheless, it should be noted that, even if sampling is accomplished from a population in which the correlation coefficient ρ is in fact equal to zero, observed values of r may be either positive or negative due to the random variation induced by sampling.

To determine whether the correlation coefficient in a particular situation is significantly different from zero (that is, whether there is a significant degree of linear correlation between the variables X and Y), we simply test the null hypothesis

$$H_0: \rho = 0.$$

Since we are interested in determining whether there is a significant linear relationship between the two variables and not whether the relationship is positive or negative, we are generally interested in testing H_0 against a two-sided alternative hypothesis. Thus, we generally test the hypotheses

$$H_0: \rho = 0 \quad \text{versus} \quad H_1: \rho \neq 0. \tag{2}$$

However, before we can test the hypotheses of (2) in the manner of past chapters, we must have available a statistic (pivotal quantity) such that

1. it is a function of the unknown parameter ρ,
2. the distribution of the statistic is known and does not depend on the true value of the parameter ρ (the form of the distribution is independent of the parameter), and
3. the statistic is a function of no other unknown.

And, since we have already indicated that we shall use r to represent the observed value of the estimator of ρ, we seek a pivotal quantity involving r.

Since the observed value r is taken from a joint normal distribution, it can be shown that the statistic for which r represents an observed value has a sampling distribution that tends to be normal. If the population from which sampling is accomplished is such that $\rho = 0$, this tendency is quite strong, because the sampling distribution of the sample correlation coefficient is symmetrical about $\rho = 0$. On the other hand, when $\rho \neq 0$, the sampling distribution of the sample correlation coefficient is skewed; hence, the tendency toward normality is not quite so strong. The sampling distribution is skewed left for negative values of ρ, skewed right for positive values of ρ, and normal when $\rho = 0$.

Since the sampling distribution of the sample correlation coefficient only tends toward the normal distribution, we do not have an exact test for a hypothesis concerning ρ unless $\rho = 0$. However, we can perform an exact test for the hypotheses of (2) by using the statistic which has as its observed value

$$t = r\sqrt{\frac{n-2}{1-r^2}}. \tag{3}$$

Fisher (14.3) has shown that the pivotal quantity for which (3) represents an observed value is distributed according to the t distribution with $(n - 2)$ degrees of freedom if, and only if, $\rho = 0$. Thus, we can test the hypotheses of (2) in a manner similar to that for any other situation where we use a t test. The essential steps for making a two-tail, 100α percent significance test that the correlation coefficient ρ of a bivariate normal distribution is equal to zero are as follows.

1. Set up the null and alternative hypotheses

$$H_0{:}\rho = 0 \quad \text{versus} \quad H_1{:}\rho \neq 0. \tag{4}$$

2. Specify the level of significance α.
3. Determine the size (n) of the sample to be taken. [This fixes the number $(n - 2)$ of degrees of freedom to be used in establishing the appropriate critical value.]
4. Determine the critical values $-t_{\alpha/2,n-2}$ and $t_{\alpha/2,n-2}$ that divide the t distribution into a critical (rejection) region and an acceptance region.
5. From a sample of size n obtain (under the assumption that the null hypothesis is true, or $\rho = 0$) one observed value of the pivotal quantity. That is, compute

$$t = r\sqrt{\frac{n-2}{1-r^2}}, \tag{5}$$

where r is obtained by computing

$$r = \frac{\sum_{i=1}^{n}(x_i - \bar{x})(y_i - \bar{y})}{\sqrt{\sum_{i=1}^{n}(x_i - \bar{x})^2 \sum_{i=1}^{n}(y_i - \bar{y})^2}} \tag{6}$$

with the x and y values being obtained from the n ordered pairs $(x_1, y_1), (x_2, y_2), \ldots, (x_n, y_n)$ representing the sample values.

6. If the value of (5) is in the critical region, reject H_0. Otherwise, accept H_0.

Example 1

An electronics firm wants to know if the number of defective components produced is correlated with the average time it takes an employee to assemble the component. A random sample of 18 employees is selected for a given work week and the number of defective components produced and average assembly time recorded in Table 14.4. Determine the estimate of the population correlation coefficient and whether there is a significant linear relationship between the two variables. Use a 5 percent significance level.

Table 14.4

Defective components produced and assembly time

Observation number i	*Number of defective items* x	*Average assembly time* y *(minutes)*
1	27	19
2	24	16
3	12	12
4	22	17
5	13	10
6	29	19
7	14	15
8	20	14
9	16	15
10	21	21
11	22	14
12	25	22
13	23	22
14	18	15
15	20	14
16	15	15
17	20	15
18	13	16

Solution

The computations required for the estimate of ρ are given in Table 14.5. Using the values of Table 14.4 and the computational formulas for $\sum_{i=1}^{n}(x_i - \bar{x})(y_i - \bar{y})$, $\sum_{i=1}^{n}(x_i - \bar{x})^2$, and $\sum_{i=1}^{n}(y_i - \bar{y})^2$, we obtain

$$r = \frac{\sum_{i=1}^{n}(x_i - \bar{x})(y_i - \bar{y})}{\sqrt{\sum_{i=1}^{n}(x_i - \bar{x})^2 \sum_{i=1}^{n}(y_i - \bar{y})^2}} = \frac{\sum_{i=1}^{n} x_i y_i - \dfrac{(\sum_{i=1}^{n} x_i)(\sum_{i=1}^{n} y_i)}{n}}{\sqrt{\left[\sum_{i=1}^{n} x_i^2 - \dfrac{(\sum_{i=1}^{n} x_i)^2}{n}\right]\left[\sum_{i=1}^{n} y_i^2 - \dfrac{(\sum_{i=1}^{n} y_i)^2}{n}\right]}}$$

$$= \frac{5{,}914 - \dfrac{(354)(291)}{18}}{\sqrt{\left[7{,}392 - \dfrac{(354)^2}{18}\right]\left[4{,}889 - \dfrac{(291)^2}{18}\right]}} = \frac{191.0}{\sqrt{(430.0)(184.5)}} = 0.678.$$

To test the hypotheses

$$H_0:\rho = 0 \quad \text{versus} \quad H_1:\rho \neq 0,$$

we first determine the critical values for $\alpha = 0.05$ and $n - 2 = 16$ degrees of freedom to be

$$-t_{0.025,\,16} = -2.120 \quad \text{and} \quad t_{0.025,\,16} = 2.120.$$

Computing the observed value of the pivotal quantity, we obtain

$$t = r\sqrt{\frac{n-2}{1-r^2}} = (0.0678)\sqrt{\frac{16}{1-(0.678)^2}} = 3.689.$$

Table 14.5

Data for calculating correlation coefficient (Example 1)

Observation number i	*Number of defective items* x	*Average time* y	xy	x^2	y^2
1	27	19	513	729	361
2	24	16	384	576	256
3	12	12	144	144	144
4	22	17	374	484	289
5	13	10	130	169	100
6	29	19	551	841	361
7	14	15	210	196	225
8	20	14	280	400	196
9	16	15	240	256	225
10	21	21	441	441	441
11	22	14	308	484	196
12	25	22	550	625	484
13	23	22	506	529	484
14	18	15	270	324	225
15	20	14	280	400	196
16	15	15	225	225	225
17	20	15	300	400	225
18	13	16	208	169	256
Total	354	291	5,914	7,392	4,889

Since the observed value of the pivotal quantity (3.689) is greater than the critical value 2.120, we reject H_0 and conclude that ρ is significantly different from zero. That is, the number of defective items produced is correlated linearly with the average time to assemble a component. □

From Example 1, the positive estimate (0.678) of the correlation coefficient indicates that the number of defective items an employee produces increases as his average time for assembling a component increases. This does not necessarily mean that the faster an employee assembles a component the fewer defective items he produces. This relationship between the two variables may be due to a third variable to which they are both related. For example, the amount of experience an employee has as an assembler could be negatively correlated with both variables—as the number of years of experience increases, the number of defective components decreases and, likewise, the average time for assembly decreases.

As we can see from this discussion, the correlation coefficient merely indicates whether there is a linear relationship between the variables, not whether one variable causes the other to change. It is this particular point many researchers fail to understand; consequently, it often happens that a perfectly valid correlation coefficient is used to "prove" a cause-and-effect relationship that does not exist. Because there is a tendency to misinterpret tests of hypotheses concerning the population correlation coefficient, we urge you to use ρ properly.

We emphasize that our discussion concerning correlation is strictly introductory. We have considered only testing the hypothesis for $\rho = 0$. An approximate test exists for other hypotheses concerning ρ (for example, $\rho = k \neq 0$), but we shall not consider it here. Discussion of this type of test and the construction of confidence intervals for ρ are given in Snedecor (14.4) and Steel and Torrie (14.5).

Problems

1. How does correlation analysis differ from regression analysis?

2. What does the correlation coefficient measure?

3. Does a small correlation coefficient value reflect no relationship between two random variables? Why?

4. Is the test for the hypotheses concerning the correlation coefficient an exact test? Why?

5. The tabulated data represent the observations for 15 days on the number of customers entering a specialty shop and the amount of sales, in hundreds of dollars, for each day.

   a. Determine the estimate of the population correlation coefficient.

   b. Determine at the 5% significance level whether the population correlation coefficient is significantly different from zero.

Number of customers (x)	*Amount of sales* (y)
30	2.6
25	3.0
20	2.1
37	3.2
31	2.9
50	3.9
29	2.6
40	3.7
42	4.3
36	3.1
24	2.7
43	3.7
36	3.6
33	3.0
37	2.9

6. The data given for ten individuals represents the weekly take-home pay, in hundreds of dollars, and the weekly dollar amount spent on entertainment.

   a. Determine the estimate of the population correlation coefficient.

   b. Determine at the 1% significance level whether the population correlation coefficient is different from zero.

Take-home pay (x)	*Spent on entertainment* (y)
1.5	10
3.3	25
2.1	20
2.3	26
1.9	18
4.2	35
3.2	12
1.7	17
2.5	27
2.3	20

7. The data for eight grocery stores represent the number of jars of peanut butter sold during a given week and the number of loaves of bread sold during the same week.

   a. Determine the estimate of the population correlation coefficient.

   b. Determine at the 10% significance level whether there is a significant linear relationship between the two variables.

Jars of peanut butter (x)	*Loaves of bread* (y)
10	40
24	85
17	50
35	110
8	28
9	80
28	60
40	90

14.8 Summary Example

To review the concepts and methods set forth in this chapter, let us consider a comprehensive example.

Example

Suppose a men's clothing manufacturer wants to know the relationship between sales revenue and advertising expenditures. This relationship is thought to be linear, with advertising expenditure the independent variable and sales revenue the dependent variable. Annual data are available for the past 13 years as given in Table 14.6. Suppose in this situation the Case B assumptions are valid. Determine estimates for the parameters β_0, β_1, and σ_e^2; estimate the regression line $\mu_{y|x}$; and construct 95 percent confidence intervals for β_0, β_1, $\mu_{y|8}$, and a future observation when advertising expenditures will be \$10,000 (that is, $x = 10$). Further, use a 5 percent significance level to determine whether there is a significant linear relationship between the means of the populations of Y values and the x values (that is, determine whether β_1 is significantly different from zero).

Table 14.6

Advertising expenditures versus sales revenues (in thousands of dollars)

Year	*Advertising expenditures* (x)	*Sales revenue* (y)
1	3	41
2	3	40
3	4	42
4	5	46
5	3	40
6	7	49
7	6	51
8	4	43
9	6	48
10	7	49
11	9	56
12	7	51
13	8	50

Solution

The values necessary for computing the estimates are given in Table 14.7. Substituting appropriate values from the table into the proper formulas for estimating $\beta_0, \beta_1, \mu_{y|x}$ and σ_e^2, we obtain

$$\hat{\beta}_1 = \frac{n\sum_{i=1}^n x_i y_i - (\sum_{i=1}^n x_i)(\sum_{i=1}^n y_i)}{n\sum_{i=1}^n x_i^2 - (\sum_{i=1}^n x_i)^2} = \frac{13(3{,}474) - (72)(606)}{13(448) - (72)^2} = \frac{1530}{640} = 2.39,$$

and

$$\hat{\beta}_0 = \bar{y} - \hat{\beta}_1\bar{x} = \frac{606}{13} - (2.39)\left(\frac{72}{13}\right) = 33.38.$$

Table 14.7

Computations for data of Table 14.6

x	y	xy	x^2	y^2
3	41	123	9	1,681
3	40	120	9	1,600
4	42	168	16	1,764
5	46	230	25	2,116
3	40	120	9	1,600
7	49	343	49	2,401
6	51	306	36	2,601
4	43	172	16	1,849
6	48	288	36	2,304
7	49	343	49	2,401
9	56	504	81	3,136
7	51	357	49	2,601
8	50	400	64	2,500
72	606	3,474	448	28,554

Thus, our estimate of the true regression line $\mu_{y|x}$ is

$$\hat{\mu}_{y|x} = \hat{\beta}_0 + \hat{\beta}_1 x = 33.38 + 2.39x.$$

Finally, we have

$$\hat{\sigma}_e^2 = \frac{1}{n-2}\left\{\sum_{i=1}^{n} y_i^2 - \frac{(\sum_{i=1}^{n} y_i)^2}{n} - \frac{\hat{\beta}_1}{n}\left[n\sum_{i=1}^{n} x_i y_i - \left(\sum_{i=1}^{n} x_i\right)\left(\sum_{i=1}^{n} y_i\right)\right]\right\}$$

$$= \frac{1}{11}\left\{28{,}554 - \frac{(606)^2}{13} - \frac{2.39}{13}[1{,}530]\right\} = \frac{1}{11}\{23.79\} = 2.16.$$

Since $\hat{\sigma}_e^2 = 2.16$, we obtain $\hat{\sigma}_e = \sqrt{2.16} = 1.47$ as our estimate of the standard error.

Substituting the appropriate values into the respective formulas for determining the 95 percent confidence intervals for β_0, β_1, $\mu_{y|8}$, and $y(10)$, we have

$$\hat{\beta}_0 \pm t_{\alpha/2, n-2}\hat{\sigma}_e\sqrt{\frac{\sum_{i=1}^{n} x_i^2}{n\sum_{i=1}^{n}(x_i - \bar{x})^2}} \Rightarrow 33.38 \pm 2.201(1.47)\sqrt{\frac{(448)}{13\left[448 - \frac{(72)^2}{13}\right]}}$$

$$\Rightarrow 33.38 \pm 2.71$$

$$\Rightarrow 30.67 < \beta_0 < 36.09.$$

Next, as a confidence interval for β_1, we have

$$\hat{\beta}_1 \pm t_{\alpha/2, n-2}\frac{\hat{\sigma}_e}{\sqrt{\sum_{i=1}^{n}(x_i - \bar{x})^2}} \Rightarrow 2.39 \pm 2.201\frac{1.47}{\sqrt{49.23}}$$

$$\Rightarrow 2.39 \pm 0.46$$

$$\Rightarrow 1.93 < \beta_1 < 2.85.$$

Now, $\hat{\mu}_{y|8} = 33.38 + 2.39(8) = 52.50$. Thus, the limits for the confidence interval for $\mu_{y|8}$ become

$$\hat{\mu}_{y|x_0} \pm t_{\alpha/2, n-2}\hat{\sigma}_e\sqrt{\frac{1}{n} + \frac{(x_0 - \bar{x})^2}{\sum_{i=1}^{n}(x_i - \bar{x})^2}} \Rightarrow 52.50 \pm 2.201(1.47)\sqrt{\frac{1}{13} + \frac{(8 - 5.54)^2}{49.23}}$$

$$\Rightarrow 52.50 \pm 1.45$$

$$\Rightarrow 51.05 < \beta_0 + \beta_1(8) < 53.95.$$

Finally, $\hat{y}(10) = 33.38 + 2.39(10) = 57.28$. Thus, the limits for the confidence interval for $y(10)$ are

$$\hat{y}(x_0) \pm t_{\alpha/2, n-2}\hat{\sigma}_e\sqrt{1 + \frac{1}{n} + \frac{(x_0 - \bar{x})^2}{\sum_{i=1}^{n}(x_i - \bar{x})^2}}$$

$$\Rightarrow 57.28 \pm 2.201(1.47)\sqrt{1 + \frac{1}{13} + \frac{(10 - 5.54)^2}{49.23}}$$

$$\Rightarrow 57.28 \pm 3.94$$

$$\Rightarrow 53.34 < y(10) < 61.22.$$

In order to determine whether there is a significant linear relationship between the two variables, we test the hypotheses

$$H_0:\beta_1 = 0 \quad \text{versus} \quad H_1:\beta_1 \neq 0.$$

Using a 5 percent significance level, we determine the critical values as

$$-t_{\alpha/2,n-2} = -t_{0.025,11} = -2.201 \quad \text{and} \quad t_{\alpha/2,n-2} = t_{0.025,11} = 2.201.$$

Computing the observed value of the appropriate pivotal quantity, we obtain

$$t = \frac{\hat{\beta}_1 - 0}{\sqrt{\dfrac{\hat{\sigma}_e^2}{\sum_{i=1}^{n}(x_i - \bar{x})^2}}} = \frac{2.39}{\sqrt{\dfrac{2.16}{49.23}}} = 11.41.$$

Comparing 11.41 with our critical values, we note that it is in the critical region; thus, we reject H_0 and conclude that there is a significant linear relationship between sales revenue and advertising expenditure. □

This example is intended to review the basic concepts, formulas, and techniques associated with linear regression analysis. In most practical situations, we would determine whether β_1 is significantly different from zero before continuing. That is, we would ordinarily test the basic assumption that there actually exists a linear relationship between the means of the Y populations and the x values prior to estimating other parameters; for, if there is strong evidence that the assumption is not valid, it is meaningless to continue with other calculations.

Finally, note once again the critical role of the Case B assumption of normally distributed errors for the regression analysis model. Without this assumption, none of the confidence intervals or tests of hypotheses of the example could be accomplished, unless possibly more stringent assumptions are made.

If there is evidence in a given situation that the error variable does not follow the normal distribution, the results based upon an assumption of normality will not be valid. In this case, another distribution must be assumed, and related tests of hypotheses and confidence intervals must be developed.

Important Terms and Concepts

- conditional mean
- confidence intervals for β_0, β_1, $\mu_{y|x}$, and $y(x)$
- correlation analysis
- correlation coefficient
- deterministic mathematical model
- estimation of β_0, β_1, σ_e^2, $\mu_{y|x}$, and $y(x)$
- future observation
- least-squares equations
- linear regression analysis
- linear regression model
- method of least squares
- negative correlation
- normal equations
- normal-theory case
- positive correlation
- probabilistic mathematical model
- product-moment correlation coefficient
- regression analysis
- regression coefficients
- tests of hypotheses concerning β_0 and β_1

References

14.1 F. A. Graybill, *An Introduction to Linear Statistical Models*, vol. 1, New York: McGraw-Hill Book Company, 1961.

14.2 A. M. Mood and F. A. Graybill, *Introduction to the Theory of Statistics*, New York: McGraw-Hill Book Company, 1963.

14.3 R. A. Fisher, "Frequency Distribution of the Values of the Correlation Coefficient in Samples from an Indefinitely Large Population," *Biometrika*, vol. 10; p. 507, 1915.

14.4 G. W. Snedecor, *Statistical Methods*, Ames, Iowa: Iowa State College Press, 1958.

14.5 R. G. D. Steel and J. H. Torrie, *Principles and Procedures of Statistics*, New York: McGraw-Hill Book Company, 1960.

Review Problem

1. A retail grocer believes there is a linear relationship between the price per pound of bananas and the weekly demand. To determine if there is such a relationship, he collected the following data for a sixteen-week period.

Price (x) cents/pound	*Demand (y) hundred pounds*	*Price (x) cents/pound*	*Demand (y) hundred pounds*
10	2.3	16	2.1
12	2.3	15	2.0
13	2.1	13	2.2
13	2.2	12	2.3
14	2.2	10	2.4
15	2.2	11	2.3
13	2.3	13	2.2
17	1.9	15	2.1

a. Determine $\hat{\beta}_0$ and $\hat{\beta}_1$.

b. Determine $\hat{\mu}_{y|x}$ for each x in the data.

c. Determine the individual deviations from the regression line $(y_i - \hat{\mu}_{y|x_i})$ for each x in the data.

d. Determine $\hat{\sigma}_e^2$ using Equation (9, Sec. 14.3).

e. Determine $\hat{\sigma}_e^2$ using Equation (11, Sec. 14.3).

f. Determine at the 5% significance level whether the slope of the regression line is significantly less than zero.

g. Determine the 90% confidence intervals for β_0, β_1, and $\mu_{y|x=10}$.

h. Determine the 90% confidence interval for a future observation when the price of bananas is 10 cents per pound.

Fifteen

Multiple Regression Analysis

15.1 Introduction

When working with linear regression analysis, we were concerned with the relationship between two variables, one a random variable and the other a mathematical (nonrandom) variable. We saw that when there is a strong linear relationship between the two variables, we can often express the dependent variable as a linear function of the independent variable. However, when no strong linear relationship exists between the two variables, adequate description and prediction may require us to express the dependent variable as a linear function of several independent variables. For example, the gasoline mileage of an automobile is a function of the horsepower rating of the engine, the total weight of the automobile, and the octane rating of the gasoline; sales revenue may be expressed as a function of expenditures for TV, radio, newspaper, and direct-mail advertising; and the breaking strength of a steel rod may be expressed as a function of the temperature and pressure to which it is subjected.

In order to establish the relationships between a dependent variable and several independent variables, we again use the technique known as regression analysis. When its use involves more than one independent variable, the technique is generally termed *multiple regression analysis.*

The situation we are about to consider is analogous to that of Chapter 14, the only difference being that we are now going to observe several (say k) characteristics for each element in the sample. We call this the *multivariate* case. For example, consider the *trivariate* situation. The sample space of a trivariate experiment can be represented by a set of ordered triples of measurements. Referring to an example of Chapter 14, consider the scores of college students on a particular achievement examination, the IQ's of the same set of students, and the grade-point average of such students. Suppose we take a sample of students and record for each the measurements associated with these three variables. If we let the IQ of a given student be represented by x_1, the score on the achievement exam by x_2, and the grade-point average by Y, we obtain from the jth student the "triple" of numbers (x_{1j}, x_{2j}, y_j). For n students in the sample, we have a sample of size n which consists of the n ordered triples

$$(x_{11}, x_{21}, y_1), (x_{12}, x_{22}, y_2), \ldots, (x_{1n}, x_{2n}, y_n).$$

It is easy to see that this reasoning can be extended to the case having k characteristics observed for each element in the sample. Then, the results of sampling n elements would be a set of n ordered k-tuples. Should a change in one (dependent) variable accompany, rather systematically, changes in several (independent) variables, we often can determine, or predict, the effect that specific changes in the independent variables will have on the dependent variable. Thus, we are interested in the relationship between the dependent variable and the independent variables that multiple regression analysis can reveal.

The logic and mechanics of multiple regression analysis parallel those of linear regression analysis as given in Chapter 14. However, the increase in the number of independent variables makes the computations laborious. Therefore it is imperative for you to be thoroughly familiar with the concepts and computations of Chapter 14 before attempting the material here.

15.2 The Multiple Linear Regression Model

We must again either know or assume a mathematical model in order to use the regression analysis technique. That is, we must express in the form of a mathematical function the relationship between the dependent variable y and the several (say k) independent variables $x_1, x_2, \ldots, x_k$. Thus, y depends upon $x_1, x_2, \ldots, x_k$ and certain unknown parameters. Again, the assumption of a functional relationship does not necessarily imply any cause-and-effect relationship between the independent variables and the dependent variable.

When variable y is a function of k independent variables (x_i; $i = 1, 2, \ldots, k$) we generally write it as

$$y = f(x_1, x_2, \ldots, x_k). \tag{1}$$

If the function is linear, then (1) can be expressed as

$$y = \beta_0 + \sum_{i=1}^{k} \beta_i x_i, \tag{2}$$

where β_0 and β_i ($i = 1, 2, \ldots, k$) are constant, unknown parameters. Since the substitution of a specific set of values for the independent variables results in a specific value of y that can be accurately determined, the model of (2) is known as a *deterministic mathematical model of several variables*. As in the case for the two-variable models of Chapter 14, deterministic mathematical models are quite adequate for representing and predicting values in many situations, especially when the error of prediction is small from a practical point of view.

When the error of prediction is more critical we use a probabilistic mathematical model of several variables to predict or explain the relationship among the variables.

Let us consider the linear probabilistic mathematical model of several variables—generally termed the *multiple linear regression model*—given by

$$Y = \beta_0 + \sum_{i=1}^{k} \beta_i x_i + e. \tag{3}$$

The assumptions commonly associated with the multiple linear regression model are:

1. The entities β_0 and β_i $(i = 1, 2, \ldots, k)$ are unknown population parameters to be estimated.
2. Y is a dependent observable variable inherently related to $x_1, x_2, \ldots, x_k$ in a linear fashion. However, because of errors in observations, observed values of Y will deviate in a random manner from the line depicting the linear relation.
3. The x_i $(i = 1, 2, \ldots, k)$ are independent mathematical (nonrandom) variables whose values can be determined without error.
4. The error variable e is a random variable that is not observable. It accounts for the failure of the deterministic linear model

$$y = \beta_0 + \sum_{i=1}^{k} \beta_i x_i$$

   to hold for all observations on the random variable Y and the mathematical variables $x_1, x_2, \ldots, x_k$.
5. The key characteristics of e are that it is normally distributed with mean zero and variance σ_e^2 and, for any two values Y_i and Y_j the associated errors (e_i and e_j) are statistically independent.

Note that as in Chapter 14, the right-hand term of (3) can, for a given set of x values (x_i; $i = 1, 2, \ldots, k$), be represented as a constant $(\beta_0 + \sum_{i=1}^{k} \beta_i x_i)$ plus a random variable (e).

In multiple regression analysis, the variable for which estimates are to be obtained is commonly called the dependent variable and is denoted by Y. The variables used as a basis for estimation are commonly designated x_i $(i = 1, 2, \ldots, k)$ and are called the independent variables. The equation (developed through multiple regression analysis) which relates the mean of the Y values to the set of x values is called the *regression of Y on the x_i $(i = 1, 2, \ldots, k)$ variables.*

A word of caution is in order: Nothing in the assumptions for the multiple linear regression model indicates that Y is linearly related to the independent variables x_i $(i = 1, 2, \ldots, k)$. It is implicit from the basic assumptions for the multiple linear regression model that there exists a population involving various values of Y for each given set of x values. Thus, for each possible set of x values, there essentially exists a distribution of possible values of Y.

For a given set of x values, we can determine the mean and variance of the random variable Y as we did in Chapter 14. That is, if we let $Y \mid x_1, x_2, \ldots, x_k$ denote the value of Y for a given set of x values and let $\mu_{y\mid x_1, x_2, \ldots, x_k}$ denote the mean of Y for a given set of values of x, we can show that

$$\mu_{y\mid x_1, x_2, \ldots, x_k} = E(Y \mid x_1, x_2, \ldots, x_k) = \beta_0 + \sum_{i=1}^{k} \beta_i x_i \tag{4}$$

and

$$\sigma^2_{y|x_1, x_2, \ldots, x_k} = \text{Var}(Y|x_1, x_2, \ldots, x_k) = \sigma_e^2. \tag{5}$$

We note that $\sigma^2_{y|x_1, x_2, \ldots, x_k}$ is independent of the x_i ($i = 1, 2, \ldots, k$). This simply means that the variance of Y, under the assumptions for the multiple linear regression model, does not depend on the values of the x_i ($i = 1, 2, \ldots, k$). Conversely, however, $\mu_{y|x_1, x_2, \ldots, x_k}$ does depend on the values of the x_i ($i = 1, 2, \ldots, k$) and consequently changes as the values of the x_i change. In fact, we draw the very important implication from (4) that, under the multiple linear regression model (3) and its related assumptions, $\mu_{y|x_1, x_2, \ldots, x_k}$ is linearly related to the x_i ($i = 1, 2, \ldots, k$).

This last statement puts in a nutshell the essence of the model assumption of multiple linear regression analysis. In other words, in multiple regression analysis, we assume a linear relationship between the expected value of the random variable Y for given values of the nonrandom variables x_i ($i = 1, 2, \ldots, k$), that is,

$$E(Y|x_1, x_2, \ldots, x_k) \quad \text{or} \quad \mu_{y|x_1, x_2, \ldots, x_k},$$

and the nonrandom variables

$$x_i \; (i = 1, 2, \ldots, k).$$

Equation (4) is termed the line of regression of Y on the variables x_i ($i = 1, 2, \ldots, k$). It represents the average relationship between the x_i ($i = 1, 2, \ldots, k$) and Y in the population. As in simple linear regression, the parameter β_0, called the y-axis intercept, represents the $E(Y)$ corresponding to $x_1 = 0$, $x_2 = 0, \ldots,$ and $x_k = 0$.

The other parameters ($\beta_1, \beta_2, \ldots, \beta_k$) are not quite so easily interpreted. We shall explain only β_1, leaving it for you to make similar interpretations for the other parameters. The parameter β_1 is called the partial slope (with respect to x_1) of the population regression line. It represents the amount by which $E(Y)$ increases when x_1 increases by one unit and $x_2, x_3, \ldots,$ and x_k are held constant. Interpretations of the other parameters are exactly parallel to that of β_1. Taken collectively, the parameters are known as the *population regression coefficients.*

We note that assumption 5, page 561, requires that the error terms be normally distributed. Thus, our assumptions here are equivalent to those for Case B of Chapter 14, page 525.

Problems

1. In multiple linear regression analysis, between what two groups of entities is the relationship assumed to be linear?

2. Why is the parameter β_i called the partial slope with respect to x_i?

3. Why must the assumptions associated with the multiple linear regression model be valid before multiple linear regression can be used to test hypotheses and establish confidence intervals?

15.3 Estimation of β_i

We are again confronted with the problem of determining estimators, which in this case are for the parameters of the multiple linear regression model,

$$Y = \beta_0 + \sum_{i=1}^{k} \beta_i x_i + e.$$

The linear relationship of interest is completely characterized by these parameters. In other words, we need to estimate the mean and variance of the Y population for each given set of x values. However, in order to estimate the population mean,

$$\mu_{y|x_1, x_2, \ldots, x_k} = \beta_0 + \sum_{i=1}^{k} \beta_i x_i, \tag{1}$$

we must estimate the parameters β_i $(i = 0, 1, \ldots, k)$. Once we find the estimator for $\mu_{y|x_1, x_2, \ldots, x_k}$, we can determine the estimator for the common variance of the populations. That is, we can determine the estimator of

$$\sigma^2_{y|x_1, x_2, \ldots, x_k} = \sigma^2_e. \tag{2}$$

Denoting the least-squares estimates of β_i $(i = 0, 1, \ldots, k)$ by $\hat{\beta}_i$ $(i = 0, 1, \ldots, k)$, we can represent the least-squares estimate of $\mu_{y|x_1, x_2, \ldots, x_k}$ by

$$\hat{\mu}_{y|x_1, x_2, \ldots, x_k} = \hat{\beta}_0 + \sum_{i=1}^{k} \hat{\beta}_i x_i. \tag{3}$$

Using this notation, we represent the deviation of the jth observation y_j from the least-squares estimate of the population mean corresponding to a given set of x values (denoted by $x_{1j}, x_{2j}, \ldots, x_{kj}$) as

$$y_j - \hat{\mu}_{y|x_{1j}, x_{2j}, \ldots, x_{kj}}, \tag{4}$$

where

$$\hat{\mu}_{y|x_{1j}, x_{2j}, \ldots, x_{kj}} = \hat{\beta}_0 + \sum_{i=1}^{k} \hat{\beta}_i x_{ij} \tag{5}$$

is the least-squares estimate of the mean of the population of Y values for the x values given by $x_{1j}, x_{2j}, \ldots, x_{kj}$. The related observations consist of the sample points $(x_{1j}, x_{2j}, \ldots, x_{kj}, y_j)$ for $j = 1, 2, \ldots, n$. Thus, we are selecting n sample units and recording $(k + 1)$ characteristics for each sample unit.

When using the method of least squares, our objective is to obtain the best-fitting plane—the one that minimizes the sum of squares of the deviations between the observed values of Y and the estimates of their corresponding population means. That is, the least-squares estimate of (3) is such that the sum of squares of deviations given by

$$SS_{\text{error}} = \sum_{j=1}^{n} (y_j - \hat{\mu}_{y|x_{1j}, x_{2j}, \ldots, x_{kj}})^2 \tag{6}$$

is the minimum sum of squares which can be obtained for any method of estimating the population means. Substituting the right-hand term of (5) into (6) in place of $\hat{\mu}_{y|x_{1j}, x_{2j}, \ldots, x_{kj}}$, we obtain

$$SS_{\text{error}} = \sum_{j=1}^{k} \left[y_j - \left(\hat{\beta}_0 + \sum_{i=1}^{k} \hat{\beta}_i x_{ij} \right) \right]^2. \tag{7}$$

Suppose we want to estimate the β_i $(i = 0, 1, \ldots, k)$ by either the maximum-likelihood method or the least-squares method. (We may use either, since, under our assumption of normally distributed errors, the results will be identical.) The derivation of the estimators results in a set of simultaneous equations known as the normal equations associated with the multiple linear regression model. The simultaneous equations required for determining from a given set of sample data the observed values $\hat{\beta}_i$ $(i = 0, 1, \ldots, k)$ of the least-squares estimators are

$$\begin{aligned}
\hat{\beta}_0 n + \hat{\beta}_1 \sum_{j=1}^{n} x_{1j} + \hat{\beta}_2 \sum_{j=1}^{n} x_{2j} + \cdots + \hat{\beta}_k \sum_{j=1}^{n} x_{kj} &= \sum_{j=1}^{n} y_j, \\
\hat{\beta}_0 \sum_{j=1}^{n} x_{1j} + \hat{\beta}_1 \sum_{j=1}^{n} x_{1j}^2 + \hat{\beta}_2 \sum_{j=1}^{n} x_{1j}x_{2j} + \cdots + \hat{\beta}_k \sum_{j=1}^{n} x_{1j}x_{kj} &= \sum_{j=1}^{n} x_{1j}y_j, \\
\hat{\beta}_0 \sum_{j=1}^{n} x_{2j} + \hat{\beta}_1 \sum_{j=1}^{n} x_{2j}x_{1j} + \hat{\beta}_2 \sum_{j=1}^{n} x_{2j}^2 + \cdots + \hat{\beta}_k \sum_{j=1}^{n} x_{2j}x_{kj} &= \sum_{j=1}^{n} x_{2j}y_j, \\
\vdots \qquad\qquad \vdots \qquad\qquad \vdots \qquad\qquad \vdots \qquad & \qquad \vdots \\
\hat{\beta}_0 \sum_{j=1}^{n} x_{kj} + \hat{\beta}_1 \sum_{j=1}^{n} x_{kj}x_{1j} + \hat{\beta}_2 \sum_{j=1}^{n} x_{kj}x_{2j} + \cdots + \hat{\beta}_k \sum_{j=1}^{n} x_{kj}^2 &= \sum_{j=1}^{n} x_{kj}y_j.
\end{aligned} \tag{8}$$

The equations of (8) represent $(k + 1)$ linear equations in $(k + 1)$ unknowns β_i $(i = 0, 1, \ldots, k)$. That is, since the sums (e.g., $\sum_{j=1}^{n} x_{1j}$) can be obtained from a given set of sample data and substituted into the equations of (8), the resulting set of simultaneous equations will have $(k + 1)$ unknowns remaining after such substitution. We obtain estimates of the parameters of the model simply by solving the resulting simultaneous equations for the $(k + 1)$ unknowns.

We encourage anyone familiar with calculus to see Graybill (15.1) for the theoretical development of the above normal equations. Those not familiar with calculus need not despair. A definite relationship exists between the $(k + 1)$ normal equations of (8) and the two normal equations of (5) and (6), Section 14.3, which gives us a rule of thumb for constructing the normal equations associated with any multiple linear regression model, no matter how many independent variables are involved. This relationship stems directly from one between the two models for linear regression and multiple linear regression. For example, the equations of (8) can be obtained merely by successively multiplying the equation for the multiple linear regression model (after inserting a subscript to represent the observation number and ignoring the error term) by 1, $x_1, x_2, \ldots, x_k$, respectively, and summing (over the number of observed values) on both sides of the resulting equations. The technique can be extended to obtain appropriate normal equations for linear models involving any number of independent variables. That is, in any multiple linear regression situation, we obtain the normal equations by successively multiplying the equation for the model (ignoring the error term) by 1 and each of the independent variables, respectively, and summing (over the number of observed values) on both sides of the resulting equations. For example, the appropriate model for a situation involving two independent variables becomes

$$Y = \beta_0 + \beta_1 x_1 + \beta_2 x_2 + e.$$

Following the described procedure, we obtain the three normal equations

$$\hat{\beta}_0 n + \hat{\beta}_1 \sum_{j=1}^{n} x_{1j} + \hat{\beta}_2 \sum_{j=1}^{n} x_{2j} = \sum_{j=1}^{n} y_j$$

$$\hat{\beta}_0 \sum_{j=1}^{n} x_{1j} + \hat{\beta}_1 \sum_{j=1}^{n} x_{1j}^2 + \hat{\beta}_2 \sum_{j=1}^{n} x_{1j}x_{2j} = \sum_{j=1}^{n} x_{1j}y_j$$

$$\hat{\beta}_0 \sum_{j=1}^{n} x_{2j} + \hat{\beta}_1 \sum_{j=1}^{n} x_{1j}x_{2j} + \hat{\beta}_2 \sum_{j=1}^{n} x_{2j}^2 = \sum_{j=1}^{n} x_{2j}y_j.$$

For $k > 2$, the problem of determining the least-squares estimates of $\hat{\beta}_i$ ($i = 0, 1, \ldots, k$) is a bit difficult, the major difficulty arising in determining the numerical values of the estimates from a given set of sample data.

Example 1

Suppose the hypothetical data in Table 15.1 came from an experiment involving 16 different automobiles. The independent variables are horsepower rating (x_1), weight (x_2), and octane rating (x_3) of the required gasoline; the dependent variable

Table 15.1

Gasoline mileage as a function of three independent variables

Independent variables			*Dependent variable*
Horsepower rating x_1	*Weight* x_2	*Octane rating* x_3	*Gasoline mileage* Y
180	2,800	90	16.0
195	3,000	90	14.9
160	2,700	80	14.1
235	3,200	110	18.5
175	2,700	100	15.3
285	3,400	110	21.9
210	3,000	100	15.2
230	3,300	100	18.2
235	3,500	100	17.0
195	2,800	90	16.8
290	3,500	110	20.0
150	2,500	80	14.8
100	2,100	80	13.0
290	3,400	110	20.3
255	3,400	100	18.2
220	2,900	100	17.1
$\sum_{j=1}^{n} x_{1j} = 3{,}405$	$\sum_{j=1}^{n} x_{2j} = 48{,}200$	$\sum_{j=1}^{n} x_{3j} = 1{,}550$	$\sum_{j=1}^{n} y_j = 271.3$

is the gasoline mileage (Y) (in miles per gallon). Suppose that the relationship between Y and the independent variables is given by the equation for the multiple linear regression model and that the assumptions of the multiple linear regression model are valid. Determine the numerical normal equations for the corresponding model.

Solution

The model for the relationship between the dependent variable and the three independent variables is

$$Y = \beta_0 + \beta_1 x_1 + \beta_2 x_2 + \beta_3 x_3 + e, \tag{9}$$

where Y stands for mileage and x_1, x_2, and x_3 represent horsepower rating, weight, and octane rating, respectively.

The normal equations for the model of (9) become

$$\begin{aligned}
\hat{\beta}_0 n + \hat{\beta}_1 \sum_{j=1}^{n} x_{1j} + \hat{\beta}_2 \sum_{j=1}^{n} x_{2j} + \hat{\beta}_3 \sum_{j=1}^{n} x_{3j} &= \sum_{j=1}^{n} y_j, \\
\hat{\beta}_0 \sum_{j=1}^{n} x_{1j} + \hat{\beta}_1 \sum_{j=1}^{n} x_{1j}^2 + \hat{\beta}_2 \sum_{j=1}^{n} x_{1j}x_{2j} + \hat{\beta}_3 \sum_{j=1}^{n} x_{1j}x_{3j} &= \sum_{j=1}^{n} x_{1j}y_j, \\
\hat{\beta}_0 \sum_{j=1}^{n} x_{2j} + \hat{\beta}_1 \sum_{j=1}^{n} x_{2j}x_{1j} + \hat{\beta}_2 \sum_{j=1}^{n} x_{2j}^2 + \hat{\beta}_3 \sum_{j=1}^{n} x_{2j}x_{3j} &= \sum_{j=1}^{n} x_{2j}y_j, \\
\hat{\beta}_0 \sum_{j=1}^{n} x_{3j} + \hat{\beta}_1 \sum_{j=1}^{n} x_{3j}x_{1j} + \hat{\beta}_2 \sum_{j=1}^{n} x_{3j}x_{2j} + \hat{\beta}_3 \sum_{j=1}^{n} x_{3j}^2 &= \sum_{j=1}^{n} x_{3j}y_j.
\end{aligned} \tag{10}$$

The appropriate sums of squares and sums of cross products are determined in Table 15.2. Substituting the appropriate values from Tables 15.1 and 15.2 into the equations of (10), we obtain the numerical normal equations

$$\begin{aligned}
\hat{\beta}_0(16) + \hat{\beta}_1(3{,}405) + \hat{\beta}_2(48{,}200) + \hat{\beta}_3(1{,}550) &= 271.3, \\
\hat{\beta}_0(3{,}405) + \hat{\beta}_1(767{,}475) + \hat{\beta}_2(10{,}563{,}000) + \hat{\beta}_3(337{,}600) &= 59{,}567.0, \\
\hat{\beta}_0(48{,}200) + \hat{\beta}_1(10{,}563{,}000) + \hat{\beta}_2(147{,}640{,}000) + \hat{\beta}_3(4{,}723{,}000) &= 829{,}530.0, \\
\hat{\beta}_0(1{,}550) + \hat{\beta}_1(337{,}600) + \hat{\beta}_2(4{,}723{,}000) + \hat{\beta}_3(151{,}900) &= 26{,}622.0.
\end{aligned} \tag{11}$$ □

The normal equations in Example 1, representing four equations in four unknowns, can be solved by conventional substitution methods. However, as often happens in multiple regression, the coefficients of the unknowns are exceptionally large and thus difficult to work with. Unless the coefficients of the unknowns are exceptionally simple, there is no easy way to solve four equations in four unknowns. However, although solving the normal equations is not easy, they can be solved by a fairly straightforward method—one that is generally used even when the above system of equations is solved through use of an electronic computer. But let us first consider ways of simplifying the coefficients in the normal equations.

Using the property

$$\sum_{j=1}^{n} (x_{ij} - \bar{x}_i) = \sum_{j=1}^{n} x_{ij} - n\bar{x}_i = \sum_{j=1}^{n} x_{ij} - \sum_{j=1}^{n} x_{ij} = 0,$$

Table 15.2

Computations for data of Table 15.1

	x_1^2	x_1x_2	x_1x_3	x_1y	x_2^2	x_2x_3	x_2y	x_3^2	x_3y	y^2
	32,400	504,000	16,200	2,880.0	7,840,000	252,000	44,800	8,100	1,440	256.00
	38,025	585,000	17,550	2,905.5	9,000,000	270,000	44,700	8,100	1,341	222.01
	25,600	432,000	12,800	2,256.0	7,290,000	216,000	38,070	6,400	1,128	198.81
	55,225	752,000	25,850	4,347.5	10,240,000	352,000	59,200	12,100	2,035	342.25
	30,625	472,500	17,500	2,677.5	7,290,000	270,000	41,310	10,000	1,530	234.09
	81,225	969,000	31,350	6,241.5	11,560,000	374,000	74,460	12,100	2,409	479.61
	44,100	630,000	21,000	3,192.0	9,000,000	300,000	45,600	10,000	1,520	231.04
	52,900	759,000	23,000	4,186.0	10,890,000	330,000	60,060	10,000	1,820	331.24
	55,225	822,500	23,500	3,995.0	12,250,000	350,000	59,500	10,000	1,700	289.00
	38,025	546,000	17,550	3,276.0	7,840,000	252,000	47,040	8,100	1,512	282.24
	84,100	1,015,000	31,900	5,800.0	12,250,000	385,000	70,000	12,100	2,200	400.00
	22,500	375,000	12,000	2,220.0	6,250,000	200,000	37,000	6,400	1,184	219.04
	10,000	210,000	8,000	1,300.0	4,410,000	168,000	27,300	6,400	1,040	169.00
	84,100	986,000	31,900	5,887.0	11,560,000	374,000	69,020	12,100	2,233	412.09
	65,025	867,000	25,500	4,641.0	11,560,000	340,000	61,880	10,000	1,820	331.24
	48,400	638,000	22,000	3,762.0	8,410,000	290,000	49,590	10,000	1,710	292.41
Total	767,475	10,563,000	337,600	59,567.0	147,640,000	4,723,000	829,530	151,900	26,622	4,690.07

we can simplify the normal equations. That is, we obtain a different set of values by the transformations

$$x'_{1j} = x_{1j} - \bar{x}_1; \qquad x'_{2j} = x_{2j} - \bar{x}_2;$$
$$x'_{3j} = x_{3j} - \bar{x}_3; \qquad y'_j = y_j - \bar{y}. \tag{12}$$

If we restate the multiple linear regression model in terms of the transformed variables, we have

$$Y' = \alpha_0 + \alpha_1 x'_1 + \alpha_2 x'_2 + \alpha_3 x'_3 + e, \tag{13}$$

where α_i $(i = 0, 1, 2, 3)$ are the appropriate coefficients for the transformed variables. Substituting the x'_i $(i = 1, 2, 3)$ values from (12) into (13), we can determine the relationship between the coefficients of models (13) and (9). Thus, (13) becomes

$$Y - \bar{Y} = \alpha_0 + \alpha_1(x_1 - \bar{x}_1) + \alpha_2(x_2 - \bar{x}_2) + \alpha_3(x_3 - \bar{x}_3) + e,$$

which implies

$$Y = (\alpha_0 + \bar{Y} - \alpha_1\bar{x}_1 - \alpha_2\bar{x}_2 - \alpha_3\bar{x}_3) + \alpha_1 x_1 + \alpha_2 x_2 + \alpha_3 x_3 + e. \tag{14}$$

Equating the equivalent models given by (9) and (14), we obtain the relationships between the coefficients of the two models. That is, in order for them to be linear models of the same dependent variable as a function of the same independent variables, the coefficients of like terms must be equal. Setting the coefficients of like terms equal, we obtain

$$\beta_1 = \alpha_1, \quad \beta_2 = \alpha_2, \quad \beta_3 = \alpha_3, \tag{15}$$

and, consequently,

$$\beta_0 = \alpha_0 + \bar{Y} - \beta_1\bar{x}_1 - \beta_2\bar{x}_2 - \beta_3\bar{x}_3. \tag{16}$$

When we substitute the x'_{ij} values obtained from (12) and the observed values of Y' into the equations of (10) and use the property $\sum_{j=1}^{n} (x_{ij} - \bar{x}_i) = 0$, the normal equations for (13) become

$$\begin{aligned}
\hat{\alpha}_0(n) + \hat{\alpha}_1(0) + \hat{\alpha}_2(0) + \hat{\alpha}_3(0) &= 0,\\
\hat{\alpha}_0(0) + \hat{\alpha}_1 \sum_{j=1}^{n} (x'_{1j})^2 + \hat{\alpha}_2 \sum_{j=1}^{n} x'_{1j}x'_{2j} + \hat{\alpha}_3 \sum_{j=1}^{n} x'_{1j}x'_{3j} &= \sum_{j=1}^{n} x'_{1j}y'_j,\\
\hat{\alpha}_0(0) + \hat{\alpha}_1 \sum_{j=1}^{n} x'_{2j}x'_{1j} + \hat{\alpha}_2 \sum_{j=1}^{n} (x'_{2j})^2 + \hat{\alpha}_3 \sum_{j=1}^{n} x'_{2j}x'_{3j} &= \sum_{j=1}^{n} x'_{2j}y'_j,\\
\hat{\alpha}_0(0) + \hat{\alpha}_1 \sum_{j=1}^{n} x'_{3j}x'_{1j} + \hat{\alpha}_2 \sum_{j=1}^{n} x'_{3j}x'_{2j} + \hat{\alpha}_3 \sum_{j=1}^{n} (x'_{3j})^2 &= \sum_{j=1}^{n} x'_{3j}y'_j.
\end{aligned} \tag{17}$$

We note from the first row that $\hat{\alpha}_0 = 0$. Further, the remaining equations represent only three equations in three unknowns. Moreover, from (15), the estimates of α_1, α_2, and α_3 are also estimates of β_1, β_2, and β_3, respectively. Thus, replacing the $\hat{\alpha}_i$ values in (17) by their corresponding $\hat{\beta}_i$ values, stating the transformed variables in terms of the original independent variables, and letting

$$s_{im} = \sum_{j=1}^{n} (x_{ij} - \bar{x}_i)(x_{mj} - \bar{x}_m), \qquad \text{for } (i = 1, 2, 3;\ m = 1, 2, 3),$$

we obtain the following simplified normal equations

$$\begin{aligned}
\hat{\beta}_1 s_{11} + \hat{\beta}_2 s_{12} + \hat{\beta}_3 s_{13} &= \sum_{j=1}^{n} (x_{1j} - \bar{x}_1)(y_j - \bar{y}), \\
\hat{\beta}_1 s_{21} + \hat{\beta}_2 s_{22} + \hat{\beta}_3 s_{23} &= \sum_{j=1}^{n} (x_{2j} - \bar{x}_2)(y_j - \bar{y}), \qquad (18) \\
\hat{\beta}_1 s_{31} + \hat{\beta}_2 s_{32} + \hat{\beta}_3 s_{33} &= \sum_{j=1}^{n} (x_{3j} - \bar{x}_3)(y_j - \bar{y}).
\end{aligned}$$

Having the estimates $\hat{\beta}_1$, $\hat{\beta}_2$, and $\hat{\beta}_3$, we can determine the estimate of $\hat{\beta}_0$ from (16). Since $\hat{\alpha}_0 = 0$, we have

$$\hat{\beta}_0 = \bar{y} - \hat{\beta}_1 \bar{x}_1 - \hat{\beta}_2 \bar{x}_2 - \hat{\beta}_3 \bar{x}_3. \qquad (19)$$

Using the same transformation on the k variables of the multiple linear regression model and letting

$$s_{im} = \sum_{j=1}^{n} (x_{ij} - \bar{x}_i)(x_{mj} - \bar{x}_m), \qquad \text{for } (i = 1, 2, \ldots, k; m = 1, 2, \ldots, k),$$

we obtain a general abbreviated set of normal equations:

$$\begin{aligned}
\hat{\beta}_1 s_{11} + \hat{\beta}_2 s_{12} + \cdots + \hat{\beta}_k s_{1k} &= \sum_{j=1}^{n} (x_{1j} - \bar{x}_1)(y_j - \bar{y}), \\
\hat{\beta}_1 s_{21} + \hat{\beta}_2 s_{22} + \cdots + \hat{\beta}_k s_{2k} &= \sum_{j=1}^{n} (x_{2j} - \bar{x}_2)(y_j - \bar{y}), \qquad (20) \\
\vdots \qquad\qquad \vdots \qquad\qquad & \quad \vdots \qquad\qquad\qquad \vdots \\
\hat{\beta}_1 s_{k1} + \hat{\beta}_2 s_{k2} + \cdots + \hat{\beta}_k s_{kk} &= \sum_{j=1}^{n} (x_{kj} - \bar{x}_k)(y_j - \bar{y}).
\end{aligned}$$

In obtaining the abbreviated set, we have transformed the system of simultaneous equations from $(k + 1)$ equations in $(k + 1)$ unknowns to an analogous but simpler system of equations. Thus, we must solve simultaneously only k equations in k unknowns and then substitute the solutions into the $(k + 1)$th equation to determine the $(k + 1)$th estimate. However, reducing the number of equations is a relatively small benefit from the procedure involving the transformations. More important is the fact that the numerical values of the corrected (adjusted for the sample means) sums of squares and cross products used in the abbreviated set of equations will be considerably smaller than the values for the uncorrected terms used in the original normal equations. Together, the two changes (fewer unknowns for which to solve and smaller values for the coefficients of the unknowns in the normal equations) will usually reduce considerably the work involved in solving the system of equations for the estimates of the unknown coefficients. Let us compare the normal equations obtained by using these changes with the original normal equations of Example 1.

Example 2

From the data of Example 1, determine the abbreviated normal equations given by (18) and the equation for estimating $\hat{\beta}_0$ through use of (19).

Solution

Using the computational formula for the corrected sum of cross products,

$$s_{im} = \sum_{j=1}^{n} (x_{ij} - \bar{x}_i)(x_{mj} - \bar{x}_m) = \sum_{j=1}^{n} x_{ij}x_{mj} - \frac{(\sum_{j=1}^{n} x_{ij})(\sum_{j=1}^{n} x_{mj})}{n},$$

and the data from Tables 15.1 and 15.2, we obtain the corrected sums of squares and cross products. (Note that $s_{im} = s_{mi}$; consequently, we need compute only one of the two terms.) That is,

$$s_{11} = \sum_{j=1}^{n} (x_{1j} - \bar{x}_1)^2 = \sum_{j=1}^{n} x_{1j}^2 - \frac{(\sum_{j=1}^{n} x_{1j})^2}{n}$$

$$= 767{,}475 - \frac{(3{,}405)^2}{16} = 42{,}848;$$

$$s_{12} = \sum_{j=1}^{n} (x_{1j} - \bar{x}_1)(x_{2j} - \bar{x}_2) = \sum_{j=1}^{n} x_{1j}x_{2j} - \frac{(\sum_{j=1}^{n} x_{1j})(\sum_{j=1}^{n} x_{2j})}{n}$$

$$= 10{,}563{,}000 - \frac{(3{,}405)(48{,}200)}{16} = 305{,}438;$$

$$s_{13} = \sum_{j=1}^{n} (x_{1j} - \bar{x}_1)(x_{3j} - \bar{x}_3) = \sum_{j=1}^{n} x_{1j}x_{3j} - \frac{(\sum_{j=1}^{n} x_{1j})(\sum_{j=1}^{n} x_{3j})}{n}$$

$$= 337{,}600 - \frac{(3{,}405)(1{,}550)}{16} = 7{,}741;$$

$$\sum_{j=1}^{n} (x_{1j} - \bar{x}_1)(y_j - \bar{y}) = \sum_{j=1}^{n} x_{1j}y_j - \frac{(\sum_{j=1}^{n} x_{1j})(\sum_{j=1}^{n} y_j)}{n}$$

$$= 59{,}567 - \frac{(3{,}405)(271.3)}{16} = 1{,}831;$$

$$s_{22} = \sum_{j=1}^{n} (x_{2j} - \bar{x}_2)^2 = \sum_{j=1}^{n} x_{2j}^2 - \frac{(\sum_{j=1}^{n} x_{2j})^2}{n}$$

$$= 147{,}640{,}000 - \frac{(48{,}200)^2}{16} = 2{,}437{,}500;$$

$$s_{23} = \sum_{j=1}^{n} (x_{2j} - \bar{x}_2)(x_{3j} - \bar{x}_3) = \sum_{j=1}^{n} x_{2j}x_{3j} - \frac{(\sum_{j=1}^{n} x_{2j})(\sum_{j=1}^{n} x_{3j})}{n}$$

$$= 4{,}723{,}000 - \frac{(48{,}200)(1{,}550)}{16} = 53{,}625;$$

$$\sum_{j=1}^{n} (x_{2j} - \bar{x}_2)(y_j - \bar{y}) = \sum_{j=1}^{n} x_{2j}y_j - \frac{(\sum_{j=1}^{n} x_{2j})(\sum_{j=1}^{n} y_j)}{n}$$

$$= 829{,}530 - \frac{(48{,}200)(271.3)}{16} = 12{,}239;$$

$$s_{33} = \sum_{j=1}^{n} (x_{3j} - \bar{x}_3)^2 = \sum_{j=1}^{n} x_{3j}^2 - \frac{(\sum_{j=1}^{n} x_{3j})^2}{n}$$

$$= 151{,}900 - \frac{(1{,}550)^2}{16} = 1{,}744;$$

and

$$\sum_{j=1}^{n} (x_{3j} - \bar{x}_3)(y_j - \bar{y}) = \sum_{j=1}^{n} x_{3j}y_j - \frac{(\sum_{j=1}^{n} x_{3j})(\sum_{j=1}^{n} y_j)}{n}$$

$$= 26{,}622 - \frac{(1{,}550)(271.3)}{16} = 340.$$

Substituting the values into their respective positions in (18), we obtain the following normal equations

$$\begin{aligned} 42{,}848\hat{\beta}_1 + 305{,}438\hat{\beta}_2 + 7{,}741\hat{\beta}_3 &= 1{,}831, \\ 305{,}438\hat{\beta}_1 + 2{,}437{,}500\hat{\beta}_2 + 53{,}625\hat{\beta}_3 &= 12{,}239, \\ 7{,}741\hat{\beta}_1 + 53{,}625\hat{\beta}_2 + 1{,}744\hat{\beta}_3 &= 340. \end{aligned} \tag{21}$$

From (19), we determine the equation for estimating β_0. That is,

$$\hat{\beta}_0 = \bar{y} - \hat{\beta}_1\bar{x}_1 - \hat{\beta}_2\bar{x}_2 - \hat{\beta}_3\bar{x}_3 = \frac{271.3}{16} - \hat{\beta}_1\frac{3{,}405}{16} - \hat{\beta}_2\frac{48{,}200}{16} - \hat{\beta}_3\frac{1{,}550}{16}$$

$$= 16.96 - 212.8\hat{\beta}_1 - 3{,}012.5\hat{\beta}_2 - 96.9\hat{\beta}_3. \quad \square$$

Comparing the abbreviated normal equations of (21) with the normal equations of (11), we note that most coefficients of $\hat{\beta}_1$, $\hat{\beta}_2$, and $\hat{\beta}_3$ have been reduced by several orders of magnitude.

From both a conceptual and a numerical standpoint, a good method for solving the normal equations is the abbreviated Doolittle method. This method, once digested, allows an easy notational base for presenting concepts associated with testing hypotheses and constructing confidence intervals in multiple linear regression. However, because of the computational manipulations posed by the method and the fact that most problems involving multiple linear regression analysis will be done on a computer, we relegate discussion of the abbreviated Doolittle method to Supplement 15.A. Nevertheless, we will need the results of either the abbreviated Doolittle method or a computer analysis for multiple linear regression analysis in order to consider confidence intervals and tests of hypotheses. Even if the results from a computer analysis are to be used, you will benefit from reading Supplement 15.A without being overly concerned about the computational manipulations.

From Supplement 15.A, we determine the estimates of the parameters in the multiple regression model. Thus, we have

$$\hat{\beta}_0 = 11.66 \qquad \hat{\beta}_1 = 0.06212$$
$$\hat{\beta}_2 = -0.003047 \qquad \hat{\beta}_3 = 0.01293.$$

Consequently, the regression equation is given by

$$\hat{\mu}_{y|x_1,x_2,x_3} = 11.66 + 0.06212x_1 - 0.003047x_2 + 0.01293x_3. \tag{22}$$

Problems

The problems of Sections 15.4, 15.5, and 15.6 require the data and results of the two problems in this section, therefore, you should retain your solutions for later use.

1. The president of a large bank believes that the weekly demand for business loans from his bank is a function of the prime interest rate and the index of regional business activity. In order to verify this relationship, the following data are obtained from bank records for twelve weeks during the past five years.

Prime interest rate (x_1) *(percent)*	*Index of regional business activity* (x_2) *(percent)*	*Weekly demand for business loans* (y) *(millions)*
6.00	90	9.0
6.25	95	9.5
6.00	98	10.2
6.50	100	9.8
6.75	104	10.1
7.00	101	9.5
7.00	104	9.7
7.25	106	9.5
7.75	105	9.0
8.00	102	8.4
8.00	98	8.0
7.50	101	9.0

a. Determine the abbreviated numerical normal equations for the corresponding model.

b. Determine the mean of the values for each variable.

2. The vice-president of marketing for a cosmetic manufacturer believes that there is a relationship between the firm's monthly sales revenue and the amount of advertising expenditures. He further believes that the expenditures for three different categories—television, radio, and newspaper—are somewhat differently related to sales revenue. Thus, he wants to determine the relationship between sales revenue and the three categories. The following data come from the last fifteen periods for which data are available.

Advertising expenditures (hundreds of dollars)			*Sales revenue (ten-thousands of dollars)*
Television (x_1)	*Radio* (x_2)	*Newspaper* (x_3)	(y)
6.4	1.2	1.0	9.3
7.0	1.4	1.1	10.4
6.8	1.6	1.3	9.8
5.7	1.9	0.8	8.6
8.2	1.7	2.0	10.7
7.1	1.4	2.3	10.3
6.3	1.8	1.4	10.1
8.4	2.3	1.9	12.2
9.1	2.2	1.4	11.8
8.5	1.7	2.1	10.9
7.7	1.4	1.5	10.3
6.9	1.9	0.9	9.8
6.3	2.0	1.4	9.9
7.2	2.3	1.1	11.2
7.4	2.4	1.0	12.0

a. Determine the numerical normal equations for the appropriate model.

b. Determine the mean of the values for each variable.

15.4 *Tests of Hypotheses Concerning the Regression Coefficients*

For the multiple linear regression model,

$$Y = \beta_0 + \sum_{i=1}^{k} \beta_i x_i + e,$$

we have considered both the method of obtaining formulas for estimating β_i ($i = 0, 1, \ldots, k$) and the technique for obtaining numerical values for these estimators. We now turn to the problem of testing hypotheses involving regression coefficients. First, however, we must know the sampling distributions for relevant estimators. Thus, we state without proof that, under the assumptions associated with the multiple linear regression model, p. 561, the variable (statistic) for which $\hat{\beta}_i$ ($i = 1, 2, \ldots, k$) represents an observed value has a sampling distribution that is normal with mean β_i and variance $c_{ii}\sigma_e^2$, where c_{ii} is the entry in the ith row and jth column of the matrix **C**, which is generally given as part of the computer printout for most multiple regression programs. It is usually called the "the inverse of the X prime X matrix" or "the inverse of the corrected-sums-of-square matrix." An explanation of the meaning of **C** and its determination by the abbreviated Doolittle method is given in detail in Supplement 15.A. Throughout the remainder of this chapter, we shall use the results from Supplement 15.A to determine the values associated with **C** for the data of Example 1, Section 15.3.

Generally, we are interested in determining whether there is any linear relationship between the group of independent variables and the dependent variable. If we conclude that there is none, we can simply estimate the means of the populations of Y values by one value that does not depend upon the values of the independent variables. That is, we can simply estimate each of the population means by $\bar{y}$.

If we conclude that there is a linear relationship between the group of independent variables and the dependent variable, we shall want to determine which of the independent variables are linearly related to the dependent variable. That is, given a significant linear relationship between the group of independent variables and the dependent variable, we want to determine which of the independent variables is contributing significantly to the linear relationship.

To determine if there is a linear relationship between the group of independent variables and the dependent variable (that is, regression is significant), we shall again resort to the analysis-of-variance technique of Chapter 13. The null hypothesis for the over-all test for a linear relationship is given by

$$H_0: \beta_i = 0 \qquad (i = 1, 2, \ldots, k).$$

In other words, we begin with the null hypothesis that all coefficients are equal to zero. The alternative hypothesis is given by

$$H_1: \beta_i \neq 0 \qquad (\text{for at least one } i;\ i = 1, 2, \ldots, k).$$

Table 15.3 presents the form of the analysis-of-variance table for testing the significance of regression.

The following partitioning of the total sum of squares is very useful for constructing the analysis-of-variance table for testing the significance of regression:

$$\sum_{j=1}^{n}(y_j - \bar{y})^2 = \sum_{i=1}^{k}\hat{\beta}_i\left[\sum_{j=1}^{n}(x_{ij} - \bar{x}_i)(y_j - \bar{y})\right] + \sum_{j=1}^{n}(y_j - \hat{\mu}_{y|x_{1j}, x_{2j}, \ldots, x_{kj}})^2. \quad (1)$$

You are encouraged to verify (1) by proper algebraic manipulation. In terms of the sums-of-squares notation, (1) becomes the identity

$$SS_{\text{total}} = SS_{\text{regrssion}} + SS_{\text{error}}. \quad (2)$$

Even though (2) is a valid identity and, thus, can be used to determine any one of the sum-of-squares terms when the remaining two are known, it is again strongly recommended that this identity be used only as a check on the computations involved in obtaining the three terms. However, when the calculation of SS_{error} is very tedious and time consuming, we may determine SS_{total} and $SS_{\text{regression}}$ by straightforward computations and use (2) to compute SS_{error}. In so doing, we must bear in mind that we have no means of checking our computations.

Table 15.3

Analysis-of-variance table for testing the significance of regression

Source	*Degrees of freedom*	*Sum of squares*	*Mean square*	*F ratio*
Regression on k variables	k	$SS_{\text{regression}} = \sum_{i=1}^{k}\hat{\beta}_i\left[\sum_{j=1}^{n}(x_{ij} - \bar{x}_i)(y_j - \bar{y})\right]$	$MS_{\text{regression}} = \frac{SS_{\text{regression}}}{k}$	$\frac{MS_{\text{regression}}}{MS_{\text{error}}}$
Error	$n - k - 1$	$SS_{\text{error}} = \sum_{j=1}^{n}(y_j - \hat{\mu}_{y\|x_{1j}, x_{2j}, \ldots, x_{kj}})^2$	$MS_{\text{error}} = \frac{SS_{\text{error}}}{n - k - 1}$	
Total	$n - 1$	$SS_{\text{total}} = \sum_{j=1}^{n}(y_j - \bar{y})^2$		

In Table 15.3, the term $SS_{\text{regression}}$ represents the observed value of a variable which, when divided by σ_e^2, has a chi-square distribution with k degrees of freedom, if and only if $\beta_i = 0$ for each $i = 1, 2, \ldots, k$. Similarly, the term SS_{error} represents the observed value of a variable which, when divided by σ_e^2, has a chi-square distribution with $(n - k - 1)$ degrees of freedom. Of course, the assumptions of Section 15.2, p. 561, must be valid for the terms to have their respective distributions. Moreover, the variables associated with $SS_{\text{regression}}$ and SS_{error} are statistically independent.

The importance of these statements is their connection with the testing of a hypothesis about the iinear relationship between the group of independent variables and the dependent variable. Using this information, we can immediately obtain a pivotal quantity to use in testing the null hypothesis

$$H_0:\beta_i = 0 \qquad (i = 1, 2, \ldots, k)$$

versus the alternative hypothesis

$$H_1:\beta_i \neq 0 \qquad \text{(for at least one } i;\ i = 1, 2, \ldots, k). \tag{3}$$

That is, since we have the observed values for two independent chi-square variables, we have the ratio

$$\frac{MS_{\text{regression}}/\sigma_e^2}{MS_{\text{error}}/\sigma_e^2} = \frac{MS_{\text{regression}}}{MS_{\text{error}}}, \tag{4}$$

which represents the observed value of a variable having an F distribution with k and $(n - k - 1)$ degrees of freedom, if and only if H_0 is true. Thus, the statistic represented by the single observation of (4) is a pivotal quantity for testing the hypotheses of (3). Let us demonstrate this by an example.

Example 1

Using the data of Table 15.1 and the associated computations, determine if a significant linear relationship exists between the three independent variables and the dependent variable. Use a 5 percent significance level.

Solution

We wish to test the hypotheses

$$H_0:\beta_i = 0 \quad (i = 1, 2, 3) \quad \text{versus} \quad H_1:\beta_i \neq 0 \quad \text{(for at least one } i;\ i = 1, 2, 3). \tag{5}$$

Since $n = 16$, $k = 3$, and $\alpha = 0.05$, we determine the critical value from Table A-5 to be $F_{0.05, 3, 12} = 3.49$. Thus, our decision rule is to accept H_0 if the F ratio of (4) is less than 3.49. Otherwise, we reject H_0.

Column 2 of Table 15.4 gives the estimates $\hat{\mu}_{y|x_1, x_2, x_3}$ (from Equation (13), Supplement 15.A) of the means of the Y populations accurate to two decimal places for the corresponding values of x_1, x_2, x_3 given initially in Table 15.1. Column 3 gives the error (or deviation) of the estimate from the observed value. Column 4 gives the squares of the error terms. The totals at the bottom of columns 3 and 4 are the sum of the errors and the sum of squares of the errors, respectively. Although the sum of errors should theoretically equal zero, we find the value 0.03 because of rounding errors. The sum of squares of the errors (9.0047) is actually the computation of the SS_{error} term of Table 15.3.

Table 15.5 gives the analysis-of-variance values (outlined in Table 15.3) for testing the hypotheses of (5). The entries in the sum-of-squares column in this table are obtained as follows.

The $SS_{\text{regression}}$ term is generally given as part of the printout for most multiple regression computer programs. However, from Equation (14) of Supplement 15.A,

Table 15.4

Computations for determining SS_{error} for data in Table 15.1

Gasoline mileage y	*Estimate* $\hat{\mu}_{y\mid x_1, x_2, x_3}$	*Deviation* $y - \hat{\mu}_{y\mid x_1, x_2, x_3}$	$(y - \hat{\mu}_{y\mid x_1, x_2, x_3})^2$
16.0	15.47	0.53	0.2809
14.9	15.80	−0.90	0.8100
14.1	14.41	−0.31	0.0961
18.5	17.93	0.57	0.3249
15.3	15.60	−0.30	0.0900
21.9	20.43	1.47	2.1609
15.2	16.86	−1.66	2.7556
18.2	17.19	1.01	1.0201
17.0	16.89	0.11	0.0121
16.8	16.41	0.39	0.1521
20.0	20.43	−0.43	0.1849
14.8	14.39	0.41	0.1681
13.0	12.51	0.49	0.2401
20.3	20.74	−0.44	0.1936
18.2	18.43	−0.23	0.0529
17.1	17.78	−0.68	0.4624
Total		0.03	9.0047

we computed this term using the abbreviated Doolittle method. Thus,

$$SS_{\text{regression}} = \sum_{i=1}^{k} \hat{\beta}_i \left[\sum_{j=1}^{n} (x_{ij} - \bar{x}_i)(y_j - \bar{y}) \right]$$
$$= (0.06212)(1{,}831) + (-0.003047)(12{,}239) + (0.01293)(340)$$
$$= 80.8457.$$

From Table 15.4, we obtain

$$SS_{\text{error}} = \sum_{j=1}^{n} (y_j - \hat{\mu}_{y\mid x_{1j}, x_{2j}, x_{3j}})^2$$
$$= 9.0047.$$

The SS_{total} term is computed in the usual manner from the equation

$$SS_{\text{total}} = \sum_{j=1}^{n} (y_j - \bar{y})^2 = \sum_{j=1}^{n} y_j^2 - \frac{(\sum_{j=1}^{n} y_j)^2}{n}.$$

Table 15.5

Analysis-of-variance table for data of Table 15.1

Source	*Degrees of freedom*	*Sum of squares*	*Mean square*	*F ratio*
Regression	3	80.8457	26.9486	35.91
Error	12	9.0047	0.7504	
Total	15	89.8394		

Therefore, from the data of Table 15.1, we have

$$SS_{\text{total}} = \sum_{j=1}^{n} y_j^2 - \frac{(\sum_{j=1}^{n} y_j)^2}{n} = 4{,}690.0700 - \frac{(271.3)^2}{16} = 89.8394.$$

Using the identity of (2) to check our computations, we note that

$$SS_{\text{regression}} + SS_{\text{error}} = 80.8457 + 9.0047 = 89.8504.$$

We have calculated SS_{total} to be 89.8394, a value that is 0.0110 less than $SS_{\text{regression}} + SS_{\text{error}}$. This difference can be attributed to both the rounding in computing the estimates of the regression coefficients and the rounding of the estimates in Table 15.4 to two decimal places.

From the sum-of-squares terms and the computations illustrated in Table 15.3, we obtain the mean square and F ratio given in Table 15.5. We note that the F ratio, the observed value of the pivotal quantity given in (4), is 35.91. Since this is greater than the critical value $F_{0.05,3,12} = 3.49$, we reject H_0. Because of the nature of the alternative hypothesis, we conclude that there is a significant linear relationship between the dependent variable and at least one of the independent variables. □

Example 1 illustrates a well-defined procedure for testing a hypothesis about the linear relationship between a group of independent variables and a dependent variable. The essential steps of the method for making a 100α percent significance test for a linear relationship are as follows:

1. Set up the null hypothesis

$$H_0: \beta_i = 0 \qquad (i = 1, 2, \ldots, k)$$

versus the alternative hypothesis (6)

$$H_1: \beta_i \neq 0 \quad \text{(for at least one } i;\ i = 1, 2, \ldots, k).$$

2. Specify the level of significance α.
3. Determine (or note) the number (k) of independent variables and the number (n) of observations to be taken from the populations of Y values. [This fixes the numbers k and $(n - k - 1)$ of degrees of freedom to be used in establishing the appropriate critical value.]
4. Determine the critical value $F_{\alpha,k,n-k-1}$ (where k and n are appropriately determined from step 3 and α is determined from step 2).
5. From a sample of size n obtain [under the assumption that the null hypothesis is true, or $\beta_i = 0$ $(i = 1, 2, \ldots, k)$] one observed value of the statistic of (4). That is, perform the analysis of variance by carrying out the calculations as illustrated in Table 15.3 to obtain

$$F = \frac{MS_{\text{regression}}}{MS_{\text{error}}}. \tag{7}$$

This value represents an observed value of an F variable with k and $(n - k - 1)$ degrees of freedom, if and only if H_0 is true.

6. If the value of (7) is in the critical region, reject H_0. Otherwise, accept H_0.

If the test indicates that regression is not significant (that is, if we accept H_0), we conclude that there is no linear relationship between the group of independent mathematical variables and the dependent random variable. However, if the result indicates that regression is significant (that is, if we reject H_0), we can conclude only that there is at least one regression coefficient that is significantly different from zero. From this test, we are unable to determine which coefficients are significantly different from zero. Fortunately, another test exists for this purpose and we consider its development.

In our remarks at the beginning of the section, we stated without proof that the variable (statistic) for which $\hat{\beta}_i$ ($i = 1, 2, \ldots, k$) represents an observed value has a sampling distribution that is normal with mean β_i and variance $c_{ii}\sigma_e^2$, where c_{ii} is the entry in the ith row and ith column of the matrix **C** from Supplement 15.A. The real importance of this is that it permits us to determine a pivotal quantity and its sampling distribution. This pivotal quantity can be used to test, for a given value of i, the hypothesis

$$H_0\!:\beta_i = (\beta_i)_0, \tag{8}$$

where $(\beta_i)_0$ represents any value β_i can assume. [Note that the hypothesis we want to test as a result of rejecting H_0 of (6) is $H_0\!:\beta_i = 0$. However, because we sometimes want to test hypotheses concerning values of β_i other than zero, we shall determine the pivotal quantity for these tests. Then, when we want to test $H_0\!: \beta_i = 0$, we can simply substitute in place of $(\beta_i)_0$ in (8).]

If we determine a pivotal quantity for a general value of i, we can use this pivotal quantity for the different values of i ($i = 1, 2, \ldots, k$) to test each hypothesis of the type given in (8). Of course, the values used in this pivotal quantity must be for the corresponding value of i.

Applying the usual transformation for converting a normal variable (in this instance, the variable for which $\hat{\beta}_i$ represents an observed value) to a standard normal variable (mean zero and variance one), we obtain an observed value of the standard normal variable Z,

$$z = \frac{\hat{\beta}_i - \beta_i}{\sigma_e\sqrt{c_{ii}}}. \tag{9}$$

However, because σ_e^2 is unknown, we cannot use (9) as a pivotal quantity for testing any hypothesis concerning β_i. Thus, we must search further for an appropriate pivotal quantity.

The MS_{error} term in Table 15.3 represents an unbiased estimate ($\hat{\sigma}_e^2$) of σ_e^2. That is,

$$\hat{\sigma}_e^2 \doteq MS_{\text{error}}. \tag{10}$$

It also represents, when divided by σ_e^2, an observed value of a chi-square variable with $(n - k - 1)$ degrees of freedom which has been divided by its degrees of freedom. Furthermore, it can be shown that the two variables for which $\hat{\beta}_i$ and $\hat{\sigma}_e^2$ represent the observed values are statistically independent. Thus, recalling the definition of a t distribution (Definition 11.1), we can construct a conceptual observed value of a t statistic by taking the ratio of an observed value of the appropriate Z variable to an observed value of the square root of an appropriate chi-square statistic divided by its degrees of freedom. That is, we can construct the following observed value

$$\frac{\dfrac{\hat{\beta}_i - \beta_i}{\sigma_e\sqrt{c_{ii}}}}{\sqrt{\dfrac{MS_{\text{error}}}{\sigma_e^2}}}$$

which, after proper factoring and canceling, reduces to

$$\frac{\hat{\beta}_i - \beta_i}{\hat{\sigma}_e\sqrt{c_{ii}}}. \tag{11}$$

The observed value given in (11) is that for a t variable with $(n - k - 1)$ degrees of freedom, and it can be determined completely from a set of data and a specified value of β_i.

To test the null hypothesis of (8), using the pivotal quantity of (11), we obtain

$$t = \frac{\hat{\beta}_i - (\beta_i)_0}{\hat{\sigma}_e\sqrt{c_{ii}}}. \tag{12}$$

Note carefully that (12) represents an observed value of a t variable with $(n - k - 1)$ degrees of freedom, if and only if $\beta_i = (\beta_i)_0$—that is, if and only if we substitute the true value of β_i into (12).

Because we are usually more interested in determining whether the coefficient is significantly different from some specific value rather than whether it is greater than or less than such value, we usually test

$$H_0\!:\beta_i = (\beta_i)_0 \quad \text{versus the two-sided alternative} \quad H_1\!:\beta_i \neq (\beta_i)_0.$$

Thus, we shall set forth the detailed steps for this case only. However, these steps can be modified to permit making a one-tail test in exactly the same manner as making a one-tail test concerning the mean of a normal population with unknown variance or a one-tail test concerning the slope of the simple linear regression line of Chapter 14. Note that the steps in the test of these general hypotheses exactly parallel those presented in Chapter 14 for testing hypotheses concerning the slope of the simple linear regression equation. Thus, the essential steps for making a two-tail, 100α percent significance test concerning β_i (for a specific value of i), when the assumptions of Section 15.2, p. 561, are valid, are as follows.

1. Set up the null and alternative hypotheses

$$H_0\!:\beta_i = (\beta_i)_0 \quad \text{versus} \quad H_1\!:\beta_i \neq (\beta_i)_0, \tag{13}$$

   where $(\beta_i)_0$ is any value β_i can assume. (As indicated previously, the most commonly tested null hypothesis of this type is $H_0\!:\beta_i = 0$.)
2. Specify the level of significance α.
3. Determine (or note) the size (n) of the sample to be taken and the number (k) of independent variables present. (This fixes the number $(n - k - 1)$ of degrees of freedom to be used in establishing the appropriate critical values.)
4. Determine the critical values $-t_{\alpha/2,n-k-1}$ and $t_{\alpha/2,n-k-1}$ that divide the t distribution with $(n - k - 1)$ degrees of freedom into two critical (rejection) regions, each of area $\alpha/2$, and an acceptance region of area $(1 - \alpha)$.

5. From a sample of size n obtain [under the assumption that H_0 is true, or $\beta_i = (\beta_i)_0$] one observed value of the t statistic given in (12). That is, compute

$$t = \frac{\hat{\beta}_i - (\beta_i)_0}{\hat{\sigma}_e\sqrt{c_{ii}}}, \tag{14}$$

where $\hat{\beta}_i$ and c_{ii} are computed by either a multiple regression computer program or the abbreviated Doolittle method and $\hat{\sigma}_e$ is computed from (10) and Table 15.3.
6. If the observed value given in (14) is in the critical region, reject H_0. Otherwise, accept H_0.

Example 2

Using the data of Table 15.1 and the associated computations, determine which of the coefficients (β_1, β_2, and β_3) of the x variables in the model of (9, Sec. 15.3),

$$Y = \beta_0 + \beta_1 x_1 + \beta_2 x_2 + \beta_3 x_3 + e,$$

are significantly different from zero. Use a 5 percent significance level.

Solution

The individual hypotheses we want to test are

$$H_0{:}\beta_1 = 0 \quad \text{versus} \quad H_1{:}\beta_1 \neq 0; \tag{15}$$

$$H_0{:}\beta_2 = 0 \quad \text{versus} \quad H_1{:}\beta_2 \neq 0; \tag{16}$$

and

$$H_0{:}\beta_3 = 0 \quad \text{versus} \quad H_1{:}\beta_3 \neq 0. \tag{17}$$

From Table 15.1, we determine $n = 16$ and $k = 3$. Thus, for $\alpha = 0.05$, we determine from Table A-4 the critical values

$$-t_{\alpha/2,n-k-1} = -t_{0.025,12} = -2.179$$

and

$$t_{\alpha/2,n-k-1} = t_{0.025,12} = 2.179.$$

To determine the observed t values given in (14), we must first determine the appropriate values of $\hat{\beta}_i$ and c_{ii} (for $i = 1, 2, 3$) and, also, $\hat{\sigma}_e$. The values of $\hat{\beta}_i$ ($i = 1, 2, 3$) are given in (12) of Supplement 15.A:

$$\hat{\beta}_1 = 0.06212, \quad \hat{\beta}_2 = -0.003047, \quad \text{and} \quad \hat{\beta}_3 = 0.01293.$$

The values for c_{ii} ($i = 1, 2, 3$) are given in (16) of Supplement 15.A:

$$c_{11} = 0.000366915,$$
$$c_{22} = 0.000003949319,$$
$$c_{33} = 0.00297449.$$

Finally, the value for $\hat{\sigma}_e$ is determined from Table 15.5 and Relation (10):

$$\hat{\sigma}_e = \sqrt{MS_{\text{error}}} = \sqrt{0.7504} = 0.87.$$

Substituting these values into Equation (14), we obtain the observed values of the pivotal quantities for testing the hypotheses of (15) through (17):

$$t = \frac{\hat{\beta}_1 - (\beta_1)_0}{\hat{\sigma}_e\sqrt{c_{11}}} = \frac{0.06212 - 0}{(0.87)\sqrt{0.000366915}} = 3.728, \tag{18}$$

$$t = \frac{\hat{\beta}_2 - (\beta_2)_0}{\hat{\sigma}_e\sqrt{c_{22}}} = \frac{-0.003047 - 0}{(0.87)\sqrt{0.000003949319}} = -1.762, \tag{19}$$

$$t = \frac{\hat{\beta}_3 - (\beta_3)_0}{\hat{\sigma}_e\sqrt{c_{33}}} = \frac{0.01293 - 0}{(0.87)\sqrt{0.00297449}} = 0.273. \tag{20}$$

Comparing these observed values with the critical values (-2.179 and 2.179) and noting that the observed value of (18) is in the critical region while the observed values of (19) and (20) are in the acceptance region, we reject $H_0:\beta_1 = 0$, and accept H_0: $\beta_2 = 0$ and $H_0:\beta_3 = 0$. □

In Example 2, where we considered model (9, Sec. 15.3) as the correct model, we determined (at the 5 percent significance level and on an individual basis) that only $\beta_1 \neq 0$. Thus, from the conclusions in this example, we have sufficient evidence to indicate that (9, Sec. 15.3) is not the correct model. In other words, since we have determined that β_2 and β_3 are equal to zero, we conclude that either x_2 or x_3 (or both) should not appear in the model. Even so, we cannot immediately eliminate both variables from our model, because the elimination of either x_2 or x_3 (or both) presents a new model for consideration. Thus, based upon the results of Example 2, we have three new model for consideration, namely,

$$Y = \beta_0 + \beta_1 x_1 + \beta_2 x_2 + e, \tag{21}$$

$$Y = \beta_0 + \beta_1 x_1 + \beta_3 x_3 + e, \tag{22}$$

or

$$Y = \beta_0 + \beta_1 x_1 + e. \tag{23}$$

Of course, the assumptions for each of the models are similar to those given in Section 15.2, p. 561.

To determine which of the models [(21), (22), or (23)] is correct, it is necessary to repeat the complete process of estimation and testing of hypotheses. Since the consideration of this "model hunting" is beyond the scope of this text, we shall pursue the problem no further. For further information see Geary and Leser (15.2) and Chapter 6 of Draper and Smith (15.3).

Problems

The problems of Sections 15.4, 15.5, and 15.6 require the results of the two problems of Supplement 15.A. If you have not solved the problems of Supplement 15.A, you may use the data given for them in the answer section to obtain the values necessary for solving the remaining problems of this chapter.

1. Use the data of Problem 1, Section 15.3, and the results of Problem 1, Supplement 15.A.
   a. Determine the estimates $\hat{\mu}_{y|x_1, x_2}$ and the deviations $(y_i - \hat{\mu}_{y|x_1, x_2})$ for each of the data points in Problem 1, Section 15.3.
   b. Determine $\sum_{i=1}^{n} (y_i - \hat{\mu}_{y|x_1, x_2})^2$ from the deviations in Part a.
   c. Determine the associated analysis-of-variance table.
   d. Determine $\hat{\sigma}_e^2$.
   e. Determine at the 5% significance level whether there is a linear relationship between the group of independent variables and the dependent variable.
   f. Determine at the 5% significance level which of the coefficients of the independent variables are significantly different from zero.
2. Use the data of Problem 2, Section 15.3, and the results of Problem 2, Supplement 15.A.
   a. Determine the estimates of $\mu_{y|x_1, x_2, x_3}$ and the deviations $(y_i - \hat{\mu}_{y|x_1, x_2, x_3})$ for each of the data points in Problem 2, Section 15.3.
   b. Determine $\sum_{i=1}^{n} (y_i - \hat{\mu}_{y|x_1, x_2, x_3})^2$ from the deviations in Part a.
   c. Determine the associated analysis-of-variance table.
   d. Determine $\hat{\sigma}_e^2$.
   e. Determine at the 5% significance level whether there is a linear relationship between the group of independent variables and the dependent variable.
   f. Determine at the 5% significance level which of the coefficients of the independent variables are significantly different from zero.

15.5 Confidence Intervals for β_i and $\mu_{y|x_1, x_2, \ldots, x_k}$

In many situations, we are interested in constructing confidence intervals for the parameters β_i $(i = 1, 2, \ldots, k)$ and $\mu_{y|x_1, x_2, \ldots, x_k}$. Before considering the derivation of the confidence intervals for these parameters, let us recall from the previous section that, when the assumptions of Section 15.2, p. 561, are valid, the least-squares estimator of β_i $(i = 1, 2, \ldots, k)$ is distributed normally with mean β_i and variance $c_{ii}\sigma_e^2$, where c_{ii} is the entry in the ith row and ith column of the matrix **C**. That is, the estimate $\hat{\beta}_i$ is the observed value of the estimator whose sampling distribution has mean

$$\mu_{\beta_i} = \beta_i \tag{1}$$

and variance

$$\sigma_{\beta_i}^2 = c_{ii}\sigma_e^2. \tag{2}$$

Solution

First, we write the corrected sums of squares and cross products in the following arrays. Let

$$\mathbf{S} = \begin{bmatrix} \sum_{j=1}^{n} (x_{1j} - \bar{x}_1)^2 & \sum_{j=1}^{n} (x_{1j} - \bar{x}_1)(x_{2j} - \bar{x}_2) & \sum_{j=1}^{n} (x_{1j} - \bar{x}_1)(x_{3j} - \bar{x}_3) \\ \sum_{j=1}^{n} (x_{2j} - \bar{x}_2)(x_{1j} - \bar{x}_1) & \sum_{j=1}^{n} (x_{2j} - \bar{x}_2)^2 & \sum_{j=1}^{n} (x_{2j} - \bar{x}_2)(x_{3j} - \bar{x}_3) \\ \sum_{j=1}^{n} (x_{3j} - \bar{x}_3)(x_{1j} - \bar{x}_1) & \sum_{j=1}^{n} (x_{3j} - \bar{x}_3)(x_{2j} - \bar{x}_2) & \sum_{j=1}^{n} (x_{3j} - \bar{x}_3)^2 \end{bmatrix} \tag{1}$$

$$= \begin{bmatrix} s_{11} & s_{12} & s_{13} \\ s_{21} & s_{22} & s_{23} \\ s_{31} & s_{32} & s_{33} \end{bmatrix} = \begin{bmatrix} 42{,}848 & 305{,}438 & 7{,}741 \\ 305{,}438 & 2{,}437{,}500 & 53{,}625 \\ 7{,}741 & 53{,}625 & 1{,}744 \end{bmatrix}$$

and let

$$\mathbf{G} = \begin{bmatrix} \sum_{j=1}^{n} (x_{1j} - \bar{x}_1)(y_j - \bar{y}) \\ \sum_{j=1}^{n} (x_{2j} - \bar{x}_2)(y_j - \bar{y}) \\ \sum_{j=1}^{n} (x_{3j} - \bar{x}_3)(y_j - \bar{y}) \end{bmatrix} = \begin{bmatrix} 1{,}831 \\ 12{,}239 \\ 340 \end{bmatrix}. \tag{2}$$

Next, we represent the unknowns by the array

$$\mathbf{B} = \begin{bmatrix} \hat{\beta}_1 \\ \hat{\beta}_2 \\ \hat{\beta}_3 \end{bmatrix}. \tag{3}$$

These rectangular arrays are known as *matrices.* We shall refer to an entry of the matrix **S** as the element s_{ij}, where s_{ij} represents the entry appearing at the intersection of the ith row and jth column. Similarly, we shall represent the entries of the matrices **B** and **G** by b_{ij} and g_{ij}, respectively.

The relationship between a set of simultaneous equations and matrices allows us to write the simultaneous equations in matrix notation. That is, the normal equations of the earlier example can be written in matrix notation as

$$\mathbf{SB} = \mathbf{G}. \tag{4}$$

If we were to solve an equation $sb = g$ for b, we would simply multiply both sides of the equation by the inverse of s, ($1/s$ if $s \neq 0$), and obtain $b = (1/s)g$. Likewise, in matrix notation, we solve the matrix equation of (4) by multiplying both sides by a matrix (if such matrix exists) known as the inverse of **S** and denoted by $\mathbf{S}^{-1}$. In so doing, we obtain

$$\mathbf{B} = \mathbf{S}^{-1}\mathbf{G}. \tag{5}$$

Actually, $\mathbf{S}^{-1}$ is simply another matrix of real numbers from which **B** can be obtained through use of Relation (5). We employ the abbreviated Doolittle method in order to find the matrix $\mathbf{S}^{-1}$. It is common in many computer analyses for the matrix $\mathbf{S}$ to be called the "X prime X" matrix, and consequently, $\mathbf{S}^{-1}$ is called the "inverse of the X prime X matrix." □

Anderson and Bancroft (15.7) recommend that, in the application of the abbreviated Doolittle method, the data be coded so that the elements of the $\mathbf{S}$ matrix are between 1 and 10 (algebraic sign is ignored). This can be accomplished by dividing the columns and their corresponding rows by 10 raised to an appropriate power. Of course, we must decode the results before using them.

Let us now return to the example and code the matrix $\mathbf{S}$ so that it conforms to the specifications of Anderson and Bancroft. We first determine the number of digits to the left of the decimal point for each of the diagonal elements (that is, for s_{ii}, $i = 1, 2, \ldots$). Next, we note whether such number is odd or even. If the number is odd, we subtract 1 from it and divide by 2 in order to obtain a new number, say r_i. If the number is even, we subtract 2 from it and divide by 2 in order to obtain r_i. Finally, for $i = 1, 2, \ldots$, we simply divide both the ith row and the ith column by 10^{r_i}.

Let us, for example, code the matrix $\mathbf{S}$ of (1). The diagonal elements are $s_{11} = 42{,}848$; $s_{22} = 2{,}437{,}500$; and $s_{33} = 1{,}744$. To determine the number r_1 for the first row and first column, we note that s_{11} has five digits to the left of the decimal point. Thus, we determine r_1 by subtracting 1 from 5 and dividing the result by 2, thus obtaining $r_1 = 2$. The divisor for the first row and the first column is given by $10^{r_1} = 10^2 = 100$. Therefore, we divide the first row and first column of $\mathbf{S}$ by 100. Similarly, we determine r_i ($i = 2, 3$) for the second and third diagonal elements (s_{22} and s_{33}) to be $r_2 = 3$ and $r_3 = 1$, respectively. Thus, the divisor for the second row and the second column is $10^{r_2} = 10^3 = 1{,}000$, while the divisor for the third row and the third column is $10^{r_3} = 10^1 = 10$. For the matrix $\mathbf{S}$ of (1), we obtain the coded matrix

$$\begin{matrix} & 100 & 1{,}000 & 10 \\ \begin{matrix}100\\1{,}000\\10\end{matrix} & \multicolumn{3}{c}{\begin{bmatrix} 4.2848 & 3.05438 & 7.741 \\ 3.05438 & 2.4375 & 5.3625 \\ 7.741 & 5.3625 & 17.44 \end{bmatrix}} \end{matrix}, \tag{6}$$

where the divisors for the rows and columns are indicated in the margins. We note that not all entries are between 1 and 10. That is, the entry 17.44 is outside of the range recommended by Anderson and Bancroft. However, we ignore this, since it is generally sufficient to have most of the entries (when sign is ignored) between 1 and 10, with all of them between 0 and 100.

When we perform the coding for $\mathbf{S}$, we must perform a similar coding on the matrix $\mathbf{G}$. Specifically, we must divide only the corresponding rows of $\mathbf{G}$ by the respective divisors for the rows of $\mathbf{S}$, to obtain the coded matrix

$$\begin{matrix}100\\1{,}000\\10\end{matrix}\begin{bmatrix} 18.31 \\ 12.239 \\ 34.0 \end{bmatrix}, \tag{7}$$

where the divisors of the rows of $\mathbf{G}$ are in the left margin.

The abbreviated Doolittle method consists of a forward and backward solution. By the forward solution, we find regression coefficients ($\hat{\beta}_i$) and the sum of squares attributable to regression, $SS_{\text{regression}}$ (a sum of squares, used in an analysis of variance concerning multiple regression). The backward solution lets us determine the inverse of the matrix obtained by coding **S**, which we can decode to determine the inverse of **S**. The inverse of **S** is needed to determine confidence intervals for the β_i coefficients and to test hypotheses concerning specific coefficients. The complete method represents a shortcut for use with a symmetric matrix **S**; thus it is not applicable for any nonsymmetric matrix.

Returning to the example, refer to Table 15.6, where each row is labeled by a letter and a subscript (R_1', R_2, r_1, etc.). Under this notational scheme, a given table entry is characterized by the letter and subscript designating its row and an additional subscript describing its column. Thus, R_{25} represents the table entry which appears in (or is understood to appear in) column 5 and row R_2.

We perform the forward and backward solutions by the abbreviated Doolittle method according to the following instructions:

1. Enter in a table (Table 15.6) relevant information from the coded matrices corresponding to **S** and **G**, that is, the respective matrices given by (6) and (7). Note that the portion of (6) below the diagonal is not included in the table. Note also that this step creates a number of rows equal to the number of simultaneous equations to be solved. For these three rows, that is, R_i' ($i = 1, 2, 3$), construct columns corresponding to those in Table 15.6 labeled E_1, E_2, E_3, and "Check." Construct E_i ($i = 1, 2, 3$) so that it is a column having a 1 in row R_i' and 0's in the other two rows. Each "Check" column entry is obtained by summing all preceding entries that appear (or are understood to appear) in the same row. The "Check" column is used to identify arithmetic errors. Although the entries R_{21}', R_{31}', and R_{32}' do not actually appear in the table, they must be included in determining the "Check" column. This can be accomplished easily because of the symmetry involved—that is, because

$$R_{21}' = R_{12}'; \quad R_{31}' = R_{13}'; \quad \text{and} \quad R_{32}' = R_{23}'.$$

For example, the second element in the check column is obtained as

$$\begin{aligned} R_{28}' &= R_{21}' + R_{22}' + R_{23}' + R_{24}' + R_{25}' + R_{26}' + R_{27}' \\ &= R_{12}' + R_{22}' + R_{23}' + R_{24}' + R_{25}' + R_{26}' + R_{27}' \\ &= 3.05438 + 2.4375 + 5.3625 + 12.239 + 0 + 1 + 0 \\ &= 24.09338. \end{aligned}$$

2. Construct the row R_1 merely by copying row R_1'. Then, construct the row r_1 by dividing each element of R_1 by the element R_{11} (4.2848). Check to see that

$$R_{18} = \sum_{j=1}^{7} R_{1j} \quad \text{and} \quad r_{18} = \sum_{j=1}^{7} r_{1j}.$$

If either relation fails to hold (with reasonable allowance for rounding error), repeat the relevant calculations. If both hold, proceed to instruction 3.

3. Construct the row R_2, where the entry for the jth column ($j = 2, 3, \ldots, 8$) of R_2 is

$$R_{2j} = R'_{2j} - R_{12}r_{1j} \qquad (j = 2, 3, \ldots, 8).$$

For example,

$$\begin{aligned} R_{24} &= R'_{24} - R_{12}r_{14} \\ &= 12.239 - (3.05438)(4.273245) \\ &= -0.813114. \end{aligned}$$

Next, obtain row r_2 by dividing each element of row R_2 by the element R_{22} (0.260216). (The elements represented by the asterisks (*) are actually zeros. However, since they are not used in subsequent computations in the abbreviated Doolittle method, they need not be entered.) If either

$$R_{28} = \sum_{j=2}^{7} R_{2j} \quad \text{or} \quad r_{28} = \sum_{j=2}^{7} r_{2j}$$

fails to hold (allowing for rounding error), repeat the process. Otherwise, proceed to instruction 4.

4. Construct the row R_3, where the entry for the jth column ($j = 3, 4, \ldots, 8$) of R_3 is

$$R_{3j} = R'_{3j} - R_{13}r_{1j} - R_{23}r_{2j} \qquad (j = 3, 4, \ldots, 8).$$

For example,

$$\begin{aligned} R_{35} &= R'_{35} - R_{13}r_{15} - R_{23}r_{25} \\ &= 0 - (7.741)(0.233383) - (-0.155601)(-2.739416) \\ &= -2.232874. \end{aligned}$$

Next, obtain row r_3 by dividing each element of row R_3 by the element R_{33} (3.361918). Perform the check

$$R_{38} \stackrel{?}{=} \sum_{j=3}^{7} R_{3j} \quad \text{and} \quad r \stackrel{?}{=} \sum_{j=3}^{7} r_{3j},$$

and repeat the calculations or continue, whichever is appropriate.

This completes the forward solution. For the backward solution, perform steps 5–7.

5. Determine the entry for the jth column ($j = 5, 6, 7$) of row C'_1, by computing $R_{15}r_{1j} + R_{25}r_{2j} + R_{35}r_{3j}$ for $j = 5, 6, 7$.
6. Determine the entry for the jth column ($j = 6, 7$) of row C'_2 by computing $R_{26}r_{2j} + R_{36}r_{3j}$ for $j = 6, 7$.
7. Determine the entry for the seventh column of row C'_3 by computing $R_{37}r_{37}$.

To minimize rounding errors in the above calculations, it is best to carry as many significant digits as feasible, preferably six or more, after coding.

The first four columns of Table 15.6 represent the forward solution, and columns 5 through 7 represent the backward solution. Let us now consider how to use these solutions to obtain the values needed for our multiple regression problem.

Arranging the relevant rows from Table 15.6 into matrices, we use the forward solution to obtain the equations that can be easily solved for the estimates $\hat{\beta}_1$, $\hat{\beta}_2$,

Table 15.6

The abbreviated Doolittle method for Equation (4). (Using coded data)

Instruction	Column / Row	(1) x_1	(2) x_2	(3) x_3	(4) y	(5) E_1	(6) E_2	(7) E_3	(8) Check
1.	R'_1	4.2848	3.05438	7.741	18.31	1.0	0	0	34.39018
	R'_2		2.4375	5.3625	12.239	0	1.0	0	24.09338
	R'_3			17.44	34.0	0	0	1.0	65.5435
2. R'_{1j}	R_1	4.2848	3.05438	7.741	18.31	1.0	0	0	34.39018
$R_{1j} \div R_{11}$	r_1	1.0	0.712840	1.806619	4.273245	0.233383	0	0	8.026088
3. $R'_{2j} - R_{12}r_{1j}$	R_2	*	0.260216	−0.155601	−0.813114	−0.712840	1.0	0	−0.421343
$R_{2j} \div R_{22}$	r_2	*	1.0	−0.597969	−3.124766	−2.739416	3.842961	0	−1.619205
4. $R'_{3j} - R_{13}r_{1j} - R_{23}r_{2j}$	R_3	*	*	3.361918	0.434594	−2.232874	0.597969	1.0	3.161603
$R_{3j} \div R_{33}$	r_3	*	*	1.0	0.129270	−0.664167	0.177865	0.297449	0.940416
5. $R_{15}r_{1j} + R_{25}r_{2j} + R_{35}r_{3j}$	C'_1					3.669150	−3.136566	−0.664166	
6. $R_{26}r_{2j} + R_{36}r_{3j}$	C'_2						3.949319	0.177865	
7. $R_{37}r_{37}$	C'_3							0.297449	

and $\hat{\beta}_3$. However, the data in Table 15.6 are coded; hence, the solutions must first be decoded. That is, let us denote by **D** the matrix which has for its rows the first 3 columns of rows r_1, r_2, and r_3:

$$\mathbf{D} = \begin{bmatrix} r_{11} & r_{12} & r_{13} \\ r_{21} & r_{22} & r_{23} \\ r_{31} & r_{32} & r_{33} \end{bmatrix}.$$

Recalling that the asterisks in Table 15.6 represent zeros, we have

$$\mathbf{D} = \begin{bmatrix} 1.0 & 0.712840 & 1.806619 \\ 0 & 1.0 & -0.597969 \\ 0 & 0 & 1.0 \end{bmatrix}$$

In order to decode the matrix **D**, we simply reverse the coding process—that is, we multiply each column by the term by which the corresponding column of the original matrix **S** was divided when we did the coding for Table 15.6. Thus, multiplying the first, second, and third columns of **D** by 100, 1,000, and 10, respectively, we obtain the matrix **T**.

$$\mathbf{T} = \begin{bmatrix} 100 & 712.840 & 18.06619 \\ 0 & 1{,}000 & -5.97969 \\ 0 & 0 & 10 \end{bmatrix}.$$

Next, we obtain the equations which can be easily solved. That is, using matrix notation, we obtain

$$\mathbf{TB} = \mathbf{r},$$

where **B** is as given in (3) and **r** is a column matrix obtained from column (4) of Table 15.6. Representing the elements of **T** by t_{ij} ($i = 1, 2, 3$; $j = 1, 2, 3$), we represent the three equations in three unknowns as

$$\begin{aligned} t_{11}\hat{\beta}_1 + t_{12}\hat{\beta}_2 + t_{13}\hat{\beta}_3 &= r_{14}, \\ t_{22}\hat{\beta}_2 + t_{23}\hat{\beta}_3 &= r_{24}, \\ t_{33}\hat{\beta}_3 &= r_{34}. \end{aligned} \tag{8}$$

The solutions to (8) are obtained by solving the last row for $\hat{\beta}_3$, substituting this value into the second row, solving for $\hat{\beta}_2$, and so forth.

$$\hat{\beta}_3 = r_{34}/t_{33},$$

$$\hat{\beta}_2 = \frac{1}{t_{22}}[r_{24} - t_{23}\hat{\beta}_3], \tag{9}$$

and

$$\hat{\beta}_1 = \frac{1}{t_{11}}[r_{14} - t_{12}\hat{\beta}_2 - t_{13}\hat{\beta}_3].$$

The resulting equations for our example become

$$\begin{aligned} 100\hat{\beta}_1 + 712.840\hat{\beta}_2 + 18.06619\hat{\beta}_3 &= 4.273245, \\ 1{,}000\hat{\beta}_2 - 5.97969\hat{\beta}_3 &= -3.124766, \\ 10\hat{\beta}_3 &= 0.129270. \end{aligned} \tag{10}$$

The solutions are

$$\hat{\beta}_3 = (0.129270)/10 = 0.012927,$$

$$\hat{\beta}_2 = \frac{1}{1{,}000}[-3.124766 + 5.97969(0.012927)] = -0.0030475, \tag{11}$$

and

$$\hat{\beta}_1 = \frac{1}{100}[4.273245 - 712.840(-0.003047) - 18.06619(0.012927)] = 0.062117.$$

Then, the value for $\hat{\beta}_0$ can be determined by (from Equation (19), Section 15.3)

$$\begin{aligned}\hat{\beta}_0 &= \bar{y} - \hat{\beta}_1\bar{x}_1 - \hat{\beta}_2\bar{x}_2 - \hat{\beta}_3\bar{x}_3 \\ &= \frac{271.3}{16} - (0.062120)\left(\frac{3{,}405}{16}\right) - (-0.003047)\left(\frac{48{,}200}{16}\right) - (0.012927)\left(\frac{1{,}550}{16}\right) \\ &= 11.663122.\end{aligned}$$

Rounding the coefficients to the first four nonzero digits gives

$$\hat{\beta}_0 = 11.66, \quad \hat{\beta}_1 = 0.06212, \quad \hat{\beta}_2 = -0.003047, \quad \text{and} \quad \hat{\beta}_3 = 0.01293. \tag{12}$$

As a check on the accuracy of the regression coefficients, we can determine whether the equations of (21), Section 15.3, are satisfied. However, generally we need check only one equation. So, checking the first equation, we determine

$$\begin{aligned}&42{,}848\hat{\beta}_1 + 305{,}438\hat{\beta}_2 + 7{,}741\hat{\beta}_3 \\ &\quad = 42{,}848(0.06212) + 305{,}438(-0.003047) + 7{,}741(0.01293) \\ &\quad = 1{,}831.14.\end{aligned}$$

This is sufficiently close to 1,831 for us to conclude that the difference is due to rounding errors.

From the estimates of (12), we obtain the regression equation,

$$\begin{aligned}\hat{\mu}_{y|x_1, x_2, x_3} &= \hat{\beta}_0 + \hat{\beta}_1 x_1 + \hat{\beta}_2 x_2 + \hat{\beta}_3 x_3 \\ &= 11.66 + 0.06212x_1 - 0.003047x_2 + 0.01293x_3,\end{aligned} \tag{13}$$

which estimates the mean of the population of Y values at the point (x_1, x_2, x_3).

Although this completes the computations immediately needed from the forward solution, let us consider one additional point. After obtaining the estimates given in (12), we can also determine, for use in an analysis of variance, the sum of squares attributable to multiple regression:

$$\begin{aligned}SS_{\text{regression}} &= \hat{\beta}_1 g_{11} + \hat{\beta}_2 g_{21} + \hat{\beta}_3 g_{31} \\ &= (0.06212)(1{,}831) + (-0.003047)(12{,}239) + (0.01293)(340) \\ &= 80.8457.\end{aligned} \tag{14}$$

In order to obtain confidence intervals and test hypotheses about the individual parameters of the multiple linear regression model, we estimate the standard deviation associated with the estimates of each parameter. For this, we use the backward solution. However, for now we shall simply determine the appropriate terms that result from the backward solution.

First, let us represent by $\mathbf{C}'$ that matrix which we obtain from Table 15.6 by selecting columns 5, 6, and 7 from rows C_1', C_2', and C_3'. That is, since the corresponding matrix in the table is symmetrical,

$$\mathbf{C}' = \begin{bmatrix} 3.669150 & -3.136566 & -0.664166 \\ -3.136566 & 3.949319 & 0.177865 \\ -0.664166 & 0.177865 & 0.297449 \end{bmatrix}. \tag{15}$$

Next, we decode $\mathbf{C}'$ to obtain the inverse of $\mathbf{S}$, by simply dividing each row and the corresponding column by the same divisor used to code $\mathbf{S}$. In other words, we divide the first row and the first column by 100, the second row and the second column by 1,000, and the third row and the third column by 10. Decoding gives us the matrix $\mathbf{C}$, which is the inverse of $\mathbf{S}$. That is, $\mathbf{C} = \mathbf{S}^{-1}$, where

$$\mathbf{C} = \begin{bmatrix} 0.000366915 & -0.00003136566 & -0.000664166 \\ -0.00003136566 & 0.000003949319 & 0.0000177865 \\ -0.000664166 & 0.0000177865 & 0.00297449 \end{bmatrix}. \tag{16}$$

One use of $\mathbf{C}$ is to check the estimates obtained by the forward solution of the abbreviated Doolittle method. That is, from (5), $\mathbf{B} = \mathbf{S}^{-1}\mathbf{G}$, we have

$$\mathbf{B} = \mathbf{CG}. \tag{17}$$

Performing the matrix multiplication of (17), we can obtain the estimates $\hat{\beta}_1$, $\hat{\beta}_2$, and $\hat{\beta}_3$. That is,

$$\begin{bmatrix} \hat{\beta}_1 \\ \hat{\beta}_2 \\ \hat{\beta}_3 \end{bmatrix} = \begin{bmatrix} 0.000366915 & -0.00003136566 & -0.000664166 \\ -0.00003136566 & 0.000003949319 & 0.0000177865 \\ -0.000664166 & 0.0000177865 & 0.00297449 \end{bmatrix} \begin{bmatrix} 1{,}831 \\ 12{,}239 \\ 340 \end{bmatrix}.$$

To obtain these estimates without matrix multiplication, we substitute the values from $\mathbf{C}$ and $\mathbf{G}$ into the equations

$$\begin{aligned} \hat{\beta}_1 &= c_{11}g_{11} + c_{12}g_{21} + c_{13}g_{31}, \\ \hat{\beta}_2 &= c_{21}g_{11} + c_{22}g_{21} + c_{23}g_{31}, \\ \hat{\beta}_3 &= c_{31}g_{11} + c_{32}g_{21} + c_{33}g_{31}. \end{aligned}$$

Thus, we obtain

$$\begin{aligned} \hat{\beta}_1 &= 0.000366915(1{,}831) - 0.00003136566(12{,}239) - 0.000664166(340) \\ &= 0.062161, \end{aligned}$$

$$\begin{aligned} \hat{\beta}_2 &= -0.00003136566(1{,}831) + 0.000003949319(12{,}239) + 0.0000177865(340) \\ &= -0.0030474, \end{aligned}$$

and

$$\begin{aligned} \hat{\beta}_3 &= -0.000664166(1{,}831) + 0.0000177865(12{,}239) + 0.00297449(340) \\ &= 0.0129276. \end{aligned}$$

Except in the last digit or two, these values agree with those of (11). The slight disagreement can be attributed to rounding error and is no cause for alarm. If the two sets of estimates had not agreed, we would need to check the values in Table 15.6 for errors. This use of $\mathbf{C}$, however, is not the primary reason for its calculation. The terms in $\mathbf{C}$ are used to determine estimates of the variances of certain estimators. □

We have considered the abbreviated Doolittle method for the case concerning three independent mathematical variables. The method may be extended to as many independent variables as necessary. Once the pattern is recognized, the procedure is repetitive and straightforward.

Problems

1. Use the normal equations obtained for Problem 1, Section 15.3, and the abbreviated Doolittle method.
   a. Determine the estimates of β_0, β_1, and β_2.
   b. Determine the regression equation for estimating $\mu_{y|x_1, x_2}$.
   c. Determine $SS_{\text{regression}}$.
   d. Determine the $\mathbf{C}$ matrix.
2. Use the normal equations obtained for Problem 2, Section 15.3, and the abbreviated Doolittle method.
   a. Determine the estimates for β_0, β_1, β_2, and β_3.
   b. Determine the regression equation for estimating $\mu_{y|x_1, x_2, x_3}$.
   c. Determine $SS_{\text{regression}}$.
   d. Determine the $\mathbf{C}$ matrix.

Important Terms and Concepts

abbreviated Doolittle method

analysis of variance for multiple linear regression

backward solution by the abbreviated Doolittle method

coding $\mathbf{S}$

confidence intervals for β_i, $\mu_{y|x_1, x_2, \ldots x_k}$, and $y(x_1, x_2, \ldots, x_k)$

decoding $\mathbf{S}$

deterministic mathematical model of several variables

estimation of β_i, σ_e^2, $\mu_{y|x_1, x_2, \ldots, x_k}$, and $y(x_1, x_2, \ldots, x_k)$

forward solution by the abbreviated Doolittle method

matrices

multiple linear regression model

multiple regression analysis

multiple regression coefficients

normal equations for the multiple linear regression model

probabilistic mathematical model of several variables

tests of hypotheses concerning a set of regression coefficients

tests of hypotheses concerning individual regression coefficients

References

15.1 F. A. Graybill, *An Introduction to Linear Statistical Models*, vol. 1, New York: McGraw-Hill Book Company, 1961.

15.2 R. C. Geary and C. E. V. Leser, "Significance Tests in Multiple Regression," *The American Statistician*, vol. 22, p. 20, 1968.

15.3 N. R. Draper and H. Smith, *Applied Regression Analysis*, New York: John Wiley & Sons, Inc., 1966.

15.4 J. Johnston, *Econometric Methods*, New York: McGraw-Hill Book Company, 1963.

15.5 M. H. Doolittle, *United States Coast and Geodetic Survey Report*, 1878.

15.6 P. S. Dwyer, *Linear Computations*, New York: John Wiley & Sons, Inc., 1951.

15.7 R. L. Anderson and T. A. Bancroft, *Statistical Theory in Research*, New York: McGraw-Hill Book Company, 1952.

Sixteen

Simple Decision Problems—Expected Value and Utility

16.1 Introduction

Statistical theory has recently become one of the most widely discussed topics in the area of decision making. Simultaneously, "Bayesian inference" has become a popular topic. Unfortunately, statistical decision theory and Bayesian inference are often discussed as though they were synonyms. However, they are not; there are many problems in statistical decision theory to which Bayesian inference is not applicable. A major objective in this chapter and the next, therefore, is to introduce the broad category of decision problems to which Bayesian inference does apply.

We begin with simple decision problems, and the two major criteria on which such decisions are based: *expected monetary value* and *expected utility*. Thus, our initial approach will assume that perfect and complete information required for an intelligent decision is available. Then, we shall attack the situation in which one must base decisions partially (at best) on sample information. In doing this, we shall take a different look at the binomial sampling distribution in order to determine a different (Bayesian) way of exploiting information from sample values. Along the way, we shall go so far as to view a parameter essentially as a random variable and look at the costs of various decision errors and their effects on terminal decisions. The aim of these two chapters is only to familiarize you with Bayesian concepts and techniques—enough for you to pursue the Bayesian area further by referring to other more specialized texts (e.g., references 16.1 and 16.2).

16.2 Bayes' Theorem and the Bayesian Approach

We discussed the essential concepts of the Bayesian approach to decision making in Section 4.10. You are encouraged to review those concepts now so that you can easily establish their relevance in our discussions of decision theory here. You will recall that we use the Bayesian approach to determine which of several possible states of nature exists in a given situation—by conducting an experiment, taking an observation, or viewing an indicator for partial information. (Of course, we implicitly assume that it is impossible to obtain complete information—otherwise we would not need the Bayesian approach.) Then, we use Bayes' theorem to combine the information collected with any other that may be available. Here, we shall consider both Bayesian techniques and interpretation of their results.

Note that in Chapter 4 we accomplished the following things in each example.

1. We somehow obtained access to a prior-probability distribution that associates probabilities with each of a set of mutually exclusive events. (In each example, we assumed this information to be given.)
2. Next, we collected sample information. That is, we observed the outcome of an indicator.
3. Finally, we used the sample information to revise our assessment of the relevant probability distribution. That is, by basing change solely on sample information, we changed our prior-probability distribution to a posterior distribution by application of Bayes' theorem.

These three steps manifest a major portion of what has come to be known as the Bayesian approach to decision making.

Let us examine an example wherein the collection of sample information provides an indicator for revising a prior-probability distribution.

Example 1

A given kind of transistor is manufactured on each of five day-shift lines, three night-shift lines, and two swing-shift lines. The fractions defective of transistors produced by day, night, and swing shifts are 0.10, 0.15, and 0.20, respectively. The identification material for one lot of transistors has been misplaced. Company officials would like to obtain as good an idea as feasible of the shift during which the lot was manufactured. However, in this case, to test a transistor is to destroy it. Thus, officials must base their decision on sample information. They take a sample of size $n = 50$ and find 8 defectives. In light of the sample information, what posterior probabilities should they attach to the possible shifts during which the lot was produced? Assume that all ten manufacturing lines are equally likely in the prior-probability sense. Assume also that the lot is large enough for us to assume that the sample of 50 items represents 50 independent trials from a Bernoulli process.

Solution

Let us identify the three shifts—day, night, and swing—as states of nature E_1, E_2, and E_3, respectively. The prior probabilities immediately become

$$P(E_1) = 0.5, \qquad P(E_2) = 0.3, \qquad P(E_3) = 0.2. \tag{1}$$

We recall from Chapter 4 that the expected value of a discrete random variable X is

$$E(X) = \sum x_i P(X = x_i), \tag{1}$$

where $\sum$ represents the summation over all possible values of X, x_i is the ith possible value of X, and $P(X = x_i)$ is the probability that the discrete random variable X takes on the value x_i. However, we must find the expected value of the monetary consequences associated with a given act. Is it proper to treat such consequences as values of a random variable? The answer, under the following reasoning, is yes. If we restrict consideration to a single act, monetary consequence can be viewed as a random variable, because it is related in a one-to-one fashion to the basic random variable of interest. For example, the basic random variable in Example 1 is "state of weather." For a single act, the occurrence of any outcome of "state of weather" is always accompanied by the occurrence of a specific monetary consequence. Hence, the probability distribution that is inherently associated with the basic random variable "state of nature" is indirectly associated with monetary consequence. Thus, monetary consequence may legitimately be viewed as a random variable. It follows immediately that we obtain the expected monetary value of an act by first multiplying each monetary consequence that can result by the probability of its occurrence and then summing these products over all possible consequences of such act.

Note from Table 16.2 that it is not intuitively obvious which of the acts (business investments) is best for the individual. If judgment is based on a single given day, any one act may be considered best. For example, for a cold wet day, act 3 is obviously best. Likewise, act 1 is best for a warm dry day, and act 2 is best for a warm wet day. Thus, the nature of the decision requires long-run consideration. It is meaningless to compare acts on the basis of any given day; hence, the criterion used must reflect long-run expected benefits. Expected monetary value is one such criterion.

Of course, if one act should give better results than the others for all states of nature, there is no need to apply decision theory. By the concept known as *dominance*, the act is said to dominate the others.

Solution for Example 1

Returning to the first example, we select the best (in the sense of expected monetary value) act by computing the expected monetary value of each possible act. The expected monetary value

$$\text{for act 1} = (0.10)(-\$20) + (0.20)(\$20) + (0.20)(-\$10) + (0.50)(\$80) = \$40;$$

$$\text{for act 2} = (0.10)(-\$15) + (0.20)(-\$5) + (0.20)(\$20) + (0.50)(\$60) = \$31.50;$$

$$\text{for act 3} = (0.10)(\$100) + (0.20)(\$60) + (0.20)(\$10) + (0.50)(\$10) = \$29.$$

Thus, we conclude that for long-run purposes, act 1 is the best when expected monetary value is our criterion. □

We could have simplified the solution by using a table. For example, the expected monetary value of act 1 is calculated in Table 16.3. Note the two columns labeled "conditional consequence" and "expected consequence." A *conditional consequence*,

Table 16.3

Calculation of expected monetary value of act 1 of Example 1

		Act 1	
Event (E_i)	*Probability* $P(E_i)$	*Conditional consequence* (\$)	*Expected consequence* (\$)
E_1 (cold $\cap$ wet)	0.10	−20	(0.10)(−20) = −2.00
E_2 (cold $\cap$ dry)	0.20	20	(0.20)(20) = 4.00
E_3 (warm $\cap$ wet)	0.20	−10	(0.20)(−10) = −2.00
E_4 (warm $\cap$ dry)	0.50	80	(0.50)(80) = 40.00
Expected monetary value of act 1			\$40.00

as the name implies, is conditional on the outcome of the basic random variable of interest and simply denotes the consequence of a given event-act combination. The *expected consequence* is simply an intermediate calculation in the determination of expected monetary value.

Note also that the possible events in Example 1 are mutually exclusive and exhaustive, a characteristic of the kind of decision problems we shall be discussing in this and the next chapter.

However, before we go too far in stereotyping the problem, let us discuss one other example designed to point out an often overlooked fact in elementary decision theory. We have been careful to point out that, when problems involve decisions under uncertainty, the decision maker cannot influence by his choice of an act an individual outcome of the basic random variable on which his decision must ultimately depend. This should by no means, however, be construed to imply that his choice cannot influence characteristics of the probability distribution of that basic random variable. This is definitely a possibility that fits into the overall framework of problems to which elementary decision theory is applicable. Let us consider an example.

Example 2

Suppose a company planning to bid on a government contract is certain that one of three outcomes will result:

1. the company will be awarded a contract that ultimately will return \$200,000 profit,
2. the company will be awarded a subcontract that ultimately will return \$75,000 in profit, or
3. the company will be awarded no contract at all.

We shall refer to these outcomes as events E_1, E_2, and E_3. The company has four feasible alternatives in preparing its proposal:

1. use only company personnel in preparing the proposal (at essentially no cost),
2. hire 50 full-time employees (average salary \$1,000 per month) for a period of two months,

3. hire 20 full-time employees (average salary of $1,000 per month) for a period of two months and engage a consulting firm (at $20,000 per month) to direct proposal efforts for the last month, or
4. engage the consulting firm to direct proposal efforts for the two months.

We shall refer to these as acts A_1, A_2, A_3, and A_4. The ultimate monetary consequence to the company as a result of each event-act combination can be calculated simply by subtracting the proposal cost associated with the appropriate act from the profit to be obtained through the occurrence of the appropriate event. For example, if the company chooses act A_3 and event E_1 occurs, the ultimate monetary return will be

$$\begin{aligned} C_{13} &= \$200{,}000 - (20)(2)(\$1{,}000) - \$20{,}000 \\ &= \$200{,}000 - \$40{,}000 - \$20{,}000 = \$140{,}000, \end{aligned}$$

where C_{13} is the monetary consequence of the simultaneous occurrence of event E_1 and act A_3. Consequences for other event-act combinations are shown as conditional consequences in Table 16.4.

Now, suppose further that, drawing from experience on past contracts of this nature and the judgment of experts in this specific contract area, company management assigns the probabilities of Table 16.4 to the various event-act combinations. For example, company officials feel that by using only their own employees (act A_1), they have a probability of only 0.3 of being awarded the contract; however, by using a management consulting firm (act A_4), they feel that they can increase this probability to 0.7. This is obviously a realistic situation in which the company is able to influence the probability distribution of the basic random variable by its choice of acts. It can enhance its chances of a successful bid by intelligent selection of the proposal work force. How, then, should a decision be made?

We note first that even the idea of dominance is not as simple as it appeared in connection with Example 1. For, if we view only the conditional consequences associated with various acts, it appears that act A_1 dominates the others in our present example. However, we cannot say that act A_1 dominates until we can say that it is better than the others under all possible states of nature. And we cannot say this before determining whether the conditional consequences for one act are associated with a different set of probabilities than corresponding conditional consequences for another act. Thus, we must look to the expected consequences. In this sense, then, the substance of dominance is that one act dominates another when the expected consequences for that act are greater than the expected consequences of the other acts for all possible states of nature. Clearly, this is not the case for the data of Table 16.4. Thus, we must again apply elementary decision theory.

The calculations required for determining the act with the greatest expected monetary value are in Table 16.4 in the columns labeled "Expected consequences." Results indicate that the company should hire the management consulting firm for two months (act A_4), assuming, of course, that expected monetary value is a good criterion and that the probabilities assigned by the company represent the true probabilities for various event-act combinations.

The last assumption of Example 2 leads us immediately to a couple of the most important questions in decision theory. How good is the probabilistic information used to reach a decision in a given problem, and can one improve such information

Table 16.4

Consequences and probabilities associated with possible event-act combinations

	Act A_1			Act A_2			Act A_3			Act A_4		
Event E_i	$P(E_i)$	*Conditional consequence* ($)	*Expected consequence* ($)	$P(E_i)$	*Conditional consequence* ($)	*Expected consequence* ($)	$P(E_i)$	*Conditional consequence* ($)	*Expected consequence* ($)	$P(E_i)$	*Conditional consequence* ($)	*Expected consequence* ($)
E_1	0.3	200,000	60,000	0.4	100,000	40,000	0.5	140,000	70,000	0.7	160,000	112,000
E_2	0.3	75,000	22,500	0.3	−25,000	−7,500	0.3	15,000	4,500	0.2	35,000	7,000
E_3	0.4	0	0	0.3	−100,000	−30,000	0.2	−60,000	−12,000	0.1	−40,000	−4,000
Expected monetary value			82,500			2,500			62,500			115,000

before embarking on an analysis that leads to a terminal decision? Until the advent of the Bayesian approach to decision making, answers to such questions were usually quite general and often ambiguous. To be sure, despite the extensive research that has been done in this area, many of the answers in specific situations are still general and ambiguous. However, due largely to the Bayesian influence, we do have some better answers now. As we pursue discussions of the Bayesian approach to decision making, we shall give some of the answers.

Indicators

Note that, in both of the previous examples involving decisions under uncertainty, we have assumed under each act a probability distribution that associates probabilities with each possible event-act combination; we have also assumed that we know (or have good information concerning) the nature of the distribution. Further, we have taken the nature of the distribution for granted. Let us now investigate a problem that represents the real essence of the so-called Bayesian approach to decision making. In it we obtain added knowledge about the probability distribution of the basic random variable for each feasible act prior to making a final decision. Note also that we are concerned in the following example with a benefit other than money.

Example 3

Suppose Mr. Morgan pilots his private plane each day to a city some distance from home and returns each evening. On any given evening, he may land at either of two airports in his home city, depending on whether or not he expects to stop at his office before going home. If he doesn't have to go to the office, his secretary meets him at the airport to help him plan his route for the next day. Thus, before taking off for home each day, he checks with his office to find out what appointments his secretary has made and let her know where to meet him, if necessary. His possible acts are:

A: land at Airport A,
B: land at Airport B.

Mr. Morgan's commuting is complicated by the fact that he doesn't always know before landing whether or not he must stop at the office. His secretary may schedule appointments after his call. If she doesn't meet him, he knows he must stop at the office. The possible states of nature (events) are

E_1: required to stop at office,
E_2: not required to stop at office.

Airport A is 10 minutes from Mr. Morgan's home and 50 minutes from his office. Airport B is 40 minutes from his home and 10 minutes from his office. His home is

45 minutes from his office. The consequences in terms of travel time from airport to home for the various event-act combinations are, therefore,

event 1-act A:	50(Airport A to office) + 45(office to home) =	95,
event 1-act B:	10(Airport B to office) + 45(office to home) =	55,
event 2-act A:	(Airport A to home)	10,
event 2-act B:	(Airport B to home)	40.

Over the past two years Mr. Morgan has found that on 60 percent of all days he has been required to stop at the office. He thus attaches a probability of 0.6 to event 1 and a probability of 0.4 to event 2. Since his objective is to choose the act that minimizes his expected travel time from airport to home, he applies simple decision theory, as shown in Table 16.5. Since his expected travel time is 12 minutes shorter (61 − 49) if he lands at Airport B, Mr. Morgan long since decided to land at Airport B when he knows he has an appointment. Thus, the analysis is restricted to times when no appointments are scheduled at the time he calls his secretary.

Table 16.5

Expected consequences for Mr. Morgan

		Consequences (travel times) associated with various acts			
		Act A (land at Airport A)		*Act B (land at Airport B)*	
Event (state of nature) E_i	$P(E_i)$	*Conditional consequence*	*Expected consequence*	*Conditional consequence*	*Expected consequence*
E_1 (appointments)	0.6	95	57	55	33
E_2 (no appointments)	0.4	10	4	40	16
Expected consequences (travel times) of acts			61		49

Quite happy that he has perhaps added (on average) 12 minutes per day to his leisure time, he proudly tells a friend of his analysis and decision. The friend (a statistician familiar with Bayesian concepts) is not very impressed. He tells Mr. Morgan that he overlooked important information that could well save him even more time, up to an additional 12 minutes per day.

The friend explains that if Mr. Morgan could know whether he had to go to the office on each day, he could plan a strategy rather than landing at the same airport every day. He could land at Airport B on those days (60 percent of all days) when he had to go to the office, and at Airport A on those days (40 percent of all days) when he could go straight home. His expected travel time would then be 0.6(55) + 0.4(10) = 37 minutes. The friend suggests that Mr. Morgan's secretary might be able to make some prediction about whether or not he will have an appointment by the time he arrives in his home city, based on what happened at the office. So for several months, Mr. Morgan asks his secretary for a prediction each day and records her responses, as well as the actual states of nature existing on his arrivals during this time. Table 16.6 shows the results of how often her response (an indicator of the true state of nature) correctly indicates the true state of nature. In this case, I represents an indicator—the result of an observation or experiment. As an example of the meaning of the joint probabilities of Table 16.6, the first entry (0.40) means that 40 percent of the time the secretary predicts appointments and there are appointments. Thus, we have the joint probabilities $P(E_i \cap I_k)$, ($i = 1, 2$ and $k = 1, 2$)

$$\begin{aligned} P(E_1 \cap I_1) &= 0.40, \qquad P(E_1 \cap I_2) = 0.20, \\ P(E_2 \cap I_1) &= 0.05, \qquad P(E_2 \cap I_2) = 0.35. \end{aligned} \tag{2}$$

These are joint probabilities in the same sense as those described in our first discussion of the Bayesian approach. The margin totals of Table 16.6 present relevant marginal probabilities of states of nature and outcomes of the indicator I. Mr. Morgan would

Table 16.6

Indicator results

Event (state of nature) E_i	*Probability that indicator is correct i.e., $P(E_i \cap I_k)$ = joint probability of the ith state of nature and the kth indicator outcome*		*Marginal probabilities of events $P(E_i)$*
	I_1 (Secretary predicts appointments)	*I_2 (Secretary predicts no appointments)*	
E_1 (appointments)	$P(E_1 \cap I_1) = 0.40$	$P(E_1 \cap I_2) = 0.20$	0.6
E_2 (no appointments)	$P(E_2 \cap I_1) = 0.05$	$P(E_2 \cap I_2) = 0.35$	0.4
Marginal probabilities of indicators $P(I_k)$	0.45	0.55	

like to use this additional information to develop an optimum strategy; by *strategy*, we mean a guideline for action by which a specific action is to be performed as a result of each possible outcome of an indicator. We shall use S to represent a strategy. For example, the four possible strategies available to Mr. Morgan are shown in Table 16.7. As an example of the interpretation of the strategies of Table 16.7, strategy S_2, (that is, $I_1 \Rightarrow A$ and $I_2 \Rightarrow B$, the adoption of action A when indicator I_1 occurs and action B when indicator I_2 occurs) means that Mr. Morgan will land at Airport A

Table 16.7

Strategies available to Mr. Morgan

Indicator outcomes	*I_1: Secretary predicts appointments*	
	I_2: Secretary predicts no appointments	
Actions	*A: Land at Airport A*	
	B: Land at Airport B	
Strategy	*Strategy description by action-indicator combination*	
S_1	$I_1 \Rightarrow A$	$I_2 \Rightarrow A$
S_2	$I_1 \Rightarrow A$	$I_2 \Rightarrow B$
S_3	$I_1 \Rightarrow B$	$I_2 \Rightarrow A$
S_4	$I_1 \Rightarrow B$	$I_2 \Rightarrow B$

if his secretary predicts appointments during their phone conversation and that he will land at Airport B if she predicts no appointments. (Note that strategy S_4 is the one he chose previously by looking only at historical data.) How should he select the optimum strategy?

We have already agreed on a decision-theory approach when we assume knowledge of probabilities associated with various event-act combinations. We change only one part of our available information by introducing the concept of the indicator. That is, we have provided a means of changing the prior probabilities associated with events E_1 and E_2 to posterior probabilities. Thus, we proceed as usual to compute expected benefits for each act. However, we shall arrive at the expected benefits for various possible acts through use of appropriate posterior (with regard to indicator outcome) probabilities rather than prior probabilities. These expected benefits will eventually lead to selection of one strategy.

First, we must determine the conditional probabilities $P(E_i|I_k)$ for $i = 1, 2$ and $k = 1, 2$—that is, the conditional probability that the ith state of nature prevails, given that the kth outcome of the indicator has occurred. (These are, in fact, posterior probabilities.) We use the simple version of Bayes' theorem to find

$$P(E_1|I_1) = \frac{P(E_1 \cap I_1)}{\sum_{i=1}^{2} P(E_i \cap I_1)} = \frac{0.40}{0.40 + 0.05} = \frac{0.40}{0.45} = \frac{8}{9}, \tag{3}$$

$$P(E_2|I_1) = \frac{P(E_2 \cap I_1)}{\sum_{i=1}^{2} P(E_i \cap I_1)} = \frac{0.05}{0.40 + 0.05} = \frac{0.05}{0.45} = \frac{1}{9}, \tag{4}$$

$$P(E_1|I_2) = \frac{P(E_1 \cap I_2)}{\sum_{i=1}^{2} P(E_i \cap I_2)} = \frac{0.20}{0.20 + 0.35} + \frac{0.20}{0.55} = \frac{4}{11}, \tag{5}$$

and

$$P(E_2|I_2) = \frac{P(E_2 \cap I_2)}{\sum_{i=1}^{2} P(E_i \cap I_2)} = \frac{0.35}{0.20 + 0.35} = \frac{0.35}{0.55} = \frac{7}{11}. \tag{6}$$

We now ask two questions. Which act (A or B) gives the least expected travel time when the secretary predicts appointments (outcome I_1 occurs)? (This is really a conditional expected travel time—conditional with respect to I and expected with respect to E_1.) Which act (A or B) gives the least expected travel time when the secretary predicts no appointments (outcome I_2 occurs)? The answers will lead us immediately to an optimum strategy based on posterior probabilities. We use the posterior probabilities of (3) through (6). Required calculations are shown in Table 16.8 for outcome I_1 of the indicator and Table 16.9 for outcome I_2 of the indicator. In Table 16.8, for example, column 1 describes the possible states of nature, column 2 gives prior probabilities, and 3 and 4 show the calculations for changing prior probabilities of events E_1 and E_2 to posterior probabilities based on outcome I_1 of the indicator. (Note that one step is now omitted because we know the appropriate joint probabilities. Ordinarily, particularly when we start to use sample information to compute posterior probabilities, such will not be the case.) Columns 5 and 7 list conditional consequences for the various event-act combinations and columns 6 and 8 show the corresponding expected consequences based on outcome I_1. Finally, the sum of column 6 represents the expected travel time Mr. Morgan will encounter as a result of always selecting act A when outcome I_1 of the indicator occurs, and the sum of column 8 represents his expected travel time as a result of always selecting act B when outcome I_1 of the

Table 16.12

Calculations for Example 3

Act A_1—One-game gamble				*Act A_2—Three-game gamble*			
Event (state of nature) E_i	$P(E_i)$	*Conditional consequence* ($)	*Expected consequence* ($)	*Event (state of nature)* E_i	$P(E_i)$	*Conditional consequence* ($)	*Expected consequence* ($)
E_1 (win)	0.6	10,000	6,000	E_1 (win)	$(0.6)^3 = 0.216$	40,000	8,640
E_2 (lose)	0.4	−11,000	−4,400	E_2 (lose)	$1 - (0.6)^3 = 0.784$	−10,000	−7,840
Expected monetary value of act A_1			1,600	Expected monetary value of act A_2			800

particular circumstances. Suppose further, that the decision maker is not sure whether the situation is such that he can use expected monetary value as the criterion, or if he should consider the utility that he attaches to the various possible consequences and, hence, make a decision on the basis of expected utility. How does he decide which criterion to use? Further, if he decides on expected utility, how should he implement the analysis?

To answer such questions for situations involving acts that can result in one of several possible outcomes, we must first be able to answer such questions for situations involving gambles that can result in only one of two possible outcomes. The question to be answered in this case is as follows: "Regardless of the probabilities attached to the two possible consequences (assuming, of course, that the probabilities total to one), will the individual be indifferent to a choice between

1. a gamble in which the two possible consequences have such probabilities of occurring, respectively, and
2. a certainty of receiving (or paying) the expected value of such gamble?"

If the answer to this question is yes, then expected monetary value is a valid guide in deciding whether to take a gamble or not. If the answer is no, expected monetary value is not a valid guide.

Example 4

Suppose there is 1 chance in 1,000 that during any given year an individual will be involved in a wreck which will ultimately cost him $50,000 in a liability suit. Is he willing to be guided by expected monetary value in deciding whether to buy liability insurance? The individual's situation is such that a $50,000 loss will ruin him financially.

Solution

The expected monetary value of his not insuring is

$$\frac{1}{1{,}000}(-\$50{,}000) + \frac{999}{1{,}000}(\$0) = -\$50.$$

However, to most of us, the threat of a $50,000 lump-sum loss is so bad that we happily pay (say) $100 per year to avoid the risk. Most people are by no means indifferent to the choice between a 1/1,000 chance of losing $50,000 and the certainty of losing $50 every year. We are willing to pay more than $50 to avoid even a small risk of the large loss. Thus, expected monetary value is obviously not a valid guide. If, on the other hand, the individual says that he will pay no more than $50 per year for insurance, he is really saying he is indifferent between the two choices. In this case, we must question him regarding other probabilities which might be attached to a wreck; we have not established that expected monetary value is a valid guide, but have simply failed to eliminate it as a valid guide. □

Example 5

Suppose the same situation as that of Example 4, except that the decision maker is a multimillion-dollar company that owns a fleet of cars. It is quite likely that, knowing that some of its cars will be involved in costly wrecks but that most of them will not, the company will draw the line at $50 per year per car for liability insurance. Further, if the probability of a wreck were different than that given in Example 4, it is likely that the company would be willing to pay no more for insurance than the expected value of the relevant gamble. In this case, we see that the company is willing to be guided by expected monetary value—because it has enough cars on the road to take advantage of the meaning of expected value.

Obviously common sense must be used in deciding whether expected monetary value is a valid guide in a gamble involving only two possible outcomes. In evaluating acts that have only two possible consequences, people act as if all dollars are of equal value as long as the best of the two consequences is not "too good" and the worst is not "too bad."

We can take this method a step further to decide whether expected monetary value is valid for evaluating an act having several possible consequences. We simply apply the same type of reasoning: we consider only the best and worst of all possible consequences of the act and then ask ourselves whether expected monetary value is a valid guide when these are the only two possible consequences. Thus, we simply apply the aforementioned question to the best and worst consequences. When we use such a question as a guide in determining whether expected monetary value is a valid criterion, we are really applying the following reasoning. If a decision maker would use expected monetary value as a criterion in choosing between an act which is certain to result in a specific dollar consequence and an act which will result in either the best or worst of all possible consequences, he should use expected monetary value as a criterion in the decision problem of interest. Otherwise, he should not.

In any real decision problem, then, the analyst first must decide through use of the above reasoning whether expected monetary value is a valid criterion. When the reasoning leads to a "no" answer, he must resort to an analysis involving expected utility. This leads us immediately to the question of how to convert absolute consequences to a measure of utility or, more generally, how to construct a *utility curve* for a given circumstance or set of circumstances. So many techniques have been set forth for constructing utility curves that it would be impossible even to mention all of them. While we are interested in developing your capability to make an analysis involving expected utilities, we do not wish to belabor the question of how best to construct a utility curve. Thus, we simply wish to present only one method for constructing a utility curve. We have selected a method which we believe will enable you to see from an intuitive standpoint the real meaning of the concept of utility.

Before we list the general steps required in an analysis involving expected utility, let us examine the meaning of the utility of a given consequence—that is, an individual's utility for that consequence. First of all, we have already indicated that a person's utility for a given amount of money depends on both his attitude toward risk and the particular circumstances surrounding an analysis. This being the case, it is meaningless to speak of "the utility to an individual" of a given consequence. We must speak of such a utility in reference to some alternative. Thus, henceforth, we shall always begin any attempt to determine a utility for a given consequence by establishing two *reference consequences*. Then, any statement of utility will be stated in relation to these reference consequences. The steps in determining the utility of a given consequence C in a given situation are:

1. Establish two reference consequences R_1 and R_2 such that $R_1 \leq C \leq R_2$.
2. Determine the value (say π_c) at which the individual is indifferent to a choice between accepting C with certainty and accepting a gamble (proposition) in which the probability of R_2 is π_c and the probability of R_1 is $1 - \pi_c$.
3. Define π_c to be the utility (to the individual) of C in terms of R_1 and R_2.

In general, the utility (to an individual) of a given consequence is the probability which makes a reference gamble involving two reference consequences exactly equivalent (at least in the mind of the individual) to the given consequence.

We generally label as a *risk avoider* an individual who consistently assigns a utility in such a way that the given consequence is less than the expected value of the corresponding reference gamble. That is, the risk avoider consistently associates a utility of π_{c_i} with the consequence C_i such that

$$(\pi_{c_i})R_2 + (1 - \pi_{c_i})R_1 > C_i \qquad (i = 1, 2, \ldots), \tag{1}$$

where C_i is any consequence in the range bounded by the reference consequences R_1 and R_2. The risk avoider essentially says that his desire (or need) to avoid consequence R_1 is so great that he will not enter a gamble involving it unless the expected value of the gamble is greater than the consequence he can receive with certainty.

On the other hand, we generally label as a *risk lover* an individual who consistently assigns a utility to a consequence in a way that the given consequence is greater than the expected value of the corresponding reference gamble. That is, the risk lover consistently associates a utility of π_{c_i} with the consequence C_i such that

$$(\pi_{c_i})R_2 + (1 - \pi_{c_i})R_1 < C_i \qquad (i = 1, 2, \ldots), \tag{2}$$

where, again, C_i is any consequence in the range bounded by the reference consequences R_1 and R_2. The risk lover essentially says that his desire (or need) to have even some chance at consequence R_2 is so great that he will enter a gamble involving it even though the expected value of the gamble is less than the consequence which he can receive with certainty.

What if an individual consistently associates a utility of π_{c_i} with the consequence C_i such that

$$(\pi_{c_i})R_2 + (1 - \pi_{c_i})R_1 = C_i \qquad (i = 1, 2, \ldots), \tag{3}$$

where C_i, R_1, and R_2 are the same as in (1)? We see immediately that this person is essentially willing to be guided by expected monetary value.

Example 6

A certain risk avoider wishes to determine the utility to him of \$5,000 in terms of the reference consequences +\$20,000 and −\$20,000. Let us denote this utility by

π_5. What is the value below which he will not assign a value of π_5? What question must he answer in order to fix π_5?

Solution

If the risk avoider were willing to be guided by expected monetary value, we know he would assign a probability (say p_5) to the reference consequence \$20,000 such that

$$p_5(20{,}000) + (1 - p_5)(-20{,}000) = 5{,}000. \tag{4}$$

In other words, he would be indifferent to a choice between a reference gamble having expected value \$5,000 and a certain consequence of \$5,000. Solving (4) for p_5, we have

$$\begin{aligned} 20{,}000p_5 - 20{,}000 + 20{,}000p_5 &= 5{,}000 \\ \Rightarrow 40{,}000p_5 &= 25{,}000 \\ \Rightarrow p_5 &= \frac{25{,}000}{40{,}000} = 0.625. \end{aligned} \tag{5}$$

We know that a risk avoider will not consider a gamble involving the reference consequences \$20,000 and $-$\$20,000 unless his probability of winning \$20,000 is greater than 0.625 and his probability of losing \$20,000 is less than $1 - 0.625 = 0.375$. Hence, we know he will associate a utility of $\pi_5 > 0.625$ with the consequence \$5,000.

In order to fix the utility, he must choose a probability π_5 that makes him indifferent to a choice between the gamble involving a probability of π_5 of winning \$20,000 [and hence a probability of $(1 - \pi_5)$ of losing \$20,000] and a certainty of receiving \$5,000. If the individual is only slightly conservative and the loss of \$20,000 will not be too damaging to him, he may assign a utility of $\pi = 0.7$ to the consequence \$5,000. On the other hand, if he is highly conservative or if the loss of \$20,000 is something he must avoid at any reasonable cost, he may attach a very high utility to the consequence \$5,000. This kind of decision can be made only by the individual involved. However, it is obvious that the expected-value calculations of (5) will provide a guideline to such a decision. □

The value $p_5 = 0.625$ in the foregoing example would obviously serve as an upper limit to the utility that a risk lover would attach to the consequence \$5,000.

It seems clear that, if we agree on a reasonable set of reference consequences, we can affect a decision analysis in a situation in which we are not willing to be guided by expected monetary value simply by assigning a utility to each possible consequence involved and choosing the act (or strategy) with the highest expected utility. However, if there are many possible consequences, this approach may be too laborious. In this case, one can construct a utility curve that shows all the utilities required in a specific analysis.

Suppose an individual must make a decision involving many possible event-act combinations and, therefore, many consequences. Suppose further that he is not willing to be guided by expected monetary value. In order to perform an analysis based on expected utility, he should construct a utility curve by use of the following general steps.

1. Determine the best possible consequence, C_B, and the worst possible consequence, C_W, involved in the analysis.
2. Establish two reference consequences, R_1 and R_2, such that
$$R_1 \leq C_W \quad \text{and} \quad R_2 \geq C_B. \tag{6}$$
3. Select a finite set of n (perhaps equally spaced) values C_i ($i = 1, 2, \ldots, n$) such that
$$R_1 \leq C_i \leq R_2 \qquad (i = 1,\ 2,\ \ldots,\ n). \tag{7}$$
4. Assign a utility π_{c_i} to each of the n values C_i. That is, for each value C_i, determine the value π_{c_i} such that the decision maker is indifferent with regard to a choice between accepting the consequence C_i with certainty and accepting a gamble (proposition) in which the probability is π_{c_i} that he will receive consequence R_2 and the probability is $(1 - \pi_{c_i})$ that he will receive consequence R_1.
5. For $i = 1, 2, \ldots, n$, plot π_{c_i} on the vertical scale and C_i on the horizontal scale.
6. Connect the points by a smooth curve, and label it the utility curve for the individual. The utility curve can now be used to assign utilities to any consequence involved in a real decision problem as long as this consequence is between R_1 and R_2 inclusive.

Naturally, the number of consequences may be too small to justify the work of constructing a utility curve. Then the analyst must apply step 4 above to each possible consequence in the decision problem. That is, he must assign utility values directly to each and then calculate the expected utility for the various alternative acts.

Example 7

Suppose a consultant (Mr. Jones) has the opportunity of bidding on two possible contracts: act A and act B. (Note also the implicit alternative of making no bid, the certain consequence of which will be \$0.) There are five contingencies (states of nature $E_1, E_2, \ldots, E_5$) that will affect the profit resulting from each possible action. The probabilities $P(E_i)$ of the various probabilistic events are shown in column 2 of Table 16.13. Columns 3 and 5 present conditional (that is, conditional on the occurrence of the possible states of nature) consequences that will accrue to Mr. Jones as a result of act A and act B, respectively. Determine the contract (if either) on which he should bid if he is willing to be guided by expected monetary value. Next, suppose he is not willing to be guided by expected monetary value. Show the reasoning (along with example answers) which he might use to make a decision based on expected utility.

Solution

In order to determine which contract (if either) to choose, Mr. Jones must answer two questions. Which act has the larger expected monetary value? Is the expected monetary value of such act greater than the value he can obtain by a no-bid? To answer the questions, he must calculate the expected monetary value of each act, as shown in columns 4 and 6 of Table 16.13. Note that the expected monetary value of both acts is positive (greater than the amount $0 which will result from a no-bid). Thus, since the expected monetary value of act *A* ($4,200) is greater than the expected monetary value of act *B* ($3,700), Mr. Jones should choose act *A* if he is willing to be guided by expected monetary value.

Table 16.13

Analysis of Mr. Jones' (a risk avoider) decision problem—decision based on expected monetary value

		Monetary consequence ($) of bidding on contracts A and B			
		Act A		*Act B*	
(1) *Event (state of nature)* E_i	(2) *Probability* $P(E_i)$	(3) *Conditional consequence*	(4) *Expected consequence*	(5) *Conditional consequence*	(6) *Expected consequence*
E_1	0.4	20,000	8,000	15,000	6,000
E_2	0.1	10,000	1,000	12,000	1,200
E_3	0.1	5,000	500	6,000	600
E_4	0.1	−8,000	−800	−5,000	−500
E_5	0.3	−15,000	−4,500	−12,000	−3,600
Expected monetary values of acts			4,200		3,700

Monetary consequences of no bid = $0

Now, suppose that Mr. Jones feels forced to be a risk avoider. How should he proceed? He has two choices. He may choose a set of reference consequences and establish a utility for each of the possible conditional consequences (Table 16.13). Or, he may use these reference consequences to establish a somewhat general utility curve that he can use in any decision problem as long as his situation and attitude do not change and none of the consequences are outside the range of the reference consequences. Suppose Mr. Jones chooses the latter route; and, noting that the best possible consequence of the problem is $20,000 and the worst possible consequence is −$15,000, he decides to base his utility values on reference consequences of −$25,000 and +$25,000. Suppose further that he decides to evaluate his utility for all consequences at $5,000 intervals over the range covered by these reference consequences. The following steps are representative of the reasoning he must go through to establish the utility curve.

1. He may first evaluate the consequence $0. To do so, he must ask himself the question "What probability (say π_0, where π represents a utility and the subscript indicates the specific consequence to which such utility refers) must I choose so that I am indifferent to a choice between receiving $0 with certainty and accepting a gamble wherein I have a probability of π_0 of winning $25,000 and a probability of $(1 - \pi_0)$ of losing $25,000?"

Being a risk avoider, he can place a lower limit on π_0 by solving for the probability (p_0) which will make the expected value of the gamble equal to the consequence being evaluated ($0). That is, he must solve the relation

$$p_0(25{,}000) + (1 - p_0)(-25{,}000) = 0$$

for p_0. He finds $p_0 = 0.5$, as one might expect. Suppose, however, that Mr. Jones is rather conservative and, finally, assigns a value of $\pi_0 = 0.80$ (Mr. Jones' utility for $0).

Note that in answering this particular question, Mr. Jones has done far more than merely evaluate one of the points on his utility curve. He has established a value against which he can eventually evaluate the expected utilities of the two acts in question. For, recall that he always has available the act of no-bidding, which has the dollar value $0.

2. He next evaluates several positive consequences through questions of the following nature. For example, if he wishes to assign a utility to the consequence +$5,000, he asks "What probability (say π_{+5}) will make me indifferent to a choice between receiving $5,000 with certainty and accepting a gamble wherein I have a probability of π_{+5} of winning $25,000 and a probability of $(1 - \pi_{+5})$ of losing $25,000?"

   You might verify that, since Mr. Jones is a risk avoider, the lowest value π_{+5} can take on according to the criterion of expected monetary value becomes the solution of

$$p_{+5}(25{,}000) + (1 - p_{+5})(-25{,}000) = 5{,}000, \tag{8}$$

   or $p_{+5} = 0.6$. However, Mr. Jones has already attached a utility of $\pi_0 = 0.80$ to the consequence $0. Certainly, then, π_{+5} will be greater than 0.80.

   Suppose he assigns a value of 0.87 to π_{+5} (Mr. Jones' utility for +$5,000). Suppose further that similar reasoning (be sure that you understand clearly the question Mr. Jones poses in each case) leads him to assign the following utilities to appropriate positive consequences:

Positive consequences	*Corresponding utility*
+$10,000	$\pi_{+10} = 0.92$
+$15,000	$\pi_{+15} = 0.96$
+$20,000	$\pi_{+20} = 0.99$

   Note that the utility which must be attached to the consequence +$25,000 is automatically $\pi_{+25} = 1.0$, since this is the only probability that will make the gamble equivalent to the *certain* consequence *even* under the criterion of expected monetary value.

3. Mr. Jones' next evaluates several *negative* consequences through questions of the following nature. For example, if he wishes to assign a utility to the consequence −$10,000, he must ask "What probability (say π_{-10}) will make me indifferent to a choice between *paying* $10,000 with certainty and accepting a gamble wherein I have a probability of π_{-10} of winning $25,000 and a probability of $(1 - \pi_{-10})$ of losing $25,000?"

Again, you should verify that the lowest value which π_{-10} can assume according to the criterion of expected monetary value becomes the solution to

$$p_{-10}(25{,}000) + (1 - p_{-10})(-25{,}000) = -10{,}000, \tag{9}$$

or $p_{-10} = 0.30$.

Suppose Mr. Jones assigns a value of 0.57 to π_{-10} and that similar reasoning leads him to assign the following utilities to appropriate *negative* consequences:

Negative consequences	*Corresponding utility*
−\$5,000	$\pi_{-5} = 0.70$
−\$15,000	$\pi_{-17} = 0.40$
−\$20,000	$\pi_{-20} = 0.22$

Note that the utility which should be attached to the consequence −\$25,000 is automatically $\pi_{-25} = 0$, since it makes the gamble in fact identical to the certain consequence.

4. Mr. Jones can now establish a utility curve by smoothing a curve through the points obtained by plotting on the horizontal axis the consequences evaluated in steps 1, 2, and 3, versus (on the vertical axis) the utilities assigned to such consequences. The resulting utility curve is denoted in Figure 16.1 as the utility

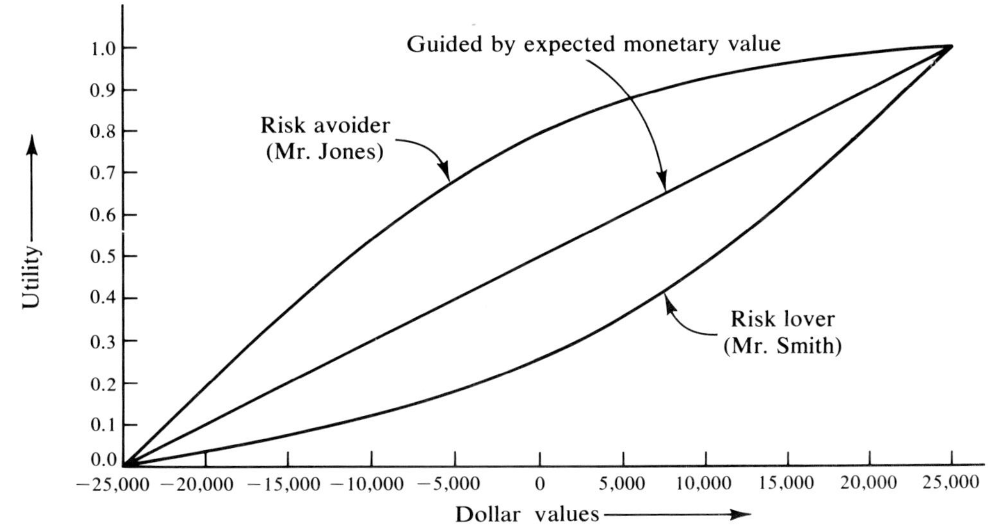

Figure 16.1 Utility curves of Example 7

curve of a risk avoider. Note that we can attach the following meaning to any point on the curve. Mr. Jones is indifferent to a choice between receiving (or paying) the corresponding amount on the horizontal axis and accepting a gamble wherein he has a probability equal to the corresponding utility shown on the vertical axis of winning \$25,000 and a probability equal to one minus the utility shown on the vertical axis of losing \$25,000.

Having now established the utility curve, Mr. Jones can implement his analysis simply by

1. replacing every possible consequence in his real decision problem (Table 16.3) by the utility which he has attached to such consequence,
2. proceeding to calculate the expected utility of each possible act, and
3. choosing the act with the largest expected utility if such expected utility is greater than his utility for \$0 (no action at all) and electing no act at all otherwise.

Required calculations are presented in Table 16.14. The utilities of columns 3 and 5 of Table 16.14 are obtained by reading directly from Mr. Jones' utility curve of Figure 16.1 the utilities corresponding to the dollar consequences involved in his decision problem. Note two very important things that happened because Mr. Jones is a risk avoider. First, since the expected utility of act *B* (0.791) is greater than the expected utility of act *A* (0.757), Contract *B* is now more attractive than Contract *A*. Recall that this represents an exact reversal of the results of a decision based on expected monetary value (Table 16.13). Second, neither contract is as attractive to Mr. Jones as a no-bid, since the utility of a no-bid $= 0.80 > 0.791 > 0.757$.

Table 16.14

Analysis of Mr. Jones' (a risk avoider) decision problem

		Utility consequences of bidding on contracts A and B			
		Act A		*Act B*	
(1) *Event (state of nature)* E_i	(2) *Probability* $P(E_i)$	(3) *Conditional consequence*	(4) *Expected consequence*	(5) *Conditional consequence*	(6) *Expected consequence*
E_1	0.4	0.99	(0.4)(0.99) = 0.396	0.96	(0.4)(0.96) = 0.384
E_2	0.1	0.92	(0.1)(0.92) = 0.092	0.93	(0.1)(0.93) = 0.093
E_3	0.1	0.87	(0.1)(0.87) = 0.087	0.88	(0.1)(0.88) = 0.088
E_4	0.1	0.62	(0.1)(0.62) = 0.062	0.70	(0.1)(0.70) = 0.070
E_5	0.3	0.40	(0.3)(0.40) = 0.120	0.52	(0.3)(0.52) = 0.156
Expected utility of acts			0.757		0.791

Utility to Mr. Jones of no bid = 0.80

Two other curves are presented in Figure 16.1. The curve labeled "Risk lover (Mr. Smith)" represents a curve resulting from our asking a risk lover specific questions similar to those answered by Mr. Jones in Example 7. The straight line represents the utility curve of a person who is willing to be guided by expected monetary value. Note once again that a person exhibiting this type of utility curve is essentially saying that he is indifferent to a choice between receiving a given consequence with certainty and accepting a gamble (involving the appropriate reference consequences) for which the expected monetary value is equal to the given consequence. □

Expected consequence ($)	*Utility*
25,000	1.00
20,000	0.98
15,000	0.95
10,000	0.90
5,000	0.79
0	0.66
−5,000	0.42
−10,000	0.00

1. Construct a utility curve.
2. Using the utility curve, which alternative will Mr. Gilbert choose?
3. Based on the utilities he placed on the different outcomes, what is Mr. Gilbert's attitude toward risk? Show explicitly how you arrive at an answer.

5. Refer to the utility curves in Figure 16.1:

a. What utility would a risk lover place on receiving $7,500 with certainty and accepting a gamble with a probability of $\pi_{7.5}$ of making $25,000 and $1 - \pi_{7.5}$ of losing $25,000?

b. To what value would a risk avoider give a utility of $\pi = 0.45$?

c. If an individual places a utility of 0.6 on the consequence $5,000, what can you say about his attitude toward risk? Explain.

d. Can you infer from Figure 16.1 what the *absolute* best alternative is for any given problem? Explain.

e. What value of P would make the expected value of making $25,000 with a probability of P_9 and losing $25,000 with a probability of $1 - P_9$ equal to the consequence $9,000? Explain.

16.5 *The General Bayesian Approach Involving Expected Consequences*

We are now ready to generalize the Bayesian approach to decision making. We shall assume for the present that "expected (material or utility) consequences" is a suitable criterion for making a decision. The essential characteristics of the general problem are as follows:

1. The analyst is repeatedly (at least conceptually) required to choose from among several (say m) acts A_j ($j = 1, 2, \ldots, m$).

2. Each act will result in a definite consequence to the decision maker. The consequence, represented by absolute material values (money, time, etc.) or by utility values, depends on a set (say n) of probabilistic events E_i ($i = 1, 2, \ldots, n$), commonly called states of nature.

3. Information is available regarding prior (with respect to some indicator) probabilities for the various events, $P(E_i)$ ($i = 1, 2, \ldots, n$).

4. There may exist an indicator I having a finite number (say p) of outcomes I_k ($k = 1, 2, \ldots, p$) that the analyst can observe before making a decision; information from it will allow the analyst to revise the prior (with respect to the indicator I) probabilities $P(E_i)$ ($i = 1, 2, \ldots, n$) to posterior (with respect to the indicator I) probabilities

$$P(E_i \mid I_k) \qquad (i = 1, 2, \ldots, n;\ k = 1, 2, \ldots, p).$$

In this case, we will be given all of the np joint probabilities

$$P(E_i \cap I_k) \qquad (i = 1, 2, \ldots, n;\ k = 1, 2, \ldots, p), \tag{1}$$

or we will be provided a means of obtaining sufficient information to compute such joint probabilities; that is, we can obtain the np conditional probabilities

$$P(I_k \mid E_i) \qquad (i = 1, 2, \ldots, n;\ k = 1, 2, \ldots, p). \tag{2}$$

5. There exists a set of strategies $S_1, S_2, \ldots$ from which the analyst must select one:
   a. If no information is available regarding an indicator, the analyst must choose from the set of acts given under point 1 of this listing. Thus, the problem is to choose one act (course of action) to be followed in all cases, and the number of possible strategies is equal to the number (m) of alternative acts available.
   b. If information is available regarding the indicator I, the analyst must select a strategy which determines the particular act (course of action) to be taken for each possible outcome of the indicator. That is, he must choose a strategy from all possible strategies of the following form:

$$\begin{aligned}
&\text{Outcome } I_1 \Rightarrow \text{one of the acts } A_1, A_2, \ldots, A_m,\\
&\text{Outcome } I_2 \Rightarrow \text{one of the acts } A_1, A_2, \ldots, A_m,\\
&\quad\vdots\\
&\text{Outcome } I_p \Rightarrow \text{one of the acts } A_1, A_2, \ldots, A_m,
\end{aligned}$$

   where, again, the symbol $\Rightarrow$ means "dictates," or "leads to." Thus, since each of the p indicator outcomes can lead to any of the m acts, the number of possible strategies (combinations of indicator outcome-act) from which the analyst must choose is

$$\underbrace{m \cdot m \cdots m}_{p \text{ times}} = (m)^p. \tag{3}$$

6. The analyst's objective is to select the strategy that will maximize or minimize, whichever is appropriate, the expected consequence. Depending on whether or not information is available on an indicator, the analyst uses one of two possible general approaches to selecting the optimum strategy.

No Indicator Available

If no indicator is available the analyst must base his decision on prior probabilities of the various possible events (states of nature). The procedure in this case is as follows:

1. Select one of the possible strategies (acts, in this case), say the sth act—that is, act A_s.

2. Compute the expected consequence of Act A_s by

$$E(A_s) = P(E_1) \cdot C_{1s} + P(E_2) \cdot C_{2s} + \cdots + P(E_n) \cdot C_{ns} = \sum_{i=1}^{n} P(E_i) C_{is}, \tag{4}$$

where C_{is} represents the consequence that will accrue if the analyst selects act A_s and if the specific event E_i occurs. (Note that we now allow the possibility that the act selected may affect the probability distribution associated with the probabilistic event of interest.)
3. Repeat steps 1 and 2 for all possible acts, that is, for $s = 1, 2, \ldots, m$.
4. Select as a strategy the act for which the expected consequence of (4) is optimum, (maximum or minimum, as desired).

Indicator I Available

If there is an indicator I, the analyst should base his decision on posterior (with respect to the indicator I) probabilities of the various possible events (states of nature). The procedure follows:

1. Select a pertinent indicator outcome, say the tth indicator outcome, or indicator outcome I_t.
2. By use of Bayes' theorem, revise the rth prior probability, $P(E_r)$, to a posterior (with respect to the indicator I) probability, $P(E_r | I_t)$; in other words, compute

$$P(E_r | I_t) = \frac{P(E_r \cap I_t)}{\sum_{i=1}^{n} P(E_i \cap I_t)} = \frac{P(E_r) P(I_t | E_r)}{\sum_{i=1}^{n} P(E_i) P(I_t | E_i)}. \tag{5}$$

3. Repeat step 2 for all possible prior probabilities, that is, for $r = 1, 2, \ldots, n$.
4. Restricting consideration to the indicator outcome I_t, select one of the possible acts, say the sth act, or Act A_s.
5. Compute

$$E(A_s | I_t) = P(E_1 | I_t) \cdot C_{1s} + P(E_2 | I_t) \cdot C_{2s} + \cdots + P(E_n | I_t) \cdot C_{ns}$$
$$= \sum_{i=1}^{n} P(E_i | I_t) \cdot C_{is}, \tag{6}$$

where $E(A_s | I_t)$ is the "conditional (with respect to the indicator I) expected consequence of act A_s given indicator outcome I_t," $P(E_i | I_t)$ is as determined in (5), and C_{is} is again the consequence that will accrue if the analyst selects act A_s and if the specific event E_i occurs.
6. Repeat steps 4 and 5 for all possible acts, that is, for $s = 1, 2, \ldots, m$.
7. Select as a partial strategy (for use whenever outcome I_t occurs) the act for which the conditional expected consequence of step 6 is optimum (maximum or minimum, as desired).
8. Repeat steps 1 through 7 for all possible indicator outcomes, that is, for $t = 1, 2, \ldots, p$. This gives the analyst a specific partial strategy to be used whenever any specific indicator outcome occurs.
9. Combine all partial strategies obtained in steps 1 through 8 to obtain an overall strategy.

The analyst may not be satisfied merely to select an optimum strategy. He may also be interested in the overall expected consequences of a strategy. Recall that any strategy based on the outcome of an indicator with p possible outcomes I_k ($k = 1, 2, \ldots, p$) is comprised of p partial strategies, one for each possible outcome of the indicator I. Recall also that associated with each partial strategy is a conditional expected consequence. For example, when the partial strategy is to let indicator outcome I_t lead to act A_s, we denote by $E(A_s | I_t)$ the conditional expected consequence of the partial strategy $I_t \Rightarrow A_s$. Then, the overall expected consequence of a strategy can be computed as

$$E(S) = \sum_{k=1}^{p} E(A_k | I_k) P(I_k), \tag{7}$$

where, for this formula only, $E(A_k | I_k)$ is the conditional expected consequence of the kth partial strategy, and $P(I_k)$ is the probability of outcome I_k occurring.

16.6 Summary

The essential characteristics of the decision problems we have been discussing are as follows. The decision maker must choose from among several alternative acts. Each act results in a definite material consequence (e.g., money), but the exact consequence depends on the states of nature that constitute the set of probabilistic events. While the decision maker may, by his choice of an act, influence the probability distribution associated with the possible states of nature (events), he has no control over or direct knowledge about which state of nature will exist at any given time. In any case, information is available on the prior probabilities for the various possible states of nature. There may exist an indicator that provides information allowing the decision maker to revise the prior (with respect to the indicator) probabilities to posterior probabilities and view an outcome before rendering a terminal decision.

The objective of the decision maker is to choose the strategy which optimizes the consequences—that is, yields the best expected benefit. If all units (e.g., dollars) of the material consequence under consideration are equally valuable to him, the expected benefit which he should choose is the one offering the maximum appropriate expected material consequence (e.g., expected monetary value). On the other hand, if all units of the material consequence are not equally valuable to the decision maker, he should convert the material consequences of his decision problem to a measure that expresses his utility for such consequences and choose the strategy that allows him to optimize expected utility—that is, the strategy yielding the greatest expected utility. The first of this two-part step can often be expedited by constructing a utility curve.

When no information is available regarding an indicator, the decision maker usually must render a terminal decision without benefit of added information. That is, he must essentially choose one of the possible acts to be employed in all cases. In this case, the strategies available to him are simply the alternative acts, and the decision maker must use available prior probabilities in the expected-value calculations in selecting an optimum strategy (act). On the other hand, when an indicator exists, he may choose one act when one outcome of the indicator occurs and a different act when another outcome occurs. In such case, a given strategy consists of a statement of the specific act to be taken under the occurrence of each possible outcome of the indicator; the decision maker should use posterior (with respect to the indicator) probabilities in the expected-value calculations leading to an optimum strategy.

Important Terms and Concepts

act	expected monetary value	risk avoider
attitude toward risk	expected utility	risk lover
Bayes' theorem	indicator	state of nature
Bayesian inference	partial strategy	statistical decision theory
conditional consequence	posterior probability	strategy
consequences	preference	terminal decision
dominance	preference scale	utility
event	prior probability	utility curve
event-act combination	reference consequence	utility scale
expected consequence	reference gamble	

References

16.1 R. Schlaifer, *Statistics for Business Decisions*, New York: McGraw-Hill Book Company, 1961.

16.2 K. Sasaki, *Statistics for Modern Decision Making*, Belmont, Calif.: Wadsworth Publishing Company, 1968.

Review Problems

1. A demolition firm stocks three grades of explosives: low, medium, and high. The firm has just moved to a new locale. During a move to a new location, the explosives, which appear identical, were packed by grade in 100-unit lots in identical, unlabeled boxes. It is known that the inventory of explosives contains low, medium, and high grades in proportions of 0.25, 0.25, and 0.50, respectively, and that the proportion defective is 0.15, 0.10, and 0.05, respectively. Company officials want to label the boxes correctly, but because the test to determine defective explosives is necessarily destructive, then decide to test a sample of 10 units from each box and make a decision based on the results. Suppose a sample is taken and one

unit is found to be defective. Using *binomial probabilities to approximate hypergeometric probabilities*, find the probability that the box contains

a. high-grade explosives,

b. medium-grade explosives,

c. low-grade explosives,

d. at least medium-grade explosives.

2. Each of six boxes contains 10 marbles. Two of them (Type A) have 6 black and 4 white marbles, three of them (Type B) have 5 black and 5 white marbles, and one (Type C) has 3 black and 7 white marbles. The boxes are selected by rolling a pair of fair dice. The dice are thrown, and if the total is 2, 3, 8, 10, or 12, a Type A box is selected; if the total is 4, 5, or 6, a Type B box is selected; and, if the total is 7, 9, or 11, a Type C box is selected.

   a. The dice are thrown, a box selected, and a white marble is withdrawn. What is the probability that a Type A box was selected?

   b. Suppose the dice are thrown, and a sample of 20 is selected (with replacement) from the box chosen. If the sample has a total of 4 black balls, what is the probability that a 7 appeared on the roll of the dice?

3. A hospital received 100 doses of a new drug from a pharmaceutical laboratory. Past experience indicates that 70 or 80 units out of the 100 will be effective. They have estimated a probability of 0.7 for the state "7 effective doses," and 0.3 for the state "8 effective doses." Ten out of the 100 doses are taken as a sample and tested; 6 are found to be effective. (Assume sampling *with* replacement.)

   a. Based on the sample information, which state is the most probable and what is its probability?

   b. Suppose an additional sample of 10 is taken and 8 are found to be effective. Does this additional sample support the results of Part a?

4. You are the president of a company that manufactures a new type of pencil sharpener and you are involved in a patent-rights suit. The party initiating the suit has set a deadline of midnight tomorrow for settlement out of court by arranging a contract for royalty payments on the patent rights. If the suit is not settled in that manner you are certain that you will be forced to stop production of the pencil sharpener, and the shut-down will cost you about $120,000. If the suit is settled out of court, the total cost per unit will increase from the present cost of $2.20 to $2.50. Your lawyers, however, assure you that you are certain to win the suit if it goes to court, and your unit costs, therefore, would remain at $2.20.

   Another factor you must consider is that a large office-supply company is opening bids for one-million pencil sharpeners next week, and you obviously cannot bid on this contract if you go to court. If you settle out of court, incurring the additional costs, your possible bids and the probabilities of winning the contract are as follows:

Bid price/unit	*Probability of obtaining the contract*
2.65	0.20
2.60	0.40
2.57	0.65
2.54	0.80

   You also have the opportunity of bidding on a large contract to be let by a foreign government at a later date. Production limitations prevent you from bidding on this contract if you should be awarded the first. To bid for this contract, you have to invest in new packing equipment. The more packing facilities you have, the more likely you will be awarded the contract. The conditional profit for this contract, exclusive of the cost of the new equipment, is $250,000 if your unit costs are $2.50, and $325,000 if your unit costs are \$2.20. The new equipment has to be amortized by the contract. Your possible investments and the probabilities of obtaining the contract at different investment levels are:

Investment	*Probability of obtaining the contract*
50,000	0.25
65,000	0.40
80,000	0.55
100,000	0.65

a. Determine the total number of alternatives available to you and list them.

b. Determine the best alternative if you choose to go to court. Use the expected monetary value as your decision criterion.

c. Determine the best alternative if you choose to settle out of court. Use the same criterion.

d. Under the same criterion, determine whether you should decide to settle out of court or fight the suit in court.

5. The table below gives the utility preferences of a particular decision maker working on Problem 4.

Dollar value (in thousands)	*Utility*
40	0
60	0.05
80	0.11
100	0.19
120	0.28
140	0.37
160	0.51
180	0.69
200	1.00

a. Construct a utility curve from the given values.

b. Using the utility curve constructed in Part a,

   1. which alternative under Act *B* would he choose?
   2. which alternative under Act *A* would he choose?
   3. would he elect to go to court?

Show all work.

Seventeen

Use of Bayes' Theorem in Binomial Sampling Problems

17.1 Introduction

In the last chapter, we considered general decision problems, criteria for making decisions, and the general approach to decision making called the Bayesian approach. Now, we shall begin applying the Bayesian approach to problems involving the binomial sampling distribution. To review the concepts associated with the binomial distribution, see Section 5.2.

The problem (involving binomial sampling) to which Bayesian inference is applicable is as follows. We know that we are sampling from a Bernoulli process (or, at least, a process with characteristics so much like those of a Bernoulli process that we may for all practical purposes assume a Bernoulli process). While we don't know the exact value of the population proportion of successes (the parameter π), we do have prior information regarding probabilities associated with the various possible values that π can assume. (Note that if we assume all possible values of π are equally likely—essentially no information—we have the same problem as that involved in classical testing of hypotheses concerning π.) We must choose among various alternative courses of action, but the consequences depend on the true value of the population proportion π. Before deciding upon the action to be taken, we can take a sample of size n from the Bernoulli process under consideration. Our objective, of course, is to select the action offering the best possible expected consequence. As before, we shall generally be concerned with maximizing expected monetary value.

There are two possible approaches. On the one hand, we may take a sample and then simply decide what action is best in light of it. In this case, as in the general problem described above, we are merely concerned with evaluating the expected monetary value of each possible act. On the other hand, we may try to determine an overall strategy before collecting any sample information. That is, we may conduct an analysis to determine the specific action to be taken under each possible outcome of the sample. In either case, we utilize Bayes' theorem.

While now considering applications of Bayesian inference in problems involving the binomial sampling distribution, we shall also expand on the problem of testing hypotheses. In particular, we shall relate the Bayesian approach to the classical techniques of decision making encountered in Chapter 9 on tests of hypotheses. You will find that the Bayesian approach to decision making allows us to answer explicitly questions for which we had only implicit (even indirect) answers in Chapter 9.

One good way to show the relationship between the classical and Bayesian approaches to decision making is to use problems involving opportunity losses associated with various event-act combinations, rather than merely absolute dollar consequences. We begin with a brief introduction to the concept of opportunity loss.

17.2 Opportunity Losses

In all previous problems, we have been concerned with absolute material consequences, usually dollar values; hence, we have always evaluated the consequence of "no action" as \$0—an evaluation that generally is inadequate in real decision problems. For clarification, suppose a person wants to evaluate a single possible action for which the consequences depend on various possible states of nature. Should the person elect "no action" when, by the true state of nature, he should take the action, he will pass up a potential profit. Thus, rather than \$0, the consequence here is the loss of potential profit. We generally refer to such losses as opportunity losses. An *opportunity loss* is defined as the difference between the consequence (cost or profit or other material value) to be realized under the decision chosen and the consequence to be realized for the best possible decision for the event under consideration.

Example 1

Suppose a company has the opportunity to bid on three possible contracts (acts A, B, and C) or to decline to bid (act D). The consequences to be derived as a result of the various acts depend on various possible probabilistic states of nature with regard to forthcoming technological breakthroughs. The absolute dollar-value consequences of the various event-act combinations are as shown in Table 17.1. Convert the table of absolute consequences to a table of opportunity losses.

Solution

The company faces a typical decision problem. It must choose an act before it can possibly know which state of nature exists (which event will occur). What if the company were positive that event E_1 would occur? Under such information, the best possible consequence would come from act B. There would be no opportunity loss. However, if the company elects act A under these circumstances, it will achieve a

Table 17.1

Consequences of various event-act combinations—Example 1

Event (state of nature) E_i	*Absolute consequences associated with various possible event-act combinations* Act A (bid contract A)	Act B (bid contract B)	Act C (bid contract C)	Act D (no bid)
E_1 (major breakthrough within one year)	\$3,000,000	\$4,000,000	\$2,000,000	\$0
E_2 (major breakthrough between one and three years)	1,000,000	0	1,500,000	0
E_3 (major breakthrough between three and five years)	−1,000,000	−500,000	0	0
E_4 (no breakthrough)	−2,000,000	−3,000,000	−2,000,000	0

consequence of only \$3,000,000 when it could have achieved \$4,000,000 through act *B*. The opportunity loss associated with this event-act combination (event E_1–act *A*) is \$1,000,000. Similar reasoning about all event-act combinations leads us immediately to the opportunity losses of Table 17.2.

Note by the relations of the values in Table 17.2 to corresponding values in Table 17.1 that the general steps required in converting absolute consequences of various possible event-act combinations to opportunity losses are as follows.

1. Select an event. Restricting consideration to this event, choose the best consequence available through any possible act.

Table 17.2

Opportunity losses—Example 1

Event (state of nature) E_i	*Opportunity losses associated with various possible event-act combinations* Act A (bid contract A)	Act B (bid contract B)	Act C (bid contract C)	Act D (no bid)
E_1 (major breakthrough within one year)	\$1,000,000	\$0	\$2,000,000	\$4,000,000
E_2 (major breakthrough between one and three years)	500,000	1,500,000	0	1,500,000
E_3 (major breakthrough between three and five years)	1,000,000	500,000	0	0
E_4 (no breakthrough)	2,000,000	3,000,000	2,000,000	0

2. Select an act. Determine the opportunity loss for the appropriate event-act combination by taking the algebraic difference between the best consequence available through any possible act under the event and the consequence available through the given act being evaluated.
3. Repeat step 2 for all possible acts.
4. Repeat steps 1 through 3 for all possible events.

You should verify that the foregoing steps, when applied to the data of Table 17.1, will produce the data of Table 17.2. □

It is very important to note (for example, event E_1–act A) that an opportunity loss may occur even when a given decision results in a positive consequence to the decision maker. Thus, in future discussions, when we use the term "loss" we shall always mean "opportunity loss," not "loss" in the usual sense of "profit" and "loss."

Problems

1. Given the table below, calculate the opportunity losses associated with each event-act combination.

	Absolute consequences associated with various possible event-act combinations (hundreds of dollars)						
Event	*Act A*	*Act B*	*Act C*	*Act D*	*Act E*	*Act F*	*Act G*
E_1	1,000	2,000	1,500	2,700	950	600	0
E_2	2,500	1,300	750	2,000	1,400	750	0
E_3	600	−150	−100	1,300	800	200	0
E_4	−150	−250	−250	−1,000	200	−100	0

2. Refer to Problem 2, Section 16.3, and construct a table of expected opportunity losses for the data given there. Show your work.

3. A manufacturing company is considering the automation of several activities in their production line. The consequences of the action depend on the disposition they make of the existing labor, and what attitude the union has on the change. The expected consequences of each different event-act combination follow.

		Consequences (savings per year in hundreds of dollars)		
(E_i)	*Union attitude*	*Retrain all present workers*	*Retrain some of present workers*	*Hire new operators*
E_1	Against	60	59	56
E_2	Indifferent	71	73	75
E_3	In favor	85	82	89

a. Construct the opportunity loss table.

b. If $P(E_1) = 0.6$, $P(E_2) = 0.3$, $P(E_3) = 0.1$, choose the act which minimizes the expected opportunity loss. Show your work.

17.3 Relation between Classical and Bayesian Approaches to Decision Analysis

Let us begin by first viewing a situation that is simple with regard to the work required to reach a decision. Suppose in our situation that sample information has been collected with no thought given to the method of analysis to be employed in reaching a decision. Of course, this is not the best procedure, but there are two very good reasons for our considering such a situation. First, this procedure is common in real-world situations. Second, an understanding of the analysis in such a situation will help you grasp the concepts regarding the selection of a strategy in more general situations.

Example 1

Since the town of Gober has no television station, its 10,000 television owners watch programs from stations in neighboring cities. Reception is poor, and the town council is being pressured for a solution. The council wishes to decide whether, under the following conditions, it should put up a receiving tower.

1. The initial cost of constructing the tower will be $500,000.
2. The five-year fixed cost of maintenance will be $250,000.
3. By charging a nominal monthly fee, the town can realize a surplus (before fixed cost) over a five-year period of $300 per customer.
4. A television station is scheduled to be built in Gober. It will be completed in five years, at which time the salvage value of the tower and related equipment will be $0.
5. While the council would like to generate a surplus (which we shall call profit) through operation of the tower, it is willing to undertake the venture if it can just break even.

Let us assume that there are only four possible values (states of nature) which π (the true proportion of the population of television owners that will eventually use the tower service) can take on, say 0.10, 0.30, 0.50, and 0.70. (There is obviously no realistic basis for such an assumption. We make it only to illustrate relevant concepts without too much complication.) Suppose further that the prior (that is, prior to obtaining sample information) probabilities associated with these possible states of nature are

Event (state of nature) E_i	*Prior probabilities* $P(E_i)$
$E_1(\pi = 0.10)$	0.19
$E_2(\pi = 0.30)$	0.79
$E_3(\pi = 0.50)$	0.01
$E_4(\pi = 0.70)$	0.01

(Prior-probability information provided to two decimal places is rather unrealistic. We assume it here merely to illustrate certain key points.)

The town council has randomly selected 50 television owners and taken them to other towns to expose them to reception made possible by towers similar to the one it is considering. In subsequent interviews, 13 of the 50 owners indicated they would take advantage of the tower service if it were available.

The major question is "Should the Gober town council construct the tower?" Answer by (a) using the classical approach to testing hypotheses, (b) using the decision-theory approach based only on prior probability information regarding the possible values of the parameter π, and (c) using the decision-theory approach based on combining the sample information with the prior-probability information.

Note that Bayesian analysis alone does not require establishing null and alternative hypotheses concerning the parameter of interest. However, since our ultimate decision here (and in countless similar situations) will depend essentially on reasoning concerning the true value of the parameter π and since establishing the null and alternative hypotheses will provide a basis for comparing the three approaches, we begin by establishing such hypotheses.

First of all, let us review the possible reasoning of a decision maker when the null hypothesis is to be tested by classical procedures. However, unless this procedure is viewed properly, you could develop an unwarranted prejudice against the classical approach to testing hypotheses and a corresponding prejudice in favor of the Bayesian approach. There is no conclusive argument that the Bayesian approach is any more theoretically sound than the classical approach, although for many, it is somewhat more comprehensible. They sometimes construe the fact that it is easy to understand and apply to mean that it is applicable to a broader range of problems. This is by no means the case!

If, prior to applying the classical approach, the analyst would avail himself of all the information regarding consequences of wrong decisions that he does in Bayesian analysis and use this information to establish his decision rule, he could accomplish the same results available through the Bayesian approach. This "if" is large, however, and really represents the "fly in the ointment" with regard to classical testing of hypotheses. Since related reasoning is so difficult, descriptions of the mechanical techniques have far over-shadowed discussions of reasoning. Even the discussions that do appear in appropriate literature are generally brief and nebulous. If the Bayesian approach does hold an edge over the classical approach, this is the reason. The body of knowledge which has come to be known as Bayesian inference provides a well-defined approach to decision making, including a specific means of combining economic considerations, prior probabilities, and sample information.

It is important to note that, when comparing the approaches in the following solution, we use the classical approach as though economic considerations and prior probabilities were not available to the analyst. Obviously, this is not true. If they are available for a Bayesian analysis, they are available for a classical analysis. However, they are usually not properly incorporated into many real-world applications of classical tests of hypotheses.

Solution

To establish the null hypothesis concerning the parameter π, the town council must simply determine the proportion of television owners who will have to subscribe to the tower service in order for the council to break even. This requires that a relationship (a function of π) be developed allowing the council to calculate the profit that will accrue for various possible values of π. The profit relation for the five-year period becomes

$$\text{Profit} = (\text{Number of customers})(\text{Profit per customer}) - \text{Related costs.} \qquad (1)$$

However, for a given proportion π of owners who subscribe, the number of customers is found by multiplying the number of television owners by π. Further, surplus (before fixed cost) per customer for the five-year period is \$300, and the related cost (\$750,000) is obtained by adding the initial construction cost (\$500,000) to the five-year fixed cost of maintaining the tower (\$250,000). Thus, we have

$$P_\pi = (10{,}000)(\pi)(300) - 750{,}000, \qquad (2)$$

where P_π represents the profit to Gober if the proportion of television owners taking advantage of the tower service is π. Solving (2), we have

$$P = 3{,}000{,}000\pi - 750{,}000. \qquad (3)$$

Now, to determine the break-even proportion (say π_B), we simply determine the value of π that will provide no profit. In other words, we set P_{π_B} equal to zero and solve for π_B:

$$\pi_B = \frac{750{,}000}{3{,}000{,}000} = 0.25. \qquad (4)$$

Since the town will just break even if $\pi = 0.25$, the council will wish to test the null hypothesis $H_0 : \pi = 0.25$.

(a) Given the null hypothesis, there are various lines of reasoning which the council might pursue in establishing the alternative hypothesis. First, it might decide that it wants to avoid the error of constructing the tower when it should not. In this case, the alternative hypothesis will be $H_1 : \pi > 0.25$. On the other hand, it might decide that it wants to avoid the error of not constructing the tower when it should. Then, the alternative hypothesis will be $H_1 : \pi < 0.25$. Let us suppose that it chooses the first, to test the null hypothesis

$$H_0 : \pi \leq 0.25 \quad \text{versus} \quad H_1 : \pi > 0.25. \qquad (5)$$

(Note that we have written the null hypothesis differently than in previous discussions involving classical testing of hypotheses. The inequality is included merely to facilitate reasoning associated with the Bayesian approach. The classical test is conducted as though the null hypothesis were $H_0: \pi = 0.25$.) Now comes the most nebulous part of the reasoning. How does the council proceed in specifying α, the probability of Type I error with which it can live? This is the question that causes so much consternation in classical testing. Regardless of the reasoning, suppose the council fixes $\alpha = 0.05$. It now knows that the probability of Type I error is 0.05. Further, it can calculate β for various possible values of π. However, unless economic considerations are incorporated into the analysis, the calculations will often be almost meaningless. Since economic considerations are often overlooked in the classical approach (as they were in Chapters 9, 10, and 11), we shall make a decision without trying to incorporate them.

Recall that our sample of size 50 produced 13 people who indicated a desire to use the tower service. Thus, our sample proportion is

$$p = \frac{13}{50} = 0.26.$$

Ignoring both the continuity correction factor, and the finite population correction factor, we test the null hypothesis of (5) by computing

$$\begin{aligned} \alpha' &= P(p > 0.26 \mid \pi = 0.25) \\ &= P\left(Z > \frac{0.26 - 0.25}{\sqrt{\frac{0.25(0.75)}{50}}}\right) \\ &\cong P(Z > 0.163) = 0.4351. \end{aligned}$$

Thus, since $\alpha' = 0.4351 > 0.05 = \alpha$, the council has no reason to reject the null hypothesis and thus will not construct the tower.

(b) Using only prior-probability information regarding the possible states of nature, we are essentially ignoring our sample information. In order to reach a decision under this approach, the Gober council must

1. Use relation (3) to calculate the profits that will accrue as a result of various event-act combinations,
2. convert such profits to opportunity losses,
3. calculate the expected opportunity loss associated with each act,
4. choose the act with the least expected opportunity loss.

First we should note that the act "construct the tower" is equivalent to rejecting the null hypotheses of (5), and the act "do not construct the tower" is equivalent to accepting the null hypothesis of (5). This will allow us to compare the forthcoming decision to that of (a) above.

We note that the profit to the council if it does not construct the tower (if it accepts H_0) is \$0, regardless of the true state of nature (true value of π), as reflected in column 2 of Table 17.3. To obtain profits associated with the act of constructing the tower (rejecting H_0), we simply evaluate Relation (3) for the various possible values of π. For example,

$$P_{0.10} = 3{,}000{,}000(0.10) - 750{,}000 = -450{,}000.$$

Other profits are obtained similarly.

Table 17.3

Profits for event-act combinations—Example 1

(1) Event (state of nature) E_i	*Profits resulting from various event-act combinations* (2) Do not construct tower (accept H_0)	(3) Construct tower (reject H_0)
$E_1(\pi = 0.10)$	0	−750,000
$E_2(\pi = 0.30)$	0	150,000
$E_3(\pi = 0.50)$	0	750,000
$E_4(\pi = 0.70)$	0	1,350,000

Table 17.4 gives the probabilities associated with the various states of nature $[P(E_i)]$, the results (conditional consequences) of converting the profits of Table 17.3 to opportunity losses, and the results of relevant expected-value calculations—that is, the expected opportunity losses associated with the two possible acts. Since the expected opportunity loss associated with the act "do not construct the tower" is less than that of "construct the tower," the council will obviously accept the null hypothesis and elect not to construct the tower. While this decision is identical to that of (a), above, it is not necessary that a decision based on incorporation of prior probabilities be the same as a corresponding decision reached by the classical approach to testing hypotheses.

(c) Now, use the decision-theory approach, combining the sample information with relevant prior-probability information. In order to reach a decision in this case, the Gober town council must, with one major exception, carry out the same steps as those of (b) above. That is, it must calculate expected opportunity losses of the two acts. However, now, the council will use posterior probabilities based on the outcome of the sample. Required calculations are shown in Table 17.5. As an example of the

Table 17.4

Opportunity losses—Example 1

		Opportunity losses associated with various event-act combinations			
		Do not construct tower (accept H_0)		Construct tower (reject H_0)	
(1) Event (state of nature) E_i	(2) Probabilities $P(E_i)$	(3) Conditional opportunity loss	(4) Expected opportunity loss	(5) Conditional opportunity loss	(6) Expected opportunity loss
$E_1(\pi = 0.10)$	0.19	$0	$0	$750,000	$142,500
$E_2(\pi = 0.30)$	0.79	150,000	118,500	0	0
$E_3(\pi = 0.50)$	0.01	750,000	7,500	0	0
$E_4(\pi = 0.70)$	0.01	1,350,000	13,500	0	0
Expected opportunity loss			$139,500		$142,500

Table 17.5

Expected consequences—Example 1

Let indicator outcome I_{13} represent the obtaining of 13 successes in 50 trials

Event (state of nature) E_i	Prior probabilities $P(E_i)$	Joint probabilities $P(E_i \cap I_{13}) = P(E_i)P(I_{13} \mid E_i)$	Posterior probabilities $P(E_i \cap I_{13})/\sum_i P(E_i \cap I_{13})$	Opportunity losses associated with various event-act combinations: Do not construct tower (accept H_0)		Construct tower (reject H_0)	
				Conditional opportunity loss	Expected opportunity loss	Conditional opportunity loss	Expected opportunity loss
$E_1(\pi = 0.10)$	0.19	(0.19)(0.0007) = 0.000133	0.00160	$0	$0	$750,000	$1,200
$E_2(\pi = 0.30)$	0.79	(0.79)(0.1050) = 0.082950	0.99836	150,000	149,754	0	0
$E_3(\pi = 0.50)$	0.01	(0.01)(0.0003) = 0.000003	0.00004	750,000	300	0	0
$E_4(\pi = 0.70)$	0.01	(0.01)(0.0000) = 0.000000	0.00000	1,350,000	0	0	0
		0.083086			$150,054		$1,200

posterior-probability calculations, consider the second row of the Table. It shows the posterior probability associated with the event $E_2(\pi = 0.30)$—that is, $P(E_2 | I_{13})$—with I_{13} representing the indicator outcome "the obtaining of 13 successes out of 50 trials," where we define as a success a customer who indicates that he will take advantagc of the service offered by the tower. We must first determine $P(E_2 \cap I_{13})$. The required calculation is

$$P(E_2 \cap I_{13}) = P(E_2)P(I_{13} | E_2). \tag{6}$$

However, we see that, in the notation of Section 5.2,

$$\begin{aligned} P(I_{13} | E_2) &= P_b(X = 13 | n = 50, \pi = 0.30) \\ &= B(13; 50, 0.30) - B(14; 50, 0.30) \\ &= 0.7771 - 0.6721 = 0.1050. \end{aligned}$$

Substituting into (6), we have

$$P(E_2 \cap I_{13}) = 0.79(0.1050) = 0.082950. \tag{7}$$

Other relevant joint probabilities are obtained similarly. We note from Table 17.5 that such joint probabilities total to 0.083086 (in the notation of Chapter 16):

$$\sum_i P(E_i \cap I_{13}) = 0.083086.$$

Finally,

$$P(E_2 | I_{13}) = \frac{P(E_2 \cap I_{13})}{\sum_i P(E_i \cap I_{13})} = \frac{0.082950}{0.083086} = 0.99836. \tag{8}$$

Other posterior probabilities are obtained similarly. All relevant posterior probabilities are employed in Table 17.5 to convert the conditional consequences of various event-act combinations to expected consequences. Adding the proper columns of Table 17.5, we obtain the expected opportunity losses (of the two acts) that result from a combination of sample information with prior-probability information. Note that, under the added information, the decision is reversed. That is, using posterior probabilities, the council finds that the expected opportunity loss associated with accepting H_0 (not constructing the tower) is greater than that associated with rejecting H_0 (constructing the tower). Thus, based on this analysis, the council will elect to construct the tower. □

Once again, you are warned against inferring too much. The example in no way implies that the Bayesian approach to decision making is superior to the classical approach. In fact, if all available relevant information is incorporated in the two approaches, they will lead to identical decisions. This is shown by Schlaifer (17.1) in a quite detailed treatment of a production example. A very important conclusion to be drawn from the example (particularly in view of the fact that the two approaches lead in this example to different decisions) is that, in our usual implementation of the classical approach to decision making, we often fail to take advantage of extremely important information. Thus, the fault lies not with the classical approach but, rather, with the applications that fail to exploit the available information fully.

Perhaps even more important is the fact that, should all relevant information be incorporated into a classical decision analysis [see Schlaifer (17.1)], the respective reasoning and required calculations are generally quite complicated and cumbersome. Thus, since the Bayesian approach gives results that are always consistent with those to which an equally detailed classical analysis would give rise, the ease of using the Bayesian analysis has made it quite popular.

Problems

1. Given the table below and a sample result of 16 successes out of 50 trials, decide whether the proposal should be accepted or rejected using the decision-theory approach based on all the information supplied. (Minimize the expected opportunity losses.) Show your work.

		Profits resulting from various event-act combinations	
State of nature E_i	$P(E_i)$	*Do not accept proposal*	*Accept proposal*
E_1 $\pi = 0.20$	0.15	\$0	\$ − 150,000
E_2 $\pi = 0.35$	0.30	0	−7,500
E_3 $\pi = 0.50$	0.45	0	85,000
E_4 $\pi = 0.75$	0.10	0	175,000

2. The Republic Banana Co. imports bananas that can be sold for \$12 per hundred pounds in ungraded lots. Market surveys indicate that if the bananas are classified by quality, Class A bananas can be sold for \$18 per hundred pounds, and Class B bananas for \$10 per hundred pounds. The estimated cost of classifying is \$1.00 per hundred pounds.

   a. At what proportion of Class A bananas is it profitable to classify the bananas?

   b. The growers estimate that they produce either 20% or 60% Class A bananas, with probabilities of 0.4 and 0.6, respectively.

      1. Construct the opportunity loss table and decide whether
         (1) the bananas should be sold in ungraded lots or
         (2) they should be classed and sold by quality.
      2. If a sample of 20 bananas yields 7 Class A bananas, which alternative should be chosen based on expected opportunity losses?

   Show all work!

3. Mr. Wolf, a destitute student working his way through college, is given the opportunity to purchase the entire lot of 1000 student tickets to a basketball game for \$1500. There are 10,000 students at the college, and he may sell the tickets for \$2.50 each. Being an astute business student, Mr. Wolf investigates the athletic office records and finds that 30% of the games have an attendance of about 500 students, 45% of the games have an attendance of about 750 students, and 25% of the games are attended by about 950 students. Not content with just this information, however, Mr. Wolf decides to gather some sample information

on the possible attendance to the coming game and he asks 50 students at random whether they would desire to purchase a ticket. He receives 31 affirmative answers.

a. Determine the null hypothesis and use the classical approach to help Mr. Wolf decide whether to purchase the tickets (let $\alpha = 0.05$).

b. Use the decision theory approach based on all the information available to help Mr. Wolf decide. Use expected opportunity losses as your decision criterion. Show all work.

17.4 The General Bayesian Approach Exemplified

In order to obtain a good feeling for the advantages of the Bayesian approach to decision making, particularly in problems involving binomial sampling, let us examine an extensive example. We begin by assuming that we know the sample size to be taken. How do we determine an optimum decision rule—that is, a critical value in terms of number of successes observed in the sample—for deciding whether to accept a given null hypothesis? Second, in the situation in which a decision has not been made with regard to either sample size or decision rule, how do we determine the optimum sample size? We answer these questions in the following example. First, we must warn you that the procedure is quite demanding and is generally used only on projects so large that the potential savings justifies extensive effort. On the other hand, when computerized, the procedure takes much less time and is not prohibitive in this respect.

Our method will be to hypothesize an example, make several assumptions relevant to the approach to be employed, and then illustrate the use of Bayesian concepts in various situations as we relax these assumptions one at a time.

Example 1 (General Information for Examples 2 and 3)

Suppose a high-school student who attended summer school for the first six weeks of the summer, wants a job the second six weeks. His best offer to date will pay \$400 for the six weeks. However, he has been offered another somewhat speculative opportunity. Mr. Ames, a friend of his father who owns a junk yard, has on hand 1,000 carburetors. He has a potential buyer whose offer expires in two days. However, Mr. Ames is willing to sell the lot to the student at a reduced rate, provided the student makes up his mind before the other offer expires. Should the student purchase the lot, he will spend the remainder of the summer testing the carburetors to determine which are repairable. Mr. Ames agrees to repurchase all that are repairable at \$5 per carburetor. The student's fixed cost (mostly test equipment) of undertaking the venture is \$100. His variable cost (purchase price, testing cost, etc.) per carburetor is \$1. The student thus must decide whether to undertake the speculative venture or accept the firm job offer which will net him \$400 certain. In essence, he must decide whether the speculative venture will pay him at least as well as the job.

Thus, a quite relevant figure is the proportion of repairable carburetors in the lot that will make the speculative venture equivalent to \$400. For the speculative venture, his earnings (as a function of π) would be

$$\$5(\pi)(1000) - \$1\,(1{,}000) - \$100 = \$5{,}000\pi - \$1{,}100. \tag{1}$$

Thus, we obtain the breakeven proportion of repairable carburetors in the lot by equating (1) to \$400 and solving for π:

$$\$5{,}000\pi_B - \$1{,}100 = \$400 \Rightarrow \pi_B = 0.30, \tag{2}$$

where π_B represents the proportion of repairable carburetors in the lot that will allow the student to break even on the speculative venture. The conclusion is that if $\pi \geq 0.30$, the student should undertake the speculative venture; if $\pi < 0.30$, he should accept the job. This information tells us immediately that, in order to make a decision, the student should test the null hypothesis

$$H_0\!:\pi = 0.30$$

versus the alternative hypothesis (3)

$$H_1\!:\pi < 0.30.$$

What will be his consequences (in terms of opportunity losses) of making an incorrect decision? (Regardless of his procedure for deciding whether to accept or reject the null hypothesis of (3), we know that there are two kinds of error which can result: Type I error and Type II error.) Recalling our original definition of opportunity loss, we arrive at a formula for opportunity loss under various possible values of π through the following reasoning.

If $\pi = 0.30$ (H_0 true) and he elects the speculative venture (accepts H_0), he will incur no opportunity loss because he will have made a correct decision. However, if $\pi = 0.30$ (H_0 true) and he elects the job (rejects H_0), his opportunity loss becomes the difference between what he could have earned under a correct decision [\$5,000π − \$1,100, as given by (1)] and what he will earn under the incorrect decision (\$400); that is,

$$\$5{,}000\pi - \$1{,}100 - \$400 = \$5{,}000\pi - \$1{,}500. \tag{4}$$

Note that this is the opportunity loss associated with Type I error.

On the other hand, if $\pi < 0.30$ (H_0 false) and he elects the job (rejects H_0), he will incur no opportunity loss because he will again have made a correct decision. However, if $\pi < 0.30$ (H_0 false) and he elects the speculative venture (accepts H_0), his opportunity loss becomes the difference between the earnings under a correct decision (\$400) and the earnings under the incorrect decision (\$5,000π − \$1,100); that is,

$$\$400 - (\$5{,}000\pi - \$1{,}100) = \$1{,}500 - \$5{,}000\pi. \tag{5}$$

Note that this is the opportunity loss associated with a Type II error.

In summary, the relations which describe the student's opportunity loss as a function of π become

$$\text{Loss of electing speculative venture} = \begin{cases} \$1{,}500 - \$5{,}000\pi & \text{if } \pi < 0.30 \\ \$0 & \text{if } \pi = 0.30 \end{cases}$$

and (6)

$$\text{Loss of electing job} = \begin{cases} \$0 & \text{if } \pi < 0.30 \\ \$5{,}000\pi - \$1{,}500 & \text{if } \pi = 0.30. \end{cases}$$

Since they depend on the value π actually assumes, the losses given by the relations of (6) are often referred to as *conditional opportunity losses*, that is, conditional with regard to the true value of π. The relations are graphed in Figure 17.1.

Suppose Mr. Ames has purchased and resold many lots of carburetors similar to the one on hand, so that he is able to tell the student the prior-probability distribution shown by Table 17.6.

We shall now analyze the student's problem.

Bayesian Analysis When Sample Information Has Been Collected

Let us begin by using the same assumptions used in the example of the previous section. That is, let us consider the binomial sampling problem in which we have

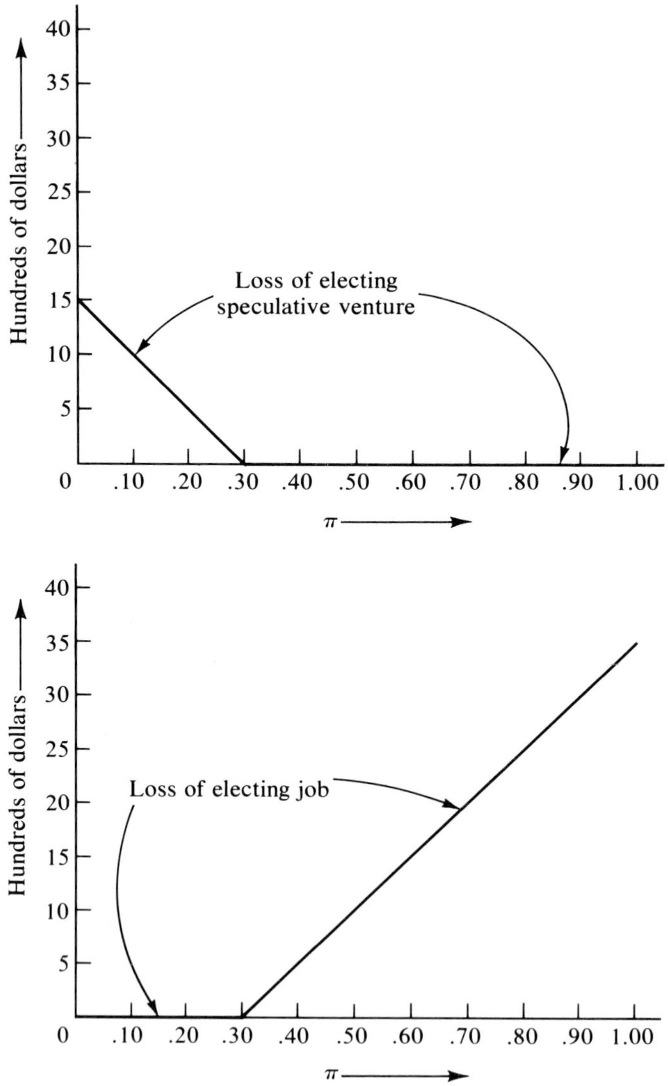

Figure 17.1 Conditional opportunity losses—Example 1

Table 17.6

Prior probabilities for Example 1

State of nature (E_i)	*Prior probability* $P(E_i)$
$E_1(\pi = 0.1)$	0.1
$E_2(\pi = 0.2)$	0.1
$E_3(\pi = 0.3)$	0.2
$E_4(\pi = 0.4)$	0.3
$E_5(\pi = 0.5)$	0.2
$E_6(\pi = 0.6)$	0.1
Total	1.0

collected sample information prior to considering the kind of analysis to be conducted. Note that we use the binomial distribution here simply to facilitate calculations. In other words, sampling (which is necessarily without replacement) from the above population does not result in a Bernoulli process in the strict sense of the word. However, we wish to emphasize the Bayesian analysis, not the details involved. We could accomplish the same type of analysis through use of probabilities involving the hypergeometric distribution (the correct one in this case). However, we shall assume that the population of size $N = 1{,}000$ is large enough to treat the probability of a success as constant from draw to draw. Under this assumption, we are justified in using the binomial distribution to calculate relevant probabilities. (In a real-world problem, we should either use the hypergeometric distribution or take into account the costs of errors due to the use of binomial probabilities to approximate hypergeometric probabilities.)

Example 2

Suppose we again have the situation involving the student and Mr. Ames. However, now Mr. Ames informs the student that he has randomly selected a sample of size $n = 20$ carburetors, tested each, and found that 5 are repairable. In light of this, as well as all the information given earlier, should the student elect the speculative venture?

Solution

The student can immediately make a decision by conducting a Bayesian analysis in which he treats the outcome of 5 successes out of 20 trials from a Bernoulli process as the relevant indicator outcome. He simply must

1. use this indicator outcome to convert the prior probabilities of Table 17.6 to posterior probabilities,
2. use the posterior probabilities to calculate the expected opportunity losses associated with the acts of accepting and rejecting the null hypothesis of (3), and
3. choose the act having the smaller expected opportunity loss.

Required calculations are shown in Table 17.7. Columns 1 through 4 show calculations for converting prior probabilities to posterior probabilities. As an example of the calculation of the posterior probabilities, we have

$$P(E_2|I_5) = \frac{P(E_2 \cap I_5)}{P(I_5)} = \frac{P(E_2)P(I_5|E_2)}{\sum_{i=1}^{6} P(E_i)P(I_5|E_i)}$$

$$= \frac{P(\pi = 0.2)P_b(X = 5|n = 20, \pi = 0.2)}{P(\pi = 0.1)P_b(X = 5|n = 20, \pi = 0.1) + \cdots + P(\pi = 0.6)P_b(X = 5|n = 20, \pi = 0.6)}$$

$$= \frac{0.01746}{0.08190} = 0.2132,$$

where we have related the notation using Bayes' theorem to that involving binomial probabilities. The conditional losses (columns 5 and 7) are obtained simply by inserting the relevant value of π into the appropriate formula of Relation (6) or by reading the appropriate value from the loss functions of Figure 17.1. For example, when $\pi = 0.2$, the conditional loss of accepting the hypothesis that $\pi \geq 0.3$ becomes

$$\$1,500 - \$5,000\pi = \$1,500 - \$5,000(0.2) = \$500.$$

We note that the expected opportunity loss (\$145.50) associated with accepting H_0 (deciding on the speculative venture) is less than the expected opportunity loss (\$175.15) associated with rejecting H_0 (deciding on the job). Thus, having considered all available relevant information, the student should elect the speculative venture. □

Now, you may note that this is the same decision the student would have reached if he had completely ignored the sample information. That is, if the student had used prior probabilities in his calculations of expected opportunity losses, he would have obtained

$$\text{Expected loss of accepting } H_0 = (0.1)(\$1,000) + (0.1)(\$500) = \$150 \tag{8}$$

and

$$\text{Expected loss of rejecting } H_0 = (0.3)(\$500) + (0.2)(\$1,000) + (0.1)(\$1,500) = \$500. \tag{9}$$

Obviously, these figures lead to the same decision we had using posterior probabilities. Does this mean that the prior probabilities completely negate the influence of the sample information? Obviously, no. The result simply means that, in this particular case, the sample information (which alone tends to imply that $\pi < 0.30$) is not conclusive enough to cause a reversal in the decision based only on prior probabilities. However, it very well could cause a reversal. Let us see how.

Suppose the sample of size $n = 20$ carburetors had resulted in only 2 repairable carburetors. What should be the student's decision when this sample information is combined with all other original information? Calculations required for reaching a decision in this case are given in Table 17.8. Note that, in light of our sample information, the prior probabilities are revised so drastically that our decision now becomes that of rejecting H_0. That is, should the sample of 20 carburetors have included only 2 repairable carburetors, the student would have elected the job rather than the speculative venture.

Table 17.7

Opportunity losses: five successes out of twenty trials—Example 2

$H_0: \pi = 0.3 \qquad H_1: \pi < 0.3$

Let indicator outcome I_5 represent the outcome "5 successes out of 20 trials" from a Bernoulli process

(1) Events (states of nature) E_i	(2) Prior probabilities $P(E_i)$	(3) Joint probabilities $P(E_i \cap I_5) = P(E_i)P(I_5 \mid E_i)$	(4) Posterior probabilities $P(E_i \cap I_5)/\sum_i P(E_i \cap I_5)$	Opportunity losses associated with various acts: Accept H_0 (6) Conditional loss	Accept H_0 (6) Expected loss	Reject H_0 (7) Conditional loss	Reject H_0 (8) Expected loss
$E_1(\pi = 0.1)$	0.1	(0.1)(0.0319) = 0.00319	0.00319/0.08190 = 0.0389	$1,000	$38.90	$0	$0
$E_2(\pi = 0.2)$	0.1	(0.1)(0.1746) = 0.01746	0.01746/0.08190 = 0.2132	500	106.60	0	0
$E_3(\pi = 0.3)$	0.2	(0.2)(0.1789) = 0.03578	0.03578/0.08190 = 0.4369	0	0	0	0
$E_4(\pi = 0.4)$	0.3	(0.3)(0.0746) = 0.02238	0.02238/0.08190 = 0.2733	0	0	500	136.65
$E_5(\pi = 0.5)$	0.2	(0.2)(0.0148) = 0.00296	0.00296/0.08190 = 0.0361	0	0	1,000	36.10
$E_6(\pi = 0.6)$	0.1	(0.1)(0.0013) = 0.00013	0.00013/0.08190 = 0.0016	0	0	1,500	2.40
		0.08190	1.0000		$145.50		$175.15

Table 17.8

Opportunity losses: two successes out of twenty trials—Example 2

$H_0: \pi = 0.3 \qquad H_1: \pi < 0.3$

Let indicator outcome I_2 represent the outcome "2 successes out of 20 trials" from a Bernoulli process

				Opportunity losses associated with various acts			
				Accept H_0		*Reject H_0*	
Events (states of nature) E_i	*Prior probabilities* $P(E_i)$	*Joint probabilities* $P(E_i \cap I_2) = P(E_i)P(I_2 \mid E_i)$	*Posterior probabilities* $P(E_i \cap I_2)/\sum_i P(E_i \cap I_2)$	*Conditional loss*	*Expected loss*	*Conditional loss*	*Expected loss*
$E_1(\pi = 0.1)$	0.1	(0.1)(0.2852) = 0.02852	0.02852/0.04876 = 0.5849	\$1,000	\$584.90	\$0	\$0
$E_2(\pi = 0.2)$	0.1	(0.1)(0.1369) = 0.01369	0.01369/0.04876 = 0.2808	500	140.40	0	0
$E_3(\pi = 0.3)$	0.2	(0.2)(0.0270) = 0.00540	0.00540/0.04876 = 0.1108	0	0	0	0
$E_4(\pi = 0.4)$	0.3	(0.3)(0.0031) = 0.00093	0.00093/0.04876 = 0.0191	0	0	500	9.55
$E_5(\pi = 0.5)$	0.2	(0.2)(0.0002) = 0.00004	0.00004/0.04876 = 0.0008	0	0	1,000	0.80
$E_6(\pi = 0.6)$	0.1	(0.1)(0.0000) = 0.00000	0.00000/0.04876 = 0.0000	0	0	1,500	0
		0.04876	1.0000		\$725.30		\$10.35

It is rather obvious that since the student should accept H_0 ($\pi = 0.3$) in light of 5 repairable carburetors out of 20, he should also accept H_0 for any number greater than 5. It is just as obvious that he should reject H_0 for any number of repairable carburetors less than 2 in the sample.

This reasoning leads to a very important question. Is there not some number such that we should reject H_0 if the number of repairable units out of 20 is less than or equal to such number and accept H_0 if the repairable units out of 20 exceed it? That is, does there not exist some general decision rule the student could use in deciding whether to accept or reject H_0? The answer is the same as when we approach the decision from a classical (rather than Bayesian) standpoint: yes. In our specific example, the number is between 2 and 4 inclusive. How can we obtain the exact value of this number? Although this question can be answered in a rather straightforward manner, the answer itself is not obvious.

Determining the Best Decision Rule for a Fixed Sample Size

Suppose our situation is similar to that of Example 2. Suppose further that we have determined the size of the sample to be taken, but not collected any sample information. And, for consistency throughout our general discussion, let us suppose further that our hypotheses are

$$H_0: \pi = \pi_B \quad \text{versus} \quad H_1: \pi < \pi_B, \tag{10}$$

where π_B is the proportion of successes in the population of interest representing the economic break-even point. That is, π_B is such that if the analyst were certain that $\pi \geq \pi_B$, he would take one action and for $\pi < \pi_B$, he would take a different action. We would like to determine the best decision rule for testing H_0 when our decision is to be based on a sample of size n. That is, if we let X represent the statistic "number of successes out of n trials from a Bernoulli process," we should like to find a value in the sample space of X to be used as a critical value in testing H_0. How do we proceed?

First, we note that common sense tells us that a sample yielding a small value of X should cause us to act as if the parameter π is small, while a sample yielding a large value of X should cause us to act as if π is large. Thus, we want a decision rule of the following nature. We must find a value (say c) such that we shall accept H_0 if our sample of size n yields a number of successes x greater than c and reject H_0 for $x \leq c$. Hereafter, we shall refer to such a decision rule simply as (n, c). Thus, for example, the decision rule (25, 8) stated in conjunction with the hypotheses of (10) would automatically imply that we are to

1. take a random sample of size 25 from the appropriate Bernoulli process;
2. accept H_0 if the number of successes is greater than 8;
3. reject H_0 if the number of successes is less than or equal to 8.

We now need to determine the value c that represents the best decision rule for a given sample size n. The general procedure can best be illustrated with an example.

Example 3

Suppose our student is very enterprising and finds that Mr. Ames quite often receives a lot of carburctors similar to thc onc now on hand, and that he will always be willing to give the student first chance at purchasing the lot under conditions identical to those outlined in Example 1. The student cannot test all lots by himself. However, he has friends who will test a lot of 1,000 carburetors for \$300. Further, the student is willing to finance a friend any time he can clear a profit of \$100 from a lot. Thus, the student believes that it will be profitable for him to purchase from Mr. Ames any lot from which a sample of size $n = 20$ indicates that he should accept the null hypothesis, given

$$H_0: \pi = 0.30 \quad \text{versus} \quad H_1: \pi < 0.30. \tag{11}$$

Note that the null hypothesis to be tested is the same as that of Example 1 because the \$300 of added costs and \$100 of required profit (total \$400) take the place of the \$400 which the student can earn on a summer job. Thus, his break-even proportion of repairable carburetors remains $\pi_B = 0.30$. Note also that there is good reason for restricting the sample to size $n = 20$, since this will ordinarily be the largest number that can be tested before Mr. Ames sells to another customer.

The decision problem now is identical to that for which a specific solution was found in Example 2. Thus, the student can take a sample and conduct an analysis similar to that of Example 2 each time Mr. Ames receives a lot of 1,000 carburetors. However, he can save considerable effort by determining the best decision rule to be used for repeatedly testing the null hypothesis of (11). In other words, if he determines a "best test" of the form $(20, c)$, he can decide on each lot by simply comparing the number of repairable carburetors in the sample with c, the critical value. What is his best decision rule?

Solution

Refer to Example 1 for information relevant to this solution. Suppose the student decides on the decision rule (20, 4), which means he will take a sample of size $n = 20$ and reject H_0 (11) if the number (x) of repairable carburetors turns out ≤ 4 and accept H_0 if $x > 4$.

Since we assume sampling from a Bernoulli process, the number of repairable carburetors (X) in the sample of size $n = 20$ which the student is to take will represent a value of a statistic having a binomial distribution. However, the binomial distribution that holds will depend on the true value of π. Thus, the probabilities of acceptance and rejection of H_0 [$P_b(X \leq 4)$ and $P_b(X > 4)$] and hence the opportunity losses from decision errors resulting from a given decision rule depend on the true value of π. Since we would like to evaluate a decision rule [e.g., (20, 4)] on the basis of opportunity losses from decision errors, we must take this fact into account.

For example, suppose $\pi = 0.2$ (that is, $\pi < \pi_B = 0.3$ and hence H_0 is false) and we are using the decision rule (20, 4). We shall make either a correct decision or a Type II error. That is, if we reject H_0 (observe $x \leq 4$), we make a correct decision. The probability of a correct decision through use of the decision rule (20, 4) when $\pi = 0.2$ is, from the binomial tables, simply

$$P_b(X \leq 4 | n = 20, \pi = 0.2) = 0.6296.$$

However, under a correct decision, the loss from a decision error is automatically \$0. On the other hand, if we accept H_0 (observe $x > 4$), we obviously make a Type II error. The probability of a decision error through use of the decision rule (20, 4) where $\pi = 0.2$ is

$$P_b(X > 4 | n = 20, \pi = 0.2) = 1 - 0.6296 = 0.3704,$$

from the binomial tables. We obtain the opportunity loss corresponding to this probability, by using (6),

$$\$1{,}500 - \$5{,}000\pi = \$1{,}500 - \$5{,}000(0.2) = \$500.$$

(We calculated all such losses for values π can assume in Table 17.7.) Finally, then, if $\pi = 0.2$, there are two possible losses (\$500 and \$0) which may result from the decision rule (20, 4), and these losses have associated with them the probabilities of occurrence 0.3704 and 0.6296, respectively. The net effect of the decision rule (20, 4) in terms of opportunity losses may be viewed simply as a problem in determining the expected value of opportunity loss under the decision rule (20, 4). The result is

$$\$500(0.3704) + \$0(0.6296) = \$185.20. \tag{12}$$

This is the amount we would expect to lose, on average, through use of the decision rule (20, 4) when the true value of π is 0.2. (Note that, if we fix the value of π, we can calculate the conditional expected opportunity loss for any decision rule by reasoning as above and implementing an expected-value calculation similar to that of Equation (12). However, since the loss due to a correct decision is always \$0, we can, determine the conditional expected loss of the decision rule merely by calculating that part of the expected loss which results from the decision error.)

Obviously, π will not necessarily take on the value 0.2. In a practical problem, it may assume any specific value between 0 and 1, inclusive. Even in our particular example, it may take on any of the values 0.1, 0.2, 0.3, 0.4, 0.5, or 0.6, by the original statement of the problem. It should be noted here, however, that when $\pi < 0.3$ (since $\pi_B = 0.3$), the decision errors resulting from a given decision rule will be Type II errors. [This statement is dictated by the directions of the inequality statements in the null and alternative hypotheses of (11).] Naturally, then, only Type I error will occur from a decision rule where $\pi \geq 0.3$.

As an illustration of the latter case, consider our use of the same decision rule (20, 4) for $\pi = 0.5$ ($\pi > \pi_B = 0.3$, and thus H_0 is true). Here, the probability of a decision error is

$$P_b(X \leq 4 | n = 20, \pi = 0.5) = 0.0059,$$

and the associated loss from (6) is

$$\$5{,}000\pi - \$1{,}500 = \$5{,}000(0.5) - \$1{,}500 = \$1{,}000.$$

The conditional expected opportunity loss of the decision rule (20, 4) for $\pi = 0.5$, is therefore,

$$\$1{,}000(0.0059) = \$5.90. \tag{13}$$

There remains only the case when π is equal to π_B (that is, $\pi = 0.3$), the break-even proportion. As always, there is no opportunity loss associated with a correct decision. The opportunity loss of an incorrect decision is also \$0:

$$\$5{,}000\pi - \$1{,}500 = \$5{,}000(0.3) - \$1{,}500 = \$0.$$

Thus, regardless of the probabilities associated with the correct decision and decision error, the conditional expected loss of any decision rule in the unique situation $\pi = \pi_B = 0.3$ is \$0.

Table 17.9 presents calculations leading to the conditional expected opportunity losses of using the decision rule (20, 4) for the various possible values π can assume. Note there that each row shows calculations paralleling those of (12) and (13).

Table 17.9

Conditional expected opportunity losses associated with the decision rule (20, 4) for $H_0: \pi = 0.3$ versus $H_1: \pi < 0.3$

Type II error losses

State of nature (E_i)	*Opportunity loss of acceptance when* H_0 *false* (\$1,500–\$5000π)	*Corresponding probability of opportunity loss* ($P_b(X > 4 \mid n = 20, \pi)$)	*Conditional expected opportunity loss (conditional on* π*)*
$E_1(\pi = 0.1)$	\$1,000	0.0432	\$1,000(0.0432) = \$43.20
$E_2(\pi = 0.2)$	500	0.3704	500(0.3704) = 185.20

Type I error losses

State of nature (E_i)	*Opportunity loss of rejection when* H_0 *true* (\$5,000π–\$1,500)	*Corresponding probability of opportunity loss* ($P_b(X \leq 4 \mid n = 20, \pi)$)	*Conditional expected opportunity loss (conditional on* π*)*
$E_3(\pi = 0.3)$	\$0	0.2375	\$0(0.2375) = \$0
$E_4(\pi = 0.4)$	500	0.0510	500(0.0510) = 25.50
$E_5(\pi = 0.5)$	1,000	0.0059	1,000(0.0059) = 5.90
$E_6(\pi = 0.6)$	1,500	0.0003	1,500(0.0003) = 0.45

Having this information regarding the decision rule, (20, 4), we now need similar information concerning other decision rules when the sample size is $n = 20$. We noted earlier that our best decision rule will be (20, 2), (20, 3), or (20, 4). However, we determined this from investigations (Example 2) that would not ordinarily be conducted in an analysis designed to choose the best decision rule. Therefore, we must proceed as though we did not have this information. We must compute information similar to that of Table 17.9 for each of the decision rules (20, 1), (20, 2), (20, 3), (20, 4) and (20, 5). Such calculations combined with the results of Table 17.9 provide entries to Table 17.10. The right-most column for each decision rule represents the conditional expected opportunity loss of that rule as a function of the value of π. The results are also presented in Figure 17.2.

Table 17.10

Conditional expected opportunity losses associated with various decision rules

$H_0: \pi = 0.3$ $\qquad$ $H_1: \pi < 0.3$

Decision rule (n, c)

			(20, 1)		(20, 2)		(20, 3)		(20, 4)		(20, 5)	
Kind of error	*Value of* π	*Conditional opportunity loss* (\$)	*Probability of loss*	*Conditional expected loss* (\$)	*Probability of loss*	*Conditional expected loss* (\$)	*Probability of loss*	*Conditional expected loss* (\$)	*Probability of loss*	*Conditional expected loss* (\$)	*Probability of loss*	*Conditional expected loss* (\$)
Type II	0.1	1,000	0.6083	608.38	0.3231	323.10	0.1330	133.00	0.0432	43.20	0.0113	11.30
	0.2	500	0.9308	465.40	0.7939	396.95	0.5886	294.30	0.3704	185.20	0.1958	97.90
Type I	0.3	0	0.0076	0	0.0355	0	0.1071	0	0.2375	0	0.4164	0
	0.4	500	0.0005	0.25	0.0036	1.80	0.0160	8.00	0.0510	25.50	0.1256	62.80
	0.5	1,000	0	0	0.0002	0.20	0.0013	1.30	0.0059	5.90	0.0207	20.70
	0.6	1,500	0	0	0	0	0	0	0.0003	0.45	0.0016	2.40

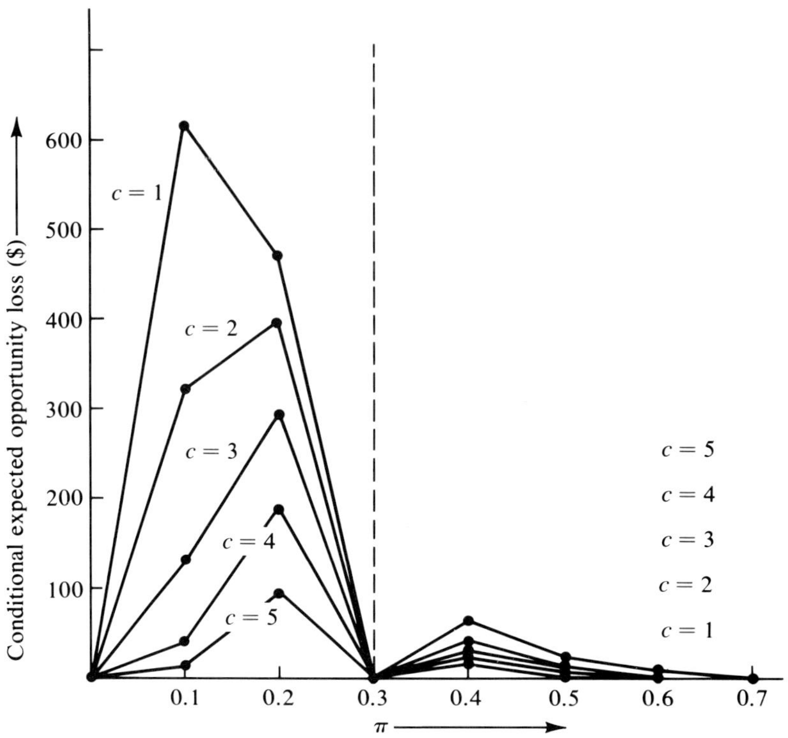

Figure 17.2 Graphs of conditional expected opportunity losses

Intuitively, the best rule in the classical sense would have minimum losses over all possible values of π. However, it is clear from Table 17.10 and Figure 17.2 that no such rule exists. For values of π such that $0 \leq \pi < 0.3$ (since $\pi_B = 0.3$), the rule (20, 5) affords minimum opportunity losses, and for values of π such that $0.3 < \pi \leq 1.0$, the rule (20, 1) affords minimum opportunity losses. Thus, we cannot choose the best rule under a strict classical definition of the term parameter. However, what has recently come to be called the modern (Bayesian) interpretation can solve our problem. We simply use the prior probabilities associated with various possible values of π.

Now the conditional expected losses (Table 17.10) associated with each decision rule can be treated as random variables whose possible values depend on the particular value π assumes. Thus, we can exploit the idea of expected losses resulting from a given decision rule over all possible values of π. Our calculations will produce a single expected loss due to incorrect decisions for each rule. The decision rule yielding the minimum expected loss for a fixed sample size will be the best decision rule of that class.

For our problem, the prior probabilities of Table 17.6 can be used to calculate the expected loss associated with each decision rule. Required calculations are shown in Table 17.11. For example, suppose we wish to calculate the expected loss associated with the decision rule (20, 2). We know that the expected loss of \$323.10 in Table 17.11 is conditional on π assuming the value 0.1. According to our prior-probability distribution, π assumes the value 0.1 with probability 0.1. Hence the "Contribution to expected loss" (Table 17.11) due to using the decision rule (20, 2) when $\pi = 0.1$ is

Table 17.11

Expected opportunity losses associated with various decision rules

$H_0: \pi = 0.3$ $H_1: \pi < 0.3$

			Decision rule (n, c)									
			(20, 1)		(20, 2)		(20, 3)		(20, 4)		(20, 5)	
Kind of error	*Value of* π	*Prior probabilities*	*Conditional expected loss* ($)	*Contribution to expected loss* ($)	*Conditional expected loss* ($)	*Contribution to expected loss* ($)	*Conditional expected loss* ($)	*Contribution to expected loss* ($)	*Conditional expected loss* ($)	*Contribution to expected loss* ($)	*Conditional expected loss* ($)	*Contribution to expected loss* ($)
Type II	0.1	0.1	608.38	60.84	323.10	32.31	133.00	13.30	43.20	4.32	11.30	1.13
	0.2	0.1	465.40	46.54	396.95	39.70	294.30	29.43	185.20	18.52	97.90	9.79
Type I	0.3	0.2	0	0	0	0	0	0	0	0	0	0
	0.4	0.3	0.25	0.08	1.80	0.54	8.00	2.40	25.50	7.65	62.80	18.84
	0.5	0.2	0	0	0.20	0.04	1.30	0.26	5.90	1.18	20.70	4.14
	0.6	0.1	0	0	0	0	0	0	0.45	0.05	2.40	0.24
Expected opportunity loss				107.46		72.59		45.39		31.72		34.14

merely (0.1)($323.10) = $32.31. Reasoning similarly for other values of π, we obtain the expected loss of the decision rule (20, 2) as

$$(0.1)(\$323.10) + (0.1)(\$396.95) + (0.2)(\$0) + (0.3)(\$1.80) + (0.2)(\$0.20) + (0.1)(\$0) = \$72.59.$$

Corresponding results for other decision rules are as shown in the Table.

From the calculations of Table 17.11, we immediately select the decision rule (20, 4) as best since, under a fixed sample size of $n = 20$, its expected loss is less than that of any other rule. Now, it might be argued that we cannot call (20, 4) the best rule because we have not investigated all possible rules for $n = 20$. However, this argument is not valid. If other rules were investigated, we would observe that as c (the critical value of the decision rule) increases from 0 to n, the expected losses from making Type I errors increase from essentially zero to very large; at the same time the expected losses from making Type II errors decrease from very large to essentially zero. Thus, as long as the prior-probability distribution assigns some nonzero probabilities to both types of error, a minimum will exist. In fact, with c increasing from zero to n, we find that, once the curve depicting expected loss (as a function of c) reaches a minimum and turns upward, it does not decrease again. Thus, for a fixed sample size n and a given prior-probability distribution, there will be only one best decision rule for minimum expected opportunity loss. The rule that yields minimum expected loss under a given prior distribution is called a *Bayesian decision rule.*

In light of all the calculations, the student of our example should set up the following operating policy. Any time Mr. Ames offers him the opportunity to purchase a lot of 1,000 carburetors, he should take a sample of size $n = 20$. Using the decision rule (20, 4), he should agree to purchase the lot if the sample yields more than 4 repairable carburetors. Otherwise, he should decline the proposition. □

Determining Optimum Sample Size

Suppose now that our student is not constrained to a sample of size $n = 20$ but there is a cost associated with sampling. How should he proceed to choose the optimum decision rule (n, c) when both n and c can be varied? Let us assume that sampling cost can be represented as a function of n (usually it is an increasing function of n). For simplicity, let us say that the sampling cost is a linear function of n, so that

$$\text{sampling cost} = a + bn, \tag{14}$$

where a represents a sampling cost that is fixed regardless of the value of n and b is the per unit sampling cost. The student now faces the problem of determining an optimal expected total-loss function over all values of n by combining, for each n, the sampling costs with the expected losses from incorrect decisions. What are the characteristics of the function leading to a general method of analysis? How is such a function determined practically?

The answer to the first question leads automatically to the answer of the second. However, to emphasize the importance of the first answer, let us consider a brute-force approach to the second. At first glance, a straightforward (though not very practical) approach would seem to involve the procedure illustrated in Example 3 in order to determine the optimum decision rule (and, hence, the expected loss from

Table 17.13

Expected opportunity loss associated with the decision rule (21, 5)

(1)	(2)	(3)	(4)	(5)	(6)	(7)
		$H_0: \pi = 0.3$		$H_1: \pi < 0.3$	[(4) · (5)]	[(3) · (6)]
Kind of error	*Value of* π	*Prior prob-abilities*	*Conditional opportunity loss* ($)	*Probability of loss*	*Conditional expected loss* ($)	*Contribution to expected loss* ($)
Type II	0.1	0.1	1,000	0.0144	14.40	1.44
	0.2	0.1	500	0.2307	115.35	11.54
Type I	0.3	0.2	0	0.3627	0	0
	0.4	0.3	500	0.0957	47.85	14.36
	0.5	0.2	1,000	0.0133	13.30	2.66
	0.6	0.1	1,500	0.0008	1.20	0.12
Expected opportunity loss						30.12

optimal expected total-loss function [which is the sum of the sampling cost function (increasing in n) and the optimum expected loss curve (decreasing in n)] will have a unique minimum. Thus, we know immediately that there exists an optimum n.

The general approach for determining the optimum sample size to be taken in deciding whether to accept or reject a null hypothesis similar to the one in Example 3 involves the following major steps.

1. Determine the sampling cost function, and plot a curve of sampling cost versus the sample size n. (It is not necessary that this function be linear in n.)
2. Use the procedure illustrated in Example 3 along with the concepts of Theorem 17.1) to plot a curve of optimum expected losses from decision errors versus the sample size n.
3. Determine the expected total-loss curve by adding the two curves of steps 1 and 2.
4. Select as an optimum sample size the value of n which corresponds to the minimum value attained by the expected total-loss curve.

Note that the optimum expected loss from decision errors for the value $n = 0$ represents a special case. It is obtained by first calculating the expected losses of respectively rejecting and accepting the null hypothesis when a decision is based on prior probabilities and then choosing the smaller of the two.

It is not always possible to determine in advance the cost of sampling. It may be necessary to modify the analysis by using an expected sampling cost based on the prior distribution of the parameter π. For a discussion of this procedure, the reader should see Schlaifer (17.1).

Finally, let us consider the correct treatment of the results of sample information collected before the analyst decides to construct a general decision rule by investigating the question of optimum sample size. We indicated earlier that in the classical approach the decision rule should always be constructed as if no sample has been taken. However, from the Bayesian point of view, any sample information should be incorporated before the optimum decision rule is determined. Thus, before an analysis is made to determine an optimum sample size, any sample information collected should be used to revise the relevant prior (to the taking of a sample) probabilities to posterior (to the taking of a sample) probabilities. Then, these posterior (to the taking of a sample) probabilities should be treated in the previously described analysis as prior (to the selecting of an optimum sample size) probabilities.

Problems

1. You have an opportunity to purchase all the programs for a forthcoming convention of 1,000 unicycle salesmen. The programs will cost you $500 and may be sold for $0.75 each.

   a. What proportion of conventioneers have to purchase a program to allow you to break even on the venture? Show all work.

   b. What would be the consequences in terms of opportunity losses of an incorrect decision? Assume the possible proportion sold to be 0, 0.1, 0.2, . . . , 1.0.

   c. Draw graphs such as those in Figure 17.1 for the results found in Part b.

2. The R. E. Tire Co. has a weekly option to purchase used tires from a car rental company. The option is for a lot of 1,000 tires at $1,500, and they have a time limit of one day in which to exercise it. A sound tire can be retreaded and sold for a profit of $5.00. Management believes that a maximum of 20 tires can be sampled within the time constraint imposed by the option agreement. Examination of previous records yields the following information:

Proportion of good tires in lots of 1000	*Frequency*
0.2	200
0.4	350
0.6	300
0.8	150
	1,000

   Determine the optimal decision rule (20, c) which will minimize the expected opportunity losses. Show all work.

17.5 The Problem Involving a Normal Sampling Distribution

While we have considered the so-called modern approach to decision making only as it applies to a probability distribution for a discrete variable, you must not infer that it cannot be applied to the case of a continuous variable. On the contrary, it is applicable in many such situations. Of particular importance is the situation where a normal sampling distribution is involved.

Suppose we face a situation similar to the general situation in which we used the binomial sampling distribution. That is, suppose we must choose only one of two possible terminal acts. However, in this case, suppose the opportunity losses associated with the possible acts depend on the unknown mean μ_x of some continuous variable X, not on the unknown population proportion π of a Bernoulli process. If we assume that the sampling distribution of the sample mean $\bar{X}$ can be treated as normal (that is, the original distribution of X is normal and/or the sample size under consideration is sufficiently large) and the variance of the population σ_x^2 can be treated as known, the modern approach to a decision almost exactly parallels that given in earlier sections of the present chapter.

Since the procedure to be used in the case of a normal sampling distribution so closely parallels the procedure for the case of binomial sampling and our objective is merely to introduce Bayesian analysis, we shall not pursue this problem in detail. However, the interested reader will find that this problem, along with other very closely related problems, is handled quite adequately in Schlaifer (17.1) and Sasaki (17.3).

Important Terms and Concepts

best decision rule
conditional opportunity losses
economic breakeven point
expected loss
expected opportunity loss
expected sampling cost
expected total-loss curve
opportunity loss
optimal expected total-loss function
optimum expected losses from decision errors
optimum-expected-loss curve
sampling cost
sampling cost function

References

17.1 R. Schlaifer, *Statistics for Business Decisions*, New York: McGraw-Hill Book Company, 1961.

17.2 Harvard University Computation Laboratory, *Tables of the Cumulative Binomial Probability Distribution*, Cambridge, Mass.: Harvard University Press, 1955.

17.3 K. Sasaki, *Statistics for Modern Business Decision Making*, Belmont, Calif.: Wadsworth Publishing Company, 1968.

Review Problems

1. Carefully study Example 1 of Section 17.3. Using the decision theory approach and minimizing expected opportunity losses, decide whether the receiving tower should be constructed based on the following data.

Number of television owners: 4,000
Cost of construction: $100,000
Cost of maintenance: $40,000 over the effective life
Fee for tower service: $100 per customer
Effective life of the tower: 5 years
Salvage value at the end of effective life: $0
Possible states of nature and prior probabilities as follows:

Event (state of nature) E_i	*Prior probability* $P(E_i)$
$E_1(\pi = 0.20)$	0.15
$E_2(\pi = 0.35)$	0.75
$E_3(\pi = 0.55)$	0.05
$E_4(\pi = 0.70)$	0.05

Sample information: 4 out of 20 owners indicated they would use the tower service if it were available.

2. Refer to Problem 3 of Section 17.3 and determine the decision rule (20, c) that would minimize Mr. Wolf's expected opportunity losses.

Eighteen

Index Numbers

18.1 Introduction

An *index number* is often considered synonymous with an indicator. It is a special type of average used to measure relative changes. The primary use of an index number is to obtain a fair representation of the average value of the many influences upon it. In other words, it represents the relative change between two or more groups of data taken from different time periods, different geographical locations, and so forth.

Today's business management views index numbers as indicators of changes in market conditions. The Consumer Price Index (CPI), is one of the best known indexes because of its wide use by companies and labor unions in collective bargaining. Increases or decreases in wages are often calculated according to some agreed-upon formula based on changes in the CPI. Among the many types of index numbers useful in business forecasting, two of the more popular indexes are the Wholesale Price Index (WPI) and the Index of Industrial Production (IIP), a quantity index.

Even though widely used, index numbers are usually based on inadequate data, and the mathematical theories on which they depend are somewhat less than rigorous. However, to say that they are intentionally misleading would be incorrect. Index numbers are actually percentages, and hence, subject to the same biased interpretation as percentages. Thus, since a biased statement using percentages (or index numbers) can be technically accurate, the correct interpretation of index numbers is of prime importance.

Since they are used mainly to describe what has happened rather than to make inferences concerning future events, index numbers are usually considered a part of descriptive statistics. So, they have the same primary utility as other descriptive measures we have considered. That is, they are used to reduce large complex sets of data to a single number or a small set of numbers that "describe" either the sets of data or the relationships among them. By an index number we mean a ratio of two numbers. For example, a salary quotient (the ratio of a new salary to a corresponding old salary) is an index number. Our main concern here is with index numbers used in business and economics.

We shall place two restrictions on our meaning of index numbers: (1) the numbers are expressed in relative terms (ratios or percentages) instead of absolute terms, and (2) the results for a group of data are combined to obtain a single representative number. However, for the resulting index number to be useful, we must combine the results in some meaningful manner. For example, we would have a difficult time justifying, let alone interpreting, an index number which is determined by combining information regarding wheat production, textbook sales, days of sunshine, turnpike traffic, and attendance at professional football games. Nevertheless, such a combination could be interpreted as an index number. Thus, for an index number to be useful, we must group characteristics that are related in some sense of common interest. For example, an index number obtained by combining prices of beef, pork, chicken, and mutton can serve as a good indicator of the relative price changes of meat. You should keep in mind that an index number is considered good or bad according to how well it measures that which it is supposed to measure.

There are four basic methods of constructing index numbers that are relatively straightforward. All are essentially the same regardless of whether we are constructing an index for prices, quantities, or values. In presenting the four basic methods, we shall present the techniques of computation in two forms. That is, we first state the formula for determining an index number by a specific method; next, we outline the steps for computing the number.

18.2 Simple Index Numbers

Before constructing index numbers for a group of characteristics, let us first consider the construction of a *simple index number*—that is, an index number for one characteristic for which the values are known for several time periods or geographical locations. We generally refer to the values as being known for different *time* periods (or simply periods), although the techniques and terminology are equally applicable for geographical locations. Thus, regardless of the method chosen for computing a series of index numbers, we shall always select a *base period*; the resulting values of the series will be given relative to the value assumed in the base period by the characteristic of interest. We shall use the term *commodity* to refer to the item for which the value of the characteristic of interest is known. Thus, for a simple index number, we consider the values of one characteristic of one commodity for several periods.

Example 1

Suppose the hypothetical data in the first two columns of Table 18.1 represent the price (p) per bushel of wheat from 1959 through 1969. Compute the simple price index numbers first using 1960 as the base period and then using 1965 as the base period.

Table 18.1

Simple price index numbers

Year	*Price of wheat (per bushel)*	*Index A* (1960 = 100)	*Index B* (1965 = 100)
1959	\$1.65	111	121
1960	1.48	100	109
1961	1.52	103	112
1962	1.43	97	105
1963	1.38	93	101
1964	1.40	95	103
1965	1.36	92	100
1966	1.32	89	97
1967	1.29	87	95
1968	1.30	88	96
1969	1.25	84	92

Solution

To determine the simple price index number for a base period of 1960, we divide the price for each period by \$1.48, the price for 1960. That is, the simple price index numbers in column 3 of Table 18.1 are determined by dividing each price by the base-year price and multiplying the result by 100. (Since index numbers are used primarily for relative comparisons, we shall omit decimal places. That is, we shall round the resulting quotient to two decimal places and express the rounded result as a percentage. And since index numbers are always percentage figures, we omit the percent sign.) For example, the wheat price index number for 1962, with 1960 as the base period, is determined as

$$\frac{\$1.43}{\$1.48}(100) = 97.$$

Similarly, the simple price index numbers for a base period of 1965 are given in column 4, where the index numbers are determined in the same manner as those in column 3, except 1965 is the base period and the appropriate divisor is thus \$1.36. Hence, the simple wheat price index number for the 1962 period when the base period is 1965 is

$$\frac{\$1.43}{\$1.36}(100) = 105.$$

The index number (97) for 1962 with 1960 as the base period means that the price of wheat for 1962 was 3 percent lower than for 1960. The number (105) for 1962 with 1965 as the base period means that the price of wheat for 1962 was 5 percent higher than it was for 1965.

We must be extremely careful in interpreting index numbers. The numerical difference between any two index numbers is not necessarily the percentage change between them. For example, using the data in Table 18.1 and comparing the index numbers for 1963 and 1967, both determined with 1960 as the base year, we note that 93 is 6 percentage points greater than 87. This does not indicate a 6 percent decrease in the price of wheat from 1963 to 1967. Rather, since the index numbers were constructed using 1960 as the base year, the 6 percent decrease is a change in the price of wheat relative to what it was in 1960. Since index numbers are constructed from a given base, they merely indicate changes relative to that base. □

Example 1 indicates our need for a standard notation. Thus, we shall use the following notation:

p_i = price of the ith commodity;
q_i = quantity of the ith commodity;
v_i = value of the ith commodity;
p_{0i} = base-period (period 0) price of the ith commodity;
q_{0i} = base-period quantity of the ith commodity;
v_{0i} = base-period value of the ith commodity;
p_{ni} = price of the ith commodity for period n, where $n = 1, 2, \ldots$;
q_{ni} = quantity of the ith commodity for period n;
v_{ni} = value of the ith commodity for period n;
I_0 = base-period index number;
I_n = index number for period n, where the base period and method of determining the index number are understood.

Note that the terms for the base period are indicated by the subscript 0 (zero), and those for the succeeding periods are indicated by the subscripts 1, 2, 3, etc. Further, the base-year index number I_0 is always 100. Finally, we see that if we limit consideration to a simple index (only one commodity involved), the subscript i is superfluous. However, for indexes involving more than one commodity, concise definitions are necessary.

Using the above notation, we determine the simple price index (involving the ith commodity) for period n by the formula

$$I_n = \frac{p_{ni}}{p_{0i}}(100). \tag{1}$$

To determine the simple index number for either quantity or value, we would simply replace p_{ni} and p_{0i} by either q_{ni} and q_{0i}, respectively, or v_{ni} and v_{0i}, respectively. Thus, the steps for constructing simple index numbers are:

1. Select the base period.
2. Divide the entry for each period by that of the base period and multiply the quotient by 100.

Problems

1. From the data on the right, determine the simple price index using as the base period

   a. 1966, b. 1968.

Year	*Price of eggs (cents/dozen)*
1966	53
1967	60
1968	64
1969	70
1970	59

2. Use the data here to determine the simple price index numbers when the base period is

   a. 1967, b. 1970.

Year	*Price of labor (dollars/hour)*
1966	3.20
1967	3.40
1968	3.67
1969	3.90
1970	4.15

18.3 Simple Aggregative Index Numbers

A *simple aggregative index number* is exactly what its name implies. It is the simple index number for an aggregate—that is, a collection of items or commodities into a group. For example, we may determine the "aggregate" price for a bushel of wheat, a bushel of oats, a bushel of corn, and a bushel of barley for each time period, and then construct the simple index number of this "aggregate" price by simply determining the ratio of the price for each period to the base-period price.

Example 1

For the data given in the first three rows of Table 18.2, construct the simple aggregative price index numbers using 1971 as the base year.

Solution

Table 18.2 also presents the appropriate totals and the corresponding index numbers for each year with a base period of 1971. The index numbers are determined by adding the prices of the three commodities for each year, dividing each total by the

Table 18.2

Simple aggregative price index numbers

Commodity i	1971 p_{0i}	1972 p_{1i}	1973 p_{2i}	1974 p_{3i}
1	\$1.50	\$1.63	\$1.70	\$1.73
2	2.10	2.15	2.22	2.26
3	5.00	4.97	5.13	5.24
Total	\$8.60	\$8.75	\$9.05	\$9.23
Index number	100	102	105	107

total for the base period 1971, and multiplying the result by 100. For the base period 1971, we obtain

$$I_0 = \frac{p_{01} + p_{02} + p_{03}}{p_{01} + p_{02} + p_{03}}(100) = \frac{\$8.60}{\$8.60}(100) = 100;$$

for 1972, the first period subsequent to the base period, we obtain

$$I_1 = \frac{p_{11} + p_{12} + p_{13}}{p_{01} + p_{02} + p_{03}}(100) = \frac{\$8.75}{\$8.60}(100) = 102;$$

for 1973, the second period, we obtain

$$I_2 = \frac{p_{21} + p_{22} + p_{23}}{p_{01} + p_{02} + p_{03}}(100) = \frac{\$9.05}{\$8.60}(100) = 105;$$

and, for 1974, the third period, we obtain

$$I_3 = \frac{p_{31} + p_{32} + p_{33}}{p_{01} + p_{02} + p_{03}}(100) = \frac{\$9.23}{\$8.60}(100) = 107. \quad \square$$

As can be surmised from Example 1, the formula for constructing nth-period simple aggregative price index numbers for k commodities is

$$I_n = \frac{\sum_{i=1}^{k} p_{ni}}{\sum_{i=1}^{k} p_{0i}}(100). \tag{1}$$

Thus, the steps for constructing simple aggregative price index numbers for k commodities are as follows.

1. Select the base period.
2. For each period, total the prices of all commodities included in the aggregate.
3. Divide each total by the base period total and multiply the result by 100.

To construct simple aggregative index numbers for quantities or values, we again simply replace p_{ni} and p_{0i} in Relation (1) by either q_{ni} and q_{0i}, respectively, or v_{ni} and v_{0i}, respectively. In the steps for constructing simple aggregative price index numbers, we simply replace prices by either quantities or values. We must, however, be careful when using (1) for constructing quantity index numbers, because it is meaningless to add, for example, bushels, tons, and crates. Relation (1) may be used for constructing quantity index numbers only if the numerator and denominator of (1) are of the same units. And, since the commodities of most interest are generally not quoted in the same units, (1) is seldom used for constructing quantity index numbers.

The simple aggregative price index number gives us a way to determine relative changes in the price of an aggregate of commodities or, as it is commonly termed, a *market basket of commodities*. However, since it assigns equal weight (that is, unity) to the absolute change in price for each commodity, the simple aggregative index number assigns equal importance to each commodity in the market basket. This particular characteristic is the primary weakness of this type of index number, because it permits any one commodity which has a substantially higher price than the

Table 18.3

Effect of a high-priced commodity on aggregative numbers

Commodity i	1971 p_{0i}	1972 p_{1i}	1973 p_{2i}	1974 p_{3i}
1	\$1.50	\$1.63	\$1.70	\$1.73
2	2.10	2.15	2.22	2.26
3	5.00	4.97	5.13	5.24
4	20.00	20.05	20.10	20.15
Total	\$28.60	\$28.80	\$29.15	\$29.38
Index number	100	101	102	103

others to dominate the index. For example, suppose, we were to include in Table 18.2 a commodity which is in the twenty-dollar range, as demonstrated in Table 18.3. We note that the simple aggregative price index numbers with a base period of 1971 are, for the three subsequent periods, 101, 102, and 103, while their respective counterparts from Table 18.2 are 102, 105, and 107. Thus, we note that the one high-priced commodity has considerably influenced the relative changes of the market-basket price. Further, the type of units in which the price of each commodity is quoted has a significant influence on the resulting index number. For example, if the price of commodity 4 of Table 18.3 represented 100 units and we restated it as a single unit price, the influence of commodity 4 would be negligible. Thus, we see that the units in which the prices are given influence the index number as a hidden weight, which often has no meaningful significance.

Problems

1. Given the following data, determine the simple aggregative price index numbers using 1972 as the base period.

Commodity	1972 p_{0i}	1973 p_{1i}	1974 p_{2i}
A	\$1.03	\$1.24	\$1.32
B	1.95	1.86	1.84
C	6.52	6.85	7.14
D	4.30	4.50	4.60

2. For the following data, determine the simple aggregative price index numbers using 1971 as the base period.

Commodity	1971 p_{0i}	1972 p_{1i}	1973 p_{2i}	1974 p_{3i}
A	\$3.91	\$4.26	\$4.10	\$4.32
B	1.56	1.64	1.72	1.74
C	2.13	2.20	2.08	2.14

3. From the following data, determine the simple aggregative price index numbers using 1970 as the base period.

Commodity	1970 p_{0i}	1971 p_{1i}	1972 p_{2i}
A	\$21.50	\$23.72	\$24.03
B	48.32	55.14	54.17
C	61.28	62.84	63.00

18.4 Simple Average of Index Numbers

The *simple average of index numbers* is, as one might expect, simply the average of a group of individual index numbers. In order to obtain the simple average index numbers for a group of commodities, we first determine the base period to be used for constructing all the index numbers. Next, we determine the simple index numbers for each commodity using its base-period price, quantity, or value (whichever is appropriate) as the divisor for each of the prices, quantities, or values of the subsequent years. Finally, we compute the average of the simple index numbers in the usual manner for each period. Consider an example.

Example 1

With 1971 as the base period, use the data of Table 18.2 to construct the simple average of the index numbers for each appropriate year.

Solution

The individual index numbers and the simple average index numbers are presented in Table 18.4, where each simple index number is determined by the relation. Thus, letting I_{ni} and p_{ni} represent the index number and price, respectively, of the ith commodity for the nth year subsequent to the base year, we have

$$I_{ni} = \frac{p_{ni}}{p_{0i}}(100). \tag{1}$$

For example, the simple index number for commodity 3 for 1973 with 1971 as the base year is determined as

$$I_{23} = \frac{p_{23}}{p_{03}}(100) = \frac{\$5.13}{\$5.00}(100) = 102.6.$$

Table 18.4

The simple average of index numbers

Commodity i	1971 $I_{0i} = \left(\frac{p_{0i}}{p_{0i}}\right)100$	1972 $I_{1i} = \left(\frac{p_{1i}}{p_{0i}}\right)100$	1973 $I_{2i} = \left(\frac{p_{2i}}{p_{0i}}\right)100$	1974 $I_{3i} = \left(\frac{p_{3i}}{p_{0i}}\right)100$
1	100	108.7	113.3	115.3
2	100	102.4	105.7	107.6
3	100	99.4	102.6	104.8
Total	300	310.5	321.6	327.7
Index number	100	104	107	109

We express the simple index numbers to one decimal place only because they are to be used for subsequent computations.

The simple average index number is determined by averaging the index numbers for each year. Thus, the index numbers are determined for the three subsequent years as follows:

$$I_1 = \frac{\sum_{i=1}^{3} I_{1i}}{3} = \frac{I_{11} + I_{12} + I_{13}}{3} = \frac{310.5}{3} = 104;$$

$$I_2 = \frac{\sum_{i=1}^{3} I_{2i}}{3} = \frac{I_{21} + I_{22} + I_{23}}{3} = \frac{321.6}{3} = 107;$$

$$I_3 = \frac{\sum_{i=1}^{3} I_{3i}}{3} = \frac{I_{31} + I_{32} + I_{33}}{3} = \frac{327.7}{3} = 109.$$

In keeping with our conventional representation, the simple average index numbers are expressed without decimal places. □

From Example 1, it can be noted that the formula for computing the simple average of index numbers for k commodities is given by

$$I_n = \frac{\sum_{i=1}^{k} I_n}{k} = \frac{\sum_{i=1}^{k}\left(\frac{p_{ni}}{p_{0i}}\right)100}{k}. \tag{2}$$

The steps for determining the simple average of index numbers as given by relation (2) are:

1. Select the base period.
2. For each period, determine the simple index numbers for each commodity by dividing the price of that commodity by its base-period price and multiplying the results by 100.
3. For a given period, add the simple index numbers for all commodities and divide by the number of commodities.

In order to construct the simple average of quantity or value index numbers, we would again simply replace p_{ni} and p_{0i} in Relation (2) by either q_{ni} and q_{0i}, respectively, or v_{ni} and v_{0i}, respectively.

Comparing the index numbers of Table 18.2 (simple aggregative index numbers for 1972 through 1974, with 1971 as the base period) with the corresponding simple average index numbers of Table 18.4, we detect notable differences. The reason for the differences lies in the fact that commodity 3 dominates the simple aggregative index numbers, while commodity 1 dominates the simple average index numbers. That is, in Table 18.2, commodity 3 is the highest-priced commodity and hence exerts the most influence on the total price of the aggregate. However, in Table 18.4, commodity 1 has the largest values for its simple index numbers and, consequently, exerts more influence on the simple average index number than the simple index number for any other commodity.

We can conclude that the simple average index numbers avoid the difficulty we encountered with simple aggregative index numbers. That is, it eliminates the undue influence created by the difference in units for which the price of a specific commodity is stated.

However, there remains a difficulty of considerable consequence—the commodities now have equal importance, a situation we may not want when we figure the relative price changes of a market basket of commodities. For example, an index number for a market basket containing one pound of beef, one loaf of bread, one quart of milk, and one pound of butter may not represent the market basket for which we wish to measure the relative changes. We may not use a pound of butter for every loaf of bread, and we may not use a pound of beef for every quart of milk, and so on. Thus, we need some means of weighting each commodity according to its importance relative to the entire market basket. This leads us to the problem of weighting index numbers.

Problems

1. Using the data of Problem 1, page 683, determine the simple average index numbers with 1972 as the base period.
2. Using the data of Problem 2, page 683, determine the simple average index numbers with 1971 as the base period.
3. Using the data of Problem 3, page 684, determine the simple average index numbers with 1970 as the base period.

18.5 *Weighted Aggregative Index Numbers*

There are many ways in which index numbers can be weighted. Here we shall consider the most commonly used weights for aggregative price indexes and quantity indexes. Later, we shall consider weights for average index numbers. Let us first consider *weighted aggregative price index numbers.* The two most commonly used weights for nth-period aggregative price indexes are base-period quantities and nth-period quantities.

For weighted aggregative price index numbers (for the nth period) determined by using *base-period quantities* as weights, we use the formula

$$I_n = \frac{\sum_{i=1}^{k} p_{ni}q_{0i}}{\sum_{i=1}^{k} p_{0i}q_{0i}}(100). \tag{1}$$

Formula (1) is generally termed the *Laspeyres' formula*, and the index numbers it produces are termed *Laspeyres-type index numbers.* In this formula, we note that the prices are weighted by the base-period quantities. This type of weighting allows us to compare the relative change in prices for each period by keeping the contents of our market basket the same as for the base period. Let us consider an example.

Example 1

Using the data given in Table 18.2, page 681, and the 1971 quantities of 1,000, 200, and 300 units for commodities 1, 2, and 3, respectively, determine the Laspeyres-type index numbers for 1972 and 1973, using 1971 as the base years.

Solution

Table 18.5 shows the computations. The numerator for each index number is determined for each year by Formula (1):

$$\sum_{i=1}^{k} p_{0i}q_{0i} = (\$1.50)(1{,}000) + (\$2.10)(200) + (\$5.00)(300) = \$3{,}420;$$

$$\sum_{i=1}^{k} p_{1i}q_{0i} = (\$1.63)(1{,}000) + (\$2.15)(200) + (\$4.97)(300) = \$3{,}551;$$

$$\sum_{i=1}^{k} p_{2i}q_{0i} = (\$1.70)(1{,}000) + (\$2.22)(200) + (\$5.13)(300) = \$3{,}683.$$

Table 18.5

Laspeyres-type index numbers

	Quantity		*Price*		*Value of base-year quantities at current-year prices*		
	1971	1971	1972	1973	1971	1972	1973
Commodity i	q_{0i}	p_{0i}	p_{1i}	p_{2i}	$p_{0i}q_{0i}$	$p_{1i}q_{0i}$	$p_{2i}q_{0i}$
1	1,000	\$1.50	\$1.63	\$1.70	\$1,500	\$1,630	\$1,700
2	200	2.10	2.15	2.22	420	430	444
3	300	5.00	4.97	5.13	1,500	1,491	1,539
Total					\$3,420	\$3,551	\$3,683
Index number					100	104	108

The denominator is the same for all terms:

$$\sum_{i=1}^{k} p_{0i} q_{0i} = \$3{,}420.$$

Dividing each of the numerators by the base-period sum, we obtain the index numbers for 1971 through 1973, with 1971 as the base period. That is,

$$I_0 = \frac{\$3{,}420}{\$3{,}420}(100) = 100,$$

$$I_1 = \frac{\$3{,}551}{\$3{,}420}(100) = 104, \quad \text{and} \quad I_2 = \frac{\$3{,}683}{\$3{,}420}(100) = 108.$$

We note that the price for the market basket containing the base-period quantities of each commodity is 4 percent higher in 1972 than it was in 1971 and 8 percent higher in 1973 than in 1971. □

From Formula (1) and the steps involved in constructing Table 18.5, we note the general steps for constructing the Laspeyres-type index numbers:

1. Select the base period.
2. Determine the base-period quantity for each commodity, where each quantity is expressed in the units for the corresponding price.
3. For each commodity, multiply the price for each period by the base-period quantity.
4. Total the products obtained in step 3 for each period.
5. Obtain the Laspeyres-type index numbers for each period by dividing each total by the base period total and then multiplying the resulting ratios by 100.

From Example 1, we see that the Laspeyres-type index numbers reflect perfectly the relative price changes of a basket of commodities only as long as the quantities for each commodity in the market basket do not change from the base-period quantities. That is, the Laspeyres-type index numbers measure the difference between an actual basket price for the nth period and an actual basket price for the base period, both prices determined by using as weights quantities associated with the base period. However, in many situations, we want to measure the difference between an actual basket price for the nth period and an actual basket price for the base period, both prices determined by using as weights quantities associated with the nth period.

We obtain the correct formula for determining an appropriate index number by slightly adjusting Formula (1). We simply replace the base-period quantities by the nth-period quantities. The resulting formula, generally known as the *Paasche formula*, is

$$I_n = \frac{\sum_{i=1}^{k} p_{ni} q_{ni}}{\sum_{i=1}^{k} p_{0i} q_{ni}}(100). \tag{2}$$

The resulting index numbers are termed *Paasche-type index numbers*.

Example 2

Using the data of Table 18.2, page 681, and the quantities of Table 18.6, determine the Paasche-type index numbers for 1972 and 1973 with 1971 as the base year.

Table 18.6

Paasche-type index numbers—initial data on quantities

Commodity i	1971 q_{0i}	1972 q_{1i}	1973 q_{2i}
1	1,000	800	700
2	200	400	300
3	300	400	500

Solution

The computations are shown in Table 18.7. The index numbers are obtained as follows:

$$I_0 = \frac{\sum_{i=1}^{k} p_{0i} q_{0i}}{\sum_{i=1}^{k} p_{0i} q_{0i}}(100) = \frac{\$3{,}420}{\$3{,}420}(100) = 100;$$

$$I_1 = \frac{\sum_{i=1}^{k} p_{1i} q_{1i}}{\sum_{i=1}^{k} p_{0i} q_{1i}}(100) = \frac{\$4{,}152}{\$4{,}040}(100) = 103;$$

$$I_2 = \frac{\sum_{i=1}^{k} p_{2i} q_{2i}}{\sum_{i=1}^{k} p_{0i} q_{2i}}(100) = \frac{\$4{,}421}{\$4{,}180}(100) = 106.$$

Since $I_1 = 103$, we conclude that the 1972 market basket quantities would cost 3 percent more than the same quantities purchased at 1971 prices. Similarly, $I_2 = 106$ implies that the cost of the 1973 market basket quantities would be 6 percent more than the same quantities purchased at 1971 prices. □

From Example 2, we note the general steps for constructing the Paasche-type price index numbers:

1. Select the base period.
2. Determine the quantity for each commodity-year combination, where the quantity for each commodity is expressed in the units for the corresponding price.
3. For each commodity, multiply the base-period price by the nth-period quantity for each period. Total these products for each period.

Table 18.7

Paasche-type index numbers

Commodity	Quantity			Price			Actual value of given-year quantities at base-year prices			Actual value in given years		
	1971	1972	1973	1971	1972	1973	1971	1972	1973	1971	1972	1973
i	q_{0i}	q_{1i}	q_{2i}	p_{0i}	p_{1i}	p_{2i}	$p_{0i}q_{0i}$	$p_{0i}q_{1i}$	$p_{0i}q_{2i}$	$p_{0i}q_{0i}$	$p_{1i}q_{1i}$	$p_{2i}q_{2i}$
1	1,000	800	700	$1.50	$1.63	$1.70	$1,500	$1,200	$1,050	$1,500	$1,304	$1,190
2	200	400	300	2.10	2.15	2.22	420	840	630	420	860	666
3	300	400	500	5.00	4.97	5.13	1,500	2,000	2,500	1,500	1,988	2,565
Total							$3,420	$4,040	$4,180	$3,420	$4,152	$4,421
Index number										100	103	106

3. Determine the simple index numbers for each commodity by dividing its price for each period by its base-period price and multiplying the result by 100.
4. Multiply the simple index numbers obtained in step 3 by the weight obtained in step 2 for each commodity. Total these products for each period.
5. Obtain the weighted average of price index numbers by dividing the total obtained in step 4 by the sum of the weights obtained in step 2.

The other index numbers obtained by using the weighted average can be determined through the above steps by simply interchanging the roles of price and quantity and/or *n*th period and base period. The interchange to be made, of course, depends upon which type of index number is desired as well as which values are to be used as weights.

Problems

1. Given the data,

	1972		1973		1974	
Commodity	q_{0i}	p_{0i}	q_{1i}	p_{1i}	q_{2i}	p_{2i}
A	10	\$1.03	12	\$1.24	14	\$1.32
B	20	1.95	20	1.86	20	1.84
C	10	6.52	7	6.85	11	7.14

a. determine the weighted average price index numbers with base-period quantities as weights for 1973 and 1974, with 1972 as the base period;

b. determine the weighted average price index numbers with *n*th-period quantities as weights for 1973 and 1974, with 1972 as the base period.

18.7 Shifting the Base of Index Numbers

We sometimes wish to shift the base of an index number. For example, we may want to compare two index numbers with different bases or change the base of an index number that has an out-of-date base. The process of changing bases is quite simple. In order to change the base of a group of index numbers, we simply divide each index number of the group by the index number corresponding to the new base year. Let us consider an example.

Example 1

Using the index numbers of Table 18.9 (where 1960 is the base year), shift the base to 1965.

Table 18.9

Index numbers

Year	1959	1960	1961	1962	1963	1964	1965	1966	1967	1968	1969
Index (1960 = 100)	111	100	103	97	93	95	92	89	87	88	84

Solution

The index numbers for the base year 1965 are given in Table 18.10. These numbers are obtained by simply dividing the index numbers in Table 18.9 (which have 1960 as a base year) by 92, the index number for 1965. For example, the index number for 1962 with a 1965 base is determined as $97/92 = 105$.

Table 18.10

Results of shifting the base of the index numbers of Table 18.9

Year	1959	1960	1961	1962	1963	1964	1965	1966	1967	1968	1969
Index (1965 = 100)	121	109	112	105	101	103	100	97	95	96	91

Note that, except for rounding errors, the index numbers in Table 18.10 are the same as those obtained in the "Index B" column of Table 18.1 through use of the individual prices. □

Problems

1. Given the following index numbers,

Year	1965	1966	1967	1968	1969	1970
Index (1966 = 100)	98	100	103	102	105	101

   a. shift the base of the index numbers from 1966 to 1969;

   b. shift the base of the index numbers from 1966 to 1970.

2. Given the following index numbers,

Year	1970	1971	1972	1973	1974	1975
Index (1970 = 100)	100	104	101	106	109	112

a. shift the base of the index numbers from 1970 to 1975;

b. shift the base of the index numbers from 1970 to 1973.

18.8 Splicing Two Overlapping Series of Index Numbers

Occasionally, new commodities or different quantities are introduced into a market basket for which we have constructed index numbers for several consecutive periods. In such a situation, we generally wish to construct index numbers using these new commodities or different quantities; thus, we have two series of index numbers measuring approximately the same thing, where we want a single continuous series of index numbers. We can splice the two overlapping series in one of two ways. We can either make the new series continuous with the old series or make the old continuous with the new, simply by using proportions in either case. That is, we use the index numbers that overlap (so to speak) in order to set up the appropriate proportions.

Example 1

Using the two series of index numbers in Table 18.11, splice the second series to the first. That is, obtain a single continuous series having the same base year as the first. Then splice the first series to the second. In other words, obtain a single continuous series having the same base year as the second of the two overlapping series.

Table 18.11

Two overlapping series of index numbers

Year	*First series of index numbers*	*Second series of index numbers*
1970	100	
1971	103	
1972	105	
1973	107	100
1974		102
1975		105

Solution

In order to splice the second series to the first, we simply establish the ratio $x_{74}:102 = 107:100$, where x_{74} represents the index number we wish to determine for 1974 with a 1970 base, the base for the first series. Expressing the ratio in terms of proportions, we obtain

$$\frac{x_{74}}{102} = \frac{107}{100}.$$

Solving for x_{74}, gives

$$x_{74} = \frac{107}{100}(102) = 109.$$

Similarly, we obtain the spliced index number for 1975 by solving

$$\frac{x_{75}}{105} = \frac{107}{100}$$

for x_{75}. Thus,

$$x_{75} = \frac{107}{100}(105) = 112.$$

These values are given in Table 18.12 under the column labeled "First series of index numbers."

Table 18.12

Results of splicing the two series of Table 18.11

	Spliced indexes	
Year	*First series of index numbers*	*Second series of index numbers*
1970	100	93
1971	103	96
1972	105	98
1973	107	100
1974	109	102
1975	112	105

In order to splice the first series to the second, we proceed similarly. That is, to obtain the index number for 1972 with 1973 as the base period, we solve

$$\frac{x_{72}}{105} = \frac{100}{107},$$

obtaining

$$x_{72} = \frac{100}{107}(105) = 98.$$

Similarly, for 1971 and 1970, we obtain, respectively,

$$x_{71} = \frac{100}{107}(103) = 96 \quad \text{and} \quad x_{70} = \frac{100}{107}(100) = 93.$$

These are the values in the last column of Table 18.12. □

Problems

1. Given the following data,

Year	*Index A*	*Index B*
1965	100	
1966	104	
1967	106	
1968	107	100
1969		102
1970		105

   a. splice Index B to Index A to obtain an index number of each year 1965 through 1970 for Index A;

   b. splice Index A to Index B to obtain an index number of each year 1965 through 1970 for Index B.

2. Given the following data,

Year	*Index A*	*Index B*
1960	100	
1961	103	
1962	105	
1963	104	
1964	104	
1965	106	100
1966		101
1967		102
1968		103
1969		105
1970		107

   a. splice Index B to Index A to obtain an index number for each year 1960 through 1970 for Index A;

   b. splice Index A to Index B to obtain an index number for each year 1960 through 1970 for Index B.

18.9 Chain Index Numbers

One defect of the weighted average or weighted aggregative index numbers is that the weights or some portion thereof (quantity or price) are fixed. Thus, when a new commodity enters the market basket, we are forced to revise our list of commodities and, consequently, our weights. To allow for such revision, we use a technique that links all of the index numbers. That is, we construct each index number so that the immediately preceding period is its base period. Index numbers constructed in this manner are referred to as *link index numbers* or *link relatives*. The link index numbers can then be multiplied together in order to establish an index number for some common base.

Example 1

For the data given in Table 18.5, page 687, use the Laspeyres' formula to construct the link index numbers for 1972 and 1973, while keeping the quantities fixed at the 1971 levels. Also, using link index numbers, construct the index numbers for 1972 and 1973 which have 1971 as the base.

Solution

Table 18.5 shows the appropriate sums for constructing the link index numbers. That is, for 1971, 1972, and 1973, respectively, we have

$$\sum_{i=1}^{k} p_{0i}q_{0i} = \$3{,}420, \tag{1}$$

$$\sum_{i=1}^{k} p_{1i}q_{0i} = \$3{,}551, \tag{2}$$

and

$$\sum_{i=1}^{k} p_{2i}q_{0i} = \$3{,}683. \tag{3}$$

Now, since each link index number is constructed with the preceding period as its base, we determine the link index numbers for 1972 and 1973 simply by using the above sums. Letting I_{n-r} represent the index number obtained for period r with period n as its base, we obtain

$$I_{71-72} = \frac{\$3{,}551}{\$3{,}420}(100) = 103.8 \quad \text{and} \quad I_{72-73} = \frac{\$3{,}683}{\$3{,}551}(100) = 103.7.$$

Multiplying these two link index numbers (expressed in decimal form rather than as percentages) and multiplying their product by 100, we obtain the index number for 1973 with 1971 as the base. That is, in our current notation, we obtain

$$I_{71-73} = I_{71-72} \cdot I_{72-73} \cdot (100) = (1.038)(1.037)(100) = 107.7,$$

which we round to 108. Note that it has the same value as I_2 of Table 18.5. We note that the value I_{71-73}, which has been obtained by use of link index numbers, is exactly the same as the index number for 1973 with 1971 as its base. That is, from (1) through (3), we obtain

$$I_{71-73} = \frac{\$3{,}551}{\$3{,}420} \cdot \frac{\$3{,}683}{\$3{,}551}(100) = \frac{\$3{,}683}{\$3{,}420}(100) = 108,$$

which is exactly the ratio for determining I_2 in Table 18.5. □

In Example 1, we constructed an index number simply by multiplying the appropriate link index numbers. This procedure can be extended to include as many consecutive link index numbers as desired. However, the base period will always be the same as the base period for the link index number which has the earliest base period. For example, beginning with a link index number for 1971 (which has a base period of 1970), we can multiply this by the one for 1972 and obtain the index number for 1972 with 1970 as the base. Next, we can multiply this resulting index number by the link index number for 1973 and obtain the index number for 1973 with 1970 as the base. We can continue this multiplication in chain fashion until we obtain the index number for the period for which the most recent link index number is determined. Index numbers obtained in this manner are occasionally termed *chain index numbers*.

Problems

1. From the data of Problem 1, page 683, and the 1972 quantities of Problem 1, page 691, use the Laspeyres' formula to determine the link index numbers for 1973 and 1974. Also, use the link index numbers to determine the index numbers for 1973 and 1974 which have 1972 as the base period.
2. From the data of Problem 2, page 683, and the 1971 quantities of Problem 2, page 691, use the Laspeyres' formula to determine the link index numbers for 1972, 1973, and 1974. Also, use the link index numbers to determine the index numbers for 1972, 1973, and 1974 which have 1971 as the base period.
3. From the data of Problem 3, page 684, and the 1970 quantities of Problem 3, page 692, use the Laspeyres' formula to determine the link index numbers for 1971 and 1972. Also, use the link index numbers to determine the index numbers for 1971 and 1972 which have 1970 as the base period.

18.10 Related Remarks

For the businessman or economist, it is very difficult to keep abreast of market conditions without encountering index numbers. Most indicators of the state of the national, state, and local environments, are expressed in the form of index numbers. Some of the indicators available to the general public are consumer prices, wholesale prices, index of weekly payrolls, foreign trade or exports and imports, construction costs, industrial production, farm production indexes, index of business activity, purchasing power of the dollar, and index of farm outputs.

Index numbers, used primarily as measures of economic change, permit management to form judgments and make decisions based on the comparison of two or more values. For example, a manager may decide to buy inventory as the result of changes in certain important index numbers. A politician may base his or her campaign platform on issues related to changes in relevant index numbers. A banker may change loan policies as the result of certain index numbers. The list of uses for index numbers is practically boundless.

We have discussed some basic methods for constructing index numbers. You should note that although the use of index numbers is very widely accepted, it is always wise to determine how an index number has been constructed before placing too much faith in it. Many variables influence the outcome of index numbers, as for example, the contents of the market basket, the choice for the base period, the method of construction, and many other conditions. This being true, the person who uses index numbers should be exceptionally careful. Since index numbers have associated with them the same pitfalls as percentages, they can be very deceiving if interpreted improperly. On the other hand, when used with proper understanding, they can be very helpful in revealing relative changes for situations where absolute changes are not so meaningful.

For a more extensive treatment of index numbers, you are referred to Bowen (18.1), Croxton and Cowden (18.2), Mudgett (18.3), and Stockton (18.4).

Important Terms and Concepts

chain index numbers
index numbers
Laspeyres' formula
Laspeyres-type index numbers
link index numbers
link relatives
market basket
Paasche's formula
Paasche-type index numbers
shifting bases of index numbers
simple aggregative index numbers
simple average of index numbers
simple index numbers
splicing index numbers
weighted aggregative index numbers
weighted average of index numbers

References

18.1 E. K. Bowen, *Statistics, With Applications in Management and Economics*, Homewood, Ill.: Richard D. Irwin, Inc., 1960.

18.2 F. E. Croxton and D. J. Cowden, *Practical Business Statistics*, 3rd ed., Englewood Cliffs, N.J.: Prentice-Hall, Inc., 1960.

18.3 B. D. Mudgett, *Index Numbers*, New York: John Wiley & Sons, Inc., 1951.

18.4 J. A. Stockton, *Introduction to Business and Economic Statistics*, 3rd ed., Cincinnati: South-Western Publishing Company, 1966.

Nineteen

Elementary Time-Series Analysis

19.1 Times Series—General

A *time series* is a set of values of a variable observed at specific time intervals. From a mathematical viewpoint, a time series can be described by the observed values $y_1, y_2, \ldots$ of a variable Y at times $t_1, t_2, \ldots$. In this sense, the variable Y is said to be a function of the variable t, that is, $Y = f(t)$. However, so that our notation will be compatible with that introduced in our study of regression analysis (Chapter 14), we shall represent the time variable by the letter x. Thus, we say that Y is a function of x, where Y is the time-series variable of interest and x is the independent variable representing time.

Typical examples of time series are annual sales of a given company, daily high temperatures for a city or region, and monthly Consumer Price Indexes (CPI).

The pictorial representation of a time series is essentially no different from many we have already discussed. It is constructed merely by plotting values of the time-series variable Y versus corresponding values of the time variable x. Figure 19.1 shows a typical time-series graph.

Although we generally think of the graph of a time series as representing the movement of a single variable over time, the shape of the resulting curve is influenced by various forces, some of which may be economic or sociological. Thus, time series in general exhibit some combination of certain characteristic movements or variations. The analysis of such a combination of movements is generally termed *time-series analysis*. While the techniques associated with time-series analysis are well defined, the techniques are almost completely descriptive and their application involves a great deal of subjectivity. Our purpose here, therefore, is merely to present a summary of the techniques and related rationale, not to justify them from a theoretical standpoint.

Classification of Time-Series Movements

We refer to the characteristic variations of time series as *components*, which we categorize into four main groups:

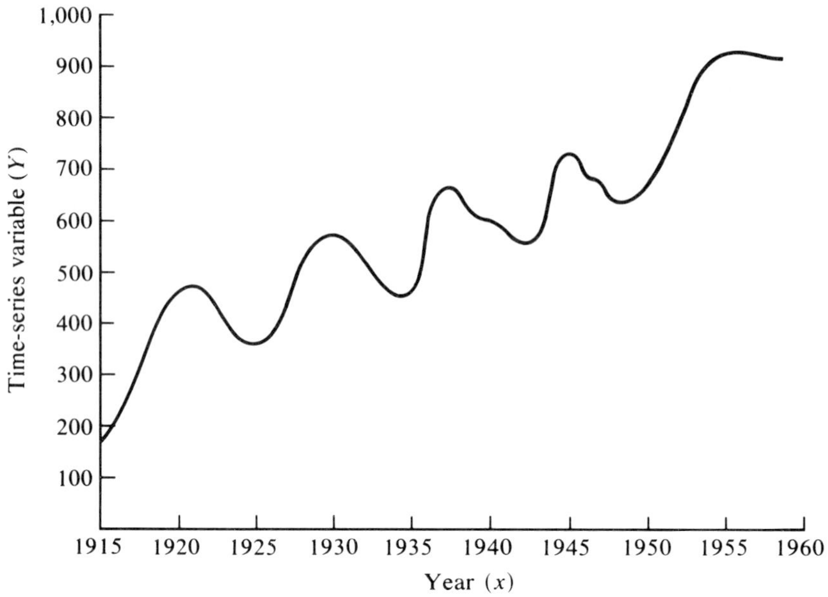

Figure 19.1 A typical time-series graph

1. *Long-term movement*, commonly called *secular variation* or *secular trend*, refers to the long-term general direction of the time-series graph. The most general manifestation of the secular movement is labeled a *trend curve*, but for many elementary time-series analyses a *trend line* will suffice for descriptive purposes.
2. *Seasonal movements* or *seasonal variations* refer to the quite similar repetitive patterns which a time-series curve exhibits in corresponding months of succeeding years. These movements can logically be traced to annually recurring events, as for example, an increase in ice cream sales during the summer and a corresponding decrease in the winter. In business and economic contexts, the seasonal movements generally show an annual periodicity. However, such variations can occur repetitively over other intervals of time such as the day, hour, or week. In this case, the movements are generally called periodic rather than seasonal. The periodicity naturally depends on the kind of data under analysis.
3. *Cyclical movements* or *cyclical variations* are long-term oscillations of the time series about the trend line or curve. These movements are often called cycles and may or may not occur with exact periodicity. For a movement to be considered cyclical in business and economic activity, we generally require that the period involved be longer than a year.
4. *Irregular* or *residual variations* refer to the random motions of a time series attributed to chance events. They essentially represent those movements not accounted for in 1, 2, and 3, above. We ordinarily view the variations resulting from such events as having an effect for only a very short period of time.

The concepts of long-term and cyclical variations are graphed in Figure 19.1, where the sketched curve represents the trend curve of the time series involved. Annual figures were used to develop the long-term trend. Thus, no seasonal movements are indicated. However, the cyclical movements about the trend are readily apparent.

The Analysis of Time Series

The essence of the principal concepts involved in time-series analysis is given in Figure 19.2. The graph of a long-term trend line is shown in a; b shows the superposition of cyclical movement on the long-term trend; and c shows the fluctuations that result from the superposition of seasonal movement upon the curve of b. Note that further fluctuations would occur if we superimposed random variation on the curve of c. The primary objectives of time-series analysis are to describe the component movements shown by curves such as those in the figure and use the information resulting from the description.

Let T, C, S, and I represent the respective forces that produce trend, cyclical, seasonal, and irregular movements. Now, the time-series variable Y is clearly a function of these four forces. Time-series analysis consists essentially of decomposing the time-series variable into these basic component movements. In order to effect this decomposition, we must make an assumption regarding the form of the functional relationship between Y and the four forces. One of the most useful assumptions of this nature is a model called the *multiplicative model of time-series analysis*,

$$Y = T \cdot C \cdot S \cdot I; \tag{1}$$

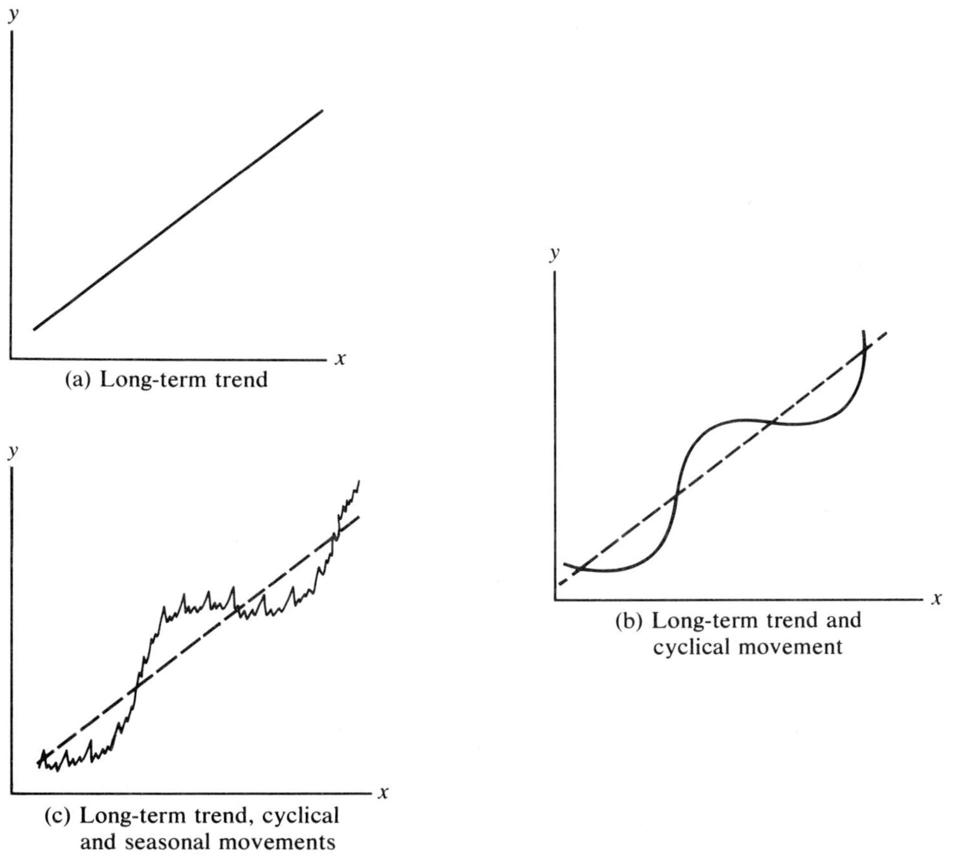

Figure 19.2 Principal concepts of time series

hence, the primary task to be accomplished in the description facet of time-series analysis is the decomposition of the model given by Equation (1) into the basic component movements T, C, S, and I. We shall study appropriate methods subsequently.

First, however, note that it is often proper to assume a model of the form

$$Y = T + C + S + I, \tag{2}$$

which is called the *additive model of time-series analysis*. Our decomposition method is based on the multiplicative model given by Equation (1). However, analogous procedures are available for the additive model. In practical time-series analysis, the proper choice of a model depends on the nature of the data.

The Use of Time-Series Analysis in Forecasting

It goes almost without saying that one of the most critical problems in any organization concerned with planning and decision making is that of forecasting. Thus, it is not surprising that much space in relevant literature is devoted to the development of forecasting techniques. One such technique uses information generated through time-series analysis.

Clearly, as soon as we make the foregoing model assumption of (1), we admit that the total variation of the relevant time series cannot be completely explained by trend and seasonal components. Yet, certain purely mechanical forecasting techniques involving time-series analysis use merely the projection of trend and adjustments for empirically determined seasonal effects. Obviously, when cyclical and residual forces are not taken into account, forecasts are in error. Thus, the analysis of trend and seasonal effects and a related projection should be viewed only as first steps to a forecast. Further refinements involving the proper management of cyclical and residual effects should be incorporated.

Now, even though there are limitations associated with formal time-series-analysis techniques, the importance of such procedures in forecasting should not be underestimated. Information developed in a well designed time-series analysis serves two very important functions. First, it provides a "ball-park" forecast that takes into account irregularities stemming from trend and seasonal movements. Second, it provides a method for adjusting the original data for these influences. Then, the resulting time series, consisting of the trend and seasonally adjusted data, often proves quite useful in the analysis and interpretation of cyclical and residual variations.

19.2 Time Series—Trend Analysis

The first problem in time-series analysis is that of trend measurement. However, in light of the discussion of Section 19.1, this particular problem reduces to that of finding a smooth curve (perhaps a straight line) that both fits the time-series data "reasonably" well and appears to represent the long-term component of change in the time-series variable. In other words, we must find a mathematical relationship which depicts Y (the time-series variable) as a function of x (the time variable). There are several ways to do this. Two of the best known are visual examination of the data and sketching of a curve, and application of the method of moving averages to smooth the data. However, neither method is theoretically based. Thus, although we shall view the method of moving averages in a later discussion devoted to seasonal variation, neither merits detailed consideration in trend analysis.

The most logical method for trend analysis is that of least squares; it is both theoretically sound and easy to apply. However, a warning is in order. Even a trend equation found through the use of mathematical procedures (for example, the method of least squares) merely represents the long-term movement of the past history of the time-series variable. It bears no necessary relationship to values the time-series variable will assume in the future. At best, the trend curve obtained through computational techniques can be interpreted as only an indication of future values to be assumed by the time-series variable. Even this is true only if the forces that interacted to produce the trend continue to have the same type of effects upon the time-series variable.

Now, except for one very minor variation, our problem is no different from the usual least-squares problem discussed in Chapter 14. The difference concerns the manner in which a time series covering an interval of time should be plotted. For example, in Figure 19.1, even though the individual observations of the time-series variable represent specific values for the whole year, they are plotted as if they represent a single point along the x-axis—the point representing the year with which the particular value is associated. Yet, each value actually corresponds to an interval (one year) of time. To resolve the problem, we adopt the rather arbitrary convention that an observed value associated with a period of time is usually plotted as though it were associated with the point of time at the middle of the time period. Thus, the observed values in a given time series are typically (but not always) plotted as though they were associated with July 1 of the respective year.

Returning to the major problem of trend analysis, we now consider how to use the least-squares method in relating the time-series variable Y to the independent time variable x. So far, we have agreed only that Y is some function of x. Obviously, the possible mathematical forms of this function are infinite. How do we settle this problem? The form of the relationship is somewhat subjective; one can never be absolutely sure that the chosen form is proper. However, there are several general forms that occur historically in time series; hence, one of these forms will usually suffice for a given situation. Fortunately, for most of the often-used forms, the method of least squares will provide the curve for which the sum of squares of deviations between the fitted curve and the observed points will be minimized.

Suppose we have n observed values on the time-series variable Y for specified time intervals. We then have a set of points $(x_1, y_1), (x_2, y_2), \ldots, (x_n, y_n)$. Our objective is to find the relationship $Y = f(x)$ that appropriately represents the trend movement. Two of the most commonly used forms for these functions are

$$Y = \beta_0 + \beta_1 x \tag{1}$$

and

$$Y = \beta_0 + \beta_1 x + \beta_2 x^2. \tag{2}$$

Equation (1) represents a straight-line relationship, while Equation (2) represents a special form of the general type curve called a parabola. However, both (1) and (2) are special cases of a polynomial. We call the relation of Equation (1) a *polynomial of the first degree* (in x) and the relation of Equation (2) a *polynomial of the second degree.* A *polynomial of the fifth degree*, for example, would be represented by the relationship

$$Y = \beta_0 + \beta_1 x + \beta_2 x^2 + \beta_3 x^3 + \beta_4 x^4 + \beta_5 x^5. \tag{3}$$

We have seen in Chapter 14 that if we wish to fit a straight line to the data given by the ordered pairs $(x_1, y_1), (x_2, y_2), \ldots, (x_n, y_n)$, the theory of least squares tells us that in order to best (in the sense of least squares) approximate the relevant coefficients (β_0 and β_1) we must solve the so-called normal equations for a least-squares straight line, or

$$\begin{aligned} \sum y &= \hat{\beta}_0 n + \hat{\beta}_1 \sum x, \\ \sum xy &= \hat{\beta}_0 \sum x + \hat{\beta}_1 \sum x^2. \end{aligned} \tag{4}$$

Through a proper description of the model involved and appropriate use of calculus, it can be shown that when we wish to fit a polynomial of the second degree, we estimate the relevant coefficients by solving the normal equations

$$\begin{aligned} \sum y &= \hat{\beta}_0 n + \hat{\beta}_1 \sum x + \hat{\beta}_2 \sum x^2, \\ \sum xy &= \hat{\beta}_0 \sum x + \hat{\beta}_1 \sum x^2 + \hat{\beta}_2 \sum x^3, \\ \sum x^2 y &= \hat{\beta}_0 \sum x^2 + \hat{\beta}_1 \sum x^3 + \hat{\beta}_2 \sum x^4. \end{aligned} \tag{5}$$

A definite relationship between the set of equations of (5) and those of (4) can be noted. It stems directly from the relationship between the two relevant models, (2) and (1), respectively. Thus, for example, the set of equations of (5) can be obtained by successively multiplying the model equation of (2) by (where the observed value of Y is substituted for Y) 1, x and x^2, respectively, and summing (over the number of observed values) on both sides of the resulting equations. Further, the technique can be extended to obtain appropriate normal equations for models involving polynomials of any degree. For example, the normal equations for a polynomial of the fifth degree are obtained by successively multiplying the model equation (3) by 1, x, x^2, x^3, x^4, and x^5, respectively, and summing on both sides of the resulting equations, that is,

$$\begin{aligned} \sum y &= \hat{\beta}_0 n + \hat{\beta}_1 \sum x + \hat{\beta}_2 \sum x^2 + \hat{\beta}_3 \sum x^3 + \hat{\beta}_4 \sum x^4 + \hat{\beta}_5 \sum x^5, \\ \sum xy &= \hat{\beta}_0 \sum x + \hat{\beta}_1 \sum x^2 + \hat{\beta}_2 \sum x^3 + \hat{\beta}_3 \sum x^4 + \hat{\beta}_4 \sum x^5 + \hat{\beta}_5 \sum x^6, \\ \sum x^2 y &= \hat{\beta}_0 \sum x^2 + \hat{\beta}_1 \sum x^3 + \hat{\beta}_2 \sum x^4 + \hat{\beta}_3 \sum x^5 + \hat{\beta}_4 \sum x^6 + \hat{\beta}_5 \sum x^7, \\ \sum x^3 y &= \hat{\beta}_0 \sum x^3 + \hat{\beta}_1 \sum x^4 + \hat{\beta}_2 \sum x^5 + \hat{\beta}_3 \sum x^6 + \hat{\beta}_4 \sum x^7 + \hat{\beta}_5 \sum x^8, \\ \sum x^4 y &= \hat{\beta}_0 \sum x^4 + \hat{\beta}_1 \sum x^5 + \hat{\beta}_2 \sum x^6 + \hat{\beta}_3 \sum x^7 + \hat{\beta}_4 \sum x^8 + \hat{\beta}_5 \sum x^9, \\ \sum x^5 y &= \hat{\beta}_0 \sum x^5 + \hat{\beta}_1 \sum x^6 + \hat{\beta}_2 \sum x^7 + \hat{\beta}_3 \sum x^8 + \hat{\beta}_4 \sum x^9 + \hat{\beta}_5 \sum x^{10}. \end{aligned} \tag{6}$$

Now, to obtain the least-squares estimate of the trend curve for a time series in which a polynomial of any degree (say k) is applicable, we need only establish and solve the appropriate normal equations for the relevant coefficients $\hat{\beta}_i$ ($i = 0, 1, \ldots, k$). Once the equation for the trend curve is obtained by the method of least squares, the trend values corresponding to the various years can be computed by simply inserting the appropriate values of x into the least-squares equation.

We have already seen from Chapter 15 that in any situation involving the determination of the least-squares equation, computations can be simplified by defining the independent variable x so that $\sum x = 0$. For, in such cases, the normal equations of (4) (two equations in two unknowns) reduce to the much simpler situation of two equations, one each in one unknown, that is,

$$\sum y = \hat{\beta}_0 n,$$
$$\sum xy = \hat{\beta}_1 \sum x^2. \tag{7}$$

Solution of these equations provides us immediately with the least-squares estimates of β_0 and β_1 in the situation where $\sum x = 0$:

$$\hat{\beta}_0 = \frac{\sum y}{n} \quad \text{and} \quad \hat{\beta}_1 = \frac{\sum xy}{\sum x^2}. \tag{8}$$

This type of simplification often comes quite easily in time-series analysis. In other words, regardless of the actual values of the time periods involved, we can represent the time periods in a time series by integers simply by choosing one of the time periods to be the base, or zero-th, period. If the observed values of a time-series variable Y occur at equally spaced time intervals (say years), we can make $\sum x = 0$ quite easily as follows:

1. When the number of observed values is odd, transform x so that the median year corresponds to $x = 0$, $x = -1$ corresponds to the year previous to the median year, $x = -2$ corresponds to the next previous year, etc.; $x = 1$ corresponds to the year following the median year, $x = 2$ corresponds to the next following year, etc.

2. When the number of observed values is even, make $x = -1$ correspond to the earlier of the two median years, $x = -3$ correspond to the next earlier year, $x = -5$ correspond to the next earlier year, etc.; then $x = 1$ corresponds to the later of the two median years, $x = 3$ corresponds to the next later year, $x = 5$ corresponds to the next later year, etc.

When data are not equally spaced, it is harder to find the transformation that gives the desired results. In such cases (as seen in Chapter 15), we define

$$x' = x - \bar{x} \quad \text{and} \quad y' = y - \bar{y},$$

and then treat x' and y' as original variables of interest. Then the least-squares line becomes

$$\hat{Y} = \bar{y} + \frac{\sum x'y'}{\sum x'^2}(x - \bar{x}). \tag{9}$$

Example 1

Suppose the sales of a company (millions of dollars) during the years 1965–1975 are as given in Table 19.1.

a. Graph the data.

b. Find the least-squares trend line by defining the independent variable so that $\sum x = 0$.

c. Find the least-squares trend line by defining the independent variable so that $x = 0$ for year 1965, $x = 1$ for year 1966, etc., and employing the transformation $x' = x - \bar{x}$ and $y' = y - \bar{y}$.

d. Use the results of each of steps b and c to estimate sales during the year 1976. Note that results are identical.

Table 19.1

Company salcs (1965–1975)

Year	*Sales (millions of dollars)*
1965	66.0
1966	85.3
1967	89.4
1968	77.3
1969	95.7
1970	104.8
1971	93.7
1972	111.4
1973	89.4
1974	115.4
1975	115.0

Solution

a. The data are graphed in Figure 19.3 along with the least-squares trend line.

b. Relevant calculations are given in Table 19.2. Having defined x so that $\sum x = 0$, we use (8) to obtain

$$\hat{\beta}_0 = \frac{\sum y}{n} = \frac{1{,}043.9}{11} = 94.9 \quad \text{and} \quad \hat{\beta}_1 = \frac{\sum xy}{\sum x^2} = \frac{433.6}{110} = 3.94.$$

Thus, the least-squares line becomes

$$\hat{Y} = 94.9 + 3.94x, \tag{10}$$

where the origin $x = 0$ is the year 1970 and the unit of x is one year.

Table 19.2

Calculations for Example 1(b)

Year	x	y	x^2	xy
1965	−5	66.0	25	−330.0
1966	−4	85.3	16	−341.2
1967	−3	89.4	9	−268.2
1968	−2	77.3	4	−154.6
1969	−1	95.7	1	−95.7
1970	0	104.8	0	0
1971	1	93.7	1	93.7
1972	2	111.4	4	222.8
1973	3	89.4	9	268.2
1974	4	115.9	16	463.6
1975	5	115.0	25	575.0
	$\sum x = 0$	$\sum y = 1{,}043.9$	$\sum x^2 = 110$	$\sum xy = 433.6$

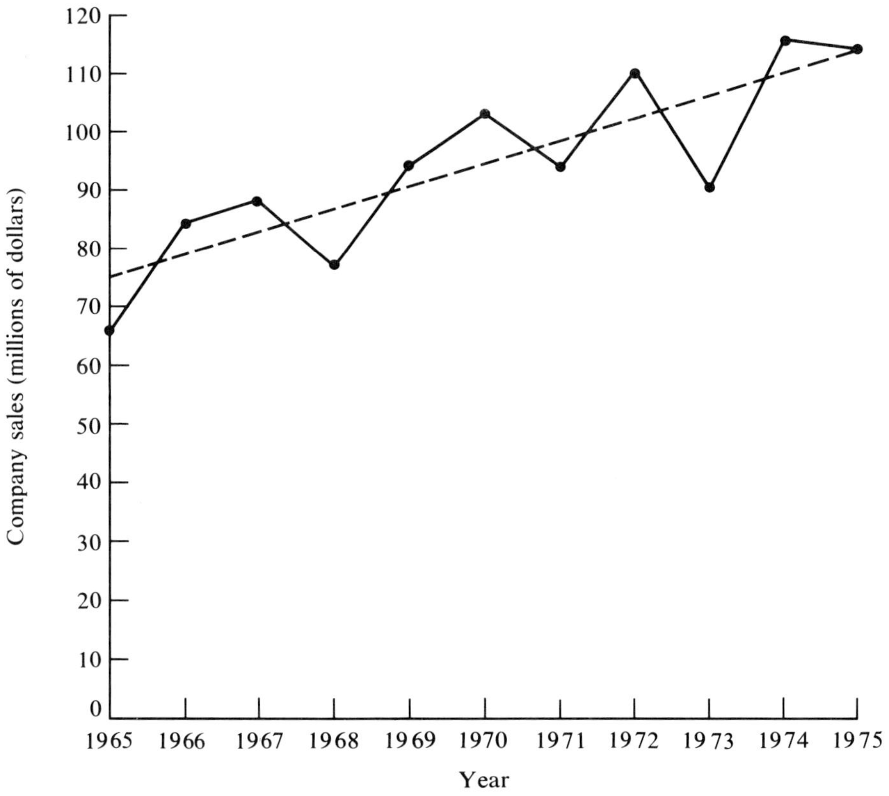

Figure 19.3 Graph of the sales of Table 19.1

c. Relevant calculations are given in Table 19.3. We use Equation (9) to obtain

$$\hat{Y} = \bar{y} + \frac{\sum x'y'}{\sum x'^2}(x - \bar{x})$$

$$\Rightarrow \hat{Y} = 94.9 + \frac{433.6}{110}(x - 5) \tag{11}$$

$$\Rightarrow \hat{Y} = 75.2 + 3.94x,$$

where the origin $x = 0$ is the year 1965 and the units of x are 1 year.

d. To obtain an estimate of sales during the year 1976 from Equations (10) and (11), we need only note the location of the origin in each case. Thus, for the result of (b), we have $x = 6$ for the year 1976, and, hence,

$$\hat{Y}_{1976} = 94.9 + 3.94x = 94.9 + 3.94(6) = 118.54.$$

On the other hand, for the result of (c), we have $x = 11$ for the year 1976. Hence,

$$\hat{Y}_{1976} = 75.2 + 3.94x = 75.2 + 3.94(11) = 118.54.$$

Table 19.3

Calculations for Example 1(c)

Year	x	y	$x' = x - \bar{x}$	$y' = y = \bar{y}$	x'^2	$x'y'$
1965	0	66.0	−5	−28.9	25	144.5
1966	1	85.3	−4	−9.6	16	38.4
1967	2	89.4	−3	−5.5	9	16.5
1968	3	77.3	−2	−17.6	4	35.2
1969	4	95.7	−1	0.8	1	−0.8
1970	5	104.8	0	9.9	0	0
1971	6	93.7	1	−1.2	1	−1.2
1972	7	111.4	2	16.5	4	33.0
1973	8	89.4	3	−5.5	9	−16.5
1974	9	115.9	4	21.0	16	84.0
1975	10	115.0	5	20.1	25	100.5
	$\sum x = 55$	$\sum y = 1{,}043.9$	$\sum x' = 0$	$\sum y' = 0$	$\sum x'^2 = 110$	$\sum x'y' = 433.6$
	$\bar{x} = 5$	$\bar{y} = 94.9$				

□

Problems

In all problems of this section, carry calculations to the same number of significant figures as in Example 1, page 710.

1. Suppose the sales of a company were as in the following table.

Year	1968	1969	1970	1971	1972	1973	1974
Sales (thousands)	\$ 10.2	14.7	20.6	23.8	27.9	32.8	38.6

a. Find the least-squares trend line by defining the independent variable so that $\sum x = 0$.

b. Find the least-squares trend line by defining the independent variable so that $x = 0$ for 1968, and using the transformation $x' = x - \bar{x}$ and $y' = y - \bar{y}$.

c. Use the results of both Part a and Part b to estimate sales for the year 1978. The results should be identical.

d. If the trend continues at its present rate, in what year would you expect sales to reach \$50,000?

2. Suppose the sales of a company for a given period were as shown in the following table.

Year	1969	1970	1971	1972	1973	1974
Sales (millions)	\$ 26	32	40	51	59	65

a. Find the least-squares trend line by defining the independent variable so that $\sum x = 0$.

b. Find the least-squares trend line by defining the independent variable so that $x = 0$ for 1969, and using the transformation $x' = x - \bar{x}$ and $y' = y - \bar{y}$.

c. From the results of each of a and b, estimate sales during the year 1978. The results should be identical.

d. Assuming the trend continued at its present rate, in what year would you expect sales to reach \$83,000,000?

3. Suppose a company's unit production (in millions of units) record for a given period is shown in the following table.

Year	1967	1968	1969	1970	1971	1972	1973	1974
Production (millions of units)	3.8	5.3	6.9	8.2	10.1	11.3	12.4	14.6

   a. Find the least-squares trend line by defining the independent variable so that $\sum x = 0$.

   b. Find the least-squares trend line by defining the independent variable so that $x = 0$ for 1967, and using the transformation $x' = x - \bar{x}$ and $y' = y - \bar{y}$.

   c. From the results of a and b, estimate production during the year 1978. The results should be identical.

   d. If the trend continues at its present rate, in what year would production reach 20,000,000 units?

4. Suppose the production of a company for a given period is shown in the following table.

Year	1965	1966	1967	1968	1969	1970	1971
Production (thousands of units)	18	30	46	65	80	100	110

   a. Find the least-squares trend line by defining the independent variable so that $\sum x = 0$.

   b. Find the least-squares trend line by defining the independent variable so that $x = 0$ for 1965, and using the transformation $x' = x - \bar{x}$ and $y' = y - \bar{y}$.

   c. From a and b, estimate production during the year 1980. The results should be identical.

   d. If the trend continues at its present rate, in what year would you expect production to reach 150,000 units?

19.3 Time Series—Seasonal Fluctuations

Recall that a primary distinction between cyclical movement and seasonal movement is that cyclical movement is not necessarily periodic, in nature, while seasonal movement is. Because of the regularity and dependability associated with a periodic fluctuation, certain time-series-analysis techniques are specifically applicable to seasonal variation. One is the method of moving averages, a technique that can be used in analyzing any periodic phenomenon.

The Moving Average

Consider a set of numbers (these may be viewed as observations on a random variable)

$$y_1, y_2, y_3, \ldots, y_n. \tag{1}$$

A *moving average of order* $k(k \leq n)$ is a set of $(n - k + 1)$ averages associated with this set of data. It is defined as

$$\frac{y_1 + y_2 + \cdots + y_k}{k}, \quad \frac{y_2 + y_3 + \cdots + y_{k+1}}{k}, \quad \frac{y_3 + y_4 + \cdots + y_{k+2}}{k}, \ldots. \tag{2}$$

The relevant sums in the numerators of Equation (2) are called *moving totals of order k*. For example, suppose we have the numbers 6, 18, 3, 15, 9, 21, 6. The moving average of order 3 becomes

$$\frac{6 + 18 + 3}{3}, \quad \frac{18 + 3 + 15}{3}, \quad \frac{3 + 15 + 9}{3}, \quad \frac{15 + 9 + 21}{3}, \quad \frac{9 + 21 + 6}{3},$$

or $\quad 9, \quad 12, \quad 9, \quad 15, \quad 12.$

We generally locate each value in the moving average at its appropriate position relative to the original data. In this case, the moving average would be shown as

Original data	6,		18,		3,		15,		9,		21,		6
Moving average of order 3			9,		12,		9,		15,		12		

Note that each number in the moving average represents the mean of the three numbers immediately above it in the original data. Further, if the moving average were of order 4, proper placement would put each moving-average value between two numbers in the original data; each number in the moving average would be the mean of the four numbers immediately above it in the original data.

In a time series, data are generally given in years, months, etc. Thus, a moving average of order k is termed, for example, a k-year moving average, a k-month moving average, and so forth. That is, we refer to 8-year moving averages, 12-month moving averages, and so forth.

The Moving-Average Method of Analysis

One can see from the above data that the moving-average process has a tendency to reduce short-duration variations. The process involving the removal of the variation is termed the *smoothing of a time series*. However, this represents only a minute aspect of the rationale for the method of moving averages. If the data demonstrate periodic variations, a moving average of order equal to the length of the period will completely dampen (eliminate) the periodic fluctuation. This feature is what makes the moving average so useful in the analysis of seasonal variation. There are, of course, certain drawbacks associated with the moving average. First, it tends sometimes to produce periodic fluctuations not present in the original data. Second, a change in the period of the moving average can substantially affect the trend shown by a graph of the moving average. Thus, in analyses involving a moving average, care in selecting the period to be used in computing the moving average is critically important.

Because of the nature of seasonal variation, the type of moving average most often used for time-series analysis is the 12-*month moving average*. For one reason, if any 12 successive monthly observations of a seasonally fluctuating time-series variable are combined by an averaging process, each month is represented once and all months are equally represented. Further, the entire period is covered exactly once. Hence, the values of a 12-month moving average obtained from averages of consecutive combinations of 12 values of the original time variable will reflect no seasonal effects. (Similar arguments can be made for moving averages involving other periods of time.)

Whereas each monthly observation is generally influenced in some degree by seasonal forces, 12-month moving averages are independent of seasonal variation. Hence, the key concept in the method we are about to look at involves the relationship between the moving average, which covers a complete period, and the original observation, which appears at the center of the time period associated with the moving average: this relationship identifies and describes the pattern of the seasonal variation.

The Ratio-to-Moving-Average Method of Analysis

Recall from Section 19.1 our model assumption that

$$Y = T \cdot S \cdot C \cdot I, \tag{3}$$

where Y represents the time-series variable and T, S, C, and I respectively stand for trend, seasonal, cyclical, and irregular movement. Now, suppose we make the very reasonable assumption that the seasonal variation S fluctuates with a 12-month period and that the shape of the variation is the same from year to year. Suppose further that we assume that the irregular variations I are independent for different periods, or years. Under such assumptions, if a 12-month moving average is determined for the values given by Equation (3), it will eliminate the S and smooth the I, and we shall have as a remainder only values involving $T \cdot C$. In other words, the resulting 12-month moving average can be represented as $T \cdot C$. We can then obtain a representation of $S \cdot I$ by taking the ratio of the original data to the proper moving average. Generally speaking, we may denote this as

$$\frac{Y}{\text{moving average}} = \frac{T \cdot S \cdot C \cdot I}{T \cdot C} = S \cdot I. \tag{4}$$

The result, since we have ostensibly removed $T \cdot C$ from the original values, should be a series of numbers that show much less variation than do the original values.

Consider the time-series data of Table 19.4. Suppose we wish to apply the ratio-to-moving-average method to this time-series data to arrive at a series of values involving only seasonal and irregular variation. We do this by constructing a table similar to Table 19.5 and performing the following steps:

1. Compute the 12-month moving total (column 4) of the original data (column 3).
2. Noting that the values of the 12-month moving total of column 4 do not correspond timewise to the original time-series values of column 3 (they are off-center, so to speak), center the values of the 12-month moving total by first computing the 2-year moving total (column 5) of the values of column 4 and then averaging the values of the 2-year moving total (dividing by 24) to obtain the centered, 12-month moving average of column 6. (Note that the values of the centered, 12-month moving average of column 6 now correspond timewise to the original observations of column 3).
3. Compute the values of column 7 by dividing the values of column 3 by the values of column 6 and multiplying by 100.

Table 19.4

Original time-series values

	Year								
Month	1965	1966	1967	1968	1969	1970	1971	1972	1973
Jan.	8,148	9,404	10,700	12,402	14,308	16,364	18,178	20,208	21,992
Feb.	7,800	9,116	10,388	11,884	13,884	15,628	17,170	19,286	21,766
Mar.	7,304	8,632	9,858	11,414	13,116	14,956	16,590	18,376	20,600
Apr.	6,992	8,168	9,370	10,892	12,618	14,158	15,870	17,734	19,684
May	6.724	7,676	8,778	10,320	11,750	13,188	14,992	16,492	18,266
June	6,588	7,480	8,598	9,994	11,474	12,938	14,640	16,176	17,800
July	6,464	7,496	8,684	9,918	11,408	12,986	14,818	16,700	18,374
Aug.	6,514	7,498	8,694	9,832	11,470	13,164	14,952	17,026	19,262
Sept.	6,874	7,886	9,088	10,362	11,980	13,488	15,164	17,296	19,474
Oct.	7,052	8,194	9,348	10,814	12,222	13,752	15,528	17,330	19,194
Nov.	7,602	8,840	9,914	11,456	13,274	14,770	16,356	18,250	19,996
Dec.	8,508	9,768	11,098	12,970	14,744	16,368	18,068	20,256	22,840
Total	86,570	100,158	114,518	132,258	152,248	171,760	192,326	215,130	239,248
Average	7,214	8,347	9,543	11,022	12,687	14,313	16,027	17,928	19,937

Table 19.5

Computations of ratios-to-moving average for values of Table 19.4

(1) *Year*	(2) *Month*	(3) *y* *Original data*	(4) *12-Month moving total*	(5) *2-Year moving total of column 4*	(6) *12-Month centered moving average*	(7) *Ratio to 12-month centered moving average*
1965	Jan.	8,148				
	Feb.	7,800				
	Mar.	7,304				
	Apr.	6,992				
	May	6,724				
	June	6,588				
			86,570			
	July	6,464		174,396	7,266.50	88.96
			87,826			
	Aug.	6,514		176,968	7,373.67	88.34
			89,142			
	Sept.	6,874		179,612	7,483.83	91.85
			90,470			
	Oct.	7,052		182,116	7,588.17	92.93
			91,646			
	Nov.	7,602		184,244	7,676.83	99.03
			92,598			
	Dec.	8,508		186,088	7,753.67	109.73
			93,490			
1966	Jan.	9,404		188,012	7,833.83	120.04
			94,522			
	Feb.	9,116		190,028	7,917.83	115.13
			95,506			
	Mar.	8,632		192,024	8,001.00	107.89
			96,518			
	Apr.	8,168		194,178	8,090.75	100.95
			97,660			

Table 19.5 (Continued)

	May	7,676		196,558	8,189.92	93.72
			98,898			
	June	7,480		199,056	8,294.00	90.19
			100,158			
	July	7,496		201,612	8,400.50	89.23
			101,454			
	Aug.	7,498		204,180	8,507.50	88.13
			102,726			
	Sept.	7,886		206,678	8,611.58	91.57
			103,952			
	Oct.	8,194		209,106	8,712.75	94.05
			105,154			
	Nov.	8,840		211,410	8,808.75	100.35
			106,256			
	Dec.	9,768		213,630	8,901.25	109.74
			107,374			
1967	Jan.	10,700		215,936	8,997.33	118.92
			108,562			
	Feb.	10,388		218,320	9,096.67	114.20
			109,758			
	Mar.	9,858		220,718	9,196.58	107.19
			110,960			
	Apr.	9,370		223,074	9,294.75	100.81
			112,114			
	May	8,778		225,302	9,387.58	93.51
			113,188			
	June	8,598		227,706	9,487.75	90.62
			114,518			
	July	8,684		230,738	9,614.08	90.33
			116,220			
	Aug.	8,694		233,936	9,747.33	89.19
			117,716			
	Sept.	9,088		236,988	9,874.50	92.04
			119,272			
	Oct.	9,348		240,066	10,002.75	93.45
			120,794			
	Nov.	9,914		243,130	10,130.42	97.86
			122,336			
	Dec.	11,098		246,068	10,252.83	108.24
			123,732			
1968	Jan.	12,402		248,698	10,362.42	119.68
			124,966			
	Feb.	11,884		251,070	10,461.25	113.60
			126,104			
	Mar.	11,414		253,482	10,561.75	108.07
			127,378			
	Apr.	10,892		256,222	10,675.92	102.02
			128,844			
	May	10,320		259,230	10,801,25	95.54
			130,386			
	June	9,994		262,644	10,943.50	91.32
			132,258			
	July	9,918		266,422	11,100.92	89.34
			134,164			
	Aug.	9,832		270,328	11,263.67	87.29
			136,164			
	Sept.	10,362		274,030	11,417,92	90.75
			137,866			
	Oct.	10,814		277,458	11,560.75	93.54
			139,592			
	Nov.	11,456		280,614	11,692.25	97.98
			141,022			
	Dec.	12,970		283,524	11,813.50	109.79
			142,502			

Table 19.5 (Continued)

(1) Year	(2) Month	(3) *y* Original data	(4) 12-Month moving total	(5) 2-Year moving total of column 4	(6) 12-Month centered moving average	(7) Ratio to 12-month centered moving average
1969	Jan.	14,308		286,494	11,937.25	119.86
			143.992			
	Feb.	13,884		289,622	12,067.58	115.05
			145,630			
	Mar.	13,116		292,878	12,203.25	107.48
			147,248			
	Apr.	12,618		295,904	12,329.33	102.34
			148,656			
	May	11,750		299,130	12,463.75	94.27
			150,474			
	June	11,474		302,722	12,613.42	90.97
			152,284			
	July	11,408		306,552	12,773.00	89.31
			154,304			
	Aug.	11,470		310,352	12,931.33	88.70
			156,048			
	Sept.	11,980		313,936	13,080.67	91.59
			157,888			
	Oct.	12,222		317,316	13,221.50	92.44
			159,428			
	Nov.	13,274		320,294	13,345.58	99.46
			160,866			
	Dec.	14,744		323,196	13,466.50	109.49
			162,330			
1970	Jan.	16,364		326,238	13,593.25	120.38
			163,908			
	Feb.	15,628		329,510	13,729.58	113.83
			165,602			
	Mar.	14,956		332,712	13,863.00	107.88
			167,110			
	Apr.	14,158		335,750	13,989.58	101.20
			168,640			
	May	13,188		338,776	14,115.67	93.43
			170,136			
	June	12,938		341,896	14,245.67	90.82
			171,760			
	July	12,986		345,334	14,388.92	90.25
			173,574			
	Aug.	13,164		348,690	14,528.75	90.61
			175,116			
	Sept.	13,488		351,866	14,661.08	92.00
			176,750			
	Oct.	13,752		355,212	14,800.50	92.92
			178,462			
	Nov.	14,770		358,728	14,947.00	98.82
			180,266			
	Dec.	16,368		362,234	15,093.08	108.45
			181,968			
1971	Jan.	18,178		365,768	15,240.33	119.28
			183,800			
	Feb.	17,170		369,388	15,391.17	111.56
			185,588			
	Mar.	16,590		372,852	15,535.50	106.79
			187,264			

Table 19.5 (Continued)

	Apr.	15,870		376,304	15,679.33	101.22
			189, 040			
	May	14,992		379,666	15,819.42	94.77
			190,626			
	June	14,640		382,952	15,956.33	91.75
			192,326			
	July	14,818		386,682	16,111.75	91.97
			194,356			
	Aug.	14,952		390,828	16,284.50	91.82
			196,472			
	Sept.	15,164		394,730	16,447.08	92.20
			198,258			
	Oct.	15,528		398,380	16,599.17	93.55
			200,122			
	Nov.	16,356		401,744	16,739.33	97.71
			201,622			
	Dec.	18,068		404,780	16,865.83	107.13
			203,158			
1972	Jan.	20,208		408,198	17,008.25	118.81
			205,040			
	Feb.	19,286		412,154	17,173.08	112.30
			207,114			
	Mar.	18,376		416,360	17,348.33	105.92
			209,246			
	Apr.	17,734		420,294	17,512.25	101.27
			211,048			
	May	16,492		423,990	17,666.25	93.35
			212,942			
	June	16,176		428,072	17,836.33	90.69
			215,130			
	July	16,700		432,044	18,001.83	92.77
			216,914			
	Aug.	17,026		436,308	18,179.50	93.65
			219,394			
	Sept.	17,296		441,012	18,375.50	94.13
			221,618			
	Oct.	17,330		445,186	18,549.42	93.43
			223,568			
	Nov.	18,250		448,910	18,704.58	97.57
			225,342			
	Dec.	20,256		452,308	18,846.17	107.48
			226,966			
1973	Jan.	21,992		455,606	18,983.58	115.85
			228,640			
	Feb.	21.766		459,516	19,146.50	113.68
			230,876			
	Mar.	20,600		463,930	19,330.42	106.57
			233,054			
	Apr.	19,684		467,972	19,498.83	100.95
			234,918			
	May	18,266		471,582	19,649.25	92.96
			236,664			
	June	17,800		475,912	19,829.67	89.76
			239,248			
	July	18,374				
	Aug.	19,262				
	Sept.	19,474				
	Oct.	19,194				
	Nov.	19,996				
	Dec.	22,840				

Note that if the assumptions associated with (3) and (4) hold, we have achieved the desired goal. The ratios-to-centered moving averages of column 7 provide the basis for subsequent analysis.

The Seasonal Index

Table 19.6 summarizes the ratios to the centered, 12-month moving averages calculated in Table 19.5. Note that there are 12 fewer entries in this table than in Table 19.4—because we have lost a year's data in taking the 12-month moving average. Despite obvious variations in the data of Table 19.6, the data (on a monthly basis) are much more nearly homogeneous than the original observations on the time-series variable (Table 19.4). The reason is, as illustrated generally in (3) and (4), that we have removed the trend and cyclical effects, perhaps even some of the irregular fluctuations. (Since the assumptions associated with (3) and (4) will not hold exactly, in most cases, we will not have removed all of the cyclical effects. Hence, we may attribute the differences among the entries of Table 19.6 to effects of seasonal, cyclical, and irregular forces.)

Now, note that the observations in any given row of Table 19.6 represent the same month. Thus, their variation certainly cannot be attributed to the effects of seasonal forces. Hence, the variation among the values for any given month must be attributed to some combination of cyclical and irregular forces. (If the model assumptions were exactly valid, these would be attributable only to irregular, or residual, forces, since we would have removed all the cyclical effects in taking the ratio to the moving average. Naturally, this comment does not hold in the case of a changing seasonal pattern (see the subsection titled "Changing Seasonal Pattern").) Finally, by "averaging" the values for each month, we can average out the effects of the irregular forces. Thus, at least conceptually, we arrive at a measure of the effects of the seasonal forces for each month of the year by applying an "averaging" process to the rows of Table 19.6.

The result of applying the "averaging" process to the values for each month is commonly known as a *seasonal index*. For example, in Table 19.6, we obtain the seasonal index for the month of March by "averaging" the eight values associated with this month. (Note that in this particular kind of problem, the median of the relevant values sometimes represents the central tendency better than does the mean.)

One common method for calculating seasonal indexes is that of determining the arithmetic mean of only a selected number of central items for a given month. This procedure is applied to the four central items in each row of Table 19.6, the four central items being obtained merely by eliminating from consideration the two largest and the two smallest observations in each row. Columns 11 and 12 give the respective totals and arithmetic means of the four central items for each month. We interpret the seasonal index of 107.4 for the month of March to mean that the March average for the data observed has been 7.4 percent higher than the average for the year centered at this month.

The averages of column 12 can be construed as seasonal indexes. However, so that each index will be based on 100 percent, we commonly make an adjustment so that the sum of the 12 seasonal indexes will equal 1,200. We do this simply by multiplying each of the averages of column 12 of Table 19.6 by a factor that is the ratio of 1,200

Table 19.6

Calculation of seasonal indexes

	Ratios to 12-month centered moving averages									*Seasonal indexes—central four months*		
Month (1)	1965 (2)	1966 (3)	1967 (4)	1968 (5)	1969 (6)	1970 (7)	1971 (8)	1972 (9)	1973 (10)	*Totals* (11)	*Averages* (12)	*Adjusted seasonal indexes* (13)
Jan.		120.0	118.9	119.7	119.9	120.4	119.3	118.8	115.9	477.8	119.5	119.7
Feb.		115.1	114.2	113.6	115.1	113.8	111.6	112.3	113.7	455.3	113.8	114.0
Mar.		107.9	107.2	108.1	107.5	107.9	106.8	105.9	106.6	429.4	107.4	107.6
Apr.		101.0	100.8	102.0	102.3	101.2	101.2	101.3	101.0	404.7	101.2	101.4
May		93.7	93.5	95.5	94.3	93.4	94.8	93.4	93.0	374.9	93.7	93.9
June		90.2	90.6	91.3	91.0	90.8	91.8	90.7	89.8	363.1	90.8	91.0
July	89.0	89.2	90.3	89.3	89.3	90.3	92.0	92.8		359.2	89.8	89.9
Aug.	88.3	88.1	89.2	87.3	88.7	90.6	91.8	93.7		356.8	89.2	89.3
Sept	91.9	91.6	92.0	90.8	91.6	92.0	92.2	94.1		367.5	91.9	92.1
Oct.	92.9	94.1	93.5	93.5	92.4	92.9	93.6	93.4		373.4	93.9	93.5
Nov.	99.0	100.4	97.9	98.0	99.5	98.8	97.7	97.6		393.7	98.4	98.6
Dec.	109.7	109.7	108.2	109.8	109.5	108.5	107.1	107.5		435.9	109.0	109.2
											1,198.0	1,200.2

to the sum of the unadjusted indexes. In the case of Table 19.6, the factor becomes 1,200/1,198 = 1.00167. Finally, the adjusted seasonal indexes of column 13 are obtained by multiplying each average of column 12 by this factor.

Use of Seasonal Indexes

Seasonal indexes similar to those computed in Table 19.6 may be used for both *decomposition* and *synthesis*. The most common use in decomposition is for obtaining *seasonally adjusted data*, or *deseasonalized data*. The procedure for obtaining deseasonalized data is simply to divide each of the individual observations of the original set of data by the appropriate monthly seasonal index. Table 19.7 presents the results of seasonally adjusting the data of Table 19.4. For example, the figures in Table 19.7

Table 19.7

Seasonally adjusted data (original data of Table 19.4 adjusted for seasonal variation)

Month	1965	1966	1967	1968	1969	1970	1971	1972	1973
Jan.	6,807	7,856	8,939	10.361	11,953	13,671	15,186	16,882	18,373
Feb.	6,842	7,996	9,112	10,425	12,179	13,709	15,061	16,918	19,093
Mar.	6,788	8,022	9,162	10,608	12,190	13,900	15,418	17,078	19,145
Apr.	6,895	8,055	9,241	10,742	12,444	13,963	15,651	17,489	19,412
May	7,161	8,175	9,348	10,990	12,513	14,045	15,966	17,563	19,453
June	7,240	8,220	9,448	10,982	12,609	14,218	16,088	17,776	19,560
July	7,190	8,338	9,660	11,032	12,690	14,445	16,483	18,576	20,438
Aug.	7,295	8,396	9,736	11,010	12,844	14,741	16,744	19,066	21,570
Sept.	7,464	8,562	9,868	11,251	13,008	14,645	16,465	18,780	21,144
Oct.	7,542	8,764	9,998	11,566	13,072	14,708	16,607	18,535	20,528
Nov.	7,710	8,966	10,055	11,619	13,462	14,980	16,588	18,509	20,280
Dec.	7,791	8,945	10,163	11,877	13,502	14,989	16,546	18,549	20,916

associated with the month of March are obtained by dividing corresponding values of Table 19.4 by the seasonal index for the month of March, 107.4 percent. When a set of data has been adjusted for seasonal variation (as in Table 19.7), the resulting data, due to the fact that the seasonal variation has been removed, reflect only trend, cyclical, and irregular effects. Referring to our original model, we see that seasonally adjusted data, since they represent the original data divided by seasonal effects, can be represented as

$$\frac{T \cdot S \cdot C \cdot I}{S} = T \cdot C \cdot I. \tag{5}$$

Thus, for decomposition purposes, removal of the seasonal fluctuations permits the analyst to determine whether there exist influences other than those associated with seasonal fluctuations. The composite effect of seasonally adjusting data is shown in Figures 19.4 and 19.5. Figure 19.4 graphs the original data of Table 19.4, while Figure 19.5 graphs the corresponding deseasonalized data.

The primary use of seasonal indexes in synthesis occurs in forecasting. Seasonal indexes can be combined with a forecast based on trend, for example, to produce

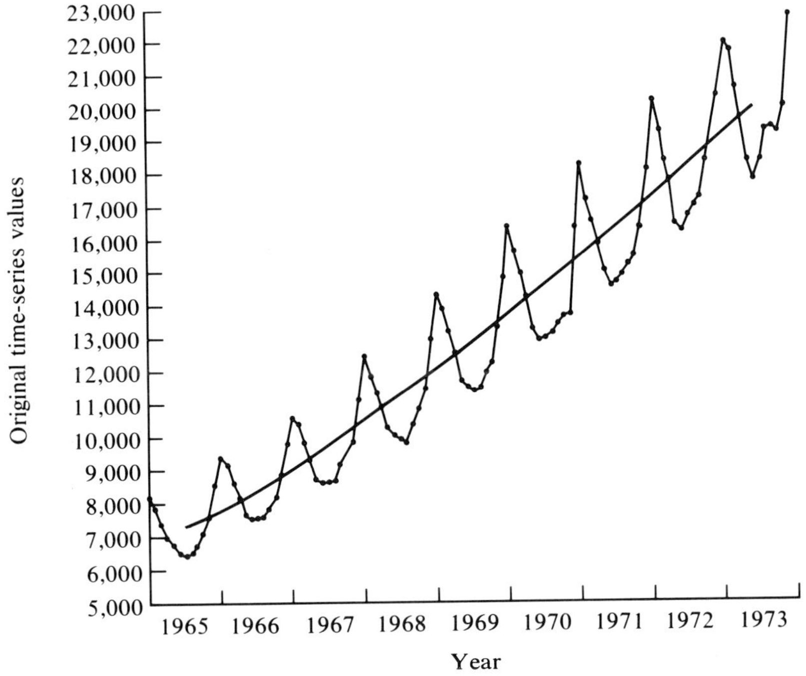

Figure 19.4 Graph of the original data of Table 19.4

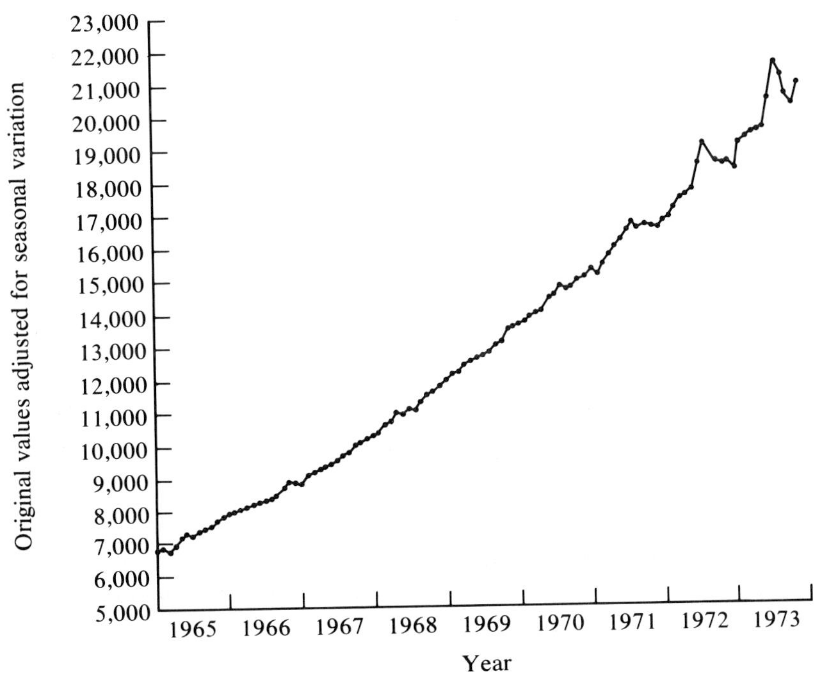

Figure 19.5 Graph of the data of Table 19.4 after deseasonalization

forecasts that reflect anticipated seasonal variations. We discuss this in more detail in Section 19.5.

Changing Seasonal Pattern

A special case of a problem involving seasonal variation occurs when the seasonal pattern is not constant over the time period under consideration. In this case, the method used for removing the seasonal variation depends on the nature of the change in the seasonal effect. If there is an abrupt change in the seasonal pattern, the analysis for seasonal variations can be conducted separately for the period before the change and the period after the change. On the other hand, if the change in seasonal pattern is of a long-term secular nature, it is impossible to represent a typical seasonal pattern for the period by an average. Thus, in either instance, the averaging process described above is not appropriate for finding the seasonal indexes—we must use a method that takes into account the change in seasonal effect. For details of this procedure, see Stockton (19.1).

Problems

1. For the data of Problem 1, Section 19.2, compute
   a. the two-year moving averages,
   b. the three-year moving averages.
2. For the data of Problem 3, Section 19.2, compute
   a. the two-year moving averages,
   b. the three-year moving averages,
   c. the four-year moving averages.

19.4 The Removal of Trend from Deseasonalized Data

We have already seen that, although many forces influence a given time-series variable, the forces can be grouped into broad categories according to effects rather than origin. In fact, the classification of such effects is reflected in the model encountered earlier

$$Y = T \cdot S \cdot C \cdot I,$$

where Y represents the times-series variable and T, S, C, and I represent, respectively, the trend, seasonal, cyclical and irregular variations that affect the variable. So far we have analyzed the time-series data for and identified the effects of seasonal variation. In the notation of the model, we have separated and identified the effects of S. Thus, we have reduced time-series data involving the effects of all four classifications of variations to a set of data reflecting only the trend, cyclical, and irregular variations. In other words, we have obtained seasonally adjusted data as

$$\frac{Y}{S} = \frac{T \cdot S \cdot C \cdot I}{S} = T \cdot C \cdot I. \tag{1}$$

Continuing this analysis, we reduce the seasonally adjusted data further by dividing it by appropriate trend values, as

$$\frac{\text{Seasonally adjusted data}}{\text{Trend}} = \frac{T \cdot C \cdot I}{T} = C \cdot I. \tag{2}$$

Now, if trend and seasonal variation were the only forces acting in a given time series, the results of Equation (2) could be represented by a horizontal line parallel to the axis on which time is measured. However, this is not usually the case; hence, quotients similar to those of column 8 of Table 19.5 are usually subject to both cyclical and residual fluctuations. The final formal step in the decomposition of time-series data, then, is the extraction from such quotients of information regarding the cyclical behavior of the time-series variable.

Consider again the original data of Table 19.4. We previously identified the effects of seasonal variation and set these forth as seasonal indexes. To prepare our investigation of the cyclical behavior, we must identify trend effects. We can do this by applying the method of least squares to the monthly data of Table 19.4. However, without a computer, this entails a prohibitive amount of work. Thus, we usually estimate the monthly trend line by first fitting a trend line to yearly data and then transforming the equation reflecting yearly trend to one reflecting monthly trend. Refer to Table 19.8. Yearly totals for the original time-series variable under consideration are given in the y column. The values labeled x are codes for the time

Table 19.8

Determination of least-squares line for totals of Table 19.4

Year	x	y	xy	x^2
1965	−4	86,570	−346,280	16
1966	−3	100,158	−300,474	9
1967	−2	114,518	−229,036	4
1968	−1	132,258	−132,258	1
1969	0	152,248	0	0
1970	1	171,760	171,760	1
1971	2	192,326	384,652	4
1972	3	215,130	645,390	9
1973	4	239,248	956,992	16
	0	1,404,216	1,150,746	60

variable. Using the method of least squares to fit a straight line to the data, we obtain, after carrying out the calculations in the xy and x^2 columns,

$$\hat{\beta}_0 = 156{,}024 \qquad \text{and} \qquad \hat{\beta}_1 = 19{,}179.1, \tag{3}$$

where $\hat{\beta}_0$ and $\hat{\beta}_1$ represent the respective estimates of the y-intercept and slope in the yearly trend equation $\hat{Y} = \hat{\beta}_0 + \hat{\beta}_1 x$. The resulting equation is

$$\hat{Y} = 156{,}024 + 19{,}179.1x, \qquad x = 0 \text{ for } 1969, \tag{4}$$

where an x unit equals 1 year. Equation (4) can be used to compute yearly trend values where, as indicated, the origin of the x-axis is at the middle of the year 1969. We must modify this equation in two ways. The transformation we use must both convert the unit of time to one month and shift the origin to the middle of the month that is to be the origin. We accomplish this by the new equation

$$\hat{Y} = \frac{156{,}024}{12} + \frac{19{,}179.1}{144}(x - 5.5) \tag{5}$$

$$\Rightarrow \hat{Y} = 12{,}269.47 + 133.19x; \qquad x = 0 \text{ at January, 1969, } x\text{-unit} = 1 \text{ month}, \tag{6}$$

where $\hat{Y}$ now represents monthly total for the time-series variable.

The rationale behind Equation (6) may not be intuitively clear. It is as follows. Equation (4) presents yearly totals of the time-series trend ($\hat{Y}$) as a function of x, where the unit in which x is measured is years. Thus, we convert $\hat{Y}$ to monthly totals simply by dividing by 12. Further, we obtain an "independent" variable having months as a unit of measure by dividing x by 12. Even after this, however, we still have an equation representing values that are centered, so to speak, between two months, that is, at the center of a year. Thus, we merely shift the origin so that the values represented by the final equation are centered at the middle of some month. In choosing the value 5.5 for the shift in Equation (5), we shift the origin from July 1, 1969, to January 15, 1969.

This procedure for converting an equation involving a time-series variable from an annual basis to a monthly basis is commonly called the *stepping down* of the equation. Note that not all time-series equations can be treated in this fashion. For such a change of scale to be valid, the data must be of the nature that the annual total tends to be approximately 12 times as large as the monthly total. A time series

Table 19.9

Monthly trend values for time series of Table 19.4

Month	1965	1966	1967	1968	1969	1970	1971	1972	1973
Jan.	5,876	7,475	9,073	10,671	12,269	13,868	15,466	17,064	18,663
Feb.	6,010	7,608	9,206	10,804	12,403	14,001	15,599	17,198	18,796
Mar.	6,143	7,741	9,339	10,938	12,536	14,134	15,732	17,331	18,929
Apr.	6,276	7,874	9,472	11,071	12,699	14,267	15,866	17,464	19,062
May	6,409	8,007	9,606	11,204	12,802	14,401	15,999	17,597	19,195
June	6,542	8,141	9,739	11,337	12,935	14,534	16,132	17,730	19,329
July	6,675	8,274	9,872	11,470	13,069	14,667	16,265	17,863	19,462
Aug.	6,809	8,407	10,005	11,604	13,202	14,800	16,398	17,997	19,595
Sept.	6,942	8,540	10,138	11,737	13,335	14,933	16,532	18,130	19,728
Oct.	7,075	8,673	10,272	11,870	13,468	15,066	16,665	18,263	19,861
Nov.	7,208	8,807	10,405	12,003	13,601	15,200	16,798	18,396	19,994
Dec.	7,341	8,940	10,538	12,136	13,735	15,333	16,931	18,529	20,128

Table 19.10

Original values of Table 19.4 adjusted for both seasonal variation and trend

Month	1965	1966	1967	1968	1969	1970	1971	1972	1973
Jan.	115.8	105.1	98.5	97.1	97.4	98.6	98.2	98.9	98.4
Feb.	113.8	105.1	99.0	96.5	98.2	97.9	96.6	98.4	101.6
Mar	110.5	103.6	98.1	97.0	97.2	98.3	98.0	98.5	101.1
Apr.	109.0	102.3	97.6	97.0	98.2	97.9	98.6	100.1	101.8
May	111.7	102.1	97.3	98.1	97.7	97.5	99.8	99.8	101.3
June	110.7	101.0	97.0	96.9	97.5	97.8	99.7	100.3	101.2
July	107.7	100.8	97.9	96.2	97.1	98.5	101.3	104.0	105.0
Aug.	107.1	99.9	97.3	94.9	97.3	99.6	102.1	105.9	110.1
Sept.	107.5	100.3	97.3	95.9	97.5	98.1	99.6	103.6	107.2
Oct.	106.6	101.0	97.3	97.4	97.1	97.6	99.7	101.5	103.5
Nov.	107.0	101.8	96.6	96.8	99.0	98.6	98.7	100.6	101.4
Dec.	106.1	100.1	96.4	96.9	98.3	97.8	97.7	100.1	103.9

of this nature is called an *aggregative time series*. Among time series of this nature are income, costs, sales, and so forth. Nonaggregative series include such variables as prices, insurance rates, population. For a more comprehensive treatment of the rationale related to the stepping down of an equation, see Stockton (19.1).

Letting $x = 0$ for January 15, 1969, we can compute monthly trend values for each month in our original time-series data by repeated use of Equation (6). The results are shown in Table 19.9. Finally, we further adjust for trend the seasonally adjusted data of Table 19.7 by dividing appropriate values by the corresponding trend values of Table 19.9. The resulting quotients are shown in Table 19.10. Note that they represent the data after adjustment for both seasonal variation and trend. The entire sequence of analysis can be seen graphically in Figures 19.4, 19.5, and 19.6 which present, respectively, the original data, the original data adjusted for seasonal variation, and the original data adjusted for both seasonal variation and trend.

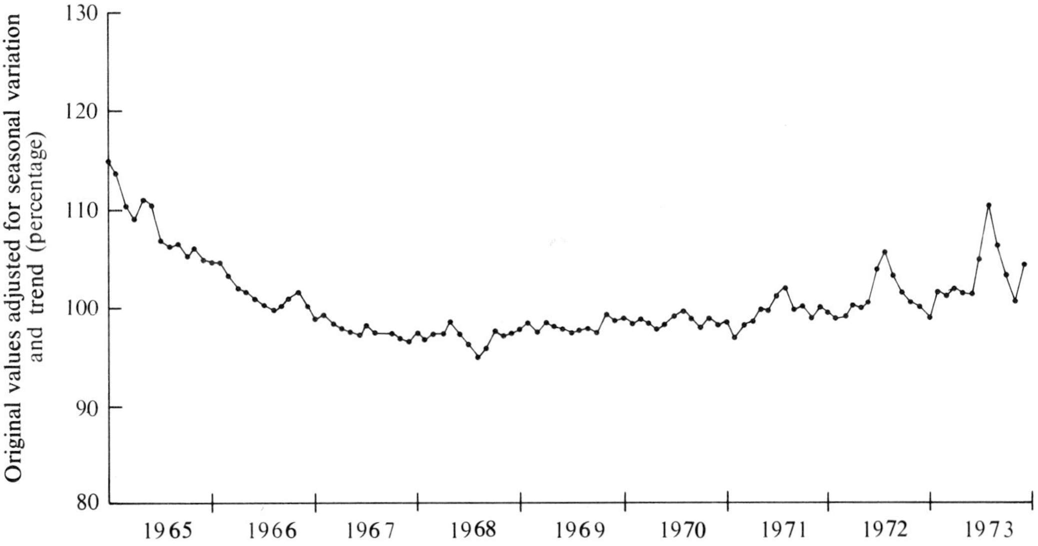

Figure 19.6 Graph of original data of Table 19.4 after adjustment for seasonal variation and trend.

Problems

1. Consider the time series data of Table A.

Table A

Original time-series values

Month	1970	1971	1972	1973	1974	1975
Jan.	3720	4296	4914	5454	6063	6597
Feb.	3555	4167	4689	5154	5787	6531
Mar.	3432	3930	4491	4980	5514	6180
Apr.	3261	3789	4254	4764	5319	5904
May	3099	3531	3951	4494	4947	5481
June	2991	3447	3876	4395	4854	5340
July	2970	3432	3897	4446	5010	5511
Aug.	2952	3444	3951	4488	5109	5778
Sept.	3114	3594	4041	4548	5190	5841
Oct.	3249	3660	4128	4656	5199	5757
Nov.	3441	3975	4425	4908	5475	5997
Dec.	3888	4428	4911	5421	6075	6852

a. Develop ratios to 12-month centered moving averages for the data. That is, perform an analysis parallelling that conducted in conjunction with Tables 19.4 and 19.5. Carry calculations to the following significant figures in constructing the table analogous to Table 19.5: column 4, whole numbers; column 5, whole numbers; column 6, two decimals; and column 7, two decimals.

b. Determine seasonal indexes for the data in Table A. That is, perform an analysis which parallels that conducted in conjunction with Table 19.6. Carry calculations to one decimal place.

c. Seasonally adjust the data of Table A. That is, perform an analysis which parallels that conducted in conjunction with Table 19.7. Carry calculations to the nearest whole number.

d. Remove the effect of trend from the deseasonalized data developed in Part c. That is, perform an analysis parallelling that conducted in conjunction with Tables 19.8 through 19.10, carrying calculations to the same number of significant figures. As an intermediate check, note that the *yearly* trend equation should be

$$\hat{Y} = 39,208.75 + 6,377.5x;$$

$x = 0$ for 1970, x-unit $= 1$ year.

19.5 *Forecasting through Use of Time-Series Analysis Results*

We have so far been primarily concerned with the breakdown of a time series into its components, or *decomposition*. We turn now to a more detailed look at the synthesis aspects of time-series analysis—the reconstruction and projection of a time series by combining its components through a procedure called *synthesis*.

The Cyclical and Residual Variation

No formal procedure has been devised for separating cyclical effects from residual effects in data of the nature of that of Table 19.9. Thus, even though they do provide a basis for further analysis of cyclical variation, the data represent the final result of *formal decomposition*. Further, this kind of limitation in decomposition carries over into synthesis, particularly with regard to forecasting. This is to say, it is very difficult, even impossible in some cases, to separate the cyclical and residual variation encountered in a specific economic time series. In fact, the very nature of irregular variations (random and nonrecurring) makes it practically impossible to forecast them (successfully) through formal analysis. And, even though the nature of cyclical variation makes it slightly more amenable to analytical forecasting, it is generally more satisfactory to arrive at predictions of cyclical behavior by

1. studying the reaction of the particular time series of interest to variations in general business and
2. basing forecasts upon related predictions concerning general business conditions.

This is due to the fact that the so-called business cycle manifests itself in some fashion in most economic time series.

Forecasting by Synthesis of Trend and Seasonal Variation

The concepts related to decomposition provide us a useful method for obtaining approximate forecasts of future values of a time-series variable. Annual forecasts can be obtained merely by extrapolation of the relevant trend line. Further, one can obtain monthly forecasts simply by multiplying monthly trend values by appropriate seasonal indexes, which depend on whether there exists a changing seasonal pattern.

While the purely mechanical forecast based on the synthesis of trend and seasonal effects often provides useful approximate forecasts, it is obviously not without limitation. This type of forecasting is completely dependent upon these assumptions:

1. The trend equation accurately reflects the effect of secular forces.
2. Such forces will continue to exert similar influences on the time-series variable.
3. Seasonal variation will behave as in the past.
4. There exist no cyclical nor residual effects.

If these assumptions are not valid, and they are generally not, more involved techniques must be used.

Even so, very good forecasts often emanate from a procedure involving a combination of an approximation forecast obtained by synthesis of trend and seasonal variation, an attempt to incorporate cyclical variations by forecasting the business cycle and taking into account the relationship of the particular economic series to such cycle, and an attempt to independently forecast irregular eventualities by analyzing specific factors and taking note of all information about anticipated effects of such forces.

Important Terms and Concepts

cyclical variation	moving average	secular trend
decomposition	ratio-to-moving average	secular variation
deseasonalized data	seasonal index	synthesis
irregular variations	seasonal variation	time series
long-term movement	seasonally adjusted data	time-series analysis
method of moving averages		trend analysis

References

19.1 J. R. Stockton, *Introduction to Business and Economic Statistics*, 5th ed., Cincinnati: South-Western Publishing Company, 1975.

19.2 F. E. Croxton, D. J. Cowden, and S. Klein, *Applied General Statistics*, 3rd ed., Englewood Cliffs, N.J.: Prentice-Hall, Inc., 1967.

Table A-1: Areas under the Normal Curve

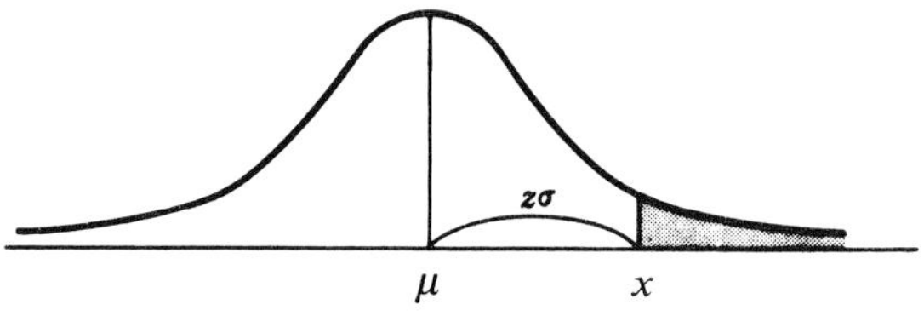

Example

$$z = \frac{x - \mu}{\sigma}$$

$P[Z > 1] = 0.1587$
$P[Z > 2] = 0.0228$

Normal Deviate z	.00	.01	.02	.03	.04	.05	.06	.07	.08	.09
0.0	.5000	.4960	.4920	.4880	.4840	.4801	.4761	.4721	.4681	.4641
0.1	.4602	.4562	.4522	.4483	.4443	.4404	.4364	.4325	.4286	.4247
0.2	.4207	.4168	.4129	.4090	.4052	.4013	.3974	.3936	.3897	.3859
0.3	.3821	.3783	.3745	.3707	.3669	.3632	.3594	.3557	.3520	.3483
0.4	.3446	.3409	.3372	.3336	.3300	.3264	.3228	.3192	.3156	.3121
0.5	.3085	.3050	.3015	.2981	.2946	.2912	.2877	.2843	.2810	.2776
0.6	.2743	.2709	.2676	.2643	.2611	.2578	.2546	.2514	.2483	.2451
0.7	.2420	.2389	.2358	.2327	.2296	.2266	.2236	.2206	.2177	.2148
0.8	.2119	.2090	.2061	.2033	.2005	.1977	.1949	.1922	.1894	.1867
0.9	.1841	.1814	.1788	.1762	.1736	.1711	.1685	.1660	.1635	.1611
1.0	.1587	.1562	.1539	.1515	.1492	.1469	.1446	.1423	.1401	.1379
1.1	.1357	.1335	.1314	.1292	.1271	.1251	.1230	.1210	.1190	.1170
1.2	.1151	.1131	.1112	.1093	.1075	.1056	.1038	.1020	.1003	.0985
1.3	.0968	.0951	.0934	.0918	.0901	.0885	.0869	.0853	.0838	.0823
1.4	.0808	.0793	.0778	.0764	.0749	.0735	.0721	.0708	.0694	.0681
1.5	.0668	.0655	.0643	.0630	.0618	.0606	.0594	.0582	.0571	.0559
1.6	.0548	.0537	.0526	.0516	.0505	.0495	.0485	.0475	.0465	.0455
1.7	.0446	.0436	.0427	.0418	.0409	.0401	.0392	.0384	.0375	.0367
1.8	.0359	.0351	.0344	.0336	.0329	.0322	.0314	.0307	.0301	.0294
1.9	.0287	.0281	.0274	.0268	.0262	.0256	.0250	.0244	.0239	.0233
2.0	.0228	.0222	.0217	.0212	.0207	.0202	.0197	.0192	.0188	.0183
2.1	.0179	.0174	.0170	.0166	.0162	.0158	.0154	.0150	.0146	.0143
2.2	.0139	.0136	.0132	.0129	.0125	.0122	.0119	.0116	.0113	.0110
2.3	.0107	.0104	.0102	.0099	.0096	.0094	.0091	.0089	.0087	.0084
2.4	.0082	.0080	.0078	.0075	.0073	.0071	.0069	.0068	.0066	.0064
2.5	.0062	.0060	.0059	.0057	.0055	.0054	.0052	.0051	.0049	.0048
2.6	.0047	.0045	.0044	.0043	.0041	.0040	.0039	.0038	.0037	.0036
2.7	.0035	.0034	.0033	.0032	.0031	.0030	.0029	.0028	.0027	.0026
2.8	.0026	.0025	.0024	.0023	.0023	.0022	.0021	.0021	.0020	.0019
2.9	.0019	.0018	.0018	.0017	.0016	.0016	.0015	.0015	.0014	.0014
3.0	.0013	.0013	.0013	.0012	.0012	.0011	.0011	.0011	.0010	.0010

Source: This table is abridged from Table II of Fisher & Yates: *Statistical Table for Biological, Agricultural and Medical Research*, published by Oliver & Boyd, Edinburgh, and by permission of the authors and publishers.

Table A-2: Ordinates of the Normal Curve

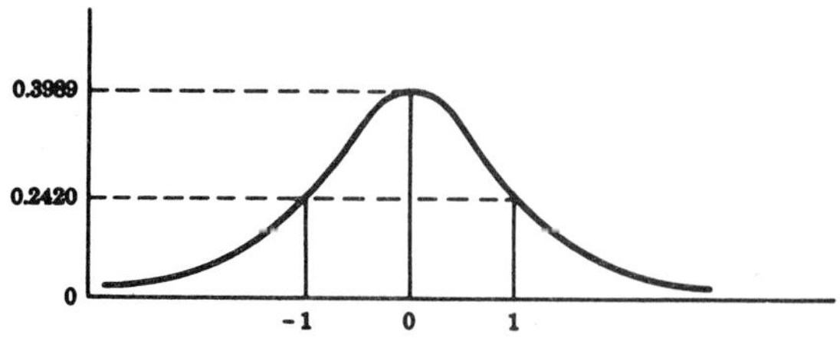

Example

$$z = \frac{x - \mu}{\sigma}$$

$z = 0$: ordinate $= 0.3989$

$z = 1$: ordinate $= 0.2420$

$z = 2$: ordinate $= 0.0540$

z	.00	.01	.02	.03	.04	.05	.06	.07	.08	.09
.0	.3989	.3989	.3989	.3988	.3986	.3984	.3982	.3980	.3977	.3973
.1	.3970	.3965	.3961	.3956	.3951	.3945	.3939	.3932	.3925	.3918
.2	.3910	.3902	.3894	.3885	.3876	.3867	.3857	.3847	.3836	.3825
.3	.3814	.3802	.3790	.3778	.3765	.3752	.3739	.3725	.3712	.3697
.4	.3683	.3668	.3653	.3637	.3621	.3605	.3589	.3572	.3555	.3538
.5	.3521	.3503	.3485	.3467	.3448	.3429	.3410	.3391	.3372	.3352
.6	.3332	.3312	.3292	.3271	.3251	.3230	.3209	.3187	.3166	.3144
.7	.3123	.3101	.3079	.3056	.3034	.3011	.2989	.2966	.2943	.2920
.8	.2897	.2874	.2850	.2827	.2803	.2780	.2756	.2732	.2709	.2685
.9	.2661	.2637	.2613	.2589	.2565	.2541	.2516	.2492	.2468	.2444
1.0	.2420	.2396	.2371	.2347	.2323	.2299	.2275	.2251	.2227	.2203
1.1	.2179	.2155	.2131	.2107	.2083	.2059	.2036	.2012	.1989	.1965
1.2	.1942	.1919	.1895	.1872	.1849	.1826	.1804	.1781	.1758	.1736
1.3	.1714	.1691	.1669	.1647	.1626	.1604	.1582	.1561	.1539	.1518
1.4	.1497	.1476	.1456	.1435	.1415	.1394	.1374	.1354	.1334	.1315
1.5	.1295	.1276	.1257	.1238	.1219	.1200	.1182	.1163	.1145	.1127
1.6	.1109	.1092	.1074	.1057	.1040	.1023	.1006	.0989	.0973	.0957
1.7	.0940	.0925	.0909	.0893	.0878	.0863	.0848	.0833	.0818	.0804
1.8	.0790	.0775	.0761	.0748	.0734	.0721	.0707	.0694	.0681	.0669
1.9	.0656	.0644	.0632	.0620	.0608	.0596	.0584	.0573	.0562	.0551
2.0	.0540	.0529	.0519	.0508	.0498	.0488	.0478	.0468	.0459	.0449
2.1	.0440	.0431	.0422	.0413	.0404	.0396	.0387	.0379	.0371	.0363
2.2	.0355	.0347	.0339	.0332	.0325	.0317	.0310	.0303	.0297	.0290
2.3	.0283	.0277	.0270	.0264	.0258	.0252	.0246	.0241	.0235	.0229
2.4	.0224	.0219	.0213	.0208	.0203	.0198	.0194	.0189	.0184	.0180
2.5	.0175	.0171	.0167	.0163	.0158	.0154	.0151	.0147	.0143	.0139
2.6	.0136	.0132	.0129	.0126	.0122	.0119	.0116	.0113	.0110	.0107
2.7	.0104	.0101	.0099	.0096	.0093	.0091	.0088	.0086	.0084	.0081
2.8	.0079	.0077	.0075	.0073	.0071	.0069	.0067	.0065	.0063	.0061
2.9	.0060	.0058	.0056	.0055	.0053	.0051	.0050	.0048	.0047	.0046
3.0	.0044	.0043	.0042	.0040	.0039	.0038	.0037	.0036	.0035	.0034
3.1	.0033	.0032	.0031	.0030	.0029	.0028	.0027	.0026	.0025	.0025
3.2	.0024	.0023	.0022	.0022	.0021	.0020	.0020	.0019	.0018	.0018
3.3	.0017	.0017	.0016	.0016	.0015	.0015	.0014	.0014	.0013	.0013
3.4	.0012	.0012	.0012	.0011	.0011	.0010	.0010	.0010	.0009	.0009
3.5	.0009	.0008	.0008	.0008	.0008	.0007	.0007	.0007	.0007	.0006
3.6	.0006	.0006	.0006	.0005	.0005	.0005	.0005	.0005	.0005	.0004
3.7	.0004	.0004	.0004	.0004	.0004	.0004	.0003	.0003	.0003	.0003
3.8	.0003	.0003	.0003	.0003	.0003	.0002	.0002	.0002	.0002	.0002
3.9	.0002	.0002	.0002	.0002	.0002	.0002	.0002	.0002	.0001	.0001

Source: This table is abridged from Table II of Fisher & Yates: *Statistical Tables for Biological, Agricultural and Medical Research*, published by Oliver & Boyd, Edinburgh, and by permission of the authors and publishers.

Table A-3: Percentage Points of the Chi-Square Distribution

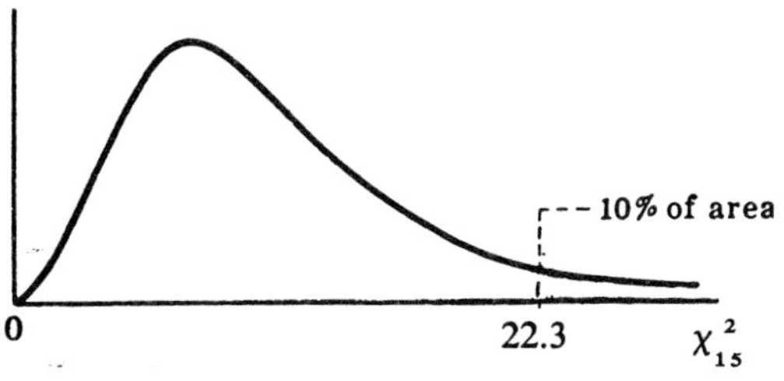

Example

For 15 degrees of freedom:

$P[\chi^2_{15} > 22.3] = 0.10$

n \ P	.995	.99	.975	.95	.90	.75	.50	.25	.10	.05	.025	.01	.005	P / n
1	0.0^4393	0.0^3157	0.0^3982	0.0^23	0.0158	0.102	0.455	1.323	2.71	3.84	5.02	6.63	7.88	1
2	0.0100	0.0201	0.0506	0.103	0.211	0.575	1.386	2.77	4.61	5.99	7.38	9.21	10.60	2
3	0.0717	0.115	0.216	0.352	0.584	1.213	2.37	4.11	6.25	7.81	9.35	11.34	12.84	3
4	0.207	0.297	0.484	0.711	1.064	1.923	3.36	5.39	7.78	9.49	11.14	13.28	14.86	4
5	0.412	0.554	0.831	1.145	1.610	2.67	4.35	6.63	9.24	11.07	12.83	15.09	16.75	5
6	0.676	0.872	1.237	1.635	2.20	3.45	5.35	7.84	10.64	12.59	14.45	16.81	18.55	6
7	0.989	1.239	1.690	2.17	2.83	4.25	6.35	9.04	12.02	14.07	16.01	18.48	20.3	7
8	1.344	1.646	2.18	2.73	3.49	5.07	7.34	10.22	13.36	15.51	17.53	20.1	22.0	8
9	1.735	2.09	2.70	3.33	4.17	5.90	8.34	11.39	14.68	16.92	19.02	21.7	23.6	9
10	2.16	2.56	3.25	3.94	4.87	6.74	9.34	12.55	15.99	18.31	20.5	23.2	25.2	10
11	2.60	3.05	3.82	4.57	5.58	7.58	10.34	13.70	17.28	19.68	21.9	24.7	26.8	11
12	3.07	3.57	4.40	5.23	6.30	8.44	11.34	14.85	18.55	21.0	23.3	26.2	28.3	12
13	3.57	4.11	5.01	5.89	7.04	9.30	12.34	15.98	19.81	22.4	24.7	27.7	29.8	13
14	4.07	4.66	5.63	6.57	7.79	10.17	13.34	17.12	21.1	23.7	26.1	29.1	31.3	14
15	4.60	5.23	6.26	7.26	8.55	11.04	14.34	18.25	22.3	25.0	27.5	30.6	32.8	15
16	5.14	5.81	6.91	7.96	9.31	11.91	15.34	19.37	23.5	26.3	28.8	32.0	34.3	16
17	5.70	6.41	7.56	8.67	10.09	12.79	16.34	20.5	24.8	27.6	30.2	33.4	35.7	17
18	6.26	7.01	8.23	9.39	10.86	13.68	17.34	21.6	26.0	28.9	31.5	34.8	37.2	18
19	6.84	7.63	8.91	10.12	11.65	14.56	18.34	22.7	27.2	30.1	32.9	36.2	38.6	19
20	7.43	8.26	9.59	10.85	12.44	15.45	19.34	23.8	28.4	31.4	34.2	37.6	40.0	20
21	8.03	8.90	10.28	11.59	13.24	16.34	20.3	24.9	29.6	32.7	35.5	38.9	41.4	21
22	8.64	9.54	10.98	12.34	14.04	17.24	21.3	26.0	30.8	33.9	36.8	40.3	42.8	22
23	9.26	10.20	11.69	13.09	14.85	18.14	22.3	27.1	32.0	35.2	38.1	41.6	44.2	23
24	9.89	10.86	12.40	13.85	15.66	19.04	23.3	28.2	33.2	36.4	39.4	43.0	45.6	24
25	10.52	11.52	13.12	14.61	16.47	19.94	24.3	29.3	34.4	37.7	40.6	44.3	46.9	25
26	11.16	12.20	13.84	15.38	17.29	20.8	25.3	30.4	35.6	38.9	41.9	45.6	48.3	26
27	11.81	12.88	14.57	16.15	18.11	21.7	26.3	31.5	36.7	40.1	43.2	47.0	49.6	27
28	12.46	13.56	15.31	16.93	18.94	22.7	27.3	32.6	37.9	41.3	44.5	48.3	51.0	28
29	13.12	14.26	16.05	17.71	19.77	23.6	28.3	33.7	39.1	42.6	45.7	49.6	52.3	29
30	13.79	14.95	16.79	18.49	20.6	24.5	29.3	34.8	40.3	43.8	47.0	50.9	53.7	30
40	20.7	22.2	24.4	26.5	29.1	33.7	39.3	45.6	51.8	55.8	59.3	63.7	66.8	40
50	28.0	29.7	32.4	34.8	37.7	42.9	49.3	56.3	63.2	67.5	71.4	76.2	79.5	50
60	35.5	37.5	40.5	43.2	46.5	52.3	59.3	67.0	74.4	79.1	83.3	88.4	92.0	60
70	43.3	45.4	48.8	51.7	55.3	61.7	69.3	77.6	85.5	90.5	95.0	100.4	104.2	70
80	51.2	53.5	57.2	60.4	64.3	71.1	79.3	88.1	96.6	101.9	106.6	112.3	116.3	80
90	59.2	61.8	65.6	69.1	73.3	80.6	89.3	98.6	107.6	113.1	118.1	124.1	128.3	90
100	67.3	70.1	74.2	77.9	82.4	90.1	99.3	109.1	118.5	124.3	129.6	135.8	140.2	100
Z_α	−2.58	−2.33	−1.96	−1.64	−1.28	−0.674	0.000	0.674	1.282	1.645	1.960	2.33	2.58	Z_α

For $n > 100$ take $\chi^2 = \frac{1}{2}(Z_\alpha + \sqrt{2n-1})^2$. Z_α is the standardized normal deviate corresponding to the α level of significance, and is shown in the bottom of the table.

Source: This table is abridged from Catherine M. Thompson, "Table of Percentage Points of the χ^2 Distribution," *Biometrika*, Vol. 32 (1941). Used by permission.

Table A-4: Percentage Points of the *t* Distribution

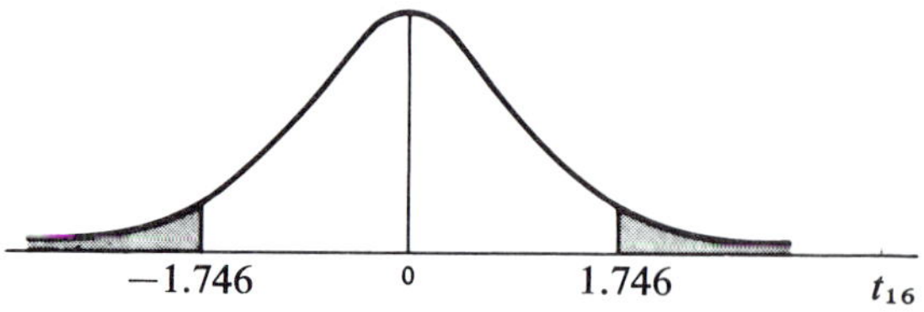

Example

For 16 degrees of freedom:

$P[t_{16} > 1.746] = 0.05$

$P[t_{16} < -1.746] = 0.05$

n \ α	.25	.20	.15	.10	.05	.025	.01	.005	.0005
1	1.000	1.376	1.963	3.078	6.314	12.706	31.821	63.657	636.619
2	.816	1.061	1.386	1.886	2.920	4.303	6.965	9.925	31.598
3	.765	.978	1.250	1.638	2.353	3.182	4.541	5.841	12.941
4	.741	.941	1.190	1.533	2.132	2.776	3.747	4.604	8.610
5	.727	.920	1.156	1.476	2.015	2.571	3.365	4.032	6.859
6	.718	.906	1.134	1.440	1.943	2.447	3.143	3.707	5.959
7	.711	.896	1.119	1.415	1.895	2.365	2.998	3.499	5.405
8	.706	.889	1.108	1.397	1.860	2.306	2.896	3.355	5.041
9	.703	.883	1.100	1.383	1.833	2.262	2.821	3.250	4.781
10	.700	.879	1.093	1.372	1.812	2.228	2.764	3.169	4.587
11	.697	.876	1.088	1.363	1.796	2.201	2.718	3.106	4.437
12	.695	.873	1.083	1.356	1.782	2.179	2.681	3.055	4.318
13	.694	.870	1.079	1.350	1.771	2.160	2.650	3.012	4.221
14	.692	.868	1.076	1.345	1.761	2.145	2.624	2.977	4.140
15	.691	.866	1.074	1.341	1.753	2.131	2.602	2.947	4.073
16	.690	.865	1.071	1.337	1.746	2.120	2.583	2.921	4.015
17	.689	.863	1.069	1.333	1.740	2.110	2.567	2.898	3.965
18	.688	.862	1.067	1.330	1.734	2.101	2.552	2.878	3.922
19	.688	.861	1.066	1.328	1.729	2.093	2.539	2.861	3.883
20	.687	.860	1.064	1.325	1.725	2.086	2.528	2.845	3.850
21	.686	.859	1.063	1.323	1.721	2.080	2.518	2.831	3.819
22	.686	.858	1.061	1.321	1.717	2.074	2.508	2.819	3.792
23	.685	.858	1.060	1.319	1.714	2.069	2.500	2.807	3.767
24	.685	.857	1.059	1.318	1.711	2.064	2.492	2.397	3.745
25	.684	.856	1.058	1.316	1.708	2.060	2.485	2.787	3.725
26	.684	.856	1.058	1.315	1.706	2.056	2.479	2.779	3.707
27	.684	.855	1.057	1.314	1.703	2.052	2.473	2.771	3.690
28	.683	.855	1.056	1.313	1.701	2.048	2.467	2.763	3.674
29	.683	.854	1.055	1.311	1.699	2.045	2.462	2.756	3.659
30	.683	.854	1.055	1.310	1.697	2.042	2.457	2.750	3.646
40	.681	.851	1.050	1.303	1.684	2.021	2.423	2.704	3.551
60	.679	.848	1.046	1.296	1.671	2.000	2.390	2.660	3.460
120	.677	.845	1.041	1.289	1.658	1.980	2.358	2.617	3.373
∞	.674	.842	1.036	1.282	1.645	1.960	2.326	2.576	3.291

Source: This table is abridged from Table III of Fisher & Yates: *Statistical Tables for Biological, Agricultural and Medical Research*, published by Oliver & Boyd, Edinburgh, and by permission of the authors and publishers.

Table A-5: Percentage Points of the *F* Distribution

Denominator *df*	Probability of a larger *F*	Numerator *df* 1	2	3	4	5	6	7	8	9
1	.100	39.86	49.50	53.59	55.83	57.24	58.20	58.91	59.44	59.86
	.050	161.4	199.5	215.7	224.6	230.2	234.0	236.8	238.9	240.5
	.025	647.8	799.5	864.2	899.6	921.8	937.1	948.2	956.7	963.3
	.010	4052	4999.5	5403	5625	5764	5859	5928	5982	6022
2	.100	8.53	9.00	9.16	9.24	9.29	9.33	9.35	9.37	9.38
	.050	18.51	19.00	19.16	19.25	19.30	19.33	19.35	19.37	19.38
	.025	38.51	39.00	39.17	39.25	39.30	39.33	39.36	39.37	39.39
	.010	98.50	99.00	99.17	99.25	99.30	99.33	99.36	99.37	99.39
3	.100	5.54	5.46	5.39	5.34	5.31	5.28	5.27	5.25	5.24
	.050	10.13	9.55	9.28	9.12	9.01	8.94	8.89	8.85	8.81
	.025	17.44	16.04	15.44	15.10	14.88	14.73	14.62	14.54	14.47
	.010	34.12	30.82	29.46	28.71	28.24	27.91	27.67	27.49	27.35
4	.100	4.54	4.32	4.19	4.11	4.05	4.01	3.98	3.95	3.94
	.050	7.71	6.94	6.59	6.39	6.26	6.16	6.09	6.04	6.00
	.025	12.22	10.65	9.98	9.60	9.36	9.20	9.07	8.98	8.90
	.010	21.20	18.00	16.69	15.98	15.52	15.21	14.98	14.80	14.66
5	.100	4.06	3.78	3.62	3.52	3.45	3.40	3.37	3.34	3.32
	.050	6.61	5.79	5.41	5.19	5.05	4.95	4.88	4.82	4.77
	.025	10.01	8.43	7.76	7.39	7.15	6.98	6.85	6.76	6.68
	.010	16.26	13.27	12.06	11.39	10.97	10.67	10.46	10.29	10.16
6	.100	3.78	3.46	3.29	3.18	3.11	3.05	3.01	2.98	2.96
	.050	5.99	5.14	4.76	4.53	4.39	4.28	4.21	4.15	4.10
	.025	8.81	7.26	6.60	6.23	5.99	5.82	5.70	5.60	5.52
	.010	13.75	10.92	9.78	9.15	8.75	8.47	8.26	8.10	7.98
7	.100	3.59	3.26	3.07	2.96	2.88	2.83	2.78	2.75	2.72
	.050	5.59	4.74	4.35	4.12	3.97	3.87	3.79	3.73	3.68
	.025	8.07	6.54	5.89	5.52	5.29	5.12	4.99	4.90	4.82
	.010	12.25	9.55	8.45	7.85	7.46	7.19	6.99	6.84	6.72
8	.100	3.46	3.11	2.92	2.81	2.73	2.67	2.62	2.59	2.56
	.050	5.32	4.46	4.07	3.84	3.69	3.58	3.50	3.44	3.39
	.025	7.57	6.06	5.42	5.05	4.82	4.65	4.53	4.43	4.36
	.010	11.26	8.65	7.59	7.01	6.63	6.37	6.18	6.03	5.91
9	.100	3.36	3.01	2.81	2.69	2.61	2.55	2.51	2.47	2.44
	.050	5.12	4.26	3.86	3.63	3.48	3.37	3.29	3.23	3.18
	.025	7.21	5.71	5.08	4.72	4.48	4.32	4.20	4.10	4.03
	.010	10.56	8.02	6.99	6.42	6.06	5.80	5.61	5.47	5.35
10	.100	3.29	2.92	2.73	2.61	2.52	2.46	2.41	2.38	2.35
	.050	4.96	4.10	3.71	3.48	3.33	3.22	3.14	3.07	3.02
	.025	6.94	5.46	4.83	4.47	4.24	4.07	3.95	3.85	3.78
	.010	10.04	7.56	6.55	5.99	5.64	5.39	5.20	5.06	4.94
11	.100	3.23	2.86	2.66	2.54	2.45	2.39	2.34	2.30	2.27
	.050	4.84	3.98	3.59	3.36	3.20	3.09	3.01	2.95	2.90
	.025	6.72	5.26	4.63	4.28	4.04	3.88	3.76	3.66	3.59
	.010	9.65	7.21	6.22	5.67	5.32	5.07	4.89	4.74	4.63
12	.100	3.18	2.81	2.61	2.48	2.39	2.33	2.28	2.24	2.21
	.050	4.75	3.89	3.49	3.26	3.11	3.00	2.91	2.85	2.80
	.025	6.55	5.10	4.47	4.12	3.89	3.73	3.61	3.51	3.44
	.010	9.33	6.93	5.95	5.41	5.06	4.82	4.64	4.50	4.39
13	.100	3.14	2.76	2.56	2.43	2.35	2.28	2.23	2.20	2.16
	.050	4.67	3.81	3.41	3.18	3.03	2.92	2.83	2.77	2.71
	.025	6.41	4.97	4.35	4.00	3.77	3.60	3.48	3.39	3.31
	.010	9.07	6.70	5.74	5.21	4.86	4.62	4.44	4.30	4.19
14	.100	3.10	2.73	2.52	2.39	2.31	2.24	2.19	2.15	2.12
	.050	4.60	3.74	3.34	3.11	2.96	2.85	2.76	2.70	2.65
	.025	6.30	4.86	4.24	3.89	3.66	3.50	3.38	3.29	3.21
	.010	8.86	6.51	5.56	5.04	4.69	4.46	4.28	4.14	4.03

Source: This table is abridged from M. Merrington and C. M. Thompson, "Tables of Percentage Points of the Inverted Beta (*F*) Distribution," *Biometrika*, Vol. 33 (1943). Used by permission.

Numerator *df*											
10	12	15	20	24	30	40	60	120	∞	*P*	*df*
60.19	60.71	61.22	61.74	62.00	62.26	62.53	62.79	63.06	63.33	.100	1
241.9	243.9	245.9	248.0	249.1	250.1	251.1	252.2	253.3	254.3	.050	
968.6	976.7	984.9	993.1	997.2	1001	1006	1010	1014	1018	.025	
6056	6106	6157	6209	6235	6261	6287	6313	6339	6366	.010	
9.39	9.41	9.42	9.44	9.45	9.46	9.47	9.47	9.48	9.49	.100	2
19.40	19.41	19.43	19.45	19.45	19.46	19.47	19.48	19.49	19.50	.050	
39.40	39.41	39.43	39.45	39.46	39.46	39.47	39.48	39.49	39.50	.025	
99.40	99.42	99.43	99.45	99.46	99.47	99.47	99.48	99.49	99.50	.010	
5.23	5.22	5.20	5.18	5.18	5.17	5.16	5.15	5.14	5.13	.100	3
8.79	8.74	8.70	8.66	8.64	8.62	8.59	8.57	8.55	8.53	.050	
14.42	14.34	14.25	14.17	14.12	14.08	14.04	13.99	13.95	13.90	.025	
27.23	27.05	26.87	26.69	26.60	26.50	26.41	26.32	26.22	26.13	.010	
3.92	3.90	3.87	3.84	3.83	3.82	3.80	3.79	3.78	3.76	.100	4
5.96	5.91	5.86	5.80	5.77	5.75	5.72	5.69	5.66	5.63	.050	
8.84	8.75	8.66	8.56	8.51	8.46	8.41	8.36	8.31	8.26	.025	
14.55	14.37	14.20	14.02	13.93	13.84	13.75	13.65	13.56	13.46	.010	
3.30	3.27	3.24	3.21	3.19	3.17	3.16	3.14	3.12	3.10	.100	5
4.74	4.68	4.62	4.56	4.53	4.50	4.46	4.43	4.40	4.36	.050	
6.62	6.52	6.43	6.33	6.28	6.23	6.18	6.12	6.07	6.02	.025	
10.05	9.89	9.72	9.55	9.47	9.38	9.29	9.20	9.11	9.02	.010	
2.94	2.90	2.87	2.84	2.82	2.80	2.78	2.76	2.74	2.72	.100	6
4.06	4.00	3.94	3.87	3.84	3.81	3.77	3.74	3.70	3.67	.050	
5.46	5.37	5.27	5.17	5.12	5.07	5.01	4.96	4.90	4.85	.025	
7.87	7.72	7.56	7.40	7.31	7.23	7.14	7.06	6.97	6.88	.010	
2.70	2.67	2.63	2.59	2.58	2.56	2.54	2.51	2.49	2.47	.100	7
3.64	3.57	3.51	3.44	3.41	3.38	3.34	3.30	3.27	3.23	.050	
4.76	4.67	4.57	4.47	4.42	4.36	4.31	4.25	4.20	4.14	.025	
6.62	6.47	6.31	6.16	6.07	5.99	5.91	5.82	5.74	5.65	.010	
2.54	2.50	2.46	2.42	2.40	2.38	2.36	2.34	2.32	2.29	.100	8
3.35	3.28	3.22	3.15	3.12	3.08	3.04	3.01	2.97	2.93	.050	
4.30	4.20	4.10	4.00	3.95	3.89	3.84	3.78	3.73	3.67	.025	
5.81	5.67	5.52	5.36	5.28	5.20	5.12	5.03	4.95	4.86	.010	
2.42	2.38	2.34	2.30	2.28	2.25	2.23	2.21	2.18	2.16	.100	9
3.14	3.07	3.01	2.94	2.90	2.86	2.83	2.79	2.75	2.71	.050	
3.96	3.87	3.77	3.67	3.61	3.56	3.51	3.45	3.39	3.33	.025	
5.26	5.11	4.96	4.81	4.73	4.65	4.57	4.48	4.40	4.31	.010	
2.32	2.28	2.24	2.20	2.18	2.16	2.13	2.11	2.08	2.06	.100	10
2.98	2.91	2.85	2.77	2.74	2.70	2.66	2.62	2.58	2.54	.050	
3.72	3.62	3.52	3.42	3.37	3.31	3.26	3.20	3.14	3.08	.025	
4.85	4.71	4.56	4.41	4.33	4.25	4.17	4.08	4.00	3.91	.010	
2.25	2.21	2.17	2.12	2.10	2.08	2.05	2.03	2.00	1.97	.100	11
2.85	2.79	2.72	2.65	2.61	2.57	2.53	2.49	2.45	2.40	.050	
3.53	3.43	3.33	3.23	3.17	3.12	3.06	3.00	2.94	2.88	.025	
4.54	4.40	4.25	4.10	4.02	3.94	3.86	3.78	3.69	3.60	.010	
2.19	2.15	2.10	2.06	2.04	2.01	1.99	1.96	1.93	1.90	.100	12
2.75	2.69	2.62	2.54	2.51	2.47	2.43	2.38	2.34	2.30	.050	
3.37	3.28	3.18	3.07	3.02	2.96	2.91	2.85	2.79	2.72	.025	
4.30	4.16	4.01	3.86	3.78	3.70	3.62	3.54	3.45	3.36	.010	
2.14	2.10	2.05	2.01	1.98	1.96	1.93	1.90	1.88	1.85	.100	13
2.67	2.60	2.53	2.46	2.42	2.38	2.34	2.30	2.25	2.21	.050	
3.25	3.15	3.05	2.95	2.89	2.84	2.78	2.72	2.66	2.60	.025	
4.10	3.96	3.82	3.66	3.59	3.51	3.43	3.34	3.25	3.17	.010	
2.10	2.05	2.01	1.96	1.94	1.91	1.89	1.86	1.83	1.80	.100	14
2.60	2.53	2.46	2.39	2.35	2.31	2.27	2.22	2.18	2.13	.050	
3.15	3.05	2.95	2.84	2.79	2.73	2.67	2.61	2.55	2.49	.025	
3.94	3.80	3.66	3.51	3.43	3.35	3.27	3.18	3.09	3.00	.010	

Denominator *df*	Probability of a larger *F*	Numerator *df* 1	2	3	4	5	6	7	8	9
15	.100	3.07	2.70	2.49	2.36	2.27	2.21	2.16	2.12	2.09
	.050	4.54	3.68	3.29	3.06	2.90	2.79	2.71	2.64	2.59
	.025	6.20	4.77	4.15	3.80	3.58	3.41	3.29	3.20	3.12
	.010	8.68	6.36	5.42	4.89	4.56	4.32	4.14	4.00	3.89
16	.100	3.05	2.67	2.46	2.33	2.24	2.18	2.13	2.09	2.06
	.050	4.49	3.63	3.24	3.01	2.85	2.74	2.66	2.59	2.54
	.025	6.12	4.69	4.08	3.73	3.50	3.34	3.22	3.12	3.05
	.010	8.53	6.23	5.29	4.77	4.44	4.20	4.03	3.89	3.78
17	.100	3.03	2.64	2.44	2.31	2.22	2.15	2.10	2.06	2.03
	.050	4.45	3.59	3.20	2.96	2.81	2.70	2.61	2.55	2.49
	.025	6.04	4.62	4.01	3.66	3.44	3.28	3.16	3.06	2.98
	.010	8.40	6.11	5.18	4.67	4.34	4.10	3.93	3.79	3.68
18	.100	3.01	2.62	2.42	2.29	2.20	2.13	2.08	2.04	2.00
	.050	4.41	3.55	3.16	2.93	2.77	2.66	2.58	2.51	2.46
	.025	5.98	4.56	3.95	3.61	3.38	3.22	3.10	3.01	2.93
	.010	8.29	6.01	5.09	4.58	4.25	4.01	3.84	3.71	3.60
19	.100	2.99	2.61	2.40	2.27	2.18	2.11	2.06	2.02	1.98
	.050	4.38	3.52	3.13	2.90	2.74	2.63	2.54	2.48	2.42
	.025	5.92	4.51	3.90	3.56	3.33	3.17	3.05	2.96	2.88
	.010	8.18	5.93	5.01	4.50	4.17	3.94	3.77	3.63	3.52
20	.100	2.97	2.59	2.38	2.25	2.16	2.09	2.04	2.00	1.96
	.050	4.35	3.49	3.10	2.87	2.71	2.60	2.51	2.45	2.39
	.025	5.87	4.46	3.86	3.51	3.29	3.13	3.01	2.91	2.84
	.010	8.10	5.85	4.94	4.43	4.10	3.87	3.70	3.56	3.46
21	.100	2.96	2.57	2.36	2.23	2.14	2.08	2.02	1.98	1.95
	.050	4.32	3.47	3.07	2.84	2.68	2.57	2.49	2.42	2.37
	.025	5.83	4.42	3.82	3.48	3.25	3.09	2.97	2.87	2.80
	.010	8.02	5.78	4.87	4.37	4.04	3.81	3.64	3.51	3.40
22	.100	2.95	2.56	2.35	2.22	2.13	2.06	2.01	1.97	1.93
	.050	4.30	3.44	3.05	2.82	2.66	2.55	2.46	2.40	2.34
	.025	5.79	4.38	3.78	3.44	3.22	3.05	2.93	2.84	2.76
	.010	7.95	5.72	4.82	4.31	3.99	3.76	3.59	3.45	3.35
23	.100	2.94	2.55	2.34	2.21	2.11	2.05	1.99	1.95	1.92
	.050	4.28	3.42	3.03	2.80	2.64	2.53	2.44	2.37	2.32
	.025	5.75	4.35	3.75	3.41	3.18	3.02	2.90	2.81	2.73
	.010	7.88	5.66	4.76	4.26	3.94	3.71	3.54	3.41	3.30
24	.100	2.93	2.54	2.33	2.19	2.10	2.04	1.98	1.94	1.91
	.050	4.26	3.40	3.01	2.78	2.62	2.51	2.42	2.36	2.30
	.025	5.72	4.32	3.72	3.38	3.15	2.99	2.87	2.78	2.70
	.010	7.82	5.61	4.72	4.22	3.90	3.67	3.50	3.36	3.26
25	.100	2.92	2.53	2.32	2.18	2.09	2.02	1.97	1.93	1.89
	.050	4.24	3.39	2.99	2.76	2.60	2.49	2.40	2.34	2.28
	.025	5.69	4.29	3.69	3.35	3.13	2.97	2.85	2.75	2.68
	.010	7.77	5.57	4.68	4.18	3.85	3.63	3.46	3.32	3.22
26	.100	2.91	2.52	2.31	2.17	2.08	2.01	1.96	1.92	1.88
	.050	4.23	3.37	2.98	2.74	2.59	2.47	2.39	2.32	2.27
	.025	5.66	4.27	3.67	3.33	3.10	2.94	2.82	2.73	2.65
	.010	7.72	5.53	4.64	4.14	3.82	3.59	3.42	3.29	3.18
27	.100	2.90	2.51	2.30	2.17	2.07	2.00	1.95	1.91	1.87
	.050	4.21	3.35	2.96	2.73	2.57	2.46	2.37	2.31	2.25
	.025	5.63	4.24	3.65	3.31	3.08	2.92	2.80	2.71	2.63
	.010	7.68	5.49	4.60	4.11	3.78	3.56	3.39	3.26	3.15
28	.100	2.89	2.50	2.29	2.16	2.06	2.00	1.94	1.90	1.87
	.050	4.20	3.34	2.95	2.71	2.56	2.45	2.36	2.29	2.24
	.025	5.61	4.22	3.63	3.29	3.06	2.90	2.78	2.69	2.61
	.010	7.64	5.45	4.57	4.07	3.75	3.53	3.36	3.23	3.12

Numerator *df*											
10	12	15	20	24	30	40	60	120	∞	*P*	*df*
2.06	2.02	1.97	1.92	1.90	1.87	1.85	1.82	1.79	1.76	.100	15
2.54	2.48	2.40	2.33	2.29	2.25	2.20	2.16	2.11	2.07	.050	
3.06	2.96	2.86	2.76	2.70	2.64	2.59	2.52	2.46	2.40	.025	
3.80	3.67	3.52	3.37	3.29	3.21	3.13	3.05	2.96	2.87	.010	
2.03	1.99	1.94	1.89	1.87	1.84	1.81	1.78	1.75	1.72	.100	16
2.49	2.42	2.35	2.28	2.24	2.19	2.15	2.11	2.06	2.01	.050	
2.99	2.89	2.79	2.68	2.63	2.57	2.51	2.45	2.38	2.32	.025	
3.69	3.55	3.41	3.26	3.18	3.10	3.02	2.93	2.84	2.75	.010	
2.00	1.96	1.91	1.86	1.84	1.81	1.78	1.75	1.72	1.69	.100	17
2.45	2.38	2.31	2.23	2.19	2.15	2.10	2.06	2.01	1.96	.050	
2.92	2.82	2.72	2.62	2.56	2.50	2.44	2.38	2.32	2.25	.025	
3.59	3.46	3.31	3.16	3.08	3.00	2.92	2.83	2.75	2.65	.010	
1.98	1.93	1.89	1.84	1.81	1.78	1.75	1.72	1.69	1.66	.100	18
2.41	2.34	2.27	2.19	2.15	2.11	2.06	2.02	1.97	1.92	.050	
2.87	2.77	2.67	2.56	2.50	2.44	2.38	2.32	2.26	2.19	.025	
3.51	3.37	3.23	3.08	3.00	2.92	2.84	2.75	2.66	2.57	.010	
1.96	1.91	1.86	1.81	1.79	1.76	1.73	1.70	1.67	1.63	.100	19
2.38	2.31	2.23	2.16	2.11	2.07	2.03	1.98	1.93	1.88	.050	
2.82	2.72	2.62	2.51	2.45	2.39	2.33	2.27	2.20	2.13	.025	
3.43	3.30	3.15	3.00	2.92	2.84	2.76	2.67	2.58	2.49	.010	
1.94	1.89	1.84	1.79	1.77	1.74	1.71	1.68	1.64	1.61	.100	20
2.35	2.28	2.20	2.12	2.08	2.04	1.99	1.95	1.90	1.84	.050	
2.77	2.68	2.57	2.46	2.41	2.35	2.29	2.22	2.16	2.09	.025	
3.37	3.23	3.09	2.94	2.86	2.78	2.69	2.61	2.52	2.42	010	
1.92	1.87	1.83	1.78	1.75	1.72	1.69	1.66	1.62	1.59	.100	21
2.32	2.25	2.18	2.10	2.05	2.01	1.96	1.92	1.87	1.81	.050	
2.73	2.64	2.53	2.42	2.37	2.31	2.25	2.18	2.11	2.04	.025	
3.31	3.17	3.03	2.88	2.80	2.72	2.64	2.55	2.46	2.36	.010	
1.90	1.86	1.81	1.76	1.73	1.70	1.67	1.64	1.60	1.57	.100	22
2.30	2.23	2.15	2.07	2.03	1.98	1.94	1.89	1.84	1.78	.050	
2.70	2.60	2.50	2.39	2.33	2.27	2.21	2.14	2.08	2.00	.025	
3.26	3.12	2.98	2.83	2.75	2.67	2.58	2.50	2.40	2.31	.010	
1.89	1.84	1.80	1.74	1.72	1.69	1.66	1.62	1.59	1.55	.100	23
2.27	2.20	2.13	2.05	2.01	1.96	1.91	1.86	1.81	1.76	.050	
2.67	2.57	2.47	2.36	2.30	2.24	2.18	2.11	2.04	1.97	.025	
3.21	3.07	2.93	2.78	2.70	2.62	2.54	2.45	2.35	2.26	.010	
1.88	1.83	1.78	1.73	1.70	1.67	1.64	1.61	1.57	1.53	.100	24
2.25	2.18	2.11	2.03	1.98	1.94	1.89	1.84	1.79	1.73	.050	
2.64	2.54	2.44	2.33	2.27	2.21	2.15	2.08	2.01	1.94	.025	
3.17	3.03	2.89	2.74	2.66	2.58	2.49	2.40	2.31	2.21	.010	
1.87	1.82	1.77	1.72	1.69	1.66	1.63	1.59	1.56	1.52	.100	25
2.24	2.16	2.09	2.01	1.96	1.92	1.87	1.82	1.77	1.71	.050	
2.61	2.51	2.41	2.30	2.24	2.18	2.12	2.05	1.98	1.91	.025	
3.13	2.99	2.85	2.70	2.62	2.54	2.45	2.36	2.27	2.17	.010	
1.86	1.81	1.76	1.71	1.68	1.65	1.61	1.58	1.54	1.50	.100	26
2.22	2.15	2.07	1.99	1.95	1.90	1.85	1.80	1.75	1.69	.050	
2.59	2.49	2.39	2.28	2.22	2.16	2.09	2.03	1.95	1.88	.025	
3.09	2.96	2.81	2.66	2.58	2.50	2.42	2.33	2.23	2.13	.010	
1.85	1.80	1.75	1.70	1.67	1.64	1.60	1.57	1.53	1.49	.100	27
2.20	2.13	2.06	1.97	1.93	1.88	1.84	1.79	1.73	1.67	.050	
2.57	2.47	2.36	2.25	2.19	2.13	2.07	2.00	1.93	1.85	.025	
3.06	2.93	2.78	2.63	2.55	2.47	2.38	2.29	2.20	2.10	.010	
1.84	1.79	1.74	1.69	1.66	1.63	1.59	1.56	1.52	1.48	.100	28
2.19	2.12	2.04	1.96	1.91	1.87	1.82	1.77	1.71	1.65	.050	
2.55	2.45	2.34	2.23	2.17	2.11	2.05	1.98	1.91	1.83	.025	
3.03	2.90	2.75	2.60	2.52	2.44	2.35	2.26	2.17	2.06	.010	

Denominator *df*	Probability of a larger *F*	Numerator *df* 1	2	3	4	5	6	7	8	9
29	.100	2.89	2.50	2.28	2.15	2.06	1.99	1.93	1.89	1.86
	.050	4.18	3.33	2.93	2.70	2.55	2.43	2.35	2.28	2.22
	.025	5.59	4.20	3.61	3.27	3.04	2.88	2.76	2.67	2.59
	.010	7.60	5.42	4.54	4.04	3.73	3.50	3.33	3.20	3.09
30	.100	2.88	2.49	2.28	2.14	2.05	1.98	1.93	1.88	1.85
	.050	4.17	3.32	2.92	2.69	2.53	2.42	2.33	2.27	2.21
	.025	5.57	4.18	3.59	3.25	3.03	2.87	2.75	2.65	2.57
	.010	7.56	5.39	4.51	4.02	3.70	3.47	3.30	3.17	3.07
40	.100	2.84	2.44	2.23	2.09	2.00	1.93	1.87	1.83	1.79
	.050	4.08	3.23	2.84	2.61	2.45	2.34	2.25	2.18	2.12
	.025	5.42	4.05	3.46	3.13	2.90	2.74	2.62	2.53	2.45
	.010	7.31	5.18	4.31	3.83	3.51	3.29	3.12	2.99	2.89
60	.100	2.79	2.39	2.18	2.04	1.95	1.87	1.82	1.77	1.74
	.050	4.00	3.15	2.76	2.53	2.37	2.25	2.17	2.10	2.04
	.025	5.29	3.93	3.34	3.01	2.79	2.63	2.51	2.41	2.33
	.010	7.08	4.98	4.13	3.65	3.34	3.12	2.95	2.82	2.72
120	.100	2.75	2.35	2.13	1.99	1.90	1.82	1.77	1.72	1.68
	.050	3.92	3.07	2.68	2.45	2.29	2.17	2.09	2.02	1.96
	.025	5.15	3.80	3.23	2.89	2.67	2.52	2.39	2.30	2.22
	.010	6.85	4.79	3.95	3.48	3.17	2.96	2.79	2.66	2.56
∞	.100	2.71	2.30	2.08	1.94	1.85	1.77	1.72	1.67	1.63
	.050	3.84	3.00	2.60	2.37	2.21	2.10	2.01	1.94	1.88
	.025	5.02	3.69	3.12	2.79	2.57	2.41	2.29	2.19	2.11
	.010	6.63	4.61	3.78	3.32	3.02	2.80	2.64	2.51	2.41

Denominator *df*	Probability of a larger *F*	Numerator *df* 10	12	15	20	24	30	40	60	120	∞
29	.100	1.83	1.78	1.73	1.68	1.65	1.62	1.58	1.55	1.51	1.47
	.050	2.18	2.10	2.03	1.94	1.90	1.85	1.81	1.75	1.70	1.64
	.025	2.53	2.43	2.32	2.21	2.15	2.09	2.03	1.96	1.89	1.81
	.010	3.00	2.87	2.73	2.57	2.49	2.41	2.33	2.23	2.14	2.03
30	.100	1.82	1.77	1.72	1.67	1.64	1.61	1.57	1.54	1.50	1.46
	.050	2.16	2.09	2.01	1.93	1.89	1.84	1.79	1.74	1.68	1.62
	.025	2.51	2.41	2.31	2.20	2.14	2.07	2.01	1.94	1.87	1.79
	.010	2.98	2.84	2.70	2.55	2.47	2.39	2.30	2.21	2.11	2.01
40	.100	1.76	1.71	1.66	1.61	1.57	1.54	1.51	1.47	1.42	1.38
	.050	2.08	2.00	1.92	1.84	1.79	1.74	1.69	1.64	1.58	1.51
	.025	2.39	2.29	2.18	2.07	2.01	1.94	1.88	1.80	1.72	1.64
	.010	2.80	2.66	2.52	2.37	2.29	2.20	2.11	2.02	1.92	1.80
60	.100	1.71	1.66	1.60	1.54	1.51	1.48	1.44	1.40	1.35	1.29
	.050	1.99	1.92	1.84	1.75	1.70	1.65	1.59	1.53	1.47	1.39
	.025	2.27	2.17	2.06	1.94	1.88	1.82	1.74	1.67	1.58	1.48
	.010	2.63	2.50	2.35	2.20	2.12	2.03	1.94	1.84	1.73	1.60
120	.100	1.65	1.60	1.55	1.48	1.45	1.41	1.37	1.32	1.26	1.19
	.050	1.91	1.83	1.75	1.66	1.61	1.55	1.50	1.43	1.35	1.25
	.025	2.16	2.05	1.94	1.82	1.76	1.69	1.61	1.53	1.43	1.31
	.010	2.47	2.34	2.19	2.03	1.95	1.86	1.76	1.66	1.53	1.38
∞	.100	1.60	1.55	1.49	1.42	1.38	1.34	1.30	1.24	1.17	1.00
	.050	1.83	1.75	1.67	1.57	1.52	1.46	1.39	1.32	1.22	1.00
	.025	2.05	1.94	1.83	1.71	1.64	1.57	1.48	1.39	1.27	1.00
	.010	2.32	2.18	2.04	1.88	1.79	1.70	1.59	1.47	1.32	1.00

Table A-6: Cumulative Binomial Distribution

$P_b(X \geq x \mid n, \pi)$

$n = 1$

x \ π	01	02	03	04	05	06	07	08	09	10
1	0100	0200	0300	0400	0500	0600	0700	0800	0900	1000

x \ π	11	12	13	14	15	16	17	18	19	20
1	1100	1200	1300	1400	1500	1600	1700	1800	1900	2000

x \ π	21	22	23	24	25	26	27	28	29	30
1	2100	2200	2300	2400	2500	2600	2700	2800	2900	3000

x \ π	31	32	33	34	35	36	37	38	39	40
1	3100	3200	3300	3400	3500	3600	3700	3800	3900	4000

x \ π	41	42	43	44	45	46	47	48	49	50
1	4100	4200	4300	4400	4500	4600	4700	4800	4900	5000

$n = 2$

x \ π	01	02	03	04	05	06	07	08	09	10
1	0199	0396	0591	0784	0975	1164	1351	1536	1719	1900
2	0001	0004	0009	0016	0025	0036	0049	0064	0081	0100

x \ π	11	12	13	14	15	16	17	18	19	20
1	2079	2256	2431	2604	2775	2944	3111	3276	3439	3600
2	0121	0144	0169	0196	0225	0256	0289	0324	0361	0400

x \ π	21	22	23	24	25	26	27	28	29	30
1	3759	3916	4071	4224	4375	4524	4671	4816	4959	5100
2	0441	0484	0529	0576	0625	0676	0729	0784	0841	0900

x \ π	31	32	33	34	35	36	37	38	39	40
1	5239	5376	5511	5644	5775	5904	6031	6156	6279	6400
2	0961	1024	1089	1156	1225	1296	1369	1444	1521	1600

x \ π	41	42	43	44	45	46	47	48	49	50
1	6519	6636	6751	6864	6975	7084	7191	7296	7399	7500
2	1681	1764	1849	1936	2025	2116	2209	2304	2401	2500

$n = 3$

x \ π	01	02	03	04	05	06	07	08	09	10
1	0297	0588	0873	1153	1426	1694	1956	2213	2464	2710
2	0003	0012	0026	0047	0073	0104	0140	0182	0228	0280
3				0001	0001	0002	0003	0005	0007	0010

x \ π	11	12	13	14	15	16	17	18	19	20
1	2950	3185	3415	3639	3859	4073	4282	4486	4686	4880
2	0336	0397	0463	0533	0608	0686	0769	0855	0946	1040
3	0013	0017	0022	0027	0034	0041	0049	0058	0069	0080

x \ π	21	22	23	24	25	26	27	28	29	30
1	5070	5254	5435	5610	5781	5948	6110	6268	6421	6570
2	1138	1239	1344	1452	1563	1676	1793	1913	2035	2160
3	0093	0106	0122	0138	0156	0176	0197	0220	0244	0270

Source: This table is abridged from "Tables of the Cumulative Binomial Probability Distribution," *Annals of the Computational Laboratory of Harvard University*, Vol. 35 (1955). Used by permission of the publisher, Harvard University Press.

x \ π	31	32	33	34	35	36	37	38	39	40
1	6715	6856	6992	7125	7254	7379	7500	7617	7730	7840
2	2287	2417	2548	2682	2818	2955	3094	3235	3377	3520
3	0298	0328	0359	0393	0429	0467	0507	0549	0593	0640

x \ π	41	42	43	44	45	46	47	48	49	50
1	7946	8049	8148	8244	8336	8425	8511	8594	8673	8750
2	3665	3810	3957	4104	4253	4401	4551	4700	4850	5000
3	0689	0741	0795	0852	0911	0973	1038	1106	1176	1250

$n = 4$

x \ π	01	02	03	04	05	06	07	08	09	10
1	0394	0776	1147	1507	1855	2193	2519	2836	3143	3439
2	0006	0023	0052	0091	0140	0199	0267	0344	0430	0523
3			0001	0002	0005	0008	0013	0019	0027	0037
4									0001	0001

x \ π	11	12	13	14	15	16	17	18	19	20
1	3726	4003	4271	4530	4780	5021	5254	5479	5695	5904
2	0624	0732	0847	0968	1095	1228	1366	1509	1656	1808
3	0049	0063	0079	0098	0120	0144	0171	0202	0235	0272
4	0001	0002	0003	0004	0005	0007	0008	0010	0013	0016

x \ π	21	22	23	24	25	26	27	28	29	30
1	6105	6298	6485	6664	6836	7001	7160	7313	7459	7599
2	1963	2122	2285	2450	2617	2787	2959	3132	3307	3483
3	0312	0356	0403	0453	0508	0566	0628	0694	0763	0837
4	0019	0023	0028	0033	0039	0046	0053	0061	0071	0081

x \ π	31	32	33	34	35	36	37	38	39	40
1	7733	7862	7985	8103	8215	8322	8425	8522	8615	8704
2	3660	3837	4015	4193	4370	4547	4724	4900	5075	5248
3	0915	0996	1082	1171	1265	1362	1464	1569	1679	1792
4	0092	0105	0119	0134	0150	0168	0187	0209	0231	0256

x \ π	41	42	43	44	45	46	47	48	49	50
1	8788	8868	8944	9017	9085	9150	9211	9269	9323	9375
2	5420	5590	5759	5926	6090	6252	6412	6569	6724	6875
3	1909	2030	2155	2283	2415	2550	2689	2831	2977	3125
4	0283	0311	0342	0375	0410	0448	0488	0531	0576	0625

$n = 5$

x \ π	01	02	03	04	05	06	07	08	09	10
1	0490	0961	1413	1846	2262	2661	3043	3409	3760	4095
2	0010	0038	0085	0148	0226	0319	0425	0544	0674	0815
3		0001	0003	0006	0012	0020	0031	0045	0063	0086
4						0001	0001	0002	0003	0005

x \ π	11	12	13	14	15	16	17	18	19	20
1	4416	4723	5016	5296	5563	5818	6061	6293	6513	6723
2	0965	1125	1292	1467	1648	1835	2027	2224	2424	2627
3	0112	0143	0179	0220	0266	0318	0375	0437	0505	0579
4	0007	0009	0013	0017	0022	0029	0036	0045	0055	0067
5				0001	0001	0001	0001	0002	0002	0003

x \ π	21	22	23	24	25	26	27	28	29	30
1	6923	7113	7293	7464	7627	7781	7927	8065	8196	8319
2	2833	3041	3251	3461	3672	3883	4093	4303	4511	4718
3	0659	0744	0836	0933	1035	1143	1257	1376	1501	1631
4	0081	0097	0114	0134	0156	0181	0208	0238	0272	0308
5	0004	0005	0006	0008	0010	0012	0014	0017	0021	0024

x \ π	31	32	33	34	35	36	37	38	39	40
1	8436	8546	8650	8748	8840	8926	9008	9084	9155	9222
2	4923	5125	5325	5522	5716	5906	6093	6276	6455	6630
3	1766	1905	2050	2199	2352	2509	2670	2835	3003	3174
4	0347	0390	0436	0486	0540	0598	0660	0726	0796	0870
5	0029	0034	0039	0045	0053	0060	0069	0079	0090	0102

x \ π	41	42	43	44	45	46	47	48	49	50
1	9285	9344	9398	9449	9497	9541	9582	9620	9655	9688
2	6801	6967	7129	7286	7438	7585	7728	7865	7998	8125
3	3349	3525	3705	3886	4069	4253	4439	4625	4813	5000
4	0949	1033	1121	1214	1312	1415	1522	1635	1753	1875
5	0116	0131	0147	0165	0185	0206	0229	0255	0282	0313

$n = 6$

x \ π	01	02	03	04	05	06	07	08	09	10
1	0585	1142	1670	2172	2649	3101	3530	3936	4321	4686
2	0015	0057	0125	0216	0328	0459	0608	0773	0952	1143
3		0002	0005	0012	0022	0038	0058	0085	0118	0159
4					0001	0002	0003	0005	0008	0013
5										0001

x \ π	11	12	13	14	15	16	17	18	19	20
1	5030	5356	5664	5954	6229	6487	6731	6960	7176	7379
2	1345	1556	1776	2003	2235	2472	2713	2956	3201	3446
3	0206	0261	0324	0395	0473	0560	0655	0759	0870	0989
4	0018	0025	0034	0045	0059	0075	0094	0116	0141	0170
5	0001	0001	0002	0003	0004	0005	0007	0010	0013	0016
6										0001

x \ π	21	22	23	24	25	26	27	28	29	30
1	7569	7748	7916	8073	8220	8358	8487	8607	8719	8824
2	3692	3937	4180	4422	4661	4896	5128	5356	5580	5798
3	1115	1250	1391	1539	1694	1856	2023	2196	2374	2557
4	0202	0239	0280	0326	0376	0431	0492	0557	0628	0705
5	0020	0025	0031	0038	0046	0056	0067	0079	0093	0109
6	0001	0001	0001	0002	0002	0003	0004	0005	0006	0007

x \ π	31	32	33	34	35	36	37	38	39	40
1	8921	9011	9095	9173	9246	9313	9375	9432	9485	9533
2	6012	6220	6422	6619	6809	6994	7172	7343	7508	7667
3	2744	2936	3130	3328	3529	3732	3937	4143	4350	4557
4	0787	0875	0969	1069	1174	1286	1404	1527	1657	1792
5	0127	0148	0170	0195	0223	0254	0288	0325	0365	0410
6	0009	0011	0013	0015	0018	0022	0026	0030	0035	0041

x \ π	41	42	43	44	45	46	47	48	49	50
1	9578	9619	9657	9692	9723	9752	9778	9802	9824	9844
2	7819	7965	8105	8238	8364	8485	8599	8707	8810	8906
3	4764	4971	5177	5382	5585	5786	5985	6180	6373	6563
4	1933	2080	2232	2390	2553	2721	2893	3070	3252	3438
5	0458	0510	0566	0627	0692	0762	0837	0917	1003	1094
6	0048	0055	0063	0073	0083	0095	0108	0122	0138	0156

$n = 7$

x \ π	01	02	03	04	05	06	07	08	09	10
1	0679	1319	1920	2486	3017	3515	3983	4422	4832	5217
2	0020	0079	0171	0294	0444	0618	0813	1026	1255	1497
3		0003	0009	0020	0038	0063	0097	0140	0193	0257
4				0001	0002	0004	0007	0012	0018	0027
5								0001	0001	0002

x \ π	11	12	13	14	15	16	17	18	19	20
1	5577	5913	6227	6521	6794	7049	7286	7507	7712	7903
2	1750	2012	2281	2556	2834	3115	3396	3677	3956	4233
3	0331	0416	0513	0620	0738	0866	1005	1154	1313	1480
4	0039	0054	0072	0094	0121	0153	0189	0231	0279	0333
5	0003	0004	0006	0009	0012	0017	0022	0029	0037	0047
6					0001	0001	0001	0002	0003	0004

x \ π	21	22	23	24	25	26	27	28	29	30
1	8080	8243	8395	8535	8665	8785	8895	8997	9090	9176
2	4506	4775	5040	5298	5551	5796	6035	6266	6490	6706
3	1657	1841	2033	2231	2436	2646	2861	3081	3304	3529
4	0394	0461	0536	0617	0706	0802	0905	1016	1134	1260
5	0058	0072	0088	0107	0129	0153	0181	0213	0248	0288
6	0005	0006	0008	0011	0013	0017	0021	0026	0031	0038
7					0001	0001	0001	0001	0002	0002

x \ π	31	32	33	34	35	36	37	38	39	40
1	9255	9328	9394	9454	9510	9560	9606	9648	9686	9720
2	6914	7113	7304	7487	7662	7828	7987	8137	8279	8414
3	3757	3987	4217	4447	4677	4906	5134	5359	5581	5801
4	1394	1534	1682	1837	1998	2167	2341	2521	2707	2898
5	0332	0380	0434	0492	0556	0625	0701	0782	0869	0963
6	0046	0055	0065	0077	0090	0105	0123	0142	0164	0188
7	0003	0003	0004	0005	0006	0008	0009	0011	0014	0016

x \ π	41	42	43	44	45	46	47	48	49	50
1	9751	9779	9805	9827	9848	9866	9883	9897	9910	9922
2	8541	8660	8772	8877	8976	9068	9153	9233	9307	9375
3	6017	6229	6436	6638	6836	7027	7213	7393	7567	7734
4	3094	3294	3498	3706	3917	4131	4346	4563	4781	5000
5	1063	1169	1282	1402	1529	1663	1803	1951	2105	2266
6	0216	0246	0279	0316	0357	0402	0451	0504	0562	0625
7	0019	0023	0027	0032	0037	0044	0051	0059	0068	0078

$n = 8$

x \ π	01	02	03	04	05	06	07	08	09	10
1	0773	1492	2163	2786	3366	3904	4404	4868	5297	5695
2	0027	0103	0223	0381	0572	0792	1035	1298	1577	1869
3	0001	0004	0013	0031	0058	0096	0147	0211	0289	0381
4			0001	0002	0004	0007	0013	0022	0034	0050
5							0001	0001	0003	0004

x \ π	11	12	13	14	15	16	17	18	19	20
1	6063	6404	6718	7008	7275	7521	7748	7956	8147	8322
2	2171	2480	2794	3111	3428	3744	4057	4366	4670	4967
3	0487	0608	0743	0891	1052	1226	1412	1608	1815	2031
4	0071	0097	0129	0168	0214	0267	0328	0397	0476	0563
5	0007	0010	0015	0021	0029	0038	0050	0065	0083	0104
6		0001	0001	0002	0002	0003	0005	0007	0009	0012
7									0001	0001

x \ π	21	22	23	24	25	26	27	28	29	30
1	8483	8630	8764	8887	8999	9101	9194	9278	9354	9424
2	5257	5538	5811	6075	6329	6573	6807	7031	7244	7447
3	2255	2486	2724	2967	3215	3465	3718	3973	4228	4482
4	0659	0765	0880	1004	1138	1281	1433	1594	1763	1941
5	0129	0158	0191	0230	0273	0322	0377	0438	0505	0580
6	0016	0021	0027	0034	0042	0052	0064	0078	0094	0113
7	0001	0002	0002	0003	0004	0005	0006	0008	0010	0013
8									0001	0001

x \ π	31	32	33	34	35	36	37	38	39	40
1	9486	9543	9594	9640	9681	9719	9752	9782	9808	9832
2	7640	7822	7994	8156	8309	8452	8586	8711	8828	8936
3	4736	4987	5236	5481	5722	5958	6189	6415	6634	6846
4	2126	2319	2519	2724	2936	3153	3374	3599	3828	4059
5	0661	0750	0846	0949	1061	1180	1307	1443	1586	1737
6	0134	0159	0187	0218	0253	0293	0336	0385	0439	0498
7	0016	0020	0024	0030	0036	0043	0051	0061	0072	0085
8	0001	0001	0001	0002	0002	0003	0004	0004	0005	0007

x \ π	41	42	43	44	45	46	47	48	49	50
1	9853	9872	9889	9903	9916	9928	9938	9947	9954	9961
2	9037	9130	9216	9295	9368	9435	9496	9552	9602	9648
3	7052	7250	7440	7624	7799	7966	8125	8276	8419	8555
4	4292	4527	4762	4996	5230	5463	5694	5922	6146	6367
5	1895	2062	2235	2416	2604	2798	2999	3205	3416	3633
6	0563	0634	0711	0794	0885	0982	1086	1198	1318	1445
7	0100	0117	0136	0157	0181	0208	0239	0272	0310	0352
8	0008	0010	0012	0014	0017	0020	0024	0028	0033	0039

$n = 9$

x \ π	01	02	03	04	05	06	07	08	09	10
1	0865	1663	2398	3075	3698	4270	4796	5278	5721	6126
2	0034	0131	0282	0478	0712	0978	1271	1583	1912	2252
3	0001	0006	0020	0045	0084	0138	0209	0298	0405	0530
4			0001	0003	0006	0013	0023	0037	0057	0083
5						0001	0002	0003	0005	0009
6										0001

x \ π	11	12	13	14	15	16	17	18	19	20
1	6496	6835	7145	7427	7684	7918	8131	8324	8499	8658
2	2599	2951	3304	3657	4005	4348	4685	5012	5330	5638
3	0672	0833	1009	1202	1409	1629	1861	2105	2357	2618
4	0117	0158	0209	0269	0339	0420	0512	0615	0730	0856
5	0014	0021	0030	0041	0056	0075	0098	0125	0158	0196
6	0001	0002	0003	0004	0006	0009	0013	0017	0023	0031
7						0001	0001	0002	0002	0003

x \ π	21	22	23	24	25	26	27	28	29	30
1	8801	8931	9048	9154	9249	9335	9411	9480	9542	9596
2	5934	6218	6491	6750	6997	7230	7452	7660	7856	8040
3	2885	3158	3434	3713	3993	4273	4552	4829	5102	5372
4	0994	1144	1304	1475	1657	1849	2050	2260	2478	2703
5	0240	0291	0350	0416	0489	0571	0662	0762	0870	0988
6	0040	0051	0065	0081	0100	0122	0149	0179	0213	0253
7	0004	0006	0008	0010	0013	0017	0022	0028	0035	0043
8			0001	0001	0001	0001	0002	0003	0003	0004

x \ π	31	32	33	34	35	36	37	38	39	40
1	9645	9689	9728	9762	9793	9820	9844	9865	9883	9899
2	8212	8372	8522	8661	8789	8908	9017	9118	9210	9295
3	5636	5894	6146	6390	6627	6856	7076	7287	7489	7682
4	2935	3173	3415	3662	3911	4163	4416	4669	4922	5174
5	1115	1252	1398	1553	1717	1890	2072	2262	2460	2666
6	0298	0348	0404	0467	0536	0612	0696	0787	0886	0994
7	0053	0064	0078	0094	0112	0133	0157	0184	0215	0250
8	0006	0007	0009	0011	0014	0017	0021	0026	0031	0038
9				0001	0001	0001	0001	0002	0002	0003

x \ π	41	42	43	44	45	46	47	48	49	50
1	9913	9926	9936	9946	9954	9961	9967	9972	9977	9980
2	9372	9442	9505	9563	9615	9662	9704	9741	9775	9805
3	7866	8039	8204	8359	8505	8642	8769	8889	8999	9102
4	5424	5670	5913	6152	6386	6614	6836	7052	7260	7461
5	2878	3097	3322	3551	3786	4024	4265	4509	4754	5000
6	1109	1233	1366	1508	1658	1817	1985	2161	2346	2539
7	0290	0334	0383	0437	0498	0564	0637	0717	0804	0898
8	0046	0055	0065	0077	0091	0107	0125	0145	0169	0195
9	0003	0004	0005	0006	0008	0009	0011	0014	0016	0020

$n = 10$

x \ π	01	02	03	04	05	06	07	08	09	10
1	0956	1829	2626	3352	4013	4614	5160	5656	6106	6513
2	0043	0162	0345	0582	0861	1176	1517	1879	2254	2639
3	0001	0009	0028	0062	0115	0188	0283	0401	0540	0702
4			0001	0004	0010	0020	0036	0058	0088	0128
5					0001	0002	0003	0006	0010	0016
6									0001	0001

x \ π	11	12	13	14	15	16	17	18	19	20
1	6882	7215	7516	7787	8031	8251	8448	8626	8784	8926
2	3028	3417	3804	4184	4557	4920	5270	5608	5932	6242
3	0884	1087	1308	1545	1798	2064	2341	2628	2922	3222
4	0178	0239	0313	0400	0500	0614	0741	0883	1039	1209
5	0025	0037	0053	0073	0099	0130	0168	0213	0266	0328
6	0003	0004	0006	0010	0014	0020	0027	0037	0049	0064
7			0001	0001	0001	0002	0003	0004	0006	0009
8									0001	0001

x \ π	21	22	23	24	25	26	27	28	29	30
1	9053	9166	9267	9357	9437	9508	9570	9626	9674	9718
2	6536	6815	7079	7327	7560	7778	7981	8170	8345	8507
3	3526	3831	4137	4442	4744	5042	5335	5622	5901	6172
4	1391	1587	1794	2012	2241	2479	2726	2979	3239	3504
5	0399	0479	0569	0670	0781	0904	1037	1181	1337	1503
6	0082	0104	0130	0161	0197	0239	0287	0342	0404	0473
7	0012	0016	0021	0027	0035	0045	0056	0070	0087	0106
8	0001	0002	0002	0003	0004	0006	0007	0010	0012	0016
9							0001	0001	0001	0001

x \ π	31	32	33	34	35	36	37	38	39	40
1	9755	9789	9818	9843	9865	9885	9902	9916	9929	9940
2	8656	8794	8920	9035	9140	9236	9323	9402	9473	9536
3	6434	6687	6930	7162	7384	7595	7794	7983	8160	8327
4	3772	4044	4316	4589	4862	5132	5400	5664	5923	6177
5	1679	1867	2064	2270	2485	2708	2939	3177	3420	3669
6	0551	0637	0732	0836	0949	1072	1205	1348	1500	1662
7	0129	0155	0185	0220	0260	0305	0356	0413	0477	0548
8	0020	0025	0032	0039	0048	0059	0071	0086	0103	0123
9	0002	0003	0003	0004	0005	0007	0009	0011	0014	0017
10								0001	0001	0001

x \ π	41	42	43	44	45	46	47	48	49	50
1	9949	9957	9964	9970	9975	9979	9983	9986	9988	9990
2	9594	9645	9691	9731	9767	9799	9827	9852	9874	9893
3	8483	8628	8764	8889	9004	9111	9209	9298	9379	9453
4	6425	6665	6898	7123	7340	7547	7745	7933	8112	8281
5	3922	4178	4436	4696	4956	5216	5474	5730	5982	6230
6	1834	2016	2207	2407	2616	2832	3057	3288	3526	3770
7	0626	0712	0806	0908	1020	1141	1271	1410	1560	1719
8	0146	0172	0202	0236	0274	0317	0366	0420	0480	0547
9	0021	0025	0031	0037	0045	0054	0065	0077	0091	0107
10	0001	0002	0002	0003	0003	0004	0005	0006	0008	0010

$n = 11$

x \ π	01	02	03	04	05	06	07	08	09	10
1	1047	1993	2847	3618	4312	4937	5499	6004	6456	6862
2	0052	0195	0413	0692	1019	1382	1772	2181	2601	3026
3	0002	0012	0037	0083	0152	0248	0370	0519	0695	0896
4			0002	0007	0016	0030	0053	0085	0129	0185
5					0001	0003	0005	0010	0017	0028
6								0001	0002	0003

x \ π	11	12	13	14	15	16	17	18	19	20
1	7225	7549	7839	8097	8327	8531	8712	8873	9015	9141
2	3452	3873	4286	4689	5078	5453	5811	6151	6474	6779
3	1120	1366	1632	1915	2212	2521	2839	3164	3494	3826
4	0256	0341	0442	0560	0694	0846	1013	1197	1397	1611
5	0042	0061	0087	0119	0159	0207	0266	0334	0413	0504
6	0005	0008	0013	0018	0027	0037	0051	0068	0090	0117
7		0001	0001	0002	0003	0005	0007	0010	0014	0020
8							0001	0001	0002	0002

x \ π	21	22	23	24	25	26	27	28	29	30
1	9252	9350	9436	9511	9578	9636	9686	9730	9769	9802
2	7065	7333	7582	7814	8029	8227	8410	8577	8730	8870
3	4158	4488	4814	5134	5448	5753	6049	6335	6610	6873
4	1840	2081	2333	2596	2867	3146	3430	3719	4011	4304
5	0607	0723	0851	0992	1146	1313	1493	1685	1888	2103
6	0148	0186	0231	0283	0343	0412	0490	0577	0674	0782
7	0027	0035	0046	0059	0076	0095	0119	0146	0179	0216
8	0003	0005	0007	0009	0012	0016	0021	0027	0034	0043
9			0001	0001	0001	0002	0002	0003	0004	0006

x \ π	31	32	33	34	35	36	37	38	39	40
1	9831	9856	9878	9896	9912	9926	9938	9948	9956	9964
2	8997	9112	9216	9310	9394	9470	9537	9597	9650	9698
3	7123	7361	7587	7799	7999	8186	8360	8522	8672	8811
4	4598	4890	5179	5464	5744	6019	6286	6545	6796	7037
5	2328	2563	2807	3059	3317	3581	3850	4122	4397	4672
6	0901	1031	1171	1324	1487	1661	1847	2043	2249	2465
7	0260	0309	0366	0430	0501	0581	0670	0768	0876	0994
8	0054	0067	0082	0101	0122	0148	0177	0210	0249	0293
9	0008	0010	0013	0016	0020	0026	0032	0039	0048	0059
10	0001	0001	0001	0002	0002	0003	0004	0005	0006	0007

x \ π	41	42	43	44	45	46	47	48	49	50
1	9970	9975	9979	9983	9986	9989	9991	9992	9994	9995
2	9739	9776	9808	9836	9861	9882	9900	9916	9930	9941
3	8938	9055	9162	9260	9348	9428	9499	9564	9622	9673
4	7269	7490	7700	7900	8089	8266	8433	8588	8733	8867
5	4948	5223	5495	5764	6029	6288	6541	6787	7026	7256
6	2690	2924	3166	3414	3669	3929	4193	4460	4729	5000
7	1121	1260	1408	1568	1738	1919	2110	2312	2523	2744
8	0343	0399	0461	0532	0610	0696	0791	0895	1009	1133
9	0072	0087	0104	0125	0148	0175	0206	0241	0282	0327
10	0009	0012	0014	0018	0022	0027	0033	0040	0049	0059
11	0001	0001	0001	0001	0002	0002	0002	0003	0004	0005

$n = 12$

x \ π	01	02	03	04	05	06	07	08	09	10
1	1136	2153	3062	3873	4596	5241	5814	6323	6775	7176
2	0062	0231	0486	0809	1184	1595	2033	2487	2948	3410
3	0002	0015	0048	0107	0196	0316	0468	0652	0866	1109
4		0001	0003	0010	0022	0043	0075	0120	0180	0256
5				0001	0002	0004	0009	0016	0027	0043
6							0001	0002	0003	0005
7										0001

x \ π	11	12	13	14	15	16	17	18	19	20
1	7530	7843	8120	8363	8578	8766	8931	9076	9202	9313
2	3867	4314	4748	5166	5565	5945	6304	6641	6957	7251
3	1377	1667	1977	2303	2642	2990	3344	3702	4060	4417
4	0351	0464	0597	0750	0922	1114	1324	1552	1795	2054
5	0065	0095	0133	0181	0239	0310	0393	0489	0600	0726
6	0009	0014	0022	0033	0046	0065	0088	0116	0151	0194
7	0001	0002	0003	0004	0007	0010	0015	0021	0029	0039
8					0001	0001	0002	0003	0004	0006
9										0001

x \ π	21	22	23	24	25	26	27	28	29	30
1	9409	9493	9566	9629	9683	9730	9771	9806	9836	9862
2	7524	7776	8009	8222	8416	8594	8755	8900	9032	9150
3	4768	5114	5450	5778	6093	6397	6687	6963	7225	7472
4	2326	2610	2904	3205	3512	3824	4137	4452	4765	5075
5	0866	1021	1192	1377	1576	1790	2016	2254	2504	2763
6	0245	0304	0374	0453	0544	0646	0760	0887	1026	1178
7	0052	0068	0089	0113	0143	0178	0219	0267	0322	0386
8	0008	0011	0016	0021	0028	0036	0047	0060	0076	0095
9	0001	0001	0002	0003	0004	0005	0007	0010	0013	0017
10						0001	0001	0001	0002	0002

x \ π	31	32	33	34	35	36	37	38	39	40
1	9884	9902	9918	9932	9943	9953	9961	9968	9973	9978
2	9256	9350	9435	9509	9576	9634	9685	9730	9770	9804
3	7704	7922	8124	8313	8487	8648	8795	8931	9054	9166
4	5381	5681	5973	6258	6533	6799	7053	7296	7528	7747
5	3032	3308	3590	3876	4167	4459	4751	5043	5332	5618
6	1343	1521	1711	1913	2127	2352	2588	2833	3087	3348
7	0458	0540	0632	0734	0846	0970	1106	1253	1411	1582
8	0118	0144	0176	0213	0255	0304	0359	0422	0493	0573
9	0022	0028	0036	0045	0056	0070	0086	0104	0127	0153
10	0003	0004	0005	0007	0008	0011	0014	0018	0022	0028
11				0001	0001	0001	0001	0002	0002	0003

x \ π	41	42	43	44	45	46	47	48	49	50
1	9982	9986	9988	9990	9992	9994	9995	9996	9997	9998
2	9834	9860	9882	9901	9917	9931	9943	9953	9961	9968
3	9267	9358	9440	9513	9579	9637	9688	9733	9773	9807
4	7953	8147	8329	8498	8655	8801	8934	9057	9168	9270
5	5899	6175	6443	6704	6956	7198	7430	7652	7862	8062
6	3616	3889	4167	4448	4731	5014	5297	5577	5855	6128
7	1765	1959	2164	2380	2607	2843	3089	3343	3604	3872
8	0662	0760	0869	0988	1117	1258	1411	1575	1751	1938
9	0183	0218	0258	0304	0356	0415	0481	0555	0638	0730
10	0035	0043	0053	0065	0079	0095	0114	0137	0163	0193
11	0004	0005	0007	0009	0011	0014	0017	0021	0026	0032
12				0001	0001	0001	0001	0001	0002	0002

$n = 13$

x \ π	01	02	03	04	05	06	07	08	09	10
1	1225	2310	3270	4118	4867	5526	6107	6617	7065	7458
2	0072	0270	0564	0932	1354	1814	2298	2794	3293	3787
3	0003	0020	0062	0135	0245	0392	0578	0799	1054	1339
4		0001	0005	0014	0031	0060	0103	0163	0242	0342
5				0001	0003	0007	0013	0024	0041	0065
6						0001	0001	0003	0005	0009
7									0001	0001

x \ π	11	12	13	14	15	16	17	18	19	20
1	7802	8102	8364	8592	8791	8963	9113	9242	9354	9450
2	4270	4738	5186	5614	6017	6396	6751	7080	7384	7664
3	1651	1985	2337	2704	3080	3463	3848	4231	4611	4983
4	0464	0609	0776	0967	1180	1414	1667	1939	2226	2527
5	0097	0139	0193	0260	0342	0438	0551	0681	0827	0991
6	0015	0024	0036	0053	0075	0104	0139	0183	0237	0300
7	0002	0003	0005	0008	0013	0019	0027	0038	0052	0070
8			0001	0001	0002	0003	0004	0006	0009	0012
9								0001	0001	0002

x \ π	21	22	23	24	25	26	27	28	29	30
1	9533	9604	9666	9718	9762	9800	9833	9860	9883	9903
2	7920	8154	8367	8559	8733	8889	9029	9154	9265	9363
3	5347	5699	6039	6364	6674	6968	7245	7505	7749	7975
4	2839	3161	3489	3822	4157	4493	4826	5155	5478	5794
5	1173	1371	1585	1816	2060	2319	2589	2870	3160	3457
6	0375	0462	0562	0675	0802	0944	1099	1270	1455	1654
7	0093	0120	0154	0195	0243	0299	0365	0440	0527	0624
8	0017	0024	0032	0043	0056	0073	0093	0118	0147	0182
9	0002	0004	0005	0007	0010	0013	0018	0024	0031	0040
10			0001	0001	0001	0002	0003	0004	0005	0007
11									0001	0001

x \ π	31	32	33	34	35	36	37	38	39	40
1	9920	9934	9945	9955	9963	9970	9975	9980	9984	9987
2	9450	9527	9594	9653	9704	9749	9787	9821	9849	9874
3	8185	8379	8557	8720	8868	9003	9125	9235	9333	9421
4	6101	6398	6683	6957	7217	7464	7698	7917	8123	8314
5	3760	4067	4376	4686	4995	5301	5603	5899	6188	6470
6	1867	2093	2331	2581	2841	3111	3388	3673	3962	4256
7	0733	0854	0988	1135	1295	1468	1654	1853	2065	2288
8	0223	0271	0326	0390	0462	0544	0635	0738	0851	0977
9	0052	0065	0082	0102	0126	0154	0187	0225	0270	0321
10	0009	0012	0015	0020	0025	0032	0040	0051	0063	0078
11	0001	0001	0002	0003	0003	0005	0006	0008	0010	0013
12							0001	0001	0001	0001

x \ π	41	42	43	44	45	46	47	48	49	50
1	9990	9992	9993	9995	9996	9997	9997	9998	9998	9999
2	9895	9912	9928	9940	9951	9960	9967	9974	9979	9983
3	9499	9569	9630	9684	9731	9772	9808	9838	9865	9888
4	8492	8656	8807	8945	9071	9185	9288	9381	9464	9539
5	6742	7003	7254	7493	7721	7935	8137	8326	8502	8666
6	4552	4849	5146	5441	5732	6019	6299	6573	6838	7095
7	2524	2770	3025	3290	3563	3842	4127	4415	4707	5000
8	1114	1264	1426	1600	1788	1988	2200	2424	2659	2905
9	0379	0446	0520	0605	0698	0803	0918	1045	1183	1334
10	0096	0117	0141	0170	0203	0242	0287	0338	0396	0461
11	0017	0021	0027	0033	0041	0051	0063	0077	0093	0112
12	0002	0002	0003	0004	0005	0007	0009	0011	0014	0017
13							0001	0001	0001	0001

$n = 14$

x \ π	01	02	03	04	05	06	07	08	09	10
1	1313	2464	3472	4353	5123	5795	6380	6888	7330	7712
2	0084	0310	0645	1059	1530	2037	2564	3100	3632	4154
3	0003	0025	0077	0167	0301	0478	0698	0958	1255	1584
4		0001	0006	0019	0042	0080	0136	0214	0315	0441
5				0002	0004	0010	0020	0035	0059	0092
6						0001	0002	0004	0008	0015
7									0001	0002

x \ π	11	12	13	14	15	16	17	18	19	20
1	8044	8330	8577	8789	8972	9129	9264	9379	9477	9560
2	4658	5141	5599	6031	6433	6807	7152	7469	7758	8021
3	1939	2315	2708	3111	3521	3932	4341	4744	5138	5519
4	0594	0774	0979	1210	1465	1742	2038	2351	2679	3018
5	0137	0196	0269	0359	0467	0594	0741	0907	1093	1298
6	0024	0038	0057	0082	0115	0157	0209	0273	0349	0439
7	0003	0006	0009	0015	0022	0032	0046	0064	0087	0116
8		0001	0001	0002	0003	0005	0008	0012	0017	0024
9						0001	0001	0002	0003	0004

x \ π	21	22	23	24	25	26	27	28	29	30
1	9631	9691	9742	9786	9822	9852	9878	9899	9917	9932
2	8259	8473	8665	8837	8990	9126	9246	9352	9444	9525
3	5887	6239	6574	6891	7189	7467	7727	7967	8188	8392
4	3366	3719	4076	4432	4787	5136	5479	5813	6137	6448
5	1523	1765	2023	2297	2585	2884	3193	3509	3832	4158
6	0543	0662	0797	0949	1117	1301	1502	1718	1949	2195
7	0152	0196	0248	0310	0383	0467	0563	0673	0796	0933
8	0033	0045	0060	0079	0103	0132	0167	0208	0257	0315
9	0006	0008	0011	0016	0022	0029	0038	0050	0065	0083
10	0001	0001	0002	0002	0003	0005	0007	0009	0012	0017
11						0001	0001	0001	0002	0002

x \ π	31	32	33	34	35	36	37	38	39	40
1	9945	9955	9963	9970	9976	9981	9984	9988	9990	9992
2	9596	9657	9710	9756	9795	9828	9857	9881	9902	9919
3	8577	8746	8899	9037	9161	9271	9370	9457	9534	9602
4	6747	7032	7301	7556	7795	8018	8226	8418	8595	8757
5	4486	4813	5138	5458	5773	6080	6378	6666	6943	7207
6	2454	2724	3006	3297	3595	3899	4208	4519	4831	5141
7	1084	1250	1431	1626	1836	2059	2296	2545	2805	3075
8	0381	0458	0545	0643	0753	0876	1012	1162	1325	1501
9	0105	0131	0163	0200	0243	0294	0353	0420	0497	0583
10	0022	0029	0037	0048	0060	0076	0095	0117	0144	0175
11	0003	0005	0006	0008	0011	0014	0019	0024	0031	0039
12		0001	0001	0001	0001	0002	0003	0003	0005	0006
13										0001

x \ π	41	42	43	44	45	46	47	48	49	50
1	9994	9995	9996	9997	9998	9998	9999	9999	9999	9999
2	9934	9946	9956	9964	9971	9977	9981	9985	9988	9991
3	9661	9713	9758	9797	9830	9858	9883	9903	9921	9935
4	8905	9039	9161	9270	9368	9455	9532	9601	9661	9713
5	7459	7697	7922	8132	8328	8510	8678	8833	8974	9102
6	5450	5754	6052	6344	6627	6900	7163	7415	7654	7880
7	3355	3643	3937	4236	4539	4843	5148	5451	5751	6047
8	1692	1896	2113	2344	2586	2840	3105	3380	3663	3953
9	0680	0789	0910	1043	1189	1348	1520	1707	1906	2120
10	0212	0255	0304	0361	0426	0500	0583	0677	0782	0898
11	0049	0061	0076	0093	0114	0139	0168	0202	0241	0287
12	0008	0010	0013	0017	0022	0027	0034	0042	0053	0065
13	0001	0001	0001	0002	0003	0003	0004	0006	0007	0009
14										0001

$n = 15$

x \ π	01	02	03	04	05	06	07	08	09	10
1	1399	2614	3667	4579	5367	6047	6633	7137	7570	7941
2	0096	0353	0730	1191	1710	2262	2832	3403	3965	4510
3	0004	0030	0094	0203	0362	0571	0829	1130	1469	1841
4		0002	0008	0024	0055	0104	0175	0273	0399	0556
5			0001	0002	0006	0014	0028	0050	0082	0127
6					0001	0001	0003	0007	0013	0022
7								0001	0002	0003

x \ π	11	12	13	14	15	16	17	18	19	20
1	8259	8530	8762	8959	9126	9269	9389	9490	9576	9648
2	5031	5524	5987	6417	6814	7179	7511	7813	8085	8329
3	2238	2654	3084	3520	3958	4392	4819	5234	5635	6020
4	0742	0959	1204	1476	1773	2092	2429	2782	3146	3518
5	0187	0265	0361	0478	0617	0778	0961	1167	1394	1642
6	0037	0057	0084	0121	0168	0227	0300	0387	0490	0611
7	0006	0010	0015	0024	0036	0052	0074	0102	0137	0181
8	0001	0001	0002	0004	0006	0010	0014	0021	0030	0042
9					0001	0001	0002	0003	0005	0008
10									0001	0001

x \ π	21	22	23	24	25	26	27	28	29	30
1	9709	9759	9802	9837	9866	9891	9911	9928	9941	9953
2	8547	8741	8913	9065	9198	9315	9417	9505	9581	9647
3	6385	6731	7055	7358	7639	7899	8137	8355	8553	8732
4	3895	4274	4650	5022	5387	5742	6086	6416	6732	7031
5	1910	2195	2495	2810	3135	3469	3810	4154	4500	4845
6	0748	0905	1079	1272	1484	1713	1958	2220	2495	2784
7	0234	0298	0374	0463	0566	0684	0817	0965	1130	1311
8	0058	0078	0104	0135	0173	0219	0274	0338	0413	0500
9	0011	0016	0023	0031	0042	0056	0073	0094	0121	0152
10	0002	0003	0004	0006	0008	0011	0015	0021	0028	0037
11			0001	0001	0001	0002	0002	0003	0005	0007
12									0001	0001

x \ π	31	32	33	34	35	36	37	38	39	40
1	9962	9969	9975	9980	9984	9988	9990	9992	9994	9995
2	9704	9752	9794	9829	9858	9883	9904	9922	9936	9948
3	8893	9038	9167	9281	9383	9472	9550	9618	9678	9729
4	7314	7580	7829	8060	8273	8469	8649	8813	8961	9095
5	5187	5523	5852	6171	6481	6778	7062	7332	7587	7827
6	3084	3393	3709	4032	4357	4684	5011	5335	5654	5968
7	1509	1722	1951	2194	2452	2722	3003	3295	3595	3902
8	0599	0711	0837	0977	1132	1302	1487	1687	1902	2131
9	0190	0236	0289	0351	0422	0504	0597	0702	0820	0950
10	0048	0062	0079	0099	0124	0154	0190	0232	0281	0338
11	0009	0012	0016	0022	0028	0037	0047	0059	0075	0093
12	0001	0002	0003	0004	0005	0006	0009	0011	0015	0019
13					0001	0001	0001	0002	0002	0003

x \ π	41	42	43	44	45	46	47	48	49	50
1	9996	9997	9998	9998	9999	9999	9999	9999	10000	10000
2	9958	9966	9973	9979	9983	9987	9990	9992	9994	9995
3	9773	9811	9843	9870	9893	9913	9929	9943	9954	9963
4	9215	9322	9417	9502	9576	9641	9697	9746	9788	9824
5	8052	8261	8454	8633	8796	8945	9080	9201	9310	9408
6	6274	6570	6856	7131	7392	7641	7875	8095	8301	8491
7	4214	4530	4847	5164	5478	5789	6095	6394	6684	6964
8	2374	2630	2898	3176	3465	3762	4065	4374	4686	5000
9	1095	1254	1427	1615	1818	2034	2265	2510	2767	3036
10	0404	0479	0565	0661	0769	0890	1024	1171	1333	1509
11	0116	0143	0174	0211	0255	0305	0363	0430	0506	0592
12	0025	0032	0040	0051	0063	0079	0097	0119	0145	0176
13	0004	0005	0007	0009	0011	0014	0018	0023	0029	0037
14			0001	0001	0001	0002	0002	0003	0004	0005

$n = 16$

x \ π	01	02	03	04	05	06	07	08	09	10
1	1485	2762	3857	4796	5599	6284	6869	7366	7789	8147
2	0109	0399	0818	1327	1892	2489	3098	3701	4289	4853
3	0005	0037	0113	0242	0429	0673	0969	1311	1694	2108
4		0002	0011	0032	0070	0132	0221	0342	0496	0684
5			0001	0003	0009	0019	0038	0068	0111	0170
6					0001	0002	0005	0010	0019	0033
7							0001	0001	0003	0005
8										0001

x \ π	11	12	13	14	15	16	17	18	19	20
1	8450	8707	8923	9105	9257	9386	9493	9582	9657	9719
2	5386	5885	6347	6773	7161	7513	7830	8115	8368	8593
3	2545	2999	3461	3926	4386	4838	5277	5698	6101	6482
4	0907	1162	1448	1763	2101	2460	2836	3223	3619	4019
5	0248	0348	0471	0618	0791	0988	1211	1458	1727	2018
6	0053	0082	0120	0171	0235	0315	0412	0527	0662	0817
7	0009	0015	0024	0038	0056	0080	0112	0153	0204	0257
8	0001	0002	0004	0007	0011	0016	0024	0036	0051	0070
9			0001	0001	0002	0003	0004	0007	0010	0015
10							0001	0001	0002	0002

x \ π	21	22	23	24	25	26	27	28	29	30
1	9770	9812	9847	9876	9900	9919	9935	9948	9958	9967
2	8791	8965	9117	9250	9365	9465	9550	9623	9686	9739
3	6839	7173	7483	7768	8029	8267	8482	8677	8851	9006
4	4418	4814	5203	5583	5950	6303	6640	6959	7260	7541
5	2327	2652	2991	3341	3698	4060	4425	4788	5147	5501
6	0992	1188	1405	1641	1897	2169	2458	2761	3077	3402
7	0342	0432	0536	0657	0796	0951	1125	1317	1526	1753
8	0095	0127	0166	0214	0271	0340	0420	0514	0621	0744
9	0021	0030	0041	0056	0075	0098	0127	0163	0206	0257
10	0004	0006	0008	0012	0016	0023	0031	0041	0055	0071
11	0001	0001	0001	0002	0003	0004	0006	0008	0011	0016
12						0001	0001	0001	0002	0003

x \ π	31	32	33	34	35	36	37	38	39	40
1	9974	9979	9984	9987	9990	9992	9994	9995	9996	9997
2	9784	9822	9854	9880	9902	9921	9936	9948	9959	9967
3	9144	9266	9374	9467	9549	9620	9681	9734	9778	9817
4	7804	8047	8270	8475	8661	8830	8982	9119	9241	9349
5	5846	6181	6504	6813	7108	7387	7649	7895	8123	8334
6	3736	4074	4416	4759	5100	5438	5770	6094	6408	6712
7	1997	2257	2531	2819	3119	3428	3746	4070	4398	4728
8	0881	1035	1205	1391	1594	1813	2048	2298	2562	2839
9	0317	0388	0470	0564	0671	0791	0926	1076	1242	1423
10	0092	0117	0148	0185	0229	0280	0341	0411	0491	0583
11	0021	0028	0037	0048	0062	0079	0100	0125	0155	0191
12	0004	0005	0007	0010	0013	0017	0023	0030	0038	0049
13		0001	0001	0001	0002	0003	0004	0005	0007	0009
14								0001	0001	0001

x \ π	41	42	43	44	45	46	47	48	49	50
1	9998	9998	9999	9999	9999	9999	10000	10000	10000	10000
2	9974	9979	9984	9987	9990	9992	9994	9995	9997	9997
3	9849	9876	9899	9918	9934	9947	9958	9966	9973	9979
4	9444	9527	9600	9664	9719	9766	9806	9840	9869	9894
5	8529	8707	8869	9015	9147	9265	9370	9463	9544	9616
6	7003	7280	7543	7792	8024	8241	8441	8626	8795	8949
7	5058	5387	5711	6029	6340	6641	6932	7210	7476	7728
8	3128	3428	3736	4051	4371	4694	5019	5343	5665	5982
9	1619	1832	2060	2302	2559	2829	3111	3405	3707	4018
10	0687	0805	0936	1081	1241	1416	1607	1814	2036	2272
11	0234	0284	0342	0409	0486	0574	0674	0786	0911	1051
12	0062	0078	0098	0121	0149	0183	0222	0268	0322	0384
13	0012	0016	0021	0027	0035	0044	0055	0069	0086	0106
14	0002	0002	0003	0004	0006	0007	0010	0013	0016	0021
15					0001	0001	0001	0001	0002	0003

$n = 17$

x \ π	01	02	03	04	05	06	07	08	09	10
1	1571	2907	4042	5004	5819	6507	7088	7577	7988	8332
2	0123	0446	0909	1465	2078	2717	3362	3995	4604	5182
3	0006	0044	0134	0286	0503	0782	1118	1503	1927	2382
4		0003	0014	0040	0088	0164	0273	0419	0603	0826
5			0001	0004	0012	0026	0051	0089	0145	0221
6					0001	0003	0007	0015	0027	0047
7							0001	0002	0004	0008
8										0001

x \ π	11	12	13	14	15	16	17	18	19	20
1	8621	8862	9063	9230	9369	9484	9579	9657	9722	9775
2	5723	6223	6682	7099	7475	7813	8113	8379	8613	8818
3	2858	3345	3836	4324	4802	5266	5711	6133	6532	6904
4	1087	1383	1710	2065	2444	2841	3251	3669	4091	4511
5	0321	0446	0598	0778	0987	1224	1487	1775	2087	2418
6	0075	0114	0166	0234	0319	0423	0548	0695	0864	1057
7	0014	0023	0037	0056	0083	0118	0163	0220	0291	0377
8	0002	0004	0007	0011	0017	0027	0039	0057	0080	0109
9		0001	0001	0002	0003	0005	0008	0012	0018	0026
10						0001	0001	0002	0003	0005
11										0001

x \ π	21	22	23	24	25	26	27	28	29	30
1	9818	9854	9882	9906	9925	9940	9953	9962	9970	9977
2	8996	9152	9285	9400	9499	9583	9654	9714	9765	9807
3	7249	7567	7859	8123	8363	8578	8771	8942	9093	9226
4	4927	5333	5728	6107	6470	6814	7137	7440	7721	7981
5	2766	3128	3500	3879	4261	4643	5023	5396	5760	6113
6	1273	1510	1770	2049	2347	2661	2989	3329	3677	4032
7	0479	0598	0736	0894	1071	1268	1485	1721	1976	2248
8	0147	0194	0251	0320	0402	0499	0611	0739	0884	1046
9	0037	0051	0070	0094	0124	0161	0206	0261	0326	0403
10	0007	0011	0016	0022	0031	0042	0057	0075	0098	0127
11	0001	0002	0003	0004	0006	0009	0013	0018	0024	0032
12				0001	0001	0002	0002	0003	0005	0007
13									0001	0001

x \ π	31	32	33	34	35	36	37	38	39	40
1	9982	9986	9989	9991	9993	9995	9996	9997	9998	9998
2	9843	9872	9896	9917	9933	9946	9957	9966	9973	9979
3	9343	9444	9532	9608	9673	9728	9775	9815	9849	9877
4	8219	8437	8634	8812	8972	9115	9241	9353	9450	9536
5	6453	6778	7087	7378	7652	7906	8142	8360	8559	8740
6	4390	4749	5105	5458	5803	6139	6465	6778	7077	7361
7	2536	2838	3153	3479	3812	4152	4495	4839	5182	5522
8	1227	1426	1642	1877	2128	2395	2676	2971	3278	3595
9	0492	0595	0712	0845	0994	1159	1341	1541	1757	1989
10	0162	0204	0254	0314	0383	0464	0557	0664	0784	0919
11	0043	0057	0074	0095	0120	0151	0189	0234	0286	0348
12	0009	0013	0017	0023	0030	0040	0051	0066	0084	0106
13	0002	0002	0003	0004	0006	0008	0011	0015	0019	0025
14				0001	0001	0001	0002	0002	0003	0005
15										0001

x \ π	41	42	43	44	45	46	47	48	49	50
1	9999	9999	9999	9999	10000	10000	10000	10000	10000	10000
2	9984	9987	9990	9992	9994	9996	9997	9998	9998	9999
3	9900	9920	9935	9948	9959	9968	9975	9980	9985	9988
4	9610	9674	9729	9776	9816	9849	9877	9901	9920	9936
5	8904	9051	9183	9301	9404	9495	9575	9644	9704	9755
6	7628	7879	8113	8330	8529	8712	8878	9028	9162	9283
7	5856	6182	6499	6805	7098	7377	7641	7890	8122	8338
8	3920	4250	4585	4921	5257	5590	5918	6239	6552	6855
9	2238	2502	2780	3072	3374	3687	4008	4335	4667	5000
10	1070	1236	1419	1618	1834	2066	2314	2577	2855	3145
11	0420	0503	0597	0705	0826	0962	1112	1279	1462	1662
12	0133	0165	0203	0248	0301	0363	0434	0517	0611	0717
13	0033	0042	0054	0069	0086	0108	0134	0165	0202	0245
14	0006	0008	0011	0014	0019	0024	0031	0040	0050	0064
15	0001	0001	0002	0002	0003	0004	0005	0007	0009	0012
16							0001	0001	0001	0001

$n = 18$

x \ π	01	02	03	04	05	06	07	08	09	10
1	1655	3049	4220	5204	6028	6717	7292	7771	8169	8499
2	0138	0495	1003	1607	2265	2945	3622	4281	4909	5497
3	0007	0052	0157	0333	0581	0898	1275	1702	2168	2662
4		0004	0018	0050	0109	0201	0333	0506	0723	0982
5			0002	0006	0015	0034	0067	0116	0186	0282
6				0001	0002	0005	0010	0021	0038	0064
7							0001	0003	0006	0012
8									0001	0002

x \ π	11	12	13	14	15	16	17	18	19	20
1	8773	8998	9185	9338	9464	9566	9651	9719	9775	9820
2	6042	6540	6992	7398	7759	8080	8362	8609	8824	9009
3	3173	3690	4206	4713	5203	5673	6119	6538	6927	7287
4	1282	1618	1986	2382	2798	3229	3669	4112	4554	4990
5	0405	0558	0743	0959	1206	1482	1787	2116	2467	2836

x \ π	11	12	13	14	15	16	17	18	19	20
6	0102	0154	0222	0310	0419	0551	0708	0889	1097	1329
7	0021	0034	0054	0081	0118	0167	0229	0306	0400	0513
8	0003	0006	0011	0017	0027	0041	0060	0086	0120	0163
9		0001	0002	0003	0005	0008	0013	0020	0029	0043
10					0001	0001	0002	0004	0006	0009
11								0001	0001	0002

x \ π	21	22	23	24	25	26	27	28	29	30
1	9856	9886	9909	9928	9944	9956	9965	9973	9979	9984
2	9169	9306	9423	9522	9605	9676	9735	9784	9824	9858
3	7616	7916	8187	8430	8647	8839	9009	9158	9288	9400
4	5414	5825	6218	6591	6943	7272	7578	7860	8119	8354
5	3220	3613	4012	4414	4813	5208	5594	5968	6329	6673
6	1586	1866	2168	2488	2825	3176	3538	3907	4281	4656
7	0645	0799	0974	1171	1390	1630	1891	2171	2469	2783
8	0217	0283	0363	0458	0569	0699	0847	1014	1200	1407
9	0060	0083	0112	0148	0193	0249	0316	0395	0488	0596
10	0014	0020	0028	0039	0054	0073	0097	0127	0164	0210
11	0003	0004	0006	0009	0012	0018	0025	0034	0046	0061
12		0001	0001	0002	0002	0003	0005	0007	0010	0014
13						0001	0001	0001	0002	0003

x \ π	31	32	33	34	35	36	37	38	39	40
1	9987	9990	9993	9994	9996	9997	9998	9998	9999	9999
2	9886	9908	9927	9942	9954	9964	9972	9978	9983	9987
3	9498	9581	9652	9713	9764	9807	9843	9873	9897	9918
4	8568	8759	8931	9083	9217	9335	9439	9528	9606	9672
5	7001	7309	7598	7866	8114	8341	8549	8737	8907	9058
6	5029	5398	5759	6111	6450	6776	7086	7379	7655	7912
7	3111	3450	3797	4151	4509	4867	5224	5576	5921	6257
8	1633	1878	2141	2421	2717	3027	3349	3681	4021	4366
9	0720	0861	1019	1196	1391	1604	1835	2084	2350	2632
10	0264	0329	0405	0494	0597	0714	0847	0997	1163	1347
11	0080	0104	0133	0169	0212	0264	0325	0397	0480	0576
12	0020	0027	0036	0047	0062	0080	0102	0130	0163	0203
13	0004	0005	0008	0011	0014	0019	0026	0034	0044	0058
14	0001	0001	0001	0002	0003	0004	0005	0007	0010	0013
15						0001	0001	0001	0002	0002

x \ π	41	42	43	44	45	46	47	48	49	50
1	9999	9999	10000	10000	10000	10000	10000	10000	10000	10000
2	9990	9992	9994	9996	9997	9998	9998	9999	9999	9999
3	9934	9948	9959	9968	9975	9981	9985	9989	9991	9993
4	9729	9777	9818	9852	9880	9904	9923	9939	9952	9962
5	9193	9313	9418	9510	9589	9658	9717	9767	9810	9846
6	8151	8372	8573	8757	8923	9072	9205	9324	9428	9519
7	6582	6895	7193	7476	7742	7991	8222	8436	8632	8811
8	4713	5062	5408	5750	6085	6412	6728	7032	7322	7597
9	2928	3236	3556	3885	4222	4562	4906	5249	5591	5927
10	1549	1768	2004	2258	2527	2812	3110	3421	3742	4073
11	0686	0811	0951	1107	1280	1470	1677	1902	2144	2403
12	0250	0307	0372	0449	0537	0638	0753	0883	1028	1189
13	0074	0094	0118	0147	0183	0225	0275	0334	0402	0481
14	0017	0022	0029	0038	0049	0063	0079	0100	0125	0154
15	0003	0004	0006	0007	0010	0013	0017	0023	0029	0038
16		0001	0001	0001	0001	0002	0003	0004	0005	0007
17									0001	0001

$n = 19$

x \ π	01	02	03	04	05	06	07	08	09	10
1	1738	3188	4394	5396	6226	6914	7481	7949	8334	8649
2	0153	0546	1100	1751	2453	3171	3879	4560	5202	5797
3	0009	0061	0183	0384	0665	1021	1439	1908	2415	2946
4		0005	0022	0061	0132	0243	0398	0602	0853	1150
5			0002	0007	0020	0044	0085	0147	0235	0352
6				0001	0002	0006	0014	0029	0051	0086
7						0001	0002	0004	0009	0017
8								0001	0001	0003

x \ π	11	12	13	14	15	16	17	18	19	20
1	8908	9119	9291	9431	9544	9636	9710	9770	9818	9856
2	6342	6835	7277	7669	8015	8318	8581	8809	9004	9171
3	3488	4032	4568	5089	5587	6059	6500	6910	7287	7631
4	1490	1867	2275	2708	3159	3620	4085	4549	5005	5449
5	0502	0685	0904	1158	1444	1762	2107	2476	2864	3267
6	0135	0202	0290	0401	0537	0700	0891	1110	1357	1631
7	0030	0048	0076	0113	0163	0228	0310	0411	0532	0676
8	0005	0009	0016	0026	0041	0061	0089	0126	0173	0233
9	0001	0002	0003	0005	0008	0014	0021	0032	0047	0067
10				0001	0001	0002	0004	0007	0010	0016
11							0001	0001	0002	0003

x \ π	21	22	23	24	25	26	27	28	29	30
1	9887	9911	9930	9946	9958	9967	9975	9981	9985	9989
2	9313	9434	9535	9619	9690	9749	9797	9837	9869	9896
3	7942	8222	8471	8692	8887	9057	9205	9333	9443	9538
4	5877	6285	6671	7032	7369	7680	7965	8224	8458	8668
5	3681	4100	4520	4936	5346	5744	6129	6498	6848	7178
6	1929	2251	2592	2950	3322	3705	4093	4484	4875	5261
7	0843	1034	1248	1487	1749	2032	2336	2657	2995	3345
8	0307	0396	0503	0629	0775	0941	1129	1338	1568	1820
9	0093	0127	0169	0222	0287	0366	0459	0568	0694	0839
10	0023	0034	0047	0066	0089	0119	0156	0202	0258	0326
11	0005	0007	0011	0016	0023	0032	0044	0060	0080	0105
12	0001	0001	0002	0003	0005	0007	0010	0015	0021	0028
13				0001	0001	0001	0002	0003	0004	0006
14									0001	0001

x \ π	31	32	33	34	35	36	37	38	39	40
1	9991	9993	9995	9996	9997	9998	9998	9999	9999	9999
2	9917	9935	9949	9960	9969	9976	9981	9986	9989	9992
3	9618	9686	9743	9791	9830	9863	9890	9913	9931	9945
4	8856	9022	9169	9297	9409	9505	9588	9659	9719	9770
5	7486	7773	8037	8280	8500	8699	8878	9038	9179	9304
6	5641	6010	6366	6707	7032	7339	7627	7895	8143	8371
7	3705	4073	4445	4818	5188	5554	5913	6261	6597	6919
8	2091	2381	2688	3010	3344	3690	4043	4401	4762	5122
9	1003	1186	1389	1612	1855	2116	2395	2691	3002	3325
10	0405	0499	0608	0733	0875	1035	1213	1410	1626	1861
11	0137	0176	0223	0280	0347	0426	0518	0625	0747	0885
12	0038	0051	0068	0089	0114	0146	0185	0231	0287	0352
13	0009	0012	0017	0023	0031	0041	0054	0070	0091	0116
14	0002	0002	0003	0005	0007	0009	0013	0017	0023	0031
15			0001	0001	0001	0002	0002	0003	0005	0006
16									0001	0001

x \ π	41	42	43	44	45	46	47	48	49	50
1	10000	10000	10000	10000	10000	10000	10000	10000	10000	10000
2	9994	9995	9996	9997	9998	9999	9999	9999	9999	10000
3	9957	9967	9974	9980	9985	9988	9991	9993	9995	9996
4	9813	9849	9878	9903	9923	9939	9952	9963	9971	9978
5	9413	9508	9590	9660	9720	9771	9814	9850	9879	9904
6	8579	8767	8937	9088	9223	9342	9446	9537	9615	9682
7	7226	7515	7787	8039	8273	8488	8684	8862	9022	9165
8	5480	5832	6176	6509	6831	7138	7430	7706	7964	8204
9	3660	4003	4353	4706	5060	5413	5762	6105	6439	6762
10	2114	2385	2672	2974	3290	3617	3954	4299	4648	5000
11	1040	1213	1404	1613	1841	2087	2351	2631	2928	3238
12	0429	0518	0621	0738	0871	1021	1187	1372	1575	1796
13	0146	0183	0227	0280	0342	0415	0500	0597	0709	0835
14	0040	0052	0067	0086	0109	0137	0171	0212	0261	0318
15	0009	0012	0016	0021	0028	0036	0046	0060	0076	0096
16	0001	0002	0003	0004	0005	0007	0010	0013	0017	0022
17				0001	0001	0001	0001	0002	0003	0004

n = 20

x \ π	01	02	03	04	05	06	07	08	09	10
1	1821	3324	4562	5580	6415	7099	7658	8113	8484	8784
2	0169	0599	1198	1897	2642	3395	4131	4831	5484	6083
3	0010	0071	0210	0439	0755	1150	1610	2121	2666	3231
4		0006	0027	0074	0159	0290	0471	0706	0993	1330
5			0003	0010	0026	0056	0107	0183	0290	0432
6				0001	0003	0009	0019	0038	0068	0113
7						0001	0003	0006	0013	0024
8								0001	0002	0004
9										0001

x \ π	11	12	13	14	15	16	17	18	19	20
1	9028	9224	9383	9510	9612	9694	9759	9811	9852	9885
2	6624	7109	7539	7916	8244	8529	8773	8982	9159	9308
3	3802	4369	4920	5450	5951	6420	6854	7252	7614	7939
4	1710	2127	2573	3041	3523	4010	4496	4974	5439	5886
5	0610	0827	1083	1375	1702	2059	2443	2849	3271	3704
6	0175	0260	0370	0507	0673	0870	1098	1356	1643	1958
7	0041	0067	0103	0153	0219	0304	0409	0537	0689	0867
8	0008	0014	0024	0038	0059	0088	0127	0177	0241	0321
9	0001	0002	0005	0008	0013	0021	0033	0049	0071	0100
10			0001	0001	0002	0004	0007	0011	0017	0026
11						0001	0001	0002	0004	0006
12									0001	0001

x \ π	21	22	23	24	25	26	27	28	29	30
1	9910	9931	9946	9959	9968	9976	9982	9986	9989	9992
2	9434	9539	9626	9698	9757	9805	9845	9877	9903	9924
3	8230	8488	8716	8915	9087	9237	9365	9474	9567	9645
4	6310	6711	7085	7431	7748	8038	8300	8534	8744	8929
5	4142	4580	5014	5439	5852	6248	6625	6981	7315	7625
6	2297	2657	3035	3427	3828	4235	4643	5048	5447	5836
7	1071	1301	1557	1838	2142	2467	2810	3169	3540	3920
8	0419	0536	0675	0835	1018	1225	1455	1707	1982	2277
9	0138	0186	0246	0320	0409	0515	0640	0784	0948	1133
10	0038	0054	0075	0103	0139	0183	0238	0305	0385	0480
11	0009	0013	0019	0028	0039	0055	0074	0100	0132	0171
12	0002	0003	0004	0006	0009	0014	0019	0027	0038	0051
13			0001	0001	0002	0003	0004	0006	0009	0013
14							0001	0001	0002	0003

x \ π	31	32	33	34	35	36	37	38	39	40
1	9994	9996	9997	9998	9998	9999	9999	9999	9999	10000
2	9940	9953	9964	9972	9979	9984	9988	9991	9993	9995
3	9711	9765	9811	9848	9879	9904	9924	9940	9953	9964
4	9092	9235	9358	9465	9556	9634	9700	9755	9802	9840
5	7911	8173	8411	8626	8818	8989	9141	9274	9390	9490
6	6213	6574	6917	7242	7546	7829	8090	8329	8547	8744
7	4305	4693	5079	5460	5834	6197	6547	6882	7200	7500
8	2591	2922	3268	3624	3990	4361	4735	5108	5478	5841
9	1340	1568	1818	2087	2376	2683	3005	3341	3688	4044
10	0591	0719	0866	1032	1218	1424	1650	1897	2163	2447
11	0220	0279	0350	0434	0532	0645	0775	0923	1090	1275
12	0069	0091	0119	0154	0196	0247	0308	0381	0466	0565
13	0018	0025	0034	0045	0060	0079	0102	0132	0167	0210
14	0004	0006	0008	0011	0015	0021	0028	0037	0049	0065
15	0001	0001	0001	0002	0003	0004	0006	0009	0012	0016
16						0001	0001	0002	0002	0003

x \ π	41	42	43	44	45	46	47	48	49	50
1	10000	10000	10000	10000	10000	10000	10000	10000	10000	10000
2	9996	9997	9998	9998	9999	9999	9999	10000	10000	10000
3	9972	9979	9984	9988	9991	9993	9995	9996	9997	9998
4	9872	9898	9920	9937	9951	9962	9971	9977	9983	9987
5	9577	9651	9714	9767	9811	9848	9879	9904	9924	9941

x \ π	41	42	43	44	45	46	47	48	49	50
6	8921	9078	9217	9340	9447	9539	9619	9687	9745	9793
7	7780	8041	8281	8501	8701	8881	9042	9186	9312	9423
8	6196	6539	6868	7183	7480	7759	8020	8261	8482	8684
9	4406	4771	5136	5499	5857	6207	6546	6873	7186	7483
10	2748	3064	3394	3736	4086	4443	4804	5166	5525	5881
11	1400	1705	1949	2210	2493	2791	3104	3432	3771	4119
12	0679	0810	0958	1123	1308	1511	1734	1977	2238	2517
13	0262	0324	0397	0482	0580	0694	0823	0969	1133	1316
14	0084	0107	0136	0172	0214	0265	0326	0397	0480	0577
15	0022	0029	0038	0050	0064	0083	0105	0133	0166	0207
16	0004	0006	0008	0011	0015	0020	0027	0035	0046	0059
17	0001	0001	0001	0002	0003	0004	0005	0007	0010	0013
18						0001	0001	0001	0001	0002

$n = 50$

x \ π	01	02	03	04	05	06	07	08	09	10
1	3950	6358	7819	8701	9231	9547	9734	9845	9910	9948
2	0894	2642	4447	5995	7206	8100	8735	9173	9468	9662
3	0138	0784	1892	3233	4595	5838	6892	7740	8395	8883
4	0016	0178	0628	1391	2396	3527	4673	5747	6697	7497
5	0001	0032	0168	0490	1036	1794	2710	3710	4723	5688
6		0005	0037	0144	0378	0776	1350	2081	2928	3839
7		0001	0007	0036	0118	0289	0583	1019	1596	2298
8			0001	0008	0032	0094	0220	0438	0768	1221
9				0001	0008	0027	0073	0167	0328	0579
10					0002	0007	0022	0056	0125	0245
11						0002	0006	0017	0043	0094
12							0001	0005	0013	0032
13								0001	0004	0010
14									0001	0003
15										0001

x \ π	11	12	13	14	15	16	17	18	19	20
1	9971	9983	9991	9995	9997	9998	9999	10000	10000	10000
2	9788	9869	9920	9951	9971	9983	9990	9994	9997	9998
3	9237	9487	9661	9779	9858	9910	9944	9965	9979	9987
4	8146	8655	9042	9330	9540	9688	9792	9863	9912	9943
5	6562	7320	7956	8472	8879	9192	9428	9601	9726	9815
6	4760	5647	6463	7186	7806	8323	8741	9071	9327	9520
7	3091	3935	4789	5616	6387	7081	7686	8199	8624	8966
8	1793	2467	3217	4010	4812	5594	6328	6996	7587	8096
9	0932	1392	1955	2605	3319	4071	4832	5576	6280	6927
10	0435	0708	1074	1537	2089	2718	3403	4122	4849	5563
11	0183	0325	0535	0824	1199	1661	2203	2813	3473	4164
12	0069	0135	0242	0402	0628	0929	1309	1768	2300	2893
13	0024	0051	0100	0179	0301	0475	0714	1022	1405	1861
14	0008	0018	0037	0073	0132	0223	0357	0544	0791	1106
15	0002	0006	0013	0027	0053	0096	0164	0266	0411	0607
16	0001	0002	0004	0009	0019	0038	0070	0120	0197	0308
17			0001	0003	0007	0014	0027	0050	0087	0144
18				0001	0008	0005	0010	0019	0036	0063
19					0001	0001	0003	0007	0013	0025
20							0001	0002	0005	0009
21								0001	0002	0003
22										0001

x \ π	21	22	23	24	25	26	27	28	29	30
1	10000	10000	10000	10000	10000	10000	10000	10000	10000	10000
2	9999	9999	10000	10000	10000	10000	10000	10000	10000	10000
3	9992	9995	9997	9998	9999	10000	10000	10000	10000	10000
4	9964	9978	9986	9992	9995	9997	9998	9999	9999	10000
5	9877	9919	9948	9967	9979	9987	9992	9995	9997	9998
6	9663	9767	9841	9893	9930	9954	9970	9981	9988	9993
7	9236	9445	9603	9720	9806	9868	9911	9941	9961	9975
8	8523	8874	9156	9377	9547	9676	9772	9842	9892	9927
9	7505	8009	8437	8794	9084	9316	9497	9635	9740	9817
10	6241	6870	7436	7934	8363	8724	9021	9260	9450	9598

x \ π	21	22	23	24	25	26	27	28	29	30
11	4864	5552	6210	6822	7378	7871	8299	8663	8965	9211
12	3533	4201	4878	5544	6184	6782	7329	7817	8244	8610
13	2383	2963	3585	4233	4890	5539	6163	6749	7287	7771
14	1490	1942	2456	3023	3630	4261	4901	5534	6145	6721
15	0862	1181	1565	2013	2519	3075	3669	4286	4912	5532
16	0462	0665	0926	1247	1631	2075	2575	3121	3703	4308
17	0229	0347	0508	0718	0983	1306	1689	2130	2623	3161
18	0105	0168	0259	0384	0551	0766	1034	1359	1741	2178
19	0045	0075	0122	0191	0287	0418	0590	0809	1080	1406
20	0018	0031	0054	0088	0139	0212	0314	0449	0626	0848
21	0006	0012	0022	0038	0063	0100	0155	0232	0338	0478
22	0002	0004	0008	0015	0026	0044	0071	0112	0170	0251
23	0001	0001	0003	0006	0010	0018	0031	0050	0080	0123
24			0001	0002	0004	0007	0012	0021	0035	0056
25				0001	0001	0002	0004	0008	0014	0024
26						0001	0002	0003	0005	0009
27								0001	0002	0003
28									0001	0001

x \ π	31	32	33	34	35	36	37	38	39	40
1	10000	10000	10000	10000	10000	10000	10000	10000	10000	10000
2	10000	10000	10000	10000	10000	10000	10000	10000	10000	10000
3	10000	10000	10000	10000	10000	10000	10000	10000	10000	10000
4	10000	10000	10000	10000	10000	10000	10000	10000	10000	10000
5	9999	9999	10000	10000	10000	10000	10000	10000	10000	10000
6	9996	9997	9998	9999	9999	10000	10000	10000	10000	10000
7	9984	9990	9994	9996	9998	9999	9999	10000	10000	10000
8	9952	9969	9980	9987	9992	9995	9997	9998	9999	9999
9	9874	9914	9942	9962	9975	9984	9990	9994	9996	9998
10	9710	9794	9856	9901	9933	9955	9971	9981	9988	9992
11	9409	9563	9683	9773	9840	9889	9924	9949	9966	9978
12	8916	9168	9371	9533	9658	9753	9825	9878	9916	9943
13	8197	8564	8873	9130	9339	9505	9635	9736	9811	9867
14	7253	7732	8157	8524	8837	9097	9310	9481	9616	9720
15	6131	6698	7223	7699	8122	8491	8805	9069	9286	9460
16	4922	5530	6120	6679	7199	7672	8094	8462	8779	9045
17	3734	4328	4931	5530	6111	6664	7179	7649	8070	8439
18	2666	3197	3760	4346	4940	5531	6105	6653	7164	7631
19	1786	2220	2703	3227	3784	4362	4949	5533	6101	6644
20	1121	1447	1826	2257	2736	3255	3805	4376	4957	5535
21	0657	0882	1156	1482	1861	2289	2764	3278	3824	4390
22	0360	0503	0685	0912	1187	1513	1890	2317	2788	3299
23	0184	0267	0379	0525	0710	0938	1214	1540	1916	2340
24	0087	0133	0196	0282	0396	0544	0730	0960	1236	1562
25	0039	0061	0094	0141	0207	0295	0411	0560	0748	0978
26	0016	0026	0042	0066	0100	0149	0216	0305	0423	0573
27	0006	0011	0018	0029	0045	0070	0106	0155	0223	0314
28	0002	0004	0007	0012	0019	0031	0048	0074	0110	0160
29	0001	0001	0002	0004	0007	0012	0020	0032	0050	0076
30			0001	0002	0003	0005	0008	0013	0021	0034
31					0001	0002	0003	0005	0008	0014
32						0001	0001	0002	0003	0005
33								0001	0001	0002
34										0001

x \ π	41	42	43	44	45	46	47	48	49	50
1	10000	10000	10000	10000	10000	10000	10000	10000	10000	10000
2	10000	10000	10000	10000	10000	10000	10000	10000	10000	10000
3	10000	10000	10000	10000	10000	10000	10000	10000	10000	10000
4	10000	10000	10000	10000	10000	10000	10000	10000	10000	10000
5	10000	10000	10000	10000	10000	10000	10000	10000	10000	10000
6	10000	10000	10000	10000	10000	10000	10000	10000	10000	10000
7	10000	10000	10000	10000	10000	10000	10000	10000	10000	10000
8	10000	10000	10000	10000	10000	10000	10000	10000	10000	10000
9	9999	9999	10000	10000	10000	10000	10000	10000	10000	10000
10	9995	9997	9998	9999	9999	10000	10000	10000	10000	10000
11	9986	9991	9994	9997	9998	9999	9999	10000	10000	10000
12	9962	9975	9984	9990	9994	9996	9998	9999	9999	10000
13	9908	9938	9958	9973	9982	9989	9993	9996	9997	9998
14	9799	9858	9902	9933	9955	9970	9981	9988	9992	9995
15	9599	9707	9789	9851	9896	9929	9952	9968	9980	9987

x \ π	41	42	43	44	45	46	47	48	49	50
16	9265	9443	9585	9696	9780	9844	9892	9926	9950	9967
17	8757	9025	9248	9429	9573	9687	9774	9839	9888	9923
18	8051	8421	8740	9010	9235	9418	9565	9680	9769	9836
19	7152	7617	8037	8406	8727	8998	9225	9410	9559	9675
20	6099	6638	7143	7608	8026	8396	8718	8991	9219	9405
21	4965	5539	6099	6635	7138	7602	8020	8391	8713	8987
22	3840	4402	4973	5543	6100	6634	7137	7599	8018	8389
23	2809	3316	3854	4412	4981	5548	6104	6636	7138	7601
24	1936	2359	2826	3331	3866	4422	4989	5554	6109	6641
25	1255	1580	1953	2375	2840	3343	3876	4431	4996	5561
26	0762	0992	1269	1593	1966	2386	2850	3352	3885	4439
27	0432	0584	0772	1003	1279	1603	1975	2395	2858	3359
28	0229	0320	0439	0591	0780	1010	1286	1609	1981	2399
29	0113	0164	0233	0325	0444	0595	0784	1013	1289	1611
30	0052	0078	0115	0166	0235	0327	0446	0596	0784	1013
31	0022	0034	0053	0079	0116	0167	0236	0327	0445	0595
32	0009	0014	0022	0035	0053	0079	0116	0166	0234	0325
33	0003	0005	0009	0014	0022	0035	0053	0078	0114	0164
34	0001	0002	0003	0005	0009	0014	0022	0034	0052	0077
35		0001	0001	0002	0003	0005	0008	0014	0021	0033
36				0001	0001	0002	0003	0005	0008	0013
37						0001	0001	0002	0003	0005
38								0001	0001	0002

n = *100*

x \ π	01	02	03	04	05	06	07	08	09	10
1	6340	8674	9524	9831	9941	9979	9993	9998	9999	10000
2	2642	5967	8054	9128	9629	9848	9940	9977	9991	9997
3	0794	3233	5802	7679	8817	9434	9742	9887	9952	9981
4	0184	1410	3528	5705	7422	8570	9256	9633	9827	9922
5	0034	0508	1821	3711	5640	7232	8368	9097	9526	9763
6	0005	0155	0808	2116	3840	5593	7086	8201	8955	9424
7	0001	0041	0312	1064	2340	3936	5557	6968	8060	8828
8		0009	0106	0475	1280	2517	4012	5529	6872	7939
9		0002	0032	0190	0631	1463	2660	4074	5506	6791
10			0009	0068	0282	0775	1620	2780	4125	5487
11			0002	0022	0115	0376	0908	1757	2882	4168
12				0007	0043	0168	0469	1028	1876	2970
13				0002	0015	0069	0224	0559	1138	1982
14					0005	0026	0099	0282	0645	1239
15					0001	0009	0041	0133	0341	0726
16						0003	0016	0058	0169	0399
17						0001	0006	0024	0078	0206
18							0002	0009	0034	0100
19							0001	0003	0014	0046
20								0001	0005	0020
21									0002	0008
22									0001	0003
23										0001

x \ π	11	12	13	14	15	16	17	18	19	20
1	10000	10000	10000	10000	10000	10000	10000	10000	10000	10000
2	9999	10000	10000	10000	10000	10000	10000	10000	10000	10000
3	9992	9997	9999	10000	10000	10000	10000	10000	10000	10000
4	9966	9985	9994	9998	9999	10000	10000	10000	10000	10000
5	9886	9947	9977	9990	9996	9998	9999	10000	10000	10000
6	9698	9848	9926	9966	9984	9993	9997	9999	10000	10000
7	9328	9633	9808	9903	9953	9978	9990	9996	9998	9999
8	8715	9239	9569	9766	9878	9939	9970	9986	9994	9997
9	7835	8614	9155	9508	9725	9853	9924	9962	9982	9991
10	6722	7743	8523	9078	9449	9684	9826	9908	9953	9977
11	5471	6663	7663	8440	9006	9393	9644	9800	9891	9943
12	4206	5458	6611	7591	8365	8939	9340	9605	9773	9874
13	3046	4239	5446	6566	7527	8297	8876	9289	9567	9747
14	2076	3114	4268	5436	6526	7469	8234	8819	9241	9531
15	1330	2160	3173	4294	5428	6490	7417	8177	8765	9196
16	0802	1414	2236	3227	4317	5420	6458	7370	8125	8715
17	0456	0874	1492	2305	3275	4338	5414	6429	7327	8077
18	0244	0511	0942	1563	2367	3319	4357	5408	6403	7288
19	0123	0282	0564	1006	1628	2424	3359	4374	5403	6379
20	0059	0147	0319	0614	1065	1689	2477	3395	4391	5398

x \ π	11	12	13	14	15	16	17	18	19	20
21	0026	0073	0172	0356	0663	1121	1745	2525	3429	4405
22	0011	0034	0088	0196	0393	0710	1174	1797	2570	3460
23	0005	0015	0042	0103	0221	0428	0754	1223	1846	2611
24	0002	0006	0020	0051	0119	0246	0462	0796	1270	1891
25	0001	0003	0009	0024	0061	0135	0271	0496	0837	1314
26		0001	0004	0011	0030	0071	0151	0295	0528	0875
27			0001	0005	0014	0035	0081	0168	0318	0558
28			0001	0002	0006	0017	0041	0091	0184	0342
29				0001	0003	0008	0020	0048	0102	0200
30					0001	0003	0009	0024	0054	0112
31						0001	0004	0011	0027	0061
32						0001	0002	0005	0013	0031
33							0001	0002	0006	0016
34								0001	0003	0007
35									0001	0003
36										0001
37										0001

x \ π	21	22	23	24	25	26	27	28	29	30
1	10000	10000	10000	10000	10000	10000	10000	10000	10000	10000
2	10000	10000	10000	10000	10000	10000	10000	10000	10000	10000
3	10000	10000	10000	10000	10000	10000	10000	10000	10000	10000
4	10000	10000	10000	10000	10000	10000	10000	10000	10000	10000
5	10000	10000	10000	10000	10000	10000	10000	10000	10000	10000
6	10000	10000	10000	10000	10000	10000	10000	10000	10000	10000
7	10000	10000	10000	10000	10000	10000	10000	10000	10000	10000
8	9999	10000	10000	10000	10000	10000	10000	10000	10000	10000
9	9996	9998	9999	10000	10000	10000	10000	10000	10000	10000
10	9989	9995	9998	9999	10000	10000	10000	10000	10000	10000
11	9971	9986	9993	9997	9999	9999	10000	10000	10000	10000
12	9933	9965	9983	9992	9996	9998	9999	10000	10000	10000
13	9857	9922	9959	9979	9990	9995	9998	9999	10000	10000
14	9721	9840	9911	9953	9975	9988	9994	9997	9999	9999
15	9496	9695	9823	9900	9946	9972	9986	9993	9997	9998
16	9153	9462	9671	9806	9889	9939	9967	9983	9992	9996
17	8668	9112	9430	9647	9789	9878	9932	9963	9981	9990
18	8032	8625	9074	9399	9624	9773	9867	9925	9959	9978
19	7252	7991	8585	9038	9370	9601	9757	9856	9918	9955
20	6358	7220	7953	8547	9005	9342	9580	9741	9846	9911
21	5394	6338	7189	7918	8512	8973	9316	9560	9726	9835
22	4419	5391	6320	7162	7886	8479	8943	9291	9540	9712
23	3488	4432	5388	6304	7136	7856	8448	8915	9267	9521
24	2649	3514	4444	5386	6289	7113	7828	8420	8889	9245
25	1933	2684	3539	4455	5383	6276	7091	7802	8393	8864
26	1355	1972	2717	3561	4465	5381	6263	7071	7778	8369
27	0911	1393	2009	2748	3583	4475	5380	6252	7053	7756
28	0588	0945	1429	2043	2776	3602	4484	5378	6242	7036
29	0364	0616	0978	1463	2075	2803	3621	4493	5377	6232
30	0216	0386	0643	1009	1495	2105	2828	3638	4501	5377
31	0123	0232	0406	0669	1038	1526	2134	2851	3654	4509
32	0067	0134	0247	0427	0693	1065	1554	2160	2873	3669
33	0035	0074	0144	0262	0446	0717	1091	1580	2184	2893
34	0018	0039	0081	0154	0276	0465	0739	1116	1605	2207
35	0009	0020	0044	0087	0164	0290	0482	0760	1139	1629
36	0004	0010	0023	0048	0094	0174	0303	0499	0780	1161
37	0002	0005	0011	0025	0052	0101	0183	0316	0515	0799
38	0001	0002	0005	0013	0027	0056	0107	0193	0328	0530
39		0001	0002	0006	0014	0030	0060	0113	0201	0340
40			0001	0003	0007	0015	0032	0064	0119	0210
41				0001	0003	0008	0017	0035	0068	0125
42				0001	0001	0004	0008	0018	0037	0072
43					0001	0002	0004	0009	0020	0040
44						0001	0002	0005	0010	0021
45							0001	0002	0005	0011
46								0001	0002	0005
47									0001	0003
48										0001
49										0001

x \ π	31	32	33	34	35	36	37	38	39	40
1	10000	10000	10000	10000	10000	10000	10000	10000	10000	10000
2	10000	10000	10000	10000	10000	10000	10000	10000	10000	10000
3	10000	10000	10000	10000	10000	10000	10000	10000	10000	10000
4	10000	10000	10000	10000	10000	10000	10000	10000	10000	10000
5	10000	10000	10000	10000	10000	10000	10000	10000	10000	10000
6	10000	10000	10000	10000	10000	10000	10000	10000	10000	10000
7	10000	10000	10000	10000	10000	10000	10000	10000	10000	10000
8	10000	10000	10000	10000	10000	10000	10000	10000	10000	10000
9	10000	10000	10000	10000	10000	10000	10000	10000	10000	10000
10	10000	10000	10000	10000	10000	10000	10000	10000	10000	10000
11	10000	10000	10000	10000	10000	10000	10000	10000	10000	10000
12	10000	10000	10000	10000	10000	10000	10000	10000	10000	10000
13	10000	10000	10000	10000	10000	10000	10000	10000	10000	10000
14	10000	10000	10000	10000	10000	10000	10000	10000	10000	10000
15	9999	10000	10000	10000	10000	10000	10000	10000	10000	10000
16	9998	9999	10000	10000	10000	10000	10000	10000	10000	10000
17	9995	9998	9999	10000	10000	10000	10000	10000	10000	10000
18	9989	9995	9997	9999	9999	10000	10000	10000	10000	10000
19	9976	9988	9994	9997	9999	9999	10000	10000	10000	10000
20	9950	9973	9986	9993	9997	9998	9999	10000	10000	10000
21	9904	9946	9971	9985	9992	9996	9998	9999	10000	10000
22	9825	9898	9942	9968	9983	9991	9996	9998	9999	10000
23	9698	9816	9891	9938	9966	9982	9991	9995	9998	9999
24	9504	9685	9806	9885	9934	9963	9980	9990	9995	9997
25	9224	9487	9672	9797	9879	9930	9961	9979	9989	9994
26	8841	9204	9471	9660	9789	9873	9926	9958	9977	9988
27	8346	8820	9185	9456	9649	9780	9867	9922	9956	9976
28	7736	8325	8800	9168	9442	9638	9773	9862	9919	9954
29	7021	7717	8305	8781	9152	9429	9628	9765	9857	9916
30	6224	7007	7699	8287	8764	9137	9417	9618	9759	9852
31	5376	6216	6994	7684	8270	8748	9123	9405	9610	9752
32	4516	5376	6209	6982	7669	8254	8733	9110	9395	9602
33	3683	4523	5375	6203	6971	7656	8240	8720	9098	9385
34	2912	3696	4530	5375	6197	6961	7643	8227	8708	9087
35	2229	2929	3708	4536	5376	6192	6953	7632	8216	8697
36	1650	2249	2946	3720	4542	5376	6188	6945	7623	8205
37	1181	1671	2268	2961	3731	4547	5377	6184	6938	7614
38	0816	1200	1690	2285	2976	3741	4553	5377	6181	6932
39	0545	0833	1218	1708	2301	2989	3750	4558	5378	6178
40	0351	0558	0849	1235	1724	2316	3001	3759	4562	5379
41	0218	0361	0571	0863	1250	1739	2330	3012	3767	4567
42	0131	0226	0371	0583	0877	1265	1753	2343	3023	3775
43	0075	0136	0233	0380	0594	0889	1278	1766	2355	3033
44	0042	0079	0141	0240	0389	0605	0901	1290	1778	2365
45	0023	0044	0082	0146	0246	0397	0614	0911	1301	1789
46	0012	0024	0046	0085	0150	0252	0405	0623	0921	1311
47	0006	0012	0025	0048	0088	0154	0257	0411	0631	0930
48	0003	0006	0013	0026	0050	0091	0158	0262	0417	0638
49	0001	0003	0007	0014	0027	0052	0094	0162	0267	0423
50	0001	0001	0003	0007	0015	0029	0054	0096	0165	0271
51		0001	0002	0003	0007	0015	0030	0055	0098	0168
52			0001	0002	0004	0008	0016	0030	0056	0100
53				0001	0002	0004	0008	0016	0031	0058
54					0001	0002	0004	0008	0017	0032
55						0001	0002	0004	0009	0017
56							0001	0002	0004	0009
57								0001	0002	0004
58									0001	0002
59										0001

x \ π	41	42	43	44	45	46	47	48	49	50
1	10000	10000	10000	10000	10000	10000	10000	10000	10000	10000
2	10000	10000	10000	10000	10000	10000	10000	10000	10000	10000
3	10000	10000	10000	10000	10000	10000	10000	10000	10000	10000
4	10000	10000	10000	10000	10000	10000	10000	10000	10000	10000
5	10000	10000	10000	10000	10000	10000	10000	10000	10000	10000
6	10000	10000	10000	10000	10000	10000	10000	10000	10000	10000
7	10000	10000	10000	10000	10000	10000	10000	10000	10000	10000
8	10000	10000	10000	10000	10000	10000	10000	10000	10000	10000
9	10000	10000	10000	10000	10000	10000	10000	10000	10000	10000
10	10000	10000	10000	10000	10000	10000	10000	10000	10000	10000
11	10000	10000	10000	10000	10000	10000	10000	10000	10000	10000
12	10000	10000	10000	10000	10000	10000	10000	10000	10000	10000
13	10000	10000	10000	10000	10000	10000	10000	10000	10000	10000
14	10000	10000	10000	10000	10000	10000	10000	10000	10000	10000
15	10000	10000	10000	10000	10000	10000	10000	10000	10000	10000
16	10000	10000	10000	10000	10000	10000	10000	10000	10000	10000
17	10000	10000	10000	10000	10000	10000	10000	10000	10000	10000
18	10000	10000	10000	10000	10000	10000	10000	10000	10000	10000
19	10000	10000	10000	10000	10000	10000	10000	10000	10000	10000
20	10000	10000	10000	10000	10000	10000	10000	10000	10000	10000
21	10000	10000	10000	10000	10000	10000	10000	10000	10000	10000
22	10000	10000	10000	10000	10000	10000	10000	10000	10000	10000
23	10000	10000	10000	10000	10000	10000	10000	10000	10000	10000
24	9999	9999	10000	10000	10000	10000	10000	10000	10000	10000
25	9997	9999	9999	10000	10000	10000	10000	10000	10000	10000
26	9994	9997	9999	9999	10000	10000	10000	10000	10000	10000
27	9987	9994	9997	9998	9999	10000	10000	10000	10000	10000
28	9975	9987	9993	9997	9998	9999	10000	10000	10000	10000
29	9952	9974	9986	9993	9996	9998	9999	10000	10000	10000
30	9913	9950	9972	9985	9992	9996	9998	9999	10000	10000
31	9848	9910	9948	9971	9985	9992	9996	9998	9999	10000
32	9746	9844	9907	9947	9970	9984	9992	9996	9998	9999
33	9594	9741	9840	9905	9945	9969	9984	9991	9996	9998
34	9376	9587	9736	9837	9902	9944	9969	9983	9991	9996
35	9078	9368	9581	9732	9834	9900	9942	9968	9983	9991
36	8687	9069	9361	9576	9728	9831	9899	9941	9967	9982
37	8196	8678	9061	9355	9571	9724	9829	9897	9941	9967
38	7606	8188	8670	9054	9349	9567	9721	9827	9896	9940
39	6927	7599	8181	8663	9049	9345	9563	9719	9825	9895
40	6176	6922	7594	8174	8657	9044	9341	9561	9717	9824
41	5380	6174	6919	7589	8169	8653	9040	9338	9558	9716
42	4571	5382	6173	6916	7585	8165	8649	9037	9335	9557
43	3782	4576	5383	6173	6913	7582	8162	8646	9035	9334
44	3041	3788	4580	5385	6172	6912	7580	8160	8645	9033
45	2375	3049	3794	4583	5387	6173	6911	7579	8159	8644
46	1799	2384	3057	3799	4587	5389	6173	6911	7579	8159
47	1320	1807	2391	3063	3804	4590	5391	6174	6912	7579
48	0938	1328	1815	2398	3069	3809	4593	5393	6176	6914
49	0644	0944	1335	1822	2404	3074	3813	4596	5395	6178
50	0428	0650	0950	1341	1827	2409	3078	3816	4599	5398
51	0275	0432	0655	0955	1346	1832	2413	3082	3819	4602
52	0170	0278	0436	0659	0960	1350	1836	2417	3084	3822
53	0102	0172	0280	0439	0662	0963	1353	1838	2419	3086
54	0059	0103	0174	0282	0441	0664	0965	1355	1840	2421
55	0033	0059	0104	0175	0284	0443	0666	0967	1356	1841
56	0017	0033	0060	0105	0176	0285	0444	0667	0967	1356
57	0009	0018	0034	0061	0106	0177	0286	0444	0667	0967
58	0004	0009	0018	0034	0061	0106	0177	0286	0444	0666
59	0002	0005	0009	0018	0034	0061	0106	0177	0285	0443
60	0001	0002	0005	0009	0018	0034	0061	0106	0177	0284
61		0001	0002	0005	0009	0018	0034	0061	0106	0176
62			0001	0002	0005	0009	0018	0034	0061	0105
63				0001	0002	0005	0009	0018	0034	0060
64					0001	0002	0005	0009	0018	0033
65						0001	0002	0005	0009	0018
66							0001	0002	0004	0009
67								0001	0002	0004
68									0001	0002
69										0001

Table A-7: Cumulative Poisson Distribution

$1000 \times P_p(X \leq k \mid \lambda)$

1,000 × probability of k or less occurrences of event that has average number of occurrences equal to λ.

λ \ k	0	1	2	3	4	5	6	7	8	9
0.02	980	1,000								
0.04	961	999	1,000							
0.06	942	998	1,000							
0.08	923	997	1,000							
0.10	905	995	1,000							
0.15	861	990	999	1,000						
0.20	819	982	999	1,000						
0.25	779	974	998	1,000						
0.30	741	963	996	1,000						
0.35	705	951	994	1,000						
0.40	670	938	992	999	1,000					
0.45	638	925	989	999	1,000					
0.50	607	910	986	998	1,000					
0.55	577	894	982	998	1,000					
0.60	549	878	977	997	1,000					
0.65	522	861	972	996	999	1,000				
0.70	497	844	966	994	999	1,000				
0.75	472	827	959	993	999	1,000				
0.80	449	809	953	991	999	1,000				
0.85	427	791	945	989	998	1,000				
0.90	407	772	937	987	998	1,000				
0.95	387	754	929	984	997	1,000				
1.00	368	736	920	981	996	999	1,000			
1.1	333	699	900	974	995	999	1,000			
1.2	301	663	879	966	992	998	1,000			
1.3	273	627	857	957	989	998	1,000			
1.4	247	592	833	946	986	997	999	1,000		
1.5	223	558	809	934	981	996	999	1,000		
1.6	202	525	783	921	976	994	999	1,000		
1.7	183	493	757	907	970	992	998	1,000		
1.8	165	463	731	891	964	990	997	999	1,000	
1.9	150	434	704	875	956	987	997	999	1,000	
2.0	135	406	677	857	947	983	995	999	1,000	

Source: This table is reproduced from E. L. Grant, *Statistical Quality Control*, Second Edition (New York: McGraw-Hill Book Company, 1952). Used by permission.

λ \ k	0	1	2	3	4	5	6	7	8	9
2.2	111	355	623	819	928	975	993	998	1,000	
2.4	091	308	570	779	904	964	988	997	999	1,000
2.6	074	267	518	736	877	951	983	995	999	1,000
2.8	061	231	469	692	848	935	976	992	998	999
3.0	050	199	423	647	815	916	966	988	996	999
3.2	041	171	380	603	781	895	955	983	994	998
3.4	033	147	340	558	744	871	942	977	992	997
3.6	027	126	303	515	706	844	927	969	988	996
3.8	022	107	269	473	668	816	909	960	984	994
4.0	018	092	238	433	629	785	889	949	979	992
4.2	015	078	210	395	590	753	867	936	972	989
4.4	012	066	185	359	551	720	844	921	964	985
4.6	010	056	163	326	513	686	818	905	955	980
4.8	008	048	143	294	476	651	791	887	944	975
5.0	007	040	125	265	440	616	762	867	932	968
5.2	006	034	109	238	406	581	732	845	918	960
5.4	005	029	095	213	373	546	702	822	903	951
5.6	004	024	082	191	342	512	670	797	886	941
5.8	003	021	072	170	313	478	638	771	867	929
6.0	002	017	062	151	285	446	606	744	847	916

λ \ k	10	11	12	13	14	15	16
2.8	1,000						
3.0	1,000						
3.2	1,000						
3.4	999	1,000					
3.6	999	1,000					
3.8	998	999	1,000				
4.0	997	999	1,000				
4.2	996	999	1,000				
4.4	994	998	999	1,000			
4.6	992	997	999	1,000			
4.8	990	996	999	1,000			
5.0	986	995	998	999	1,000		
5.2	982	993	997	999	1,000		
5.4	977	990	996	999	1,000		
5.6	972	988	995	998	999	1,000	
5.8	965	984	993	997	999	1,000	
6.0	957	980	991	996	999	999	1,000

λ \ k	0	1	2	3	4	5	6	7	8	9
6.2	002	015	054	134	259	414	574	716	826	902
6.4	002	012	046	119	235	384	542	687	803	886
6.6	001	010	040	105	213	355	511	658	780	869
6.8	001	009	034	093	192	327	480	628	755	850
7.0	001	007	030	082	173	301	450	599	729	830
7.2	001	006	025	072	156	276	420	569	703	810
7.4	001	005	022	063	140	253	392	539	676	788
7.6	001	004	019	055	125	231	365	510	648	765
7.8	000	004	016	048	112	210	338	481	620	741
8.0	000	003	014	042	100	191	313	453	593	717
8.5	000	002	009	030	074	150	256	386	523	653
9.0	000	001	006	021	055	116	207	324	456	587
9.5	000	001	004	015	040	089	165	269	392	522
10.0	000	000	003	010	029	067	130	220	333	458

λ \ k	10	11	12	13	14	15	16	17	18	19
6.2	949	975	989	995	998	999	1,000			
6.4	939	969	986	994	997	999	1,000			
6.6	927	963	982	992	997	999	999	1,000		
6.8	915	955	978	990	996	998	999	1,000		
7.0	901	947	973	987	994	998	999	1,000		
7.2	887	937	967	984	993	997	999	999	1,000	
7.4	871	926	961	980	991	996	998	999	1,000	
7.6	854	915	954	976	989	995	998	999	1,000	
7.8	835	902	945	971	986	993	997	999	1,000	
8.0	816	888	936	966	983	992	996	998	999	1,000
8.5	763	849	909	949	973	986	993	997	999	999
9.0	706	803	876	926	959	978	989	995	998	999
9.5	645	752	836	898	940	967	982	991	996	998
10.0	583	697	792	864	917	951	973	986	993	997

λ \ k	20	21	22
8.5	1,000		
9.0	1,000		
9.5	999	1,000	
10.0	998	999	1,000

λ \ k	0	1	2	3	4	5	6	7	8	9
10.5	000	000	002	007	021	050	102	179	279	397
11.0	000	000	001	005	015	038	079	143	232	341
11.5	000	000	001	003	011	028	060	114	191	289
12.0	000	000	001	002	008	020	046	090	155	242
12.5	000	000	000	002	005	015	035	070	125	201
13.0	000	000	000	001	004	011	026	054	100	166
13.5	000	000	000	001	003	008	019	041	079	135
14.0	000	000	000	000	002	006	014	032	062	109
14.5	000	000	000	000	001	004	010	024	048	088
15.0	000	000	000	000	001	003	008	018	037	070

λ \ k	10	11	12	13	14	15	16	17	18	19
10.5	521	639	742	825	888	932	960	978	988	994
11.0	460	579	689	781	854	907	944	968	982	991
11.5	402	520	633	733	815	878	924	954	974	986
12.0	347	462	576	682	772	844	899	937	963	979
12.5	297	406	519	628	725	806	869	916	948	969
13.0	252	353	463	573	675	764	835	890	930	957
13.5	211	304	409	518	623	718	798	861	908	942
14.0	176	260	358	464	570	669	756	827	883	923
14.5	145	220	311	413	518	619	711	790	853	901
15.0	118	185	268	363	466	568	664	749	819	875

λ \ k	20	21	22	23	24	25	26	27	28	29
10.5	997	999	999	1,000						
11.0	995	998	999	1,000						
11.5	992	996	998	999	1,000					
12.0	988	994	997	999	999	1,000				
12.5	983	991	995	998	999	999	1,000			
13.0	975	986	992	996	998	999	1,000			
13.5	965	980	989	994	997	998	999	1,000		
14.0	952	971	983	991	995	997	999	999	1,000	
14.5	936	960	976	986	992	996	998	999	999	1,000
15.0	917	947	967	981	989	994	997	998	999	1,000

λ \ k	4	5	6	7	8	9	10	11	12	13
16	000	001	004	010	022	043	077	127	193	275
17	000	001	002	005	013	026	049	085	135	201
18	000	000	001	003	007	015	030	055	092	143
19	000	000	001	002	004	009	018	035	061	098
20	000	000	000	001	002	005	011	021	039	066
21	000	000	000	000	001	003	006	013	025	043
22	000	000	000	000	001	002	004	008	015	028
23	000	000	000	000	000	001	002	004	009	017
24	000	000	000	000	000	000	001	003	005	011
25	000	000	000	000	000	000	001	001	003	006
	14	15	16	17	18	19	20	21	22	23
16	368	467	566	659	742	812	868	911	942	963
17	281	371	468	564	655	736	805	861	905	937
18	208	287	375	469	562	651	731	799	855	899
19	150	215	292	378	469	561	647	725	793	849
20	105	157	221	297	381	470	559	644	721	787
21	072	111	163	227	302	384	471	558	640	716
22	048	077	117	169	232	306	387	472	556	637
23	031	052	082	123	175	238	310	389	472	555
24	020	034	056	087	128	180	243	314	392	473
25	012	022	038	060	092	134	185	247	318	394
	24	25	26	27	28	29	30	31	32	33
16	978	987	993	996	998	999	999	1,000		
17	959	975	985	991	995	997	999	999	1,000	
18	932	955	972	983	990	994	997	998	999	1,000
19	893	927	951	969	980	988	993	996	998	999
20	843	888	922	948	966	978	987	992	995	997
21	782	838	883	917	944	963	976	985	991	994
22	712	777	832	877	913	940	959	973	983	989
23	635	708	772	827	873	908	936	956	971	981
24	554	632	704	768	823	868	904	932	953	969
25	473	553	629	700	763	818	863	900	929	950
	34	35	36	37	38	39	40	41	42	43
19	999	1,000								
20	999	999	1,000							
21	997	998	999	999	1,000					
22	994	996	998	999	999	1,000				
23	988	993	996	997	999	999	1,000			
24	979	987	992	995	997	998	999	999	1,000	
25	966	978	985	991	994	997	998	999	999	1,000

Table A-8: Binomial Coefficients

n	$\binom{n}{0}$	$\binom{n}{1}$	$\binom{n}{2}$	$\binom{n}{3}$	$\binom{n}{4}$	$\binom{n}{5}$	$\binom{n}{6}$	$\binom{n}{7}$	$\binom{n}{8}$	$\binom{n}{9}$	$\binom{n}{10}$
0	1										
1	1	1									
2	1	2	1								
3	1	3	3	1							
4	1	4	6	4	1						
5	1	5	10	10	5	1					
6	1	6	15	20	15	6	1				
7	1	7	21	35	35	21	7	1			
8	1	8	28	56	70	56	28	8	1		
9	1	9	36	84	126	126	84	36	9	1	
10	1	10	45	120	210	252	210	120	45	10	1
11	1	11	55	165	330	462	462	330	165	55	11
12	1	12	66	220	495	792	924	792	495	220	66
13	1	13	78	286	715	1287	1716	1716	1287	715	286
14	1	14	91	364	1001	2002	3003	3432	3003	2002	1001
15	1	15	105	455	1365	3003	5005	6435	6435	5005	3003
16	1	16	120	560	1820	4368	8008	11440	12870	11440	8008
17	1	17	136	680	2380	6188	12376	19448	24310	24310	19448
18	1	18	153	816	3060	8568	18564	31824	43758	48620	43758
19	1	19	171	969	3876	11628	27132	50388	75582	92378	92378
20	1	20	190	1140	4845	15504	38760	77520	125970	167960	184756

$$C_r^n = \binom{n}{r} = \frac{n!}{r!(n-r)!} = \binom{n}{n-r}, \quad \binom{n}{0} = 1.$$

$$(q+p)^n = q^n + C_1^n p q^{n-1} + \cdots + C_r^n p^r q^{n-r} + \cdots + p^n.$$

Source: This table is reproduced from *Handbook of Probability and Statistics with Tables* by R. S. Burington and D. C. May. Second Edition Copyright © 1970 by McGraw-Hill, Inc. Used with permission of McGraw-Hill Book Company.

Table A-9: Common Logarithms of Numbers

N	0	1	2	3	4	5	6	7	8	9	Proportional parts 1	2	3	4	5
10	0000	0043	0086	0128	0170	0212	0253	0294	0334	0374	4	8	12	17	21
11	0414	0453	0492	0531	0569	0607	0645	0682	0719	0755	4	8	11	15	19
12	0792	0828	0864	0899	0934	0969	1004	1038	1072	1106	3	7	10	14	17
13	1139	1173	1206	1239	1271	1303	1335	1367	1399	1430	3	6	10	13	16
14	1461	1492	1523	1553	1584	1614	1644	1673	1703	1732	3	6	9	12	15
15	1761	1790	1818	1847	1875	1903	1931	1959	1987	2014	3	6	8	11	14
16	2041	2068	2095	2122	2148	2175	2201	2227	2253	2279	3	5	8	11	13
17	2304	2330	2355	2380	2405	2430	2455	2480	2504	2529	2	5	7	10	12
18	2553	2577	2601	2625	2648	2672	2695	2718	2742	2765	2	5	7	9	12
19	2788	2810	2833	2856	2878	2900	2923	2945	2967	2989	2	4	7	9	11
20	3010	3032	3054	3075	3096	3118	3139	3160	3181	3201	2	4	6	8	11
21	3222	3243	3263	3284	3304	3324	3345	3365	3385	3404	2	4	6	8	10
22	3424	3444	3464	3483	3502	3522	3541	3560	3579	3598	2	4	6	8	10
23	3617	3636	3655	3674	3692	3711	3729	3747	3766	3784	2	4	6	7	9
24	3802	3820	3838	3856	3874	3892	3909	3927	3945	3962	2	4	5	7	9
25	3979	3997	4014	4031	4048	4065	4082	4099	4116	4133	2	4	5	7	9
26	4150	4166	4183	4200	4216	4232	4249	4265	4281	4298	2	3	5	7	8
27	4314	4330	4346	4362	4378	4393	4409	4425	4440	4456	2	3	5	6	8
28	4472	4487	4502	4518	4533	4548	4564	4579	4594	4609	2	3	5	6	8
29	4624	4639	4654	4669	4683	4698	4713	4728	4742	4757	1	3	4	6	7
30	4771	4786	4800	4814	4829	4843	4857	4871	4886	4900	1	3	4	6	7
31	4914	4928	4942	4955	4969	4983	4997	5011	5024	5038	1	3	4	5	7
32	5051	5065	5079	5092	5105	5119	5132	5145	5159	5172	1	3	4	5	7
33	5185	5198	5211	5224	5237	5250	5263	5276	5289	5302	1	3	4	5	7
34	5315	5328	5340	5353	5366	5378	5391	5403	5416	5428	1	2	4	5	6
35	5441	5453	5465	5478	5490	5502	5514	5527	5539	5551	1	2	4	5	6
36	5563	5575	5587	5599	5611	5623	5635	5647	5658	5670	1	2	4	5	6
37	5682	5694	5705	5717	5729	5740	5752	5763	5775	5786	1	2	4	5	6
38	5798	5809	5821	5832	5843	5855	5866	5877	5888	5899	1	2	3	5	6
39	5911	5922	5933	5944	5955	5966	5977	5988	5999	6010	1	2	3	4	5
40	6021	6031	6042	6053	6064	6075	6085	6096	6107	6117	1	2	3	4	5
41	6128	6138	6149	6160	6170	6180	6191	6201	6212	6222	1	2	3	4	5
42	6232	6243	6253	6263	6274	6284	6294	6304	6314	6325	1	2	3	4	5
43	6335	6345	6355	6365	6375	6385	6395	6405	6415	6425	1	2	3	4	5
44	6435	6444	6454	6464	6474	6484	6493	6503	6513	6522	1	2	3	4	5
45	6532	6542	6551	6561	6571	6580	6590	6599	6609	6618	1	2	3	4	5
46	6628	6637	6646	6656	6665	6675	6684	6693	6702	6712	1	2	3	4	5
47	6721	6730	6739	6749	6758	6767	6776	6785	6794	6803	1	2	3	4	5
48	6812	6821	6830	6839	6848	6857	6866	6875	6884	6893	1	2	3	4	5
49	6902	6911	6920	6928	6937	6946	6955	6964	6972	6981	1	2	3	4	4
50	6990	6998	7007	7016	7024	7033	7042	7050	7059	7067	1	2	3	3	4
51	7076	7084	7093	7101	7110	7118	7126	7135	7143	7152	1	2	3	3	4
52	7160	7168	7177	7185	7193	7202	7210	7218	7226	7235	1	2	3	3	4
53	7243	7251	7259	7267	7275	7284	7292	7300	7308	7316	1	2	2	3	4
54	7324	7332	7340	7348	7356	7364	7372	7380	7388	7396	1	2	2	3	4
N	0	1	2	3	4	5	6	7	8	9	1	2	3	4	5

N	0	1	2	3	4	5	6	7	8	9	Proportional parts				
											1	2	3	4	5
55	7404	7412	7419	7427	7435	7443	7451	7459	7466	7474	1	2	2	3	4
56	7482	7490	7497	7505	7513	7520	7528	7536	7543	7551	1	2	2	3	4
57	7559	7566	7574	7582	7589	7597	7604	7612	7619	7627	1	1	2	3	4
58	7634	7642	7649	7657	7664	7672	7679	7686	7694	7701	1	1	2	3	4
59	7709	7716	7723	7731	7738	7745	7752	7760	7767	7774	1	1	2	3	4
60	7782	7789	7796	7803	7810	7818	7825	7832	7839	7846	1	1	2	3	4
61	7853	7860	7868	7875	7882	7889	7896	7903	7910	7917	1	1	2	3	3
62	7924	7931	7938	7945	7952	7959	7966	7973	7980	7987	1	1	2	3	3
63	7993	8000	8007	8014	8021	8028	8035	8041	8048	8055	1	1	2	3	3
64	8062	8069	8075	8082	8089	8096	8102	8109	8116	8122	1	1	2	3	3
65	8129	8136	8142	8149	8156	8162	8169	8176	8182	8189	1	1	2	3	3
66	8195	8202	8209	8215	8222	8228	8235	8241	8248	8254	1	1	2	3	3
67	8261	8267	8274	8280	8287	8293	8299	8306	8312	8319	1	1	2	3	3
68	8325	8331	8338	8344	8351	8357	8363	8370	8376	8382	1	1	2	3	3
69	8388	8395	8401	8407	8414	8420	8426	8432	8439	8445	1	1	2	3	3
70	8451	8457	8463	8470	8476	8482	8488	8494	8500	8506	1	1	2	3	3
71	8513	8519	8525	8531	8537	8543	8549	8555	8561	8567	1	1	2	3	3
72	8573	8579	8585	8591	8597	8603	8609	8615	8621	8627	1	1	2	3	3
73	8633	8639	8645	8651	8657	8663	8669	8675	8681	8686	1	1	2	2	3
74	8692	8698	8704	8710	8716	8722	8727	8733	8739	8745	1	1	2	2	3
75	8751	8756	8762	8768	8774	8779	8785	8791	8797	8802	1	1	2	2	3
76	8808	8814	8820	8825	8831	8837	8842	8848	8854	8859	1	1	2	2	3
77	8865	8871	8876	8882	8887	8893	8899	8904	8910	8915	1	1	2	2	3
78	8921	8927	8932	8938	8943	8949	8954	8960	8965	8971	1	1	2	2	3
79	8976	8982	8987	8993	8998	9004	9009	9015	9020	9025	1	1	2	2	3
80	9031	9036	9042	9047	9053	9058	9063	9069	9074	9079	1	1	2	2	3
81	9085	9090	9096	9101	9106	9112	9117	9122	9128	9133	1	1	2	2	3
82	9138	9143	9149	9154	9159	9165	9170	9175	9180	9186	1	1	2	2	3
83	9191	9196	9201	9206	9212	9217	9222	9227	9232	9238	1	1	2	2	3
84	9243	9248	9253	9258	9263	9269	9274	9279	9284	9289	1	1	2	2	3
85	9294	9299	9304	9309	9315	9320	9325	9330	9335	9340	1	1	2	2	3
86	9345	9350	9355	9360	9365	9370	9375	9380	9385	9390	1	1	2	2	3
87	9395	9400	9405	9410	9415	9420	9425	9430	9435	9440	1	1	2	2	3
88	9445	9450	9455	9460	9465	9469	9474	9479	9484	9489	0	1	1	2	2
89	9494	9499	9504	9509	9513	9518	9523	9528	9533	9538	0	1	1	2	2
90	9542	9547	9552	9557	9562	9566	9571	9576	9581	9586	0	1	1	2	2
91	9590	9595	9600	9605	9609	9614	9619	9624	9628	9633	0	1	1	2	2
92	9638	9643	9647	9652	9657	9661	9666	9671	9675	9680	0	1	1	2	2
93	9685	9689	9694	9699	9703	9708	9713	9717	9722	9727	0	1	1	2	2
94	9731	9736	9741	9745	9750	9754	9759	9763	9768	9773	0	1	1	2	2
95	9777	9782	9786	9791	9795	9800	9805	9809	9814	9818	0	1	1	2	2
96	9823	9827	9832	9836	9841	9845	9850	9854	9859	9863	0	1	1	2	2
97	9868	9872	9877	9881	9886	9890	9894	9899	9903	9908	0	1	1	2	2
98	9912	9917	9921	9926	9930	9934	9939	9943	9948	9952	0	1	1	2	2
99	9956	9961	9965	9969	9974	9978	9983	9987	9991	9996	0	1	1	2	2
N	0	1	2	3	4	5	6	7	8	9	1	2	3	4	5

Table A-10: Squares, Square-Roots, and Reciprocals

n	n^2	$\sqrt{n}$	$\sqrt{10n}$	$1000/n$
1	1	1.0000	3.1623	1000.0
2	4	1.4142	4.4721	500.00
3	9	1.7321	5.4772	333.33
4	16	2.0000	6.3246	250.00
5	25	2.2361	7.0711	200.00
6	36	2.4495	7.7460	166.67
7	49	2.6458	8.3666	142.86
8	64	2.8284	8.9443	125.00
9	81	3.0000	9.4868	111.11
10	100	3.1623	10.000	100.00
11	121	3.3166	10.488	90.909
12	144	3.4641	10.954	83.333
13	169	3.6056	11.402	76.923
14	196	3.7417	11.832	71.429
15	225	3.8730	12.247	66.667
16	256	4.0000	12.649	62.500
17	289	4.1231	13.038	58.824
18	324	4.2426	13.416	55.556
19	361	4.3589	13.784	52.632
20	400	4.4721	14.142	50.000
21	441	4.5826	14.491	47.619
22	484	4.6904	14.832	45.455
23	529	4.7958	15.166	43.478
24	576	4.8990	15.492	41.667
25	625	5.0000	15.811	40.000
26	676	5.0990	16.125	38.462
27	729	5.1962	16.432	37.037
28	784	5.2915	16.733	35.714
29	841	5.3852	17.029	34.483
30	900	5.4772	17.321	33.333
31	961	5.5678	17.607	32.258
32	1 024	5.6569	17.889	31.250
33	1 089	5.7446	18.166	30.303
34	1 156	5.8310	18.439	29.412
35	1 225	5.9161	18.708	28.571
36	1 296	6.0000	18.974	27.778
37	1 369	6.0828	19.235	27.027
38	1 444	6.1644	19.494	26.316
39	1 521	6.2450	19.748	25.641
40	1 600	6.3246	20.000	25.000
41	1 681	6.4031	20.248	24.390
42	1 764	6.4807	20.494	23.810
43	1 849	6.5574	20.736	23.256
44	1 936	6.6333	20.976	22.727
45	2 025	6.7082	21.213	22.222
46	2 116	6.7823	21.448	21.739
47	2 209	6.8557	21.679	21.277
48	2 304	6.9282	21.909	20.833
49	2 401	7.0000	22.136	20.408

n	n^2	$\sqrt{n}$	$\sqrt{10n}$	$1000/n$
50	2 500	7.0711	22.361	20.000
51	2 601	7.1414	22.583	19.608
52	2 704	7.2111	22.804	19.231
53	2 809	7.2801	23.022	18.868
54	2 916	7.3485	23.238	18.519
55	3 025	7.4162	23.452	18.182
56	3 136	7.4833	23.664	17.857
57	3 249	7.5498	23.875	17.544
58	3 364	7.6158	24.083	17.241
59	3 481	7.6811	24.290	16.949
60	3 600	7.7460	24.495	16.667
61	3 721	7.8103	24.698	16.393
62	3 844	7.8740	24.900	16.129
63	3 969	7.9373	25.100	15.873
64	4 096	8.0000	25.298	15.625
65	4 225	8.0623	25.495	15.385
66	4 356	8.1240	25.690	15.152
67	4 489	8.1854	25.884	14.925
68	4 624	8.2462	26.077	14.706
69	4 761	8.3066	26.268	14.493
70	4 900	8.3666	26.458	14.286
71	5 041	8.4262	26.646	14.085
72	5 184	8.4853	26.833	13.889
73	5 329	8.5440	27.019	13.699
74	5 476	8.6023	27.203	13.514
75	5 625	8.6603	27.386	13.333
76	5 776	8.7178	27.568	13.158
77	5 929	8.7750	27.749	12.987
78	6 084	8.8318	27.928	12.821
79	6 241	8.8882	28.107	12.658
80	6 400	8.9443	28.284	12.500
81	6 561	9.0000	28.461	12.346
82	6 724	9.0554	28.636	12.195
83	6 889	9.1104	28.810	12.048
84	7 056	9.1652	28.983	11.905
85	7 225	9.2195	29.155	11.765
86	7 396	9.2736	29.326	11.628
87	7 569	9.3274	29.496	11.494
88	7 744	9.3808	29.665	11.364
89	7 921	9.4340	29.833	11.236
90	8 100	9.4868	30.000	11.111
91	8 281	9.5394	30.166	10.989
92	8 464	9.5917	30.332	10.870
93	8 649	9.6437	30.496	10.753
94	8 836	9.6954	30.659	10.638
95	9 025	9.7468	30.822	10.526
96	9 216	9.7980	30.984	10.417
97	9 409	9.8489	31.145	10.309
98	9 604	9.8995	31.305	10.204
99	9 801	9.9499	31.464	10.101

n	n^2	$\sqrt{n}$	$\sqrt{10n}$	$1000/n$
100	10 000	10.000	31.623	10.000
101	10 201	10.050	31.781	9.9010
102	10 404	10.100	31.937	9.8039
103	10 609	10.149	32.094	9.7087
104	10 816	10.198	32.249	9.6154
105	11 025	10.247	32.404	9.5238
106	11 236	10.296	32.558	9.4340
107	11 449	10.344	32.711	9.3458
108	11 664	10.392	32.863	9.2593
109	11 881	10.440	33.015	9.1743
110	12 100	10.488	33.166	9.0909
111	12 321	10.536	33.317	9.0090
112	12 544	10.583	33.466	8.9286
113	12 769	10.630	33.615	8.8496
114	12 996	10.677	33.764	8.7719
115	13 225	10.724	33.912	8.6957
116	13 456	10.770	34.059	8.6207
117	13 689	10.817	34.205	8.5470
118	13 924	10.863	34.351	8.4746
119	14 161	10.909	34.496	8.4034
120	14 400	10.954	34.641	8.3333
121	14 641	11.000	34.785	8.2645
122	14 884	11.045	34.929	8.1967
123	15 129	11.091	35.071	8.1301
124	15 376	11.136	35.214	8.0645
125	15 625	11.180	35.355	8.0000
126	15 876	11.225	35.496	7.9365
127	16 129	11.269	35.637	7.8740
128	16 384	11.314	35.777	7.8125
129	16 641	11.358	35.917	7.7519
130	16 900	11.402	36.056	7.6923
131	17 161	11.446	36.194	7.6336
132	17 424	11.489	36.332	7.5758
133	17 689	11.533	36.469	7.5188
134	17 956	11.576	36.606	7.4627
135	18 225	11.619	36.742	7.4074
136	18 496	11.662	36.878	7.3529
137	18 769	11.705	37.014	7.2993
138	19 044	11.747	37.148	7.2464
139	19 321	11.790	37.283	7.1942
140	19 600	11.832	37.417	7.1429
141	19 881	11.874	37.550	7.0922
142	20 164	11.916	37.683	7.0423
143	20 449	11.958	37.815	6.9930
144	20 736	12.000	37.947	6.9444
145	21 025	12.042	38.079	6.8966
146	21 316	12.083	38.210	6.8493
147	21 609	12.124	38.341	6.8027
148	21 904	12.166	38.471	6.7568
149	22 201	12.207	38.601	6.7114

n	n^2	$\sqrt{n}$	$\sqrt{10n}$	$1000/n$
150	22 500	12.247	38.730	6.6667
151	22 801	12.288	38.859	6.6225
152	23 104	12.329	38.987	6.5789
153	23 409	12.369	39.115	6.5359
154	23 716	12.410	39.243	6.4935
155	24 025	12.450	39.370	6.4516
156	24 336	12.490	39.497	6.4103
157	24 649	12.530	39.623	6.3694
158	24 964	12.570	39.749	6.3291
159	25 281	12.610	39.875	6.2893
160	25 600	12.649	40.000	6.2500
161	25 921	12.689	40.125	6.2112
162	26 244	12.728	40.249	6.1728
163	26 569	12.767	40.373	6.1350
164	26 896	12.806	40.497	6.0976
165	27 225	12.845	40.620	6.0606
166	27 556	12.884	40.743	6.0241
167	27 889	12.923	40.866	5.9880
168	28 224	12.961	40.988	5.9524
169	28 561	13.000	41.110	5.9172
170	28 900	13.038	41.231	5.8824
171	29 241	13.077	41.352	5.8480
172	29 584	13.115	41.473	5.8140
173	29 929	13.153	41.593	5.7803
174	30 276	13.191	41.713	5.7471
175	30 625	13.229	41.833	5.7143
176	30 976	13.267	41.952	5.6818
177	31 329	13.304	42.071	5.6497
178	31 684	13.342	42.190	5.6180
179	32 041	13.379	42.308	5.5866
180	32 400	13.416	42.426	5.5556
181	32 761	13.454	42.544	5.5249
182	33 124	13.491	42.661	5.4945
183	33 489	13.528	42.779	5.4645
184	33 856	13.565	42.895	5.4348
185	34 225	13.601	43.012	5.4054
186	34 596	13.638	43.128	5.3763
187	34 969	13.675	43.244	5.3476
188	35 344	13.711	43.359	5.3191
189	35 721	13.748	43.474	5.2910
190	36 100	13.784	43.589	5.2632
191	36 481	13.820	43.704	5.2356
192	36 864	13.856	43.818	5.2083
193	37 249	13.892	43.932	5.1813
194	37 636	13.928	44.045	5.1546
195	38 025	13.964	44.159	5.1282
196	38 416	14.000	44.272	5.1020
197	38 809	14.036	44.385	5.0761
198	39 204	14.071	44.497	5.0505
199	39 601	14.107	44.609	5.0251

n	n^2	$\sqrt{n}$	$\sqrt{10n}$	$1000/n$
200	40 000	14.142	44.721	5.0000
201	40 401	14.177	44.833	4.9751
202	40 804	14.213	44.944	4.9505
203	41 209	14.248	45.056	4.9261
204	41 616	14.283	45.166	4.9020
205	42 025	14.318	45.277	4.8780
206	42 436	14.353	45.387	4.8544
207	42 849	14.387	45.497	4.8309
208	43 264	14.422	45.607	4.8077
209	43 681	14.457	45.717	4.7847
210	44 100	14.491	45.826	4.7619
211	44 521	14.526	45.935	4.7393
212	44 944	14.560	46.043	4.7170
213	45 369	14.595	46.152	4.6948
214	45 796	14.629	46.260	4.6729
215	46 225	14.663	46.368	4.6512
216	46 656	14.697	46.476	4.6296
217	47 089	14.731	46.583	4.6083
218	47 524	14.765	46.690	4.5872
219	47 961	14.799	46.797	4.5662
220	48 400	14.832	46.904	4.5455
221	48 841	14.866	47.011	4.5249
222	49 284	14.900	47.117	4.5045
223	49 729	14.933	47.223	4.4843
224	50 176	14.967	47.329	4.4643
225	50 625	15.000	47.434	4.4444
226	51 076	15.033	47.539	4.4248
227	51 529	15.067	47.645	4.4053
228	51 984	15.100	47.749	4.3860
229	52 441	15.133	47.854	4.3668
230	52 900	15.166	47.958	4.3478
231	53 361	15.199	48.062	4.3290
232	53 824	15.232	48.166	4.3103
233	54 289	15.264	48.270	4.2918
234	54 756	15.297	48.374	4.2735
235	55 225	15.330	48.477	4.2553
236	55 696	15.362	48.580	4.2373
237	56 169	15.395	48.683	4.2194
238	56 644	15.427	48.785	4.2017
239	57 121	15.460	48.888	4.1841
240	57 600	15.492	48.990	4.1667
241	58 081	15.524	49.092	4.1494
242	58 564	15.556	49.194	4.1322
243	59 049	15.588	49.295	4.1152
244	59 536	15.621	49.396	4.0984
245	60 025	15.652	49.497	4.0816
246	60 516	15.684	49.598	4.0650
247	61 009	15.716	49.699	4.0486
248	61 504	15.748	49.800	4.0323
249	62 001	15.780	49.900	4.0161
250	62 500	15.811	50.000	4.0000
251	63 001	15.843	50.100	3.9841
252	63 504	15.875	50.200	3.9683
253	64 009	15.906	50.299	3.9526
254	64 516	15.937	50.398	3.9370
255	65 025	15.969	50.498	3.9216
256	65 536	16.000	50.596	3.9063
257	66 049	16.031	50.695	3.8911
258	66 564	16.062	50.794	3.8760
259	67 081	16.093	50.892	3.8610
260	67 600	16.125	50.990	3.8462
261	68 121	16.155	51.088	3.8314
262	68 644	16.186	51.186	3.8168
263	69 169	16.217	51.284	3.8023
264	69 696	16.248	51.381	3.7879
265	70 225	16.279	51.478	3.7736
266	70 756	16.310	51.575	3.7594
267	71 289	16.340	51.672	3.7453
268	71 824	16.371	51.769	3.7313
269	72 361	16.401	51.865	3.7175
270	72 900	16.432	51.962	3.7037
271	73 441	16.462	52.058	3.6900
272	73 984	16.492	52.154	3.6765
273	74 529	16.523	52.249	3.6630
274	75 076	16.553	52.345	3.6496
275	75 625	16.583	52.440	3.6364
276	76 176	16.613	52.536	3.6232
277	76 729	16.643	52.631	3.6101
278	77 284	16.673	52.726	3.5971
279	77 841	16.703	52.820	3.5842
280	78 400	16.733	52.915	3.5714
281	78 961	16.763	53.009	3.5587
282	79 524	16.793	53.104	3.5461
283	80 089	16.823	53.198	3.5336
284	80 656	16.852	53.292	3.5211
285	81 225	16.882	53.385	3.5088
286	81 796	16.912	53.479	3.4965
287	82 369	16.941	53.572	3.4843
288	82 944	16.971	53.666	3.4722
289	83 521	17.000	53.759	3.4602
290	84 100	17.029	53.852	3.4483
291	84 681	17.059	53.944	3.4364
292	85 264	17.088	54.037	3.4247
293	85 849	17.117	54.129	3.4130
294	86 436	17.146	54.222	3.4014
295	87 025	17.176	54.314	3.3898
296	87 616	17.205	54.406	3.3784
297	88 209	17.234	54.498	3.3670
298	88 804	17.263	54.589	3.3557
299	89 401	17.292	54.681	3.3445
300	90 000	17.321	54.772	3.3333
301	90 601	17.349	54.863	3.3223
302	91 204	17.378	54.955	3.3113
303	91 809	17.407	55.045	3.3003
304	92 416	17.436	55.136	3.2895
305	93 025	17.464	55.227	3.2787
306	93 636	17.493	55.317	3.2680
307	94 249	17.521	55.408	3.2573
308	94 864	17.550	55.498	3.2468
309	95 481	17.578	55.588	3.2362
310	96 100	17.607	55.678	3.2258
311	96 721	17.635	55.767	3.2154
312	97 344	17.664	55.857	3.2051
313	97 969	17.692	55.946	3.1949
314	98 596	17.720	56.036	3.1847
315	99 225	17.748	56.125	3.1746
316	99 856	17.776	56.214	3.1646
317	100 489	17.804	56.303	3.1546
318	101 124	17.833	56.391	3.1447
319	101 761	17.861	56.480	3.1348
320	102 400	17.889	56.569	3.1250
321	103 041	17.916	56.657	3.1153
322	103 684	17.944	56.745	3.1056
323	104 329	17.972	56.833	3.0960
324	104 976	18.000	56.921	3.0864
325	105 625	18.028	57.009	3.0769
326	106 276	18.055	57.096	3.0675
327	106 929	18.083	57.184	3.0581
328	107 584	18.111	57.271	3.0488
329	108 241	18.138	57.359	3.0395
330	108 900	18.166	57.446	3.0303
331	109 561	18.193	57.533	3.0211
332	110 224	18.221	57.619	3.0120
333	110 889	18.248	57.706	3.0030
334	111 556	18.276	57.793	2.9940
335	112 225	18.303	57.879	2.9851
336	112 896	18.330	57.966	2.9762
337	113 569	18.358	58.052	2.9674
338	114 244	18.385	58.138	2.9586
339	114 921	18.412	58.224	2.9499
340	115 600	18.439	58.310	2.9412
341	116 281	18.466	58.395	2.9326
342	116 964	18.493	58.481	2.9240
343	117 649	18.520	58.566	2.9155
344	118 336	18.547	58.652	2.9070
345	119 025	18.574	58.737	2.8986
346	119 716	18.601	58.822	2.8902
347	120 409	18.628	58.907	2.8818
348	121 104	18.655	58.992	2.8736
349	121 801	18.682	59.076	2.8653
350	122 500	18.708	59.161	2.8571
351	123 201	18.735	59.245	2.8490
352	123 904	18.762	59.330	2.8409
353	124 609	18.788	59.414	2.8329
354	125 316	18.815	59.498	2.8249
355	126 025	18.841	59.582	2.8169
356	126 736	18.868	59.666	2.8090
357	127 449	18.894	59.749	2.8011
358	128 164	18.921	59.833	2.7933
359	128 881	18.947	59.917	2.7855
360	129 600	18.974	60.000	2.7778
361	130 321	19.000	60.083	2.7701
362	131 044	19.026	60.166	2.7624
363	131 769	19.053	60.249	2.7548
364	132 496	19.079	60.332	2.7473
365	133 225	19.105	60.415	2.7397
366	133 956	19.131	60.498	2.7322
367	134 689	19.157	60.581	2.7248
368	135 424	19.183	60.663	2.7174
369	136 161	19.209	60.745	2.7100
370	136 900	19.235	60.828	2.7027
371	137 641	19.261	60.910	2.6954
372	138 384	19.287	60.992	2.6882
373	139 129	19.313	61.074	2.6810
374	139 876	19.339	61.156	2.6738
375	140 625	19.365	61.237	2.6667
376	141 376	19.391	61.319	2.6596
377	142 129	19.416	61.400	2.6525
378	142 884	19.442	61.482	2.6455
379	143 641	19.468	61.563	2.6385
380	144 400	19.494	61.644	2.6316
381	145 161	19.519	61.725	2.6247
382	145 924	19.545	61.806	2.6178
383	146 689	19.570	61.887	2.6110
384	147 456	19.596	61.968	2.6042
385	148 225	19.621	62.048	2.5974
386	148 996	19.647	62.129	2.5907
387	149 769	19.672	62.209	2.5840
388	150 544	19.698	62.290	2.5773
389	151 321	19.723	62.370	2.5707
390	152 100	19.748	62.450	2.5641
391	152 881	19.774	62.530	2.5575
392	153 664	19.799	62.610	2.5510
393	154 449	19.824	62.690	2.5445
394	155 236	19.849	62.769	2.5381
395	156 025	19.875	62.849	2.5316
396	156 816	19.900	62.929	2.5253
397	157 609	19.925	63.008	2.5189
398	158 404	19.950	63.087	2.5126
399	159 201	19.975	63.166	2.5063

n	n^2	$\sqrt{n}$	$\sqrt{10n}$	$1000/n$
400	160 000	20.000	63.246	2.5000
401	160 801	20.025	63.325	2.4938
402	161 604	20.050	63.403	2.4876
403	162 409	20.075	63.482	2.4814
404	163 216	20.100	63.561	2.4752
405	164 025	20.125	63.640	2.4691
406	164 836	20.149	63.718	2.4631
407	165 649	20.174	63.797	2.4570
408	166 464	20.199	63.875	2.4510
409	167 281	20.224	63.953	2.4450
410	168 100	20.248	64.031	2.4390
411	168 921	20.273	64.109	2.4331
412	169 744	20.298	64.187	2.4272
413	170 569	20.322	64.265	2.4213
414	171 396	20.347	64.343	2.4155
415	172 225	20.372	64.420	2.4096
416	173 056	20.396	64.498	2.4038
417	173 889	20.421	64.576	2.3981
418	174 724	20.445	64.653	2.3923
419	175 561	20.469	64.730	2.3866
420	176 400	20.494	64.807	2.3810
421	177 241	20.518	64.885	2.3753
422	178 084	20.543	64.962	2.3697
423	178 929	20.567	65.038	2.3641
424	179 776	20.591	65.115	2.3585
425	180 625	20.616	65.192	2.3529
426	181 476	20.640	65.269	2.3474
427	182 329	20.664	65.345	2.3419
428	183 184	20.688	65.422	2.3364
429	184 041	20.712	65.498	2.3310
430	184 900	20.736	65.574	2.3256
431	185 761	20.761	65.651	2.3202
432	186 624	20.785	65.727	2.3148
433	187 489	20.809	65.803	2.3095
434	188 356	20.833	65.879	2.3041
435	189 225	20.857	65.955	2.2989
436	190 096	20.881	66.030	2.2936
437	190 969	20.905	66.106	2.2883
438	191 844	20.928	66.182	2.2831
439	192 721	20.952	66.257	2.2779
440	193 600	20.976	66.333	2.2727
441	194 481	21.000	66.408	2.2676
442	195 364	21.024	66.483	2.2624
443	196 249	21.048	66.558	2.2573
444	197 136	21.071	66.633	2.2523
445	198 025	21.095	66.708	2.2472
446	198 916	21.119	66.783	2.2422
447	199 809	21.142	66.858	2.2371
448	200 704	21.166	66.933	2.2321
449	201 601	21.190	67.007	2.2272

n	n^2	$\sqrt{n}$	$\sqrt{10n}$	$1000/n$
450	202 500	21.213	67.082	2.2222
451	203 401	21.237	67.157	2.2173
452	204 304	21.260	67.231	2.2124
453	205 209	21.284	67.305	2.2075
454	206 116	21.307	67.380	2.2026
455	207 025	21.331	67.454	2.1978
456	207 936	21.354	67.528	2.1930
457	208 849	21.378	67.602	2.1882
458	209 764	21.401	67.676	2.1834
459	210 681	21.424	67.750	2.1786
460	211 600	21.448	67.823	2.1739
461	212 521	21.471	67.897	2.1692
462	213 444	21.494	67.971	2.1645
463	214 369	21.517	68.044	2.1598
464	215 296	21.541	68.118	2.1552
465	216 225	21.564	68.191	2.1505
466	217 156	21.587	68.264	2.1459
467	218 089	21.610	68.337	2.1413
468	219 024	21.633	68.411	2.1368
469	219 961	21.656	68.484	2.1322
470	220 900	21.679	68.557	2.1277
471	221 841	21.703	68.629	2.1231
472	222 784	21.726	68.702	2.1186
473	223 729	21.749	68.775	2.1142
474	224 676	21.772	68.848	2.1097
475	225 625	21.794	68.920	2.1053
476	226 576	21.817	68.993	2.1008
477	227 529	21.840	69.065	2.0964
478	228 484	21.863	69.138	2.0921
479	229 441	21.886	69.210	2.0877
480	230 400	21.909	69.282	2.0833
481	231 361	21.932	69.354	2.0790
482	232 324	21.955	69.426	2.0747
483	233 289	21.977	69.498	2.0704
484	234 256	22.000	69.570	2.0661
485	235 225	22.023	69.642	2.0619
486	236 196	22.045	69.714	2.0576
487	237 169	22.068	69.785	2.0534
488	238 144	22.091	69.857	2.0492
489	239 121	22.113	69.929	2.0450
490	240 100	22.136	70.000	2.0408
491	241 081	22.159	70.071	2.0367
492	242 064	22.181	70.143	2.0325
493	243 049	22.204	70.214	2.0284
494	244 036	22.226	70.285	2.0243
495	245 025	22.249	70.356	2.0202
496	246 016	22.271	70.427	2.0161
497	247 009	22.294	70.498	2.0121
498	248 004	22.316	70.569	2.0080
499	249 001	22.338	70.640	2.0040

n	n^2	$\sqrt{n}$	$\sqrt{10n}$	$1000/n$
500	250 000	22.361	70.711	2.0000
501	251 001	22.383	70.781	1.9960
502	252 004	22.405	70.852	1.9920
503	253 009	22.428	70.922	1.9881
504	254 016	22.450	70.993	1.9841
505	255 025	22.472	71.063	1.9802
506	256 036	22.494	71.134	1.9763
507	257 049	22.517	71.204	1.9724
508	258 064	22.539	71.274	1.9685
509	259 081	22.561	71.344	1.9646
510	260 100	22.583	71.414	1.9608
511	261 121	22.605	71.484	1.9569
512	262 144	22.627	71.554	1.9531
513	263 169	22.650	71.624	1.9493
514	264 196	22.672	71.694	1.9455
515	265 225	22.694	71.764	1.9417
516	266 256	22.716	71.833	1.9380
517	267 289	22.738	71.903	1.9342
518	268 324	22.760	71.972	1.9305
519	269 361	22.782	72.042	1.9268
520	270 400	22.804	72.111	1.9231
521	271 441	22.825	72.180	1.9194
522	272 484	22.847	72.250	1.9157
523	273 529	22.869	72.319	1.9120
524	274 576	22.891	72.388	1.9084
525	275 625	22.913	72.457	1.9048
526	276 676	22.935	72.526	1.9011
527	277 729	22.956	72.595	1.8975
528	278 784	22.978	72.664	1.8939
529	279 841	23.000	72.732	1.8904
530	280 900	23.022	72.801	1.8868
531	281 961	23.043	72.870	1.8832
532	283 024	23.065	72.938	1.8797
533	284 089	23.087	73.007	1.8762
534	285 156	23.108	73.075	1.8727
535	286 225	23.130	73.144	1.8692
536	287 296	23.152	73.212	1.8657
537	288 369	23.173	73.280	1.8622
538	289 444	23.195	73.348	1.8587
539	290 521	23.216	73.417	1.8553
540	291 600	23.238	73.485	1.8519
541	292 681	23.259	73.553	1.8484
542	293 764	23.281	73.621	1.8450
543	294 849	23.302	73.689	1.8416
544	295 936	23.324	73.756	1.8382
545	297 025	23.345	73.824	1.8349
546	298 116	23.367	73.892	1.8315
547	299 209	23.388	73.959	1.8282
548	300 304	23.409	74.027	1.8248
549	301 401	23.431	74.095	1.8215

n	n^2	$\sqrt{n}$	$\sqrt{10n}$	$1000/n$
550	302 500	23.452	74.162	1.8182
551	303 601	23.473	74.229	1.8149
552	304 704	23.495	74.297	1.8116
553	305 809	23.516	74.364	1.8083
554	306 916	23.537	74.431	1.8051
555	308 025	23.558	74.498	1.8018
556	309 136	23.580	74.565	1.7986
557	310 249	23.601	74.632	1.7953
558	311 364	23.622	74.699	1.7921
559	312 481	23.643	74.766	1.7889
560	313 600	23.664	74.833	1.7857
561	314 721	23.685	74.900	1.7825
562	315 844	23.707	74.967	1.7794
563	316 969	23.728	75.033	1.7762
564	318 096	23.749	75.100	1.7731
565	319 225	23.770	75.166	1.7699
566	320 356	23.791	75.233	1.7668
567	321 489	23.812	75.299	1.7637
568	322 624	23.833	75.366	1.7606
569	323 761	23.854	75.432	1.7575
570	324 900	23.875	75.498	1.7544
571	326 041	23.896	75.565	1.7513
572	327 184	23.917	75.631	1.7483
573	328 329	23.937	75.697	1.7452
574	329 476	23.958	75.763	1.7422
575	330 625	23.979	75.829	1.7391
576	331 776	24.000	75.895	1.7361
577	332 929	24.021	75.961	1.7331
578	334 084	24.042	76.026	1.7301
579	335 241	24.062	76.092	1.7271
580	336 400	24.083	76.158	1.7241
581	337 561	24.104	76.223	1.7212
582	338 724	24.125	76.289	1.7182
583	339 889	24.145	76.354	1.7153
584	341 056	24.166	76.420	1.7123
585	342 225	24.187	76.485	1.7094
586	343 396	24.207	76.551	1.7065
587	344 569	24.228	76.616	1.7036
588	345 744	24.249	76.681	1.7007
589	346 921	24.269	76.746	1.6978
590	348 100	24.290	76.811	1.6949
591	349 281	24.310	76.877	1.6920
592	350 464	24.331	76.942	1.6892
593	351 649	24.352	77.006	1.6863
594	352 836	24.372	77.071	1.6835
595	354 025	24.393	77.136	1.6807
596	355 216	24.413	77.201	1.6779
597	356 409	24.434	77.266	1.6750
598	357 604	24.454	77.330	1.6722
599	358 801	24.474	77.395	1.6694

n	n^2	$\sqrt{n}$	$\sqrt{10n}$	$1000/n$
600	360 000	24.495	77.460	1.6667
601	361 201	24.515	77.524	1.6639
602	362 404	24.536	77.589	1.6611
603	363 609	24.556	77.653	1.6584
604	364 816	24.576	77.717	1.6556
605	366 025	24.597	77.782	1.6529
606	367 236	24.617	77.846	1.6502
607	368 449	24.637	77.910	1.6474
608	369 664	24.658	77.974	1.6447
609	370 881	24.678	78.038	1.6420
610	372 100	24.698	78.103	1.6393
611	373 321	24.718	78.166	1.6367
612	374 544	24.739	78.230	1.6340
613	375 769	24.759	78.294	1.6313
614	376 996	24.779	78.358	1.6287
615	378 225	24.799	78.422	1.6260
616	379 456	24.819	78.486	1.6234
617	380 689	24.839	78.549	1.6207
618	381 924	24.860	78.613	1.6181
619	383 161	24.880	78.677	1.6155
620	384 400	24.900	78.740	1.6129
621	385 641	24.920	78.804	1.6103
622	386 884	24.940	78.867	1.6077
623	388 129	24.960	78.930	1.6051
624	389 376	24.980	78.994	1.6026
625	390 625	25.000	79.057	1.6000
626	391 876	25.020	79.120	1.5974
627	393 129	25.040	79.183	1.5949
628	394 384	25.060	79.246	1.5924
629	395 641	25.080	79.310	1.5898
630	396 900	25.100	79.373	1.5873
631	398 161	25.120	79.436	1.5848
632	399 424	25.140	79.498	1.5823
633	400 689	25.159	79.561	1.5798
634	401 956	25.179	79.624	1.5773
635	403 225	25.199	79.687	1.5748
636	404 496	25.219	79.750	1.5723
637	405 769	25.239	79.812	1.5699
638	407 044	25.259	79.875	1.5674
639	408 321	25.278	79.937	1.5649
640	409 600	25.298	80.000	1.5625
641	410 881	25.318	80.062	1.5601
642	412 164	25.338	80.125	1.5576
643	413 449	25.357	80.187	1.5552
644	414 736	25.377	80.250	1.5528
645	416 025	25.397	80.312	1.5504
646	417 316	25.417	80.374	1.5480
647	418 609	25.436	80.436	1.5456
648	419 904	25.456	80.498	1.5432
649	421 201	25.475	80.561	1.5408

n	n^2	$\sqrt{n}$	$\sqrt{10n}$	$1000/n$
650	422 500	25.495	80.623	1.5385
651	423 801	25.515	80.685	1.5361
652	425 104	25.534	80.747	1.5337
653	426 409	25.554	80.808	1.5314
654	427 716	25.573	80.870	1.5291
655	429 025	25.593	80.932	1.5267
656	430 336	25.613	80.994	1.5244
657	431 649	25.632	81.056	1.5221
658	432 964	25.652	81.117	1.5198
659	434 281	25.671	81.179	1.5175
660	435 600	25.690	81.240	1.5152
661	436 921	25.710	81.302	1.5129
662	438 244	25.729	81.363	1.5106
663	439 569	25.749	81.425	1.5083
664	440 896	25.768	81.486	1.5060
665	442 225	25.788	81.548	1.5038
666	443 556	25.807	81.609	1.5015
667	444 889	25.826	81.670	1.4993
668	446 224	25.846	81.731	1.4970
669	447 561	25.865	81.792	1.4948
670	448 900	25.884	81.854	1.4925
671	450 241	25.904	81.915	1.4903
672	451 584	25.923	81.976	1.4881
673	452 929	25.942	82.037	1.4859
674	454 276	25.962	82.098	1.4837
675	455 625	25.981	82.158	1.4815
676	456 976	26.000	82.219	1.4793
677	458 329	26.019	82.280	1.4771
678	459 684	26.038	82.341	1.4749
679	461 041	26.058	82.401	1.4728
680	462 400	26.077	82.462	1.4706
681	463 761	26.096	82.523	1.4684
682	465 124	26.115	82.583	1.4663
683	466 489	26.134	82.644	1.4641
684	467 856	26.153	82.704	1.4620
685	469 225	26.173	82.765	1.4599
686	470 596	26.192	82.825	1.4577
687	471 969	26.211	82.885	1.4556
688	473 344	26.230	82.946	1.4535
689	474 721	26.249	83.006	1.4514
690	476 100	26.268	83.066	1.4493
691	477 481	26.287	83.126	1.4472
692	478 864	26.306	83.187	1.4451
693	480 249	26.325	83.247	1.4430
694	481 636	26.344	83.307	1.4409
695	483 025	26.363	83.367	1.4388
696	484 416	26.382	83.427	1.4368
697	485 809	26.401	83.487	1.4347
698	487 204	26.420	83.546	1.4327
699	488 601	26.439	83.606	1.4306

n	n^2	$\sqrt{n}$	$\sqrt{10n}$	$1000/n$
700	490 000	26.458	83.666	1.4286
701	491 401	26.476	83.726	1.4265
702	492 804	26.495	83.785	1.4245
703	494 209	26.514	83.845	1.4225
704	495 616	26.533	83.905	1.4205
705	497 025	26.552	83.964	1.4184
706	498 436	26.571	84.024	1.4164
707	499 849	26.589	84.083	1.4144
708	501 264	26.608	84.143	1.4124
709	502 681	26.627	84.202	1.4104
710	504 100	26.646	84.262	1.4085
711	505 521	26.665	84.321	1.4065
712	506 944	26.683	84.380	1.4045
713	508 369	26.702	84.439	1.4025
714	509 796	26.721	84.499	1.4006
715	511 225	26.739	84.558	1.3986
716	512 656	26.758	84.617	1.3966
717	514 089	26.777	84.676	1.3947
718	515 524	26.796	84.735	1.3928
719	516 961	26.814	84.794	1.3908
720	518 400	26.833	84.853	1.3889
721	519 841	26.851	84.912	1.3870
722	521 284	26.870	84.971	1.3850
723	522 729	26.889	85.029	1.3831
724	524 176	26.907	85.088	1.3812
725	525 625	26.926	85.147	1.3793
726	527 076	26.944	85.206	1.3774
727	528 529	26.963	85.264	1.3755
728	529 984	26.981	85.323	1.3736
729	531 441	27.000	85.382	1.3717
730	532 900	27.019	85.440	1.3699
731	534 361	27.037	85.499	1.3680
732	535 824	27.056	85.557	1.3661
733	537 289	27.074	85.615	1.3643
734	538 756	27.092	85.674	1.3624
735	540 225	27.111	85.732	1.3605
736	541 696	27.129	85.790	1.3587
737	543 169	27.148	85.849	1.3569
738	544 644	27.166	85.907	1.3550
739	546 121	27.185	85.965	1.3532
740	547 600	27.203	86.023	1.3514
741	549 081	27.221	86.081	1.3495
742	550 564	27.240	86.139	1.3477
743	552 049	27.258	86.197	1.3459
744	553 536	27.276	86.255	1.3441
745	555 025	27.295	86.313	1.3423
746	556 516	27.313	86.371	1.3405
747	558 009	27.331	86.429	1.3387
748	559 504	27.350	86.487	1.3369
749	561 001	27.368	86.545	1.3351

n	n^2	$\sqrt{n}$	$\sqrt{10n}$	$1000/n$
750	562 500	27.386	86.603	1.3333
751	564 001	27.404	86.660	1.3316
752	565 504	27.423	86.718	1.3298
753	567 009	27.441	86.776	1.3280
754	568 516	27.459	86.833	1.3263
755	570 025	27.477	86.891	1.3245
756	571 536	27.495	86.948	1.3228
757	573 049	27.514	87.006	1.3210
758	574 564	27.532	87.063	1.3193
759	576 081	27.550	87.121	1.3175
760	577 600	27.568	87.178	1.3158
761	579 121	27.586	87.235	1.3141
762	580 644	27.604	87.293	1.3123
763	582 169	27.622	87.350	1.3106
764	583 696	27.641	87.407	1.3089
765	585 225	27.659	87.464	1.3072
766	586 756	27.677	87.521	1.3055
767	588 289	27.695	87.579	1.3038
768	589 824	27.713	87.636	1.3021
769	591 361	27.731	87.693	1.3004
770	592 900	27.749	87.750	1.2987
771	594 441	27.767	87.807	1.2970
772	595 984	27.785	87.864	1.2953
773	597 529	27.803	87.920	1.2937
774	599 076	27.821	87.977	1.2920
775	600 625	27.839	88.034	1.2903
776	602 176	27.857	88.091	1.2887
777	603 729	27.875	88.148	1.2870
778	605 284	27.893	88.204	1.2853
779	606 841	27.911	88.261	1.2837
780	608 400	27.928	88.318	1.2821
781	609 961	27.946	88.374	1.2804
782	611 524	27.964	88.431	1.2788
783	613 089	27.982	88.487	1.2771
784	614 656	28.000	88.544	1.2755
785	616 225	28.018	88.600	1.2739
786	617 796	28.036	88.657	1.2723
787	619 369	28.054	88.713	1.2706
788	620 944	28.071	88.769	1.2690
789	622 521	28.089	88.826	1.2674
790	624 100	28.107	88.882	1.2658
791	625 681	28.125	88.938	1.2642
792	627 264	28.142	88.994	1.2626
793	628 849	28.160	89.051	1.2610
794	630 436	28.178	89.107	1.2594
795	632 025	28.196	89.163	1.2579
796	633 616	28.213	89.219	1.2563
797	635 209	28.231	89.275	1.2547
798	636 804	28.249	89.331	1.2531
799	638 401	28.267	89.387	1.2516

n	n^2	$\sqrt{n}$	$\sqrt{10n}$	$1000/n$
800	640 000	28.284	89.443	1.2500
801	641 601	28.302	89.499	1.2484
802	643 204	28.320	89.554	1.2469
803	644 809	28.337	89.610	1.2453
804	646 416	28.355	89.666	1.2438
805	648 025	28.373	89.722	1.2422
806	649 636	28.390	89.778	1.2407
807	651 249	28.408	89.833	1.2392
808	652 864	28.425	89.889	1.2376
809	654 481	28.443	89.944	1.2361
810	656 100	28.461	90.000	1.2346
811	657 721	28.478	90.056	1.2330
812	659 344	28.496	90.111	1.2315
813	660 969	28.513	90.167	1.2300
814	662 596	28.531	90.222	1.2285
815	664 225	28.548	90.277	1.2270
816	665 856	28.566	90.333	1.2255
817	667 489	28.583	90.388	1.2240
818	669 124	28.601	90.443	1.2225
819	670 761	28.618	90.499	1.2210
820	672 400	28.636	90.554	1.2195
821	674 041	28.653	90.609	1.2180
822	675 684	28.671	90.664	1.2165
823	677 329	28.688	90.719	1.2151
824	678 976	28.705	90.774	1.2136
825	680 625	28.723	90.830	1.2121
826	682 276	28.740	90.885	1.2107
827	683 929	28.758	90.940	1.2092
828	685 584	28.775	90.995	1.2077
829	687 241	28.792	91.049	1.2063
830	688 900	28.810	91.104	1.2048
831	690 561	28.827	91.159	1.2034
832	692 224	28.844	91.214	1.2019
833	693 889	28.862	91.269	1.2005
834	695 556	28.879	91.324	1.1990
835	697 225	28.896	91.378	1.1976
836	698 896	28.914	91.433	1.1962
837	700 569	28.931	91.488	1.1947
838	702 244	28.948	91.542	1.1933
839	703 921	28.966	91.597	1.1919
840	705 600	28.983	91.652	1.1905
841	707 281	29.000	91.706	1.1891
842	708 964	29.017	91.761	1.1876
843	710 649	29.034	91.815	1.1862
844	712 336	29.052	91.869	1.1848
845	714 025	29.069	91.924	1.1834
846	715 716	29.086	91.978	1.1820
847	717 409	29.103	92.033	1.1806
848	719 104	29.120	92.087	1.1792
849	720 801	29.138	92.141	1.1779

n	n^2	$\sqrt{n}$	$\sqrt{10n}$	$1000/n$
850	722 500	29.155	92.195	1.1765
851	724 201	29.172	92.250	1.1751
852	725 904	29.189	92.304	1.1737
853	727 609	29.206	92.358	1.1723
854	729 316	29.223	92.412	1.1710
855	731 025	29.240	92.466	1.1696
856	732 736	29.257	92.520	1.1682
857	734 449	29.275	92.574	1.1669
858	736 164	29.292	92.628	1.1655
859	737 881	29.309	92.682	1.1641
860	739 600	29.326	92.736	1.1628
861	741 321	29.343	92.790	1.1614
862	743 044	29.360	92.844	1.1601
863	744 769	29.377	92.898	1.1587
864	746 496	29.394	92.952	1.1574
865	748 225	29.411	93.005	1.1561
866	749 956	29.428	93.059	1.1547
867	751 689	29.445	93.113	1.1534
868	753 424	29.462	93.167	1.1521
869	755 161	29.479	93.220	1.1507
870	756 900	29.496	93.274	1.1494
871	758 641	29.513	93.327	1.1481
872	760 384	29.530	93.381	1.1468
873	762 129	29.547	93.434	1.1455
874	763 876	29.563	93.488	1.1442
875	765 625	29.580	93.541	1.1429
876	767 376	29.597	93.595	1.1416
877	769 129	29.614	93.648	1.1403
878	770 884	29.631	93.702	1.1390
879	772 641	29.648	93.755	1.1377
880	774 400	29.665	93.808	1.1364
881	776 161	29.682	93.862	1.1351
882	777 924	29.698	93.915	1.1338
883	779 689	29.715	93.968	1.1325
884	781 456	29.732	94.021	1.1312
885	783 225	29.749	94.074	1.1299
886	784 996	29.766	94.128	1.1287
887	786 769	29.783	94.181	1.1274
888	788 544	29.799	94.234	1.1261
889	790 321	29.816	94.287	1.1249
890	792 100	29.833	94.340	1.1236
891	793 881	29.850	94.393	1.1223
892	795 664	29.866	94.446	1.1211
893	797 449	29.883	94.499	1.1198
894	799 236	29.900	94.552	1.1186
895	801 025	29.917	94.604	1.1173
896	802 816	29.933	94.657	1.1161
897	804 609	29.950	94.710	1.1148
898	806 404	29.967	94.763	1.1136
899	808 201	29.983	94.816	1.1123

n	n^2	$\sqrt{n}$	$\sqrt{10n}$	$1000/n$
900	810 000	30.000	94.868	1.1111
901	811 801	30.017	94.921	1.1099
902	813 604	30.033	94.974	1.1086
903	815 409	30.050	95.026	1.1074
904	817 216	30.067	95.079	1.1062
905	819 025	30.083	95.131	1.1050
906	820 836	30.100	95.184	1.1038
907	822 649	30.116	95.237	1.1025
908	824 464	30.133	95.289	1.1013
909	826 281	30.150	95.341	1.1001
910	828 100	30.166	95.394	1.0989
911	829 921	30.183	95.446	1.0977
912	831 744	30.199	95.499	1.0965
913	833 569	30.216	95.551	1.0953
914	835 396	30.232	95.603	1.0941
915	837 225	30.249	95.656	1.0929
916	839 056	30.265	95.708	1.0917
917	840 889	30.282	95.760	1.0905
918	842 724	30.299	95.812	1.0893
919	844 561	30.315	95.864	1.0881
920	846 400	30.332	95.917	1.0870
921	848 241	30.348	95.969	1.0858
922	850 084	30.364	96.021	1.0846
923	851 929	30.381	96.073	1.0834
924	853 776	30.397	96.125	1.0823
925	855 625	30.414	96.177	1.0811
926	857 476	30.430	96.229	1.0799
927	859 329	30.447	96.281	1.0787
928	861 184	30.463	96.333	1.0776
929	863 041	30.480	96.385	1.0764
930	864 900	30.496	96.437	1.0753
931	866 761	30.512	96.488	1.0741
932	868 624	30.529	96.540	1.0730
933	870 489	30.545	96.592	1.0718
934	872 356	30.561	96.644	1.0707
935	874 225	30.578	96.695	1.0695
936	876 096	30.594	96.747	1.0684
937	877 969	30.610	96.799	1.0672
938	879 844	30.627	96.850	1.0661
939	881 721	30.643	96.902	1.0650
940	883 600	30.659	96.954	1.0638
941	885 481	30.676	97.005	1.0627
942	887 364	30.692	97.057	1.0616
943	889 249	30.708	97.108	1.0604
944	891 136	30.725	97.160	1.0593
945	893 025	30.741	97.211	1.0582
946	894 916	30.757	97.263	1.0571
947	896 809	30.773	97.314	1.0560
948	898 704	30.790	97.365	1.0549
949	900 601	30.806	97.417	1.0537

n	n^2	$\sqrt{n}$	$\sqrt{10n}$	$1000/n$
950	902 500	30.822	97.468	1.0526
951	904 401	30.838	97.519	1.0515
952	906 304	30.855	97.570	1.0504
953	908 209	30.871	97.622	1.0493
954	910 116	30.887	97.673	1.0482
955	912 025	30.903	97.724	1.0471
956	913 936	30.919	97.775	1.0460
957	915 849	30.935	97.826	1.0449
958	917 764	30.952	97.877	1.0438
959	919 681	30.968	97.929	1.0428
960	921 600	30.984	97.980	1.0417
961	923 521	31.000	98.031	1.0406
962	925 444	31.016	98.082	1.0395
963	927 369	31.032	98.133	1.0384
964	929 296	31.048	98.184	1.0373
965	931 225	31.064	98.234	1.0363
966	933 156	31.081	98.285	1.0352
967	935 089	31.097	98.336	1.0341
968	937 024	31.113	98.387	1.0331
969	938 961	31.129	98.438	1.0320
970	940 900	31.145	98.489	1.0309
971	942 841	31.161	98.539	1.0299
972	944 784	31.177	98.590	1.0288
973	946 729	31.193	98.641	1.0277
974	948 676	31.209	98.691	1.0267
975	950 625	31.225	98.742	1.0256
976	952 576	31.241	98.793	1.0246
977	954 529	31.257	98.843	1.0235
978	956 484	31.273	98.894	1.0225
979	958 441	31.289	98.944	1.0215
980	960 400	31.305	98.995	1.0204
981	962 361	31.321	99.045	1.0194
982	964 324	31.337	99.096	1.0183
983	966 289	31.353	99.146	1.0173
984	968 256	31.369	99.197	1.0163
985	970 225	31.385	99.247	1.0152
986	972 196	31.401	99.298	1.0142
987	974 169	31.417	99.348	1.0132
988	976 144	31.432	99.398	1.0121
989	978 121	31.448	99.448	1.0111
990	980 100	31.464	99.499	1.0101
991	982 081	31.480	99.549	1.0091
992	984 064	31.496	99.599	1.0081
993	986 049	31.512	99.649	1.0070
994	988 036	31.528	99.700	1.0060
995	990 025	31.544	99.750	1.0050
996	992 016	31.559	99.800	1.0040
997	994 009	31.575	99.850	1.0030
998	996 004	31.591	99.900	1.0020
999	998 001	31.607	99.950	1.0010

Table A-11: Critical Values for the Kolmogorov-Smirnov Test of Goodness of Fit

Sample Size (n)	Significance Level .20	.15	.10	.05	.01
1	.900	.925	.950	.975	.995
2	.684	.726	.776	.842	.929
3	.565	.597	.642	.708	.829
4	.494	.525	.564	.624	.734
5	.446	.474	.510	.563	.669
6	.410	.436	.470	.521	.618
7	.381	.405	.438	.486	.577
8	.358	.381	.411	.457	.543
9	.339	.360	.388	.432	.514
10	.322	.342	.368	.409	.486
11	.307	.326	.352	.391	.468
12	.295	.313	.338	.375	.450
13	.284	.302	.325	.361	.433
14	.274	.292	.314	.349	.418
15	.266	.283	.304	.338	.404
16	.258	.274	.295	.328	.391
17	.250	.266	.286	.318	.380
18	.244	.259	.278	.309	.370
19	.237	.252	.272	.301	.361
20	.231	.246	.264	.294	.352
25	.21	.22	.24	.264	.32
30	.19	.20	.22	.242	.29
35	.18	.19	.21	.23	.27
40				.21	.25
50				.19	.23
60				.17	.21
70				.16	.19
80				.15	.18
90				.14	
100				.14	
Asymptotic Formula:	$\frac{1.07}{\sqrt{n}}$	$\frac{1.14}{\sqrt{n}}$	$\frac{1.22}{\sqrt{n}}$	$\frac{1.36}{\sqrt{n}}$	$\frac{1.63}{\sqrt{n}}$

Source: Massey, F. J., Jr. 1951. "The Kolmogorov-Smirnov test for goodness of fit," *Journal of American Statistical Association*, 46, 70, with the permission of the author and publisher.

Answers to Selected Problems

Chapter 2

Section 2.2

5. **a.** 6, **b.** 0.08, **c.** Frequencies: 3, 3, 6, 5, 3, 4. **6.** 8. **7.** 200.

10. **a.** Frequencies: 8, 10, 1, 0, 2; **b.** Frequencies: 3, 5, 6, 4, 3.

12. Relative frequencies: 3/24, 3/24, 6/24, 5/24, 3/24, 4/24.

Section 2.3

1. **a.** Relative frequencies: 0.08, 0.38, 0.16, 0.24, 0.14;
   b. Less-than cumulative frequencies: 0, 4, 23, 31, 43, 50;
   c. Greater-than-or-equal-to cumulative frequencies: 50, 46, 27, 19, 7, 0;
   d. Less-than cumulative frequencies: 0.00, 0.08, 0.46, 0.62, 0.86, 1.00;
   e. Greater-than-or-equal-to cumulative relative frequencies: 1.00, 0.92, 0.54, 0.38, 0.14, 0.00.

3. **a.** Less-than-or-equal-to cumulative frequencies: 0, 4, 23, 31, 43, 50;
   b. Greater-than cumulative frequencies: 50, 46, 27, 19, 7, 0.

Section 2.4

8. **a.** 83, **b.** 46%, **c.** 38, **d.** 91%, **e.** 82,
   f. 45%, **g.** 90, **h.** 89.5, **i.** 88.5, **j.** 92.5.

12. **a.** Relative frequencies: 0.12, 0.32, 0.24, 0.32.

14. **a.** 3200, **b.** 12.5%, **c.** 3.182, **d.** 1200, **e.** 3.224.

Chapter 3

Section 3.2

1. 42. 3. 20. 5. 17. 7. 24.6. 9. 286. 11. a. 123.

13. a. 40. 15. 89.

Section 3.3

1. a. 0.22, b. 0.325, c. 0.231. 3. 1.17.

Section 3.4

1. a. 8, b. 2.4, c. 7.6, d. 7.6, e. 2.8.

3. a. 13, b. 4, c. 20, d. 20, e. 4.5. 8. 6300.0.

Section 3.5

3. $\bar{x} = 6$, $s_x^2 = 4$; a. -2, 4; b. 42, 196; c. 0.5, 1/36; d. 1/7, 4/49.

Section 3.6

1. a. 252, b. 92.

Section 3.7

1. a. 19.3, b. 19.3, d. 18.63, e. 17-18, f. 4.16.

3. a. 15.0, 11.0, 13.0; b. 10.0, 2.0; c. 10.0. 5. 1.73. 7. 6.

Section 3.8

1. a. 127.0, b. 127.0, c. 127.0, f. 125.3, g. 115-119, h. 10.0, i. 0.08.

3. 40.0. 5. 8.1, 26.1, $s_b > s_a$.

Section 3.9

1. a. 4.05, b. 3.00, c. 2.00, e. Positive.

Chapter 4

Section 4.2

1. a. True, c. True, e. True, g. False, i. True, k. True, m. True, o. False, q. True, s. True.

6. a. 13, b. 1, c. 7, d. 16, e. 33, f. 21, g. 27, h. 34.

9. a. 60, c. 20.

Section 4.3

1. 20. 5. a. 495, b. 19,958,400. 7. 2,494,800.

9. a. 20!/5!5!5!5!, b. 20!/4!4!4!4!4!, c. 20!/10!10!.

Section 4.5

3. a. $P(E_1) = 0.20$, $P(E_2) = 0.16$, $P(E_3) = 0.17$, $P(E_4) = 0.14$, $P(E_5) = 0.18$, $P(E_6) = 0.15$.

Section 4.6

3. b. 0.59, d. 0.03, f. 0.50, h. 0.90.

Section 4.7

5. 5/6. 7. a. 1/5, b. 4/15, c. 4/15, d. 4/15.

9. a. 49/225, b. 56/225, c. 56/225, d. 64/225.

Section 4.8

1. a. {1, 2, 3, 4, 5, 6}, b. {6}, c. {1, 3, 5, 6, 7}, d. {1, 3}, e. ϕ, f. {1, 2, 3, 4, 5, 6, 7}, g. {1, 3, 6}, h. {1, 3, 5, 6}, i. No, j. Yes.

3. a. Yes, b. Yes, c. Yes, d. No, e. No.

5. a. 1/36, b. 1/18, c. 1/12, d. 1/9, e. 5/36, f. 1/6, g. 5/36, h. 1/9, i. 1/12, j. 1/18, k. 1/36, l. 1/9, m. 2/9, n. 1/6, o. 2/9, p. 5/18, q. 1/3, r. 2/3.

Section 4.10

1. a. $P(L|F) = 5/16$, $P(M|F) = 10/16$, $P(U|F) = 1/16$, $P(L|A) = 5/26$
 $P(M|A) = 15/26$, $P(U|A) = 6/26$;
 b. $P(F|L) = 3/5$, $P(F|M) = 1/2$, $P(F|U) = 1/5$, $P(A|L) = 2/5$,
 $P(A|M) = 1/2$, $P(A|U) = 4/5$.

3. 0.57. 5. 2/3.

Section 4.11

1. a.

Outcome	2	3	4	5	6	7	8	9	10	11	12
Probability	1/36	2/36	3/36	4/36	5/36	6/36	5/36	4/36	3/36	2/36	1/36

 b. 1/6, c. 1/2, d. 1/9, e. 5/9.

3. a.

Value (X)	2	3	4	5	6	7	8	9	10
Probability	1/25	2/25	3/25	4/25	5/25	4/25	3/25	2/25	1/25

 b. 3/5, c. 12/25, d. 6/25, e. 21/25.

5. a.

X	0	1	2	3
$P(X)$	1/8	3/8	3/8	1/8

b.

X	0	1	2	3
$P(X)$	2/22	9/22	9/22	2/22

c.

X	0	1	2	3
$P(X)$	8/27	12/27	6/27	1/27

d.

X	0	1	2	3
$P(X)$	14/55	28/55	12/55	1/55

e.

X	0	1	2	3
$P(X)$	125/216	25/72	5/72	1/216

f.

X	0	1	2	3
$f(X)$	6/11	9/22	1/22	6/11

g.

X	0	1	2	3
$P(X)$	1/8	3/8	3/8	1/8

h.

X	0	1	2	3
$f(X)$	1/11	9/22	9/22	1/11

9.

X	3	4	5	6	7	8	9	10
$P(X \leq x)$	1/216	4/216	10/216	20/216	35/216	56/216	81/216	108/216

11	12	13	14	15	16	17	18
135/216	160/216	181/216	196/216	206/216	212/216	215/216	216/216

 a. 3/8, b. 7/27.

Section 4.12

3. 101,750. 5. -0.07. 7. a. 12.44, b. 5.46, c. 1.19.

Section 4.13

3. a. 0.25, b. 0.4375, c. 0.3125, d. 0, e. 0, f. Yes.

Chapter 5

Section 5.2

5. a. 0.113, b. 0.113,

c.

x	0	1	2	3	4	5
$P(X = x)$	0.050	0.206	0.337	0.276	0.113	0.018

7. 0.00856. 9. a. 0.0146, b. 0.8982, c. 1.0000, d. 0.9283, e. 0.0000, f. 0.0539, g. 0.9917, h. 0.9115.

Section 5.3

3. a. 0.762, b. 0.009, c. 0.003, d. 0.081.

5. a. 0.014, b. 0.758, c. 0.662. 7. 0.423.

8. a. 0.947, b. 0.801, c. 0.544, d. 0.693, e. 0.577.

Section 5.4

3. a. 0.2398, b. 0.0000, c. 0.0001, d. 0.2499, e. 0.0012.

5. a. 0.0468, b. 0.9487, c. 0.0513, d. 0.8213.

7. a. 0.2048, b. 0.0512, c. 0.2627, d. 0.9421.

Chapter 6

Section 6.2

1.

x	85	90	95	100	105	110	115
$f(x)$	0.00088	0.01080	0.04840	0.07978	0.04840	0.01080	0.00088

3. a. 5, 10; b. -100, 5; c. 800, 15; d. 0, 1; e. 0, 1.

Section 6.3

1. a. 0.1359, b. 0.8185, c. 0.1574, d. 0.8400.

3. a. 0.1574, b. 0.8400, c. 0.1574, d. 0.8185.

Section 6.4

1. a. 0.2358, b. 0.0143, c. 0.0405, d. 0.4235, e. 1.11, f. 1.779, g. 0.227, h. 2.637.

3. a. 0.9126, b. 0.0023, c. 0.7773, d. 0.0234, e. 0.9840, f. 0.6338, g. -0.495, h. 1.664, i. -1.23, j. 0.458.

5. a. 0.0287, b. 0.5512, c. 0.1314, d. 0.1513, e. 0.0372, f. 0.9589.

7. a. 1.35, b. -0.53, c. -2.21, d. -1.74, e. -0.13, f. 1.15.

9. a. -2.30, b. 0.06, c. -0.16, d. 0.42, e. -0.77, f. 2.85.

11. a. 0.1359, b. 0.8185, c. 0.1574, d. 0.8400.

Section 6.5

1. a. 0.0359, b. 0.9974, c. 0.1587, d. 0.9452, e. 0.1104, f. 0.5138, g. 0.0409.

3. 0.9544.

5. a. 8.38%, b. 98.64%, c. 33.00%, d. 98.64%, e. 15.10%, f. 8.60%, g. 79.05%.

7. 0.7980.

Section 6.6

1. a. 85.65, b. 71.46, c. 47.70, d. 85.65, e. 82.95, f. 69.45, g. 69.225, h. 71.52, i. 51.00, j. 93.00.

3. 144. 5. 117. 7. -0.4456. 9. 25,500.

Section 6.7

1. a. 67.36%, b. 2.56%, c. 63.66%, d. 37.57%.

3. a. 223.5, b. 231.5, c. 199.04, d. 196.8.

Section 6.8

1. a. 0.9332, b. 0.3538, c. 0.0733, d. 0.0183.

Section 6.9

1. a. 0.0189, b. 0.2874, c. 0.1176, d. 0.2649.

Section 6.10

1. a. $R \sim N(910, 32{,}005)$, b. $S \sim N(-757, 14{,}520)$, c. $T \sim N(-457, 67{,}400)$.

3. 410. 5. 0.3721.

Chapter 7

Section 7.2

7. a. $\mu_x = 10.8$, $X_{\text{med}} = 10$;

b.

Possible sample	Mean	Median
6, 8, 10	8.0	8.0
6, 8, 14	9.3	8.0
6, 8, 16	10.0	8.0
6, 10, 14	10.0	10.0
6, 10, 16	10.7	10.0
6, 14, 16	12.0	14.0
8, 10, 14	10.7	10.0
8, 10, 16	11.3	10.0
8, 14, 16	12.7	14.0
10, 14, 16	13.3	14.0

Section 7.3

1. a. Sampling distribution of sample mean deviation from mean

Sample space	Probability
0.0	5/25
1.5	8/25
3.0	6/25
4.5	4/25
6.0	2/25

b. Sampling distribution of sample range

Sample space	Probability
0	5/25
3	8/25
6	6/25
9	4/25
12	2/25

3. a. 10^4, 0.0001; b. ${}_{11}C_3$, 1/165; c. ${}_{15}C_8$, 1/6435; d. 20^{10}, $1/20^{10}$.

5. Sampling distribution of the "sample range"

a. Sampling with replacement

Sample space	Probability
0	5/25
4	8/25
8	6/25
12	4/25
16	2/25
Total	1.0

b. Sampling without replacement

Sample space	Probability
4	4/10
8	3/10
12	2/10
16	1/10
Total	1.0

7. $\mu_r = 6.40$, $\sigma_r^2 = 23.04$.

Section 7.4

1. $\mu_x = 44/8 = 5.500$, $\sigma_x^2 = 42.000/8 = 5.250$, $\sigma_x^2/n = 5.25/2 = 2.625$, $\mu_{\bar{x}} = 352.000/64 = 5.500 = \mu_x$, $\sigma_{\bar{x}}^2 = 168/64 = 2.625 = \sigma_x^2/n$.

Section 7.5

1. a. 0.9452, b. 0.2483, c. 114.28. 3. 0.0302. 5. 0.9872.

Section 7.6

1. a. 0.9099, b. 0.7752, c. 0.8966, d. 0.7680.

3. a. 148.467, b. 1199.395, c. 30.057, d. 242.009. 5. 0.9200.

Section 7.7

1. c. 0.9332. 3. 0.1844. 5. a. 0.8786, b. 0.8786.

Chapter 8

Section 8.2

3. $E(R - \sigma_x^2)^2 = 6.4$, $E(H - \sigma_x^2)^2 = 16.6$, R is better.

Section 8.3

3.

	Sample space	8	12	13	16	17	18	20
$N = 7,\ n = 1$	Probability	1/7	1/7	1/7	1/7	1/7	1/7	1/7
$N = 7,\ n = 3$	Probability		5/35	8/35	9/35	8/35	5/35	
$N = 7,\ n = 5$	Probability			6/21	9/21	6/21		
$N = 7,\ n = 7$	Probability				1			

Section 8.5

3. a. 0.4532, b. 1537, c. 20.56, d. 68. 5. 0.0456.

7. $\alpha = 2P\left(Z > \dfrac{e\sqrt{n(N-1)}}{\sigma_x\sqrt{N-n}}\right)$, $e = \dfrac{z_{\alpha/2}\sigma_x\sqrt{N-n}}{\sqrt{n(N-1)}}$, $n = \dfrac{\sigma_x^2 N}{\dfrac{e^2(N-1)}{z_{\alpha/2}^2} + \sigma_x^2}$.

9. 3.61. 11. a. 9604, b. 97.

Section 8.6

3. $P_F(474.80 < \mu_x < 525.20) = 0.95$. 5. 312.

9. $P_F(-25.92 < (\mu_A - \mu_B) < -4.08) = 0.80$.

11. $P\left(\overline{X} - z_{\alpha/2}\sqrt{\dfrac{\sigma_x^2}{n} \cdot \dfrac{N-n}{N-1}} < \mu_x < \overline{X} + z_{\alpha/2}\sqrt{\dfrac{\sigma_x^2}{n} \cdot \dfrac{N-n}{N-1}}\right) = 1 - \alpha$.

Chapter 9

Section 9.2

5. c. 5/10. 7. a. $\alpha = 0.4013$, $\beta = 0.0401$, b. $x^* = 48.58$, $\beta = 0.3613$, c. $x^* = 44.872$, $\alpha = 0.2364$.

Section 9.3

1. a. $\overline{x} > \overline{x}^* = 195.8875$, accept H_0, c. $\alpha' = 0.0548$, d. $\beta = 0.0093$, e. $\overline{x}^* = 194.1125$, $\alpha = 0.0093$.

3. a. $\overline{x}^* = 59.358$, reject H_0, b. $z = -3.4 < -1.284 = z^*$, reject H_0, c. $\alpha' \cong 0 > \alpha$, reject H_0, d. $\beta(59) = 0.2380$, e. $\overline{x}^* = 59.8225$, $P(Z < 0.3550) = 0.3661$.

5. a. $\overline{x}^* = 86.410$, reject H_0, b. $z = 1.8 > 1.282 = z^*$.

7. a. $\overline{x}^* = 608.225$, accept H_0, b. $z = 1.6 < 1.645 = z^*$, c. $\alpha' = 0.548 < \alpha$, d. $\beta(615) = 0.0877$, e. $\overline{x}^* = 606.775$, $\alpha = 0.0877$.

9. a.

μ_x	590	595	598	600	602	604	606
$\beta(\mu_x)$	1.000	0.9960	0.9796	0.9500	0.8935	0.8009	0.6718
μ_x	608	610	612	615	620		
$\beta(\mu_x)$	0.5180	0.3613	0.2251	0.0877	0.0092		

11. b. $\bar{x} = 4.15 < 4.1645 = \bar{x}^*$, accept H_0.

13. a. $\bar{x}_2^* = 122.94$, accept H_0; b. $z = 1.933 < 1.96 = z^*$, accept H_0;

c. $\alpha' = 0.0266 < \alpha/2$, accept H_0;

d.

μ_x	114.0	115.5	117.0	118.5	120.0	121.5	123.0	124.5	126.0
$\beta(\mu_x)$	0.0207	0.1492	0.4840	0.8300	0.9500	0.8300	0.4840	0.1492	0.0207

15. a. $H_0:\mu_x = 1.75$, $H_1:\mu_x \neq 1.75$;

b. $\bar{x}^* = 1.706 < \bar{x} = 1.73 < 1.794 = \bar{x}_2^*$, accept H_0;

c. $\beta(1.85) = 0.0045$.

Section 9.5

3. a. $p = \dfrac{296}{400} < 0.7671 = p^*$; reject $H_0:\pi = 0.80$;

b. $z = -3.0 < -1.645 = z^*$, reject H_0;

c. $\alpha' = 0.0013 < 0.05 = \alpha$, reject H_0; d. $\beta(0.76) = 0.3696$;

e. $p^* = 0.7951$, $\alpha = 0.4032$.

5. a. $H_0:\pi = 0.10$, $H_1:\pi > 0.10$, $z = 2.133 < 2.327 = z^*$, accept H_0;

b. $\beta(0.20) = 0.3997$; c. $p^* = 0.3164$, $\pi \cong 0$.

7. a. $p = 0.82 < 0.827 = p_2^*$, accept H_0;

b. $z = 1.781 < 1.96 = z_2^*$, accept H_0;

c. $\alpha' = 0.0374 > 0.025 = \alpha/2$, accept H_0; d. $\beta(0.62) = 0.1147$.

Section 9.6

3. $n \cong 701$, $\bar{x}^* = 608.7868$.

5. $n \cong 105$, $p^* = 0.7501$.

Section 9.7

1. a. $z = -1.887 < -1.645 = z^*$, reject H_0;

b. $\alpha' = 0.0296 < 0.05 = \alpha$, reject H_0; d. $\beta(-20) = 0.0370$;

e. $z^* = 1.282$, $\alpha = 0.0158$.

3. a. $z = 0.79 < z^* = 1.28$, accept H_0;

b. $\alpha' = 0.2148 > 0.10 = \alpha$, accept H_0;

c. $\beta(4.0) = 0.0606$; d. $z^* = -1.645$, $\alpha = 0.1170$.

5. a. $H_0:\mu_1 - \mu_2 = 0$, $H_1:\mu_1 - \mu_2 > 0$;
   b. $z = 4.11 > 1.28 = z^*$, reject H_0; c. $\beta(0.30) = 0.0023$;
   d. $z^* = 1.645$, $\alpha = 0.0069$.

7. a. $H_0:\mu_1 - \mu_2 = 0$, $H_1:\mu_1 - \mu_2 \neq 0$;
   b. $z = 4.11 > 1.645 = z^*$, reject H_0; c. $\beta(0.30) = 0.0068$.

Section 9.8

1. a. $H_0:\pi = 0.20$, $H_1:\pi > 0.20$; b. Rejection region: $Z > z^* = 1.645$;
   c. $H_0:\pi = 0.20$, $H_1:\pi < 0.20$; d. Rejection region: $Z < z^* = -1.29$;
   e.

π	0.04	0.08	0.12	0.16	0.20	0.24	0.28	0.32	0.36
$\beta(\pi)$	1.0	1.0	1.0	0.9981	0.9505	0.7291	0.3745	0.1251	0.0250

3. a. $H_0:\mu_x = 8000$, $H_1:\mu_x > 8000$; b. Rejection region: $Z > z^* - 1.88$;
   c. $H_0:\mu_x = 8000$, $H_1:\mu_x < 8000$; d. Rejection region: $Z < z^* = -1.88$;
   e.

μ_x	7600	7700	7800	7900	8000	8100	8200	8300	8400
$\beta(\mu_x)$	1.00	1.00	1.00	0.9980	0.9699	0.8106	0.4522	0.1314	0.0170

Chapter 10

Section 10.2

1. a. 0.10, b. 0.10, c. 0.225, d. 5.01, e. 63.7, f. 31.5.

3. a. 0.81, b. 0.472, c. 9.64, d. 64.3. 5. 0.90.

Section 10.3

3. Since $18.75 > \chi^2_{0.90,24} = 15.66$, accept $H_0:\sigma_x^2 = 200$.

5. Since $19.29 < \chi^2_{0.10,14} = 21.1$, accept $H_0:\sigma_x^2 = 63$.

7. Since $\chi^2_{0.975,23} = 11.69 < 12.00 < 38.1 = \chi^2_{0.025,23}$, accept $H_0:\sigma_x^2 = 100$.

Section 10.4

1. a. $P_F(13.7 < \sigma_x^2 < 71.6) = 0.98$; b. $P_F(119.0 < \sigma_x^2 < 286.3) = 0.90$;
   c. $P_F(7.76 < \sigma_x^2 < 80.0) = 0.99$.

3. $P_F(0.48 < \sigma_x^2 < 1.81) = 0.98$.

Section 10.5

1. Since $\sum_{i=1}^{k} \frac{(f_i - e_i)^2}{e_i} = 12.17 > 6.25 = \chi^2_{0.10,3}$, reject H_0.

3. Since $\sum_{i=1}^{k} \frac{(f_i - e_i)^2}{e_i} = 8.30 < 25.7 = \chi^2_{0.02,13}$, accept H_0.

Section 10.6

1. Since $\sum_{i=1}^{4}\sum_{j=1}^{3} \frac{(n_{ij} - e_{ij})^2}{e_{ij}} = 24.58 > 15.24 = \chi^2_{0.02,6}$, reject H_0.

3. Since $\sum_{i=1}^{3}\sum_{j=1}^{2} \frac{(n_{ij} - e_{ij})^2}{e_{ij}} = 8.45 > 4.61 = \chi^2_{0.10,2}$, reject H_0.

Chapter 11

Section 11.2

3. a. 0.01, b. 0.005, c. 0.90, d. 0.995, e. 0.195, f. 0.945,
   g. 0.245, h. 1.761, i. -1.717, j. 2.110.

Section 11.3

3. Since $\frac{\bar{x} - \mu_0}{\hat{\sigma}_x/\sqrt{n}} = -1.25 > -1.711 = -t_{0.05,24}$, accept H_0.

5. $\frac{\bar{x} - \mu_0}{\hat{\sigma}_x/\sqrt{n}} = -10.0 < -1.711 = -t_{0.05,24}$, reject H_0.

7. Since $\frac{\bar{x} - \mu_0}{\hat{\sigma}_x/\sqrt{n}} = 2.00 < 2.485 = t_{0.01,25}$, accept H_0.

9. Since $\frac{\bar{x} - \mu_0}{\hat{\sigma}_x/\sqrt{n}} = -10.0 < -2.064 = -t_{0.025,24}$, reject H_0.

11. Since $\dfrac{\bar{x} - \mu_0}{\hat{\sigma}_x/\sqrt{n}} = 2.00 > 1.708 = t_{0.05,25}$, reject H_0.

Section 11.4

1. $P_F(188.16 < \mu_x < 201.84) = 0.90$. 3. $P_F(9.375 < \mu_x < 9.625) = 0.98$.

5. $P_F(2.396 < \mu_x < 2.844) = 0.999$.

Section 11.5

1. 57.2. 3. $s_1^2 = 5.3$, $s_2^2 = 7.3$, $\hat{\sigma}^2 = 6.8$.

Section 11.6

1. Since $\dfrac{(\bar{x}_1 - \bar{x}_2) - (\mu_1 - \mu_2)_0}{\hat{\sigma}\sqrt{\dfrac{1}{n_1} + \dfrac{1}{n_2}}} = 1.62 > 1.684 = t_{0.05,38}$, accept H_0.

3. Since $\dfrac{(\bar{x}_1 - \bar{x}_2) - (\mu_1 - \mu_2)_0}{\hat{\sigma}\sqrt{\dfrac{1}{n_1} + \dfrac{1}{n_2}}} = 1.48 > 1.316 = t_{0.10,25}$, reject H_0.

5. Since $-t_{0.025,8} = -2.048 < \dfrac{(\bar{x}_1 - \bar{x}_2) - (\mu_1 - \mu_2)_0}{\hat{\sigma}\sqrt{\dfrac{1}{n_1} + \dfrac{1}{n_2}}} = -1.90 < 2.048 = t_{0.025,8}$, accept H_0.

7. Since $-t_{0.005,40} = -2.704 < \dfrac{(\bar{x}_1 - \bar{x}_2) - (\mu_1 - \mu_2)_0}{\hat{\sigma}\sqrt{\dfrac{1}{n_1} + \dfrac{1}{n_2}}} = -2.68 < 2.704 = t_{0.005,40}$, accept H_0.

9. Since $\dfrac{(\bar{x}_1 - \bar{x}_2) - (\mu_1 - \mu_2)_0}{\hat{\sigma}\sqrt{\dfrac{1}{n_1} + \dfrac{1}{n_2}}} = 1.50 > 1.315 = t_{0.10,26}$, reject H_0.

Section 11.7

1. $P_F(-0.115 < \mu_1 - \mu_2 < 6.115) = 0.90$.

3. $P_F(-4.486 < \mu_1 - \mu_2 < 1.886) = 0.98$.

5. $P_F(-7.741 < \mu_1 - \mu_2 < 0.271) = 0.95$.

7. $P_F(-3.299 < \mu_1 - \mu_2 < -0.301) = 0.80$.

Chapter 12

Section 12.2

1. $\bar{x} = 78.8$, $\hat{\sigma}_x^2 = 221.87$.

Section 12.3

5. $\bar{x} = 6.96$, $\hat{\sigma}_x^2 = 0.009606$.

Chapter 13

Section 13.1

3. a. 0.010, b. 0.050, c. 0.100, d. 0.950, e. 0.950, f. 0.975, g. 2.85, h. 5.18, i. 3.72, j. 2.85, k. 5.18, l. 3.72.

5. a. 0.157, b. 0.215, c. 0.279, d. 0.375, e. 0.249, f. 0.314, g. 0.382, h. 0.476.

Section 13.3

1. Since calculated $F = 2.50 > 2.42 = F_{0.05,9,19}$, reject H_0.

3. Since calculated $F = 1.33 < 3.54 \cong F_{0.01,19,14}$, accept H_0.

5. Since calculated $F = 1.44 < 3.03 \cong F_{0.05,14,9}$, accept H_0.

Section 13.4

5. a.

Source	Degrees of freedom	Sum of squares	Mean square	F ratio
Among	4	61.92	15.48	5.17
Within	15	44.88	2.99	
Total	19	106.80		

b. Absolute difference between means

Absolute difference between means	LSD
$\lvert \bar{x}_{1\cdot} - \bar{x}_{2\cdot} \rvert = 3.7$*	2.69
$\lvert \bar{x}_{1\cdot} - \bar{x}_{3\cdot} \rvert = 2.9$*	2.69
$\lvert \bar{x}_{1\cdot} - \bar{x}_{4\cdot} \rvert = 0.0$	2.80
$\lvert \bar{x}_{1\cdot} - \bar{x}_{5\cdot} \rvert = 4.4$*	3.03
$\lvert \bar{x}_{2\cdot} - \bar{x}_{3\cdot} \rvert = 0.8$	2.32
$\lvert \bar{x}_{2\cdot} - \bar{x}_{4\cdot} \rvert = 3.7$*	2.47
$\lvert \bar{x}_{2\cdot} - \bar{x}_{5\cdot} \rvert = 0.7$	2.69
$\lvert \bar{x}_{3\cdot} - \bar{x}_{4\cdot} \rvert = 2.9$*	2.47
$\lvert \bar{x}_{3\cdot} - \bar{x}_{5\cdot} \rvert = 1.5$	2.69
$\lvert \bar{x}_{4\cdot} - \bar{x}_{5\cdot} \rvert = 4.4$*	2.80

*Significant difference.

7. a. $\hat{\sigma}_e^2 = 2.99$, $\hat{\sigma}_\tau^2 = 3.27$, $\hat{\sigma}_x^2 = 6.26$.

Section 13.5

5. a.

Source	Degrees of freedom	Sum of squares	Mean square	F ratio
Plants	2	12.11	6.06	4.21
Fertilizers	2	7.11	3.56	2.47
Interaction	4	20.22	5.06	3.51
Error	9	13.00	1.44	
Total	17	52.44		

b. Interaction not significant;

c. At the 1% significance level, no significant effects for either of the main effects.

Chapter 14

Section 14.3

3. a. 2.088, 0.343; b. 2.50, 2.53, 2.43, 2.50, 2.91, 2.67, 2.74, 2.88, 2.67, 2.60, 2.64, 2.53;

c. -0.02, -0.08, 0.02, 0.03, -0.01, -0.08, 0.06, 0.00, 0.06, 0.05, -0.07, 0.05; d. 0.00317; e. 0.00315.

5. a. 16.227, 0.0478; b. 17.1, 17.0, 17.0, 17.1, 17.1, 17.2, 17.1, 17.1, 17.2, 17.2;

c. 0.0, 0.2, 0.0, -0.3, -0.1, -0.1, 0.1, 0.2, 0.0, 0.1;

d. 0.0263; e. 0.0213.

Section 14.4

5. Since observed $t = 4.33 > 1.895 = t_{0.05,7}$, reject H_0.

Section 14.5

5. $P_F(10.130 < \beta_0 < 10.942) = 0.90$, $P_F = (1.810 < \beta_1 < 4.630) = 0.90$, $P_F(20.858 < \mu_{y|x=3.5} < 22.754) = 0.90$.

Section 14.6

1. $P_F(2.639 < y(2.0) < 2.909) = 0.95$.

3. $P_F(16.606 < y(17.5) < 17.522) = 0.98$.

Section 14.7

5. **a.** 0.86, **b.** Since observed $t = 6.05 > 2.145 = t_{0.025,13}$, reject H_0.

7. **a.** 0.73, **b.** Since observed $t = 2.61 > 1.943 = t_{0.05,6}$, reject H_0.

Chapter 15

Section 15.3

1. **a.**

$$5.75\hat{\beta}_1 + 20.50\hat{\beta}_2 = -3.725,$$
$$20.50\hat{\beta}_1 + 230.67\hat{\beta}_2 = 5.367;$$

b. $\overline{x}_1 = 7.00$, $\overline{x}_2 = 100.33$, $\overline{y} = 9.31$.

Section 15.4

1. **a.**

$\hat{\mu}_{y\|x_1,x_2}$	$y_i - \hat{\mu}_{y\|x_1,x_2}$
9.16	-0.16
9.48	0.02
10.10	0.10
9.81	-0.01
10.01	0.09
9.39	0.11
9.74	-0.04
9.71	-0.21
9.06	-0.06
8.44	-0.04
7.96	0.04
8.85	0.15

b. 0.1313.

c.

Source	Degrees of freedom	Sum of squares	Mean square	F ratio
Regression	2	4.6192	2.3096	158.19
Error	9	0.1313	0.0146	
Total	11	4.7505		

d. 0.0146 **e.** Since observed $F = 158.19 > 4.26 = F_{0.05,2,9}$, reject H_0.

f. Since observed $t = -28.99 < -2.26 = -t_{0.025,9}$, reject H_0.

Section 15.5

1. **a.** $P_F(-1.1531 < \beta_1 < -0.9861) = 0.95$, $P_F(0.0966 < \beta_2 < 0.14) = 0.95$;

   b. $P_F(9.7153 < \mu_{y|x_1=6.5,x_2=100} < 9.8963) = 0.95$.

Section 15.6

1. $P_F[9.52 < y(6.5, 100) < 10.09] = 0.95$.

Supplement 15.A

1. **a.** 4.9282, -1.0696, 0.1183; **b.** $\hat{\mu}_{y|x_1,x_2} = 4.9282 - 1.0696x_1 + 0.1183x_2$,

   c. 4.6192 **d.** $C = \begin{bmatrix} 0.0933 & 0.0063 \\ 0.0063 & 0.0063 \end{bmatrix}$.

2. **a.** 3.231, 0.659, 1.224, 0.176;

   b. $\hat{\mu}_{y|x_1,x_2,x_3} = 3.231 + 0.659x_1 + 1.224x_2 + 0.176x_3$, **c.** 11.414,

   d. $C = \begin{bmatrix} 0.139812 & -0.131795 & -0.176132 \\ -0.131795 & 0.644013 & 0.234946 \\ -0.176132 & 0.234946 & 0.560263 \end{bmatrix}$

Chapter 16

Section 16.2

1. **a.** $P(L|F) = 5/16$, $P(M|F) = 10/16$, $P(U|F) = 1/16$,

   $P(L|A) = 5/26$, $P(M|A) = 15/26$, $P(U|A) = 6/26$;

   b. $P(F|L) = 3/5$, $P(F|M) = 1/2$, $P(F|U) = 1/5$, $P(A|L) = 2/5$,

   $P(A|M) = 1/2$, $P(A|U) = 4/5$.

3. 0.57. 5. 2/3.

Section 16.3

1. a. -139.04, b. 0.37.

3.

Consequence	1	2	3
Expected value	240	310	265

Choose alternative 2.

5. a.

Consequence	A	B
Expected time	60.5	57.5

b. (3) 56.0.

Section 16.4

1. a. 1.00 3. 0.44 5. a. 0.48, b. -15,000, e. 9000.

Chapter 17

Section 17.2

1.

	Opportunity losses						
Event	*A*	*B*	*C*	*D*	*E*	*F*	*G*
E_1	1700	700	1200	0	1750	2100	2700
E_2	0	1200	1750	500	1100	1750	2500
E_3	700	1450	1400	0	500	1100	1300
E_4	350	450	450	1200	0	300	200

3. Choose alternative *A*—retrain all present workers.
Expected opportunity loss = 1.6.

Section 17.3

1.

				Opportunity losses			
				Accept		Reject	
Event	Prior	Joint	Posterior	Conditional	Expected	Conditional	Expected
$\pi = 0.20$	0.15	0.0025	0.067	0	0	150,000	10,050.00
$\pi = 0.35$	0.30	0.0326	0.879	0	0	7,500	6,592.50
$\pi = 0.50$	0.45	0.0020	0.054	85,000	4590	0	0
$\pi = 0.75$	0.10	0.0000	0	175,000	0		
			1.000		4590		16,642.50

Decision: accept the proposal.

3. a. $H_0: \pi \geq 0.60$,

b. Define $E_1: \pi = 0.50$; $E_2: \pi = 0.75$; $E_3: \pi = 0.95$;
I_{31}: indicator outcome of 31 affirmative responses.

$H_0: \pi \geq 0.60$

				Opportunity losses			
				Buy tickets (Accept)		Do not buy (Reject)	
Event	Prior	Joint	Posterior	Conditional	Expected	Conditional	Expected
E_1	0.30	0.0081	0.547	250	136.75	0	0.00
E_2	0.45	0.0067	0.453	0	0.00	375	169.88
E_3	0.25	0.0000	0.000	0	0.00	875	0.00
			1.000		136.75		169.88

Decision: purchase the tickets (accept H_0).

Section 17.4

1. a. $\pi_b = 0.67$, $H_0: \pi \geq 0.67$;

b.

	Opportunity loss	
Event	Accept H_0	Reject H_0
$\pi = 0.0$	500	0
$\pi = 0.1$	425	0
$\pi = 0.2$	350	0
$\pi = 0.3$	275	0
$\pi = 0.4$	200	0
$\pi = 0.5$	125	0
$\pi = 0.6$	50	0
$\pi = 0.7$	0	25
$\pi = 0.8$	0	100
$\pi = 0.9$	0	175
$\pi = 1.0$	0	250

Chapter 18

Section 18.2

1.

Year	1966	1967	1968	1969	1970
a. 1966 = 100	100	113	121	132	111
b. 1968 = 100	83	94	100	109	92

Section 18.3

1. $I_0 = 100.0$, $I_1 = 104.7$, $I_2 = 108.0$.

3. $I_0 = 100.0$, $I_1 = 108.1$, $I_2 = 107.7$.

Section 18.4

1. $I_0 = 100.0$, $I_1 = 106.4$, $I_2 = 109.9$.

3. $I_0 = 100.0$, $I_2 = 109.0$, $I_3 = 108.9$.

Section 18.5

1. a. $I_0 = 100.0$, $I_1 = 103.9$, $I_2 = 106.5$;
   b. $I_0 = 100.0$, $I_1 = 103.9$, $I_2 = 107.0$.

3. a. $I_0 = 100.0$, $I_1 = 109.2$, $I_2 = 109.0$;
   b. $I_0 = 100.0$, $I_1 = 109.4$, $I_3 = 110.7$.

Section 18.6

1. a. $I_1 = 104.1$, $I_2 = 106.6$; b. $I_1 = 104.8$, $I_2 = 108.6$.

Section 18.7

1.

Year	1965	1966	1967	1968	1969	1970
a. Index (1969 = 100)	93.3	95.2	98.1	97.1	100.6	96.2
b. Index (1970 = 100)	97.0	99.0	102.0	101.0	104.0	100.0

Section 18.8

1.

Year	1965	1966	1967	1968	1969	1970
a. Index *A*	100.0	104.0	106.0	107.0	109.1	112.4
b. Index *B*	93.5	97.2	99.1	100.0	102.0	105.0

Section 18.9

1. $I_{72-73} = 103.9$, $I_{73-74} = 102.5$, $I_{72-74} = 106.5$.

3. $I_{70-71} = 109.2$, $I_{71-72} = 99.7$, $I_{70-72} = 108.9$.

Chapter 19

Section 19.2

1. a. $\hat{Y} = 24.1 + 4.60x$, $x = 0$ in 1971, x-unit = 1 year;

   b. $\hat{Y} = 10.3 + 4.60x$, $x = 0$ in 1968, x-unit = 1 year;

   c. $\hat{Y}(1978) = 56.3$; d. 1976.

3. a. $\hat{Y} = 9.1 + 0.75x$, $x = 0$ between 1970 and 1971, x-unit = 1/2 year;

   b. $\hat{Y} = 3.8 + 1.5x$, $x = 0$ in 1967, x-unit = 1 year;

   c. $\hat{Y}(1978) = 20.3$; d. 1977.

Section 19.3

1. a. Two-year: 12.45, 17.65, 22.20, 25.85, 30.35, 35.70;

   b. Three-year: 15.17, 19.70, 24.10, 28.17, 33.10.

Section 19.4

1. a. and b.

Calculation of seasonal indexes

Month (1)	1970 (2)	1971 (3)	1972 (4)	1973 (5)	1974 (6)	1975 (7)	Totals (8)	Averages (9)	Adjusted seasonal indexes (10)
Jan.		119.9	120.5	119.3	118.8	115.8	594.3	118.9	119.1
Feb.		115.0	113.8	111.6	112.3	113.7	566.4	113.3	113.5
March		107.3	108.0	106.8	105.9	106.6	534.6	106.9	107.1
April		102.4	101.4	101.3	101.2	100.9	507.2	101.4	101.5
May		94.4	93.3	94.7	93.3	93.0	468.7	93.7	93.8
June		91.1	90.7	91.8	90.7	89.8	454.1	90.8	90.9
July	89.2	89.5	90.3	92.0	92.8		453.8	90.8	90.9
Aug.	87.4	88.7	90.7	91.9	93.7		452.4	90.5	90.6
Sept.	90.9	91.5	91.9	92.2	94.2		460.7	92.1	92.2
Oct.	93.7	92.2	92.9	93.5	93.4		465.7	93.1	93.2
Nov.	98.1	99.2	98.7	97.7	97.6		491.3	98.3	98.4
Dec.	109.7	109.6	108.5	107.1	107.5		542.4	108.5	108.7
								1198.3	1199.9

Adjustment factor = 1200/1198.3 = 1.00142.

c. Seasonally adjusted data (original data of Table A adjusted for seasonal variation)

Month	1970	1971	1972	1973	1974	1975
Jan.	3,123	3,607	4,126	4,579	5,091	5,539
Feb.	3,132	3,671	4,131	4,541	5,099	5,754
March	3,204	3,669	4,193	4,650	5,148	5,770
April	3,213	3,733	4,191	4,694	5,240	5,817
May	3,304	3,764	4,212	4,791	5,274	5,843
June	3,290	3,792	4,264	4,835	5,340	5,875
July	3,267	3,776	4,287	4,891	5,512	6,063
Aug.	3,258	3,801	4,361	4,954	5,639	6,377
Sept.	3,377	3,898	4,383	4,933	5,629	6,335
Oct.	3,486	3,927	4,429	4,996	5,578	6,177
Nov.	3,497	4,040	4,497	4,988	5,564	6,095
Dec.	3,577	4,074	4,518	4,987	5,592	6,304

d. Original values of Table A adjusted for both seasonal variation and trend

Month	1970	1971	1972	1973	1974	1975
Jan.	103.3	101.5	101.0	99.2	98.9	97.5
Feb.	102.1	102.0	100.0	97.4	98.2	100.5
March	103.0	100.7	100.4	98.8	98.3	100.0
April	101.8	101.2	99.3	98.8	99.2	100.1
May	103.2	100.9	98.8	99.9	99.0	99.7
June	101.4	100.4	99.0	99.9	99.4	99.5
July	99.3	98.8	98.5	100.1	101.8	102.0
Aug.	97.7	98.3	99.2	100.5	103.3	106.4
Sept.	100.0	99.7	98.7	99.2	102.3	105.0
Oct.	101.9	99.3	98.8	99.6	100.5	101.6
Nov.	100.9	101.1	99.3	98.6	99.5	99.5
Dec.	101.9	100.8	98.8	97.7	99.2	102.2

Index